T0145350

Advances in Intelligent Systems and Computing

Volume 837

Series editor

Janusz Kacprzyk, Polish Academy of Sciences, Warsaw, Poland
e-mail: kacprzyk@ibspan.waw.pl

The series "Advances in Intelligent Systems and Computing" contains publications on theory, applications, and design methods of Intelligent Systems and Intelligent Computing. Virtually all disciplines such as engineering, natural sciences, computer and information science, ICT, economics, business, e-commerce, environment, healthcare, life science are covered. The list of topics spans all the areas of modern intelligent systems and computing such as: computational intelligence, soft computing including neural networks, fuzzy systems, evolutionary computing and the fusion of these paradigms, social intelligence, ambient intelligence, computational neuroscience, artificial life, virtual worlds and society, cognitive science and systems, Perception and Vision, DNA and immune based systems, self-organizing and adaptive systems, e-Learning and teaching, human-centered and human-centric computing, recommender systems, intelligent control, robotics and mechatronics including human-machine teaming, knowledge-based paradigms, learning paradigms, machine ethics, intelligent data analysis, knowledge management, intelligent agents, intelligent decision making and support, intelligent network security, trust management, interactive entertainment, Web intelligence and multimedia.

The publications within "Advances in Intelligent Systems and Computing" are primarily proceedings of important conferences, symposia and congresses. They cover significant recent developments in the field, both of a foundational and applicable character. An important characteristic feature of the series is the short publication time and world-wide distribution. This permits a rapid and broad dissemination of research results.

More information about this series at http://www.springer.com/series/11156

Radek Matoušek
Editor

Recent Advances in Soft Computing

Proceedings of 23rd International Conference on Soft Computing (MENDEL 2017) Held in Brno, Czech Republic, June 20–22, 2017

Springer

Editor
Radek Matoušek
Department of Applied Computer Science,
Institute of Automation and Computer
Science, Faculty of Mechanical
Engineering
Brno University of Technology
Brno, Czech Republic

ISSN 2194-5357 ISSN 2194-5365 (electronic)
Advances in Intelligent Systems and Computing
ISBN 978-3-319-97887-1 ISBN 978-3-319-97888-8 (eBook)
https://doi.org/10.1007/978-3-319-97888-8

Library of Congress Control Number: 2018950189

This Springer imprint is published by the registered company Springer Nature Switzerland AG
The registered company address is: Gewerbestrasse 11, 6330 Cham, Switzerland

Preface

This proceedings book of the Mendel conference (http://www.mendel-conference.org) contains a collection of selected accepted papers which have been presented at this event in June 2017. The Mendel conference was held in the second largest city in the Czech Republic—Brno (http://www.brno.cz/en)—which is a well-known university city. The Mendel conference was established in 1995 and is named after the scientist and Augustinian priest Gregor J. Mendel who discovered the famous Laws of Heredity.

The main aim of the Mendel conference is to create a regular possibility for students, academics, and researchers to exchange their ideas on novel research methods as well as to establish new friendships on a yearly basis. The scope of the conference includes many areas of soft computing including: *Genetic Algorithms, Genetic Programming, Grammatical Evolution, Differential Evolution, Evolutionary Strategies, Hybrid and Distributed Algorithms, Probabilistic Metaheuristics, Swarm Intelligence, Ant Colonies, Artificial Immune Systems, Computational Intelligence, Evolvable Hardware, Chemical Evolution, Fuzzy Logic, Bayesian methods, Neural Networks, Data mining, Multi-Agent Systems, Artificial Life, Self-organization, Chaos, Complexity, Fractals, Image Processing, Computer Vision, Control Design, Robotics, Motion Planning, Decision making, Metaheuristic Optimization Algorithms, Intelligent Control, Bio-Inspired Robots, Computer Vision, and Intelligent Image Processing*.

Soft computing is a formal area of computer science and an important part in the field of artificial intelligence. Professor Lotfi A. Zadeh introduced the first definition of soft computing in the early 1990s: "Soft computing principles differs from hard (conventional) computing in that, unlike hard computing, it is tolerant of imprecision, uncertainty, partial truth, and approximation." The role model for soft computing is the human mind and its cognitive abilities. The guiding principle of soft computing can be specified as follows: Exploit the tolerance for imprecision, uncertainty, partial truth, and approximation to achieve tractability and robustness at a low solution cost.

The main constituents of soft computing include fuzzy logic, neural computing, evolutionary computation, machine learning, and probabilistic reasoning, whereby probabilistic reasoning contains belief networks as well as chaos theory. It is important to say that soft computing is not a random mixture of solution approaches. Rather, it is a collection of methodologies in which each part contributes a distinct way to address a certain problem in its specific domain. In this point of view, the set of soft computing methodologies can be seen as complementary rather than competitive. Furthermore, soft computing is an important component for the emerging field of contemporary artificial intelligence.

Image processing is a complex process, in which image processing routines and domain-dependent interpretation steps often alternate. In many cases, image processing has to be extensively intelligent regarding the tolerance of imprecision and uncertainty. A typical application of intelligent image processing is computer vision in robotics.

This proceedings book contains three chapters which present recent advances in soft computing, including intelligent image processing and bio-inspired robotics. The accepted selection of papers was rigorously reviewed in order to maintain the high quality of the conference. Based on the topics of accepted papers, the proceeding book consists of two chapters: Chapter 1: *Evolutionary Computing, Swarm intelligence, Metaheuristics, and Optimization*, Chapter 2: *Neural Networks, Machine Learning, Self-organization, Fuzzy Systems, and Advanced Statistics*.

We would like to thank the members of the International Program Committees and Reviewers for their hard work. We believe that Mendel conference represents a high standard conference in the domain of soft computing. Mendel 2017 enjoyed outstanding keynote lectures by distinguished guest speakers: Thomas Bäck, the Netherlands; Marco Castellani, UK; Swagatam Das, India; Carola Doerr, France; Shengxiang Yang, UK; Krzysztof Krawiec, Poland; and Jiri Sima, Czech Republic.

Particular thanks go to the conference organizers and main sponsors as well. In 2017, the conference is organized under the auspices of the rector of Brno University of Technology with support of WU Vienna University of Economics and Business, and University of Vaasa. The conference sponsors are Humusoft Ltd. (international reseller and developer for MathWorks, Inc., USA), B&R automation Ltd. (multi-national company, specialized in factory and process automation software), and Autocont Ltd. (private Czech company that operates successfully in the area of ICT).

We would like to thank all contributing authors, as well as the members of the International Program Committees, the Local Organizing Committee, and the Executive Organizing Committee, namely Ronald Hochreiter and Lars Nolle for their hard and highly valuable work. Their work has definitely contributed to the success of the Mendel 2017 conference.

Radek Matoušek

Organization

International Program Committee and Reviewers Board

Alex Gammerman	Royal Holloway, University of London, UK
Camelia Chira	Babes-Bolyai University, Romania
Conor Ryan	University of Limerick, Ireland
Daniela Zaharie	West University of Timisoara, Romania
Eva Volná	Universitas Ostraviensis, Czech Republic
Halina Kwasnicka	Wroclaw University of Technology, Poland
Hans-Georg Beyer	Vorarlberg University of Applied Sciences, Austria
Hana Druckmullerova	Brno University of Technology, Czech Rep.
Ivan Sekaj	Slovak University of Technology, Slovakia
Ivan Zelinka	Technical University of Ostrava, Czech Rep.
Janez Brest	University of Maribor, Slovenia
Jaromír Kukal	Czech Technical University in Prague, Czech Republic
Jerry Mendel	University of Southern California, USA
Jiri Pospichal	Slovak University of Technology, Slovakia
Jirí Bila	Czech Technical University in Prague, Czech Republic
Josef Tvrdik	Universitas Ostraviensis, Czech Republic
Jouni Lampinen	University of Vaasa, Finland
Katerina Mouralova	Brno University of Technology, Czech Republic
Lars Nolle	Jade University of Applied Science, Germany
Luis Gacogne	LIP6 Universite Paris VI, France
Lukáš Sekanina	Brno University of Technology, Czech Republic
Michael O'Neill	University College Dublin, Ireland
Michael Wagenknecht	University Applied Sciences, Zittau, Germany
Miloslav Druckmuller	Brno University of Technology, Czech Republic
Napoleon Reyes	Massey University, New Zealand
Olaru Adrian	University Politehnica of Bucharest, Romania
Patric Suganthan	Nanyang Technological University, Singapore
Pavel Popela	Brno University of Technology, Czech Republic

Riccardo Poli	University of Essex, UK
Roman Senkerik	Tomas Bata University in Zlín, Czech Republic
Ronald Hochreiter	WU Vienna University, Austria
Salem Abdel-Badeh	Ain Shams University, Egypt
Tomoharu Nakashima	Osaka Prefecture University, Japan
Urszula Boryczka	University of Silesia, Poland
William Langdon	University College London, UK
Xin-She Yang	National Physical Laboratory UK, UK
Zuzana Oplatkova	Tomas Bata University in Zlín, Czech Republic

Executive Organizing Committee and Local Organizers

Radek Matoušek (Chair and Volume Editor-in-Chief), Czech Republic
Jouni Lampinen (Co-chair and Associate Adviser), Finland
Ronald Hochreiter (Co-chair and Associate Adviser), Austria
Lars Nolle (Co-chair and Associate Adviser), Germany

Local Organizers

Jiří Dvořák (Peer Review Manager)
Ladislav Dobrovský (Technical Review Manager)
Jitka Pavlíková (Senior Organizer)
Daniel Zuth (Junior Organizer)
Petr Šoustek (Junior Organizer)
Tomas Marada (Junior Organizer)

Contents

Evolutionary Computing, Swarm Intelligence, Metaheuristics, Optimization

Solving Sudoku's by Evolutionary Algorithms with Pre-processing

Pedro Redondo Amil[1] and Timo Mantere[2](✉)

[1] University of Córdoba, C2, Rabanales Campus, C.P. 14071 Córdoba, Spain
[2] University of Vaasa, PO Box 700, 65101 Vaasa, Finland
timan@uva.fi

Abstract. This paper handles the popular Sudoku puzzle and studies how to improve evolutionary algorithm solving by first pre-processing Sudoku solving with the most common known solving methods. We found that the pre-processing solves some of the easiest Sudoku's so we do not even need other methods. With more difficult Sudoku's the pre-processing reduce the positions needed to solve dramatically, which means that evolutionary algorithm finds the solution much faster than without the pre-processing.

Keywords: Ant colony optimization · Cultural algorithms · Genetic algorithms Hybrid algorithms · Puzzle solving · Sudoku

1 Introduction

In this paper we will report our newest developments with solving Sudoku's with evolutionary algorithms. This research project started as a hobby back in 2006 when we published our first Sudoku paper [1]. Since then we have published other papers, and got many citations to them and still many people inquire us of our Sudoku papers and software. Therefore, we felt that now is time to update our Sudoku experiments.

This time we decided to program the most common Sudoku solving methods and use them to pre-process Sudoku's before applying evolutionary algorithms to solve them. We also add some literature and background information that we have not discussed in our earlier Sudoku papers.

According to Wikipedia [2], Sudoku, originally called Number Place, is a logic-based, combinatorial number-placement puzzle (Fig. 1). The objective is to fill a 9×9 grid with digits so that each column, each row, and each of the nine 3×3 sub grids that compose the grid (also called "boxes", "blocks", "regions", or "sub squares") contains all of the digits from 1 to 9. The puzzle setter provides a partially completed grid, which for a well-posed puzzle has a unique solution.

We will try to solve the Sudoku with different algorithms: Genetic Algorithms, Cultural Algorithms and Genetic Algorithm/Ant colony optimization hybrid.

Genetic algorithms (GAs) [3] are computer based optimization methods that use the Darwinian evolution [4] of nature as a model and inspiration.

Cultural algorithms (CAs) were introduced by Reynolds [5], they are a branch of evolutionary computation methods, where algorithms include knowledge component,

R. Matoušek (Ed.): MENDEL 2017, AISC 837, pp. 3–15, 2019.
https://doi.org/10.1007/978-3-319-97888-8_1

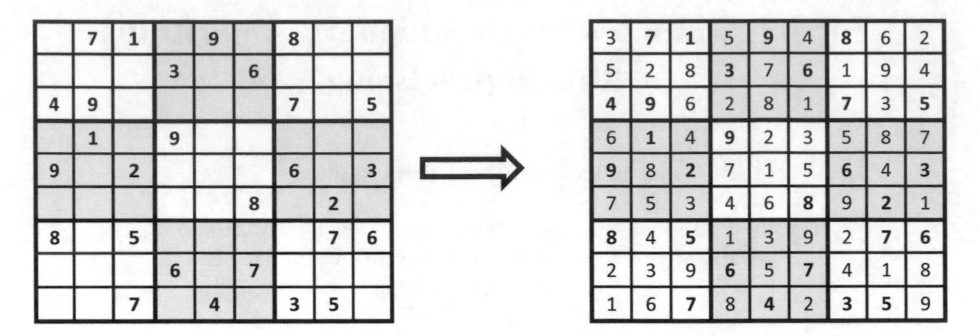

Fig. 1. Sudoku and its solution

belief space. Cultural algorithm is usually an extension of conventional genetic algorithm that has an added belief space. Ant colony optimization algorithm (ACO) [6] in turn is a probabilistic technique for solving computational problems, which can be reduced to finding good paths through graphs, ant colony also gather cultural knowledge.

Hybrid algorithm [7, 8] refers to the combination of two or more algorithms. We will use here the Genetic Algorithms/Ant colony optimization hybrid algorithm.

The pre-processing consists in filling sure number in different squares of the Sudoku through several methods, we will use this pre-processing with every algorithm that we have described before.

The objective of this study is to test if Genetic algorithms, Cultural algorithms and Genetic algorithm/Ant colony optimization hybrid algorithm are efficient methods for solving Sudoku puzzles; some Sudoku's generate by our Genetic Algorithm and some of them from a newspaper or website.

The document is organized as follows. In Sect. 1, we have included the objectives of this study. In Sect. 2 we describe the Sudoku. In Sect. 3 we have defined the different algorithms. In Sect. 4 we explain the result of the experiments and finally in the Sect. 5 we say our conclusions.

2 Sudoku

Sudoku puzzles are related to Latin squares [9], which were developed by the 18th century Swiss mathematician Leonhard Euler. Latin squares (Fig. 2) are square-grids of size $n \times n$ where each of the character (numbers, letters, symbols…) from 1 through n appears in every column and in every row precisely once (rank n Latin squares).

Magic squares [10] are square grids that are filled with (not necessarily different) numbers such that the numbers in each row and column add to up to the same sum.

In the late 19th century a Paris-based daily newspaper, Le Siècle (discontinued) published a partially completed 9×9 magic square that had 3×3 sub grids. The object of the game was to fill out the magic square such that the numbers in the grids also sum to the same number as in the rows and columns.

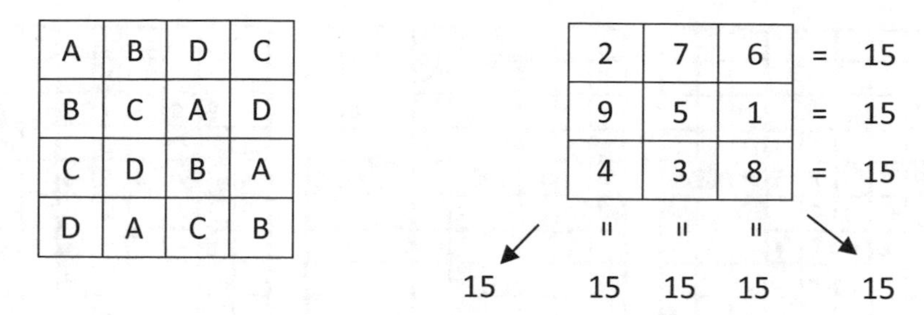

Fig. 2. Examples of Latin square (left) and magic square (right)

The name of the game is of Japanese origin, the word "SuDoku" is abbreviation of "Suuji wa dokushin ni kagiru" that it means, "The digits must remain single". It was not until 1986 when the Japanese company Nikoli, Inc. started to publish a version of the Sudoku at the suggestion of its president, Mr. Maki Kaji. He gave the game its now famous name. Almost two decades passed before (near the end of 2004) The Times newspaper in London started to publish Sudoku as its daily puzzle due to the effort of Wayne Gould, who spend many years to develop a computer program that generates Sudoku puzzles. By 2005, it became an international hit.

A complete Sudoku solution may be arrived at in more than one ways, as we can start from any of the given clues that are distributed over the sub grids of a given Sudoku (to be solved). There is no known technique that we surveyed to determine how many different starting squares (the first square that we can fill) there are. Removal of a single number given may generate another Sudoku different for which other solutions may exist and the solution is no longer unique.

The puzzles require logic, sometimes intricate, to solve but no formal mathematics is required. However, the puzzles lead naturally to certain mathematical questions. For example: Which puzzles have solutions and which do not? If a puzzle has a solution, is it unique? A couple of years ago it was proven that there has to be minimum of 17 givens in order to Sudoku have unique solution [11]. With the experiments of our Sudoku puzzle generator (yet unpublished paper), we have observed that there are several Sudoku's that do have two or more valid solutions even if the number of givens is 17 or more.

There could be Sudoku's with multiple solutions, but those that are published in newspapers usually have only one unique solution. Therefore, we can say that, a subset of Sudoku puzzles may have one and only one valid solution, but in general, a Sudoku might have two or more solutions as well. Sudoku is a combinatorial optimization problem [12], where each row, column, and 3×3 sub squares of the problem must have each integer from 1 to 9 once and only once. This means that the sum of the numbers in each row and each column of the solution are equal to 45 and the multiply of the numbers in each row and each column of the solution are equal to 9! (362880).

2.1 Pre-processing the Sudoku

To solve the Sudoku, we can pre-process it before applying some algorithm. The pre-processing consists in filling sure number in different squares of the Sudoku through several methods: full house, naked singles, hidden singles, lone rangers and naked pairs.

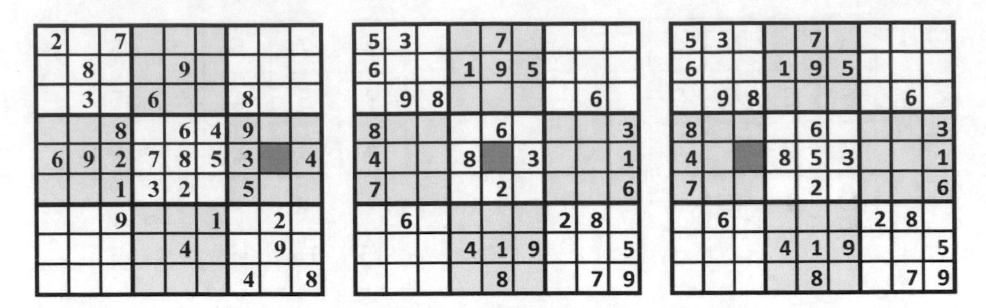

Fig. 3. Examples of full house (left), naked singles (middle) and hidden singles (right)

We used only these methods [13] because the many of the rest of methods take more time than to solve the Sudoku, because they perform several checks.

After applying the pre-processing, we use the different evolutionary algorithms to solve rest of the Sudoku, if it was not already solved.

The "Full house" method [3] consists in filling an empty cell where its value can be deduced when its row, column or 3 × 3 block has eight numbers, so we can get the last remaining number. In this example (Fig. 3), we check the central row and we know that the only number missing from this row is {1}, so we can fill this square (marked dark) with this number.

The "Naked singles" method [3] consist in filling an empty cell with the number not present in the set union of those in the same row, column and 3 × 3 sub block. This technique is applicable only when there is only one missing number. In this example (Fig. 3), we check the central square (dark square) and we know that the set from its row is {1, 3, 4, 8}, the set from its column is {1, 2, 6, 7, 8, 9} and the set from its 3 × 3 sub block is {2, 3, 6, 8}. We get a unique set joining all the sets and the result is {1, 2, 3, 4, 6, 7, 8, 9}. The only number missing from the set is {5}, so we can fill this square with this number.

The "Hidden singles" method [14] consists in filling an empty square if there is a value that can be placed in only one cell of a row, column, or 3 × 3 sub block. Then, this cell has to hold that value. In this example (Fig. 3), we check the fourth, and sixth rows and we see that the number {6} is in the fourth and sixth row, but it is not in the fifth one. This number must appear in the fifth row and in the left central 3 × 3 sub block. Since number {6} is already in first and second column, there is only one empty cell there where it fits (dark square), so it must hold this value, so we can fill this square with the number {6}.

The "Lone rangers" method [14] consists in filling an empty square that seems to have more than one possible candidate. Whereas a careful observation across the possible candidate lists of row, column or 3 × 3 sub block neighbours of that empty square reveals the exact candidate, because one of the values appear just in one cell of the row, column or 3 × 3 sub block. In this example (Fig. 4), we check the third row and we see that the number {6} only appears in one square of this row (with circle) like a candidate, so we can fill this square with this number.

The "Naked pairs" method [14] consists in removing candidates from the candidate list of squares. If two cells in a row, column or 3 × 3 sub block have the same pair of

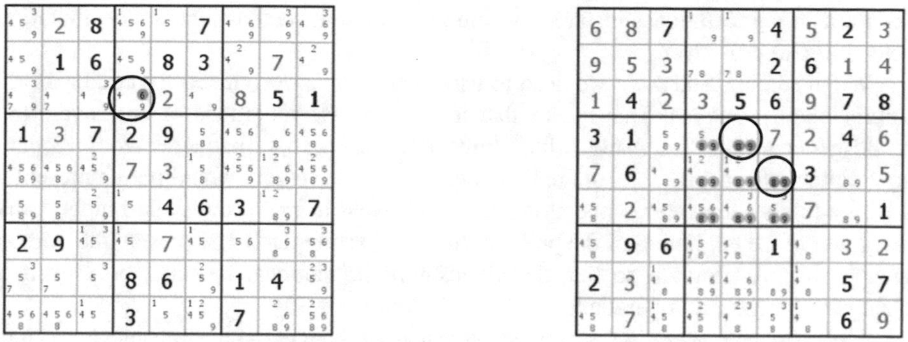

Fig. 4. Examples of lone rangers (left), and naked pairs (right)

candidates then they cannot be housed in any other cell of the same row, column, or 3×3 sub block, and therefore they can be removed from their candidate lists. In this example (Fig. 4), we check the central square and we see that two squares have the same candidates (marked with circle) that they are {8, 9}, then we can delete these numbers from the candidate lists of the other cells in the same sub block. Subsequently, another technique might possibly be applied to deduce the value of a cell. For instance, by the "Naked single", so we get the number {5} is the number of the first square of this 3×3 square because it is the only candidate number.

3 The Algorithms Used

We used different algorithms for solving the Sudoku's. The first one is a Genetic Algorithm [3], using crossover and mutation operators. The second one is a method that pre-process the Sudoku before applying the Genetic Algorithm, filling the squares that have only one possible value. The third one is a Cultural Algorithm [5]. The fourth one is a method that pre-process the Sudoku before applying Cultural Algorithm. The fifth one is a hybrid algorithm of Genetic Algorithm with the ant colony optimization [6]. The sixth one is a method that pre-process the Sudoku before applying the hybrid algorithm.

The algorithms used are the same as in papers [15, 16] where we have explained them with details. The algorithms used in [16] were used as a base and pre-processing methods were programmed to the source codes used in that study. Also all GA, CA and GA/ACO results without pre-processing are obtained with algorithms presented in [16].

3.1 Genetic Algorithms

Genetic Algorithms (GAs) [3] are computer based optimization methods that use the Darwinian evolution [4] of nature as a model and inspiration. It can be defined also like a search heuristic that mimics the process of natural selection.

This heuristic (also sometimes called a metaheuristic) is routinely used to generate useful solutions to optimization and search problems. Genetic algorithms belong to the larger class of evolutionary algorithms (EA), which generate solutions to optimization problems using techniques inspired by natural evolution, such as inheritance, mutation, selection and crossover.

When solving Sudoku's we need to use a GA that is designated for combinatorial optimization problems. That means that it will not use mutations or crossovers that could generate illegal situations, like: rows, columns, and sub squares would contain some integer from {1, …, 9} more than once, or some integers would not be present at all. In addition, the genetic operators are not allowed to move the static numbers that are given in the beginning of the problem (givens), so we need to represent the Sudoku puzzles in GA program so that the givens will be static and cannot be moved or changed with genetic operations.

The crossover operation is applied so that it exchanges whole sub blocks of nine numbers between individuals, thus, the crossover point cannot be inside a building block. The mutations contrarily are applied only inside a sub block.

We use the technique of Swap mutation. In swap mutation, the values in two positions are exchanged. In this example (Fig. 5), we change the numbers 2 and 5. Each time mutation is applied inside the sub block (3×3 blocks), the array of givens is referred. If it is illegal to change that position, we randomly reselect the positions and we recheck until legal positions are found. We are using the mutation strategy of swap mutation with an overall mutation probability of 0.12. In swap mutation, the values in two positions are exchanged.

Fig. 5. Swap mutation

3.2 Cultural Algorithms

Cultural algorithms (CAs) were introduced by Reynolds [5], they are a branch of evolutionary computation methods, where algorithms includes knowledge component, belief space. A CA is usually an extension of conventional GA that has added belief space.

The belief space of a cultural algorithm is divided into distinct categories. These categories represent different domains of knowledge collected from the search space. Belief space could collect, *e.g.* domain specific knowledge, situational knowledge, temporal knowledge, or spatial knowledge. Belief space acts as an extra reproduction component that somehow affects to the generation of new individuals.

If we compare our CA with GA, the main difference is that we added a belief space model, which in this case is simple: it is a $9 \times 9 \times 9$ cube, where the first two dimensions correspond to the positions of a Sudoku puzzle, and the third dimension represents the nine possible digits for each location.

After each generation, the belief space is updated if:

(1) The fitness value of best individual is 2 (two numbers in wrong positions).
(2) The best individual is not identical with the individual that updated the belief space the previous time.

The belief space is updated so that the value of the digit that appears in the best Sudoku solution is incremented by one in the belief space. This model also means that the belief space is updated only with near-optimal solutions.

The belief space collects information from the near-optimal solutions (only two numbers in "wrong" positions), and this information is used only in the population re-initialization process. When the population is reinitialized, positions that have only one non-zero digit value in the belief space are considered as givens. These include the actual givens of the problem, but also so called hidden givens that the belief space have learned, that is, those positions that always contain the same digit in the near-optimal solutions. This also connects the reinitialized population to the previous generations, since the new initial population is not formed freely.

3.3 GA/ACO Hybrid Algorithm

In computer science and operations research, the Ant Colony Optimization algorithm (ACO) is a probabilistic technique for solving computational problems, which can be reduced to finding good paths through graphs, initially proposed by Marco Dorigo 1992 in his PhD thesis [6]. The original idea was to search for an optimal path in a graph based on the behaviour of ants seeking a path between their colony and a source of food. The method has since diversified to solve a wider class of numerical problems, and simulating on various aspects of the behaviour of ants. The basic idea is that ants leave trail of pheromone, which is stronger, if the path is good.

In our Genetic Algorithm/Ant colony optimization hybrid algorithm, the GA part acts as a heuristic global searcher that generates new paths and most of the new individuals, the GA also finds the initial population for this hybrid. The ACO part acts like a greedy local searcher that tries quickly to converge into the strongest path. In addition, ACO collects cultural information stored in pheromone trails and it inserts new individuals generated by ACO, that is, by weighted random generator, where weights are proportionate to the pheromone strength. The ACO parts act in this hybrid algorithm somewhat similarly as a belief space in CAs.

In our algorithm, we generate the initial generation randomly. After each generation, the old pheromone paths are weakened 10%. Then the best individuals update the pheromone trails by adding strength value.

3.4 Fitness Function

To design a fitness function that would aid a GAs, CAs or hybrid algorithms search is often difficult in combinatorial problems. In this case, we decided to use to a simple fitness function that penalizes different constraint violations differently.

The condition that every 3×3 sub block contains a number from 1 to 9 is guaranteed intrinsically, because of the chosen solution encoding. Penalty functions are used to evaluate the other conditions. Each row and column of the Sudoku solution must contain each number $\{1, \ldots, 9\}$ once and only once. This can be transformed to a set of inequality constraints.

Fitness function is combined of three different rules (Eqs. 1–3). The first part (1) requires that all digits {1, …, 9} must be present in each row and column, otherwise penalty P_x is added to the fitness value.

$$P_x = \sum_{i=1}^{8}\sum_{j=1}^{8}\sum_{ii=i+1}^{9}\sum_{jj=j+1}^{9}\left[\left(x_{i,j} \equiv x_{ii,j}\right) + \left(x_{i,j} \equiv x_{i,j}\right)\right] \tag{1}$$

```
if Best(generation(i))=Best(generation(i-1)) then Value(Best)+= 1;    (2)
```

$$P_g = \sum_{i=1}^{9}\sum_{j=1}^{9}\left(x_{ij} \equiv g_{ij}\right) \tag{3}$$

Equation (1) count the missing digits in each row (x_i) and column (x_j). In the optimal situation, all digits appear in all the row and column sets and value of this fitness function part becomes zero. The second part of the fitness function (2) is for aging the best individual, it increments its fitness value by one on each generation, when it remains the best. This is easy to perform since our sorting algorithm only sorts indexes, if the same individual remains best, also, the index to it remains same.

The third component (3) of the fitness function requires that the same digit (x_{ij}) as some given (g_{ij}) must not appear in the same row or column as that given. Otherwise, penalty P_g is added.

Note, due to lack of space we do not explain all the details about the old versions of our GA, CA and GA/ACO Sudoku solvers in this paper. Instead, we are trying to summarize the main details needed in order to follow this paper separately from our previous papers. Those who are interested to learn more can are encouraged to read [15, 16]. Our genetic encoding is also explained in our Sudoku webpage [17]. Algorithms in this paper are identical to those in [16, 17] with the exception of added preprocessing part.

4 Experiments

We test five Sudoku puzzles taken from the newspaper Helsingin Sanomat marked with their difficulty rating 1–5 stars, where 1 star is the easiest. We test four Sudoku's taken from newspaper Aamulehti, they are marked with difficulty ratings: Easy, Challenging, Difficult, and Super difficult and we also tested three Sudoku's made with Genetic Algorithms from the Sudoku webpage [17] marked with difficulty ratings: Easy (GA-E), Medium (GA-M), Hard (GA-H).

We tried to solve each of the ten Sudoku puzzles 100 times. The stopping condition was the optimal solution found, which was always found with maximum seven hundred thousand trials (fitness function calls) for GA and CA, and four million trials for GA/ACO hybrid algorithm.

To compare which method is better, we will use the generations required to find the solution and the time spent in seconds. For this reason, we will get from each execution of the code the minimum, maximum and average of generations and the minimum, maximum and average of the time spent.

We done every test with a laptop with Windows 8 and the following hardware: Intel core i5-4200U 1.6 GHz with turbo boost up to 2.6 GHz, 4 GB DDR3 of Memory ram and 750 GB of capacity of Hard disk.

4.1 Genetic Algorithm

The results of the GA are given in Table 1 where the columns (left to right) stand for: difficulty rating, number of givens, minimum, maximum and average of the generations and minimum, maximum and average of the time required (in seconds) to solution.

Table 1 shows that the difficulty ratings of Sudoku's correlate with their GAs hardness. Therefore, the more difficult Sudoku's for a human solver seem to be also the more difficult for the GAs. The Easy from Aamulehti was the most easiest to solve in general. In addition, the average of the generations and the time needed to find the solution increased somewhat monotonically with the rating. The puzzles Difficult and Super Difficult of Aamulehti were much more difficult than any of the puzzles in Helsingin sanomat.

The results with pre-processed GA, also are given in Table 1, are obtained by first pre-processing 5 times (with 5 times the most of the Sudoku's are solved or it is not possible to find any more sure number) and later we use the Genetic Algorithms to solve the Sudoku's.

With respect to the results obtained without pre-processing, we can see that if we use the pre-processing of the Sudoku, we reduce drastically the generations needed to solve all Sudoku's. We reduce drastically the time spent also because the average of the time of every test is less than 1 s. For the Sudoku's that we put inside of the Easy and Medium group, only the pre-processing was enough to solve the Sudoku, so their generations needed and the time spent is very similar in all of them.

Table 1. Results of GA and GA with pre-processing.

Difficulty		Genetic algorithm						GA with pre-processing					
rating	Givens	Min. gen	Max. gen	Avg. gene	Min. time	Max. tim	Avg. Tim	Min. gen	Max. gen	Avg. gene	Min. time	Max. tim	Avg. Time
1 star	33	14	95	23	0.036	0.246	0.062	1	1	1	0.022	0.045	0.035
2stars	30	24	2217	355	0.059	5.326	0.836	1	1	1	0.023	0.046	0.035
3 stars	28	36	1838	405	0.084	4.162	0.927	1	1	1	0.023	0.046	0.035
4 stars	28	39	6426	1139	0.088	15.016	2.701	8	363	92	0.025	1.099	0.289
5 stars	30	59	14805	2837	0.165	35.248	6.831	1	234	20	0.023	1.955	0.166
Easy	36	9	33	19	0.026	0.088	0.052	1	1	1	0.022	0.046	0.035
Challengi	25	38	3595	628	0.082	8.903	1.425	1	1	1	0.023	0.046	0.035
Difficult	23	88	44168	6093	0.197	102.103	13.890	6	379	55	0.059	3.629	0.538
Super dif	22	115	30977	7814	0.250	66.997	17.672	9	213	62	0.054	1.482	0.395
GA-E	32	15	291	74	0.038	0.700	0.187	1	1	1	0.022	0.045	0.035
GA-M	29	35	4436	893	0.083	10.360	2.110	1	1	1	0.023	0.046	0.035
GA-H	27	50	27143	4989	0.118	63.442	12.193	4	654	149	0.026	4.241	0.935
Avg. Gen+sum tim		43.50	11335.33	2105.75	1.226	312.591	58.886	2.92	154.17	32.08	0.345	12.726	2.568

In the other hand, when we solve Sudoku's that we consider inside of the level Hard, the pre-processing basically changes their difficulty to Easy level, so when we solve those Sudoku's with the Genetic Algorithms, we do not need many generations.

With respect to the Sudoku's with 4 stars, after the pre-processing, it is still relatively difficult to solve, but easier than the 5 stars, difficult and super difficult Sudoku's.

4.2 Cultural Algorithm

The results given in Table 2 are using Cultural Algorithms. The columns used are the same than the Table 1. The results obtain with this algorithm are very similar to the results obtain with the Genetic Algorithms; in the most of the Easy and Medium Sudoku's this algorithm is a little bit better than the GAs one, but in the most of the hard Sudoku's the GA is a little bit better that CA. Both algorithms have similar average of the generation, average of the time spent, minimum generations and minimum time but the CA has better maximum generations and maximum time spent than the GA.

The results with CA and pre-processing are also given in Table 2. They are obtained using first the pre-processing 5 times and after that CA to solve rest of the Sudoku. The columns used are the same than the Table 1. The results obtained with this algorithm are a little bit better than the results with the pre-processed GA, but there is not a significant difference.

Table 2. Results of CA and CA with pre-processing.

Difficulty		Cultural algorithm						CA with pre-processing					
rating	Givens	Min. gene	Max. gene	Avg. gene	Min. time	Max. tim	Avg. Time	Min. gene	Max. gen	Avg. gene	Min. time	Max. tim	Avg. Time
1 star	33	10	68	25	0.027	0.173	0.065	1	1	1	0.031	0.042	0.035
2stars	30	24	1240	362	0.056	2.878	0.841	1	1	1	0.032	0.038	0.035
3 stars	28	31	2411	445	0.071	5.390	1.005	1	1	1	0.032	0.039	0.035
4 stars	28	34	5810	908	0.081	13.333	2.132	9	362	87	0.027	1.06	0.259
5 stars	30	31	9129	2301	0.076	21.542	5.422	1	208	18	0.008	1.709	0.142
Easy	36	10	30	18	0.028	0.076	0.049	1	1	1	0.031	0.039	0.035
Challengi	25	40	10266	792	0.087	22.629	1.762	1	1	1	0.032	0.038	0.035
Difficult	23	80	26094	5829	0.170	56.193	12.651	6	230	51	0.039	2.124	0.486
Super dif	22	116	34207	8603	0.250	72.975	18.421	7	253	59	0.065	2.387	0.552
GA-E	32	19	329	84	0.047	0.781	0.204	1	1	1	0.031	0.039	0.035
GA-M	29	28	4059	779	0.064	9.394	1.815	1	1	1	0.032	0.04	0.035
GA-H	27	39	24083	5331	0.090	55.619	12.272	8	796	163	0.048	4.886	1.024
Avg. Gen+sum tim		38.50	9810.50	2123.08	1.047	260.983	56.639	3.17	154.67	32.08	0.408	12.441	2.708

4.3 Hybrid Algorithm of GA and Ant Colony Optimization

The results showed in Table 3 are got by using a hybrid algorithm of Genetic algorithms and Ant colony optimization algorithm. The columns are same as in Table 1.

The results show that the average of the generations needed is less than with the GA or CA with every Sudoku tried. On the other hand, the average time spent is longer than with GA or CA, but it is not a great difference. GA/ACO hybrid uses more time per one trial than GA or CA. This might still be considered the best method without pre-processing.

The results showed in Table 3 (right) are using first the pre-processing 5 times and after that, we use a hybrid algorithm of Genetic Algorithms and Ant colony optimization algorithm. The columns used are the same as the Table 1.

Now, with pre-processing, we also obtain smaller average of the generations than with pre-processed GA or CA, but again we obtain longer average of the time spent than those algorithms for the Sudoku's that we consider as Hard level. For the rest of the Sudoku's (Easy and Medium level), we need the same number of generations but more time.

Table 3. Results of GA/ACO without and with pre-processing.

Difficulty rating	Givens	Genetic algorithm/Ant colony hybrid Min. gen	Max. gen	Avg. gene	Min. time	Max. tim	Avg. Time	GA/ACO hybrid with pre-processing Min. gen	Max. gen	Avg. gene	Min. time	Max. tim	Avg. Time
1 star	33	10	50	22	0.043	0.309	0.082	1	1	1	0.04	0.061	0.044
2stars	30	28	2199	199	0.098	7.531	0.794	1	1	1	0.037	0.051	0.043
3 stars	28	37	3280	347	0.135	12.458	1.235	1	1	1	0.038	0.061	0.046
4 stars	28	38	6438	905	0.132	22.925	3.391	6	229	71	0.027	1.431	0.399
5 stars	30	42	6571	1515	0.141	22.636	5.345	1	11	4	0.009	0.103	0.036
Easy	36	9	31	17	0.032	0.165	0.059	1	1	1	0.038	0.05	0.043
Challengi	25	41	3720	581	0.153	11.617	1.939	1	1	1	0.037	0.046	0.042
Difficult	23	77	17673	5104	0.269	67.874	13.040	6	221	35	0.043	3.877	0.534
Super dif	22	123	24199	6867	0.394	85.914	15.529	10	375	52	0.056	3.905	0.564
GA-E	32	15	338	65	0.055	1.353	0.343	1	1	1	0.055	0.13	0.082
GA-M	29	34	3865	606	0.128	17.392	2.240	1	1	1	0.05	0.108	0.071
GA-H	27	51	15073	4136	0.172	69.384	16.755	7	554	91	0.05	4.88	0.887
Avg. Gen ı sum tim		42.08	6953.08	1697.00	1.752	319.558	60.752	3.08	116.42	21.67	0.480	14.703	2.791

In the end of each Tables 1–3, we have calculated the average of the generations and the sum of the time spent to solve the different Sudoku's. According to these results, we can say that the best algorithm (without using the pre-processing) is GA/ACO hybrid algorithm because this algorithm need fewer generations to solve every level of difficulty but this algorithm has a problem. This problem is that it needs more time to solve the Sudoku but it is not a great difference.

The next best algorithm is GA because needs less generations to solve every level of difficulty and it needs less time also in every level of difficulty except in the Hard level, although the difference between GA and CA is not as high as the difference between GA/ACO and GA. Therefore, the worst algorithm studied was CA.

About the pre-processing, there are not differences between solving Sudoku's with GA, CA or GA/ACO in Easy or Medium level. For Hard level, GA/ACO is the best algorithm with slightly more time than GA but less time spent than CA. GA uses the same generations as CA but it needs slightly less time (thanks to no belief space updates) to solve the Sudoku's. So, GA is a little bit better than CA when we use pre-processing.

5 Conclusions

We studied if Sudoku puzzles can be solved with a combinatorial Genetic Algorithms, Cultural Algorithms and a hybrid algorithm with Genetic Algorithms and Ant Colony optimization. The results show that all these methods can solve Sudoku puzzles, but with slightly different effectivity. We find out that the hybrid algorithm GA/ACO is the most efficient with every Sudoku's, even more efficient than GAs or CAs. In this study we for the first time applied pre-processing of Sudoku's by first applying the most well known Sudoku solving methods. When we apply the pre-processing we obtain really good results and get a really efficient algorithm for solving Sudoku's, where first the easy parts of Sudoku is solved and then heuristic evolutionary algorithm is applied to solve rest instead of brute force used e.g. in SudokuExplainer.

The other goal was to study if difficulty ratings given for Sudoku puzzles (human difficulty) are consistent with their difficulty for our algorithms. The answer to that question seems to be positive: those Sudoku's that have a higher difficulty rating proved more difficult also for our algorithms. This also means that our algorithms can be used for testing the difficulty of a new Sudoku puzzle: Easy, Medium and Hard depending of the algorithm we used for testing and the generations needed for solving Sudoku.

We are using the same algorithms as here with Sudoku for other combinatorial optimization purposes, e.g. evenly loading the vehicle etc. technical balancing problems. We found Sudoku solving as a good benchmark problem for creating the better and more efficient algorithm. With real world problems, we have found that evaluating and comparing our algorithms and their evolution versions sometimes more uncertain than with exactly defined combinatorial balancing problem as Sudoku.

References

1. Mantere, T., Koljonen, J.: Solving and rating Sudoku puzzles with genetic algorithms. In: Hyvönen, E., et al. (eds.) Proceedings of the 12th Finnish Artificial Conference STeP 2006, Espoo, Finland, 26–27 October, pp. 86–92 (2006)
2. Wikipedia: Sudoku. http://en.wikipedia.org/wiki/Sudoku
3. Holland, J.: Adaptation in Natural and Artificial Systems. The MIT Press, Cambridge (1992)
4. Darwin, C.: The Origin of Species: By Means of Natural Selection or the Preservation of Favoured Races in the Struggle for Life. Oxford University Press, London (1859)
5. Reynolds, R.G.: An overview of cultural algorithms. In: Advances in Evolutionary Computation, McGraw Hill Press, New York (1999)
6. Colorni, A., Dorigo M., Maniezzo, V.: Distributed optimization by ant colonies. In: actes de la première conférence européenne sur la vie artificielle, pp. 134–142. Elsevier, Paris (1991)
7. Lozano, M., García-Martínez, C.: Hybrid metaheuristics with evolutionary algorithms specializing in intensification and diversification: overview and progress report. Comput. Oper. Res. **37**(3), 481–497 (2010)
8. Blum, C., Puchinger, J., Raidl, G., Roli, A.: Hybrid metaheuristics in combinatorial optimization: a survey. Appl. Soft Comput. **11**(6), 4135–4151 (2011)
9. Wikipedia: Latin square. http://en.wikipedia.org/wiki/Latin_square
10. Wikipedia: Magic square. http://en.wikipedia.org/wiki/Magic_square

11. McGuire, G., Tugemann, B., Civario, G.: There is no 16-clue Sudoku: solving the Sudoku minimum number of clues problem via hitting set enumeration. Exp. Math. **23**(2), 190–217 (2014)
12. Lawler, E.L., Lentra, J.K., Rinnooy, A.H.G., Shmoys, D.B. (eds.): The Traveling Salesman Problem – A Guided Tour of Combinatorial Optimization. Wiley, New York (1985)
13. Saha, S., Kumar, R.: Unifying heuristics and evolutionary computing for solving and rating Sudoku puzzles. Communicated 0 (2013)
14. Sudokuwiki: Sudoku. http://www.sudokuwiki.org/sudoku.htm
15. Mantere, T., Koljonen J.: Ant colony optimization and a hybrid genetic algorithm for Sudoku solving. In: MENDEL 2009 – 15th International Conference on Soft Computing, Brno, Czech Republic, 24–26 June, pp. 41–48 (2009)
16. Mantere T.: Improved ant colony genetic algorithm hybrid for Sudoku solving. In: Proceedings of the 2013 3rd World Congress on Information and Communication Technologies (WICT 2013), Hanoi, Vietnam, 15–18 December, pp. 276–281 (2013)
17. Mantere, T., Koljonen, J.: Sudoku page. http://lipas.uwasa.fi/~timan/sudoku/

Weak Ties and How to Find Them

Iveta Dirgová Luptáková, Marek Šimon, and Jiří Pospíchal(✉)

Faculty of Natural Sciences, University of SS. Cyril and Methodius in Trnava,
917 01 Trnava, Slovak Republic
{iveta.dirgova,marek.simon}@ucm.sk,
jiri.pospichal@gmail.com

Abstract. This paper aims to obtain a better characterization of "weak ties", edges between communities in a network. Community detection has been, for the last decade, one of the major topics in network science, with applications ranging from e-commerce in social networks or web page characterizations, up to control and engineering problems. There are many methods, which characterize, how well a network is split into communities or clusters, and a set of methods, based on a wide range set of principles, have been designed to find such communities. In a network, where nodes and edges are characterized only by their topological properties, communities are characterized only by being densely internally connected by edges, while edges between communities are scarce. Usually, the most convenient method how to find the in-between edges is to use the community detection methods to find the communities and from them the edges between them. However, the sole characterization of these edges, "weak ties", is missing. When such characterization is mentioned, typically it is based on high value of edge betweenness, i.e. number of shortest paths going through the edge. While this is mostly valid for networks characterized by a high modularity value, it is very often misleading. Search for a community detection resulting in maximum modularity value is an NP complete problem, and since the selection of weak ties characterizes the communities, their selection is an NP complete problem as well. One can only hope to find a good heuristic. In our paper, we show, to what extent the high edge betweenness characterization is misleading and we design a method, which better characterizes the "weak tie" edges.

Keywords: Weak ties · Heuristic · Networks · Edge betweenness
Community detection · Modularity

1 Introduction

In the study of networks [1, 15], small world property, scale free characteristic of vertex degrees, or community/cluster structure are among the most studied subjects. A number of partitioning, clustering or community detection methods can provide the separation of nodes of the network into (mostly studied) disjunctive subsets. The formal division of methods into partitioning, clustering or community detection is not generally agreed upon. Mostly, partitioning is understood to require split of the set of nodes into partitions of roughly the same size, and clustering often requires preceding knowledge of

R. Matoušek (Ed.): MENDEL 2017, AISC 837, pp. 16–26, 2019.
https://doi.org/10.1007/978-3-319-97888-8_2

required number of clusters. Community detection should ideally be able to find both small as well as large, densely connected subsets of nodes, with few edges in-between, while the number of the communities should naturally result from the hidden inner parameters of the community detection method.

The term "weak tie" comes from mathematical sociology [10], where nodes in a network represent people and edges interpersonal ties. Strong ties connect the cliques of friends, while the connection between two people belonging to different clusters is called a weak tie. When all other characteristics of people or their connection are stripped away, what remains is a graph (i.e. network) composed of nodes connected by edges. The weak ties are then edges between clusters/cliques/communities.

How to characterize, which division of a set of nodes into subsets characterizes better the division of network into communities? Such a question, similarly to the problem of definition of an ideal split of a network into clusters, does not have a generally accepted answer.

However, there is a most often accepted measure of quality of division of a network into communities, which is called modularity [14]. Modularity is the fraction of the edges that connect nodes within the given communities minus the expected fraction if the edges were distributed at random. Even though this measure sometimes neglects very small communities and has other negative issues, it is used in a number of practical applications.

Usage of community detection generally ranges from WWW page recommendations, detection of communities in Facebook for e-commerce, or brain analysis, up to control and engineering applications like truss structure manufacturing [5], preventing blackout in power systems [13], water distribution [9], image segmentation for finding cracks in bridges [20], or security [6, 11, 12, 18].

Once the division of the network is made, its modularity can be calculated easily. Unfortunately, the maximization of modularity, i.e. finding the best division, is an NP complete problem [3]. Therefore, a number of heuristic methods are used for community detection, based on a wide range of principles, from Edge betweenness, designed by Girvan and Newman [8], Louvain algorithm introduced by Blondel et al. [2] iteratively connecting communities from single nodes, up to Optimal by Brandes et al. [4] which maximizes modularity measure using Integer Linear Programming. The last algorithm has exponential time complexity, but in practice can easily handle up to 50 nodes, therefore we shall use it in our further computations.

Optimal selection of weak ties, i.e. in-between community edges, can be done by using Optimal [4] community detection algorithm and selecting edges between communities. This approach is however limited to smaller graphs. All other approaches are suboptimal.

If we would like to find, if an edge is likely a week tie, without using a community detection algorithm, the approach mostly suggested so far is to calculate its edge betweenness [15], which is the number of shortest paths going through the edge. This characteristic is however far from perfect. Recently, efforts in this direction have been made e.g. in [7, 16, 18]. In our paper, we aim to measure a quality of using edge betweenness as weak tie edge characteristic, and try to improve this characteristic.

2 Where Maximum Edge Betweenness Fails to Match Weak Ties

Here we try to ascertain, how good or bad is the heuristic, which urges to look for weak ties determined by the optimum modularity maximization among edges with the highest betweenness value.

As simple test cases, we used all 1253 undirected non-isomorphic graphs with up to seven vertices from the book [17]. From these graphs, 941 were connected and can be optimally split into two or more communities, according to the Optimum community detection method [4]. For all of these graphs, we evaluated their edge betweenness and created a sequence of decreasing values of their edge betweenness. Then we selected the weak ties, i.e. edges connecting nodes from different communities determined by the Optimum method, took their values of edge betweenness and calculated the average of their rank within the previously ordered decreasing edge betweenness sequence.

The results can be seen in Fig. 1, where we plotted the mean rank of weak ties against the modularity values of the processed graphs.

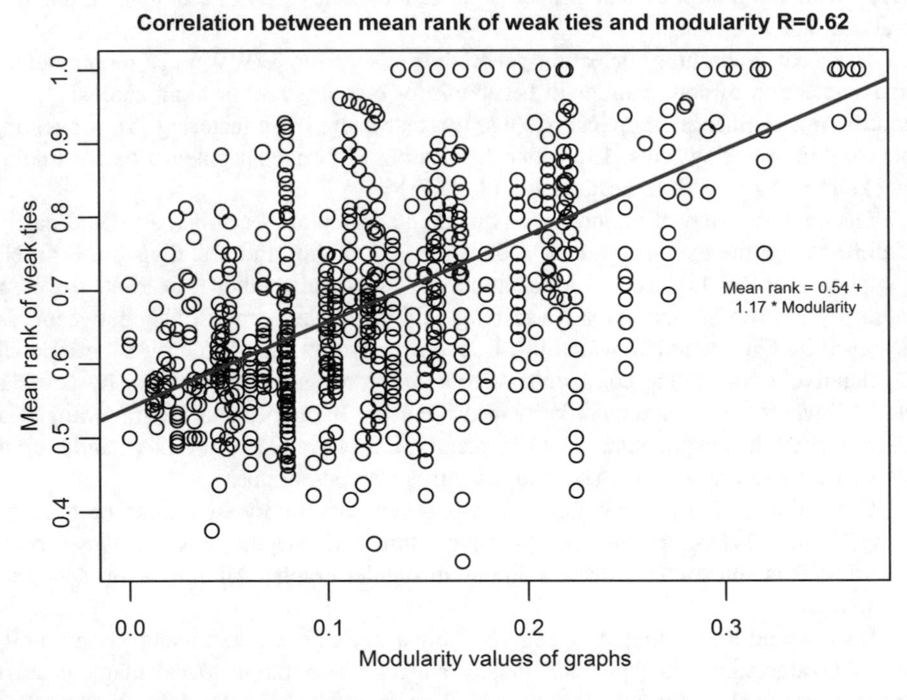

Fig. 1. Mean rank of weak ties among all edges ordered decreasingly according to their betweenness correlates with modularity value of the graph. While correlation coefficient $r = 0.62$ is good, weak ties of graphs with lower modularity values can have quite low edge betweenness.

The scatter plot can be approximated by a red linear regression line shown in the Fig. 1, defined as Eq. (1):

$$\text{Mean rank} = 0.54 + 1.17 * \text{Modularity} \tag{1}$$

with correlation coefficient $r_{xy} = 0.62$. The plot shows, that for graphs with the highest values of modularity (i.e. with dense clusters connected by a few edges) the weak ties are among the edges with the highest betweenness values. This shows that the old heuristic how to find weak ties among other edges works very well for graphs with well-defined modules. However, there are much more graphs with lower values of modularity, and for them the heuristic is inefficient. Moreover, to know the modularity value of the graph means to know the modules, and therefore weak ties. So, when for an input graph we do not know the modularity, we cannot well guess, if the high edge betweenness heuristic will work well or not.

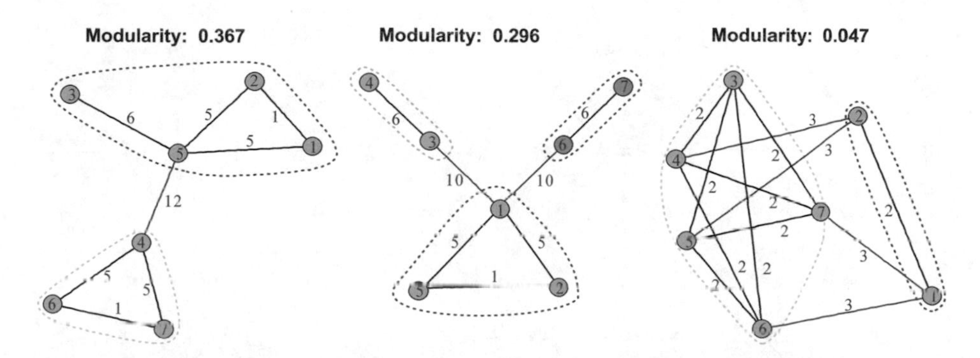

Fig. 2. Examples of graphs, where the highest values of edge betweenness (shown next to the edges) match weak ties enhanced by red color. Dotted lines show communities. The first graph has a bridge edge; the other two graphs have more of weak ties. The first graph has the highest modularity value, while the last graph has the lowest modularity value.

The Fig. 2 shows examples, where the edges with high value of betweenness exactly match the weak ties. The first two graphs exhibit quite a high modularity value, but the last graph shows, that the heuristic occasionally works for a graph with low value of modularity as well.

On the other hand, Fig. 3 shows graphs, where none of the edges with a high value of edge betweenness match the weak ties (i.e. none of the red edges have the highest edge betweenness value). Although such graphs are relatively seldom, and most of the shown graphs in the figure have a low modularity value, it is clear, that finding weak ties by high edge betweenness values occasionally fails. Mostly, a few weak ties have a high edge betweenness value, while the rest of weak ties have an average edge betweenness value.

3 How to Improve the Edge Betweenness Heuristic

A weak tie edge, which connects communities, should likely be part of long paths containing many edges. Therefore, if we delete such an edge, if it is a bridge (i.e. its deletion disconnects the network), those paths which went through the edge no longer exist and if each path is long, edge betweenness value decreases for all the edges which were part of the path. Therefore, if we sum the decreases of all the edge betweenness

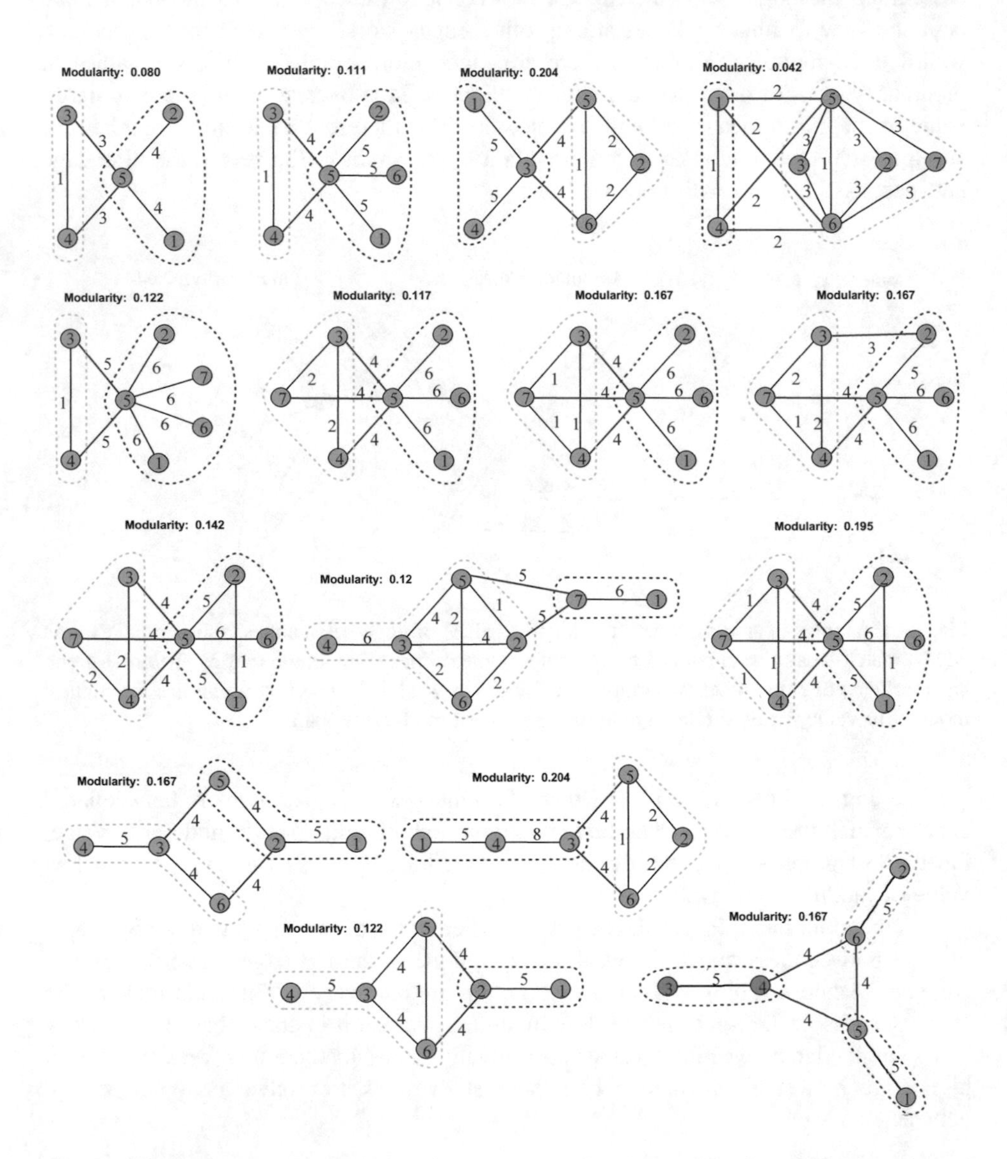

Fig. 3. All graphs up to 7 vertices, where for m weak ties, none of the first m highest values of edge betweenness (shown next to the edges) match them. As in the Fig. 2, dotted lines encompass communities and weak ties are denoted by a red edge color.

after the edge deletion, the absolute value of the sum should be for a weak tie greater than for a regular edge with the same betweenness value. On the other hand, if the weak tie edge is not a bridge, its deletion forces all the shortest paths going through this edge to be redirected through other weak ties. Therefore a deletion of a weak tie, which is not a bridge, should increase edge betweenness of other edges, where the greatest increase is likely to occur in a few of the remaining weak ties. An example of such behavior can be seen in the first graph in Fig. 3, which is in the details shown in Fig. 4.

Let us have a set $\boldsymbol{E}$ of edges of the graph A in Fig. 4 specified for couples of vertices as follows: $\boldsymbol{E} = \{a = \{1, 5\}, b = \{2, 5\}, c = \{3, 4\}, d = \{3, 5\}, e = \{4, 5\}\}$. If we calculate a difference in edge betweenness values for each edge, after we remove each edge to get subgraphs (B)–(F) from Fig. 4, we can arrive at the matrix of differences. For each edge defined at the heads of columns in the matrix of **e**dge **b**etweenness **d**ifferences ***EBD*** in Eq. (2), the values in the columns specify the change in the edge betweenness for edges denoted at the heads of rows.

EBD =

	Deleted edge				
Change in e.b. for edge	*a*	*b*	*c*	*d*	*e*
a	-4	-1	0	0	0
b	-1	-4	0	0	0
c	0	0	-1	3	3
d	-1	-1	1	-3	3
e	-1	-1	1	3	-3

(2)

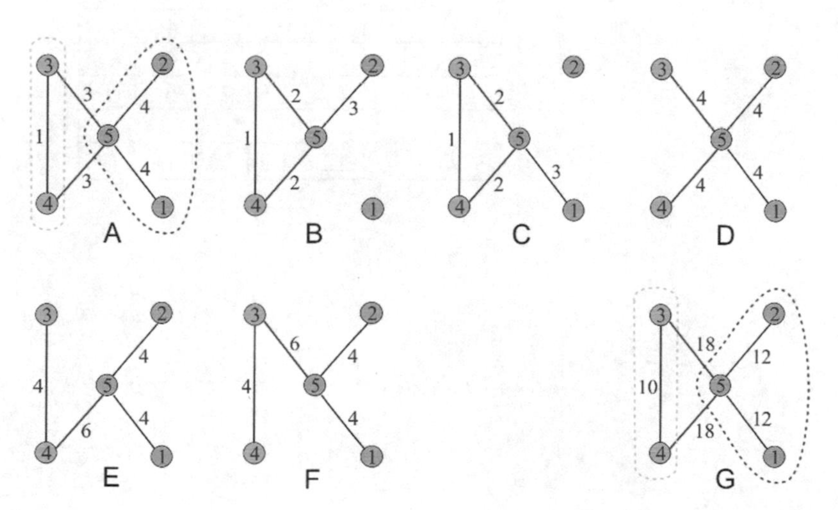

Fig. 4. (A) First of the graphs from Fig. 3, edges with the highest values of edge betweenness (shown next to the edges) do not match weak ties. (B)–(F) All the subgraphs of the graph (A) with one edge deleted, with their new edge betweenness values. (G) Graph (A) with correlated edge betweenness values, calculated by Eq. (3). Maximum values now correspond to weak ties.

The matrix ***EBD*** is not symmetric. For example, deleting edge *e* will increase edge betweenness of edge *c* by 3, but deleting edge *c* will increase edge betweenness of edge *e* only by 1. Since it would be ideal to measure the mutual influence of edges, we decided to add the transposed matrix to obtain a symmetric resulting matrix, ***EBD*** + *t*(***EBD***).

If there is a bridge edge as a weak tie (i.e. one edge separates two clusters or communities), its removal would only decrease edge betweenness values of the other edges and the values of its row/column in the matrix ***EBD*** + *t*(***EBD***) would only be negative. On the other hand, if there are more weak tie edges, removal of one of them would increase the number of shortest paths going through other weak tie edges, so the nondiagonal entries in the ***EBD*** + *t*(***EBD***) would be positive. Since we do not know in advance, if we are looking for a bridge or several weak tie edges, we decided to take the absolute values of the changes, *abs*(***EBD*** + *t*(***EBD***)). In order to find out most influential edges as far as shortest paths are concerned, we simply sum up the columns of the matrix *abs*(***EBD*** + *t*(***EBD***)) and select the edge/edges with the maximum value of the sum. Thus, we determine for an edge with index *j* its **c**orrelated **e**dge **b**etweenness ceb_j.

$$ceb_j = \sum_{i=0}^{|E|} abs(\boldsymbol{EBD} + \boldsymbol{EBD}^t)_{ij} \tag{3}$$

$abs(\boldsymbol{EBD} + \boldsymbol{EBD}^t) =$ (4)

	Deleted edge				
Change in e.b. for edge	*a*	*b*	*c*	*d*	*e*
a	8	2	0	0	0
b	2	8	0	0	0
c	0	0	2	4	4
d	1	1	4	6	6
e	1	1	4	6	6
$ceb_j = \sum$	12	12	10	14	14

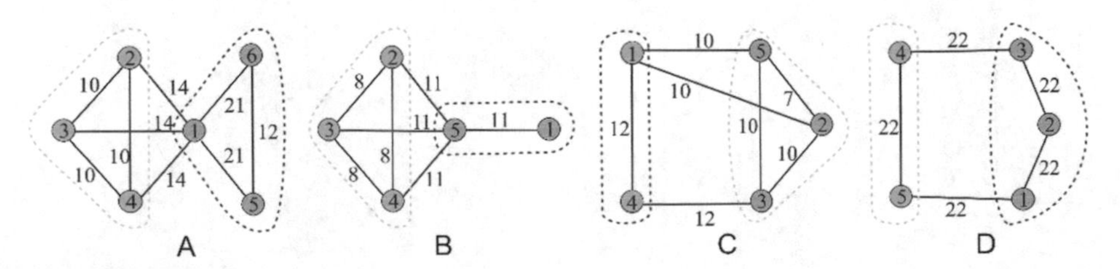

Fig. 5. Examples of failures of maximum correlated edge betweenness to match weak ties. The computations were adjusted to fix the problems of graphs A and B, the error in C remains, and the error in D is actually an artificial problem caused by our failure to take into account the automorphism of studied graphs.

The values of *ceb* from bottom row in the Eq. (4) are shown in the graph (G) in Fig. 4. It is clear, that unlike the edge betweenness values in the graph (A), the maximum values of correlated edge betweenness exactly correspond to weak tie edges.

Figure 5 shows, that even though the correlated edge betweenness values calculated by Eq. (3) provided improvement in many cases, like in the graph A from Fig. 4, where it was possible to match maximum of these values with weak ties, there remain graphs, for which this improvement failed. The graph A in Fig. 5 shows, that while the weak ties have *ceb* values 14, edges $\{1, 5\}$ and $\{1, 6\}$ has its value equal to 21. Since we observed, that similar failures often occur, where two edges are connected by a node with a degree 2, in such cases we reduced the values of their mutual correlation in the ECB matrix by half. This heuristic was supplemented by another heuristic, that the edges, where one vertex is of degree one, has its *ceb* value zeroed. This solves the case of graph B in Fig. 5 and similar other cases. The problem of graph C from Fig. 5 remains unsolved. It could be solved by determining the set of weak tie edges by Eq. (5). The equation basically states, that in a set of weak ties most couples of edges should be positively correlated (by deleting one weak tie the edge betweenness of others should increase), so a subset of edges corresponding weak ties should have the greatest sum of the corresponding submatrix of EBD, without the diagonal. However, to find such a subset, we would have to examine in the worst case $2^{|E|}$ subsets, which would increase the complexity exponentially.

$$\boldsymbol{E}_w = \underset{\boldsymbol{E}_w \subset E}{max} \sum\nolimits_{i \neq j; e_i, e_j \in \boldsymbol{E}_w}^{|\boldsymbol{E}_w|} (\boldsymbol{EBD} + \boldsymbol{EBD}^t)_{ij}, \boldsymbol{EBD}_{ij}, \boldsymbol{EBD}^t_{ij} \geq 0 \quad (5)$$

In such a case, one could use the Optimum community detection algorithm [4] instead of looking for weak ties separately. Finally, the problem with the graph D in Fig. 5 is in its isomorphism; whatever we do, the edges will be evaluated equally. Since there are 5 edges, each of them would have a rank equal to 3 and an average rank 3/5 = 0.6. There are graphs with similar problems regarding automorphism, that lower the average values of mean rank for the test set of all graphs, and there is not any way to improve those results, other than using an entirely different measure than rank.

4 Experimental Results of the Improved Correlated Edge Betweenness

The correlated edge betweenness defined in Eq. (3) together with further described improving heuristics from the preceding Sect. 3 were applied to the same set of 941 graphs as in Fig. 1. The results are shown in Fig. 6; the scatter plot can be approximated by a red linear regression line shown in the Fig. 6, defined as Eq. (6):

$$\text{Mean rank} = 0.6 + 1.24 * \text{Modularity} \quad (6)$$

the correlation coefficient r_{xy} is slightly improved from 0.62 to 0.65. The average mean rank improved more substantially. This is shown in Fig. 7 in violin plots of mean rank of weak ties among all edges. The first violin plot shows the results from Fig. 1 for

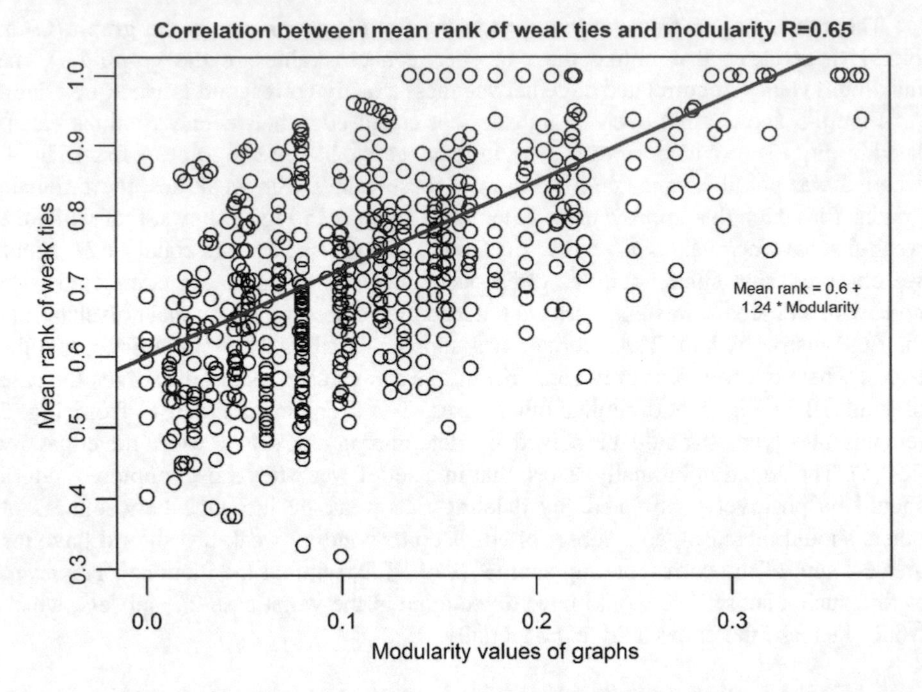

Fig. 6. Mean rank of weak ties among all edges ordered decreasingly according to their adjusted correlated edge betweenness correlates with the modularity values of the graphs. Correlation coefficient $r = 0.65$ is a slight improvement compared to Fig. 1.

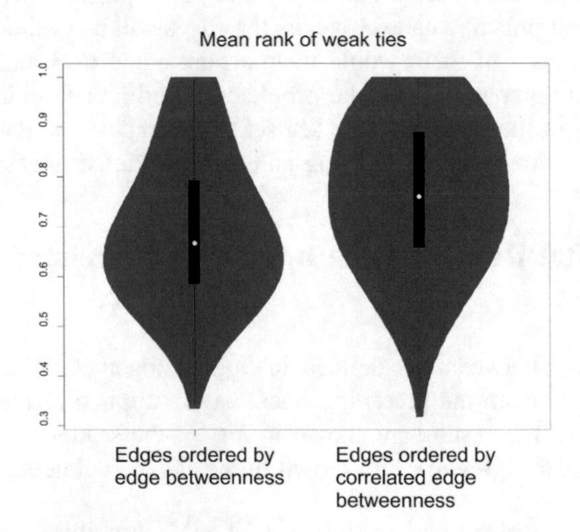

Fig. 7. Violin plots of mean rank of weak ties among all edges, first plot is calculated for edges ordered decreasingly according to edge betweenness with average value 0.695, the second plot is for the same computation for correlated edge betweenness with average value 0.764.

straight edge betweenness with average value 0.695, while the second violin plot is calculated for correlated edge betweenness with the resulting average value 0.764.

5 Conclusions

The edge betweenness heuristic for finding weak ties proved correct only for graphs or networks with a high modularity value. For networks with less clearly defined communities, this heuristic does not provide very good answers. Our correlated edge betweenness heuristic improved the rank values results by nearly ten percent. However, it comes at a cost of calculating edge betweenness of the network $|E|$ times instead of once. Since the edge betweenness calculation has a complexity $O(|V||E|)$, our improvement of the heuristic comes with a substantially increased complexity $O(|V||E|^2)$. Possible application of our improved heuristic might be a new community detection algorithm for smaller sparse networks. The question, if the computational cost is worth the improvement must be answered for each application individually.

Acknowledgement. This research was supported by a grant SK-SRB-2016-0003 of Slovak Research and Development Agency.

References

1. Barabási, A.-L., Pósfai, M.: Network Science. Cambridge University Press, Cambridge (2016). See also http://barabasi.com/networksciencebook. Accessed 7 Apr 2017
2. Blondel, V.D., Guillaume, J.-L., Lambiotte, R., Lefebvre, E.: Fast unfolding of communities in large networks. J. Stat. Mech: Theory Exp. **2008**, P10008 (2008)
3. Brandes, U., Delling, D., Gaertler, M., Görke, R., Hoefer, M., Nikoloski, Z., Wagner, D.: On modularity-NP-completeness and beyond. Faculty of Informatics, Universität Karlsruhe (TH), Technical report a 2006-19 (2006)
4. Brandes, U., Delling, D., Gaertler, M., Görke, R., Hoefer, M., Nikoloski, Z., Wagner, D.: On modularity clustering. IEEE Trans. Knowl. Data Eng. **20**(2), 172–188 (2008)
5. Cao, H., Mo, R., Wan, N., Shang, F., Li, C., Zhang, D.: A subassembly identification method for truss structures manufacturing based on community detection. Assem. Autom. **35** (3), 249–258 (2015)
6. Duijn, P.A., Klerks, P.P.: Social network analysis applied to criminal networks: recent developments in Dutch law enforcement. In: Masys, A.J. (ed.) Networks and Network Analysis for Defence and Security, pp. 121–159. Springer, Heidelberg (2014)
7. Friedman, E.J., Landsberg, A.S., Owen, J., Hsieh, W., Kam, L., Mukherjee, P.: Edge correlations in spatial networks. J. Complex Netw. **4**(1), 1–14 (2016)
8. Girvan, M., Newman, M.E.: Community structure in social and biological networks. Proc. Natl. Acad. Sci. **99**, 7821–7826 (2002)
9. Giustolisi, O., Ridolfi, L.: A novel infrastructure modularity index for the segmentation of water distribution networks. Water Resour. Res. **50**(10), 7648–7661 (2014)
10. Granovetter, M.S.: The strength of weak ties. Amer. J. Sociol. **78**(6), 1360–1380 (1973)
11. Huraj, L., Siládi, V.: Authorization through trust chains in ad hoc grids. In: Proceedings of the 2009 Euro American Conference on Telematics and Information Systems: New Opportunities to Increase Digital Citizenship, p. 13. ACM (2009)

12. Korytar, M., Gabriska, D.: Integrated security levels and analysis of their implications to the maintenance. J. Appl. Math. Stat. Inf. **10**(2), 33–42 (2014)
13. Liu, Y., Liu, T., Li, Q., Hu, X.: Power system black-start recovery subsystems partition based on improved CNM community detection algorithm. In: Shao F., Shu W., Tian, T. (eds.) Proceedings of the 2015 International Conference on Electric, Electronic and Control Engineering (ICEECE 2015), pp. 183–189. CRC Press (2015). Chapter 33
14. Newman, M.E.J.: Modularity and community structure in networks. Proc. Natl. Acad. Sci. U.S.A. **103**(23), 8577–8696 (2006)
15. Newman, M.E.J.: Networks: An Introduction. Oxford University Press, Oxford (2010)
16. Qian, Y., Li, Y., Zhang, M., Ma, G., Lu, F.: Quantifying edge significance on maintaining global connectivity. Sci. Rep. **7**, 45380 (2017)
17. Read, R.C., Wilson, R.J.: An Atlas of Graphs (Mathematics). Oxford University Press, Oxford (2005)
18. Schaub, M.T., Lehmann, J., Yaliraki, S.N., Barahona, M.: Structure of complex networks: quantifying edge-to-edge relations by failure-induced flow redistribution. Netw. Sci. **2**(1), 66–89 (2014)
19. Šimon, M., Huraj, L., Čerňanský, M.: Performance evaluations of IPTables firewall solutions under DDoS attacks. J. Appl. Math. Stat. Inf. **11**(2), 35–45 (2015)
20. Yeum, C.M., Dyke, S.J.: Vision-based automated crack detection for bridge inspection. Comput. Aided Civ. Infrastruct. Eng. **30**(10), 759–770 (2015)

Robustness and Evolutionary Dynamic Optimisation of Airport Security Schedules

Darren M. Chitty(✉), Shengxiang Yang, and Mario Gongora

Centre for Computational Intelligence (CCI),
School of Computer Science and Informatics,
De Montfort University, The Gateway, Leicester LE1 9BH, UK
darrenchitty@googlemail.com

Abstract. Reducing security lane operations whilst minimising passenger waiting times in unforseen circumstances is important for airports. Evolutionary methods can design optimised schedules but these tend to *over-fit* passenger arrival forecasts resulting in lengthy waiting times for unforeseen events. Dynamic re-optimisation can mitigate for this issue but security lane schedules are an example of a *constrained* problem due to the *human* element preventing major modifications. This paper postulates that for dynamic re-optimisation to be more effective in *constrained* circumstances consideration of schedule *robustness* is required. To reduce *over-fitting* a simple methodology for evolving more robust schedules is investigated. Random delays are introduced into forecasts of passenger arrivals to better reflect actuality and a range of these randomly perturbed forecasts are used to evaluate schedules. These steps reduced passenger waiting times for actual events for both static and dynamic policies with minimal increases in security operations.

Keywords: Airport security lane scheduling
Robust dynamic optimisation · Evolutionary algorithm

1 Introduction

Airports face pressures to reduce their operational costs whilst improving passenger experiences, a conflicting objective. A key operational cost is the manning of security lanes but reducing security lane operations has the net effect of increasing passenger waiting times and thereby dissatisfaction. Consequently, it is a problem area suited to optimisation to derive security lane schedules that reduce both waiting times and lane operations. Evolutionary methods have been frequently used to solve scheduling problems such as this. However, in uncertain environments such as passenger arrivals, there is a tendency for optimised schedules to *over-fit* predictions of future events with poor performance for unforeseen events. Real-time schedule modification can mitigate for this issue

R. Matoušek (Ed.): MENDEL 2017, AISC 837, pp. 27–39, 2019.
https://doi.org/10.1007/978-3-319-97888-8_3

and the authors of this work successfully demonstrated evolutionary dynamic re-optimisation methods to improve security lane schedules at airports [3].

However, within a *constrained* environment such as staff scheduling there are limitations to the modification of schedules restricting the effectiveness of dynamic approaches. Therefore, it is postulated that the *robustness* of original schedules must be of consideration for *constrained* environments to reduce *over-fitting* and improve the performance of both static and dynamic scheduling policies. To raise the robustness of security lane schedules two steps will be considered. Firstly, modifying the supplied forecasts of passenger arrivals by increasing arrival variability to better reflect actual events. The second step then uses this methodology to generate a range of varying forecast scenarios which candidate schedules can be tested upon to measure their robustness.

The paper is laid out as follows: Sect. 2 provides an overview of related works to airport security lane optimisation. Section 3 describes the security lane optimisaton problem and the evolutionary design of security lane schedules for both static and dynamic policies. Section 4 investigates methods to improve the robustness of schedules and contrast the performance of derived schedules with non-robust methods for unforeseen actual events. Finally, Sect. 5 draws conclusions and presents options for future work.

2 Related Work

There is limited literature associated with the optimisation of airport security lane schedules to reduce passenger waiting times. Soukour *et al.* [16] used a memetic algorithm merged with an evolutionary algorithm to assign security staff concentrating on reducing over and undertime and raising staff satisfaction. However, the security lane problem is similar to optimising airport check-in desks to minimise passenger delays and the degree to which desks are open. Wang and Chun [17] used a GA for optimal counter assignment for check-in desks. Chun and Mak [4] used simulation and search heuristics to determine the optimal check-in desk allocation that reduces the time desks are open and acceptable queue lengths for Hong Kong Airport. Bruno and Genovese [2] proposed a number of optimisation models for the check-in service balancing operational costs with passenger waiting times for Naples airport. Araujo and Repolho [1] present a new methodology to optimise the check-in desk allocation problem of maintaining a service level whilst reducing operational costs. Three phases are used whereby the first optimises the number of desks based upon [2], the second uses simulation to test the service level and the third uses an optimisation model to solve an adjacent desk constraint. Integer programming is used to solve both a common and dedicated desk problem. Mota [11] uses an evolutionary algorithm and a simulation approach to establish the allocation and opening times of check-in desks to reduce passenger waiting times.

The dynamic optimisation of check-in desks has been investigated by Parlar *et al.* [14,15] with regards the optimal opening of desks to minimise a monetary cost determined as the financial cost of waiting passengers and the cost of open check-in desks and aircraft delays solved using dynamic programming for a single flight scenario. A static policy was recommended as a dynamic policy was found to suffer from the curse of dimensionality [14]. Hsu *et al.* [8] investigated the dynamic allocation of check-in facilities and passengers to desks defined as a Sequential Stochastic Assignment Problem and solved using binary integer programming with positive results. Nandhini *et al.* [12] investigated the dynamic optimisation of check-in desks to minimise the conflicting objectives of resource allocation and passenger waiting times using a GA.

With regards robustness and evolutionary dynamic scheduling there have been some significant works within the related area of job shop scheduling. Jensen [10] considers a robustness measure for evolved schedules based on the performance of other schedules within the neighbourhood since a modified schedule will likely be one of these. The measure was found to outperform alternatives from the literature. Hart *et al.* [5] consider the use of an Artificial Immune System (AIS) to design robust dynamic job shop schedules by generating sets of schedules such that if a reschedule is required another from the set can be used. More generally for evolutionary methods Paenke *et al.* [13] use a similar approach to achieve robustness as used in this paper by averaging over a range of scenarios. However, the authors use a fitness approximation approach rather than pure simulation to reduce computational cost.

3 Dynamic Optimisation of Security Lane Schedules

3.1 The Security Lane Optimisation Problem

Passengers travelling by air are required to pass through stringent security checks such as hand baggage searches and passing through metal detectors etc. with a number of available security lanes for processing passengers. Security checks are staff intensive and cannot be compromised as maintaining security is paramount. One aspect of security that is open to optimisation is the schedules of opening these security lanes. Clearly, minimising passenger waiting times at security reduces passenger dissatisfaction. Thus, opening all security lanes will achieve this but to the expense of the airport but alternatively, closing lanes will increase passenger waiting times and hence increase dissatisfaction. Therefore, it can be considered that the problem is multi-objective in nature, minimising waiting times and minimising security operations are mutually exclusive objectives. However, passenger demand will ebb and flow and therefore the problem becomes the design of a schedule that ensures low passenger waiting times at peak times and lower security lane opening hours at times of low passenger demand. Figure 1 demonstrates the ebb and flow of passenger demand during a 24 h period at an airport for four exemplar problems with regards actual passenger flow data and a supplied generalised forecast of passenger arrivals.

3.2 Evolutionary Optimisation of Security Lane Schedules

The optimisation objective of the security lane problem is to simultaneously reduce passenger waiting times whilst also minimising the degree to which

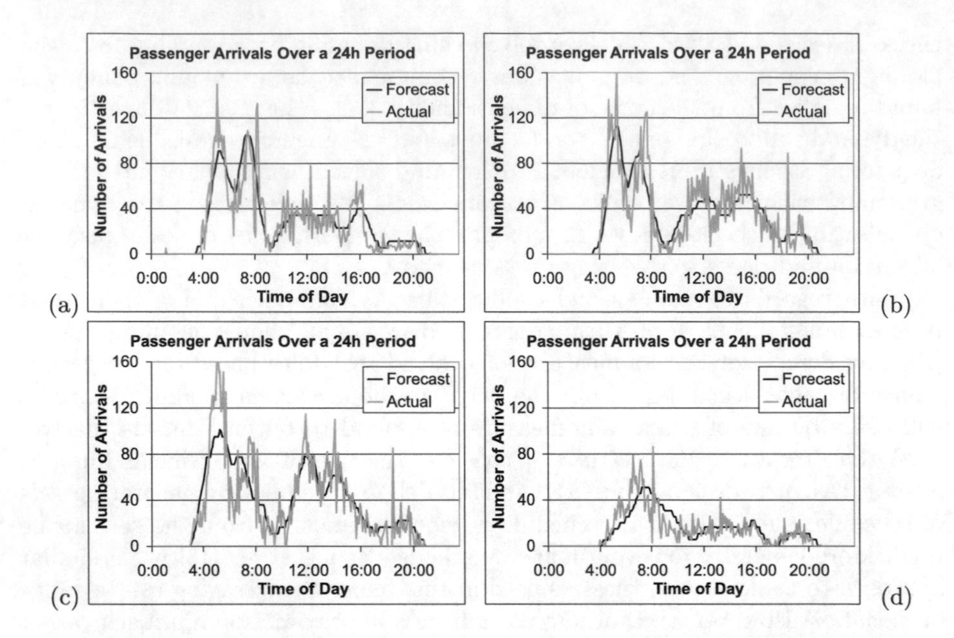

Fig. 1. The forecast arrivals of passengers at security and a set of actual passenger arrivals over a 24 h period for the exemplar problems labelled (a) F_PAXflow_2425, (b) F_PAXflow_2428, (c) F_PAXflow_2501 and (d) F_PAXflow_21113 respectively.

security lanes are open hence reducing costs. The two objectives are mutually exclusive hence the problem is multi-objective. The primary objective is to minimise passenger waiting times as defined by:

$$\text{minimise} \qquad f_1 = \max_{i \in \{1,\ldots,m\}} (W_i), \tag{1}$$

where W_i is the waiting time experienced by the ith passenger at the security queue and m is the number of passengers that arrive over the time period.

The secondary objective is to minimise the degree of time to which security lanes are open over the stated time period, defined as follows:

$$\text{minimise} \qquad f_2 = \sum_{i=1}^{i \leq n} S_i, \tag{2}$$

where S_i is the time for which the ith security lane shift lasts and n is the number of shifts within the schedule.

Essentially, the key objective is to minimise the maximum passenger waiting time experienced by a single passenger across the whole time period. Therefore, the multi-objective problem can be simplified to finding the lowest maximum waiting time experienced by a passenger with the fewest hours of security lane operation. To derive the optimal security lane operational schedule an evolutionary approach can be used by deploying a Genetic Algorithm (GA) [7] whereby a candidate schedule is represented as a set of shifts defined by a start and finish time with shifts restricted to being between two and four hours in length and each gene represents a shift. Since a set of shifts constituting a schedule can be variable in nature, a variable GA approach is used [6]. Two point crossover swaps subsets of shifts between two candidate solutions with these subsets being of differing size. Mutation consists of either swapping a subset of shifts with a random replacement set or a low probability bitwise mutation of starting and finishing times of shifts. In terms of fitness selection, a candidate schedule with a lower maximum passenger waiting time is considered the fitter. If the times are identical then the schedule with the lower degree of lane operation is considered the fitter. A simulation based approach is used to measure passenger waiting times. Passengers are simulated arriving at security defined by the passenger flow forecast and enter a queue operating in a First In First Out (FIFO) manner. Open security lanes take passengers from this queue and process them which is defined as randomly taking between 15 and 21 s per passenger. Ten simulations are used to account passenger processing variance.

However, as evidenced in Fig. 1, passenger arrivals will often not reflect the predicted forecast with bad weather or road traffic accidents causing changes to passenger arrivals. With schedules optimised to the forecast this will likely cause significant queues with security lanes not being open, these schedules essentially *over-fit* the forecast and by minimising lane opening hours there is no spare capacity. To address this issue a dynamic re-optimisaton approach can be used to improve these optimised schedules by modifying the shifts. In fact, resource managers often alter schedules to suit demand known as real-time shift updating [9]. However, there are constraints with this policy in that shifts due to their human component may only have their start time brought forward or pushed back by up to an hour and similarly for the finish time with shifts restricted to being between two and four hours in length. To dynamically modify security lane schedules a re-optimisation is performed every hour using the same aforementioned evolutionary approach. Forecast passenger flow is used for simulated future arrivals and actual passenger events are represented purely by the current passengers in the queue which could be much larger than expected. Further details of the approach can be found in [3].

Table 1. GA parameters used throughout unless otherwise stated.

Population size	100
Max generations	2,000
Tournament size	7
Crossover probability	0.9
Mutation probability	0.1
Primary fitness measure	Minimisation of max. passenger waiting time
Secondary fitness measure	Minimisation of total lane opening time

To establish the effectiveness of an evolutionary approach to the design of security lane schedules for both static and dynamic policies, experiments are

Table 2. The average maximum waiting times and the scheduled total lane opening time using actual passenger arrivals for a range of available lanes for the optimal static and the dynamically re-optimised schedules. Results averaged over 25 evolved schedules and 10 simulations with varying passenger processing times.

Problem	Max. lanes	Max. wait experienced over 24 h period (in hours)		Schedule total lane opening (in hours)	
		Static	Dynamic	Static	Dynamic
F_PAXflow_2425	4	2.70 ± 0.26	1.80 ± 0.32	38.14 ± 1.07	42.88 ± 1.55
	5	2.39 ± 0.13	1.90 ± 0.29	42.04 ± 2.06	45.97 ± 2.72
	6	2.39 ± 0.15	1.95 ± 0.26	46.74 ± 2.48	50.67 ± 2.72
	7	2.29 ± 0.45	1.77 ± 0.45	60.06 ± 2.40	65.38 ± 2.46
	8	2.33 ± 0.32	1.85 ± 0.35	64.08 ± 3.17	70.48 ± 3.68
F_PAXflow_2428	4	2.27 ± 1.17	0.80 ± 0.09	41.02 ± 0.78	44.03 ± 1.61
	5	1.04 ± 0.21	0.69 ± 0.13	44.48 ± 1.68	47.90 ± 2.06
	6	0.45 ± 0.02	0.43 ± 0.05	51.68 ± 2.34	53.80 ± 2.58
	7	0.26 ± 0.00	0.23 ± 0.03	64.10 ± 2.70	66.18 ± 2.89
	8	0.25 ± 0.01	0.22 ± 0.04	69.88 ± 2.47	72.67 ± 3.22
F_PAXflow_2501	4	3.20 ± 0.86	1.33 ± 0.44	45.12 ± 1.00	48.30 ± 1.70
	5	0.78 ± 0.08	0.73 ± 0.01	48.82 ± 1.81	52.70 ± 1.72
	6	0.43 ± 0.01	0.43 ± 0.01	65.22 ± 1.70	67.70 ± 1.87
	7	0.23 ± 0.00	0.23 ± 0.00	73.34 ± 3.23	76.70 ± 3.57
	8	0.20 ± 0.02	0.17 ± 0.02	83.34 ± 3.07	87.66 ± 3.60
F_PAXflow_21113	4	0.14 ± 0.00	0.14 ± 0.00	35.20 ± 1.45	36.87 ± 1.88
	5	0.12 ± 0.01	0.12 ± 0.01	44.64 ± 1.78	47.38 ± 2.07
	6	0.12 ± 0.02	0.12 ± 0.02	53.56 ± 2.21	56.21 ± 2.69
	7	0.12 ± 0.02	0.12 ± 0.02	60.08 ± 2.83	63.05 ± 3.35
	8	0.10 ± 0.01	0.10 ± 0.01	66.82 ± 4.66	70.91 ± 5.07

conducted for the four exemplar problems with results averaged over 25 random runs. Initial schedules are evolved using the forecast passenger flow information. These are then tested against the actual passenger flow information. Moreover, the dynamic re-optimisation of these schedules are also tested against actual passenger flow events. The parameters used by the GA are shown in Table 1.

The results in terms of maximum passenger waiting times and total security lane opening hours from evolving schedules and dynamically re-optimising them throughout the given time period are shown in Table 2. It is clear to see that there are considerable maximum passenger waiting times over several hours in length for the static schedules for actual arrival events. Dynamic re-optimisation of these schedules results in significant reductions in these maximum waiting times. This demonstrates how the static schedules have *over-fit* the forecast in terms of the pattern of lane opening hours matching projected peaks in passenger demand. Deviations from this forecast results in significant passenger delays. Clearly, the dynamic approach is highly effective in mitigating for the *over-fitting* issue but it could be considered that there is a limit to the degree to which the technique can improve passenger experiences as a result of the constrained nature of modifying schedules.

4 Evolving More Robust Security Lane Schedules

4.1 Considering a More Realistic Forecast of Passenger Arrivals

A problem with the optimisation of the airport security lane schedules can be considered that the supplied projected forecast of passenger arrivals is too smooth when compared to actual events as evidenced in Fig. 1. Moreover, from the results in Table 2 it was observed that the static schedules have unacceptably lengthy passenger queueing times most likely a result of the evolved schedules *over-fitting* this smooth forecast. Consequently, it is considered that a more realistic forecast could improve results. A simple methodology to achieve this is to take each passenger's forecast arrival time at security and add a random time penalty causing a passenger to be either later or earlier than predicted. A normal (gaussian) distribution with a mean of zero and a standard deviation of fifteen minutes is used to generate penalties. The new forecasts for each exemplar problem are shown in Fig. 2 alongside the actual passenger arrival events data whereby it can be observed that the forecasts are now more variable than previously although not mirroring actual events which is to be expected.

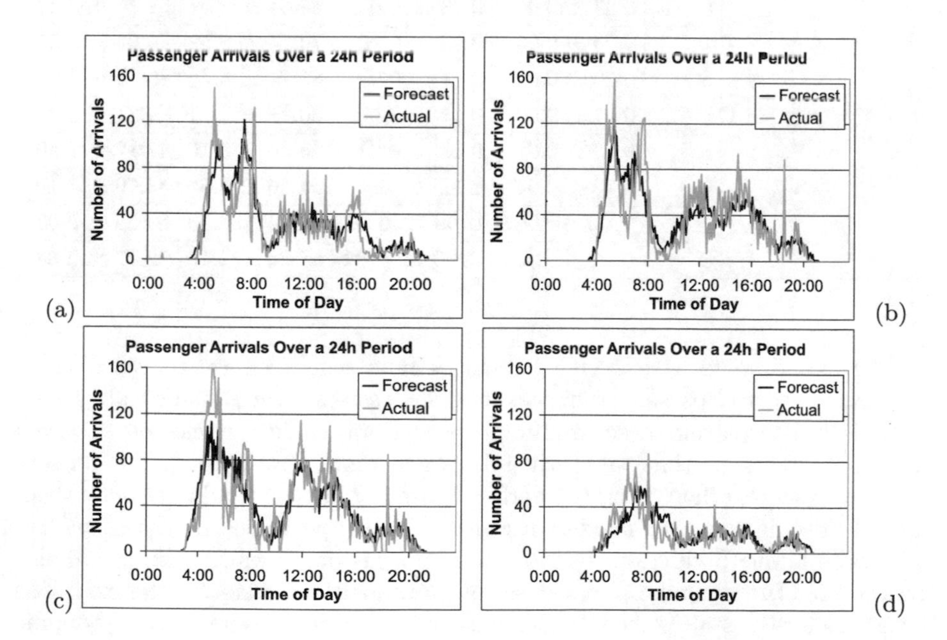

Fig. 2. Randomly perturbed forecast arrivals of passengers at security and a set of actual passenger arrivals during a given 24 h period for the exemplar problems labelled (a) F_PAXflow_2425, (b) F_PAXflow_2428, (c) F_PAXflow_2501 and (d) F_PAXflow_21113 respectively.

Table 3. The average maximum waiting times and lane opening hours using the actual passenger volumes for static and dynamically re-optimised schedules evolved using the randomly modified forecast information. Results averaged over 25 evolved schedules and 10 simulations with varying passenger processing times. Bold values indicate improvements in waiting times over those in Table 2

Problem	Max. lanes	Max. wait experienced over 24 h period (in hours)		Schedule total lane opening (in hours)	
		Static	Dynamic	Static	Dynamic
F_PAXflow_2425	4	**2.36 ± 0.18**	**1.44 ± 0.01**	38.88 ± 1.71	42.97 ± 1.80
	5	**1.90 ± 0.14**	**1.43 ± 0.01**	43.64 ± 1.60	46.18 ± 1.92
	6	**1.72 ± 0.52**	**1.29 ± 0.35**	54.28 ± 3.39	57.26 ± 3.39
	7	**1.60 ± 0.61**	**1.23 ± 0.42**	56.70 ± 2.54	60.16 ± 2.77
	8	**1.84 ± 0.25**	**1.41 ± 0.10**	66.74 ± 3.78	71.11 ± 4.08
F_PAXflow_2428	4	**1.32 ± 0.28**	0.82 ± 0.06	42.11 ± 1.27	45.48 ± 1.41
	5	**0.78 ± 0.13**	**0.60 ± 0.10**	45.98 ± 1.48	48.89 ± 1.83
	6	0.45 ± 0.02	0.43 ± 0.04	54.02 ± 2.02	56.65 ± 2.33
	7	0.26 ± 0.01	0.24 ± 0.03	64.78 ± 2.63	66.78 ± 3.13
	8	**0.22 ± 0.01**	**0.21 ± 0.02**	70.41 ± 3.34	72.94 ± 3.64
F_PAXflow_2501	4	**2.50 ± 0.73**	**1.25 ± 0.04**	46.36 ± 0.78	49.73 ± 1.34
	5	**0.75 ± 0.02**	0.73 ± 0.01	52.02 ± 1.29	55.71 ± 1.41
	6	0.43 ± 0.01	0.43 ± 0.01	65.34 ± 3.02	67.76 ± 3.44
	7	0.23 ± 0.00	0.23 ± 0.00	70.74 ± 2.52	74.04 ± 2.57
	8	0.20 ± 0.02	0.17 ± 0.02	81.74 ± 3.45	86.35 ± 4.02
F_PAXflow_21113	4	0.17 ± 0.00	0.17 ± 0.00	36.78 ± 1.61	39.31 ± 1.82
	5	0.13 ± 0.01	0.13 ± 0.01	44.88 ± 2.16	47.12 ± 2.50
	6	0.12 ± 0.01	0.12 ± 0.01	51.56 ± 2.61	53.60 ± 3.19
	7	**0.09 ± 0.01**	**0.09 ± 0.01**	62.14 ± 2.93	65.58 ± 3.07
	8	**0.09 ± 0.01**	**0.09 ± 0.01**	68.86 ± 2.15	72.77 ± 2.88

The experiments with evolving security lane schedules are repeated using this new forecast of passenger arrivals with the results from actual events shown in Table 3. Comparing these results to those from Table 2, it can be observed that in many cases the static schedules have much lower maximum passenger waiting times when using the updated forecast that is randomly perturbed. Indeed, in some cases the maximum waiting time experienced by a passenger is up to 50% lower. Reductions in the maximum passenger waiting times are also observed for the dynamic re-optimisation policy although not to the extent as those seen for the static schedules demonstrating the effectiveness of the dynamic re-optimisation for sub-optimal schedules. However, an obvious reason that passenger waiting times are improved from using the updated forecast is that there has been an increase in security lane opening hours and inspection of Table 3 and comparing to Table 2 bears this out. However, in most cases from using the updated forecast information, the increase in opening hours is relatively small. The increases though get larger as the number of available lanes increases.

4.2 Using Multiple Randomly Manipulated Predictions

The current methodology has been to use a single forecast of passenger arrivals to evolve initial static security lane schedules that minimise both passenger waiting times and security lane opening hours. A problem of these optimised schedules is that they can *over-fit* these forecasts. Evolutionary dynamic re-optimisation though has been shown to successfully mitigate for this issue to some degree. However, dynamic re-optimsation is limited in its success due to

Table 4. The average maximum waiting times and lane opening hours using the actual passenger volumes for the static and dynamically re-optimised schedules evolved using a range of randomly modified forecasts. Results averaged over 25 evolved schedules and 10 simulations with varying passenger processing times. Bold values indicate improvements in waiting times over those in Table 3

Problem	Max. lanes	Max. wait experienced over 24 h period (in hours)		Schedule total lane opening (in hours)	
		Static	Dynamic	Static	Dynamic
F_PAXflow_2425	4	2.40 ± 0.14	1.44 ± 0.01	38.77 ± 0.82	43.28 ± 1.30
	5	**1.78 ± 0.31**	**1.39 ± 0.13**	44.98 ± 1.75	47.56 ± 2.03
	6	1.79 ± 0.45	1.34 ± 0.31	59.40 ± 2.08	63.66 ± 2.51
	7	1.74 ± 0.52	1.30 ± 0.37	67.46 ± 2.28	71.93 ± 2.50
	8	1.92 ± 0.10	1.43 ± 0.00	77.76 ± 3.59	82.93 ± 4.18
F_PAXflow_2428	4	1.34 ± 0.28	**0.77 ± 0.07**	42.56 ± 1.09	46.10 ± 1.66
	5	**0.70 ± 0.12**	0.60 ± 0.09	47.36 ± 1.55	50.74 ± 1.96
	6	**0.22 ± 0.00**	**0.22 ± 0.01**	62.98 ± 1.62	65.19 ± 1.89
	7	**0.21 ± 0.02**	**0.19 ± 0.03**	74.48 ± 2.19	77.60 ± 2.67
	8	**0.21 ± 0.02**	**0.19 ± 0.03**	82.60 ± 2.19	85.88 ± 2.90
F_PAXflow_2501	4	**1.85 ± 0.28**	1.25 ± 0.01	47.02 ± 1.22	50.50 ± 1.72
	5	0.76 ± 0.02	0.73 ± 0.01	52.77 ± 1.80	56.54 ± 1.97
	6	0.42 ± 0.00	**0.42 ± 0.00**	69.32 ± 1.46	72.23 ± 1.82
	7	0.23 ± 0.00	0.23 ± 0.00	79.79 ± 1.97	83.57 ± 2.41
	8	**0.16 ± 0.02**	**0.15 ± 0.02**	91.36 ± 3.47	96.00 ± 3.98
F_PAXflow_21113	4	**0.12 ± 0.01**	**0.12 ± 0.01**	42.34 ± 1.17	43.80 ± 1.55
	5	**0.09 ± 0.00**	0.09 ± 0.00	51.36 ± 1.78	53.07 ± 2.02
	6	**0.08 ± 0.00**	**0.08 ± 0.00**	59.90 ± 2.05	62.97 ± 2.67
	7	**0.07 ± 0.00**	**0.07 ± 0.00**	68.84 ± 2.09	71.43 ± 2.86
	8	**0.07 ± 0.00**	**0.07 ± 0.00**	78.60 ± 4.28	82.25 ± 4.86

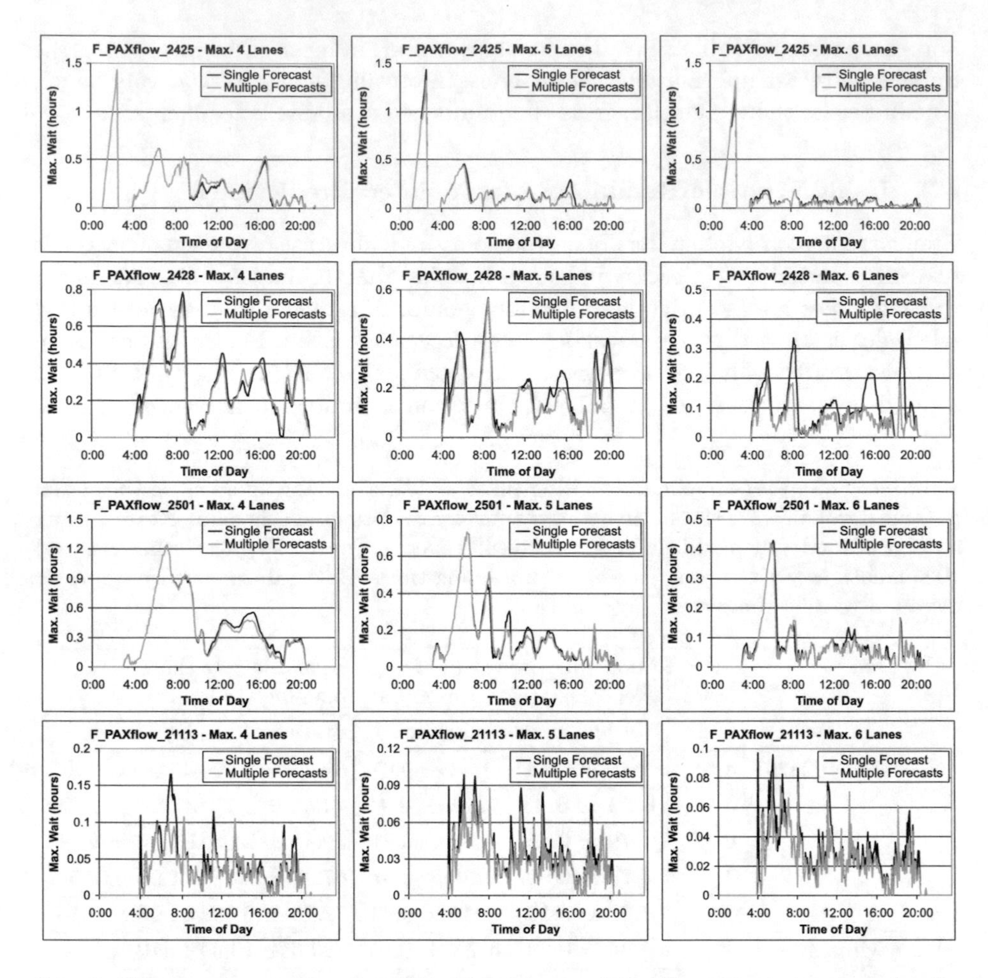

Fig. 3. The average passenger waiting time at security at five minute intervals for the dynamically re-optimised schedules using a single forecast and a range of forecasts.

the constrained nature of schedules, a schedule *unfit* for actual events cannot be greatly improved. Using a more accurate forecast that better reflects actual events has provided improved results for both static and dynamic policies.

Consequently, it is considered that further steps could be taken to reduce this *over-fitting* of a forecast of passenger arrivals by ensuring the initial schedules have a greater degree of robustness. Indeed, the approach from the previous section of generating a new forecast by applying random time penalties to projected passenger arrival times could be extended. Instead of doing this once, a range of randomly perturbed forecasts can be generated to prevent the security lane schedules matching a single prediction of passenger events. A set of ten forecasts can be generated using the approach from the previous section. Candidate schedules can then be tested against each forecast using the simulation based approach as described previously. The maximum passenger waiting time experienced is then derived as the average across these ten forecasts. This approach should derive schedules that are more robust to unexpected passenger arrivals as evolved schedules do not fit a single forecast.

The experiments from previously are repeated but now using ten differing randomly perturbed forecasts with the results shown in Table 4. By comparing these results to those from Table 3 it can be observed that the use of multiple forecasts to reduce *over-fitting* has provided generally better results in terms of the maximum passenger waiting time for both static and dynamic schedules although some increases occur especially for the F_PAXflow_2425 problem. However, these are caused by the highly unusual nature of passengers arriving very early in the day when no lanes are open which is difficult to mitigate for. A similar situation can be observed with the evolutionary dynamic re-optimisation of the static schedules. Clearly, as a result of the constraints on schedules, if the original schedule is poor then the dynamic re-optimisation approach will only be able to achieve minor improvements in passenger waiting times. Greater detail is shown in Fig. 3 for the dynamic policy in terms of average passenger waiting times throughout the time period. Although the waiting times are mixed between using a single and multiple forecasts it can be observed that the use of multiple forecasts does reduce passenger waiting times in the main. Of additional note is that in contrast to Table 3, the degree of security lane openings has increased for both static and dynamic policies in some cases over 10% although this is with greater lane availability whereby passenger waiting times are lower. Clearly though, more robust schedules will have an increase in security lane opening hours in any case. The original methodology of using a single forecast meant the evolved schedules reduced security lane operations to fit only this forecast.

5 Conclusion

This paper has considered the role of robustness in evolutionary dynamic re-optimisation of schedules in *constrained* environments. It was demonstrated that when evolving optimised schedules for airport security lane problem for minimising the conflicting objectives of waiting times and lane operations, schedules would *over-fit* forecasts and perform badly for unforeseen passenger arrival events with lengthy delays. Dynamic re-optimisation of schedules was shown to mitigate for this issue. However, modification of schedules involving human operators is an example of a *constrained* environment in that scheduled shifts cannot be changed to a great degree. Consequently, this paper hypothesised that *robustness* must be considered when optimising schedules to improve schedule performance in unforeseen circumstances. Two simple measures were used to facilitate robustness, the generation of forecasts of passenger arrivals with randomised delays to better reflect actual events and using a range of these scenarios to evaluate candidate schedules upon. Experiments demonstrated significant reductions in waiting times when introducing random delays into the forecasts of passenger arrivals for both static and dynamic policies highlighting the importance of improving forecast accuracies. Further experiments with robustness demonstrated reductions in passenger waiting times for both static and dynamic re-optimisation policies over a non-robust approach in a majority of cases with minimal increases in security lane operations. However, the degree of the reductions is much larger for static schedules rather than an evolutionary dynamic re-optimisation policy. This demonstrates the effectiveness of evolutionary dynamic re-optimisation of security lane schedules but that some consideration of robustness will benefit the approach.

Further work should consider improved methods of generating forecasts of passenger arrivals perhaps using Bayesian methods to apply penalties taking into consideration the time of day whereby delays are more likely to happen and also be lengthier such as at rush hour. Furthermore, schedule flexibility could be incorporated into the initial evolution of schedules to improve the evolutionary dynamic re-optimisation methodology.

Acknowledgment. This work was supported by the Engineering and Physical Sciences Research Council (EPSRC) of U.K. under grant EP/K001310/1.

References

1. Araujo, G.E., Repolho, H.M.: Optimizing the airport check-in counter allocation problem. J. Transp. Lit. **9**(4), 15–19 (2015)
2. Bruno, G., Genovese, A.: A mathematical model for the optimization of the airport check-in service problem. Electron. Notes Discret. Math. **36**, 703–710 (2010)
3. Chitty, D.M., Gongora, M., Yang, S.: Evolutionary dynamic optimisation of airport security lane schedules. In: 2016 IEEE Symposium Series on Computational Intelligence (SSCI), pp. 1–8, December 2016
4. Chun, H.W., Mak, R.W.T.: Intelligent resource simulation for an airport check-in counter allocation system. IEEE Trans. Syst. Man Cybern. Part C (Appl. Rev.) **29**(3), 325–335 (1999)
5. Hart, E., Ross, P., Nelson, J.: Producing robust schedules via an artificial immune system. In: Evolutionary Computation Proceedings 1998 IEEE World Congress on Computational Intelligence, pp. 464–469. IEEE (1998)
6. Harvey, I.: Species adaptation genetic algorithms: a basis for a continuing SAGA. In: Toward a Practice of Autonomous Systems: Proceedings of the First European Conference on Artificial Life, pp. 346–354 (1992)
7. Holland, J.H.: Adaptation in natural and artificial systems: an introductory analysis with applications to biology, control, and artificial intelligence. University of Michigan Press (1975)
8. Hsu, C.I., Chao, C.C., Shih, K.Y.: Dynamic allocation of check-in facilities and dynamic assignment of passengers at air terminals. Comput. Ind. Eng. **63**(2), 410–417 (2012)
9. Hur, D., Mabert, V.A., Bretthauer, K.M.: Real-time work schedule adjustment decisions: an investigation and evaluation. Prod. Oper. Manag. **13**(4), 322–339 (2004)
10. Jensen, M.T.: Generating robust and flexible job shop schedules using genetic algorithms. IEEE Trans. Evol. comput. **7**(3), 275–288 (2003)
11. Mota, M.A.M., Zuniga, C.: A simulation-evolutionary approach for the allocation of check-in desks in airport terminals. In: Proceeding of the 4th International Air Transport and Operations Symposium, ATOS 2013 (2013)

12. Nandhini, M., Palanivel, K., Oruganti, S.: Optimization of airport check-in service scheduling (2012)
13. Paenke, I., Branke, J., Jin, Y.: Efficient search for robust solutions by means of evolutionary algorithms and fitness approximation. IEEE Trans. Evol. Comput. **10**(4), 405–420 (2006)
14. Parlar, M., Rodrigues, B., Sharafali, M.: On the allocation of exclusive-use counters for airport check-in queues: static vs. dynamic policies. Opsearch **50**(3), 433–453 (2013)
15. Parlar, M., Sharafali, M.: Dynamic allocation of airline check-in counters: a queueing optimization approach. Manag. Sci. **54**(8), 1410–1424 (2008)
16. Soukour, A.A., Devendeville, L., Lucet, C., Moukrim, A.: A memetic algorithm for staff scheduling problem in airport security service. Expert Syst. Appl. **40**(18), 7504–7512 (2013)
17. Wang, B., Chun, H.: Check-in counter allocation using genetic algorithm. In: Proceeding of the 7th Conference on Artificial Intelligence and Expert Systems Applications (1995)

Enhanced Archive for SHADE

Adam Viktorin(✉), Roman Senkerik, Michal Pluhacek, and Tomas Kadavy

Faculty of Applied Informatics, Tomas Bata University in Zlin, T. G. Masaryka 5555, 760 01 Zlin, Czech Republic
{aviktorin, senkerik, pluhacek, kadavy}@fai.utb.cz

Abstract. This research paper analyses an external archive of inferior solutions used in Success-History based Adaptive Differential Evolution (SHADE) and its variant with a linear decrease in population size L-SHADE. A novel implementation of an archive is proposed and compared to the original one on CEC2015 benchmark set of test functions for two distinctive dimensionality settings. The proposed archive implementation is referred to as Enhanced Archive (EA) and therefore two Differential Evolution (DE) variants are titled EA-SHADE and EA-L-SHADE. The results on CEC2015 benchmark set are analyzed and discussed.

Keywords: Differential Evolution · External archive · JADE · SHADE L-SHADE

1 Introduction

Differential Evolution (DE) is, for more than two decades, one of the best performing optimization algorithms for numerical optimization. The algorithm was proposed by Storn and Price in 1995 [1] and since then, it is thoroughly studied and further developed by researchers all over the world. The progress of DE development and its variants was adroitly summarized in [2, 3] and most recently in updated survey [4].

The canonical version of DE by Storn and Price relies on three control parameters – population size NP, scaling factor F and crossover rate CR. Other adjustable parameters are stopping criterion (e.g. a total number of generations, a total number of objective function evaluations, precision, variance in the population, etc.), mutation strategy (for canonical DE usually “rand/1”) and crossover (for canonical DE usually “binary”). The setting of the three control parameters has a significant impact on the performance of the algorithm and therefore leads to time-consuming fine-tuning. Therefore, a new branch of adaptive DE algorithms emerged, where the tuning of control parameters is done during evolution process and relies on predefined mechanisms. The main goal of these algorithms is to achieve robustness and comparable performance without exhausting fine-tuning. Examples of such algorithms are jDE [5], SDE [6], SaDE [7], MDE_pBX [8] and JADE [9]. The JADE algorithm was proposed in 2009 and along with the adaptation mechanisms and a new mutation strategy “current-to-*p*best/1”, it also implemented an external archive of inferior solutions A. This archive stores individuals from previous generations, which were replaced by their better products.

R. Matoušek (Ed.): MENDEL 2017, AISC 837, pp. 40–55, 2019.
https://doi.org/10.1007/978-3-319-97888-8_4

Individuals from an archive are used in approximately 50% of mutations and serve to maintain diversity and to help to overcome the premature convergence.

JADE algorithm created a foundation for more recent adaptive DE variant, Success-History based Adaptive Differential Evolution (SHADE), which was proposed by Tanabe and Fukunaga in 2013 [10]. SHADE uses the same implementation of an archive as JADE and also the same mutation strategy, but proposes new historical memories for F and CR values and different adaptation strategy for these parameters.

The SHADE algorithm was one of the competitors in CEC2013 competition on real parameter single-objective optimization [11], where it was ranked 3rd and the next year, its variant with a linear decrease in population size L-SHADE [12] was ranked 1st in CEC2014 competition [13]. Therefore, these two variants (SHADE and L-SHADE) were used in this paper as representatives of state of art DE variants with an external archive and also to test the influence of population reduction to the archive use.

This paper proposes a new implementation of an archive – Enhanced Archive (EA). The basic idea of EA is that the discarding of possibly beneficial trial individuals in selection should be addressed by an archive, which would hold the best of these individuals and leave them for future use in a mutation step. The original implementation and EA were both tested on CEC2015 benchmark set [14] and the beneficial use of an archive was analyzed. Also, obtained values of benchmark test functions are compared for pairs of SHADE, EA-SHADE and L-SHADE, EA-L-SHADE algorithms.

2 Differential Evolution

The DE algorithm is initialized with a random population of individuals $\boldsymbol{P}$, that represent solutions of the optimization problem. The population size NP is set by the user along with other control parameters – scaling factor F and crossover rate CR.

In continuous optimization, each individual is composed of a vector $\boldsymbol{x}$ of length D, which is a dimensionality (number of optimized attributes) of the problem, and each vector component represents a value of the corresponding attribute, and of objective function value $f(\boldsymbol{x})$.

For each individual in a population, three mutually different individuals are selected for mutation of vectors and resulting mutated vector $\boldsymbol{v}$ is combined with the original vector $\boldsymbol{x}$ in crossover step. The objective function value $f(\boldsymbol{u})$ of the resulting trial vector $\boldsymbol{u}$ is evaluated and compared to that of the original individual. When the quality (objective function value) of the trial individual is better, it is placed into the next generation, otherwise, the original individual is placed there. This step is called selection. The process is repeated until the stopping criterion is met (e.g. the maximum number of objective function evaluations, the maximum number of generations, the low bound for diversity between objective function values in population).

The following sections describe four steps of DE: Initialization, mutation, crossover and selection.

2.1 Initialization

As aforementioned, the initial population $\boldsymbol{P}$, of size NP, of individuals is randomly generated. For this purpose, the individual vector $\boldsymbol{x}_i$ components are generated by Random Number Generator (RNG) with uniform distribution from the range which is specified for the problem by *lower* and *upper* bound (1).

$$x_{j,i} = U\left[lower_j, upper_j\right] \text{ for } j = 1, \ldots, D \tag{1}$$

where i is the index of a current individual, j is the index of current attribute and D is the dimensionality of the problem.

In the initialization phase, a scaling factor value F and crossover value CR has to be assigned as well. The typical range for F value is [0, 2] and for CR, it is [0, 1].

2.2 Mutation

In the mutation step, three mutually different individuals $\boldsymbol{x}_{r1}$, $\boldsymbol{x}_{r2}$, $\boldsymbol{x}_{r3}$ from a population are randomly selected and combined in mutation according to the mutation strategy. The original mutation strategy of canonical DE is “rand/1” and is depicted in (2).

$$\boldsymbol{v}_i = \boldsymbol{x}_{r1} + F(\boldsymbol{x}_{r2} - \boldsymbol{x}_{r3}) \tag{2}$$

where $r1 \neq r2 \neq r3 \neq i$, F is the scaling factor value and $\boldsymbol{v}_i$ is the resulting mutated vector.

2.3 Crossover

In the crossover step, mutated vector $\boldsymbol{v}_i$ is combined with the original vector $\boldsymbol{x}_i$ and produces trial vector $\boldsymbol{u}_i$. The binary crossover (3) is used in canonical DE.

$$u_{j,i} = \begin{cases} v_{j,i} & \text{if } U[0,\ 1] \leq CR \text{ or } j = j_{rand} \\ x_{j,i} & \text{otherwise} \end{cases} \tag{3}$$

where CR is the used crossover rate value and j_{rand} is an index of an attribute that has to be from the mutated vector $\boldsymbol{v}_i$ (ensures generation of a vector with at least one new component).

2.4 Selection

The selection step ensures, that the optimization progress will lead to better solutions because it allows only individuals of better or at least equal objective function value to proceed into next generation $G + 1$ (4).

$$\boldsymbol{x}_{i,G+1} = \begin{cases} \boldsymbol{u}_{i,G} & \text{if } f(\boldsymbol{u}_{i,G}) \leq f(\boldsymbol{x}_{i,G}) \\ \boldsymbol{x}_{i,G} & \text{otherwise} \end{cases} \tag{4}$$

where G is the index of current generation.

The whole DE algorithm is depicted in pseudo-code below.

Algorithm pseudo-code 1: DE

```
1.  Set NP, CR, F and stopping criterion;
2.  G = 0, x_best = {};
3.  Randomly initialize (1) population P = (x_1,G,…,x_NP,G);
4.  P_new = {}, x_best = best from population P;
5.  while stopping criterion not met
6.    for i = 1 to NP do
7.      x_i,G = P[i];
8.      v_i,G by mutation (2);
9.      u_i,G by crossover (3);
10.     if f(u_i,G) < f(x_i,G) then
11.       x_i,G+1 = u_i,G;
12.     else
13.       x_i,G+1 = x_i,G;
14.     end
15.     x_i,G+1 → P_new;
16.   end
17.   P = P_new, P_new = {}, x_best = best from population P;
18. end
19. return x_best as the best found solution
```

3 SHADE, L-SHADE and Enhanced Archive

In SHADE, the only control parameter that can be set by the user is population size *NP*, other two (*F*, *CR*) are adapted to the given optimization task, a new parameter *H* is introduced, which determines the size of *F* and *CR* value memories. The initialization step of the SHADE is, therefore, similar to DE. Mutation, however, is completely different because of the used strategy "current-to-*p*best/1" and the fact, that it uses different scaling factor value F_i for each individual. Mutation strategy also works with a new feature – external archive of inferior solutions. This archive holds individuals from previous generations, that were outperformed in selection step. The size of the archive retains the same size as the size of the population by randomly discarding its contents whenever the size overflows *NP*.

Crossover is still binary, but similarly to the mutation and scaling factor values, crossover rate value CR_i is also different for each individual.

The selection step is the same and therefore following sections describe only different aspects of initialization, mutation and crossover. The last section is devoted to the proposed EA and its built into the SHADE algorithm.

3.1 Initialization

As aforementioned, initial population $\boldsymbol{P}$ is randomly generated as in DE, but additional memories for F and CR values are initialized as well. Both memories have the same size H and are equally initialized, the memory for CR values is titled $\boldsymbol{M}_{CR}$ and the memory for F is titled $\boldsymbol{M}_F$. Their initialization is depicted in (5).

$$M_{CR,i} = M_{F,i} = 0.5 \text{ for } i = 1, \ldots, H \tag{5}$$

Also, the external archive of inferior solutions $\boldsymbol{A}$ is initialized. Since there are no solutions so far, it is initialized empty $\boldsymbol{A} = \emptyset$ and its maximum size is set to NP.

3.2 Mutation

Mutation strategy "current-to-*p*best/1" was introduced in and unlike "rand/1", it combines four mutually different vectors, therefore $pbest \neq r1 \neq r2 \neq i$ (6).

$$\boldsymbol{v}_i = \boldsymbol{x}_i + F_i\left(\boldsymbol{x}_{pbest} - \boldsymbol{x}_i\right) + F_i(\boldsymbol{x}_{r1} - \boldsymbol{x}_{r2}) \tag{6}$$

where $\boldsymbol{x}_{pbest}$ is randomly selected from the best $NP \times p$ best individuals in the current population. The p value is randomly generated for each mutation by RNG with uniform distribution from the range $[p_{min}, 0.2]$. Where $p_{min} = 2/NP$. Vector $\boldsymbol{x}_{r1}$ is randomly selected from the current population and vector $\boldsymbol{x}_{r2}$ is randomly selected from the union of current population $\boldsymbol{P}$ and archive $\boldsymbol{A}$. The scaling factor value F_i is given by (7).

$$F_i = C\left[M_{F,r}, 0.1\right] \tag{7}$$

where $M_{F,r}$ is a randomly selected value (by index r) from $\boldsymbol{M}_F$ memory and C stands for Cauchy distribution, therefore the F_i value is generated from the Cauchy distribution with location parameter value $M_{F,r}$ and scale parameter value 0.1. If the generated value $F_i > 1$, it is truncated to 1 and if it is $F_i \leq 0$, it is generated again by (7).

3.3 Crossover

Crossover is the same as in (3), but the CR value is changed to CR_i, which is generated separately for each individual (8). The value is generated from the Gaussian distribution with mean parameter value of $M_{CR,r}$, which is randomly selected (by the same index r as in mutation) from $\boldsymbol{M}_{CR}$ memory and standard deviation value of 0.1.

$$CR_i = N\left[M_{CR,r}, 0.1\right] \tag{8}$$

3.4 Historical Memory Updates

Historical memories $\boldsymbol{M}_F$ and $\boldsymbol{M}_{CR}$ are initialized according to (5), but its components change during the evolution. These memories serve to hold successful values of F and CR used in mutation and crossover steps. Successful in terms of producing trial individual better than the original individual. During one generation, these successful values are stored in corresponding arrays $\boldsymbol{S}_F$ and $\boldsymbol{S}_{CR}$. After each generation, one cell of $\boldsymbol{M}_F$ and $\boldsymbol{M}_{CR}$ memories is updated. This cell is given by the index k, which starts at 1 and increases by 1 after each generation. When it overflows the size limit of memories H, it is again set to 1. The new value of k-th cell for $\boldsymbol{M}_F$ is calculated by (9) and for $\boldsymbol{M}_{CR}$ by (10).

$$M_{F,k} = \begin{cases} \text{mean}_{WL}(\boldsymbol{S}_F) & \text{if } \boldsymbol{S}_F \neq \emptyset \\ M_{F,k} & \text{otherwise} \end{cases} \tag{9}$$

$$M_{CR,k} = \begin{cases} \text{mean}_{WA}(\boldsymbol{S}_{CR}) & \text{if } \boldsymbol{S}_{CR} \neq \emptyset \\ M_{CR,k} & \text{otherwise} \end{cases} \tag{10}$$

where $\text{mean}_{WL}()$ and $\text{mean}_{WA}()$ are weighted Lehmer (11) and weighted arithmetic (12) means correspondingly.

$$\text{mean}_{WL}(\boldsymbol{S}_F) = \frac{\sum_{k=1}^{|\boldsymbol{S}_F|} w_k \cdot S_{F,k}^2}{\sum_{k=1}^{|\boldsymbol{S}_F|} w_k \cdot S_{F,k}} \tag{11}$$

$$\text{mean}_{WA}(\boldsymbol{S}_{CR}) = \sum_{k=1}^{|\boldsymbol{S}_{CR}|} w_k \cdot S_{CR,k} \tag{12}$$

where the weight vector $\boldsymbol{w}$ is given by (13) and is based on the improvement in objective function value between trial and original individuals.

$$w_k = \frac{\text{abs}\left(f\left(\boldsymbol{u}_{k,G}\right) - f\left(\boldsymbol{x}_{k,G}\right)\right)}{\sum_{m=1}^{|\boldsymbol{S}_{CR}|} \text{abs}\left(f\left(\boldsymbol{u}_{m,G}\right) - f\left(\boldsymbol{x}_{m,G}\right)\right)} \tag{13}$$

And since both arrays $\boldsymbol{S}_F$ and $\boldsymbol{S}_{CR}$ have the same size, it is arbitrary which size will be used for the upper boundary for m in (13). Complete SHADE algorithm is depicted in pseudo-code below.

```
Algorithm pseudo-code 2: SHADE
1.  Set NP, H and stopping criterion;
2.  G = 0, x_best = {}, k = 1, p_min = 2/NP, A = Ø;
3.  Randomly initialize (1) population P = (x_1,G,…,x_NP,G);
4.  Set M_F and M_CR according to (5);
5.  P_new = {}, x_best = best from population P;
6.  while stopping criterion not met
7.    S_F = Ø, S_CR = Ø;
8.    for i = 1 to NP do
9.      x_i,G = P[i];
10.     r = U[1, H], p_i = U[p_min, 0.2];
11.     Set F_i by (7) and CR_i by (8);
12.     v_i,G by mutation (6);
13.     u_i,G by crossover (3);
14.     if f(u_i,G) < f(x_i,G) then
15.       x_i,G+1 = u_i,G, x_i,G → A, F_i → S_F, CR_i → S_CR;
16.     else
17.       x_i,G+1 = x_i,G;
18.     end
19.     if |A|>NP then randomly delete an ind. from A;
20.     x_i,G+1 → P_new;
21.   end
22.   if S_F ≠ Ø and S_CR ≠ Ø then
23.     Update M_F,k (9) and M_CR,k (10), k++;
24.     if k > H then k = 1, end;
25.   end
26.   P = P_new, P_new = {}, x_best = best from population P;
27. end
28. return x_best as the best found solution
```

3.5 Linear Decrease in Population Size

Linear decrease in population size was introduced to SHADE in [12] to improve the performance. The basic idea is to reduce the population size to promote exploitation in later phases of the evolution. Therefore, a new formula to estimate the population size was formed (14) and is calculated after each generation. Whenever the new population size NP_{new} is smaller than current population size NP, the population is sorted according to the objective function value and the worst $NP - NP_{new}$ individuals are discarded. The size of an external archive is reduced as well.

$$NP_{new} = \text{round}\left(NP_{init} - \frac{FES}{MAXFES} * \left(NP_{init} - NP_f\right)\right) \tag{14}$$

where NP_{init} is the initial population size, NP_f is the end population size. *FES* and *MAXFES* are objective function number evaluations and a maximum number of objective function evaluations respectively. The pseudo-code of the L-SHADE algorithm is depicted below.

Algorithm pseudo-code 3: L-SHADE

1. Set NP, H and stopping criterion;
2. NP = NP_{init}, $G = 0$, $x_{best} = \{\}$, $k = 1$, $p_{min} = 2/NP$, $\mathbf{A} = \emptyset$;
3. Randomly initialize (1) population $\mathbf{P} = (\mathbf{x}_{1,G}, \ldots, \mathbf{x}_{NP,G})$;
4. Set $\mathbf{M}_F$ *and* $\mathbf{M}_{CR}$ according to (5);
5. $\mathbf{P}_{new} = \{\}$, $\mathbf{x}_{best}$ = best from population $\mathbf{P}$;
6. **while** stopping criterion not met
7. $\mathbf{S}_F = \emptyset$, $\mathbf{S}_{CR} = \emptyset$;
8.   **for** $i = 1$ to NP **do**
9. $\mathbf{x}_{i,G} = \mathbf{P}[i]$;
10. $r = U[1, H]$, $p_i = U[p_{min}, 0.2]$;
11.   Set F_i by (7) and CR_i by (8);
12. $\mathbf{v}_{i,G}$ by mutation (6);
13. $\mathbf{u}_{i,G}$ by crossover (3);
14.   **if** $f(\mathbf{u}_{i,G}) < f(\mathbf{x}_{i,G})$ **then**
15. $\mathbf{x}_{i,G+1} = \mathbf{u}_{i,G}$, $\mathbf{x}_{i,G} \rightarrow \mathbf{A}$, $F_i \rightarrow \mathbf{S}_F$, $CR_i \rightarrow \mathbf{S}_{CR}$;
16.   **else**
17. $\mathbf{x}_{i,G+1} = \mathbf{x}_{i,G}$;
18.   **end**
19.   **if** $|\mathbf{A}|>NP$ **then** randomly delete an ind. from $\mathbf{A}$;
20. $\mathbf{x}_{i,G+1} \rightarrow \mathbf{P}_{new}$;
21.   **end**
22.   Calculate NP_{new} according to (14);
23.   **if** $NP_{new} < NP$ **then**
24.   Sort individuals in $\mathbf{P}$ according to their objective function value and remove $NP - NP_{new}$ worst ones;
25. $NP = NP_{new}$;
26.   **end**
27.   **if** $|\mathbf{A}| > NP$ **then** delete $|\mathbf{A}| - NP$ random individuals from $\mathbf{A}$;
28.   **if** $\mathbf{S}_F \neq \emptyset$ **and** $\mathbf{S}_{CR} \neq \emptyset$ **then**
29.   Update $\mathbf{M}_{F,k}$ (9) and $\mathbf{M}_{CR,k}$ (10), k++;
30.   **if** $k > H$ **then** $k = 1$, **end**;
31.   **end**
32. $\mathbf{P} = \mathbf{P}_{new}$, $\mathbf{P}_{new} = \{\}$, $\mathbf{x}_{best}$ = best from population $\mathbf{P}$;
33. **end**
34. **return** $\mathbf{x}_{best}$ as the best found solution

3.6 Enhanced Archive

The original archive of inferior solutions $\boldsymbol{A}$ in SHADE and L-SHADE is filled during the elitism step and contains the original individuals, that had worse objective function values than trial individuals produced from them. The maximum size of an archive is NP and wherever it overflows, a random individual is removed. Because individuals from an archive are used in at least 50% (depends on the implementation) of mutations, it is interesting to analyze whether or not these individuals help to improve the performance of an algorithm and whether this particular implementation of an archive is optimal.

During the preliminary testing, results shown, that the frequency of successful uses of individuals from an archive is quite low and that there might be a possibility of implementing more beneficial archive solution. Therefore, the EA was proposed as an alternative which was inspired from the field of discrete optimization. The basic idea is, that trial individuals unsuccessful in elitism should not be discarded as bad solutions, but the best of them should be stored in an archive in order to provide promising search directions. This is done by placing only unsuccessful trial individuals in an archive and when it overflows the maximum size, the worst individual (in terms of objective function value) is discarded. Thus, the original SHADE and L-SHADE algorithms were partially changed to implement this novel archive and the changes in pseudo-code are depicted below.

Algorithm pseudo-code 4: Changes in SHADE

14. **if** $f(\boldsymbol{u}_{i,G}) < f(\boldsymbol{x}_{i,G})$ **then**
15. $\boldsymbol{x}_{i,G+1} = \boldsymbol{u}_{i,G}$, ~~$\boldsymbol{x}_{i,G} \rightarrow \boldsymbol{A}$~~, $F_i \rightarrow \boldsymbol{S}_F$, $CR_i \rightarrow \boldsymbol{S}_{CR}$;
16. **else**
17. $\boldsymbol{x}_{i,G+1} = \boldsymbol{x}_{i,G}$, $\boldsymbol{u}_{i,G} \rightarrow \boldsymbol{A}$;
18. **end**
19. **if** $|\boldsymbol{A}|>NP$ **then** delete the worst individual from $\boldsymbol{A}$;
20. $\boldsymbol{x}_{i,G+1} \rightarrow \boldsymbol{P}_{new}$;

Algorithm pseudo-code 5: Changes in L-SHADE

14. **if** $f(\boldsymbol{u}_{i,G}) < f(\boldsymbol{x}_{i,G})$ **then**
15. $\boldsymbol{x}_{i,G+1} = \boldsymbol{u}_{i,G}$, ~~$\boldsymbol{x}_{i,G} \rightarrow \boldsymbol{A}$~~, $F_i \rightarrow \boldsymbol{S}_F$, $CR_i \rightarrow \boldsymbol{S}_{CR}$;
16. **else**
17. $\boldsymbol{x}_{i,G+1} = \boldsymbol{x}_{i,G}$, $\boldsymbol{u}_{i,G} \rightarrow \boldsymbol{A}$;
18. **end**
19. **if** $|\boldsymbol{A}|>NP$ **then** delete the worst individual from $\boldsymbol{A}$;
20. $\boldsymbol{x}_{i,G+1} \rightarrow \boldsymbol{P}_{new}$;
21. **end**
22. Calculate NP_{new} according to (14);
23. **if** $NP_{new} < NP$ **then**
24. Sort individuals in $\boldsymbol{P}$ according to their objective function value and remove $NP - NP_{new}$ worst ones;
25. $NP = NP_{new}$;
26. **end**
27. **if** $|\boldsymbol{A}| > NP$ **then** delete $|\boldsymbol{A}| - NP$ worst individuals from $\boldsymbol{A}$;

4 Experimental Setting

All experiments were in accordance with CEC2015 benchmark test set, therefore, 51 independent runes were performed on 15 test functions with the stopping criterion of a maximum number of objective function evaluations set to $10{,}000 \times D$, where D is the dimension setting. Both pairs of algorithms (SHADE, EA-SHADE and L-SHADE, EA-L-SHADE) were tested in $10D$ and $30D$. In order to analyze archive use, the total number of usages of an archive was recorded as well as the total number of successful archive usages (mutation with archive individual produced trial individual that succeeded in elitism) for each algorithm and run.

5 Results

In order to find out whether there is a significant difference in performance of the algorithm with the original external archive and the algorithm with EA, Tables 1, 2, 3, and 4 were created. Each table contains median and mean values of both algorithms and the last column depicts the significant differences evaluated by Wilcoxon rank sum test with the significance level of 5%. When the algorithm with original archive provides significantly better results, "–" sign is used, when the algorithm with EA performs significantly better, "+" sign is used, and when there is no significant difference in the performance, "=" sign is used.

Table 1. SHADE vs. EA-SHADE results on CEC2015 in 10D.

10D	SHADE		EA-SHADE		
f	Median	Mean	Median	Mean	Diff.
1	0.00E+00	0.00E+00	0.00E+00	0.00E+00	–
2	0.00E+00	0.00E+00	0.00E+00	0.00E+00	=
3	2.00E+01	1.72E+01	2.00E+01	1.84E+01	=
4	3.11E+00	3.24E+00	3.16E+00	3.22E+00	=
5	3.35E+01	6.02E+01	2.69E+01	5.47E+01	=
6	4.16E−01	3.30E+00	4.16E−01	4.58E−01	+
7	2.18E−01	2.49E−01	1.98E−01	2.05E−01	+
8	3.15E−01	2.94E−01	3.18E-01	2.85E−01	=
9	1.00E+02	1.00E+02	1.00E+02	1.00E+02	=
10	2.17E+02	2.17E+02	2.17E+02	2.17E+02	=
11	2.94E+00	1.07E+02	3.95E+00	1.25E+02	–
12	1.02E+02	1.02E+02	1.02E+02	1.02E+02	=
13	2.77E+01	2.75E+01	2.81E+01	2.81E+01	–
14	6.68E+03	4.92E+03	2.94E+03	4.50E+03	=
15	1.00E+02	1.00E+02	1.00E+02	1.00E+02	=

Table 2. SHADE vs. EA-SHADE results on CEC2015 in 30D.

30D	SHADE		EA-SHADE		
f	Median	Mean	Median	Mean	Diff.
1	1.01E−01	2.84E+00	2.15E+00	1.75E+01	−
2	0.00E+00	0.00E+00	0.00E+00	0.00E+00	=
3	2.02E+01	2.02E+01	2.02E+01	2.02E+01	−
4	2.34E+01	2.35E+01	2.33E+01	2.39E+01	=
5	1.68E+03	1.64E+03	1.69E+03	1.67E+03	=
6	7.34E+02	7.61E+02	6.88E+02	7.21E+02	=
7	7.40E+00	7.45E+00	7.27E+00	7.30E+00	=
8	1.65E+02	1.62E+02	9.12E+01	1.05E+02	+
9	1.03E+02	1.03E+02	1.03E+02	1.03E+02	−
10	6.65E+02	6.77E+02	7.93E+02	7.45E+02	=
11	4.15E+02	4.29E+02	4.05E+02	4.16E+02	=
12	1.05E+02	1.05E+02	1.05E+02	1.05E+02	=
13	9.91E+01	9.86E+01	9.94E+01	9.83E+01	=
14	3.31E+04	3.25E+04	3.20E+04	3.24E+04	=
15	1.00E+02	1.00E+02	1.00E+02	1.00E+02	=

Table 3. L-SHADE vs. EA-L-SHADE results on CEC2015 in 10D.

10D	L-SHADE		EA-L-SHADE		
f	Median	Mean	Median	Mean	Diff.
1	0.00E+00	0.00E+00	0.00E+00	0.00E+00	=
2	0.00E+00	0.00E+00	0.00E+00	0.00E+00	=
3	2.00E+01	1.76E+01	2.00E+01	1.87E+01	=
4	2.99E+00	3.08E+00	2.99E+00	2.70E+00	+
5	1.88E+01	3.55E+01	1.82E+01	3.99E+01	=
6	4.16E−01	7.71E−01	3.68E−01	4.05E−01	+
7	6.45E−02	9.63E−02	4.43E−02	5.82E−02	+
8	3.55E−01	3.72E−01	3.20E−01	3.43E−01	=
9	1.00E+02	1.00E+02	1.00E+02	1.00E+02	=
10	2.17E+02	2.17E+02	2.17E+02	2.17E+02	=
11	3.00E+02	1.90E+02	3.81E+00	1.08E+02	+
12	1.01E+02	1.01E+02	1.01E+02	1.01E+02	=
13	2.69E+01	2.68E+01	2.75E+01	2.70E+01	=
14	2.94E+03	4.50E+03	2.94E+03	4.50E+03	=
15	1.00E+02	1.00E+02	1.00E+02	1.00E+02	=

The comparison of the archive usage can be seen in Tables 5, 6, 7, and 8, where there is a mean value of successful archive usages as a percentage of all usages and Pearson's correlation coefficient for pairs of the obtained value and successful archive usage percentage. Higher values of mean successful archive usages are highlighted by bold font and correlation values for the better performing algorithm are highlighted by bold font as well.

Table 4. L-SHADE vs. EA-L-SHADE results on CEC2015 in 30D.

30D	L-SHADE		EA-L-SHADE		
f	Median	Mean	Median	Mean	Diff.
1	4.70E−01	3.34E+00	5.46E+00	2.67E+01	−
2	0.00E+00	0.00E+00	0.00E+00	0.00E+00	=
3	2.01E+01	2.01E+01	2.01E+01	2.01E+01	=
4	2.19E+01	2.22E+01	2.49E+01	2.42E+01	−
5	1.55E+03	1.56E+03	1.57E+03	1.54E+03	=
6	7.87E+02	7.56E+02	6.81E+02	7.03E+02	=
7	7.04E+00	6.99E+00	6.77E+00	6.52E+00	+
8	1.69E+02	1.80E+02	1.30E+02	1.26E+02	+
9	1.03E+02	1.03E+02	1.03E+02	1.03E+02	−
10	7.20E+02	6.95E+02	6.30E+02	6.90E+02	=
11	4.02E+02	4.24E+02	4.08E+02	4.11E+02	=
12	1.05E+02	1.05E+02	1.05E+02	1.05E+02	=
13	9.69E+01	9.65E+01	9.80E+01	9.75E+01	=
14	3.31E+04	3.27E+04	3.24E+04	3.25E+04	=
15	1.00E+02	1.00E+02	1.00E+02	1.00E+02	=

Table 5. SHADE vs. EA-SHADE archive usage results on CEC2015 in 10D.

10D	SHADE		EA-SHADE	
f	Mean good %	Cor.	Mean good %	Cor.
1	9.68E+00	–	**1.07E+01**	–
2	1.12E+01	–	**1.22E+01**	–
3	1.54E+00	−9.40E–02	**1.56E+00**	−1.59E−01
4	4.79E+00	1.85E−01	**4.94E+00**	−3.10E−01
5	3.93E+00	−3.49E−01	**4.03E+00**	2.90E−02
6	6.04E+00	2.03E−01	**9.27E+00**	**−1.80E−02**
7	**6.26E+00**	−4.76E−01	6.18E+00	**−3.54E−01**
8	5.75E+00	6.52E−01	**8.08E+00**	7.00E−01
9	2.43E+00	−1.40E−02	**2.49E+00**	−6.60E−02
10	9.22E+00	2.32E–01	**1.07E+01**	4.00E−03
11	6.08E+00	**9.98E–01**	**7.34E+00**	9.99E−01
12	2.26E+00	8.00E−03	**2.32E+00**	−2.70E−01
13	4.17E+00	**−1.29E−01**	**4.26E+00**	−2.90E−02
14	1.47E+01	8.57E−01	**1.53E+01**	6.46E−01
15	1.55E+01	–	**1.71E+01**	–

Table 6. SHADE vs. EA-SHADE archive usage results on CEC2015 in 30D.

30D	SHADE		EA-SHADE	
f	Mean good %	Cor.	Mean good %	Cor.
1	**3.66E+01**	**1.30E−01**	3.25E+01	1.24E−01
2	1.32E+01	–	**1.47E+01**	–
3	1.31E+00	**7.40E−02**	**1.33E+00**	−2.30E−01
4	4.66E+00	−5.44E−01	**4.75E+00**	−2.62E−01
5	3.57E+00	−1.01E−01	**3.60E+00**	−2.30E−02
6	1.71E+01	2.35E−01	**1.79E+01**	−1.57E−01
7	9.15E+00	4.10E−02	**9.88E+00**	−7.39E−01
8	**2.52E+01**	1.72E−01	2.34E+01	**1.90E−02**
9	1.56E+00	**1.92E−01**	**1.57E+00**	−3.10E−02
10	2.18E+01	−1.87E−01	**2.36E+01**	−2.38E−01
11	9.93E+00	−1.96E−01	**1.02E+01**	2.04E−01
12	1.20E+00	−2.77E−01	**1.21E+00**	−4.90E−02
13	3.01E+00	−9.00E−02	**3.06E+00**	−1.13E−01
14	1.34E+01	−3.42E−01	**1.36E+01**	−2.91E−01
15	9.84E+00	–	**1.06E+01**	–

Table 7. L-SHADE vs. EA-L-SHADE archive usage results on CEC2015 in 10D.

10D	L-SHADE		EA-L-SHADE	
f	Mean good %	Cor.	Mean good %	Cor.
1	8.02E+00	–	**9.02E+00**	–
2	9.37E+00	–	**1.03E+01**	–
3	1.37E+00	−2.81E−01	**1.39E+00**	-2.78E−01
4	4.47E+00	−2.83E−01	**4.71E+00**	**-1.47E–01**
5	3.63E+00	−2.10E−02	**3.69E+00**	–2.85E–01
6	5.84E+00	−2.13E−01	**6.87E+00**	**5.20E–02**
7	**7.00E+00**	−6.17E−01	6.92E+00	**−3.72E−01**
8	7.89E+00	5.32E−01	**9.45E+00**	3.10E−01
9	2.22E+00	1.04E−01	**2.26E+00**	−8.00E−02
10	6.60E+00	–1.49E−01	**7.90E+00**	3.20E−02
11	**6.76E+00**	9.98E−01	5.47E+00	**9.99E−01**
12	2.07E+00	–2.29E−01	**2.11E+00**	−4.04E−01
13	4.18E+00	2.78E–01	**4.22E+00**	2.15E−01
14	1.21E+01	5.38E−01	**1.25E+01**	5.54E−01
15	1.28E+01	–	**1.43E+01**	–

Table 8. L-SHADE vs. EA-L-SHADE archive usage results on CEC2015 in 30D.

30D	L-SHADE		EA-L-SHADE	
f	Mean good %	Cor.	Mean good %	Cor.
1	**3.71E+01**	**1.34E-01**	3.37E+01	2.87E−01
2	1.25E+01	–	**1.38E+01**	–
3	1.19E+00	5.40E−02	**1.21E+00**	1.09E−01
4	4.31E+00	**-5.59E−01**	**4.43E+00**	−3.92E−01
5	3.22E+00	-2.82E−01	**3.31E+00**	−1.07E−01
6	2.62E+01	1.91E−01	**3.05E+01**	−5.10E−02
7	9.43E+00	2.09E−01	**1.09E+01**	**−7.79E−01**
8	**2.72E+01**	1.34E−01	2.57E+01	**1.52E−01**
9	**1.44E+00**	**1.48E−01**	**1.44E+00**	–2.00E−03
10	2.95E+01	9.50E−02	**2.98E+01**	−2.16E−01
11	8.67E+00	−2.36E−01	**9.53E+00**	2.96E−01
12	**1.13E+00**	−3.14E−01	**1.13E+00**	−2.80E−02
13	2.86E+00	−4.50E−02	**2.93E+00**	−2.54E−01
14	1.23E+01	−2.86E−01	**1.44E+01**	−1.28E−01
15	9.11E+00	–	**9.83E+00**	–

6 Results Discussion

As can be seen in Tables 1 and 2, EA-SHADE performs significantly better on two functions in 10*D* and on one function in 30*D*, whereas canonical SHADE performs better on two functions and three functions respectively. Therefore, it can be concluded that EA implementation does not improve the robustness of the method, but is rather comparable to the original archive implementation. However, interesting points are in Tables 5 and 6, where the archive usage is depicted. There, negative correlation value suggests that it is beneficial for the algorithm to use individuals from an archive. The score of negative correlations is 5:7 in 10*D* and 7:10 in 30*D* which is both in favor for the EA implementation. Moreover, in 30*D*, significant differences are only on functions where the correlation is positive and, therefore, it seems that it might be better to not use an archive at all. In 10*D*, the situation is completely opposite, significant differences in results are on functions with negative correlations, which suggests that increasing exploration ability of the algorithm via archive is beneficial in lower dimensional objective space, whereas, in higher dimensional space, the exploitation is more suitable.

In the case of the L-SHADE algorithm, it can be seen in Tables 3 and 4 that in 10*D*, the EA version of the algorithm performs significantly better on four test functions. In 30*D*, it performs better on three test functions and worse on three also. The difference in performance is higher than in the case of SHADE. This might be due to the decreasing size of an archive along with the population. Also, negative correlation scores in Tables 7 and 8 are different. It is 7:6 in 10*D* and 6:9 in 30*D*, which slightly favors EA in higher dimensional objective space. Another interesting aspect is that in

10*D* and 30*D*, the score of positive and negative correlations on functions with significantly different results is 2:2 and 3:2 respectively.

Obtained results imply that proposed EA implementation supports exploration more than the original implementation. Higher exploration ability is suitable for lower dimensional objective spaces, but still can be beneficial for some functions (e.g. f7 and f8 in Table 4) in higher dimensions. Overall, it can be concluded, that EA is more suitable for the L-SHADE algorithm.

The variation in results confirms the famous "no free lunch" theorem [15], thus, it might be useful to invent an adaptive behavior for an external archive of inferior solutions.

7 Conclusion

This paper presented a novel implementation of an external archive, titled EA, in SHADE and L-SHADE algorithms. The proposed version was compared to the original archive implementation on CEC2015 benchmark set of test functions.

Obtained results imply that the proposed variant supports exploration more than the original version, which led to better results in lower dimensional space. The implementation of EA is more suitable for the L-SHADE algorithm because of the linear decrease of both, population and archive sizes. Overall, the successful generation of trial individuals was higher for the EA, which suggests better use of the archive than the original version.

Future work will be focused on the analysis of exploration/exploitation abilities of EA in the L-SHADE algorithm and an adaptive version of an archive will be developed.

Acknowledgements. This work was supported by Grant Agency of the Czech Republic – GACR P103/15/06700S, further by the Ministry of Education, Youth and Sports of the Czech Republic within the National Sustainability Programme Project no. LO1303 (MSMT-7778/2014). Also by the European Regional Development Fund under the Project CEBIA-Tech no. CZ.1.05/2.1.00/03.0089 and by Internal Grant Agency of Tomas Bata University under the Projects no. IGA/CebiaTech/2017/004.

References

1. Storn, R., Price, K.: Differential Evolution-A Simple and Efficient Adaptive Scheme for Global Optimization Over Continuous Spaces, vol. 3. ICSI, Berkeley (1995)
2. Neri, F., Tirronen, V.: Recent advances in differential evolution: a survey and experimental analysis. Artif. Intell. Rev. **33**(1–2), 61–106 (2010)
3. Das, S., Suganthan, P.N.: Differential evolution: a survey of the state-of-the-art. Evol. Comput. IEEE Trans. **15**(1), 4–31 (2011)
4. Das, S., Mullick, S.S., Suganthan, P.N.: Recent advances in differential evolution–an updated survey. Swarm Evol. Comput. **27**, 1–30 (2016)

5. Brest, J., Greiner, S., Bošković, B., Mernik, M., Zumer, V.: Self-adapting control parameters in differential evolution: a comparative study on numerical benchmark problems. Evol. Comput. IEEE Trans. **10**(6), 646–657 (2006)
6. Omran, M.G., Salman, A., Engelbrecht, A.P.: Self-adaptive differential evolution. In: Computational Intelligence and Security, pp. 192–199. Springer, Heidelberg (2005)
7. Qin, A.K., Huang, V.L., Suganthan, P.N.: Differential evolution algorithm with strategy adaptation for global numerical optimization. Evol. Comput. IEEE Trans. **13**(2), 398–417 (2009)
8. Islam, S.M., Das, S., Ghosh, S., Roy, S., Suganthan, P.N.: An adaptive differential evolution algorithm with novel mutation and crossover strategies for global numerical optimization. IEEE Trans. Syst. Man Cybern. Part B (Cybern.) **42**(2), 482–500 (2012)
9. Zhang, J., Sanderson, A.C.: JADE: adaptive differential evolution with optional external archive. Evol. Comput. IEEE Trans. **13**(5), 945–958 (2009)
10. Tanabe, R., Fukunaga, A.: Success-history based parameter adaptation for differential evolution. In: IEEE Congress on 2013 Evolutionary Computation (CEC), pp. 71–78. IEEE, June 2013
11. Liang, J.J., Qu, B.Y., Suganthan, P.N., Hernández-Díaz, A.G.: Problem definitions and evaluation criteria for the CEC 2013 special session on real-parameter optimization. Computational Intelligence Laboratory, Zhengzhou University, Zhengzhou, China and Nanyang Technological University, Singapore, Technical report, 201212 (2013)
12. Tanabe, R., Fukunaga, A.S.: Improving the search performance of SHADE using linear population size reduction. In: IEEE Congress on 2014 Evolutionary Computation (CEC), pp. 1658–1665. IEEE, 2014 July
13. Liang, J.J., Qu, B.Y., Suganthan, P.N.: Problem definitions and evaluation criteria for the CEC 2014 special session and competition on single objective real-parameter numerical optimization. Computational Intelligence Laboratory, Zhengzhou University, Zhengzhou China and Technical report, Nanyang Technological University, Singapore (2013)
14. Liang, J.J., Qu, B.Y., Suganthan, P.N., Chen, Q.: Problem definitions and evaluation criteria for the CEC 2015 competition on learning-based real-parameter single objective optimization. Technical report 201411A, Computational Intelligence Laboratory, Zhengzhou University, Zhengzhou China and Technical report, Nanyang Technological University, Singapore (2014)
15. Wolpert, D.H., Macready, W.G.: No free lunch theorems for optimization. Evol. Comput. IEEE Trans. **1**(1), 67–82 (1997)

Adaptation of Sport Training Plans by Swarm Intelligence

Iztok Fister Jr.[1](✉), Andres Iglesias[3], Eneko Osaba[2], Uroš Mlakar[1], Janez Brest[1], and Iztok Fister[1]

[1] Faculty of electrical engineering and computer science, University of Maribor, Smetanova 17, 2000 Maribor, Slovenia
iztok.fister1@um.si
[2] University of Deusto, Av. Universidades 24, 48007 Bilbao, Spain
[3] University of Cantabria, E.T.S.I. Caminos, Canales y Puertos, Avda. de los Castros, s/n, 39005 Santander, Spain

Abstract. Automatic planning of sport training sessions with Swarm Intelligence algorithms has been proposed recently in the scientific literature that influences the sports training process in practice dramatically. These algorithms are capable of generating sophisticated training plans based on an archive of the existing sports training sessions. In recent years, training plans have been generated for various sport disciplines, like road cycling, mountain biking, running. These plans have also been verified by professional sport trainers confirming that the proposed training plans correspond with the theory of sports training. Unfortunately, not enough devotion has been given to adapting the generated sports training plans due to the changing conditions that may occur frequently during their realization and causes a break in continuity of the sports training process. For instance, athletes involved in the training process can become ill or injured. These facts imply disruption of the systematic increase of the athlete's capacity. In this paper, therefore, we propose a novel solution that is capable of adapting training plans due to the absence of an athlete from the training process.

Keywords: Artificial sport trainer · Multisport · Sport training
Swarm intelligence · Running

1 Introduction

During the process of sports training, an athlete's body adapts to loading that causes the raising of the athlete's capacity due to external loading or adjustment to specific environmental conditions [2]. Indeed, external loading is determined by sports trainers prescribing their trainees the corresponding sports training plans. The sports training plans include a sequence of sports training sessions for a specific period of sports training (also a cycle). Thus, each sports training session is determined by exercise type, duration, intensity and frequency.

R. Matoušek (Ed.): MENDEL 2017, AISC 837, pp. 56–67, 2019.
https://doi.org/10.1007/978-3-319-97888-8_5

Actually, the athlete's body undergoes three kinds of adaptation during the sports training process, i.e., physical, intellectual and emotional adaptation. However, all three kinds refer to a psycho-physiological adaptation of the athlete in the sports training process. This process has been led by the sports trainers until now, who affect the normal realization of the sports training plan with adapting the sports training plans according to the current athlete's performance.

Usually, the normal realization of the sports training plan can be disturbed due to an athlete's injury, illness, travel during the competition season, or even some other psychological problems. The mentioned factors imply disruption of the systematic increase of loading by an athlete in sports training. Consequently, this means a break in continuity of the physiological adaptation process, where the effects of loading are lost, especially when the intervals between training exercises are too long.

In more detail, injuries are one of the more complex disturbing factors that may disable the athletes from the normal training process for a longer time. When injuries have been acquired by athletes, they usually have to seek medical advice. However, sports trainers are also involved in the recovery process. Actually, sports trainers, along with medical experts, decide how to continue with the sports training process, i.e., what is the best way to continue treatment and what will be the starting steps for recovery. The major sports injuries are as follows: Various bone fractures, head injuries [7], strains, etc. On the other hand, there are also some minor sports injuries that affect the athlete's sports training for some days only. In this group, we count different abrasions, burns, blisters and so on. Not only injuries, but also illness, is very unpredictable disturbing factor that depends not only on sports involvement, but also on the other areas of a human's life. For example, in the winter season there is a lot of flu that may infect the athletes as well. This infection can prevent them from the sports training process from one day to several weeks. Automatic planning of the sport training sessions has been become a very popular research topic. In recent years, the concept of an Artificial Sport Trainer (AST) has been proposed by Fister et al. in [4]. The aim of the AST is to automate tasks of the real Sport Trainer, especially in those phases of the sports training where a lot of data are generated by using the wearable mobile devices and the data analysis cannot be performed without the help of digital computers. In a nutshell, an AST is able to plan sport training sessions for short, as well as long, periods of sports training.

Adaptation of sports training plans using an AST has not been studied deeply, until recently. In this paper, we present the problem of adapting the training plans. The Particle Swarm Optimization (PSO) algorithm [6] is one of the first members of Swarm-Intelligence (SI) based algorithms that was proposed back in 1997. The remainder of the paper is organized as follows. Firstly, we expose a problem theoretically, while secondly we present a solution and results.

2 A Problem Description

In the case of an athlete's deviance from the planned sport training sessions, an athlete has basically the following scenarios with which to catch up missed sports training sessions:

- Increase the intensity of the next training session that is planned in the schedule: In this case, the intensity of the training sessions that are still waiting in the schedule are enhanced. Usually, we cannot enhance them too much, because athletes can be injured again. On the other hand, they can easily become over-trained.
- Prolonged schedule: In this case, we increase the number of days in the schedule.
- Generate a totally new training plan: In this case, we generate a new training plan for the remaining cycle.
- Perform more training sessions in one day: This is a very uncommon situation because, performing more sports training sessions in one day, demands from the recovered unfit athlete too much effort. Although performing more training sessions in one day is normally for professional or some other semi-professional athletes, this is untypical for the amateur athletes, who go to work each day.
- Nothing with missed training: The athlete just continues with the proposed training plan as if nothing has happened. However, this is good solution only for athletes without any competition goals.

However, there is no universal recipe for which of the mentioned scenarios is the more appropriate for adapting their sports plans in the case of the sports training session loss. Indeed, all athletes are unique and possess their own characteristics. Therefore, the adaptation of sports training plans demands a perfect knowledge about the psycho-physical characteristics of the athlete in training. In line with this, this adaptation seems to be a much harder problem than generating a training plan from scratch. In the remainder of the paper, we present a mathematical description of a problem along with one case study. In our case, we proceed from the first scenario, where increasing the intensity of the future training sessions is taken into account.

3 Materials and Methods

This section acquaints readers with the background information needed for understanding the subject in the continuation of the paper. In line with this, the sports training plan generated by the AST is discussed, that serves as a benchmark for testing the proposed algorithm for adapting the sports training plan. Additionally, the basics are described of the Particle Swarm Optimization (PSO) algorithm underlying the proposed algorithm. Finally, the proposed PSO for adapting the sports training plans is illustrated.

3.1 Test Sports Training Plan

A sports training plan generated by an AST [4] serves as a test instance for adaptation. Although this supports various nature-inspired algorithms, this instance was generated with the PSO algorithm. Let us notice that this training plan is intended for a semi-professional runner, who prepares himself for 10 km long-distance runs and half-marathons (21.1 km). The test sports training plan was generated for a cycle of 50 days, while the total training load indicator of this sports training plan amounts to TRIMP = 386484.0. The training load indicator TRIMP is simply the product of TRIMP $= TD \cdot HR$, where the HR denotes

Table 1. Proposed training plan for semi-professional runner generated by the AST.

Training sessions	*HR*	*TD*	TRIMP	Training sessions	*HR*	*TD*	TRIMP
1	142	65	9230	26	105	29	3045
2	156	64	9984	27	142	65	9230
3	141	91	12831	28	141	55	7755
4	125	38	4750	29	125	45	5625
5	141	74	10434	30	162	51	8262
6	158	47	7426	31	136	31	4216
7	128	36	4608	32	126	38	4788
8	128	26	3328	33	169	65	10985
9	142	65	9230	34	128	75	9600
10	141	55	7755	35	158	45	7110
11	141	91	12831	36	121	41	4961
12	115	28	3220	37	141	52	7332
13	142	65	9230	38	133	71	9443
14	158	51	8058	39	142	65	9230
15	129	45	5805	40	115	28	3220
16	125	45	5625	41	125	38	4750
17	128	75	9600	42	138	87	12006
18	142	65	9230	43	169	65	10985
19	132	29	3828	44	141	71	10011
20	115	28	3220	45	138	87	12006
21	125	29	3625	46	162	41	6642
22	141	87	12267	47	121	26	3146
23	141	55	7755	48	129	84	10836
24	128	75	9600	49	141	54	7614
25	157	65	10205	50	141	71	10011
Total	136.84	55.76	193675	Total	137.96	55.2	192809

the intensity and the *TD* the duration of the sports training session [1]. This measure is devoted for measuring the internal loading, where the effective values of the intensity and duration are measured using wearable mobile devices.

Let us imagine that the athlete who is training according to the schedule presented in Table 1 becomes injured and absent from training after conducting 50% of the total schedule (training sessions 1–25 in Table 1). Then, the injury causes an absence of 10 days (20% of the schedule), which means that 30% of the whole schedule remains yet unfinished. The border between both schedules is denoted with two double line between 35 and 36 training days in Table 1. In the sense of the training load indicator TRIMP, the exposed facts can be expressed as follows (Fig. 1):

- 50% of completed sports training sessions in the schedule is 25 days (Total TRIMP: 193675.0),
- 20% of failed sports training sessions in the schedule is 10 days (Total TRIMP: 70616.0), and
- 30% of unfinished sports training sessions in the schedule is 15 days (Total TRIMP: 122193.0).

It means that the athlete has realized the first 25 days of the prescribed sports training plan successfully, while he was disturbed for the next 10 days. In summary, 15 days of training remain for him to realize, where he must catch up the training sessions in the 10 days that were not performed due to his illness. Catching up the whole TRIMP for these 10 days is impossible, because of exhaustion, over-training or any injuries. However, an athlete can conduct training sessions that are in the schedule, while the intensity of these training sessions can be enhanced.

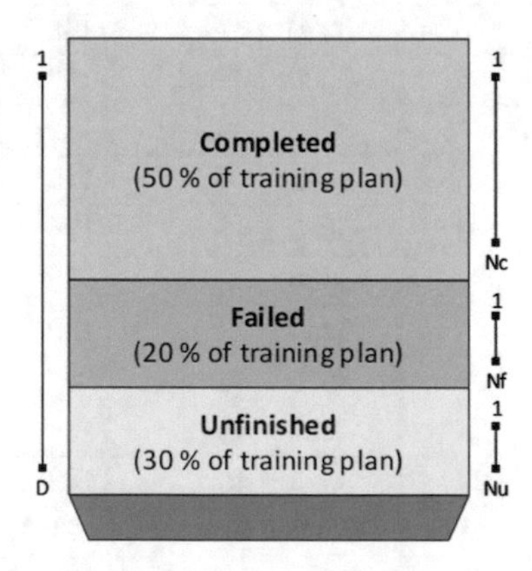

Fig. 1. Dividing the sports training plan into various types of sports sessions by adaptation, where the unfinished training sessions must be adapted with regard to the failed ones.

The sports training cycle of length D in Fig. 1 is divided into: The completed sports training sessions of dimension Nc, the failed sports training sessions of dimension Nf, and the unfinished sports training sessions of dimension Nu.

3.2 Particle Swarm Optimization

The PSO algorithm maintains a population of solutions, where each solution is represented as a real-valued vector $\mathbf{x} = (x_{i,1}, \ldots, q_{i,D})^T$ for $i = 1, \ldots, Np$ and $j = 1, \ldots, D$, and the parameter Np denotes the population size and the parameter D the problem dimension. This algorithm explores the new solutions by moving the particles throughout the search space in the direction of the current best solution. In addition to the current population $\mathbf{x}_i^{(t)}$ for $i = 1, \ldots, Np$, also the local best solutions $\mathbf{p}_i^{(t)}$ for $i = 1, \ldots, Np$ are maintained denoting the best i-th solution found. Finally, the best solution in the population $\mathbf{g}^{(t)}$ is determined in each generation. The new particle position is generated according to Eq. (1):

$$\begin{aligned} \mathbf{v}_i^{(t+1)} &= \mathbf{v}_i^{(t)} + C_1 U(0,1)(\mathbf{p}_i^{(t)} - \mathbf{x}_i^{(t)}) + C_2 U(0,1)(\mathbf{g}^{(t)} - \mathbf{x}_i^{(t)}), \\ \mathbf{x}_i^{(t+1)} &= \mathbf{x}_i^{(t)} + \mathbf{v}_i^{(t+1)}, \end{aligned} \tag{1}$$

where $U(0, 1)$ denotes a random value drawn from the uniform distribution in interval $[0, 1]$, and C_1 and C_2 are learning factors. The pseudo-code of the original PSO algorithm is illustrated in Algorithm 1.

Algorithm 1. Pseudocode of the basic Particle Swarm Optimization algorithm

Input: PSO population of particles $\mathbf{x_i} = (x_{i1}, \ldots, x_{iD})^T$ for $i = 1 \ldots Np$, MAX_FE.
Output: The best solution $\mathbf{x}_{best}$ and its corresponding value $f_{min} = \min(f(\mathbf{x}))$.

```
 1: init_particles;
 2: eval = 0;
 3: while termination_condition_not_meet do
 4:    for i = 1 to Np do
 5:       f_i = evaluate_the_new_solution(x_i);
 6:       eval = eval + 1;
 7:       if f_i ≤ pBest_i then
 8:          p_i = x_i; pBest_i = f_i; // save the local best solution
 9:       end if
10:       if f_i ≤ f_min then
11:          x_best = x_i; f_min = f_i; // save the global best solution
12:       end if
13:       x_i = generate_new_solution(x_i);
14:    end for
15: end while
```

In the next subsection, the proposed PSO algorithm is presented for adaptation of training plans.

The PSO for Adaptation of Training Plans. The algorithm for adapting sports training sessions relates to the results of the algorithm for planning sports training sessions based on existing sports activities proposed by Fister et al. in [3]. Actually, the solution in this algorithm is presented as a real-valued vector:

$$\mathbf{x}_i^{(t)} = (x_{i,1}^{(t)}, \ldots, x_{i,D}^{(t)})^T, \quad \text{for } i = 1, \ldots, Np. \tag{2}$$

Additionally, a permutation of clusters $\mathbf{\Pi}_i^{(t)}$ is assigned to each solution $\mathbf{x}_i^{(t)}$, as follows:

$$\mathbf{\Pi}_i^{(t)} = (C_{\pi_{i,1}}^{(t)}, \ldots, C_{\pi_{i,D}}^{(t)})^T, \quad \text{for } i = 1, \ldots, Np, \tag{3}$$

where $\pi_{i,j}$ for $j = 1, \ldots, D$ denotes a permutation of a cluster set $C = \{C_1, \ldots, C_m\}$, where m is the number of clusters. Thus, each cluster $C_{\pi_{i,j}}^{(t)} = \{s_{\pi_{i,j},1}^{(t)}, \ldots, s_{\pi_{i,j},n_{\pi_{i,j}}}^{(t)}\}$, where $n_{\pi_{i,j}}$ is the cluster size, combines sports training sessions $s_{\pi_{i,j},k}^{(t)}$ of the similar internal loading measured by the training load indicator TRIMP. The specific sports training session $s_{\pi_{i,j},k}^{(t)}$ is selected from the cluster $C_{\pi_{i,j}}^{(t)}$ by calculating the index k according to the following equation:

$$k_j = \left\lceil x_{i,j}^{(t)} \cdot n_{\pi_{i,j}} \right\rceil. \tag{4}$$

Each sports training activity specified in the cluster $C_{\pi_{i,j}}^{(t)}$ is a couple $s_{\pi_{i,j},k_j}^{(t)} = \langle TD_{\pi_{i,j},k_j}^{(t)}, HR_{\pi_{i,j},k_j}^{(t)} \rangle$, where $TD_{\pi_{i,j},k_j}^{(t)}$ denotes the duration and $HR_{\pi_{i,j},k_j}^{(t)}$ the average heart rate of the $x_{i,j}^{(t)}$-th element of the corresponding solution. The fitness function of the algorithm for planning sports training sessions based on existing sports activities is expressed as follows:

$$f^*(\mathbf{x}_i) = \max \sum_{j=1}^{D} TD_{\pi_{i,j},k_j}^{(t)} \cdot HR_{\pi_{i,j},k_j}^{(t)}, \tag{5}$$

subject to

$$\begin{aligned} TD_{\pi_{i,j},k_j}^{(\min)} \leq TD_{\pi_{i,j},k_j}^{(t)} \leq TD_{\pi_{i,j},k_j}^{(\max)}, \text{ and} \\ HR_{\pi_{i,j},k_j}^{(\min)} \leq HR_{\pi_{i,j},k_j}^{(t)} \leq HR_{\pi_{i,j},k_j}^{(\max)}, \end{aligned} \tag{6}$$

where $TD_{\pi_{i,j},k_j}^{(\min)}$ and $HR_{\pi_{i,j},k_j}^{(\min)}$, and $TD_{\pi_{i,j},k_j}^{(\max)}$ and $HR_{\pi_{i,j},k_j}^{(\max)}$ denote the minimum and maximum values of variables $TD_{\pi_{i,j},k_j}^{(t)}$ and $HR_{\pi_{i,j},k_j}^{(t)}$, respectively. The main modifications of the original PSO algorithm for adapting the sports training plans encompass the following components:

- Representation of individuals,
- Initialization of solutions (function 'init_particles' in Algorithm 1),
- Fitness function evaluation (function 'evaluate_the_new_solution' in Algorithm 1).

Individual in the PSO algorithm for adapting the training plans is represented as a real-valued vector $\mathbf{y}_i^{(t)}$ for $i = 1, \ldots, Np$ and $j = 1, \ldots, 2 \cdot Nu + 1$, where the parameter Nu denotes the number of days that need to be adapted. In fact, two elements of the solution vector $y_{i,2*j}^{(t)}$ and $y_{i,2*j+1}^{(t)}$ are reserved for each day denoting the adapted duration $TD_{i,j}^{(t)}$ and average heart rate $HR_{i,j}^{(t)}$ for the corresponding sports training session $\langle TD_{i,j}^{(t)}, HR_{i,j}^{(t)} \rangle$, respectively.

Fitness function maximizes the training load indicator TRIMP, as follows:

$$f^*(\mathbf{y}_i^{(t)}) = \min \left| \sum_{j=1}^{Nc} TD_{\pi_{i,j},k_j}^{(t)} \cdot HR_{\pi_{i,j},k_j}^{(t)} + \sum_{j=1}^{Nu} y_{i,j*2}^{(t)} \cdot y_{i,j*2+1}^{(t)} - f^*(\mathbf{x}_i^{(t)}) \right|, \quad (7)$$

subject to

$$\begin{aligned} \widetilde{TD}_{i,j}^{(\min)} &\leq y_{i,2\cdot j}^{(t)} \leq \widetilde{TD}_{i,j}^{(\max)}, \\ \widetilde{HR}_{i,j}^{(\min)} &\leq y_{i,2\cdot j+1}^{(t)} \leq \widetilde{HR}_{i,j}^{(\max)}, \end{aligned} \quad (8)$$

where $\widetilde{TD}_{i,j}^{(\min)}$ and $\widetilde{TD}_{i,j}^{(\max)}$ are by 60%, enhanced values of $TD_{i,j}^{(\min)}$ and $TD_{i,j}^{(\max)}$, and $\widetilde{HR}_{i,j}^{(\min)}$ and $\widetilde{HR}_{i,j}^{(\max)}$ are by 15%, enhanced values of $HR_{i,j}^{(\min)}$ and $HR_{i,j}^{(\max)}$, respectively.

In Eq. (7), the first term denotes the training load of complete training sessions according to Eq. (5), the second term the training load of the adapted training load, while the $f^*(\mathbf{x}_i^{(t)})$ is the training load of the proposed sports training plan. The purpose of the fitness function $f^*(\mathbf{y}_i^{(t)})$ is to adapt the remaining training sessions so that the difference between the proposed and real sports training plan is as small as possible.

4 Experiments and Results

The goal of the experimental work was to show that the proposed PSO algorithm is appropriate for adapting the sports training plans, when the normal realization of the sports training plans is disturbed due to an athlete's illness. In our case, the illness prevented the athlete from training for ten days. From the sports trainer standpoint, this means that the realization of the sports training plans was deteriorated by 20%. The PSO algorithm for adapting the sports training plans was implemented in Python programming language using no special external libraries. The task of the algorithm was to adapt the training plan of duration 15 days for the semi-professional runner that missed 10 sports training sessions due to illness. The parameter settings of the PSO as presented in Table 2 were used during the experimental work. Three different runs of the proposed PSO algorithm were analyzed in detail.

Test problems as illustrated in Sect. 3.1 were used in our experimental work. The results were measured according the minimum value of the difference

Table 2. Parameter setting for the PSO algorithm

Parameter	Variable	Value
Population size	NP	100
Maximal number of fitness function evaluations	MAX_FE	50000
Dimension of the problem	D	30
Velocity coefficients	C_1, C_2	2.0

between the training load indicator TRIMP of the proposed and adapted sports training plans. The results of the optimization obtained by the PSO algorithm for adaptation of sports training plans are presented in Tables 3 and 4. Actually, Table 3 depicts the reference training as proposed by the algorithm for generating the training plans based on the existing sports activities, and the results of the adapting the sports training sessions in the first run. Moreover, Table 4 depicts the results of adapting as proposed in runs two and three.

Table 3. Results of the reference training and the run 1.

Reference training	HR	TD	TRIMP	Adapted training 1	HR	TD	TRIMP
1	121.00	41.00	4961.00	1	105.00	29.45	3092.04
2	141.00	52.00	7332.00	2	154.30	103.64	15991.99
3	133.00	71.00	9443.00	3	130.66	75.01	9800.94
4	142.00	65.00	9230.00	4	136.04	32.01	4354.44
5	115.00	28.00	3220.00	5	154.75	80.85	12512.51
6	125.00	38.00	4750.00	6	111.74	27.59	3083.23
7	138.00	87.00	12006.00	7	124.41	45.02	5600.95
8	169.00	65.00	10985.00	8	183.54	101.23	18579.72
9	141.00	71.00	10011.00	9	142.41	101.37	14435.17
10	138.00	87.00	12006.00	10	127.19	61.31	7797.60
11	162.00	41.00	6642.00	11	114.67	59.32	6802.42
12	121.00	26.00	3146.00	12	158.10	49.13	7768.27
13	129.00	84.00	10836.00	13	135.22	112.37	15193.61
14	141.00	54.00	7614.00	14	161.53	73.36	11850.48
15	141.00	71.00	10011.00	15	125.46	28.39	3561.54
Summary	137.13	58.73	**122193.00**	Summary	137.67	65.34	**140424.91**

As can be seen from Tables 3 and 4, the internal training load of the adapted sports training plans 1 to 3 increases from 140424.91 to 148583.51. All adapted training plans increase the intensity of the adapted sports training plan significantly, although the days lost cannot be caught up. Actually, the realization of the sports training plan can be increased from 68.30% (TRIMP=264291) by

Table 4. Results of the run 2 and run 3.

Adapted training 2	HR	TD	TRIMP	Adapted training 3	HR	TD	TRIMP
1	114.66	29.05	3331.37	1	104.99	29.28	3074.43
2	143.01	68.10	9738.62	2	146.26	103.47	15133.47
3	132.01	87.18	11508.11	3	138.70	67.38	9345.47
4	121.38	71.49	8676.65	4	139.59	56.20	7845.18
5	185.67	80.43	14932.78	5	153.08	80.03	12250.38
6	147.35	30.04	4426.45	6	123.09	31.50	3877.42
7	132.54	49.10	6507.57	7	135.56	22.68	3074.11
8	193.07	70.20	13553.10	8	183.07	103.57	18959.61
9	139.37	119.52	16656.97	9	145.80	98.21	14319.49
10	166.55	70.48	11739.26	10	139.28	62.54	8710.66
11	138.75	43.07	5975.89	11	128.89	61.09	7873.57
12	159.46	82.76	13197.50	12	160.68	82.53	13260.56
13	125.75	69.26	8708.94	13	138.88	74.00	10277.32
14	144.12	103.91	14975.78	14	155.43	103.47	16083.29
15	111.95	32.22	3607.20	15	115.05	39.10	4498.55
Summary	143.71	67.12	**147536.18**	Summary	140.56	67.67	**148583.51**

the reference sports training plan to 74.95% (TRIMP–289681.55) in the case of the most intensively adapted sports training plan 3. The results according to adapted sports training sessions obtained in three different runs with regard to the proposed sports training plan are presented in Fig. 2, from which it can be seen that the adapted sports training plans obtained in runs 1–3 increase the intensity load indicator TRIMP significantly compared with the reference sports training plan.

The summary results of the PSO algorithm for adapting the sports training plans are illustrated in Table 5, that compares the results of planning the reference sports training sessions with the results of adapting the sports training plans obtained in three different runs. This comparison is performed in the sense of calculating the difference between the adapted training load obtained in three independent runs and the training load obtained by the reference sports training plan.

Table 5. Summary results of the adapting the sports training plans.

Training	$\overline{HR}$	$\overline{TD}$	TRIMP	$\Delta\overline{HR}$	$\Delta\overline{TD}$	ΔTRIMP
Reference	137.13	58.73	122193.00	0.00(0.00%)	0.00(0.00%)	0.00(0.00%)
Run 1	137.67	65.34	140424.91	0.53(0.39%)	6.60(11.24%)	18231.91(14.92%)
Run 2	143.71	67.12	147536.18	6.57(4.79%)	8.39(14.28%)	25343.18(20.74%)
Run 3	140.56	67.67	148583.51	3.42(2.50%)	8.94(15.22%)	26390.51(21.69%)

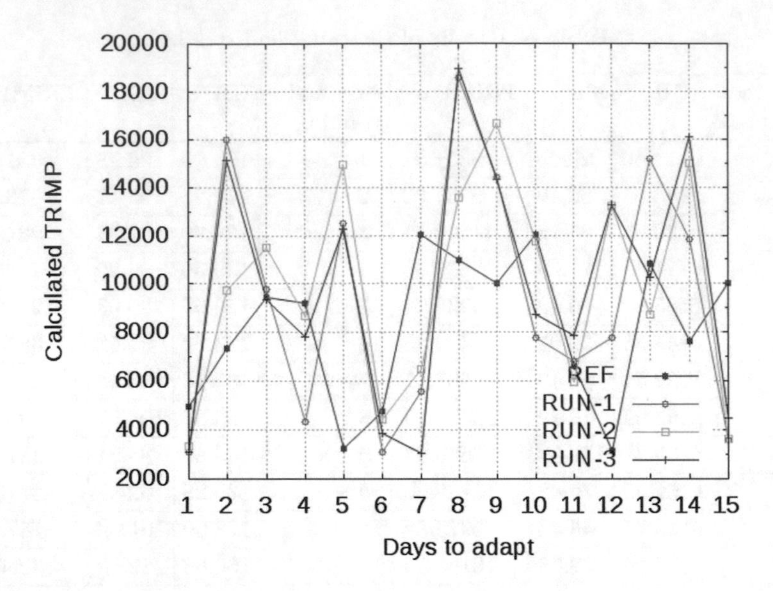

Fig. 2. Graphical presentation of the results obtained by adapting the sports training plans.

Interestingly, the internal loading can be increased by raising either the intensity (HR) or duration (TD) of the corresponding sports training plan. In the first adapted training plan, the average heart rate is increased only by 0.39%, while the duration even by 11.24%. However, this is in accordance with the boundary constraints that enable increasing the intensity by, at most, 15% and the duration by, at most, 60%. The second run proposes raising the intensity by 4.79% and duration by 14.28%, while the third run, in contrast, raises the intensity only by 2.50%, but the duration even by 15.22% that, in summary, means increasing the sports training load by 21.69%. This is a characteristic of the training load TRIMP, where the increasing of this coefficient can be performed either by increasing the average heart rate or the duration of the sports training session.

5 Conclusion

Adaptation of sport training plans is a very important task of real sport trainers, especially for those who are not very experienced [8]. Dealing with this is very hard, because adaptation takes place usually when athletes are injured or ill. Because of that, they are absent from sports training and, therefore, not able to perform sports training sessions according to the schedule. This paper proposed a simple yet effective solution for adaptation of sports training plans based on the Particle Swarm Optimization algorithm. The results presented in this paper confirmed that it is possible to adapt the sports training plans that were initially generated by an AST efficiently. In fact, there are still really big opportunities for improvement of this solution. Firstly, results should be more aligned to the pure theory of sport training while, secondly, other ways of adapting the sports training plans should be explored. Additionally, it would be interesting to see the performance of various population nature-inspired algorithms [5] in solving this problem.

References

1. Banister, E.: Modeling elite athletic performance. Physiol. Test. Elite Athletes **403**, 403–424 (1991)
2. Dick, F.W.: Sports Training Principles: An Introduction to Sports Science, 6th edn. Bloomsbury Sport, London (2015)
3. Fister Jr., I., Fister, I.: Generating the training plans based on existing sports activities using swarm intelligence. In: Nakamatsu, K., Patnaik, S., Yang, X.S. (eds.) Nature-Inspired Computing and Optimization: Theory and Applications, pp. 79–94. Springer International Publishing, Switzerland (2017)
4. Fister Jr., I., Ljubič, K., Suganthan, P.N., Perc, M., Fister, I.: Computational intelligence in sports: challenges and opportunities within a new research domain. Appl. Math. Comput. **262**, 178–186 (2015)
5. Fister Jr., I., Yang, X.-S., Fister, I., Brest, J., Fister, D.: A brief review of nature-inspired algorithms for optimization. Elektroteh. Vestn. **80**(3), 116–122 (2013)
6. Kennedy, J., Eberhart, R.: Particle swarm optimization. In: 1995 Proceedings of the IEEE International Conference on Neural Networks, vol. 4, pp. 1942–1948. IEEE (1995)
7. Omalu, B.I., DeKosky, S.T., Minster, R.L., Kamboh, M.I., Hamilton, R.L., Wecht, C.H.: Chronic traumatic encephalopathy in a national football league player. Neurosurgery **57**(1), 128–134 (2005)
8. Søvik, M.L.: Evaluating the implementation of the empowering coaching program in Norway. Ph.D thesis (2017)

On the Influence of Localisation and Communication Error on the Behaviour of a Swarm of Autonomous Underwater Vehicles

Christoph Tholen(✉), Lars Nolle, and Jens Werner

Department of Engineering Science, Jade University of Applied Sciences, Friedrich-Paffrath-Straße 101, 26389 Wilhelmshaven, Germany
{christoph.tholen, lars.nolle, jens.werner}@jade-hs.de

Abstract. The long term goal of this research is to develop a swarm of autonomous underwater vehicles (AUVs), which can be used to locate submarine sources of interest, like dumped radioactive waste or ammunition. The overall search strategy of the swarm is based on particle swarm optimisation (PSO). Standard PSO relies on correct localisation and timely communication in order to be able to converge towards the global optimum. However, underwater communication is slow and unreliable and the exact localisation of an AUV is difficult. Therefore, this paper presents an empirical study of the effect of communication and localisation error on the convergence capabilities of PSO. A simulation based on cellular automata is presented and a model of communication and localisation error is incorporated into the PSO. It is shown that both types of errors have a negative effect on the performance of the search, with localisation error having the greater contribution.

Keywords: Submarine groundwater discharge
Fluorescent Dissolved Organic Matter · Particle swarm optimisation
Cellular automata · Communication error · Localisation error

1 Introduction

The long term aim of this project is to develop a flexible and low cost environmental marine observatory. This observatory will be based on a small swarm of autonomous underwater vehicles (AUV). AUVs can be used for the exploration of intermediate size areas and the precise measurement of environmental parameters, for example salinity, nutrients or oxygen or to locate submarine sources of interest, e.g. dumped waste or ammunition [1]. An example for the hazard detection with sensor systems is oil and pollution detection using airplanes [2]. One disadvantage of such systems is, that they are only capable of observing the surface of the sea. To investigate submarine systems, AUVs could be used instead. Because AUVs have a limited payload and energy capacity, it is proposed to use a swarm of collaborative AUVs to carry out the search.

Another application for such an observatory, for example, is the search for submarine groundwater discharges (SGD) in coastal waters. SGD consist of an inflow of fresh groundwater and recirculated seawater from the sea floor into the ocean [3] (Fig. 1). The localisation of SGDs is a technical effort; however marine scientists are

R. Matoušek (Ed.): MENDEL 2017, AISC 837, pp. 68–79, 2019.
https://doi.org/10.1007/978-3-319-97888-8_6

interested in their localisation, because the SGDs discharge continuously nutrients into the coastal environment [4, 5].

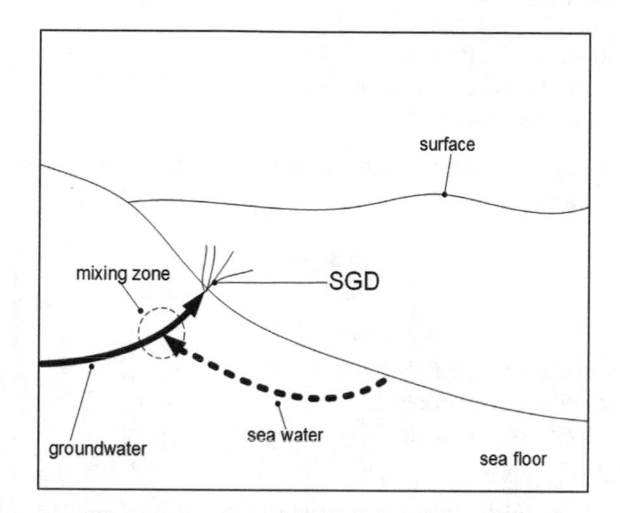

Fig. 1. Submarine Groundwater Discharge, modified after [6]

In order to coordinate the search of a swarm of collaborating AUVs, an overall search strategy has to be employed [7]. In this research, particle swarm optimisation (PSO) will be used for controlling the behaviour of the swarm.

2 Particle Swarm Optimisation

PSO is modelled on the behaviour of social entities, which collaborate to achieve a common goal [8, 9]. Examples of such behaviour in nature are fish schools or flocks of birds. Each individual of the swarm searches for itself, following a certain strategy. This is called the cognitive aspect of the search. The individuals are also influenced by the other swarm members. This is referred to as social aspect.

The PSO starts with random initialisation of the positions and the velocities of each swarm particle in the n-dimensional search space. After that, all particles of the swarm move through the search space with an adjustable velocity. The velocity of a particle is based on its current fitness value, the best solution found so far by this particle and the best solution found so far by the whole swarm (1):

$$\vec{v}_{i+1} = \vec{v}_i \cdot \varpi + r_1 \cdot c_1(\vec{p}_{best} - \vec{p}_i) + r_2 \cdot c_2(\vec{g}_{best} - \vec{p}_i) \tag{1}$$

Where:

$\vec{v}_{i+1}$: new velocity of a particle
$\vec{v}_i$: current velocity of a particle
ϖ: inertia weight

c_1: cognitive scaling factor
c_2: social scaling factor
r_1: random number from range [0, 1]
r_2: random number from range [0, 1]
$\vec{p}_i$: current position of a particle
$\vec{p}_{best}$: best known position of a particle
$\vec{g}_{best}$: best known position of the swarm

After calculating the new velocity of the particle, his new position $\vec{p}_i$ can be calculated as follows:

$$\vec{p}_{i+1} = \vec{p}_i + \vec{v}_{i+1} \cdot \Delta t \tag{2}$$

Where:

$\vec{p}_{i+1}$: new position of a particle
$\vec{p}_i$: current position of a particle
$\vec{v}_{i+1}$: new velocity of a particle
Δt: time step (one unit)

In (2), Δt, which always has the constant value of one unit, is multiplied to the velocity vector $\vec{v}_{i+1}$ in order to get consistency in the physical units [7].

3 Simulations

In this research a computer simulation based on a cellular automaton [10] is used, to simulate the distribution of discharged fresh- and recirculating seawater in a coastal environment. Fluorescent Dissolved Organic Matter (FDOM) is utilised to trace the distribution of the discharged water. The measurement of the FDOM-level is based on an optical method. Optical methods are highly suitable for detecting spatiotemporal patterns of relevant biogeochemical parameters, like dissolved organic matter or nutrients [11, 12].

3.1 Simulation Based on Cellular Automata

For the testing and fine-tuning of the search algorithm a dynamical simulation of submarine groundwater discharge in coastal water, based on a cellular automaton, was developed in recent work [13, 14].

The implemented simulation covers an area of 400 m × 400 m. This area is divided into 160,000 symmetric cells. The edge length of each cell is 1 m. Each cell has an x- and a y-position as well as an FDOM-level. Furthermore, the simulation uses rectangle shaped cells and the von-Neumann neighbourhood [15].

The simulation models the coastal water near the island of Spiekeroog, which is one of the East Frisian Islands, off the North Sea coast of Germany. The dynamic model contains two springs (SGDs) with a constant inflow rate of water and a constant FDOM level. The springs are located at position A = (200 m/200 m) and B = (100 m/300 m) (see Table 1). The parameters of the two springs differ; hence there is a global optimum and a local optimum. Furthermore, the simulation includes an obstacle (sandbank).

Table 1. Parameters of springs

	Spring A	Spring B
Position	(200 m/200 m)	(100 m/300 m)
Flow rate (m^3/iteration)	0.05	0.025
FDOM-level (arbitrary units)	100	50
Comment	Global optimum	Local optimum

The simulation was allowed to run for 5,000 iterations. Over the time the FDOM-level of each cell was logged. Figure 2 provides the FDOM-distribution after 50, 500, 1,000, 2,000, 3,000 and 5,000 iterations. The contour lines show the FDOM-level in steps of 0.1 units.

As it can be seen in Fig. 2, the simulation yields a symmetric distribution around both springs, while the inflow of spring A is much bigger than the inflow of spring B. The FDOM-level increases significantly only in the vicinity of the springs, while in all other areas the FDOM-level increases only marginally. This makes it a difficult search problem since no gradient information can be gathered in these areas.

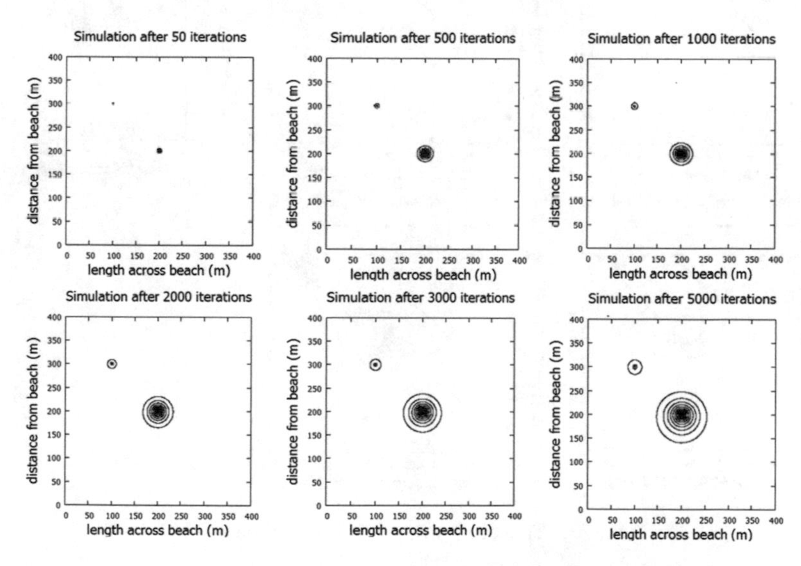

Fig. 2. Distribution of FDOM after different numbers of iterations

The simulation developed can be used as a dynamic test environment for direct search algorithms. Here, one iteration represents a time step of 1.7 s. To stabilise the behaviour of the AUVs, they update their velocities after 10 iterations, hence they move with the same velocity and direction for 17 s. In order to determine the effect of localisation and communication error on the convergence behaviour of PSO, four sets of experiments have been carried out; the first using standard PSO, the second using PSO with communication error only, the third using PSO with localisation error only, and the last one using PSO with both error sources.

3.2 Basic Particle Swarm Optimisation

For the experiments, three AUVs with a maximum velocity of 0.8 m/s were used. The cognitive scaling and the social scaling parameters were set to 2.0 according to [8]. The inertia weight ϖ determines the contribution rate of a particle's previous velocity to its velocity at the current time step. It is common to lower this value over time [9]. However, it was set to the constant value of 1.0 because of the time and energy constraints of an AUV.

The swarm operates in a dynamic simulated environment. Figure 3 shows the trajectories for a typical search run over time, in which the swarm finds the global optimum.

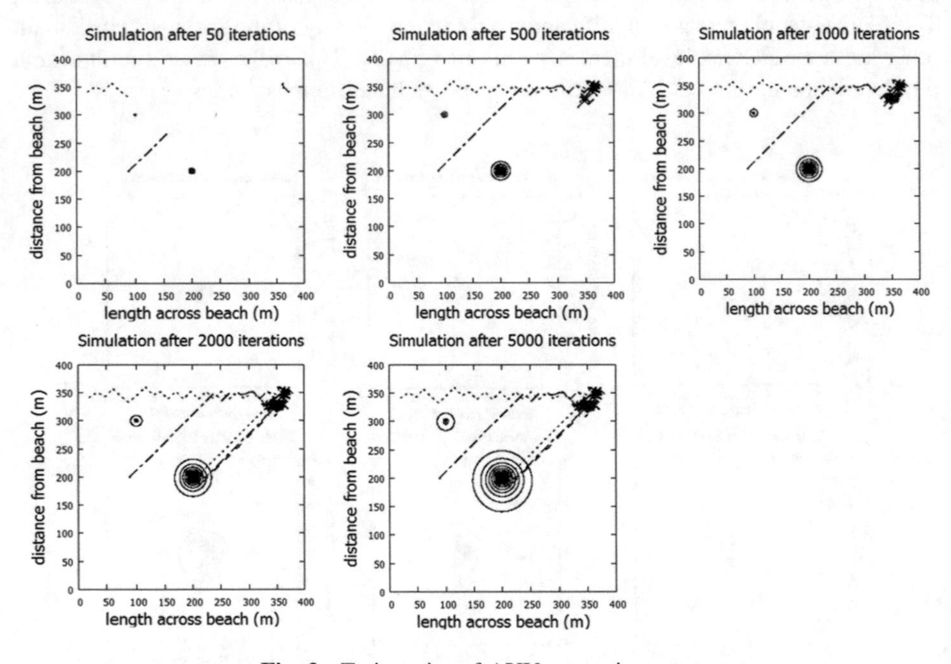

Fig. 3. Trajectories of AUVs over time steps

1,000 experiments were carried out in order to evaluate the performance of the basic PSO algorithm. From Fig. 4 it can be seen that the standard PSO has found the global optimum in 68.7% of the experiments. In 27.7% of the experiments it converged towards the local optimum and in 3.6% it did not converge at all.

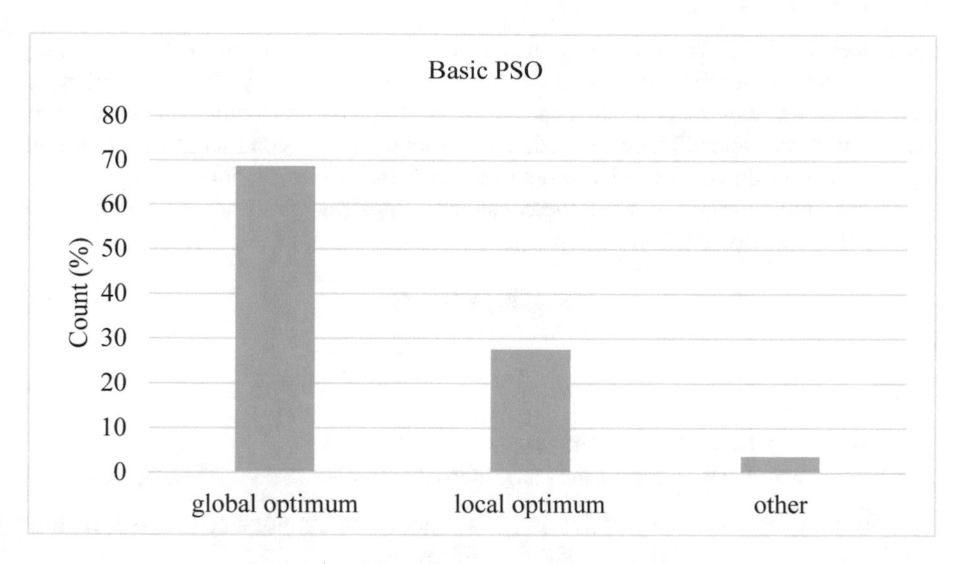

Fig. 4. Clustering of search results for basic PSO

Figure 5 shows examples of typical trajectories of the AUV´s for the three possible results, i.e. found global optimum (a), local optimum (b) or did not find an optimum (c). It can be seen, that in most cases, when the swarm finds the local optimum, one of the AUVs starts its search near by the local optimum (Fig. 5b).

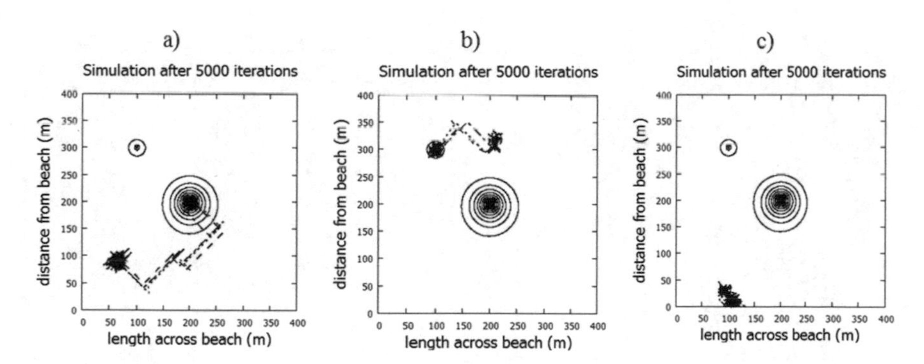

Fig. 5. Basic PSO converge against global optimum (a), local optimum (b) and without convergence against an optimum (c)

3.3 Particle Swarm Optimisation with Communication Error

Underwater communication suffers from various detrimental physical effects like propagation loss, absorption loss, multi-path propagation and temperature and pressure dependent sound velocity [16]. In a real underwater environment the reliability of a communication link is also influenced by the momentary three-dimensional orientation of each communication system (i.e. transmitting and receiving AUV) as the acoustic transducers exhibit a directional response pattern. Commercial acoustic modems apply spread spectrum techniques and OFDM (orthogonal frequency division multiplex) modulation schemes in order to cope with the frequency selective impairments of multi-path propagation. These methods allow maintaining a communication link with high probability on cost of reduced data rate (automatic rate adaption).

In the simulation of this work the considered impairments were restricted onto the propagation loss L_P as given by Eq. (3):

$$L_P = 20\log(d/d_0). \tag{3}$$

Where:

L_P: path loss
d: distance between two AUVs
d_0: reference distance for normalising the path loss.

Based on this loss a probability p_{com} for successful communication in each time step between any two AUVs is derived as following:

$$p_{com} = 100\% - L_P. \tag{4}$$

Where:

p_{com}: probability of successful communication.

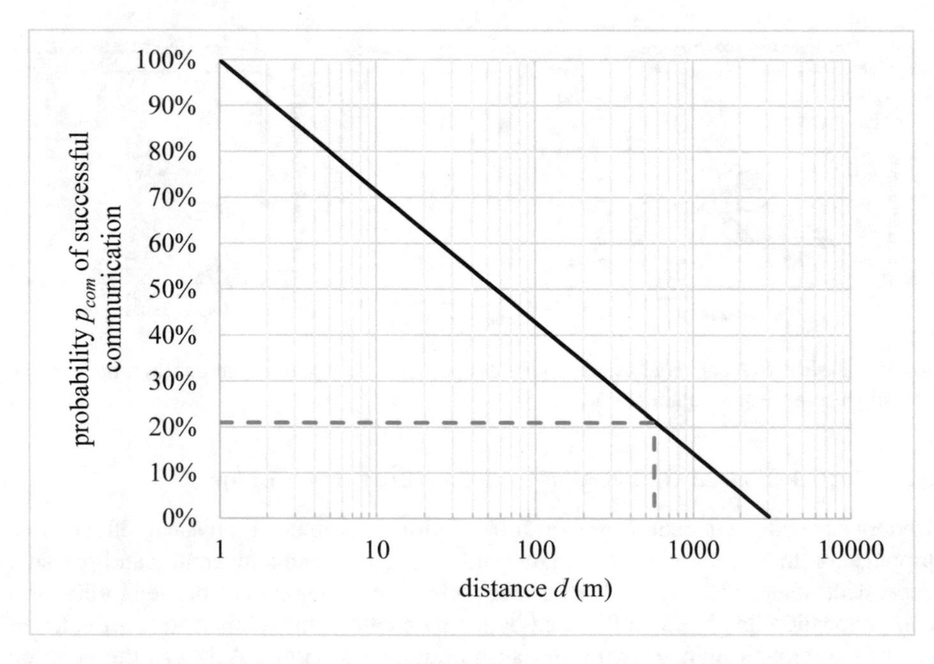

Fig. 6. Communication error function (d_0 = 70 m)

As it can be seen in Fig. 6 the maximum communication range is limited to approximately 3000 m. For the given search area of 400 m by 400 m the lowest possible value for p_{com} is 21% for the diagonally largest distance of 565.7 m.

1,000 experiments were carried out using PSO with communication error. It can be seen from Fig. 7 that the algorithm converged towards the global optimum in 58.8% of the runs. In 35.4% of the experiments the algorithm converged towards the local optimum, whereas in 5.8% it did not converge.

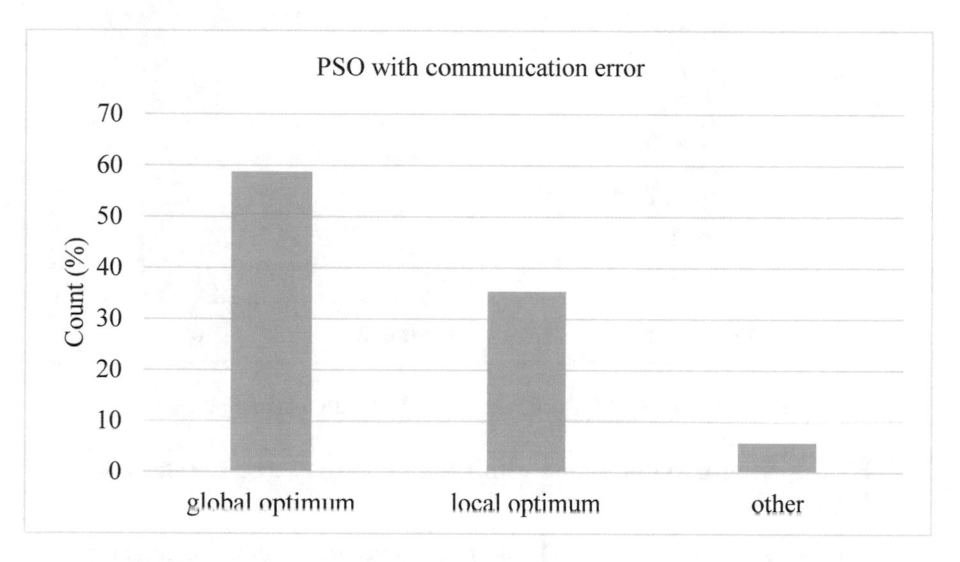

Fig. 7. Clustering of search results for PSO with communication error

3.4 Particle Swarm Optimisation with Localisation Error

The localisation of a submerged vehicle is difficult, because signals of global navigation satellite systems (GNSS) cannot be received underwater. Here, inertial navigation systems (INS) [17, 18] can be used instead, which continuously calculate the position, orientation and velocity of a moving object using dead reckoning without requiring external references.

An alternative approach is using acoustic based positioning systems, such as short baseline (SBL) and ultra short baseline (USBL) [19], which utilize acoustic distance and direction measurements for localisation. The disadvantage is that external acoustic sources are needed.

Another possibility for exact localisation is to allow the AUV to periodically surface and to update the position using GNSS [20, 21].

All the above methods are error prone. The magnitude of the error depends on the methods used as well as on the environmental conditions [20].

In this set of experiments, the localisation error was modelled using a Gaussian probability distribution with a standard deviation of 0.05 m. Again, 1,000 experiments were carried out, in which the global optimum was found in 39.6% of the searches.

In 12.8% of the runs PSO converged towards the local optimum and in 47.6% the algorithm did not converge (Fig. 8).

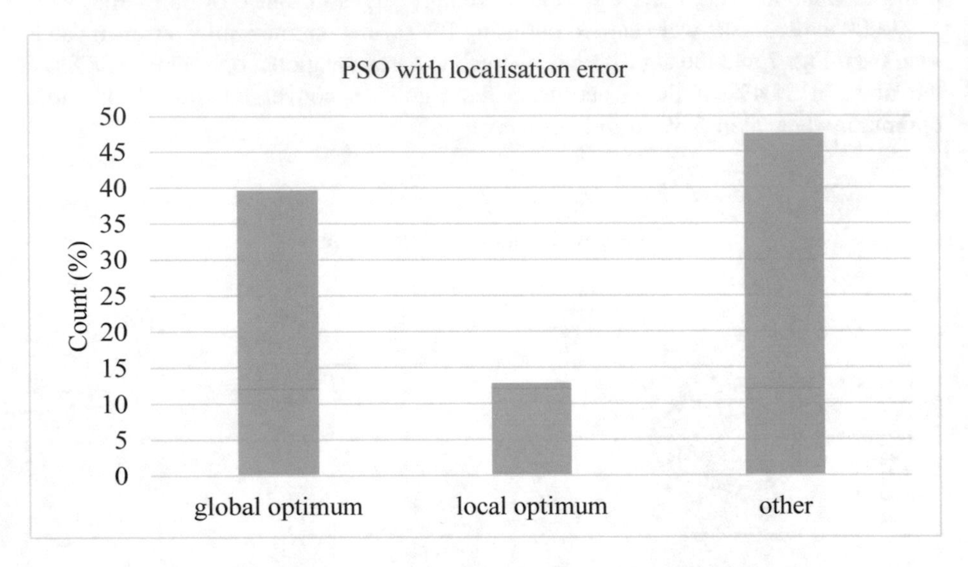

Fig. 8. Clustering of search results for PSO with localisation error

3.5 Particle Swarm Optimisation with Communication and Localisation Error

In the last set of experiments, both, the communication and the localisation error model, described above, were used in 1,000 simulations. It can be seen from Fig. 9 that the algorithm found the global optimum in 46.1% of the runs.

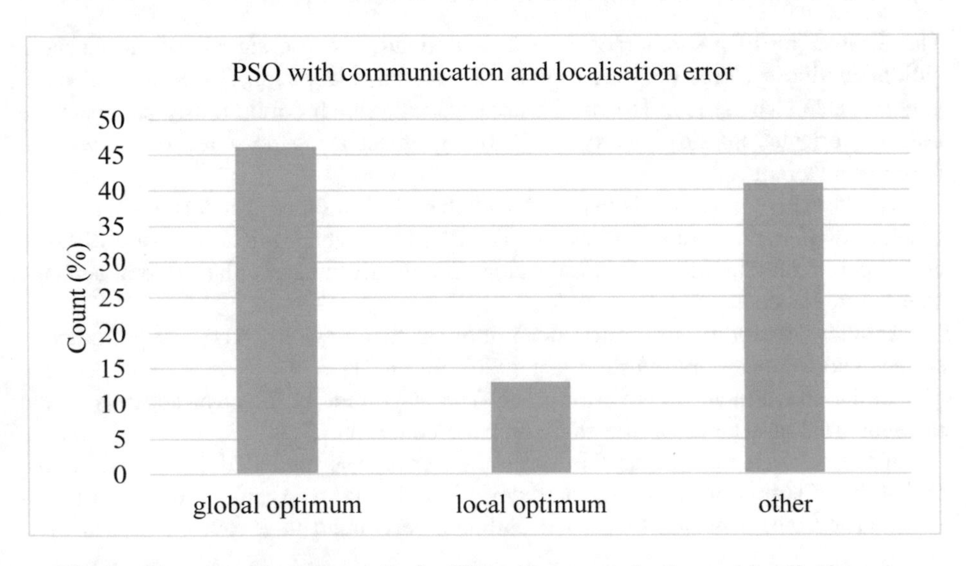

Fig. 9. Clustering of search results for PSO with communication and localisation error

The local optimum was found in 13% of the searches and in 40.9% of the simulation the algorithm did not converge.

4 Discussion

Figure 10 shows a comparison of the results obtained from the above experiments. It can be seen that both error types have a negative effect on the performance of the PSO algorithm: if communication error is introduced, the probability of PSO to find the global optimum drops from 68.7% to 58.8%. A similar effect can be observed if localisation error is applied. In this case, the probability of finding the global optimum drops from 68.7% to 39.6%. If both error types are applied, the ability to find the global optimum drops down to 46.1%, which is in the same order of magnitude as for localisation error alone. The probability of not converging increases from 3.7% for standard PSO to 5.8% when communication error is introduced. If the localisation error is included, the probability increases from 3.7% to 57.5%. When both errors are used, the probability increases to 40.9%.

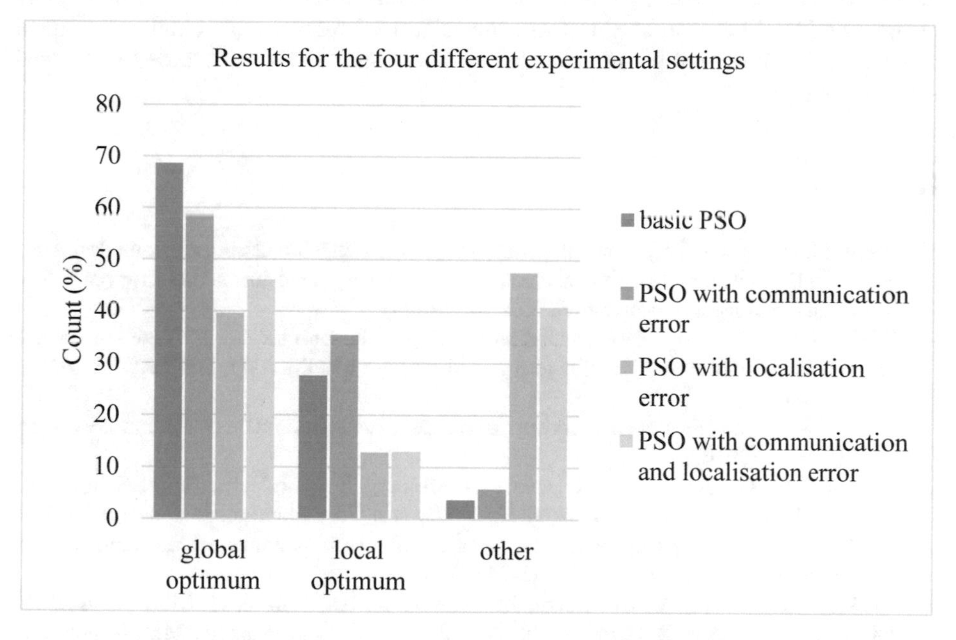

Fig. 10. Clustering of search results for the different search settings

This behaviour can be explained by the fact that in PSO, when an AUV finds a new global best known position, it sends this information to the other vehicles; if an AUV misses this update due to communication errors, it will still operate on the old global best known position and hence potentially perform optimisation in the area of a local optimum. This decreases the ability to find the global optimum.

When the localisation is error prone, the AUVs operate on false position information and hence the search is misguided, resulting in poor overall performance of PSO. When both error sources are combined, the latter effect seems to have a greater influence on the search capabilities than the communication error itself.

5 Conclusion and Future Work

The aim of this research was to investigate the effect of communication and localisation error on the behaviour of a small swarm of AUVs using a PSO algorithm for the search of submarine groundwater discharges. A dynamic underwater simulation based on a cellular automaton was used as test environment for the different search runs.

From the experiments it can be seen that both, communication and localisation errors, have a negative influence on the behaviour of the search algorithm. As depicted in Fig. 10, the localisation error has the greater negative influence on the capability of finding the global optimum.

In the next phase of this research it is proposed to investigate the behaviour of other search algorithms, for example hill climbing, random walk or Lévy flight [22] in the developed simulation. Also, the rule base of the simulation itself should be fine-tune by using real FDOM measurements from the island of Spiekeroog. Finally, the most promising algorithm will be implemented on a swarm of AUVs and tested in the real environment.

References

1. Zielinski, O., Busch, J.A., Cembella, A.D., Daly, K.L., Engelbrektsson, J., Hannides, A.K., Schmidt, H.: Detecting marine hazardous substances and organisms: sensors for pollutants, toxins and pathogens. Ocean Sci. **5**, 329–349 (2009)
2. Zielinski, O.: Airborne pollution surveillance using multi-sensor systems – new sensors and algorithms for improved oil spill detection and polluter identification. Sea Technol. **44**(10), 28–32 (2003)
3. Moore, W.S.: The effect of submarine groundwater discharge on the ocean. Ann. Rev. Mar. Sci. **2**, 59–88 (2010)
4. Nelson, C.E., Donahue, M.J., Dulaiova, H., Goldberg, S.J., La Valle, F.F., Lubarsky, K., Miyano, J., Richardson, C., Silbiger, N.J., Thomas, F.I.: Fluorescent dissolved organic matter as a multivariate biogeochemical tracer of submarine groundwater discharge in coral reef ecosystems. Mar. Chem. **177**, 232–243 (2015)
5. Beck, M., Reckhardt, A., Amelsberg, J., Bartholomä, A., Brumsack, H.J., Cypionka, H., Dittmar, T., Engelen, B., Greskowiak, J., Hillebrand, H., Holtappels, M., Neuholz, R., Köster, J., Kuypers, M.M.M., Massmann, G., Meier, D., Niggemann, J., Paffrath, R., Pahnke, K., Rovo, S., Striebel, M., Vandieken, V., Wehrmann, A., Zielinski, I.: The drivers of biogeochemistry in beach ecosystems: a crossshore transect from the dunes to the low water line. Mar. Chem. **190**, 35–50 (2017)
6. Evans, T.B., Wilson, A.M.: Groundwater transport and the freshwater–saltwater interface below sandy beaches. J. Hydrol. **538**, 563–573 (2016)

7. Nolle, L.: On a search strategy for collaborating autonomous underwater vehicles. In: Mendel 2015, 21st International Conference on Soft Computing, Brno, CZ, pp. 159–164 (2015)
8. Kennedy, J., Eberhart, R.C., Shi, Y.: Swarm Intelligence. Academic Press, Cambridge (2001)
9. Bansal, J.C., Singh, P.K., Saraswat, M., Verma, A., Jadon, S.S., Abraham, A.: Inertia weight strategies in particle swarm optimization. In: Third World Congress on Nature and Biologically Inspired Computing, Salamanca, Spain, 19–21 October, pp. 633–640 (2011)
10. Wolfram, S.: Universality and complexity in cellular automata. Physica D **10**(1–2), 1–35 (1984)
11. Moore, C., Barnard, A., Fietzek, P., Lewis, M.R., Sosik, H.M., White, S., Zielinski, O.: Optical tools for ocean monitoring and research. Ocean Sci. **5**, 661–684 (2009)
12. Zielinski, O., Voß, D., Saworski, B., Fiedler, B., Körtzinger, A.: Computation of nitrate concentrations in turbid coastal waters using an in situ ultraviolet spectrophotometer. J. Sea Res. **65**, 456–460 (2011)
13. Nolle, L., Thormählen, H., Musa, H.: Simulation of submarine groundwater discharge of dissolved organic matter using cellular automata. In: 30st European Conference on Modelling and Simulation ECMS 2016, pp. 265–269 (2016)
14. Tholen, C., Nolle, L., Zielinski, O.: On the effect of neighborhood schemes and cell shape on the behaviour of cellular automata applied to the simulation of submarine groundwater discharge. In: 31th European Conference on Modelling and Simulation ECMS 2017, pp. 255–261 (2017)
15. Jiménez, A., Posadas, A.M., Marfil, J.M.: A probabilistic seismic hazard model based on cellular automata and information theory. Nonlinear Processes Geophys. **12**(3), 381–396 (2005)
16. Lurton, X.: An introduction to underwater acoustics: principles and applications. New York Springer, London (2002)
17. Levinson, E., Ter Horst, J., Willcocks, M.: The next generation marine inertial navigator is here now. In: 1994 Position Location and Navigation Symposium, Las Vegas, NV, pp. 121–127. IEEE (1994)
18. Curey, R.K., Ash, M.E., Thielman, L.O., Barker, C.H.: Proposed IEEE inertial systems terminology standard and other inertial sensor standards. In: PLANS 2004. Position Location and Navigation Symposium, pp. 83–90 (2004)
19. Rigby, P., Pizarro, O., Williams, S.B.: Towards geo-referenced AUV navigation through fusion of USBL and DVL measurements. In: OCEANS 2006, Boston, MA, pp. 1–6 (2006)
20. Dunbabin, M., Roberts, J., Usher, K., Winstanley, G., Corke, P.: A hybrid AUV design for shallow water reef navigation. In: 2005 IEEE International Conference on Robotics and Automation, pp. 2105–2110 (2005)
21. Dunbabin, M., Corke, P., Vasilescu, I., Rus, D.: Data muling over underwater wireless sensor networks using an autonomous underwater vehicle. In: 2006 IEEE International Conference on Robotics and Automation, pp. 2091–2098 (2006)
22. Edwards, A.M., Phillips, R.A., Watkins, N.W., Freeman, M.P., Murphy, E.J., Afanasyev, V., Buldyrev, S.V., da Luz, M.G.E., Raposo, E.P., Stanley, H.E., Viswanathan, G.M.: Revisiting levy flight search patterns of wandering albatrosses, bumblebees and deer. Nature **449**, 1044–1049 (2007)

Spam Detection Using Linear Genetic Programming

Clyde Meli[1(✉)], Vitezslav Nezval[1],
Zuzana Kominkova Oplatkova[2], and Victor Buttigieg[3]

[1] Department of Computer Information Systems,
University of Malta, Msida, Malta
clyde.meli@um.edu.mt, vnez@cis.um.edu.mt
[2] Department of Informatics and Artificial Intelligence,
Tomas Bata University, Zlín, Czech Republic
kominkovaoplatkova@fai.utb.cz
[3] Department of Communications and Computer Engineering,
University of Malta, Msida, Malta
victor.buttigieg@um.edu.mt

Abstract. Spam refers to unsolicited bulk email. Many algorithms have been applied to the spam detection problem and many programs have been developed. The problem is an adversarial one and an ongoing fight against spammers. We prove that reliable Spam detection is an NP-complete problem, by mapping email spams to metamorphic viruses and applying Spinellis's [30] proof of NP-completeness of metamorphic viruses. Using a number of features extracted from the SpamAssassin Data set, a linear genetic programming (LGP) system called Gagenes LGP (or GLGP) has been implemented. The system has been shown to give 99.83% accuracy, higher than Awad et al.'s [3] result with the Naïve Bayes algorithm. GLGP's recall and precision are higher than Awad et al.'s, and GLGP's Accuracy is also higher than the reported results by Lai and Tsai [19].

Keywords: Spam detection · NP-complete · Identification · Security
Linear genetic programming

1 Introduction

The paper deals with the ongoing problem of spam detection. Spam is unsolicited bulk email (UBE), which are emails sent to a large number of recipients against their wishes. The TREC 2005 conference [8] defined spam as "unsolicited, unwanted email that was sent indiscriminately, directly or indirectly, by a sender having no current relationship with the recipient." The total yearly cost worldwide of dealing with spam was estimated by Rao and Riley to be of 20 billion dollars [28]. Such costs include the time wasted going through irrelevant advertisements and missing important emails which end up in the spam folder. It also includes costs of server hardware which have to support about five times more disk capacity than would be needed without the existence of spam, as

R. Matoušek (Ed.): MENDEL 2017, AISC 837, pp. 80–92, 2019.
https://doi.org/10.1007/978-3-319-97888-8_7

well as the labour cost of anti-spam support staff. This estimate ignores help-desk support for spam.

The evolution of spam is discussed in the literature for instance, Hunt and Carpinter [15], Almeida and Yamakami [1].

To defend against the onslaught of spam in our email boxes, we use spam filters. Spammers continually attempt to improve the way they send such spams. Email spams include these days emails with image spams, emails with malicious attachments and ransomware. The latter can look innocuous and the same as a normal email, however, they contain a very malicious attachment or link to a compromised website ready to infect your PC. Indeed, botnets are one of the main sources of spam [17]. Symantec in their 2016 Internet Security Threat Report [37] comment that the spam problem is not going away. Today we find blogs with the problem of blog spam [18], web spam misleading search engines [25] and social spam on social networks [20]. This paper will focus on email spam only. In this paper, we will prove that there exist spam emails whose reliable detection is NP-complete [9] in complexity and therefore the problem of reliable email spam detection is NP-complete. This will proceed by showing the equivalency of spam emails and virus malware. Spinellis [30] proved that reliable metamorphic virus detection is NP-complete, thus it will be shown that reliable email spam detection is similarly NP-complete. Further to our initial work in [22], we have expanded on the original set of features and utilised the SpamAssassin dataset[1]. We show that the use of linear genetic programming can reliably detect email spam. A number of innovative features have been used.

1.1 Defence Against Spam

Spam filters are software which detects spam. Various solutions to tackle the spam problem have been suggested. They range from legislation e.g. USA Can Spam Act 2004 and EU Directive 2002/58/EC, email challenge response systems to software filters. The effectiveness of such filters is essentially how well they keep legitimate messages out of their spam folder. Software filters can be divided into two. Those not based on Machine Learning, such as rule-based filters, blacklists, greylisting [13] and signatures, and those based on Machine Learning, such as naïve Bayes [10, 11], SVM [32], Neural Networks [29] and our work on LGP using Reverse Polish Notation (RPN) [22].

In our earlier work [22], we showed theoretically that RPN expressions are more expressive than naïve Bayes. Similarly in [23], it is shown that they are more expressive than SVMs. By the use of a set of features which also includes URL-based ones, our LGP system can obtain higher accuracy compared to the literature.

In the following section, we establish that the reliable detection of instances of such spams is a problem of NP-complete complexity.

[1] Available at http://csmining.org/index.php/spam-assassin-datasets.html

2 Detection Complexity and Addendum

The following proof maps metamorphic viruses to email spam. Metamorphic computer viruses can generate modified but logically equivalent copies of themselves through editing, translating and rewriting of their own code.

Theorem 1: Email Spams may be mapped to Metamorphic Viruses.

Proof. Clearly, email spam, like computer viruses, is a moving target – it changes with the times.

Recall that spammers often purposely misspell words. If we group a finite number (N) of the possible misspellings and modifications of every word in an original (spam) message, we can consider a finite possible number of modifications of the original spam email.

Every spam email may be modified by spammers to evade detection. Consider the detection of that group of spams as follows. Consider an M-term Boolean formula S corresponding to all these finite possibilities defined as follows (as similarly defined in Spinellis [30] for viruses, but in our case, also for email spam):

$$S = \left(x_{a_{1,1}} \vee x_{a_{1,2}} \vee \neg x_{a_{1,3}} \vee \ldots\right) \wedge \left(x_{a_{2,1}} \vee x_{a_{2,2}} \vee \neg x_{a_{2,3}} \vee \ldots\right) \wedge \left(x_{a_{3,1}} \ldots\right) \wedge \ldots \quad (1)$$

$$\vdots$$

$$0 \leq a_{i,j} < N \quad (2)$$

where

$a_{i,j}$ are all indices to potential modifications of words in an email and
$x_{a_{i,j}}$ is a particular modification to a word in an email
$a_{i,1}, a_{i,2}, .., a_{i,j}$ refer to the presence of a group of words (1 = present, 0 = not present)
$x_{a_{i,1}}, x_{a_{i,2}}, .., x_{a_{i,j}}$ refer to the particular group of words themselves
An email spam can be defined as a triple (*f*, *s*, *c*) where
f is an email modification function or map
s is a Boolean variable indicating that the spam matches (or is a "solution" to) S, and
c is an integer encoding the candidate values for S.

Thus if we have a spam such as 'Purchase your Viagra' represented by the triple (*f*, *s*, *c*), then we might have a new triple (*f*, *s*', *c*') which represents 'Purchase your Via4gra'.

For a typical NP-complete problem with M terms, there are M variables with each having 1 of b values. The number of candidate solutions, in this case, is equal to b^M, which is all of the possible values which can be taken.

In general, there are M terms and N possible values per individual word. Thus, the number of candidate values is equal to N^M. Here candidate values would represent all the possible modifications – misspellings, swapping characters or inserting other characters (up to a limit) – to a sentence made up of words in a dictionary. A simplified example (using a word instead of a sentence) for the dictionary word 'viagra' would be the following candidate values {'viagra', 'vi0gra', 'via4gra', 'via5gra', 'via3gra', 'viag4ra'}. A real-world example would have much more values of course.

A metamorphic self-replicating virus generates the next virus in a sequence and may be defined (by Spinellis [30]) as a triple (*ff*, *ss*, *cc*) where

ff is a virus replication and processing function or map

ss is a Boolean variable which indicates whether an instance of the virus has found a "solution" to S and

cc is an integer encoding the candidate values for S.

Now consider a particular virus triple and the email spam triple such that there are the same number of candidate values. Since *ss* and *cc* are of the same type and number, then we can say the virus triple maps to the email triple. Thus, it is proved that Email Spams may be mapped to Metamorphic Viruses.

Theorem 2: Email Spam detection is NP.

Proof. Clearly, the problem is in NP as there is no P algorithm to detect spam. It is known to be an open-ended problem.

The adversarial nature of the problem means that it is an intractable problem with no completely reasonable polynomial time algorithm. Indeed, it is intractable since the email protocol made several wrong assumptions. It assumed that every email user and server is trustworthy, every message is wanted by the recipient and that no one will abuse the service.

Theorem 3: Email Spam detection is NP-complete.

Proof. Minimising the number of misclassified documents (even in just two categories), which is theoretically the same as finding a separating hyper-plane to do this classification, is an NP-complete problem, called "Open Hemisphere" by Garey and Johnson [5, 9].

It has been proved above that the problem of email spam detection is in NP, and that Email Spams may be mapped to Metamorphic Viruses.

Thus, since Spinellis proved that Metamorphic Viruses are NP-complete, then it follows that email spam detection too is NP-complete.

Thus, a spam detector D for a particular group of spams GS (all made of possible simple modifications (misspellings, additions of digits, etc.) to a "progenitor" spam words such as viagra) can be used to solve the satisfiability problem [30]. The latter is known to be NP-complete.

3 Methodology and Features

This work utilises the SpamAssassin Corpus[2] which consists of 4150 ham emails and 1896 spam emails (31% spam ratio). It was preferred due to its medium size. The dataset was split into five equally sized segments for the purpose of cross-validation.

[2] As updated by http://csmining.org.

Four of the five were taken as the training set, and the rest was split in half as a validation set and the other half as a testing set. Every generation, the current segment is switched, but the validation and testing sets are always kept the same. This was based on a technique by [16] to avoid overfitting.

Figure 1 shows the overall architecture of the Gagenes LGP (GLGP) system.

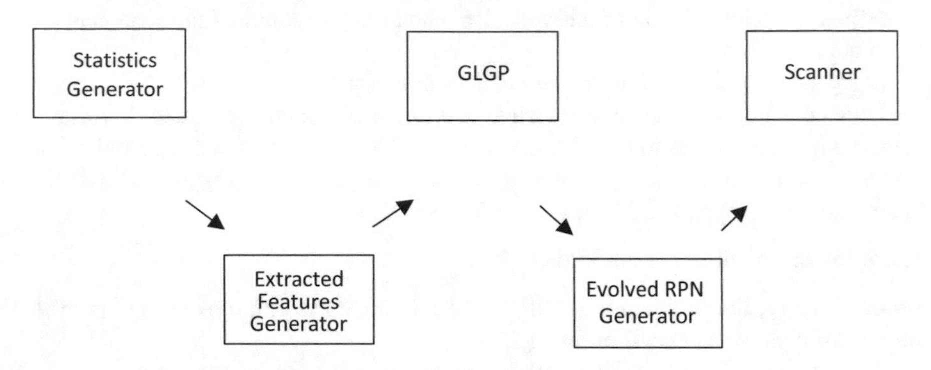

Fig. 1. LGP system architecture (figure from [23])

HTML comments are not taken to be token separators, meaning that any word may have an HTML comment in its middle. The reason for this is that spammers sometimes insert HTML comments in the middle of a word, known as Fake Taps [7]. It is assumed that alphanumeric characters, dashes, apostrophes, the euro symbol and dollar signs are part of tokens, and every other symbol that may remain is taken to be a token separator. It was chosen to ignore tokens that are all digits. The reason for this being that the original proposal by Graham [11] called "A Plan for Spam" as well as Stuart et al's Neural Network Classifier [31] ignore digits too. Stop-words are words filtered out before or after processing of text. They were not removed by the tokenizer since it was not expected [2] to have a such a great effect on the system's accuracy. Any MIME-Encoded emails encountered by GLGP when processing emails are decoded and the result replaces the original message body. Some emails use this encoding to evade detection.

Following the tokenization process, a number of features are evaluated for each email message. Feature detector values (returned by the detectors working on a particular email) are called *registers* by the expression evaluator. Operators can be +, −, *, /(division), sine and cosine. Operands are either constants or registers. There are a fixed number of registers, as many as the features which are used. Similarly, there are a fixed number of ephemeral random constants (ERC) used, similar to Brameier's original LGP [6]. This set of constants is generated randomly within a preset range at the instantiation of the chromosome, except for the first two which are always 0 and 1.

Operands and operators are combined using RPN [36] expressions. The latter are encoded using a proposed RPN-LGP Block Representation which restricts the chromosome to only valid RPN expressions.

Single-point crossover was used with elitism. Bit level Mutation was implemented together with Macro-Mutation as in [6]. Holland's Proportional Fitness Selection [4] was used.

We encountered the LGP Hurdle problem [35]. The hurdle problem occurs when it is easy to classify a bulk of fitness cases for one class, but at a cost of reducing accuracy for learning to recognize the other class. This creates a strong selection pressure. In our case, we solved it by introducing a new algorithm we are calling the First Subexpression Algorithm. It resembles the use of multiple output registers in LGP, as well as Multi Expression Programming (MEP) [26, 27]. The first subexpression result of the evaluated RPN expression is used as an output register. The first subexpression is part (or all) of the first block. The algorithm is as follows:

gpresult=v_i .... the complete evaluated value of the expression *i*

gpsub=the evaluated value of the first subexpression of expression *i*

ht=ham threshold for RPN

st=spam threshold for RPN

Training Phase

When Training Spam: For a classifier i, if gpresult ≥ st then i∈Spam

When Training Ham: For a classifier i, if gpsub ≤ ht then i∈Ham

The features used are grouped into four main categories: subject features, priority and content-type headers, message-body headers and URL-based headers. Some of the features used are lexical and syntactical, such as Yule's Measure [33]. A subset of the features are called Meli's Measure [23]. They are based on modifications of Yule's Characteristic K [24].

The features are summarised in Tables 1, 2, 3 and 4. Some of these features find their primary basis in Sahami et al. [21]. In [21] they defined specific features, such as phrasal features, e.g., the appearance of specific phrases such as "FREE" or "GRATIS", as well as other features such as sender domains, whether the email was sent via a mailing list.

The URL-based features (Table 3) are traditionally used in Fast Flux detection [12, 14] and are being proposed as being useful in email spam detection.

Evaluation is done using standard spam filter performance measures accuracy, precision, recall, F1 and TCR [15, 34].

[3] All these features are found and explained further in my Ph.D. thesis [23]

[4] URL features (Table 4) are practically numbered as also being part of message body features

Table 1. Subject-based features (table based on [23])[3]

Feature number	Feature name	Description
1	NOOFALPHANOVOWELS	No of alphabetic words without any Vowels in subject
2	NOOFSTRANGEWORDS	No of strange words in subject
3	NOOFSENTENCES	No of sentences in subject
4	AVGWORDSPERSENTENCE	Average words per sentence in subject
5	NOOFATLEAST2SUBJECT	No of alphabetic words with at least 2 characters from B, C, M, F, V, G, R in subject
6	NOOFATLEAST2BSUBJECT	No of alphabetic words with at least 2 characters from C, E, A, S, M, X, L in subject
7	NOOFWORDS15	No of words at least 15 characters long in subject
8	NOOFSTRANGEWORDSPUNCDIGIT	No of words with no Maltese or English alphanumeric characters in subject
9	NOOFSOMECAPS	No of some capitals in subject
10	NOOFALLCAPS	No of all capitals in subject
11	NOOFBINREP3	Occurrence of a character in a word repeated at least 3 times in succession (Binary Feature. Yes = 1, No = 0)
12	NOOFINTREP3	No of words in subject header with the repetitive occurrence of a character in a word at least 3 times in succession
13	NOOFMORETHAN12	More than 12 words in subject
14	YULES	Yule's measure in subject
15	HAPAXLEGOMENA	Number of Hapaxlegomena in subject
16	HAPAXDISLEGOMENA	Number of Hapaxdislegomena in subject
17	SIMPSONS	Simpson's in subject
18	HONORER	Honore's Score in subject
19	CMSIN	CM (SIN) in subject
20	ZIPFSLAW	Zipf's Law in subject

Table 2. Priority and content-type header features (table based on [23])

Feature number	Feature name	Feature description
21	NOOFALPHANOVOWELS	No of alphabetic words without any Vowels in PCT
22	NOOFSTRANGEWORDS	No of strange words in PCT
23	CHECKPRIORITY	Check priority
24	YULES	Yule's measure in PCT
25	HAPAXLEGOMENA	Number of Hapax legomena in PCT
26	HAPAXDISLOGEMENA	Number of Hapaxdislegomena in PCT
27	SIMPSONS	Simpson's in PCT
28	HONORE	Honore's score in PCT
29	CMSIN	Meli's sin in PCT
30	NOOFSENTENCES	No of sentences in PCT
31	AVGWORDSPERSENTENCE	Average words per sentence in PCT

Table 3. URL-based features (table based on [23])

Feature number	Feature name	Description
57	WOT	Web of trust
58	SBL	Safe browsing lookup by google
59	DNSTTLFLUX	TTL flux
60	DNSARECORDS	DNS A records
61	WHOIS	Whois
62	DNSAVGTTL	DNS TTL Avg
63	DNSSTDDEVTTL	DNS TTL Stddev
64	DNSPERCENTNUMCHARS	DNS numeric percentage
65	PERCENTLENLMS	Domain name longest meaningful substring

Table 4. Message body features (table based on [23])[4]

Feature Number	Feature name	Feature name
32	NOOFALPHANOVOWELS	Number of alphabetic words without any Vowels
33	NOOFSTRANGEWORDS	Number of strange words
34	NOOFSENTENCES	Number of sentences
35	AVGWORDSPERSENTENCE	Average words per sentence
36	NOOFATLEAST2	Number of alphabetic words with at least 2 characters from B, C, M, F, V, G, R
37	NOOFATLEAST2B	Number of alphabetic words with at least 2 characters from C, E, A, S, M, X, L
38	NOOFWORDS15	Number of words at least 15 characters long
39	NOOFSTRANGEWORDSPUNCDIGITS	Numbers of words with no Maltese or English alphanumeric characters
40	NOOFSOMECAPS	Number of some capitals
41	NOOFALLCAPS	Number of all capitals
42	NOOFBINREP3	Occurrence of a character in a word repeated at least 3 times in succession (Binary Feature. Yes = 1, No = 0)
43	NOOFINTREP3	Number of words in the body which have the occurrence of a character in a word repeated at least 3 times in succession
44	NOOFMORETHAN12	More than 12 words
45	YULES	Yule's measure

(*continued*)

Table 4. (*continued*)

Feature Number	Feature name	Feature name
46	HAPAXLEGOMENA	Number of Hapax legomena
47	HAPAXDISLEGOMENA	Number of Hapaxdislegomena
48	SIMPSONS	Simpson's Rule
49	HONORER	Honore's Score
50	CMSIN	Meli's Measure (SIN)
51	ZIPFSLAW	Zipf's Law
52	HTMLTAGS	Html tags present in message
53	HYPERLINKS	Hyperlinks
54	IMGCLICKS	Html contains a clickable image
55	HTMLWHITE	Html contains white text
56	ISTOOLONG	Checks if a message is overly long
66	CMPLAIN1	Meli's (PLAIN1)
67	CMPLAIN2	Meli's (PLAIN2)
68	CMCOS	Meli's (COS)
69	CMTAN	Meli's (TAN)
70	CMLOG1	Meli's (LOG1)
71	CMLOG2	Meli's (LOG2)
72	CMSHANNON	Meli's (SHANNON ENTROPY)

4 Results

Experimental parameters are given in Table 5.

Table 5. Experimental parameters (table based on [23])

Parameter	Setting
Population size	60
Mutation probability	0.1
Crossover probability	0.7
Copy probability	0.1
Ham threshold	−1 (no uncertainty)
Spam threshold	0
Maximum generations	100
Instruction set	{+,−,*,/,*sin*,*cos*}
Register set (features)	$\{x_0, x_{k-1}\}$ (*k* inputs)
Constant set	{0,1,...} (each real-valued constant[a] ranges between −1.5 to 3.5)
Number of runs	30

[a]These real-valued constants are a fixed set and are "write-protected" as [6] calls it. Except for 0 and 1 which are always present in the set, they are only initialised and randomly generated at the initialisation of a chromosome, within the stated range i.e. −1.5 to 3.5

Training was performed as follows:

1st Generation: (For every chromosome) Training on 100 spams and 100 hams from the 1st Segment. Validation Set is used for validation of the best chromosome.

2nd to 4th Generations: Training on 100 spams and 100 hams from the corresponding Segment (2nd Segment for the 2nd Generation, …). Validation Set is used for validation of the best chromosome.

5th Generation: Training on all four Training Segments i.e. all the training data (which is unbalanced). Any issue with unbalanced classification is helped by having the other generations working on a balanced subset. Validation Set is used for validation of the best chromosome.

From the 6th Generation onwards the 1st, 2nd, … Segments are used again. On the 10th Generation, all four segments are used again, and termination can occur. This process goes on repeatedly until 100 generations are reached.

The chart in Fig. 2 shows the median maximum training and testing fitness and any overfitting, over 30 runs with the full (entire) feature set. Overfitting can be seen to reduce from the 70th generation approximately, towards the 100th generation where the difference between training and testing fitnesses stabilises towards small values close to zero.

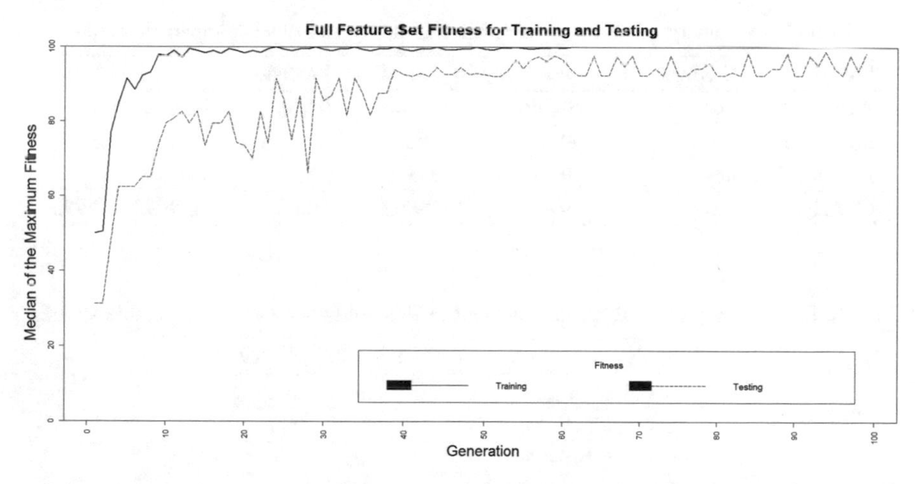

Fig. 2. Median of the maximum fitness values calculated over 30 runs as it varies across generations for both training and test sets when using the full feature set and using the RPN first subexpression algorithm (figure from [23])

Spam Filter Accuracy was subsequently evaluated. The largest Testing Fitness was of 99.9173. This corresponds to 99.9173% accuracy, 0.989418 Precision, 0.98421 Recall and 0.98681 F1-score. The average Testing Fitness at Generation 100 corresponded to 88.5785% accuracy. This was calculated by averaging the Testing accuracy output over 30 runs for all the techniques investigated, and taking the largest value reached. We assume that FALSE POSITIVES are as costly as FALSE NEGATIVES and thus Sahami's λ [21] is equal to 1 (i.e. TCR1) or 9 (i.e. TCR9).

In Table 6 one can see a combined overview of averages, medians and maximum values obtained for precision, recall, F1 and accuracy measures at Generation 100 over 30 runs.

Table 6. Comparison with averages and median

	Precision	Recall	F1	Accuracy
Average	0.999173	0.823245	0.902718	0.857904
Median	0.999173	0.807032	0.892883	0.982644
Maximum	0.999173	0.999173	0.999173	0.998347

The highest precision reached was 0.999173, and this was the same as the highest recall reached.

Tables 7 and 8 show the main best results obtained by GLGP using the full feature set. They compare well with the results in the literature on the same dataset. GLGP's Accuracy is higher than Awad et al.'s [3] result with the Naïve Bayes algorithm. GLGP's recall and precision are higher than Awad et al.'s, and GLGP's Accuracy is also higher than the reported results by Lai and Tsai [19].

Table 7. Summary of results obtained in this study and other comparable studies

Study	Stopwords kept	Accuracy %	Recall %	Precision %	F1	WAcc
Awad et al.	no	99.46	98.46	99.66	n/a	n/a
Lai and Tsai	no	91.78	n/a	n/a	n/a	n/a
Lai and Tsai	yes	91.71	n/a	n/a	n/a	n/a
GLGP	**yes**	**99.83**	**99.92**	**99.92**	**0.9992**	**0.9992**

Table 8. Summary of TCR results obtained in this study and other comparable studies

Study	Stopwords kept	TCR1	TCR9
Awad et al.	no	n/a	n/a
Lai and Tsai	no	n/a	n/a
Lai and Tsai	yes	n/a	n/a
GLGP	**yes**	**414**	**46**

5 Conclusion

In this paper, a proof that perfect email detection is NP-complete has been demonstrated. Our LGP system has been successfully implemented using the entire feature set. It was tested on the SpamAssassin dataset. Its results are comparable to state-of-the-art methods. Its accuracy, recall and precision are higher than Awad et al.'s [3] result with the Naïve Bayes algorithm. Also, its accuracy is higher than Lai and Tsai's [19] accuracy.

In the future, we will investigate further the Meli's Measure features. We will also be looking at other datasets and potentially applying the GLGP system to other problem areas.

Acknowledgements. This work was supported by the Ministry of Education, Youth and Sports of the Czech Republic within the National Sustainability Programme project No. LO1303 (MSMT-7778/2014) and also by the European Regional Development Fund under the project CEBIA-Tech No. CZ.1.05/2.1.00/03.0089 and further it was supported by Grant Agency of the Czech Republic—GACR P103/15/06700S.

This research has in part been carried out using computational facilities procured through the European Regional Development Fund, Project ERDF-076 'Refurbishing the Signal Processing Laboratory within the Department of CCE', University of Malta.

References

1. Almeida, T.A., Yamakami, A.: Advances in spam filtering techniques. In: Computational Intelligence for Privacy and Security, pp. 199–214. Springer, Heidelberg (2012)
2. Androutsopoulos, I., et al.: An experimental comparison of naive bayesian and keyword-based anti-spam filtering with personal e-mail messages. In: Proceedings of the 23rd Annual International ACM SIGIR Conference on Research and Development in Information Retrieval, pp. 160–167. ACM, New York (2000)
3. Awad, W.A., ELseuofi, S.M.: Machine learning methods for e-mail classification. Int. J. Comput. Appl. **16**(1), 0975–8887 (2011)
4. Blickle, T., Thiele, L.: A Comparison of selection schemes used in genetic algorithms. Gloriastrasse 35, CH-8092 Zurich: Swiss Federal Institute of Technology (ETH) Zurich, Computer Engineering and Communications Networks Lab (TIK (1995)
5. Borodin, Y., et al.: Live and learn from mistakes: a lightweight system for document classification. Inf. Process. Manag. **49**(1), 83–98 (2013)
6. Brameier, M.: On linear genetic programming. Fachbereich Informatik, Universität Dortmund (2004)
7. Cid, I., et al.: The impact of noise in spam filtering: a case study. In: Perner, P. (ed.) Advances in Data Mining. Medical Applications, E-Commerce, Marketing, and Theoretical Aspects, pp. 228–241. Springer, Heidelberg (2008)
8. Cormack, G.V., Lynam, T.: TREC 2005 spam track overview. In: The Fourteenth Text REtrieval Conference (TREC 2005) Proceedings (2005)
9. Garey, M.R., Johnson, D.S.: Computers and Intractability: A Guide to the Theory of NP-Completeness. W. H. Freeman, San Francisco (1979)
10. Graham, P.: Better Bayesian Filtering. http://www.paulgraham.com/better.html
11. Graham, P.: A Plan for Spam. http://www.paulgraham.com/spam.html
12. Gržinić, T., et al.: CROFlux—Passive DNS method for detecting fast-flux domains. In: 2014 37th International Convention on Information and Communication Technology, Electronics and Microelectronics (MIPRO), pp. 1376–1380 (2014)
13. Harris, E.: The Next Step in the Spam Control War: Greylisting. http://projects.puremagic.com/greylisting/whitepaper.html
14. Holz, T., et al.: Measuring and detecting fast-flux service networks. In: 15th Network and Distributed System Security Symposium (NDSS) (2008)
15. Hunt, R., Carpinter, J.: Current and new developments in spam filtering. In: 2006 14th IEEE International Conference on Networks, pp. 1–6 (2006)

16. Gonçalves, I.: Controlling Overfitting in Genetic Programming. CISUG (2011)
17. Juknius, J., Čenys, A.: Intelligent botnet attacks in modern Information warfare. In: 15th International Conference on Information and Software Technology, pp. 37–39 (2009)
18. Kolari, P., et al.: Detecting spam blogs: a machine learning approach. In: Proceedings of the National Conference on Artificial Intelligence, p. 1351. AAAI Press/MIT Press, Menlo Park/Cambridge 1999 (2006)
19. Lai, C.-C., Tsai, M.-C.: An empirical performance comparison of machine learning methods for spam e-mail categorization. In: Fourth International Conference on Hybrid Intelligent Systems, HIS 2004, pp. 44–48 IEEE (2004)
20. Lee, K., et al.: Uncovering social spammers: social honeypots + machine learning. In: Proceedings of the 33rd International ACM SIGIR Conference on Research and Development in Information Retrieval, pp. 435–442 ACM, New York (2010)
21. Sahami, M., et al.: A Bayesian approach to filtering junk e-mail. In: Proceedings of AAAI-98 Workshop on Learning for Text Categorization (1998)
22. Meli, C., Oplatkova, Z.K.: SPAM detection: Naïve Bayesian classification and RPN expression-based LGP approaches compared. In: Software Engineering Perspectives and Application in Intelligent Systems, pp. 399–411. Springer, Heidelberg (2016)
23. Meli, C.: Application and improvement of genetic algorithms and genetic programming towards the fight against spam and other internet malware. Submitted Ph.D. thesis, University of Malta, Malta (2017)
24. Miranda-García, A., Calle-Martín, J.: Yule's characteristic K revisited. Lang. Resour. Eval. **39**(4), 287–294 (2005)
25. Ntoulas, A., et al.: Detecting spam web pages through content analysis. In: Proceedings of the 15th International Conference on World Wide Web, pp. 83–92. ACM, New York (2006)
26. Oltean, M., Grosan, C.: Evolving evolutionary algorithms using multi expression programming. In: ECAL, pp. 651–658 (2003)
27. Oltean, M., Dumitrescu, D.: Multi expression programming. Babes-Bolyai University (2002)
28. Rao, J.M., Reiley, D.H.: The economics of spam. J. Econ. Perspect. **26**(3), 87–110 (2012)
29. Ruan, G., Tan, Y.: A three-layer back-propagation neural network for spam detection using artificial immune concentration. Soft. Comput. **14**(2), 139–150 (2009)
30. Spinellis, D.: Reliable identification of bounded-length viruses is NP-complete. IEEE Trans. Inf. Theory **49**(1), 280–284 (2003)
31. Stuart, I., et al.: A neural network classifier for junk e-mail. In: Document Analysis Systems VI, pp. 442–450. Springer, Heidelberg (2004)
32. Wang, Z.-Q., et al.: An efficient SVM-based spam filtering algorithm. In: 2006 International Conference on Machine Learning and Cybernetics, pp. 3682–3686. IEEE (2006)
33. Yule, G.U.: On sentence- length as a statistical characteristic of style in prose: with application to two cases of disputed authorship. Biometrika **30**(3–4), 363–390 (1939)
34. Zhang, L., et al.: An evaluation of statistical spam filtering. Techniques **3**(4), 243–269 (2004)
35. Zhang, M., Fogelberg, C.G.: Genetic programming for image recognition: an LGP approach. In: EvoWorkshops 2007, pp. 340–350. Springer, Heidelberg (2007)
36. RPN, An Introduction To Reverse Polish Notation. http://h41111.www4.hp.com/calculators/uk/en/articles/rpn.html
37. Symantec Internet Security Report (2016). https://resource.elq.symantec.com/LP=2899

Optimization of Protein Mixture Filtration

Dagmar Janáčová[1(✉)], Vladimír Vašek[1], Pavel Mokrejš[1], Karel Kolomazník[1], and Ján Piteľ[2]

[1] Tomas Bata University, nám. T.G.M. 5555, 760 01 Zlín, Czech Republic
janacova@utb.cz
[2] FMT, TU of Košice, Bayerova 1, 08001 Prešov, Slovakia
jan.pitel@tuke.sk

Abstract. In this paper there is described filtering process for separating reaction mixture after enzymatic hydrolysis to process the chromium tanning waste [1]. Filtration process of this mixture is very complicated because it is case of mixture filtration with compressible cake. Successful process strongly depends on mathematical describing of filtration, calculating optimal values of pressure difference, specific resistant of filtration cake and temperature maintenance which is connected with viscosity change. The mathematic model of filtration with compressible cake we verified in laboratory conditions on special filtration device developed on our department. Industrial filtration differs from that in laboratory merely in the quantity of material being processed and in the necessity of processing it as cheaply as possible. In order to increase capacity while keeping filter dimensions not particularly large, the filter area per volume unit of plant gets increased as well as the difference between pressures before and behind the filtering screen or filter cake. A variety of different filter types are used and they depend on whether the particle content in liquid is low or high, whether the filter operates intermittently, semicontinuously or continuously, and on how the pressure difference is produced. Liquid flows through the cake and screen under the influence of a difference between pressures before the cake and behind the screen. We solved problems connected with filtration of reaction mixture after enzymatic hydrolysis.

Keywords: Filtration process · Optimization · Compressible filtration cake

1 Filtration Process of Reaction Mixture

Filtration belongs to the most common cases practically making use of phenomena taking place during the flow of liquid through a layer of granular material. Its industrial application is similar to laboratory filtration performed on filter paper in a filter funnel. The objective is to separate solid particles from the liquid in which they are dispersed. This separation is achieved during filtration by forcing the liquid containing solid particles through a porous barrier. Solid particles get caught in pores of the barrier and gradually build up a layer on it. Liquid passes through the particle layer and porous barrier and leaves, gradually freed of its solid phase.

A suspension of solid particles in liquid is brought to the filter, and filtrate and filter cake are drawn off [1, 3]. Let us designate: m_F - mass of filtrate, m_S - mass of

R. Matoušek (Ed.): MENDEL 2017, AISC 837, pp. 93–103, 2019.
https://doi.org/10.1007/978-3-319-97888-8_8

suspension, m_K - mass of filter cake, w_F - mass fraction of solid phase concentration in filtrate, w_S - mass fraction of solid phase in suspension, w_K - mass fraction of solid phase in cake. Balance relationships are balances of whole flows as well as balances of both phases. If the filter operates efficiently, value w_F is very small and negligible in relation to the others and the quantity of filtrate can then be determined from the quantity of suspension and concentrations of solid particles in suspension and filter cake (see Fig. 1). Balance relationships are balances of whole flows as well as balances of both phases.

$$m_s = m_F + m_K \tag{1}$$

On the Fig. 1 is shown easier technology scheme of filtration hydrolytic mixture [1].

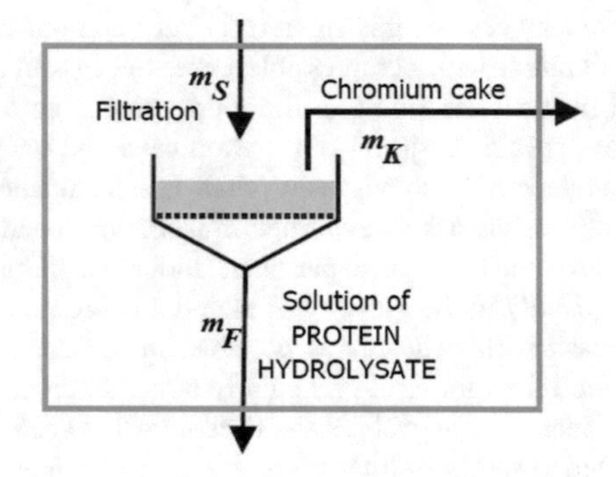

Fig. 1. Scheme of filtration process.

$$w_S \cdot m_S = w_F \cdot m_F + w_K \cdot m_K \tag{2}$$

after adjustment:

$$m_F = m_S \cdot \left(1 - \frac{w_S}{w_K}\right) \tag{3}$$

$$p_1 - p_2 = \rho \cdot \left(\frac{\lambda \cdot h \cdot A \cdot v_F^2}{8 \cdot \varepsilon^3} + g \cdot h\right) \tag{4}$$

where: p_1 - pressure before cake, p_2 - pressure behind cake, ρ - density, v_F - velocity, λ - friction coefficient, ε -porosity of layer, g - gravitational constant, h - thickness of cake, A - surface density of articles. Term $g{\cdot}h$ usually has no significance in filtration as the hydrostatic pressure produced by height of suspension level is generally negligible when compared with a pressure difference produced by industrial means [1, 4].

Provided the particles are sufficiently firm and do not undergo deformation, porosity remains constant throughout filtration and the filter cake is regarded as incompressible. Constant C and cake properties, i.e. porosity and surface density, then combine into a single constant:

$$\alpha = \frac{C \cdot A^2}{\varepsilon^3} \quad (5)$$

after simplification:

$$p_1 - p_2 = \alpha \cdot h \cdot \mu \cdot v_F \quad (6)$$

Practical account is taken of the difference between p_1 and p_3. When passing through this screen, filtrate changes its pressure from p_2 to p_3 because the screen offers resistance to its flow

$$p_2 - p_3 = R \cdot v_F \quad (7)$$

Combining equations for filter cake and screen, we obtain the filtration equation:

$$p_1 - p_3 = (\alpha \cdot h \cdot \mu + R) \cdot v_F \quad (8)$$

The flow of filtrate through cake is described as a flow of liquid through a layer of granular material [3]. This equation is used more practically in the form containing volume of filtrate:

$$\frac{dV_F}{d\tau} = \frac{(p_1 - p_3) \cdot S^2}{\beta \cdot \mu \cdot (V + V_{EKV})} \quad (9)$$

where

$$\beta = \alpha \cdot \tau \cdot \frac{V_K}{V_F} \quad (10)$$

and V_{EKV} is the resistance of filter screen expressed by means of equivalent thickness of filter cake. Integration of (7) gives:

$$V^2 + V_{EKV} \cdot V = 2 \cdot K \cdot \tau \quad (11)$$

where

$$K = \frac{(p_1 - p_2) \cdot S^2}{\beta \cdot \mu} \quad (12)$$

With a constant rate of filtrate flow, the solution of (9) is:

$$V^2 + V_{EKV} \cdot V = K \cdot \tau \quad (13)$$

Accordingly, the deformation of particles has to be taken into account as this affects particle porosity. In that case specific resistance of the filter cake is not constant. On passage through the cake, filtrate pressure drops from value p_1 to p_2. Deformation of the cake is caused by a resistive force when the stream flows past particles in a layer placed on a firm screen. The intensity of resistance F'_R equals the difference in pressures $(p_1 - p_2)$ multiplied by flow cross-section, S - area of filter [5]:

$$F'_R = (p_1 - p_2) \cdot S \tag{14}$$

As pressure p decreases when going from the interface with suspension to interface with screen, resistive force in the same direction increases and this is the force causing deformation of particles. This deformation then affects value of α and/or β. A certain value is found only across a differential section of the cake between h' and $h' + dh'$ described in articles [1, 6]. Integration across the whole thickness of cake and substitutions $dp' = -dp$ and $dh = -dh'$ give:

$$p_1 - p_2 = \langle \alpha \rangle \cdot \mu \cdot v_F \cdot h \tag{15}$$

where $\langle \alpha \rangle$ is the mean value of α. The mean value of $\alpha(\beta)$ is determined from the dependence on p'. It is found by filtration tests in an apparatus where value of p' can be altered. This dependence is usually expressed in the form as follows:

$$\alpha = \alpha_0 \cdot \left[1 + \beta (p')^{\gamma}\right] \tag{16}$$

Assuming that $\Delta p = p_s$, The course of pressure differences was broken down into seventeen time intervals. In single intervals, the dependence of pressure differences was approximated by means of polynomial equations of first to fourth degree. Their integration in relevant limits and subsequent division by length of corresponding interval provided mean integral values.

$$p_S = \frac{1}{\tau_1 - \tau_2} \int_{\tau_1}^{\tau_2} f(\tau) d\tau \tag{17}$$

and that β_i, μ, Δp and S_f^2 are constant within individual intervals and can be substituted by K_i, Constant K enabled us to determine:

$$\beta_i = \frac{\Delta p \cdot S_f^2}{\mu \cdot K_i} \tag{18}$$

The dependence was approximated by a dx potential

$$\beta = 1.76 \cdot 10^{14} \cdot e^{7.7275.10^5 \cdot p_s} \tag{19}$$

The exponential approximation was replaced by parameter β in the general equation of filtration and dx is the parameter that measures the resistance of the filter cake.

2 Experimental Part

In laboratory conditions we verified recycling technology of leather shavings treatment. The laboratory filtration model (see Fig. 2) was designed to meet following requirements [3]: constant temperature of 30–80 °C in filtering space, filter underpressure recording (0.6 kPa), filtrate volume 5 litres. The model consists of parts as follows [1]: temperature-controlled heated filter funnel, filtration flask, underpressure measurement, level height measurement, unified signal conversion, data storage. The filter funnel of circular shape consists of two main parts, the funnel proper and filter screen. Both parts were made of duralumin, which is light and conducts heat well [2, 3]. The basic dimensions of the funnel are inside diameter of 0.165 m and funnel edge of 0.08 m height. It can be filled at a time with 1.7 litres of the filtered suspension.

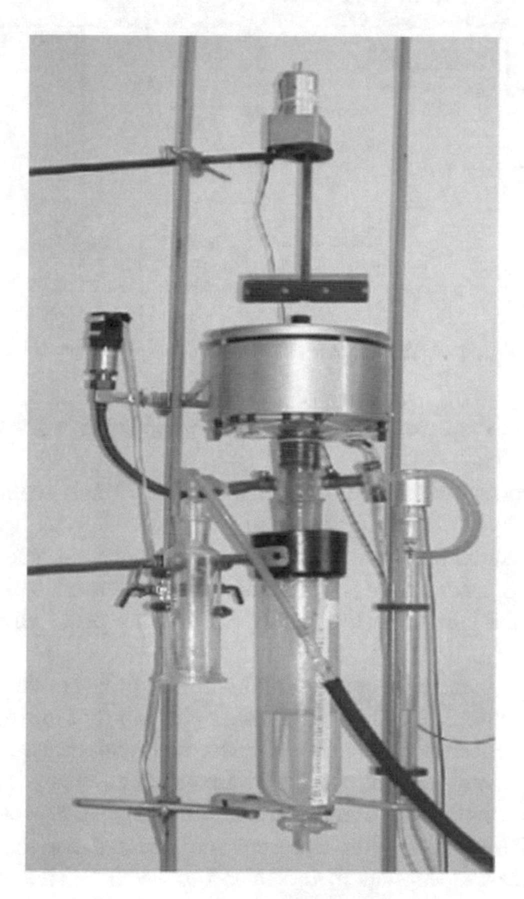

Fig. 2. Laboratory filtering equipment.

The funnel wall contain heating elements. The second, suitably shaped packing is mounted between the filter funnel and mouth of filtration flask [3, 4]. Filtration was performed while the funnel was heated to 50 °C (see Fig. 3).

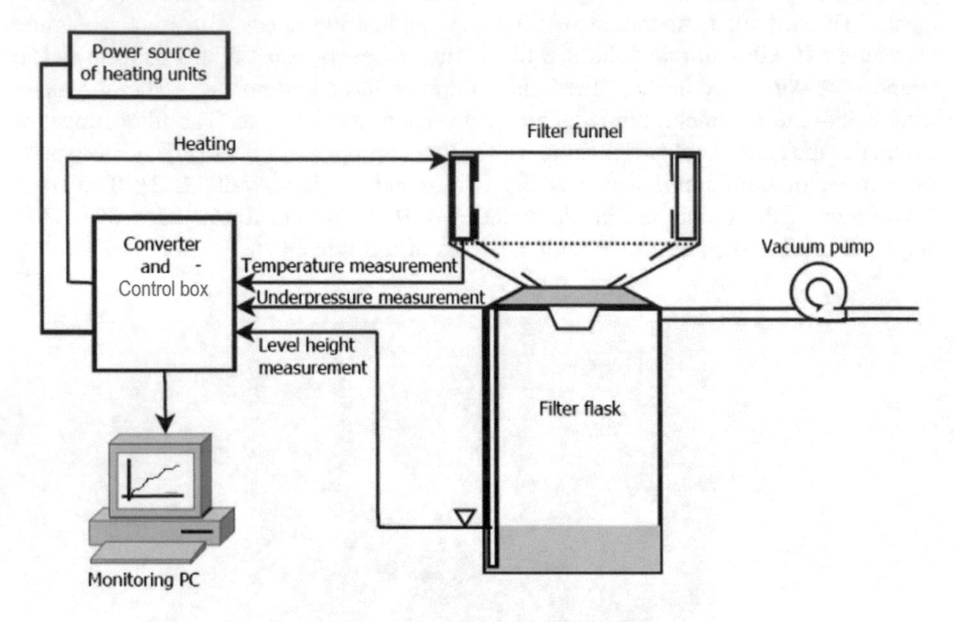

Fig. 3. Laboratory arrangement of filtration process.

Properties of the suspension were as follows: density $\rho = 2500\ \mathrm{kg{\cdot}m^{-3}}$, dynamic viscosity $\mu = 0.001\ \mathrm{Pa{\cdot}s}$.

Primary filtration proceeded for the first 100 s without external underpressure, merely under own hydrostatic pressure. After 100 s, underpressure was produced by the vacuum pump. At the time when an underpressure decrease characterizing stoppage of filtration appeared, the value of underpressure was increased. This was repeated until the cake cracked and filtration stopped. An analogous procedure was employed in secondary filtration.

Primary filtration is the term applied to filtration in progress during the first stage of enzymatic dechromation, secondary filtration applied to that during the second stage. The pressure difference means value of produced underpressure or value at the beginning of hydrostatic pressure and both terms are interchangeable.

Four stages are obvious in primary filtration. The first (0–100 s) without underpressure, second (100–460 s) with underpressure exerted for the first time. Further underpressure increases occurred in the third (460–920 s) and fourth (920–2490 s) stages. A drop in underpressure follows stage four - the already mentioned filter cake rupture.

The calculation of parameters p_s and β was repeated for all time intervals. Values calculated are listed in the following Table 1.

Table 1. Calculated values of pressure p_s (see Eq. 17) and parameter β (see Eq. 18).

Interval No.	$\tau_1 [s]$	$\tau_2 [s]$	$\tau_1 - \tau_2$ [s]	p_s [Pa]	β [m^{-2}]
I	0	100	100	0	5.86E+14
II	100	200	100	10 802	2.02E+14
III	200	300	100	20 464	2.75E+14
IV	300	400	100	23 993	2.08E+15
V	400	460	60	24 202	3.32E+15
VI	460	600	140	34 248	2.74E+15
VII	600	700	100	43 850	2.22E+15
VIII	700	800	100	48 841	3.18E+15
IX	800	920	120	53 665	1.16E+16
X	920	1 020	100	62 995	1.88E+16
XI	1 020	1 120	100	73 155	4.18E+15
XII	1 120	1 220	100	78 431	3.18E+15
XIII	1 220	1 310	90	80 628	8.77E+15
XIV	1 310	1 400	90	82 235	8.95E+15
XV	1 400	1 600	200	83 193	9.74E+16
XVI	1 600	1 800	200	83 791	9.81E+16
XVII	1 800	2 490	690	84 498	3.41E+17

The course of pressure differences during primary filtration of first-stage suspension is depicted (see Fig. 4).

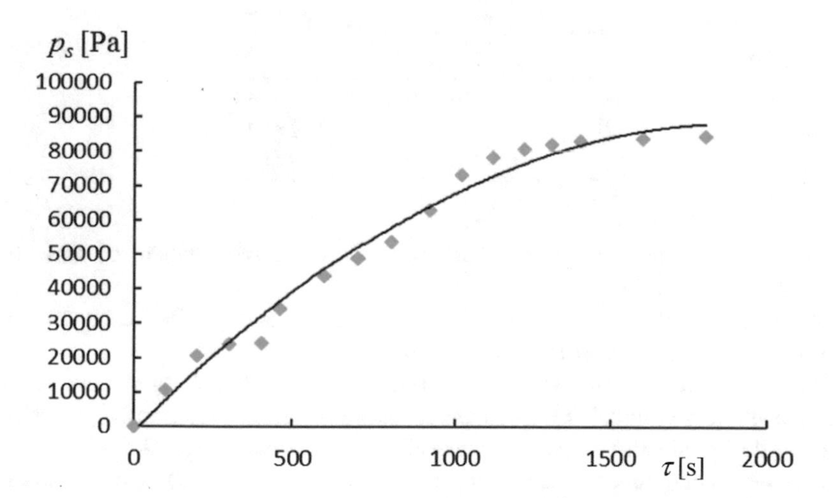

Fig. 4. Course of underpressure during primary filtration

Parameter β characterizes filter cake resistance. If this does not change when pressure differences during filtration change, the cake in question is incompressible; on the contrary, it is compressible if resistance changes (Fig. 5).

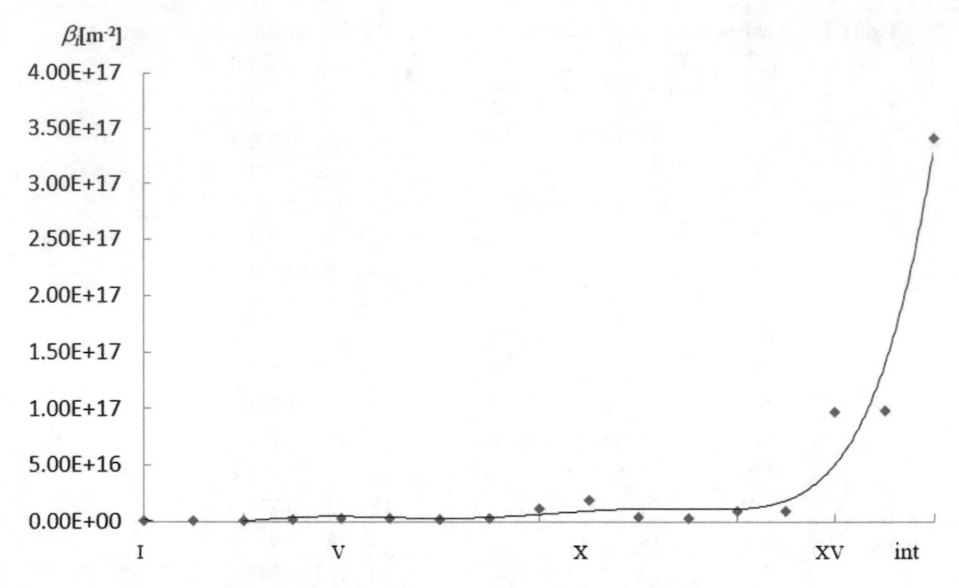

Fig. 5. Dependence of parameter β on filtering interval

The procedure in secondary filtration of hydrolysate was the same as in primary filtration. Calculated values are presented in the Table 2 as follows.

Table 2. Calculated values for primary filtration.

Interval No.	τ_1 [s]	τ_2 [s]	p_s [Pa]	β [m^{-2}]
I	0	50	1 962	4.3E+12
II	50	80	10 802	2.4E+12
III	80	300	26 813	5.6E+12

This signifies that the specific resistance of filter is independent of pressure difference $\Delta p = 12940\, Pa$.

We also observed dependence of the resistivity of the filter cake on number decanting cycles. The result is follows (Table 3).

The process having started, filtration runs without external cause only due to its own hydrostatic pressure. When the defined sampling period has elapsed, constant *K* is and the value of inductive force is increased by a difference defined in advance. With this first incrementation, the source of force changes from pressure to underpressure. Constant *K* is again calculated after this period, namely by recording changes in height of level.

The preceding value of *K* is compared with the present. If it fails to comply with the specified condition, underpressure is increased again, and *K* is newly calculated after the time difference has elapsed. The cycle is terminated upon condition being met. A further condition follows, comparing actual underpressure with the value of optimal

Table 3. Calculation of specific resistance β.

Measuring interval	V [m^3]	K [$m^4 \cdot s^{-1}$]	β [m^{-2}]
Temperature 20 °C			
I.	1.82E−05	8.32E−12	3.99E+15
II.	1.70E−04	7.25E−10	2.63E+13
III.	2.13E−04	1.13E−09	1.65E+13
IV.	3.04E−04	2.31E−09	7.37E+12
V.	2.30E−04	1.47E−09	1.18E+13
Temperature 30 °C			
I.	2.49E−04	1.55E−09	1.07E+13
II.	1.94E−04	9.46E−10	1.71E+13
III.	2.46E−04	1.57E−09	1.04E+13
Temperature 40 °C			
I:	2.12E−04	1.13E−09	7.72E+12
II:	3.72E−04	3.55E−09	2.45E+12
III:	2.37E−04	1.41E−09	5.94E+12
Temperature 50 °C			
I:	1.62E−04	6.74E−10	1.17E+13
II:	1.88E−04	8.88E−10	9.36E+12
III:	7.29E−05	1.33E−11	5.93E+12

underpressure found in identifying filtration. Control is finished upon reaching accordance within a certain tolerance. If values of underpressures differ, the value found during process modelling is set up and control is finished. Parameter x of the first condition is chosen a priori and has an influence on underpressure being possibly exceeded (see Fig. 6).

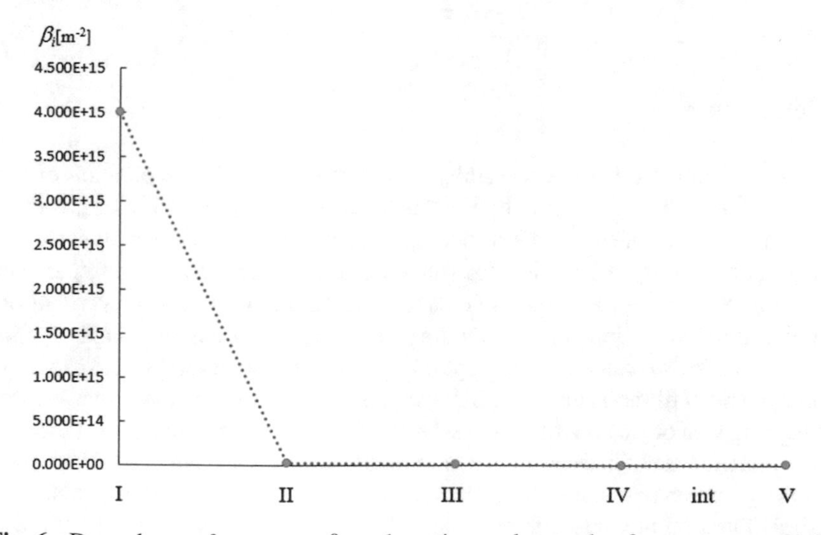

Fig. 6. Dependence of parameter β on decanting cycles number for temperature 20 °C

Curve on the Fig. 7 indicates partial filtration procedure. In the case of two-stage hydrolysis because vacuum filter was less than 5 m^2 which was insufficient for filtering almost 14 metric tons of heterogeneous mixture during one working cycle.

Filtration of a similar volume of suspension proceeded much faster (about 22 times). Values of the cake specific resistance varied merely within the same order; the cake, therefore, was further considered to be incompressible. The general expression for calculating optimal values of pressure difference we used for control of primary filtration of suspension.

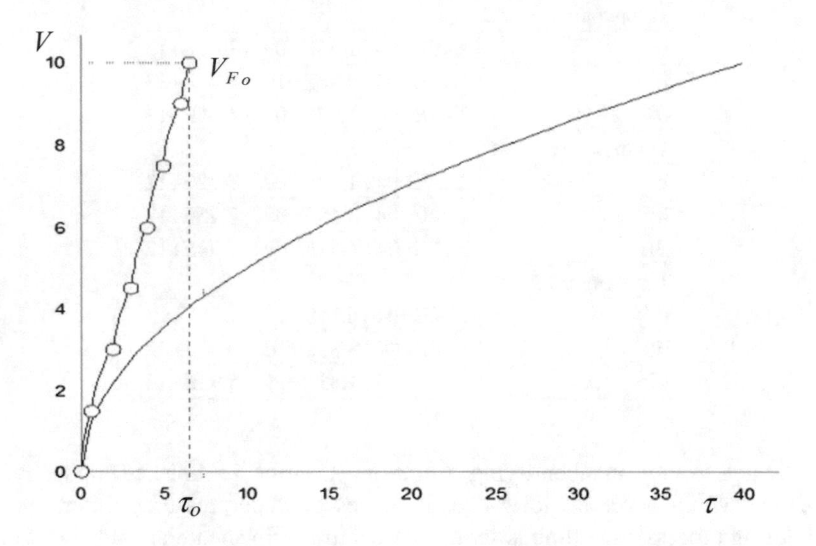

Fig. 7. Comparing of partial and total filtration time

where τ_0 - operation time of cycle (maximal permitted technological time of process).

3 Conclusion

In this contribution we have successfully solved the problem of separation of heterogeneous mixtures after enzymatic hydrolysis in the case of production of gelatine protein. By using controlled filtration during her constant speed, we have increased production of gelatine protein. We determined optimal underpressure and parameter β for filtration process with compressible cake. The filtration process we verified on our developed filtration equipment in laboratory conditions. For another suspensions with compressible cake we can determine optimal parameters: underpressure, specific resistanceβ and time of filtration also. The key element of the whole technology was filtration as the filtering area of vacuum filter was less than 5 m^2 which was insufficient for filtering almost 14 metric tons of heterogeneous mixture during one working cycle. We overcame this issue by gradual partial filtration of reaction mixture fractions by the necessity to find an optimal filtration pressure difference in order to achieve a maximal rate of filtration.

Acknowledgement. This work was supported by the Ministry of Education, Youth and Sports of the Czech Republic within the National Sustainability Programme Project No. LO1303 (MSMT-7778/2014) and also by the European Regional Development Fund under the project CEBIA-Tech No. CZ.1.05/2.1.00/03.0089.

References

1. Kolomaznik, K.: Recovery and recycling of chromium and protein from liquid and solid wastes from leather industry, ERB 3512 PL 940974 (COP 974) (1996)
2. Janacova, D., Kolomaznik, K., Vasek, V.: Optimization of liquids consumption at washing. In: 6th International Carpathian Control Conference 2005. Miskolc-Lilafuered, Hungary (2005). ISBN 963-661-644-2
3. Vasek, V., Kolomaznik, K., Janacova, D.: WSEAS Trans. Syst. **5**(11) (2006). ISSN 2651-2654
4. Vasek, L., Dolinay, V.: Int. J. Math. Models Methods Appl. Sci. **4**(4), 240–248 (2010)
5. Božek, P.: Proceedings of the International Conference on Systems Science, ICSS 2013. Advances in Intelligent Systems and Computing, Wroclaw, Poland, pp. 229–240 (2013). ISSN 2194-5357
6. More, M., Liska, O.: IEEE 11th International Symposium on Applied Machine Intelligence and Informatics (SAMI). IEEE (2013)

Accelerated Genetic Programming

Vladimír Hlaváč(✉)

Faculty of Mechanical Engineering, Czech Technical University in Prague,
Technická 4, 166 07 Prague 6, Czech Republic
hlavac@fs.cvut.cz

Abstract. Symbolic regression by the genetic programming is one of the options for obtaining a mathematical model for known data of output dependencies on inputs. Compared to neural networks (MLP), they can find a model in the form of a relatively simple mathematical relationship. The disadvantage is their computational difficulty. The following text describes several algorithm adjustments to enable acceleration and wider usage of the genetic programming. The performance of the resulting program was verified by several test functions containing several percent of the noise. The results are presented in graphs. The application is available at www.zpp.wz.cz/g.

Keywords: Symbolic regression · Genetic programming
Exponencionated gradient descent · Constant evaluation

1 Introduction

Searching for the functional dependence of measured data was mentioned as one of the basic applications of the genetic programming already in the [1]. If data are generated by a computer for a simple function, it can be recovered by this way. The problem is with real data. For example, in the case of a harmonic signal, it is necessary to determine its amplitude, period and phase in addition to the sinus function. This leads to the introduction of constants into genetic programming.

Constants can be generated with a limited variability together with the searched function if certain constants are available in the primitive function dictionary. The program can create other constants by their combination. Changing the values of constants can then be added as another form of the mutations. This method is referred to as a variation of constants and often gives very fast results with accuracy that is sufficient in a number of cases. A similar solution is already described, for example, in [2] and is used in [3]. This solution typically results in a more complex record of the function tree and the use of a greater number of constants as compared to the variant where the constants are evaluated exactly.

The possibility to add the exact value of constants to the tree record is described, for example, in [4], constant evaluation by using the method of differential evolution is used in [5], in the [6–8] evolutionary strategies are used. All of these approaches lead to a separate one-dimensional array of constants for each of the generated functions. The constants have to be determined separately in another way; the referred cases used

R. Matoušek (Ed.): MENDEL 2017, AISC 837, pp. 104–112, 2019.
https://doi.org/10.1007/978-3-319-97888-8_9

another evolutionary algorithm. The evaluation of the appropriate values of the constants for the proposed function then represents the most time-consuming part of the genetic programming algorithm. The solution in these examples is to use more powerful computers, so for example [5] describes the use of parallel grammar evolution, which implies use of multiple computers, and [7] uses a computing cluster. The availability of similar resources then limits the use of this method by ordinary users.

In this work, several partial modifications of the genetic programming algorithm have been proposed to allow a faster evolution of the genetic programming population. Subsequently, an application has been created to demonstrate these improvements. The version presented here assumes the searched function with two independent variables, in the form

$$z = f(x, y) \tag{1}$$

The fitness function[1] is then defined either as the sum of squares of errors (2) or as the sum of the absolute values of the errors in the specified points of the function (values of data, recorded in the file, is denoted as G(x,y)). It is possible to switch between these two variants using the user interface. The fitness function further includes penalization of the individual record (genome) length. It can be set to zero to disable this penalty. However, it is desirable to search for reasonably complex functions.

$$F_j = \sum_{i-1}^{n} \left(G(x_i, y_i) - f_j(x_i, y_i)\right)^2 = \sum_{i=1}^{n} \left(z_i - f_j(x_i, y_i)\right)^2 \tag{2}$$

The last chapter describes the results obtained on proposed test functions.

2 Performed Modifications of the Genetic Programming

2.1 Record of the Tree of a Function

The tree of a function (individual) can be directly rewritten into the prefix notation. In the described program, each node and leaf of the tree is represented by one byte. The symbols representing primitive functions, constants or variables do not repeat between groups, so no further markings or translation (as in grammatical evolution) is required. A record of a function written in the genome contains no unused symbol, nor is reused during evaluation of the function. To ensure that the record of a function is unambiguous and readable, a check and correction function is called after each tree modification (generation, child creation, mutation) so that any errors in the tree record will not slow down the evaluation of the fitness function. The value of the function at the point is still evaluated by a recursive program. For the next step of an algorithm improvement, a stack structure should be implemented instead. The evaluation of the

[1] [9] uses term "cost function", when looking for minimum

function, recorded as a tree of the individual, is a key point of making a speed application, since this function has to be calculated at all the given points to evaluate the fitness function.

2.2 Recording Constants in the Tree

The understanding of the meaning of the calculated constants was changed. In the record of the function they are not represented as leaves, but some nodes or leaves of the tree are marked as including a constant. At those positions, the resulting value during the evaluation is multiplied by the constant. The constants, evaluated by this program, are stored in a row of a two-dimensional array of real numbers, where each of individual has its own row. Since only ASCII symbols are used to encode the genome record, the node (or leaf), including the multiplication by a constant, is indicated by switching the seventh bit to one. An example of this representation is on a Fig. 1, where the places marked as multiplication by the constant are shown by doubling the circle of the node or leaf. It is necessary to emphasize that in the set of the basic symbols are still constants 1, 2 and π. The program can then generate any constant at the point of the leaf by using one of these symbols and marking it as a place for multiplying by an evaluated constant.

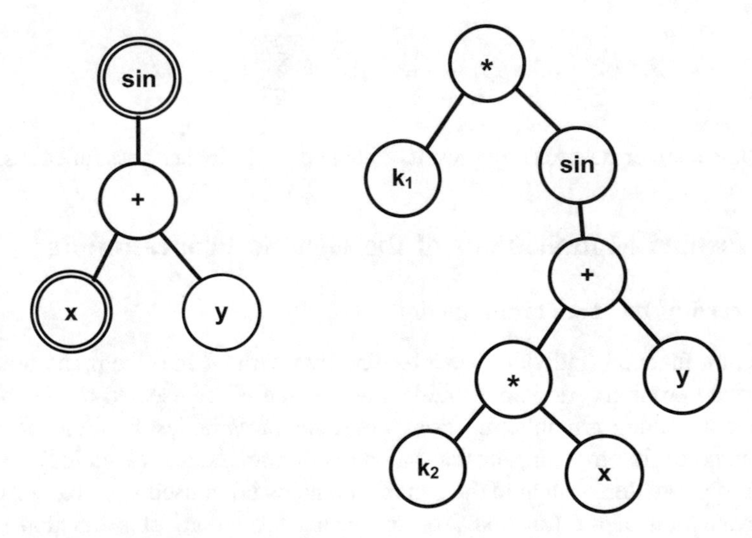

Fig. 1. On the left is a shortened tree representation of (4). On the right is the same function after translation to the standard genetic programming tree.

The maximum amount of the multiplicative constants per individual is limited to ten. If (accidentally) an individual with more constants in the tree record is created, some of them are randomly deleted, till the limit of ten is satisfied. Probability of marking of a tree node or leaf as containing a constant can be set, but only for the case of a new individual is being generated.

2.3 Enumerating Values of the Constants

Constants are calculated by a gradient method. Since these are multiplicative constants, the exponentionated gradient method (EGD, (3)) [10] was chosen. The gradient is approximated at a given point by a differential of one third of an anticipated step. The method is applied with a decreasing approximation step.

$$k_{t+1}^{(i)} = k_t^{(i)} \cdot e^{-\nabla_\gamma (f(x,y,\gamma_t))\gamma_t^{(i)}} \tag{3}$$

The gradient method is significantly faster than commonly used evolutionary techniques, but this is still the most time-consuming part of the algorithm. To avoid deceleration, only a few (typically ten) steps of the gradient method are performed and the function is included in the population in this state. If it survives to the next generation, a few next steps of gradient method are done in each of the next generation for this individual. A similar method using a successive step is mentioned in [11] as a lifetime evolution. The gradient method is generally restarted in a different (random) point of the EGD algorithm sequence, then where it was interrupted, and another six to ten steps are done. If the function in any genetic programming cycle is no longer improving, the individual is marked as final, so no other gradient method is called in the next cycles in order to improve this individual.

Separately, the possibility of negative constants is treated (the method does not allow to change the sign of the calculated constant). The problem of the local minima, which is the main disadvantage of gradient methods, is solved by the main cycle of the genetic programming by the method of variation of constants, which is implemented in the way corresponding to another method of recording constants to the genome in this program. With this particular mutation, a mutant clone is always added to the end of the population, and immediately the first ten steps of the constant evaluation are done. The original individual stays in the population unmodified.

A very compact way of recording the tree of the individual to the memory allows easy operations with the genome of an individual (in realization of mutations and crossover) as well as a quick evaluation of function values in the required points. The probabilities of selecting primitive functions as well as leaf objects (constants, variables), and the probability of marking a place where a constant should be placed can be changed during program execution.

2.4 A New Type of Mutation

Because of the new organization of recording constants into the tree, the constants are firmly tied to the object, the result of which multiplies. This required to propose a new mutation, consisting in a remarking the space for multiplying the constant to another location of the tree (to another object, that is, a leaf or a node). The program tries to find a place where the presence of a constant is indicated by the seventh bit and cancels the mark. In another random place, it adds it. Because the function "xor 80_H" is used, it may be canceled instead (if previously present). If it does not find it in the first case, then it can add two (program after a determined number of attempts to guess the place with constant uses the last tested place as the target of the mutation).

A roulette selection was used in this program, which allows choosing the evolutionary pressure and accelerating the progress of evolution. It is also possible to evaluate the current variation coefficient of the population and use it to control the level of mutations. These may be reduced or increased up to three times the preset level.

3 Data

The program assumes data in the form of a text file, where there are three values in each row, the variable x, y, and the required function value at this point. Three functions have been proposed for testing:

$$f(x,y) = k_1 \sin(k_2 x + y) \tag{4}$$

$$f(x,y) = k_1 - \ln(k_2 x + \sqrt{y}) \tag{5}$$

$$f(x,y) = k_2 . \exp(k_1 x + y^3) \tag{6}$$

Constant values were $k_1 = 3.71$ and $k_2 = 1.73$. Training data were modified with a Gaussian noise. Because of this, it is not possible to achieve a zero value of the fitness function. The calculated minimal errors (theoretical minimal value of the fitness function (2), it there are no overfitting) for each function with a 5% of noise (as the standard deviation of the actual value of the function) were 0.23, $1.92.10^{-4}$, and 0.106 (evaluated by comparing the original function without noise and the original values of constants with the training data, including noise). The fitness function was calculated as the sum of squares of errors, increased by a penalization of the individual genome length.

The values of the function were evaluated in 300 randomly generated points in a preselected range.

4 Testing

The program was run on an office PC (MS Windows 8.1 operating system, Intel Pentium CPU G3260 3.30 GHz). The graphs show the fitness function of the best individual (thin line) and the population average (dashed thicker line). The population consisted of 130 individuals. 20 mutations and 70 crossovers per generation have been executed (two offspring per crossover). Relatively small population [12] is compensated by higher level of mutations. Level of mutations grows with diversity loss [13].

To protect against frozen population, limited lifetime [14] is implemented. During tests, lifetime was set to 11.

The program has been tested in about 40 runs with a different set of data, including higher and lower level of noise. For an illustration, progress of the fitness function for three runs for each of testing function has been recorded and is presented on the Figs. 2 to 4. To be comparable, results for data with 5% of noise is present. The running time is on the horizontal axis, where the fitness value is on the vertical axis. Fitness is calculated as a sum of squares of errors. The best value in each of the cycles is drawn with a solid line, while the average of the population is drawn as a dashed line (Figs. 2, 3 and 4).

Fig. 2. The fitness function progress, data generated using (4). The first test has been interrupted while the correct solution has been found.

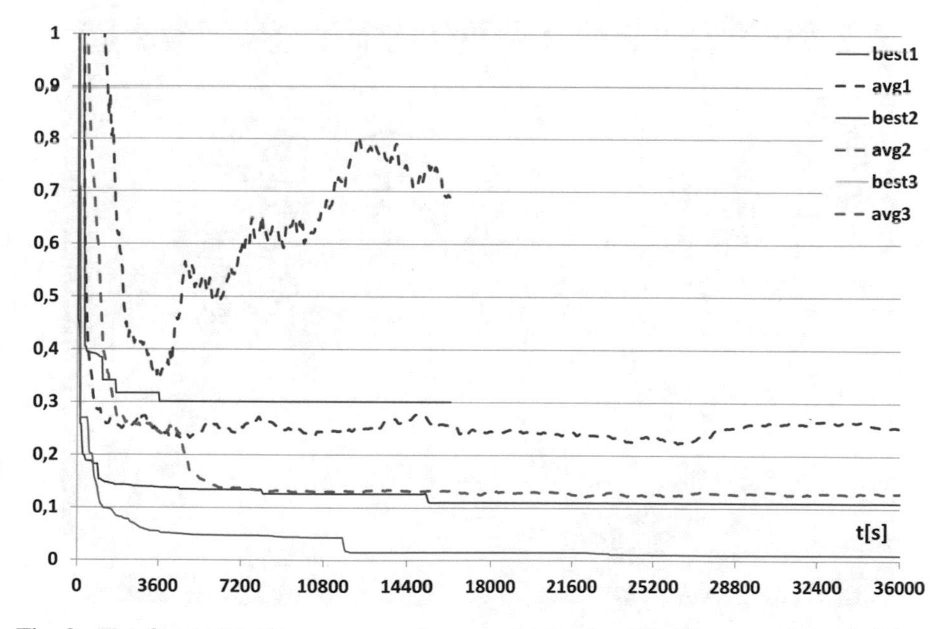

Fig. 3. The fitness function progress, data generated using (5). The second test has been interrupted while freezing.

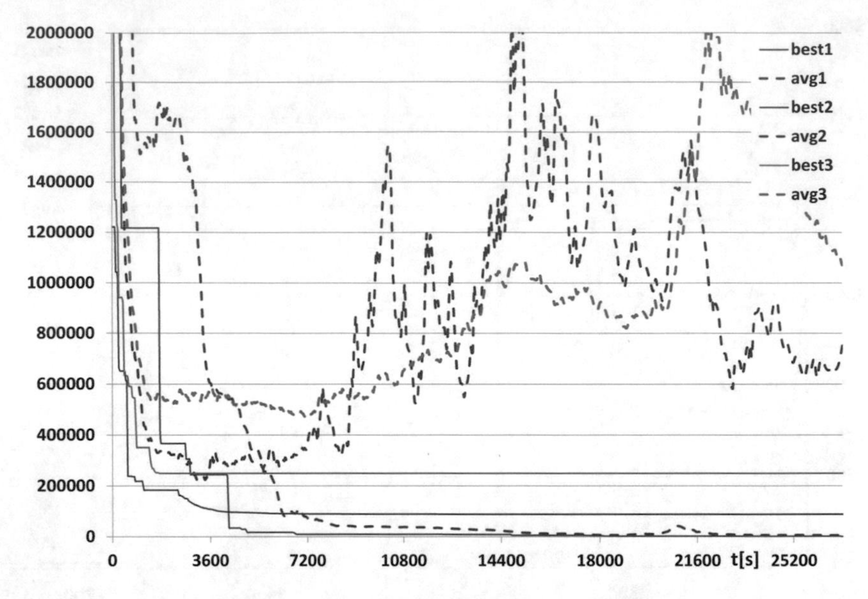

Fig. 4. The fitness function progress, data generated using (6). Most of the errors were for the biggest x values of sample data. Growing average is caused by increasing the level of mutations, used by the program in the case of no progress of the best individual.

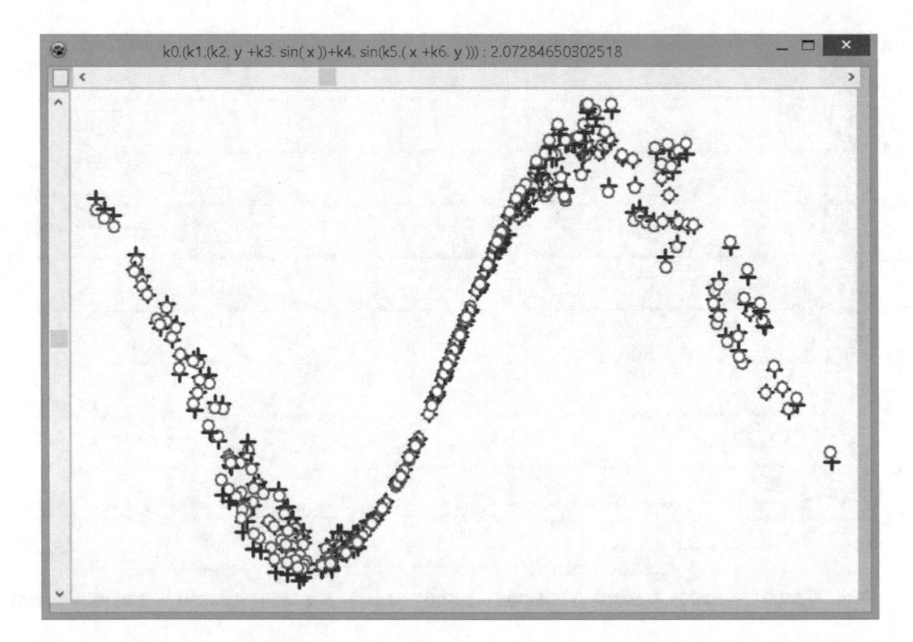

Fig. 5. The resulting graph of the evolved function (data from (4)) displayed in the application. Sample data as red crosses, interpolated data as circles. 3D mode, rotated 31° around the vertical axis and 6° around the horizontal axis. Scrollbars allow rotation of the graph. The proposed function is written in the window title (above the graph).

5 Conclusion

Total time in the range of hours allows the program to be used on standard PC computers. For better results (simpler function), it is enough to run within the order of twenty hours, which can be done by running evolution throughout the whole night, when computers are often unused. The program takes up a minimal amount of memory, allowing it to be run in a background. If shorter genome of proposed function is required, it is possible to set a higher level of mutations and a greater penalization of the length of the genome. The application used allows the generated function to be saved as a function definition for the C language (text including array of constants).

The program showed a poor applicability for functions, which can be easily interpolated by linear interpolation, and a better applicability for complicated or harmonic functions. It allows genetic programming being used as a method of symbolic regression in the cases, where actual physical description is not known, and where neural networks are now used to generate a model of a system.

References

1. Koza, J.R.: Genetic Programming: On the Programming of Computers by Means of Natural Selection. MIT Press, Cambridge (1992). ISBN 978-0262111706
2. Banzhaf, W., et al.: Genetic Programming, an Introduction. Morgan Kaufmann Publishers, Inc., San Francisco (1998). ISBN 978-1-55860-510-7
3. Barmpalexisa, P., Kachrimanisa, K., Tsakonasb, A., Georgarakis, E.: Symbolic regression via genetic programming in the optimization of a controlled release pharmaceutical formulation. In: Chemo metrics and Intelligent Laboratory Systems, vol. 107, no. 1, pp. 75–82. Elsevier, May 2011
4. Zelinka, I.: Analytic programming by means of soma algorithm. In: Proceedings of 8th International Conference on Soft Computing Mendel 2002, Brno, Czech Republic, pp. 93–101 (2002). ISBN 80-214-2135-5
5. Weisser R., Ošmera, P., Matoušek, R.: Transplant evolution with modified schema of differential evolution: optimization structure of controllers. In: International Conference on Soft Computing MENDEL, Brno (2010)
6. Brandejsky, T.: Genetic programming algorithm with constants pre-optimization of modified candidates of new population. In: Mendel 2004, Brno, pp. 34–38 (2004)
7. Brandejsky, T.: Small populations in GPA-ES algorithm. In: Mendel 2013, pp. 31–36. ISBN 978-802144755-4
8. Brandejsky, T.: Influence of random number generators on GPA-ES algorithm efficiency. In: Advances in Intelligent Systems and Computing, vol. 576, pp. 26–33. Springer International Publishing AG (2017)
9. Roupec, J.: Advanced genetic algorithms for engineering design problems. Eng. Mech. **17** (5–6), 407–417 (2011)
10. Kivinen, J., Warmuth, M.K.: Exponentiated gradient versus gradient descent for linear predictors. Inf. Comput. **132**(1), 1–63 (1997). https://doi.org/10.1006/inco.1996.2612
11. Azad, R.M.A., Ryan, C.: A simple approach to lifetime learning in genetic programming-based symbolic regression. Evol. Comput. **22**(2), 287–317 (2014)

12. Castelli, M., et al.: The influence of population size in geometric semantic GP. In: Swarm and Evolutionary Computation, vol. 32, pp. 110–120. Elsevier BV (2017). ISSN 2210-6502
13. Affenzeller, M., et al.: Dynamic observation of genotypic and phenotypic diversity for different symbolic regression gp variants. In: Proceedings of the Genetic and Evolutionary Computation Conference Companion, GECCO 2017, pp. 1553–1558. ACM, New York (2017)
14. Ošmera, P., Roupec, J.: Limited lifetime genetic algorithms in comparison with sexual reproduction based GAs. In: Proceedings of MENDEL 2000, Brno, Czech Republic, pp. 118–126 (2000)

Hybrid Multi-objective PSO for Filter-Based Feature Selection

Uroš Mlakar(✉), Iztok Fister Jr., Janez Brest, and Iztok Fister

Faculty of Electrical Engineering and Computer Science,
Univeristy of Maribor, Maribor, Slovenia
uros.mlakar@um.si
http://labraj.feri.um.si/en/Uros

Abstract. This paper proposes a novel filter-based multi-objective particle swarm optimization (PSO) algorithm for feature selection, based on information theory. The PSO is enhanced with clustering and crowding features, which enable the algorithm to maintain a diverse set of solutions throughout the optimization process. Two objectives based on mutual information are used for selecting the optimal features, where the first aims to maximize relevance of features to the class labels, while the second to minimize the redundancy among the selected features. The proposed method is tested on four datasets, giving promising results when compared to multi-objective PSO, and multi-objective Bat algorithm.

Keywords: Feature selection · Information theory
Particle swarm optimization · Mutual information · Crowding
Clustering

1 Introduction

Nowadays many databases, which are used in either supervised or unsupervised machine learning suffer the "curse of dimensionality" [5]. This means that the databases grow in both the number of data samples, and the number of features in each sample. This problem can become troublesome in a task like classification, because a presence of unimportant or highly correlated features can decrease the classification accuracy [5]. Given an original set of features D consisting of n elements, the goal of feature selection is to select a subset of features F with m elements such that $m < n$ and $F \subset D$. There are three common approaches to feature selection found in literature, based on the evolution criteria, namely: (i) wrapper-based methods, (ii) filter-based methods, and (iii) embedded methods. Wrapper-based methods use a learning algorithm to determine the fitness of each feature subset. Wrappers-based feature selection methods usually achieve better results, but are computationally inefficient. The filter-based methods use simple measures for ranking the feature subsets, where the feature selection is actually a preprocessing step. These methods are faster and independent of a learning algorithm. Researchers have used many different criteria for filter-based

R. Matoušek (Ed.): MENDEL 2017, AISC 837, pp. 113–123, 2019.
https://doi.org/10.1007/978-3-319-97888-8_10

methods such as distance measures, dependency measures, and the other [5]. Finally, embedded methods combine the advantages of filter- and wrapper-based methods. Here, a learning algorithm is employed, which takes advantage of its own feature selection process, and simultaneously calculates the classification accuracy.

Since the search space for feature selection is 2^n, where n denotes the number of features, the exhaustive search for finding an optimal subset of features by enumerating all possible solutions is impractical due to their time complexity. Therefore a computationally cheap method, with a good global search ability (i.e. exploration) is needed for efficient solving of the feature selection problem. This characteristic exhibit all nature-inspired population-based algorithms, such as swarm intelligence (SI) based algorithms (e.g., particle swarm optimization (PSO) [5,17], cuckoo search [11,13]), or evolutionary algorithms (EAs) (e.g., genetic algorithms (GA) [1,16]), which have been thoroughly studied and applied to the feature selection problem. Compared with other algorithms, the PSO has attracted a lot of attention, because of its easy implementation, the computational efficiency, the fewer number of control parameters, and the fast convergence rate.

Many PSO algorithms for feature selection are wrapper-based approaches that treat the feature selection as a single-objective problem, where the optimization is performed with regard to one objective only. Usually, the classifier error rate is considered during the evolutionary process as the objective. In general, feature selection can be treated as a multi-objective (MO) problem, where two conflicting objectives are minimized, with one being the size of the selected feature subset, and the other the classification error rate, considering the selected features [12]. This is a common approach for wrapper-based methods, where relatively high classification results are obtained, but at the greater computational cost.

In order to search for an optimal subset of features in reasonable time, the novel filter-based multi-objective PSO enhanced with clustering and crowding features (CCMOPSO) is proposed. Instead of using the feature subset size and classifier accuracy as objective measures, we use information theory for defining the objectives to be minimized for optimal subset selection. The first objective is used for measuring the relevance among the selected features and the class labels, while the other for minimizing the redundancy in the selected feature subset.

The PSO is known for its fast convergence. This characteristic is particularly helpful when solving problems. In MO optimization, however, it is desired to maintain a diverse set of solutions during the whole optimization process. With this in mind, we enhance the PSO with clustering and crowding mechanisms that are experienced features for maintaining the population diversity. By using clustering, the solutions in the PSO population are divided in several promising regions/clusters in each generation of the optimization process. Crowding assures that all solutions within the new population are compared with their neighboring solutions from the original population, where the better fitted solution is preserved in the new generation.

The structure of the paper is as follows. In Sect. 2 related work is discussed. Section 3 describes the proposed CCMOPSO algorithm. The experimental environment is detailed Sect. 4, then results are presented in Sect. 5. The paper is concluded in Sect. 6 with future work directions.

2 Related Work

Several single- and multi-objective PSO algorithms for feature selection have been proposed recently. The former search for a solution regarding to one objective, i.e., the classification error rate of the selected features subset. Usually, they converge towards a local optimum very fast. The latter consider feature selection as multi-objective problem and search for solutions according to two objectives: the minimum number of used features, and the minimum classification error of the selected feature subset. Both types of algorithms are briefly reviewed in this section.

2.1 Single-objective Methods

Cervante et al. [5] proposed a binary PSO (BPSO) filter-based feature selection method, which utilizes the mutual information shared between the features. The search for the best feature subset is designed through a fitness function, where redundancy and relevance are considered. A decision tree (DT) classifier is used to evaluate the selected feature subsets tested on four databases. They reported a similar or even higher classification accuracy, with a smaller number of used features in almost all cases.

Azevedo et al. [2] proposed a feature selection method for personal identification in keystroke dynamics systems. Their method is wrapper-based PSO, where the Support Vector Machine (SVM) is used as a classifier. They compare the performance of the proposed method to the GA, where their method gives superior results with regard to: the classification error, processing time, and feature reduction rate.

Mohemmed et al. [14] used the PSO within an Adaboost framework for face detection. The PSO is used for finding the best features and thresholds of weak classifier simultaneously. Their methodology reduces the training time, and improves the classification rate of the model.

2.2 Multi-objective Methods

Xue et al. [17] performed a study on multi-objective PSO for feature selection. They compare the efficiency of two algorithms, where the first incorporates non-dominated sorting, while the second applies the idea of crowding, mutation, and dominance to search for the Pareto front. Both algorithms achieve comparable or even better results than the other known multi-objective algorithms, such as SPEA2, NSGAII and PAES [9].

Zhang et al. [20] proposed an improved PSO for feature selection by introducing two new operators. The first is used to extend the exploration capability of the population, and the second introduces a local learning strategy designed to exploit the areas with sparse solutions in the search space. Also their method is supplemented with an archive, and the crowding distance.

3 The Proposed Algorithm

In this section the proposed multi-objective PSO for feature selection enhanced with crowding and clustering (CCMOPSO) is described that utilizes two objectives which are based on mutual information (MI). The MI is a measure used in information theory [15] that is one of the more important theories capable of capturing the relevance between features and class labels [5]. It is also computationally efficient, and therefore desirable for deployment in real life applications.

3.1 Information Theory

Developed by Shannon [15], information theory provides a way to measure the information of entropy and mutual information. Entropy can be viewed as an uncertainty of a discrete random variable X. Thus, the uncertainty X can be measured using entropy $H(X)$, as follows:

$$H(X) = -\sum_{x \in X} p(x) \cdot \log_2 p(x), \tag{1}$$

where $p(x)$ is the probability density function of X.

The probability density function of discrete random variables X and Y denoted as joint entropy $H(X, Y)$ can be formulated as:

$$H(X, Y) = -\sum_{x \in X, y \in Y} p(x, y) \cdot \log_2 p(x, y). \tag{2}$$

When the value of the discrete random variable X given random variable Y, the uncertainty can be measured using the conditional entropy $H(X|Y)$:

$$H(X|Y) = -\sum_{x \in X, y \in Y} p(x, y) \cdot \log_2 p(x|y), \tag{3}$$

where the $p(x|y)$ is the posterior probability of X given Y. The conditional entropy $H(X|Y)$ of two variables is zero, if one completely depends on the other. Consequently, $H(X|Y) = H(X)$ means that knowing one variable will do nothing to observe the other.

The information shared between two random variables is mutual information, and can be interpreted as follows: given a variable X, how much information can be gained about variable Y. Mutual information $I(X;Y)$ can be formally written as:

$$
\begin{aligned}
I(X;Y) &= H(X) - H(X|Y) \\
&= H(Y) - H(Y|X) \\
&= - \sum_{x \in X, y \in Y} p(x,y) \cdot \log_2 \frac{p(x,y)}{p(x) \cdot p(y)}
\end{aligned}
\tag{4}
$$

3.2 Particle Swarm Optimization

The Particle Swarm Optimization (PSO) is a computational algorithm, developed by Kennedy and Eberhart in 1995 [7]. It belongs to the Swarm Intelligence (SI) family of algorithms, since it mimics the social behavior of birds flocking. The algorithm maintains a population of solutions, where each solution is denoted as a particle in the search space that has a current position and a velocity. The position of a particle i is denoted by a vector $\mathbf{x_i} = (x_{i,1}, x_{i,2}, \ldots, x_{i,n})$, where n is the dimensionality of the problem. The velocity of the same particle is represented by a vector $\mathbf{v_i} = (v_{i,1}, v_{i,2}, \ldots, v_{i,n})$, which can be limited by a predefined maximum velocity v_{max}. Each particle holds the best previous solution denoted as the *pbest*, while the global best solution is *gbest*. The *pbest* guide the search process by updating the position and velocity of each particle in the population, by employing the following equations:

$$
\mathbf{x}_i^{t+1} = \mathbf{x}_i^t + \mathbf{v}_i^{t+1}, \tag{5}
$$

$$
\mathbf{v}_i^{t+1} = v_i^t + c_1 * r_1 * (\mathbf{pbest}_i - \mathbf{x}_i^t) + c_2 * r_2 * (\mathbf{gbest} - \mathbf{x}_i^t), \tag{6}
$$

where t denotes the the current iteration, c_1 and c_2 are acceleration constants, while r_1 and r_2 are uniformly distributed values within the range $[0, 1]$.

3.3 The Proposed Multi-objective PSO for Feature Selection Enhanced with Clustering and Crowding

The PSO algorithm usually converges fast towards the solution in the population. Although useful in many problems, this might become an obstacle in feature selection, since the problem is multi-objective in nature and therefore a diversity of population is needed during the whole optimization process. With this in mind, we propose a novel multi-objective PSO, extended with clustering and crowding (CCMOPSO). Clustering and crowding are features implemented within several evolutionary computation algorithms [3,6,20]. Actually, these features enable discovering the more promising regions in the search space, and consequently maintaining more good solutions in the population.

Before discussing the CCMOPSO, the concept of multi-objective optimization (MOOP) needs to be explained. In MOOP, a number of objectives need to be either minimized or maximized, which can be mathematically expressed as:

$$
\min/\max \quad F(\mathbf{x}) = [f_1(\mathbf{x}), f_2(\mathbf{x}), \ldots, f_k(\mathbf{x})], \tag{7}
$$

where $\mathbf{x}$ is a vector of decision variables, k is the number of functions to be minimized/maximized. Additionally there are also several constraints, which need to be satisfied during optimization:

$$g_i(\mathbf{x}) \leq 0, \quad i = 1, 2, \ldots, m, \tag{8}$$

and

$$h_i(\mathbf{x}) = 0, \quad i = 1, 2, \ldots, l, \tag{9}$$

where m and l are the number inequality and equality constraints respectively.

The quality of a solution in MOOP is determined in terms of trade-offs between conflicting objectives that in other words is called domination. As a result, the solution $\mathbf{x}$ dominates the solution $\mathbf{y}$, when:

$$\forall i : f_i(\mathbf{x}) \leq f_i(\mathbf{y}) \quad and \quad \exists j : f_j(\mathbf{x}) \leq f_j(\mathbf{y}), \quad i, j \in 1, 2, \ldots, k. \tag{10}$$

The solution of a MO algorithm is a set of non-dominated solutions called a Pareto Front (PF).

As was mentioned in the beginning of this section, the PSO is enhanced with clustering and crowding. Hereinafter, both mechanism are described in detail:

- **Clustering**: In each generation t of the search process, the population is divided into more sub-populations (clusters), which are assigned to different regions of the search space. Each cluster C_j consists of M particles, and is supplied by the user. Before the creation of clusters, all non-dominated solutions (ND^t) from the swarm are extracted. These particles represent the centers of clusters. A randomly selected particle from the ND^t is assigned to the cluster C_j, then an additional $M - 1$ nearest particles are added. If the number of clusters exceeds the number of particles in ND^t, then for the remaining clusters, just M nearest particles to a randomly selected particle from ND^t are selected. We should stress, that one particle cannot occur in two clusters at the same time.
- **Crowding**: After the clusters are created, the positions of the particles are updated according to Eqs. 5 and 6. Afterwards, each particle from both clusters is evaluated and compared with the nearest particle in its neighborhood. However, the better particle becomes a member of the swarm.

The CCMOPSO uses an archive of solutions, where all non-dominated ones are stored after performing a generation. This archive is also used, when the position of the particles is updated within the clusters, i.e., the random solution from the archive is taken instead of using the $\mathbf{gbest}_i$ in Eq. (6). The CCMOPSO algorithm is presented in Algorithm 1.

Two objectives are defined in the CCMOPSO since it is the MO algorithm. Thus, the first objective is responsible for evaluating the relevance between features and the class labels. In feature selection, the features may have some kind of interactions, thus a combination of several individually good features may not be the best combination for good classification accuracy. An attention must be paid to the redundancy among the selected features, where the goal is to select that subset of good features, where minimal redundancy of features interactions (the second objective) is achieved, but at the same time maximizing the relevance to the class labels. Eqs. (11) and (12) denote the objectives, used in this study:

$$Rel = -1 \sum_{x \in X} I(x; c), \tag{11}$$

$$Red = \sum_{x_i, x_j \in X} I(x_i, x_j), \tag{12}$$

where X is the set of selected features, and c the class label. Variable Rel is reserved for calculating the mutual information between each feature and the class labels. In simple terms, variable Rel is used as a measure of reporting each features importance to the classification. The Red on the other hand, calculates the mutual information among all selected features. It is used for measuring the redundancy contained within the selected feature subset. Both objectives need to be minimized during the optimization process.

Algorithm 1. CCMOPSO algorithm

Input: Population of particles $\mathbf{x}_i = (x_{i,1}, \ldots, x_{i,D})^T$ for $i = 1 \ldots NP$, MAX_G.
Output: Archive of non-dominated particles A.

1: *generate_particles*();
2: $g = 1$;
3: **while** *termination_condition_not_meet* **do**
4: $C = makeClusters(\mathbf{x})$
5: **for** $j = 1$ **to** $size(clusters)$ **do**
6: $NS = NS \cup moveParticles(C_j)$
7: **end for**
8: $\mathbf{x} = Crowding(NS)$
9: $A = A \cup non\_dom\_sort(\mathbf{x})$
10: $A = non\_dom\_sort(A)$
11: $g = g + 1$
12: **end while**

4 Experimental Environment

In this study four benchmark datasets are taken from the UCI machine learning repository [10]. All datasets have different numbers of instances, features, and classes. The used datasets are listed in Table 1. The goal of the experimental work is two-fold. On one hand, the value of parameter M, and its impact on the results is inspected, while the second part of the experiments is devoted to comparing the CCMOPSO with single- and multi-objective evolutionary computation algorithms.

Table 1. Datasets used in the study.

Dataset	Instances	Attributes	Classes
Chess	3196	36	2
Splice	3190	61	3
Mushroom	8124	22	2
Spect	267	22	2

Each dataset was split randomly into two sets: 70% of instances were considered for training, while the other 30% for testing purposes. The algorithm is firstly run on the training set to obtain the archive of non-dominated solutions, where each solutions corresponds to a subset of optimal features with minimal redundancy and maximal relevance. Because the proposed algorithm is a filter-based, each solution from the archive is evaluated using the test set by employing a learning algorithm. Each subset of features is thus extracted from the training set, then fed to a machine learning algorithm. The same subset of features are also extracted from the test set, and then tested on the model of training. For this study, the decision tree (DT) classifier is selected. Since each solution in the archive holds a different number of selected features, the results of solutions having the same number of features are averaged.

The following parameters were used for testing: population size of all algorithms was set to 30, and the algorithm was run for 300 generations. The inertia weights of the PSO algorithms were set to 0.7, while the acceleration parameters c_1 and c_2 to 1.49618. For the bat algorithm the parameters from [8] were taken. All algorithms were run 30 times, then an average Pareto front of those runs was calculated. Since the CCMOPSO has an additional parameter M, that is responsible for the cluster sizes, both values recommended by [4] were tested ($M = 5$ and $M = 10$).

In the remainder of the paper, the results of experimental work are illustrated.

5 Results

The proposed CCMOPSO was compared with the classical multi-objective PSO, described in [2], multi-objective Bat algorithm (MOBA) [19], the PSO [7], and Bat algorithm (BA) [18]. The results of comparison are presented in Fig. 1, where the discovered average Pareto fronts denote the quality of the results obtained by the specific algorithm. The numbers in brackets depict the classification accuracy, when all features are used.

The Figures show that both CCMOPSO variants give better results when compared to MOBA and MOPSO on all datasets. They are able to locate a minimal optimal subset of solutions, with relatively low classification error. Both comparing single-objective algorithms (i.e. PSO and BA) provide fairly good results, since they are able to find a small number of features, which lead to good classification accuracy. Compared to CCMOPSO, they provide comparable results for a smaller number of features, while the CCMOPSO is better when the number of features increases.

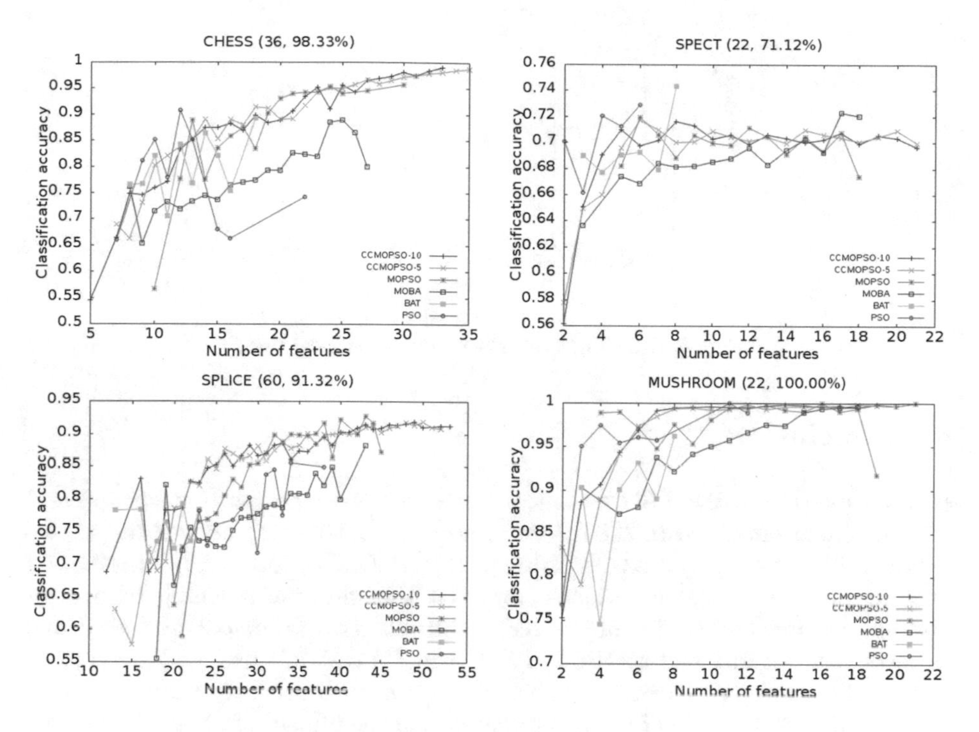

Fig. 1. Results for $Np = 30$ and $M \in \{5, 10\}$

It seems that the value of M does not have a great impact on the obtained results, yet the CCMOPSO with $M = 10$ does provide a small improvement over $M = 5$. In order to assess the results statistically, a Friedman statistical test was conducted at significance level 0.05. The Friedman test was used to check if each algorithm in this study is significantly better than the others. The measures used for the test were the classification accuracies according to the number of selected features. If an algorithm was not able to find the feature set with a said number of features, a classification accuracy of zero was used. The results of the Friedman tests are presented in Fig. 2, where squares represent average ranks and the lines represent confidence intervals. The lower the rank, the better is the observed algorithm. When the intervals of any two algorithms do not overlap, they are significantly different between each other.

According to the Friedman statistical test, both CCMOPSO algorithms outperformed all comparing algorithms. Although single-objective algorithms variants managed to find some promising feature subsets, they were no match for the multi-objective ones.

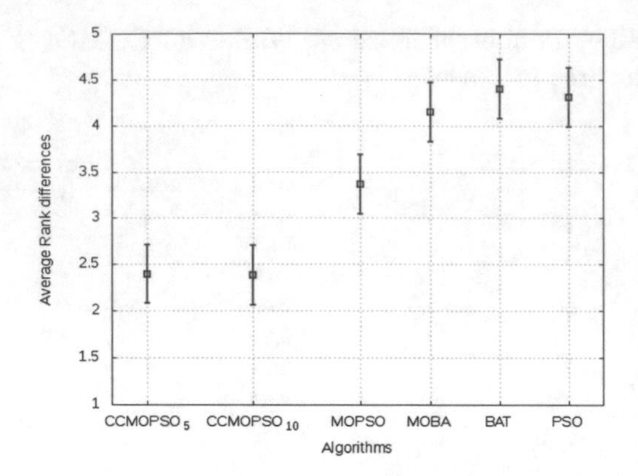

Fig. 2. Results of the Friedman statistical test.

6 Conclusion

A novel multi-objective PSO for feature selection enhanced with clustering and crowding was presented in this paper, where both features improve the exploration ability of the proposed algorithm CCMOPSO. The CCMOPSO was tested on four publicly available datasets from the UCI machine learning repository obtaining significantly different (better) results than other comparing algorithms in tests (i.e., MOBA and MOPSO, PSO, and BA). For future work, we intent to test the proposed algorithm CCMOPSO on several other datasets, and also make a large-scale study of input parameters on the impact of selected features.

References

1. Alexandridis, A., Patrinos, P., Sarimveis, H., Tsekouras, G.: A two-stage evolutionary algorithm for variable selection in the development of RBF neural network models. Chemom. Intell. Lab. Syst. **75**(2), 149–162 (2005)
2. Azevedo, G.L.F.B.G., Cavalcanti, G.D.C., Carvalho Filho, E.C.B.: An approach to feature selection for keystroke dynamics systems based on PSO and feature weighting. In: 2007 IEEE Congress on Evolutionary Computation, pp. 3577–3584, September 2007
3. Bošković, B., Brest, J.: Differential evolution for protein folding optimization based on a three-dimensional AB off-lattice model. J. Mol. Model. **22**(10), 252 (2016)
4. Bošković, B., Brest, J.: Clustering and differential evolution for multimodal optimization. In: 2012 IEEE Congress on Evolutionary Computation (CEC). IEEE (2012)
5. Cervante, L., Xue, B., Zhang, M., Shang, L.: Binary particle swarm optimisation for feature selection: a filter based approach. In: 2012 IEEE Congress on Evolutionary Computation (CEC), pp. 1–8. IEEE (2012)
6. Dara, S., Banka, H., Annavarapu, C.S.R.: A rough based hybrid binary PSO algorithm for flat feature selection and classification in gene expression data. Ann. Data Sci. **4**, 1–20 (2017)

7. Eberhart, R., Kennedy, J.: A new optimizer using particle swarm theory. In: 1995 Proceedings of the Sixth International Symposium on Micro Machine and Human Science, MHS 1995, pp. 39–43. IEEE (1995)
8. Fister Jr., I., Mlakar, U., Yang, X.-S., Fister, I.: Parameterless bat algorithm and its performance study. In: Nature-Inspired Computation in Engineering, pp. 267–276. Springer (2016)
9. Kalyanmoy, D.: Multi-objective Optimization Using Evolutionary Algorithms. Wiley, Hoboken (2001)
10. Lichman, M.: UCI machine learning repository (2013)
11. Mlakar, U., Fister Jr., I., Fister, I.: Hybrid self-adaptive cuckoo search for global optimization. Swarm Evol. Comput. **29**, 47–72 (2016)
12. Mlakar, U., Fister, I., Brest, J., Potočnik, B.: Multi-objective differential evolution for feature selection in facial expression recognition systems. Expert Syst. Appl. **89**, 129–137 (2017)
13. Mlakar, U., Zorman, M., Fister Jr., I., Fister, I.: Modified binary cuckoo search for association rule mining. J. Intell. Fuzzy Syst. **31**(6), 4319–4330 (2017)
14. Mohemmed, A.W., Zhang, M., Johnston, M.: Particle swarm optimization based adaboost for face detection. In: 2009 IEEE Congress on Evolutionary Computation, pp. 2494–2501, May 2009
15. Shannon, C., Weaver, W.: The Mathematical Theory of Communication. The University Illinois Press, Urbana (1949)
16. Sun, Y., Babbs, C.F., Delp, E.J.: A comparison of feature selection methods for the detection of breast cancers in mammograms: adaptive sequential floating search vs. genetic algorithm. In: 2005 27th Annual International Conference of the Engineering in Medicine and Biology Society, IEEE-EMBS 2005, pp. 6532–6535. IEEE (2006)
17. Xue, B., Zhang, M., Browne, W.N.: Particle swarm optimization for feature selection in classification: a multi-objective approach. IEEE Trans. Cybern. **43**(6), 1656–1671 (2013)
18. Yang, X.-S.: A new metaheuristic bat-inspired algorithm. In: Nature Inspired Cooperative Strategies for Optimization (NICSO 2010), pp. 65–74 (2010)
19. Yang, X.-S.: Bat algorithm for multi-objective optimisation. Int. J. Bio-Inspired Comput. **3**(5), 267–274 (2011)
20. Zhang, Y., Gong, D., Sun, X., Guo, Y.: A PSO-based multi-objective multi-label feature selection method in classification. Sci. Rep. **7**, 376 (2017)

Scalability of GPA-ES Algorithm

Tomas Brandejsky(✉) and Roman Divis

University of Pardubice, CS Legies Square 565,
53002 Pardubice, Czech Republic
{Tomas.Brandejsky,Roman.Divis}@upce.cz

Abstract. This contribution is dedicated to analysis of GPA-ES scalability both from analytical and experimental viewpoint. The paper tries to identify limits of the algorithm, sources of possible limitations and practical ways of its application. The paper is focused to standard implementation of GPA-ES algorithm and tests were provided on task of symbolical regression of Lorenz attractor system. After introduction onto the problem, Amdahl's law is discussed in Sect. 2, characterization of GPA-ES algorithm in Sect. 3, analysis if its computational complexity in Sect. 4, experiment results analysis to determine computing time dependency on number of threads both HW and SW in Sect. 5 and analysis of influence of population size in Sect. 6. Presented results underlines that except one anomaly all experiments concludes applicability of Amdahl's law onto GPA-ES algorithm even if it is implemented on multi-core and many-core systems, which are out of original scope of this law.

Keywords: Genetic programming algorithm · Efficiency · Scalability
Parallel algorithm · Amdahl's law

1 Introduction

Applicability of any symbolic regression algorithm depends on its ability to solve complex problems. Because the power of single processor is limited, the most significant is the dependence of solution time on data size. In the case of symbolic regression there are two data sets – training data one and GPA individual population. GPA-ES algorithm brings third data vectors – ES populations representing solution parameters while GPA population in this algorithm represents pure solution structure. Evolutionary algorithms in common are parallel, because they implement Natural selection, which is scalable (in the Nature). Paradoxically, this might be reason why scalability of these algorithms as not studied as extensively as one should expect, because there exists large amount of different algorithms and implementations. There it is possible to name works as e.g. [1, 2] about complexity of simple genetic algorithms or e.g. [3–5] about complexity of genetic programming algorithms. Scalability depends also on suitability of the algorithm to implementation on multi-core or multi-processor systems. Unfortunately, in this moment we have many different architectures, especially multi-core and many-core systems. It is probable that each of them will have different requirements to SW implementation of the algorithm.

R. Matoušek (Ed.): MENDEL 2017, AISC 837, pp. 124–133, 2019.
https://doi.org/10.1007/978-3-319-97888-8_11

Presented algorithm starts with description if Amdahl's law in Sect. 2, characterization of GPA-ES algorithm in Sect. 3, analysis if its computational complexity in Sect. 4, experiment results analysis to determine computing time dependency on number of threads both HW and SW in Sect. 5 and analysis of influence of population size in Sect. 6.

2 Amdahl's Law Describing Speed-Up of Parallel Algorithm

Amdahl's law describes possible speed-up of given algorithm if it is executed on n-processor (in original), or n-core processor today. It was published in [6]. This law was in the form (1):

$$Spedupparallel(f,n) = \frac{1}{(1-f)+\frac{f}{n}} \tag{1}$$

where
f represents parallelizable fraction of program
n is the number of processors

It assumes that program is infinitely parallelizable with no overhead. In the reality, starting of any thread consumes some time, all threads do not need to take equal time and it complicates situation and decreases reachable speed-up. Also real algorithms cannot be divided into any number of threads but only to limited number, as it will be discussed in Sect. 4. Examples of some significant modifications of Amdahl's law for various parallel system architectures are discussed e.g. in work [7]. Novel Hyper Threading (HT) processors bring new problems, which multiplies its computational power only if each of parallely executed threads in the given time used different type of instruction executed by different module of processor core (e.g. Integer and Float Point ones). It is also need to distinguish SW and HW threads because modern operating systems allow to run many threads on small number of processor cores. Such parallelization has not positive influence to computing efficiency.

3 GPA-ES Algorithm Description

GPA-ES algorithm was derived from standard Genetic Programming Algorithm by adding parameter optimization cycle in the form of Evolutionary Strategy into each gene evaluation step to optimise solution parameters.

This hierarchical algorithm was introduced in [8] and its main idea of separate optimization of individual's parameters is based on observation [9]. Standard GP algorithms sometimes eliminate good structures because of their parameters are estimated wrong. Separate optimization by nested (evolutionary) algorithm eliminates this problem and concluding blowing – producing of overcomplicated solutions. These overcomplicated structures are eliminated two ways. First of them is comparing of structures with well fitted parameters. Thus the information about residual error contains more information about suitability of individual structure to training data set than in standard GPA. Second way is given by limitation of nested ES algorithm evolutionary steps. This limitation increases chance of overcomplicated structures with many constants to be well fitted and thus they are eliminated by insufficient fitness magnitude.

Above written facts also means that standard GPA-ES algorithm was optimised to applications in symbolic regression domain. Because symbolic regression of non-trivial problems require more powerful computer systems than single thread single core one, in was implemented using OpenMP library [10]. On the other hand it is true that original ideas were formulated when novel hierarchical algorithm of system design was formulated because of three was idea of multi-layer algorithm where different layers works on different levels of abstraction (or detailness).

From the practical implementation viewpoint in means that to each GPA individual (algebraic structure represented by graph of functions/operator) there it is related table of ES individuals representing population of parameters – constants to be optimised, as it is illustrated by Fig. 1 for 2 of n GPA individuals. While on the top of this figure there are sketched tree-like representations of algebraic expressions – individuals of GPA, bottom half of the figure sketches arrays of constant magnitudes which are subject of separated ES optimization. For each GPA individual there exist separate array of ES individuals to be optimized.

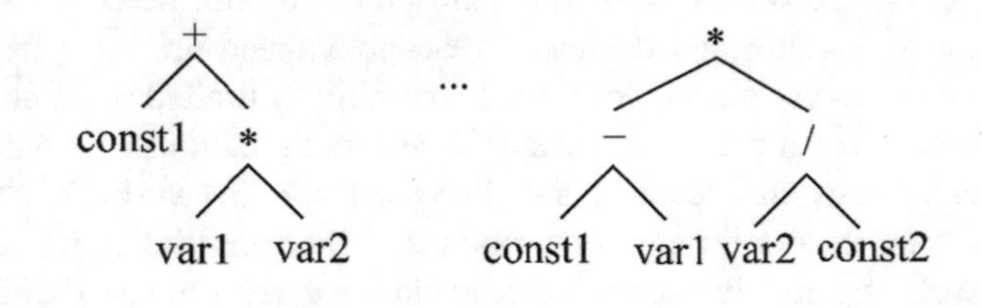

ESIndividualNo,	Const1
1,	17
2,	12
...	
m0,	23

ESIndividualNo,	Const1,	Const2
1,	173,	1414
2,	94,	1523
...		
mn,	10,	28

Fig. 1. Structure of GPA-ES algorithm data

The GPA-ES algorithm is outlined in the following Algorithm 1 list and study of its complexity is subject of the following Sect. 4.

4 GPA-ES Algorithm Complexity

When the evaluation loop of GPA is parallelized, each ES population of parameters is optimized in parallel with others. Natural limit of parallelization of inner GPA-ES cycle is set by size of GPA population because today implementation of this algorithm does not support application of more processors than the number of GPA population individuals as it will be discussed latter.

Algorithm 1. Studied GP algorithm with parameter pre-optimization.

```
1) FOR ALL individuals DO Initialize() END FOR;
2) FOR ALL individuals DO Evaluate()=>fitness END FOR;
3) Sort(individuals);
4) IF Terminal_condition() THEN STOP END IF;
5) FOR ALL individuals DO
   SELECT Rand() OF
      CASE a DO Mutate()=> new_individuals;
      CASE b DO Symmetric_crossover() => new_individuals;
      CASE c DO One_point_crossover() => new_individuals;
      CASE d DO Re-generating() => new_individuals;
   END SELECT;
   END FOR;
6) FOR ALL individuals DO Evaluate => new_fitness END FOR;
7) FOR ALL individuals DO IF new_fitness<fitness THEN
      individual = new_individual;
      fitness = new_fitness;
   END IF;
8) GOTO 3);
```

For each individual of the GPA population, special copy of the ES object is created and executed. Because no communication between different ES solving optimization of different structure parameters is required, execution runs extremely efficiently without necessity of any synchronization or communication.

Analysing computational complexity of GPA-ES algorithm it is possible to expect that it will be expressed in the form (2)

$$O(GPAES) = f(n, m, l, k, p, q), \tag{2}$$

where

n is number of individuals in GPA population
m is number of individuals in ES population
l is complexity of structures created by GPA
k is average number of constants in GPA genes
p is number of GPA evolution cycles
q is number of ES evolution cycles in each GPA evolution step

From the practical viewpoint, complexity is represented as the function of m, n, p and q only, because parameters k and l depends on above listed parameters and solved problem data and it is impossible to express them analytically. This complexity estimation of GPA-ES algorithm outlined as Algorithm 1 presumes that computational complexity is determined by building blocks of GPA and ES algorithms consisting of initial population, evaluation, sorting and evolutionary operations discussed below:

GPA population initialization average complexity is given by (3) because it is need to initialize each of n individuals and three of functions with complexity given by two initial parameters max_depth and min_depth describing maximal and minimal initial tree depth respectively. Average depth of initial trees depend on the used random number generator distribution (on the probability of terminal building block occurrence). Because the magnitudes which is in the initialization time determined by two magnitudes max_depth and min_depth are constant as well as probability distribution of used pRNG, the resulting complexity of GPA population initialization is function of n, as it is described by (3):

$$O(GPAinit) = n \quad (3)$$

GPA population evaluation consist of execution of special copy of ES algorithm. Without study of this optimization algorithm it is only possible to specify, that for each GPA individual one copy of ES algorithm with specific structures is stared, as it is outlined on (4). This dependency is analogical to previous case (3):

$$O(GPAeval) = n * O(ES) \quad (4)$$

Ending condition evaluation complexity is constant because its computational complexity does not depend on any parameter.

$$O(GPAend) = 1 \quad (5)$$

Individual sorting in GPA population is in the worst case expressed by (6):

$$O(GPAsort) = n^2 \quad (6)$$

GPA-ES algorithm uses four different structural evolutionary operators – symmetric crossover, one-point-crossover, mutation and generating on totally new individuals. They have similar complexities which might be represented as (7):

$$O(GPAoperators) = n * (3/4 * 2 + 1/4)^{((l-2)/2)} \approx n \quad (7)$$

The expression (7) might be explained as follows: We reason n individuals. Complexity of evolutionary operation on each individual depends on complexity of single operation (which is equal to 1) multiplied by number of tree nodes to be scanned. Because in the actual implementation there is probability of binary operator occurrence 3 times bigger than probability of unary one, tree has 1 root node and lowest nodes must be terminals, for average depth l we obtain expression $(3/4 * 2 + 1/4)^{((l-2)/2)}$ which is constant for reasoned arguments m, n, p and q, thus can be reduced to n in (7).

Number of GPA evolution cycles is constrained, but this limit typically much exceeds real number of cycles, thus it does not need to be considered.

Nested parameter optimization algorithm ES population initialization complexity is expressed as (8) because it needs only to generate array of constant vectors. Length of this error is given by constant representing length of ES algorithm populations. Each vector contains number of constant equals to number used in corresponding GPA tree representing particular individual. To each GPA individual specific copy of ES algorithm is applied.

$$O(ESinit) = m * k = m \tag{8}$$

ES population evaluation complexity depends on size of ES population and the complexity of functions/individuals to be optimized, it means on the number of nodes of representing graph. It does not depend directly on number of constants, because one fitness function and single fitness is evaluated for whole vector of constants (9). The average number of constants is ½ of terminal symbols number and depends on complexity of GPA individuals representing tree, which was discussed with relation o Eq. (7). From the analogical reasons it also might be reduced to simple m:

$$O(ESeval) = m * (3/4 * 2 + 1/4)^{(l-1)/2} = m \tag{9}$$

ES termination condition complexity is analogically as in the GPA case expressed as (10):

$$O(ESterm) = 1 \tag{10}$$

Population ordering complexity depends on applied sorting algorithm and in the worst case it is expressed as (11):

$$O(ESsort) = m^2 \tag{11}$$

Evolutionary strategy uses so-called intelligent crossover to produce new individuals with complexity (12):

$$O(ESnew) = m * k = m \tag{12}$$

ES algorithm uses fixed number of evolutionary cycles, because there is no reason e.g. to check if the given precision was reached (this precision can be reached only in the final cycle of outer GP algorithm.

In case of the hardware threads and parallelization of evaluation cycle of GPA only, the relationship (2) substituting (3–12) changes to (13):

$$O(GPAES) = n * O(ES) + n^2 = n\left(m + 1 + m^2\right) + n^2 = nm^2 + n^2 \tag{13}$$

5 HW and SW Threads Influence Checking Experiments

Single processor single core systems are in one moment capable to run only one program thread. Situation from the computing capacity complicates when system is equipped by many processor or modern processors with many cores, because each processor pro processor core is capable to run separate thread simultaneously.

Tests were provided on two different kinds of systems. Tests on small core number systems (2 to 6 cores) were provided clusters of standard PCs. Tests on many-core systems were executed on system equipped with 64-core Intel Xeon Phi 7210.

Turbo-boost technology of Intel Core processors was identified as source of unexpected increase of grid computational power of single-thread execution. Also there might be observed on Fig. 2 unexpected execution times for numbers of SW threads exceeding magnitude 5. Such behaviour occurs when number of SW threads exceeds number of HW threads and in used cluster it was caused by application of processors with 2 to 6 HW threads only.

Many core processors as Xeon Phi bring another uncertainty into algorithm scalability evaluation. They support hyper threading not only of two threads on single processor core, but of four ones. Problem is that results computation times look like if pair of processor share some processor modules, because if processor evaluates 64 identical threads, it has nearly half of computational power presented if it executes two different groups of 32 identical tasks.

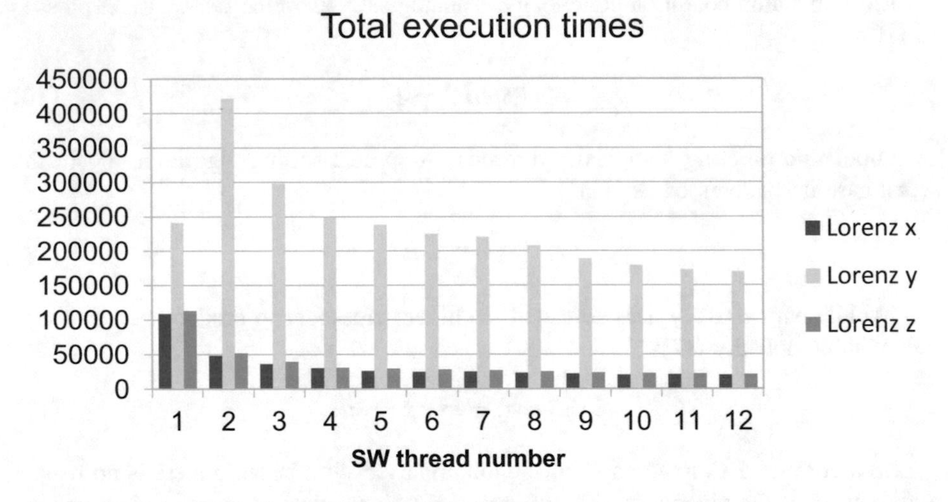

Fig. 2. Total execution times of 2000 Lorenz attractor system symbolic regression experiments execution on multi-core processors cluster with GPA population size 50 individuals for different numbers of SW threads. Execution times were evaluated independently for symbolic regression of x, y and z Lorenz attractor system coordinate equations independently.

6 Population Size Influence Checking Experiments

Experiments on multi-core Intel Core processor grid produced results summarised in Fig. 2. Each experiments consisted of symbolic regression of Lorenz attractor system equations from dataset of 500 samples. Due to dependency on pseudo random number generator seed magnitudes there was evaluated series of 2000 experiments, each of them with unique pRNG seed magnitude. The size of GPA population was 50, the size of ES populations was 30. Number of evolution cycles was unconstrained for GPA and 30 for ES. Lorenz attractor system was used for comparability with other experiments.

There exist standard GP benchmarks, some of them are oriented to symbolic regression problem. As it was published in [11], if there is sufficient number of experiments and if we solve symbolical regression problems with small residual error, average results depends only on statistical properties of initial population generator and applied evolutionary operators and it is possible to estimate them by application of Markov chain.

Each experiment was run on independent computer to prevent mutual affect of pRNGs. During experiments there was find out, that implementations of pRNG C++ library <random> are not implement thread safe way. The only positive exception was found in OpenSuse Linux distribution in GNU C++ compiler 4.8.5, but also this implementation is thread safe only until number of SW threads is less or equal to number of processor cores. Other tested compilers did not offer thread safety. They was Mageia Linux with GCC versions 4.8.2 and 4.9.2 and CentOS Linux with Intel C++ compiler version 17.0.0 on Xeon Phi systems.

GPA-ES algorithm runs in parallel only ES part. GPA part works as single thread because it represents only mall part of computing effort. Maximal number of executing HW threads must then be less or equal to number of SW threads – to number of individuals in ES population. Results of all presented experiments on manycore system correspond to previous experiments on multi-core processors. Numbers of SW threads on many-core system (64, 128 and 256) were chosen to maximize utilization of Intel Xeon Phi processor. The rest parameters of GPA-ES algorithm remain the same.

The 50 individual multi-core experiments correspond to previously mentioned experiments. There is visible influence of processor optimization oriented to maximal single thread application power. Even more significant is this situation in the case of 10 individual multicore system experiments which were chosen with respect to former small population experiments were introduced in the work [12]. They pointed that the use of small populations (e.g. 10 individuals) tends to extremely large numbers of evaluation cycles – populations and their variance, but also to much smaller (10–1000 times) number of fitness function evaluations and concluding decrease of computing time. This observation corresponds to results of above presented analysis of GPAES algorithm complexity. Problem of application of small populations and related increase of evolutionary pressure is that today processor development goes to increase of processor core number and thus against this observations.

7 Conclusion

Presented results points good scalability of GPA-ES algorithm in situation where the GPA population size is greater or equal to number of HW threads. They also shows that running of threads has appreciable overheads and that significance of this effect decreases with solved problem complexity. Dependence on specific function is well illustrated by presented problems where Amdahl's law was confirmed even for many-core systems with four times hyper threading. The only observed anomaly for single thread execution cannot conclude non applicability of Amdahl's law because it was not observed for z coordinate, for x coordinate it was observed only for small population and only for x-coordinate it was observed for both smaller and larger populations (Figs. 3 and 4).

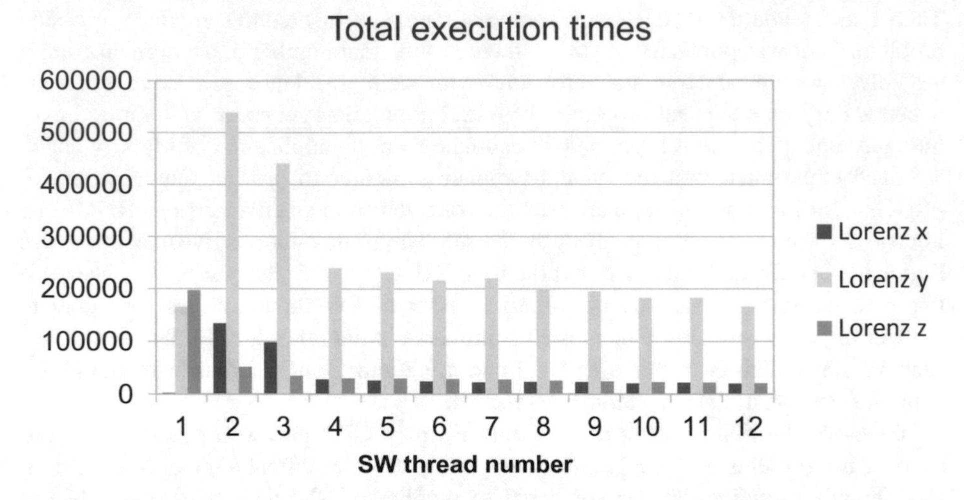

Fig. 3. Total execution time of 2000 Lorenz attractor system symbolic regression experiments execution on multi-core processors cluster with GPA population size of 10 individuals for different numbers of SW threads. Execution times were evaluated independently for symbolic regression of x, y and z Lorenz attractor system coordinate equations independently.

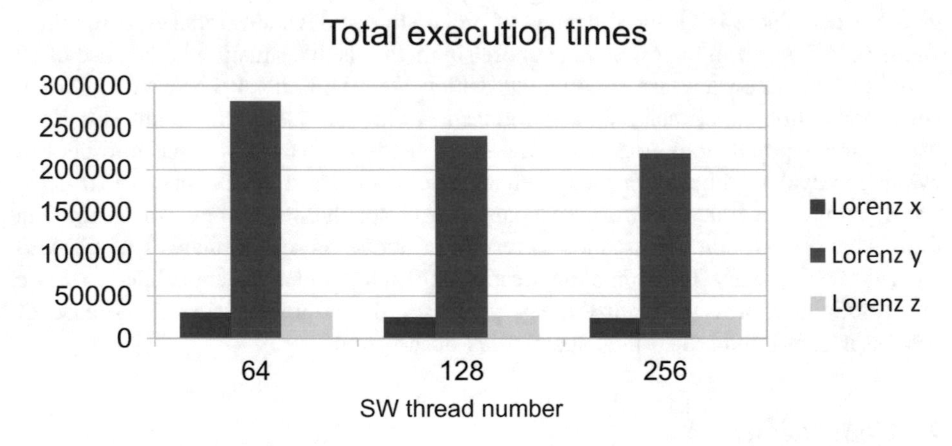

Fig. 4. Total execution time of 2000 Lorenz attractor system symbolic regression experiments execution on many-core processor Xeon Phi with GPA population sizes of 64, 128 and 256 individuals and same numbers of SW threads. Execution times were evaluated independently for symbolic regression of x, y and z Lorenz attractor system coordinate equations independently.

Acknowledgements. Computational resources were provided by the CESNET LM2015042 and the CERIT Scientific Cloud LM2015085, provided under the programme “Projects of Large Research, Development, and Innovations Infrastructures”.

Computational resources were supplied by the Ministry of Education, Youth and Sports of the Czech Republic under the Projects CESNET (Project No. LM2015042) and CERIT-Scientific Cloud (Project No. LM2015085) provided within the program Projects of Large Research, Development and Innovations Infrastructures.

References

1. Thierens, D.: Scalability problems of simple genetic algorithms. Evol. Comput. **7**(4), 331–352 (1999)
2. Liu, Y.Y., Wang, S.: A scalable parallel genetic algorithm for the generalized assignment problem. **46**, 98–119 (2015). Article no. 2184
3. Folino, G., Pizzuti, C., Spezzano, G.: A scalable cellular implementation of parallel genetic programming. IEEE Trans. Evol. Comput. **7**, 37–53 (2003)
4. Virgolin, M., Alderliesten, T., Witteveen, C., Bosman, A.N.P.: Scalable genetic programming by gene-pool optimal mixing and input-space entropy-based building-block learning, vol. 1 (2017). https://doi.org/10.1145/3071178.3071287
5. Christensen, S.: Towards scalable genetic programming. Ph.D. Study, Ottawa-Carleton Institute for Computer Science (2006)
6. Amdahl, G.M.: Validity of the single-processor approach to achieving large scale computing capabilities. In: AFIPS Conference Proceedings, pp. 483–485 (1967)
7. Hill, M.D., Marty, M.R.: Amdahl's Law in the Multicore Era. http://research.cs.wisc.edu/multifacet/papers/tr1593_amdahl_multicore.pdf. Accessed 1 June 2016
8. Brandejsky, T.: Multi-layered evolutionary system suitable to symbolic model regression. In: Proceedings of the NAUN/IEEE.AM International Conferences, 2nd International Conference on Applied Informatics and Computing Theory, Praha, 26–28 September 2011, pp. 222–225. WSEAS Press, Athens (2011). ISBN 978-1-61804-038-1
9. Brandejsky, T.: Genetic programming algorithm with parameters preoptimization - problem of structure quality measuring. In: Osmera, P. (ed.) Mendel 2005. 11th International Conference on Soft Computing, Brno, 15–17 June 2005, pp. 138–144. Brno University of Technology, Brno (2005). ISBN 80-214-2961-5
10. Chandra, R., Menon, R., Dagu, L., Kohr, D., Maydan, D., McDonald, J.: Parallel Programming in OpenMP. Morgan Kaufmann (2000). ISBN: 9781558606715
11. Brandejsky, T., Zelinka, I.: Specific behaviour of GPA-ES evolutionary system observed in deterministic chaos regression. In: Zelinka, I. et al. (eds.) Nostradamus: Modern Methods of Prediction, Modeling and Analysis of Nonlinear Systems. Advances in Intelligent Systems and Computing, Nostradamus, Ostrava, 05–07 September 2012, pp. 73–81. Springer, Heidelberg (2013). ISSN 2194-5357, ISBN 978-3-642-33226-5
12. Brandejsky, T.: Small populations in GPA-ES algorithm In: MENDEL 2013. Brno University of Technology, Brno, Faculty of Mechanical Engineering, pp. 31–36 (2013). ISSN 1803-3814, ISBN 978-80-214-4755-4

Comparison of Three Novelty Approaches to Constants (Ks) Handling in Analytic Programming Powered by SHADE

Zuzana Kominkova Oplatkova(✉), Adam Viktorin, and Roman Senkerik

Faculty of Applied Informatics, Tomas Bata University in Zlin, Nam. T. G. Masaryka 5555, 76001 Zlin, Czech Republic
{oplatkova, aviktorin, senkerik}@fai.utb.cz

Abstract. This research deals with the comparison of three novelty approaches for constant estimation in analytic programming (AP) powered by Success-history based Differential evolution (SHADE). AP is a tool for symbolic regression tasks which enables to synthesise an analytical solution based on the required behaviour of the system. This paper offers another strategy to already known and used by the AP from the very beginning and approaches published recently in 2016. This paper compares these procedures and the discussion also includes nonlinear fitting and metaevolutionary approach. As the main evolutionary algorithm, a differential algorithm in the version SHADE for the main process of AP is used. The proposed comparison is performed out on quintic, sextic, Sine 3 and Sine 4 benchmark problems.

Keywords: SHADE · Analytic programming · Constant handling

1 Introduction

Analytic Programming (AP) [17] is a tool of symbolic regression which uses techniques from the area of evolutionary computation techniques (EVT). The basic case of a regression represents a process in which the measured data is fitted and a suitable mathematical formula is obtained in an analytical way. This process is widely known for mathematicians. They use this process when a need arises for a mathematical model of unknown data, i.e. the relation between input and output values. Classical regression usually requires to select an expected type of the model in advance and a suitable method is applied for the coefficient estimation of this proposed model. Compared to that, symbolic regression in the context of EVT means to build a complex formula from basic operators defined by users. The final shape of the expression is managed to breed via evolutionary optimisation algorithms.

Initially, John Koza proposed the idea of symbolic regression done by means of a computer in Genetic Programming (GP) [1, 6, 7]. The other approaches are e.g. Grammatical Evolution (GE) developed by Ryan [9] and some others included Analytic Programming [17]. The symbolic regression can be used for different tasks: data approximation, design of electronic circuits, optimal trajectory for robots, classical

R. Matoušek (Ed.): MENDEL 2017, AISC 837, pp. 134–145, 2019.
https://doi.org/10.1007/978-3-319-97888-8_12

neural networks and pseudo neural networks synthesis and many other applications [1–7, 9, 10, 16, 17]. The results and usage depend on the user-defined set of operators and their possible combinations and nesting into themselves.

This paper deals with the comparison of three different strategies for constants (coefficients) estimation - direct encoding in the individuals (extended individuals [15] and a special handling with an individual [14] and a stance proposed in this paper similar to [14]. It extends the preliminary study [5] and provides results for more benchmark functions and for another strategy of the used evolutionary algorithm. The previous approaches for constant handling - nonlinear fitting [17] and metaevolutionary approach [17] are discussed too.

2 Analytic Programming

Basic principles of the AP were developed in 2001 [17–19].

The substance of AP lies in a special set, general functional set(GFS), of mathematical objects (set of functions, operators and constants or independent variables. The latter two are usually called terminals) and operations. This set is called a general functional set (GFS) due to its variability of the content since various functions and terminals can be mixed here. The structure of GFS is created by subsets of functions according to the number of their arguments. For example, GFS_{all} is a set of all functions, operators and terminals, GFS_{3arg} is a subset containing functions with only three arguments, GFS_{0arg} represents only terminals, etc. The subset structure presence in GFS is of vital importance for AP. It is used to avoid a synthesis of pathological programs, i.e. programs containing functions without arguments, etc. The content of GFS is dependent only on the user, [10, 17, 18].

The second part of the AP core is a procedure of mathematical operations, which are used for the program synthesis. These operations are used for transformation of an individual into a suitable program. Mathematically stated, it is a mapping from an individual domain into a program domain. This mapping consists of two main parts. The first part is called discrete set handling (DSH) (Fig. 1) [8, 17] and the second one stands for security procedures which do not allow synthesising pathological programs. The method of DSH, when used, allows handling arbitrary objects including non-numeric objects like linguistic terms {hot, cold, dark…}, logic terms (True, False) or other user defined functions.

AP needs some evolutionary algorithm [8, 11, 12, 17] that consists of a population of individuals for its run. Individuals in the population consist of integer parameters, i.e. an individual is an integer index pointing into GFS. The creation of the program can be schematically observed in Fig. 2.

An example of the process of the final complex formula synthesis (according to the Fig. 2) follows.

The number 1 in the position of the first parameter means that the operator plus (+) from GFS_{all} is used (the end of the individual is far enough). Because the operator + must have at least two arguments, the next two index pointers 6 (sin from GFS) and 7 (cos from GFS) are dedicated to this operator as its arguments. The two functions, sin and cos, are one-argument functions; therefore the next unused pointers 8 (tan

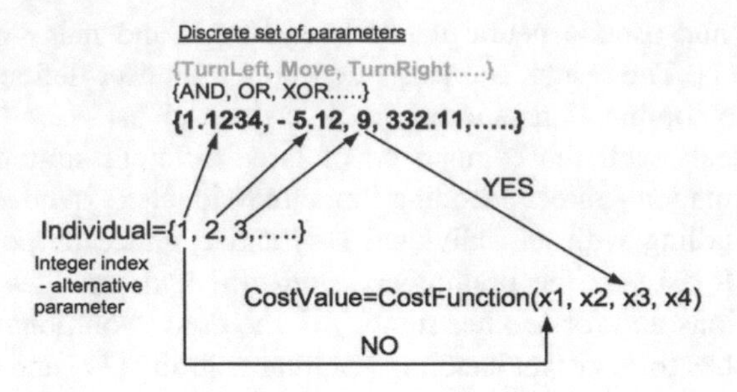

Fig. 1. Discrete set handling

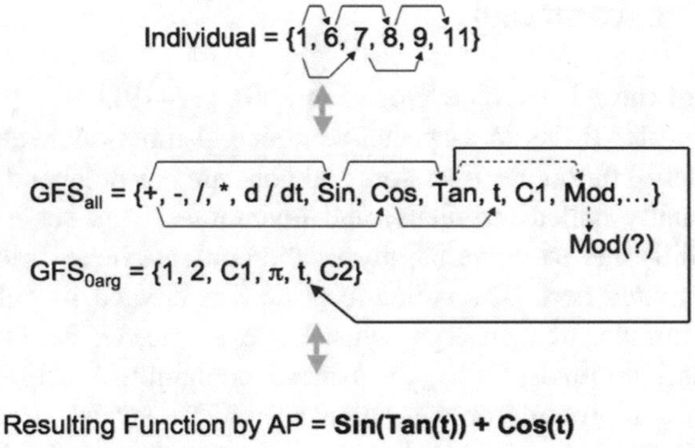

Individual = {1, 6, 7, 8, 9, 11}

GFS$_{all}$ = {+, -, /, *, d / dt, Sin, Cos, Tan, t, C1, Mod,...}

Mod(?)

GFS$_{0arg}$ = {1, 2, C1, π, t, C2}

Resulting Function by AP = **Sin(Tan(t)) + Cos(t)**

Fig. 2. Main principles of AP

from GFS) and 9 (*t* from GFS) are dedicated to the sin and cos functions. As an argument of cos, the variable *t* is used, and this part of the resulting function is closed (*t* has zero arguments) in its AP development. The one-argument function tan remains, and there is one unused pointer 11, which stands for Mod in GFS_{all}. The modulo operator needs two arguments but the individual in the example has no other indices (pointers, arguments). In this case, it is necessary to employ security procedures and jump to the subset with GFS_{0arg}. The function tan is mapped on *t* from GFS_{0arg} which is in the 11th position, cyclically from the beginning. The detailed description is represented in [10, 17–19].

3 Analytic Programming - Versions

The above-described version is the basic one AP_{basic} [19] - without constant estimation. Such approach is used for tasks like logic circuit design where numerical coefficients are not usually used. It can also be applied to a pre-generated set of numerical values as in genetic programming where e.g. 4000 random numerical values of constants are selected. They are used as standard terminals like variable x.

When a constant estimation is necessary, e.g. in data approximation or pseudo neural network synthesis, etc., firstly general approach is applied which is different from genetic programming technique where all constants (e.g. 4000 random generated values) were part of the nonterminal and terminal sets.

AP uses the constant K [19] which is indexed during the evolution (1) – (3). The K is a terminal, i.e. GFS_{0arg}. So it is used as a standard terminal, e.g. similar to variable x in the evolutionary process (1). When K is needed, a proper index is assigned – K_1, K_2, … K_n (2). Numeric values of indexed Ks are estimated (3) via different techniques - AP_{nf} (nonlinear fitting package in Mathematica) [19], AP_{meta}, (metaevolutionary approach with a second/slave evolutionary algorithm) [10, 19]. The paper deals with the comparison of three novel direct approaches AP_{extend} (extended individual - a part of the individual is used for the main process of AP and the rest of it for constant estimation) [15], $AP_{direct1}$ (the part behind decimal point determines the K from the selected range) [14] and $AP_{direct2}$ (new proposed approach in this paper - the whole value determines the K from the selected range). The authors expect that the novel approaches will decrease the computation time for the final analytic solution synthesis.

$$\frac{x^2 + K}{\pi^K} \tag{1}$$

$$\frac{x^2 + K_1}{\pi^{K_2}} \tag{2}$$

$$\frac{x^2 + 3.156}{\pi^{90.78}} \tag{3}$$

3.1 AP_{nf} - Nonlinear Fitting Version

The estimation of constants K has been done via a package for nonlinear fitting in Wolfram Mathematica environment (www.wolfram.com). The used function was FindFit, which includes different methods. Documentation refers to Conjugate Gradient, Gradient, Levenberg-Marquardt, Newton, NMinimize and Quasi-Newton. These techniques belong to a group of iterative algorithms. Thus, each constant estimation of the particularly found model is not only one step evaluation but many iterations are needed. Therefore, cost function evaluations for AP_{nf} were not interpreted correctly in previous publications. Authors think that nonlinear fitting case is a specific approach of AP_{meta} (metaevolutionary approach), except the second slave algorithm does not need to be only an evolutionary algorithm. The above-mentioned function FindFit selects its computation method automatically in Mathematica and usually around 5000 iterations are necessary in the cases of sextic, quintic, Sine 3 and Sine 4 problems. For the suggested model, the nonlinear fitting tries to find the best constants (coefficients). The final cost value comes from the performed nonlinear fitting process - the error between required and actual obtained model. Above mentioned methods might be even faster and more precise for this particular task.

3.2 AP_{meta} - Metaevolutionary Version

Generally, metaevolution means the evolution of evolution. In AP_{meta}, the metaevolution means that one evolutionary algorithm drives the main process of symbolic regression and the second is used for the constant estimation. This meta approach of analytic programming is used when the constants are not possible to estimate by AP_{nf} because of the character of the problem.

AP_{meta} is a time-consuming process and the number of cost function evaluations, which is one of the comparable factors, is usually very high. This fact is given by two evolutionary procedures (Fig. 3).

$$EA_{master} \Rightarrow program \Rightarrow K_{indexing} \Rightarrow EA_{slave} \Rightarrow K_{estimation} \Rightarrow final \cdot solution$$

Fig. 3. Schema of AP procedures

EA_{master} is the main evolutionary algorithm for AP, EA_{slave} is the second evolutionary algorithm inside AP (or a nonlinear fitting (NF) method adopted in Mathematica environment because it is an iterative process). Thus, the number of cost function evaluation (CFE) is given by (4).

$$CFE = EA_{master} * EA_{slave} \tag{4}$$

The following three approaches were developed to find a suitable constant estimation which will decrease the number of cost function evaluations to (5).

$$CFE = EA_{master} \tag{5}$$

3.3 $AP_{extended}$ - Extended Individual

The constant handling technique with an extended individual was introduced in [15]. The individual used in AP has an extended part which is used for the evolution of constant values.

The important task was to determine what the correct size of an extension is (6).

$$k = l - floor((l - 1)/(\max\_arg)) \tag{6}$$

where k is the maximum number of constants that can appear in the synthesised program (extension) of length l and max_arg is the maximum number of arguments needed by functions in GFS. Also, the *floor()* is a common floor round function. The final individual dimensionality (length) will be $k + l$ and the example might be:

- Program length $l = 10$
- GFS: $\{+, -, *, /, sin, cos, x, k\}$
- GFS maximum argument $max\_arg = 2$
- Extension size $k = 10 - floor((10-1)/2) = 6$
- Dimensionality of the extended individual $k + l = 16$.

This means, that the EA will work with individuals of length 16, but only first 10 features will be used for indexing into the GFS and the rest will be used as constant values.

It is worthwhile to note that only features which are going to be mapped to GFS are rounded and the rest is omitted (not rounded). An example can be viewed in Fig. 4. Individual features in bold are the constant values.

$$Individual = \left\{ \begin{array}{l} 5.08, 1.64, 6.72, 1.09, 6.20, \\ 1.28, \mathbf{0.07}, \mathbf{3.99}, \mathbf{5.27}, \mathbf{2.64} \end{array} \right\}$$
$$Rounded\ individual = \{5, 2, 7, 1, 6, 1\}$$
$$GFS_{all} = \{+, -, *, /, sin, cos, x, k\}$$
$$\textbf{Program: } \cos(k1 * (x - k2))$$
$$\textbf{Replaced: } \cos(0.07 * (x - 3.99))$$

Fig. 4. Principles of $AP_{extended}$

3.4 $AP_{direct1}$ - Direct Encoding of K in the Individual 1

This constant handling technique was introduced in [14]. It works with a direct encoding in the individual. It is based on a part behind a decimal point which determines the value proportionally from the selected range of K.

The part behind decimal point is obtained from (7).

$$ind_K = |ind - ind_f| \quad (7)$$

where $ind = \{x_1, x_2, x_3 \ldots x_n\}$ and $ind_f = \{floor(x_1), floor(x_2), floor(x_3) \ldots floor(x_n)\}$

The decimal values in ind_K are in the interval <0,1> . The corresponding K is then computed easily from (8).

$$K = ind_K * |rangeK_{max} - rangeK_{min}| + rangeK_{min} \quad (8)$$

The mapping is done via the standard procedure in AP_{basic} and a general approach of K indexation. When K is needed, the value in the corresponding position from (8) is directly used.

3.5 $AP_{direct2}$ - Direct Encoding of K in the Individual 2

Later authors analysed that $AP_{direct1}$ might have troubles during the run of the algorithm. Expected problematic issues are connected with the neighbourhood of arguments which are responsible for K estimation. Since they are dependent only on the decimal part of the argument regardless the integer part of the value, two points placed on the opposite sides of the coordinate system can be neighbours from ind_K point of view. It does not help the evolutionary optimisation process which expects for a successful performance that two points lie next to each other physically in the coordinate system.

This new approach was proposed in [5] and it is based on the previous and above-described $AP_{direct1}$ [14]. The difference is in the different computation of ind_K (9). It takes the value of the not rounded individual as the proportional part in respect of length of all components in GFS_{All}.

$$ind_K = \frac{ind}{Dim(GFS_{All})} \tag{9}$$

where $ind = \{x_1, x_2, x_3 \ldots x_n\}$ and $Dim(GFS_{All})$ means the number of all non-terminals and terminals used in AP. For instance, if $GFS_{All} = \{+, -, /, *, x, K\}$, the $Dim(GFS_{All}) = 6$ and the valid range for arguments in the individual is in the interval <1, 6> .

4 Used Evolutionary Algorithm - Success-History Based Adaptive Differential Evolution

Differential evolution (DE) and its strategy Success-History Based Adaptive Differential Evolution (SHADE) belong to the group of population based evolutionary algorithms. SHADE wins the competitions held during the world congress on evolutionary computation CEC. Therefore, this strategy was selected for the purpose of this paper.

DE algorithm has three control parameters – population size *NP*, crossover rate *CR* and scaling factor *F*. In the canonical form of DE, those three parameters are static and depend on the user setting. Other important features of DE algorithm are mutation strategy and crossover strategy. Canonical DE uses "rand/1/bin" mutation strategy and binomial crossover. SHADE algorithm, on the other hand, uses only two control parameters – population size *NP* and size of historical memories *H*. *F* and *CR* parameters are automatically adapted based on the evolutionary process. Also, the mutation strategy is different than that of canonical DE. Novel mutation strategy used in SHADE is called "current-to-*p*best/1". For a detailed description on feature constraint correction, update of historical memories and external archive handling in SHADE see [13].

5 Problem Design

These above-mentioned strategies of $AP_{extended}$, $AP_{direct1}$ and $AP_{direct2}$ were applied on standard benchmark tests - approximation of polynomial expression - quintic (10), sextic (11), Sine 3 (12) and Sine 4 (13) problems.

$$x^5 - 2x^3 + x \tag{10}$$

$$x^6 - 2x^4 + x^2 \tag{11}$$

$$\sin(x) + \sin(2x) + \sin(3x) \tag{12}$$

$$\sin(x) + \sin(2x) + \sin(3x) + \sin(4x) \tag{13}$$

6 Results and Discussion

The paper compares $AP_{extended}$, $AP_{direct1}$ and $AP_{direct2}$ strategies powered by SHADE. The setting was based on previous experiences of authors in this field (Table 1).

Table 1. SHADE settings.

Parameter	Quintic and sextic	Sine 3 and Sine 4
Population	50	100
Generations	4000	4000
CFE	200 000	400 000
Historical memory size	50	50
Dimensions (Direct 1 and Direct 2)	40	80
Dimensions (Extended)	60 (40 program + 20 constants)	120 (80 program + 40 constants)

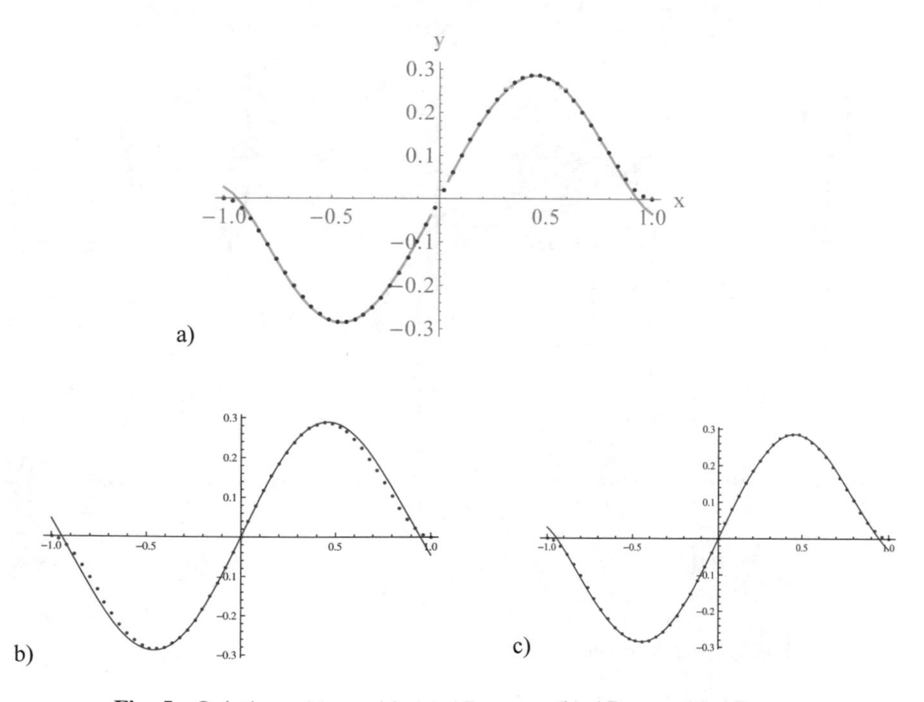

Fig. 5. Quintic problem with (a) $AP_{extended}$, (b) $AP_{direct1}$, (c) $AP_{direct2}$

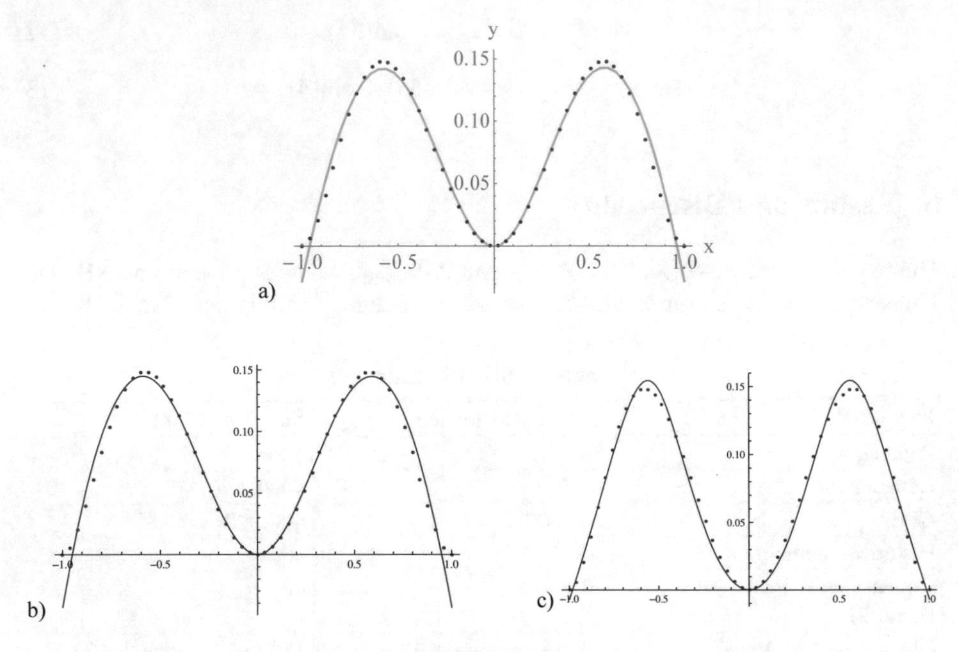

Fig. 6. Sextic problem with (a) $AP_{extended}$, (b) $AP_{direct1}$, (c) $AP_{direct2}$

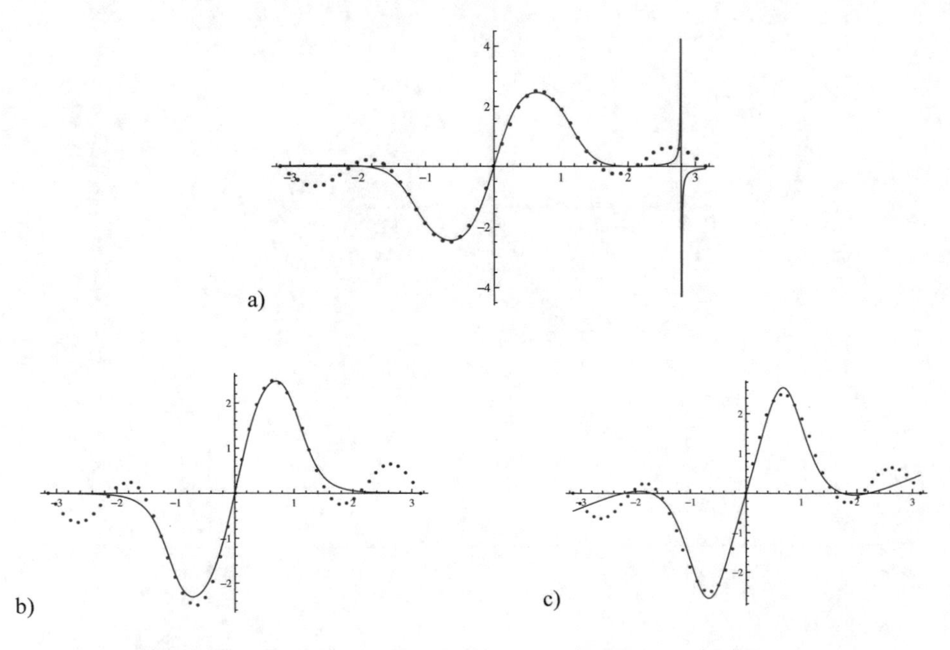

Fig. 7. Sine 3 problem with (a) $AP_{extended}$, (b) $AP_{direct1}$, (c) $AP_{direct2}$

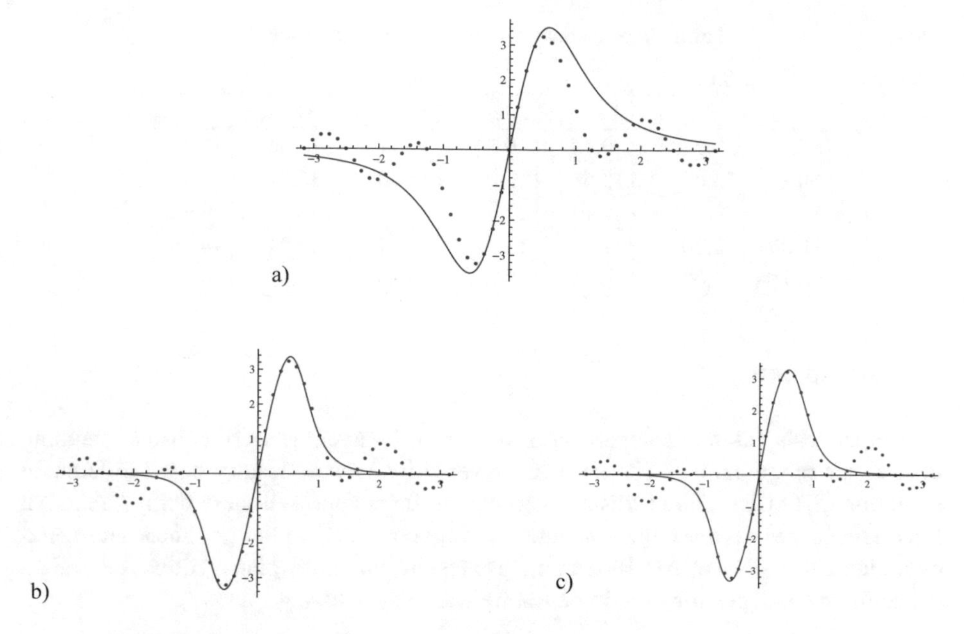

Fig. 8. Sine 4 problem with (a) $AP_{extended}$, (b) $AP_{direct1}$, (c) $AP_{direct2}$

The authors has already discussed the problem of incorrectly interpreted number of cost function evaluations (CFE) in the case of AP_{nf} [5]. For instance, the quintic or sextic problems were discovered in the interval <500, 18 000> of CFE. As mentioned in Sect. 3.1, the correct statement of the cost function evaluations should be <500, 18 000> * cca 5000 iterations which is equal to <2 500 000, 90 000 000> . In the light of these values, our setting in Table 1. means the significant reduction of computation time.

All simulations were performed 30 times out. Examples of the found results for benchmark problem are depicted in Fig. 5 (quintic), Fig. 6 (sextic), Fig. 7 (sine 3) and Fig. 8 (sine 4).

Tables 2 and 3 show statistical measures from the performed simulations.

Table 2. Statistical results for quintic and sextic.

	Quintic			Sextic		
	$AP_{extended}$	$AP_{direct1}$	$AP_{direct2}$	$AP_{extended}$	$AP_{direct1}$	$AP_{direct2}$
Min	0.21	0.44	0.17	0.24	0.29	0.21
Max	6.36	1.57	1.36	2.37	0.96	1.02
Mean	1.60	1.08	0.97	0.75	0.71	0.65
Median	0.61	1.12	1.02	0.62	0.78	0.67
St. Dev	2.19	0.30	0.26	0.54	0.20	0.22

Table 3. Statistical results for Sine 3 and Sine 4.

	Sine 3			Sine 4		
	$AP_{extended}$	$AP_{direct1}$	$AP_{direct2}$	$AP_{extended}$	$AP_{direct1}$	$AP_{direct2}$
Min	9.74	10.76	9.86	13.11	13.86	13.51
Max	33.00	13.61	13.04	15.80	15.77	15.91
Mean	12.21	12.10	11.71	14.37	14.89	14.60
Median	11.30	12.11	11.81	14.51	14.91	14.55
St. Dev	4.08	0.74	0.88	0.80	0.58	0.70

7 Conclusion

The paper deals with the comparison of three novelty approaches to constant handling in Analytic programming. The AP was powered by Success-history based differential evolution (SHADE). The results showed that all three approaches are comparable. All three approaches decrease the computation time compared to the nonlinear and meta-evolutionary versions of AP. Based on these results, the future plans will be focused on usage these strategies for pseudo neural networks synthesis.

Acknowledgement. This work was supported by the Ministry of Education, Youth and Sports of the Czech Republic within the National Sustainability Programme project No. LO1303 (MSMT-7778/2014) and also by the European Regional Development Fund under the project CEBIA-Tech No. CZ.1.05/2.1.00/03.0089, further it was supported by Grant Agency of the Czech Republic—GACR P103/15/06700S and by Internal Grant Agency of Tomas Bata University in Zlin under the project No. IGA/CebiaTech/2017/004.

References

1. Back, T., Fogel, D.B., Michalewicz, Z.: Handbook of Evolutionary Algorithms. Oxford University Press, Oxford (1997). ISBN 0750303921
2. Eiben, A.E., Michalewicz, Z., Schoenauer, M., Smith J.E.: Parameter control in evolutionary algorithms, pp. 19–46. Springer (2007)
3. Oplatkova, Z.K., Senkerik, R.: Evolutionary synthesis of complex structures - pseudo neural networks for the task of iris dataset classification. In: Nostradamus 2013: Prediction, Modeling and Analysis of Complex Systems, pp. 211–220. Springer, Heidelberg (2013). ISSN 2194-5357, 978-3-319-00541-6
4. Oplatkova, Z.K., Senkerik, R.: Control law and pseudo neural networks synthesized by evolutionary symbolic regression technique. In: Al-Begain, K., Bargiela, A.: Seminal Contributions to Modelling and Simulation - Part of the series Simulation Foundations, Methods and Applications, pp. 91–113 (2016). https://doi.org/10.1007/978-3-319-33786-9_9. ISBN 978-3-319-33785-2
5. Oplatkova, Z.K., Viktorin, A., Senkerik, R., Urbanek, T.: Different approaches for constant estimation in analytic programming. In: 31st European Conference on Modelling and Simulation, pp. 326–332 (2017). ISBN 978-0-9932440-4-9. ISSN 2522-2422
6. Koza, J.R., et al.: Genetic Programming III: Darwinian Invention and problem Solving. Morgan Kaufmann Publisher, Burlington (1999). ISBN 1-55860-543-6

7. Koza, J.R.: Genetic Programming. MIT Press, Cambridge (1998). ISBN 0-262-11189-6
8. Lampinen, J., Zelinka, I.: New ideas in optimization – mechanical engineering design optimization by differential evolution, vol. 1. McGraw-hill, London (1999). 20 p, ISBN 007-709506-5
9. O'Neill, M., Ryan, C.: Grammatical Evolution: Evolutionary Automatic Programming in an Arbitrary Language. Kluwer Academic Publishers, Dordrecht (2003). ISBN 1402074441
10. Oplatkova, Z.: Metaevolution: Synthesis of Optimization Algorithms by means of Symbolic Regression and Evolutionary Algorithms. Lambert Academic Publishing, Saarbrücken (2009)
11. Price, K., Storn, R.M., Lampinen, J.A.: Differential Evolution: A Practical Approach to Global Optimization. Natural Computing Series, 1st edn. Springer, Heidelberg (2005)
12. Price, K., Storn, R.: Differential evolution homepage (2001). http://www.icsi.berkeley.edu/~storn/code.html. Accessed 29 Feb 2012
13. Tanabe, R., Fukunaga, A.: Success-history based parameter adaptation for differential evolution. In: 2013 IEEE Congress on Evolutionary Computation (CEC), pp. 71–78. IEEE (2013)
14. Urbanek, T., Prokopova, Z., Silhavy, R., Kuncar, A.: New approach of constant resolving of analytical programming. In: 30th European Conference on Modelling and Simulation, pp. 231–236 (2016). ISBN 978-0-9932440-2-5
15. Viktorin, A., Pluhacek, M., Oplatkova, Z.K., Senkerik, R.: Analytical programming with extended individuals. In: 30th European Conference on Modelling and Simulation, pp. 237–244 (2016). ISBN 978-0-9932440-2-5
16. Volna, E., Kotyrba, M., Jarusek, R.: Multi-classifier based on Elliott wave's recognition. Comput. Math. Appl. **66**(2), 213–225 (2013)
17. Zelinka, I., et al.: Analytical programming - a novel approach for evolutionary synthesis of symbolic structures. In: Kita, E. (ed.) Evolutionary Algorithms, InTech (2011). ISBN 978-953-307-171-8
18. Zelinka, I., Varacha, P., Oplatkova, Z.: Evolutionary synthesis of neural network. In: Mendel 2006 – 12th International Conference on Softcomputing, Brno, Czech Republic, 31 May – 2 June 2006, pp. 25– 31 (2006). ISBN 80-214-3195-4
19. Zelinka, I., Oplatkova, Z., Nolle, L.: Boolean symmetry function synthesis by means of arbitrary evolutionary algorithms-comparative study. Int. J. Simul. Syst. Sci. Technol. **6**(9), 44–56 (2005). ISSN 1473-8031

Proposal of a New Swarm Optimization Method Inspired in Bison Behavior

Anezka Kazikova(✉), Michal Pluhacek, Roman Senkerik, and Adam Viktorin

Faculty of Applied Informatics, Tomas Bata University in Zlin, T.G. Masaryka 5555, 760 01 Zlin, Czech Republic
{kazikova, pluhacek, senkerik, aviktorin}@fai.utb.cz

Abstract. This paper proposes a new swarm optimization algorithm inspired by bison behavior. The algorithm mimics two survival mechanisms of the bison herds: swarming into the circle of the strongest individuals and exploring the search space via organized run throughout the optimization process. The proposed algorithm is compared to the Particle Swarm Optimization and the Cuckoo Search algorithms on four benchmark functions.

Keywords: Bison algorithm · Bison · Optimization · Swarm intelligence

1 Introduction

The collective intelligence phenomenon is intriguing: without an actual leader, swarms can make intelligent decisions for hunting, mating and surviving. Simulating such behavior patterns is successfully used in the optimization field [1, 2]. Mating fireflies in the Firefly Algorithm [2, 3], cuckoos laying eggs in the Cuckoo Search [2], hunting wolfs in the Grey Wolf Optimizer [4], communication within bee colonies in the Artificial Bee Colony algorithm [4] and the Particle Swarm Optimization [6] could make a typical and intensively utilized examples. The African Buffalo Optimization algorithm simulates exploitation and exploration techniques influenced by the African buffalos' communication skills and the common knowledge of global and individual best position [7]. Even though African buffalos and bison are related species, the proposed Bison Algorithm is set on a different principle.

Many optimization algorithms start with individuals scattered all over the search space but end with the whole population set in a little space around the found solution. Algorithms tend to neglect the exploration phase in higher iterations to exploit the local optimum. However, what if we optimize function with the global optimum in such a small decreasing neighborhood, that most optimization algorithms would be stuck in the wrong local optimum? A possible solution is to use an algorithm that would not stop exploring the search space even when assuming that the global optimum has been already found.

R. Matoušek (Ed.): MENDEL 2017, AISC 837, pp. 146–156, 2019.
https://doi.org/10.1007/978-3-319-97888-8_13

2 Bison Algorithm

The bison herd makes the perfect example of the swarming behavior. When the herd is in danger, bison form a circle of strong individuals, making the weaker ones squeeze inside. Bison are also great runners, constantly on the move. They can easily outrun men with their maximum speed of 56 km per hour, keeping this velocity for half an hour [8]. These two particular behavior patterns are used in the proposed optimization algorithm.

2.1 Bison Algorithm: The Main Loop

The algorithm combines both exploitation and exploration techniques. The stronger group exploits by swarming closer to the center of several strongest individuals, and the weaker group explores by running through the search space, preventing the population from being stuck in a local optimum. The main loop of the Bison Algorithm is explained in source code below (Algorithm 1).

Algorithm 1. Main loop

```
1. Objective function: f(x) = (x1, x2, ..., xd)
2. Initialize the swarming group with random position
   Initialize the running group around the worst
   swarming individual
3. for every migration round do
4.   strongest bison swarm
5.   weaker bison run
6.   sort population by the objective function value
7. end for
```

2.2 Swarming Behavior

The swarming movement is the exploitation part of the motion. It starts with the computation of the elite bison center. Each from the swarming group (including the elites) is then offered to move closer to the computed center (1). The movement happens in the direction to the center of a random length (2). The length of motion is affected by the overstep parameter which determines, how much can individuals exceed the center, similarly to the definition of *PathLength* in SOMA algorithm [9, 10]. Individuals move to the offered position only if it improves their fitness value. The swarming technique is explained in source code below this paragraph (Algorithm 2). Figure 1 presents the actual implementation of the swarming behavior on the Rastrigin function.

$$\boldsymbol{x_{new}} = \boldsymbol{x_{old}} + \boldsymbol{direction} * \boldsymbol{random}(0, \boldsymbol{overstep})_{\boldsymbol{D}} \tag{1}$$

$$\boldsymbol{direction} = \boldsymbol{center} - \boldsymbol{x} \tag{2}$$

Where *overstep* $\in \mathbb{R}$ allowing individuals to overstep the current computed center point and *direction* is a vector to the current center of elite bisons.

Algorithm 2. Swarming behavior

```
1. Determine the way of the center computation and the
   overstep parameters
2. Compute the center of elite bison group
   (arithmetic / weighted / ranked)
3. Compute a new possible position for each swarming
   individual (1)
4. Place the out-bounded individuals onto the hy-
   persphere
5. If f(x_new) < f(x_old) then move to the new position
```

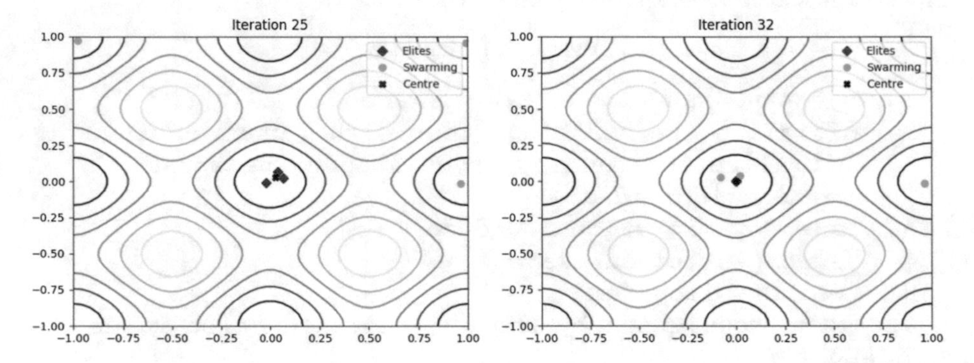

Fig. 1. Swarming movement

2.3 Running Behavior

Like any radical change of the herd direction might get tricky in the real world, proposed running movement is based on one specific leading run direction. This vector is proportional to the search space size and is randomly chosen during the initialization (3) and only slightly differs in every iteration (4). Adjusted running vector is then added to the current individual's position regardless its value (5). As a result, the running group runs through the search space more or less in the same direction.

$$\boldsymbol{run\,direction} = \boldsymbol{random}\left(\frac{\boldsymbol{up\,bound} - \boldsymbol{low\,bound}}{45}, \frac{\boldsymbol{up\,bound} - \boldsymbol{low\,bound}}{15}\right)_{\boldsymbol{dim}} \tag{3}$$

$$\boldsymbol{run\,direction} = \boldsymbol{run\,direction} * \boldsymbol{random}(0.9, 1.1)_{\boldsymbol{dim}} \tag{4}$$

$$x_{new} = x_{old} + \boldsymbol{run\,direction} \tag{5}$$

After each iteration, the population is sorted by its objective function value. The strongest individuals join the swarming group, while the weaker individuals proceed on running. When a swarming individual joins the running group, its position is recalculated to be close to the herd itself.

The whole process is outlined as a pseudo code beneath (Algorithm 3). The movement of the running group is shown in Fig. 2.

Algorithm 3. Running behavior

```
1.  Initialize the run direction vector at the beginning
    of the algorithm (3)
2.  At iteration start adjust the run direction (4)
3.  Move all running individuals to their new positions
    (5)
4.  Place out-bounded individuals into the hypersphere
```

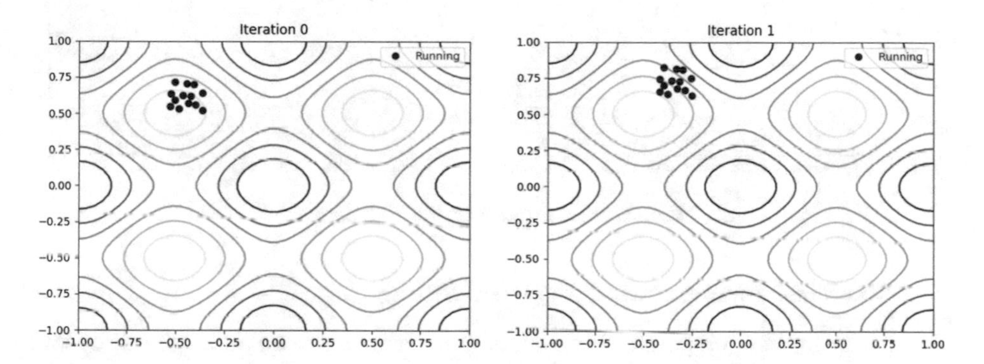

Fig. 2. Running movement

2.4 Out of the Boundaries Behavior

The Bison algorithm considers the search space a hyper-sphere. When it crosses the boundaries in any dimension, the position is just recalculated to a corresponding position on the other side. This approach is beneficial to the algorithm design; no boundaries allow the stable leading running vector. Therefore, the search space can be scanned systematically throughout the optimization process. Figure 3 shows the movement of the whole bison herd. In addition to the running, swarming and elite groups it also displays a trace of the running group from the previous five iterations.

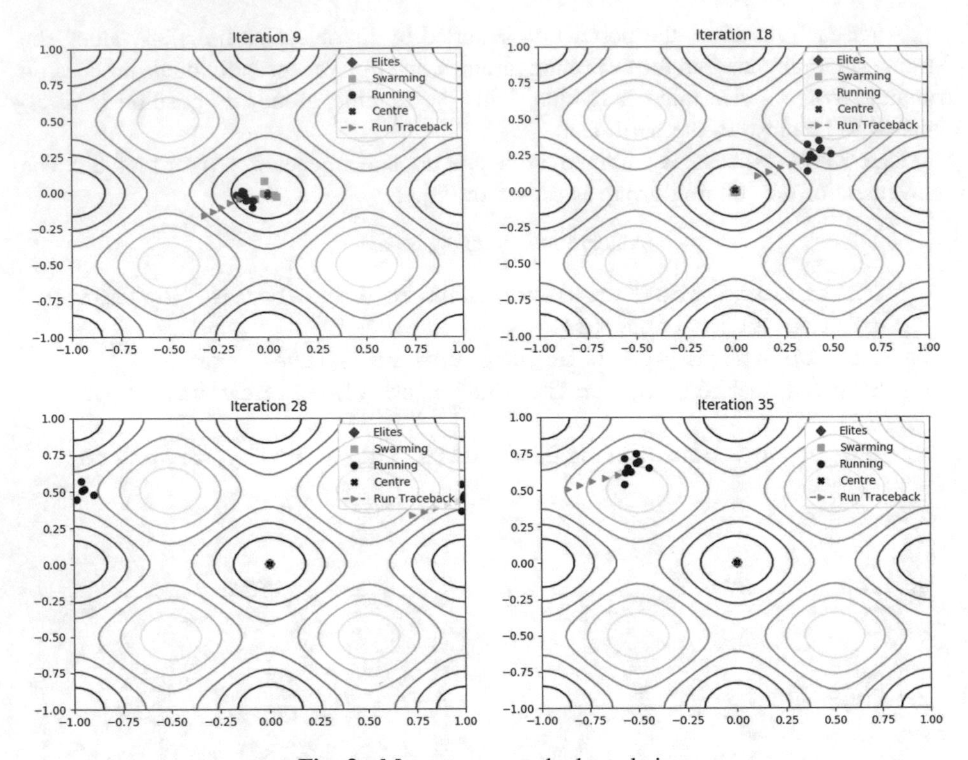

Fig. 3. Movement over the boundaries

2.5 Experiment and Control Parameters Setting

The design of the proposed Bison algorithm came from an idea: fewer control parameters mean better usability. That is why population parameters are set to be dimension based (Table 1). If dimension based approach is not appropriate, the population size can be configured directly, leaving the other internal parameters proportional to the population size.

The dimension based population size approach leaves the Bison algorithm with two configurable parameters: the way of the center computation and the possible overstep of the swarming movement. Both parameters were tested below.

Four different simple static test functions were utilized within this research: the 2nd De Jong's function (Rosenbrock's function) (6), Rastrigin's function (7), Schwefel's function (8) and Easom function (9), where dim represents the dimension.

The 2nd De Jong's Function (Rosenbrock's Valley)

$$f(x) = \sum_{i=1}^{\dim - 1} 100\left(x_i^2 - x_{i+1}\right)^2 + \left(1 - x_i\right)^2 \tag{6}$$

Table 1. Parameters of the Bison algorithm

Parameter	Value
Internal parameters	
Population	10 · dimension
Number of elite bison	Dimension + 1
Number of swarming bison	5 · number of elite bison
Configurable parameters	
Center computation	Arithmetic/weighted/ranked
Overstep	0 … no movement 1 … maximum length of the movement is to the center 3.5 – 4.1 … recommended values

Function minimum: Position for E_n: $(x_1, x_2 \ldots x_n) = (1, 1, \ldots, 1)$. Value for E_n: $y = 0$.

Rastrigin's Function

$$f(x) = 10\,\mathrm{dim} + \sum_{i=1}^{\mathrm{dim}} x_i^2 - 10\cos(2\pi x_i) \tag{7}$$

Function minimum: Position for E_n: $(x_1, x_2 \ldots x_n) = (0, 0, \ldots, 0)$. Value for E_n: $y = 0$.

Schwefel's Function

$$f(x) = 418.9829 - \sum_{i=1}^{\mathrm{dim}} -x_i \sin(\sqrt{|x|}) \tag{8}$$

Function minimum: Position for E_n: $(x_1, x_2 \ldots x_n) = (420.9687, 420.9687, \ldots, 420.9687)$. Value for E_n: $y = 0$.

Easom Function

$$f(x) = -\prod_{i=1}^{\mathrm{dim}} \cos(x_i) \cdot \sum_{i=1}^{\mathrm{dim}} -(x_i - \pi)^2 \tag{9}$$

Function minimum: Position for E_n: $(x_1, x_2 \ldots x_n) = (\pi, \pi, \ldots, \pi)$. Value for E_n: $y = -dim + 1$.

Three approaches of the center computation are proposed:

- Weighted center: minding the quality of elite bison by the actual value (10) and (11):

$$weight(\boldsymbol{x}) = -f(\boldsymbol{x}) + \sum\nolimits_{i=1}^{|elites|} f(\boldsymbol{elites}_i) \tag{10}$$

$$\boldsymbol{weighted\ center} = \sum\nolimits_{i=1}^{|elites|} \frac{weight(\boldsymbol{elites}_i) * \boldsymbol{elites}_i}{\sum\nolimits_{j=1}^{|elites|} weight(\boldsymbol{elites}_i)} \tag{11}$$

- Ranked center: computing the center by the order of elites quality (12) and (13):

$$\boldsymbol{weight} = (10, 20, \ldots 10 * |elites|) \tag{12}$$

$$\boldsymbol{ranked\ center} = \sum\nolimits_{i=1}^{|elites|} \frac{weight_i * \boldsymbol{elites}_i}{\sum\nolimits_{j=1}^{|elites|} weight_i} \tag{13}$$

- Arithmetic center: not minding the objective function value at all (14):

$$\boldsymbol{arithmetic\ center} = \sum\nolimits_{i=1}^{|elites|} \frac{\boldsymbol{elites}_i}{|elites|} \tag{14}$$

Where *elites* represents the number of elite bison parameter, x is the individual's position and $f(x)$ is the objective function value.

Table 2 reports the statistical results (average and standard deviation) of benchmark functions based on the center computation parameter. Figure 4 shows the mean convergence of 30 runs of the Bison Algorithm on four 10 dimensional functions. While the convergence curve in Fig. 5 seems to be independent of the center computation in Rastrigin and Rosenbrock functions, in Schwefel and Easom functions the difference is notable. The arithmetic and the weighted center outperformed the ranked computation in most of the tested scenarios.

Table 2. Average results and standard deviation of the benchmark functions by the center computation configuration

	Arithmetic center		Weighted center		Ranked center	
	avg	*std*	*avg*	*std*	*avg*	*std*
Rastrigin 10d	**7.022**	5.670	9.385	9.385	22.047	5.879
Rosenbrock 10d	2.989	1.480	2.998	1.079	**2.994**	1.354
Schwefel 10d	1692.512	142.722	**1667.497**	145.045	1683.962	159.121
Easom 10d	**−5.291**	3.550	−4.867	3.623	−3.109	2.768

The overstep value was examined on a set of values from 0.5 to 10.0. Testing involved 100 runs, three 10 dimensional test functions (Rosenbrock, Schwefel and Rastrigin) and arithmetic and weighted center computation. Table 3 shows the selection of the finest results of Rosenbrock function with the arithmetic center. The overstep values 3.5, 3.7 and 4.1 proved the most efficient results even in other tested functions, therefore using one of these values would be recommended.

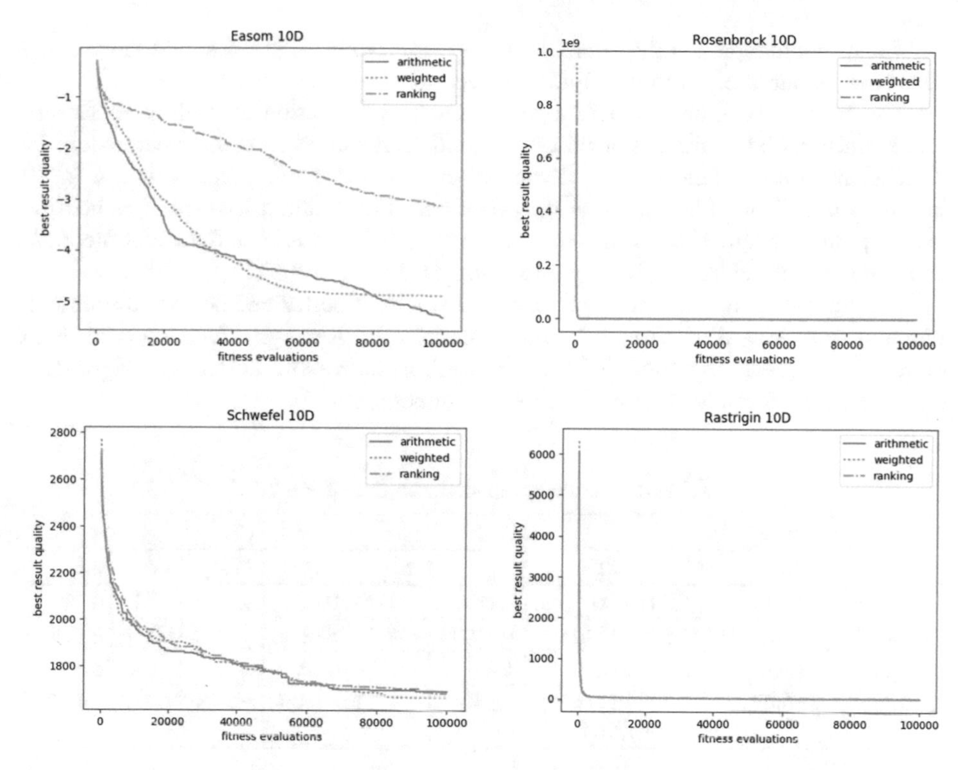

Fig. 4. Average convergence by the center computation configuration

Table 3. The Bison Algorithm results by the overstep value

Overstep	Best result	Average result	Last population deviation	Best result deviation
3.4	1.276	3.265	1.085	2.295
3.5	**1.243**	3.296	1.767	2.306
3.7	1.310	3.219	4.098	2.037
4.1	1.993	**3.011**	9.493	1.194
4.5	3.142	3.973	33.858	1.041

3 Results and Discussion

The proposed algorithm was compared to the Particle Swarm Optimization algorithm and the Cuckoo Search (from now on referred as PSO and CS) on 10D and 50D nontrivial multimodal benchmark functions. All algorithms were tested on 30 runs, each run having 10 000 · *dim* evaluations. Both PSO and CS had 50 population members, other PSO parameters were set to: $v_{max} = 6$, $w_{max} = 0.9$, $w_{min} = 0.2$, $c_1 = 2$, $c_2 = 2$. CS discovery rate of finding alien eggs was $pa = 0.25$ as in EvoloPy toolbox on GitHub [11], which has been utilized for several simulations. The Bison Algorithm was tested with the default recommended configuration, arithmetic center computation and overstep parameter set to 3.5.

Tables 4 and 5 present the results overview for 30 runs. Figure 5 shows the average convergence curve of 30 runs of four selected functions.

The Bison Algorithm outperformed the other optimization algorithms in Rosenbrock function and 10 dimensional Easom function. It should be pointed out, that in 10 dimensional Easom function, the Bison Algorithm found the exact optimum in 20 instances out of 30 (Fig. 6), while the other tested algorithms have not reached the exact optimum at all. The CS showed substantially better results in Schwefel function, and PSO performed best in Rastrigin function (Tables 4 and 5).

Judging the Bison Algorithm performance solely, it performed best in Rosenbrock function and low dimensional Easom function. In Rastrigin function and high dimensional Easom function it was comparably successful to the CS algorithm. However, it performed worst in the Schwefel function.

Table 4. Results comparison on dimension 10

	Bison			PSO			CS		
	avg	*std*	*min*	*avg*	*std*	*min*	*avg*	*std*	*min*
Rastrigin	7.70	6.83	1.00	**2.42**	1.74	0.00	9.77	2.75	4.32
Schwefel	1672.32	131.32	1323.14	1667.24	348.20	888.30	**398.56**	60.56	293.71
Rosenbrock	3.03	2.21	0.05	4.43	5.21	0.08	4.25	1.80	1.12
Easom	−6.67	3.34	−9.000 (20x)	−2.80	1.54	−8.00	−3.46	0.53	−4.77

Table 5. Results comparison on dimension 50

	Bison			PSO			CS		
	avg	*std*	*min*	*avg*	*std*	*min*	*avg*	*std*	*min*
Rastrigin	114.36	78.88	10.95	**65.34**	16.05	41.79	170.05	20.88	120.29
Schwefel	14658.37	415.4	13450.33	9909.39	1247.68	6878.38	**5192.11**	280.65	4705.3
Rosenbrock	**44.40**	10.09	41.87	68.32	32.83	7.30	53.15	25.11	26.33
Easom	−3.03	0.31	−3.65	**−12.47**	5.63	−24.00	−4.81	0.463	−6.00

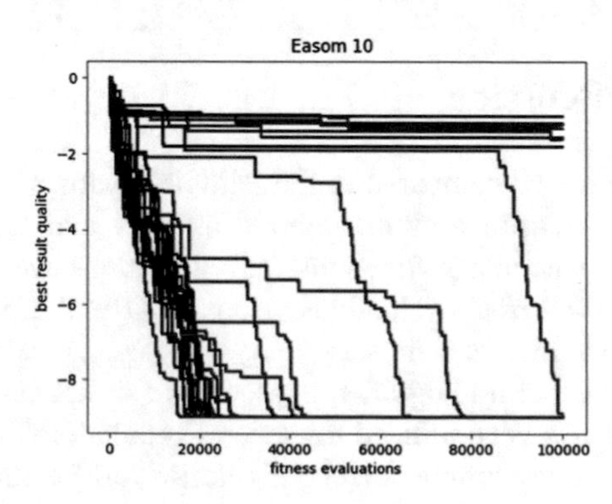

Fig. 5. Convergence of the Bison algorithm on 30 runs in Easom 10D functions

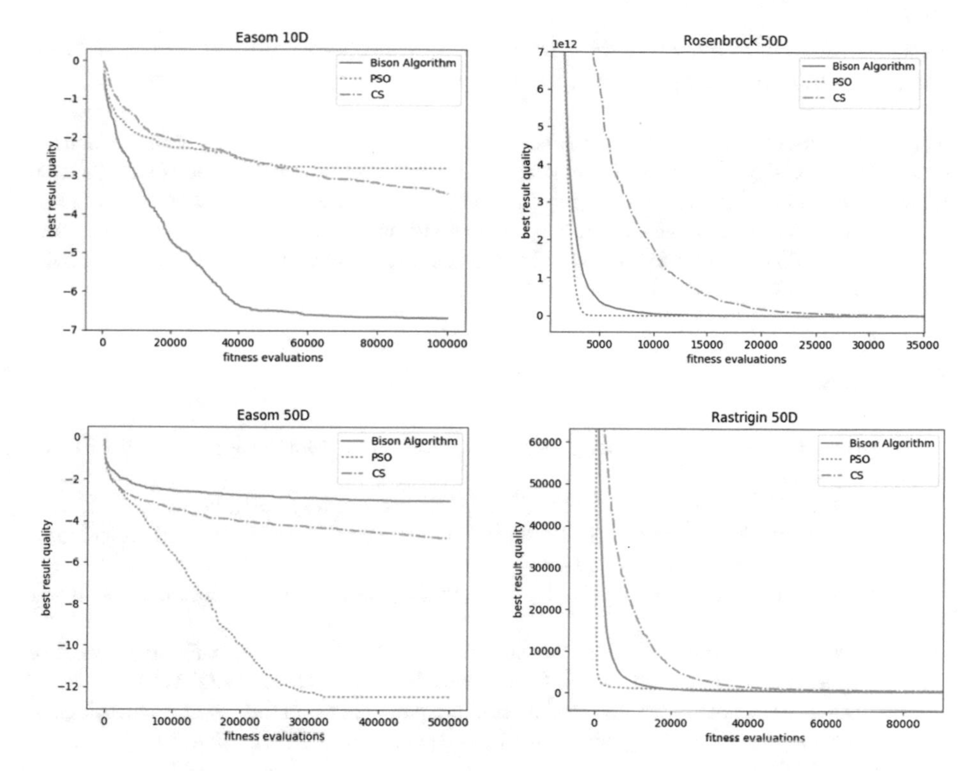

Fig. 6. Average convergence of tested optimization algorithms

4 Conclusion

In this paper, a new optimization algorithm has been proposed. It emphasizes the exploration of the search space through the whole optimization process assuming that this approach might be useful for functions with unexpected global optimum with small (or none) decreasing neighborhood. The algorithm was tested on four multimodal functions, while one of the functions – the Easom function – fulfilled the narrow decreasing neighborhood description.

The Bison Algorithm proved to be able to solve nontrivial multimodal functions. It showed comparable results on some functions, best results on Rosenbrock function and even worst results on the Schwefel function. However, the outcome of testing the Easom function was ambiguous as in the low dimensions it performed the best and in the high dimensions the worst.

The testing process brought up several questions about the Bison algorithm to be examined in later research. First – there were no multiple tests of appropriate parameter settings including the leading run direction vector and the population distribution. The algorithm could perform better with different settings (though this is also a problem dependent issue). Second, that results in some functions quite often find the right optimum position in most dimensions, while not in others – this could be fixed by adjusting the algorithm with a technique searching in just some particular dimensions.

This behavior could empower the performance of the algorithm and prevent it from completely missing the optimum in some dimensions while finding it in others.

Acknowledgements. This work was supported by Grant Agency of the Czech Republic – GACR P103/15/06700S, further by the Ministry of Education, Youth and Sports of the Czech Republic within the National Sustainability Programme Project no. LO1303 (MSMT-7778/2014. Also by the European Regional Development Fund under the Project CEBIA-Tech no. CZ.1.05/2.1.00/03.0089 and by Internal Grant Agency of Tomas Bata University under the Projects no. IGA/CebiaTech/2017/004.

References

1. Kennedy, J., Eberhart, R.C., Shi, Y.: Swarm Intelligence. Morgan Kaufmann Publishers, Burlington (2001)
2. Yang, X.S.: Nature-Inspired Metaheuristic Algorithms. Luniver Press, Frome (2010)
3. Fister, I., Fister Jr., I., Yang, X.S., Brest, J.: A comprehensive review of firefly algorithms. Swarm Evol. Comput. **13**, 34–46 (2013)
4. Mirjalili, S., Mirjalili, S.M., Lewis, A.: Gray wolf optimizer. Adv. Eng. Softw. **69**, 46–61 (2014)
5. Rajasekhar, A., Lynn, N., Das, S., Suganthan, P.N.: Computing with the collective intelligence of honey bees–a survey. Swarm Evol. Comput. **32**, 25–48 (2017)
6. Eberhart, R., Kennedy, J.: A new optimizer using particle swarm theory. In: Proceedings of the Sixth International Symposium on Micro Machine and Human Science MHS 1995, pp. 39–43 (1995)
7. Odili, J.B., Kahar, M.N.M.: African buffalo optimization. Int. J. Softw. Eng. Comput. Syst. **2** (1), 28–50 (2016)
8. Berman, R.: American Bison (Nature Watch). Lerner Publications, Minneapolis (2008)
9. Zelinka, I.: SOMA—self-organizing migrating algorithm. In: New Optimization Techniques in Engineering, pp. 167–217. Springer, Heidelberg (2004)
10. Zelinka, I.: SOMA—self-organizing migrating algorithm. In: Self-Organizing Migrating Algorithm, pp. 3–49. Springer, Heidelberg (2016)
11. Faris, H.: EvoloPy GitHub repository (2017). https://github.com/7ossam81/EvoloPy. Accessed 1 May 2017

Stochastic Heuristics for Knapsack Problems

Miloš Šeda[1(✉)] and Pavel Šeda[2]

[1] Faculty of Mechanical Engineering, Brno University of Technology, Technická 2, 616 69 Brno, Czech Republic
seda@fme.vutbr.cz

[2] IBM Global Services Delivery Center, Technická 21, 616 00 Brno, Czech Republic
pavelseda@email.cz

Abstract. In this paper, we introduce knapsack problem formulations, discuss their time complexity and propose their representation and solution based on the instance size. First, deterministic methods are briefly summarized. They can be applied to small-size tasks with a single constraint. However, because of NP-completeness of the problem, more complex problem instances must be solved by means of heuristic techniques to achieve an approximation of the exact solution in a reasonable amount of time. The problem representations and parameter settings for a genetic algorithm and simulated annealing frameworks are shown.

Keywords: Knapsack problem · Dynamic programming
Branch and bound method · Heuristic · Genetic algorithm · Simulated annealing

1 Introduction

The Knapsack Problem (KP) is a problem of combinatorial optimisation that, in spite of its very simple formulation, belongs to the class of NP-complete problems [1] so that, in general, a precise solution cannot be found in a reasonable time.

A set of n items is given of which some are to be packed in a knapsack with capacity of C units. Item i has a value of v_i and uses up to w_i of capacity. We try to maximise the total value of the packed items subject to capacity constraint. If a binary decision variable x_i specifies whether or not item i is included in the knapsack, then the problem known as the 0-1 *knapsack problem* (0-1 KP) may be formally expressed as follows:

$$\text{Maximise} \sum_{i=1}^{n} v_i x_i \tag{1}$$

subject to

$$\sum_{i=1}^{n} w_i x_i \leq C \tag{2}$$

$$x_i \in \{0, 1\},\ i = 1, \ldots, n \tag{3}$$

R. Matoušek (Ed.): MENDEL 2017, AISC 837, pp. 157–166, 2019.
https://doi.org/10.1007/978-3-319-97888-8_14

This problem has many practical applications:

- Capital budgeting. Here, C represents an available sum of money, w_i are investment requirements, and v_i the current values of these investments.
- Healthcare. Patients need surgeries, but the number of surgeons is limited. This number corresponds to the capacity while weight w_i represents the complexity and v_i the need of an operation.
- Resource-constrained project scheduling problem. A classical task of project management is to create the network graph of a project and, on the basis of knowledge or an estimation of time of activities, to determine the critical activities that could cause a project delay. Each activity draws on certain resources, e.g. financial resources, natural resources (energy, water, material, etc.), labour, managerial skills. The solution of this task is well known in the case of limited resources being not taken into consideration. However, in real situations, available capacities of resources are constrained, and the question is which activities may be carried out in the near future and which of them must be delayed.

2 Problem Solving

The knapsack problem may be solved in many ways. The simplest one is to generate all possible choices of items and to determine the optimal solution among them. Their number for n items is given by the binomial theorem as follows:

$$\binom{n}{1}+\binom{n}{2}+\cdots+\binom{n}{n-2}+\binom{n}{n-1}+\binom{n}{n}=(1+1)^n-\binom{n}{0}=2^n-1=O(2^n) \tag{4}$$

Unfortunately, this enumerating approach for large instances is impracticable because of the huge amount of computation time.

By $f_k(y)$, let us denote the maximal possible value using only the first k items when the capacity limit is y.

$$f_i(y)=\max\left\{\sum_{i=1}^{n} v_i x_i \,\middle|\, \sum_{i=1}^{n} w_i x_i \le y,\ x_i \in \{0,1\},\ i=1,\ldots,k,\ 0\le k\le n,\ 0\le y\le C\right\} \tag{5}$$

Using dynamic programming approach, we can derive

$$f_k(y)=\begin{cases}\max\{f_{k-1}(y), f_{k-1}(y-w_k)+v_k\}, & \text{if } y\ge w_k\\ f_{k-1}(y), & \text{if } y<w_k\end{cases} \tag{6}$$

If we increase the capacity of the knapsack and the set of objects that can be used to fill the knapsack when $y = C$ and $k = n$, then $F_n(C)$ corresponds to the maximum value of a knapsack with the full capacity using all of the items.

The major drawback of the dynamic programming procedure for solving the knapsack problem is that the time complexity depends not only on the number of items but also on the capacity C of the knapsack. Since it runs in $O(nC)$ time, it seems to be quite good for solving problem instances with the capacity parameter C and n not too high. However, the use of the dynamic programming procedure approach is based on the assumption that the capacity parameter C and weights w_i are non-negative integers. Fractional values may be converted to integers via multiplying by the common denominator. On the other hand, if all weights are integers, we may increment the parameter y in each step by the greatest common divisor. The drawback of the dynamic programming procedure is that, with growing n and C, memory requirements grow as well and, for values higher than 1000, the arrays $F_k(y)$ and $i_k(y)$ each have more than a million items, which causes problems with memory management or even, in the static representation of these arrays, the corresponding programme cannot be compiled.

The most efficient deterministic approach uses branch and bound method [5], which restricts the growth of the search tree.

Depending on the precision of the upper bounds, much enumeration may be avoided (the lower the upper bounds, the faster the solution is found). Let the items be numbered so that

$$\frac{v_1}{w_1} \geq \frac{v_2}{w_2} \geq \cdots \geq \frac{v_n}{w_n}, \tag{7}$$

i.e., the unit values of the items form a non-increasing sequence. Using *QuickSort, HeapSort*, or *MergeSort*, the time complexity of this step is $O(n \log_2 n)$.

We place items in the knapsack according to this non-increasing sequence. Let $x_1, x_2, \ldots, x_p$ be fixed values of 0 or 1 and

$$M_k = \{\mathbf{x} | \mathbf{x} \in M,\ x_j = \xi_j,\ \xi_j \in \{0,1\},\ j = 1, \ldots, p\} \tag{8}$$

where M is a set of feasible solutions. If

$$(\exists q)(p < q \leq n) : \sum_{j=p+1}^{q-1} w_j \leq C - \sum_{j=1}^{p} w_j \xi_j < \sum_{j=p+1}^{q} w_j, \tag{9}$$

then the upper bound for M_k can be determined as follows:

$$U_{\mathrm{B}}(M_k) = \sum_{j=1}^{p} v_j \xi_j + \sum_{j=p+1}^{q-1} v_j + \frac{v_q}{w_q} \left(C - \sum_{j=1}^{p} w_j \xi_j + \sum_{j=p+1}^{q-1} w_j \right) \tag{10}$$

However, the branch and bound method and optimisation toolboxes, which implement it for solving integer programming problems, cannot find the optimum in a reasonable amount of time for large instances (e.g. for $n \geq 50$). In this case, heuristic methods [2, 3, 6–9, 12] must be used.

Two of them, genetic algorithm and simulated annealing, have been implemented and recommendations of their parameter settings are presented, based on many tests with various sets of possible operators (selection, crossover, mutation, etc.).

2.1 Genetic Algorithm

Since the principles of GA's are well-known, we will only deal with GA parameter settings for the problems to be studied. Now we describe the general settings and the problem-oriented setting used in our application.

Individuals in the population (*chromosomes*) are represented by binary strings of length n, where a value of 0 or 1 at bit i (*gene*) implies that $x_i = 0$ or 1 in the solution, respectively.

The *population size* is usually set in the range [50, 200]. In our programme, implemented in Java, 200 individuals in the population were used because using 50 individuals led to a reduction chromosome diversity and premature convergence.

Initial population is obtained by generating random strings of 0s and 1s in the following way: First, all bits in all strings are set to 0 and, then, for each of the strings, randomly selected bits are set to 1 until the solutions (represented by strings) are feasible.

The *fitness function* corresponds to the objective function to be maximised or minimised, here, it is maximised.

Three most commonly used methods of *selection* of two parents for *reproduction*, roulette selection, ranking selection, and tournament selection were tested.

As to *crossover*, uniform crossover, one-point and two-point crossover operators were implemented.

Mutation was set to 5, 10 and 15%, exchange mutation, shift mutation, and mutation inspired by the well-known *Lin-2-Opt change operator* usually used for solving the travelling salesman problem [4] were implemented.

In *replacement* operation two randomly selected individuals with below-average fitness were replaced by the children generated.

Termination of a GA was controlled by specifying a maximum number of generations t_{max}, e.g. $t_{max} \leq 10000$.

The chromosome is represented by an n-bit binary string S where n is the number of items in the KP. A value of 1 for bit i implies that item i is in the solution and 0 otherwise.

$$f(S) = \sum_{i=1}^{m} S_i \tag{11}$$

The binary representation causes problems with generating infeasible chromosomes such as in initial population, crossover, and mutation operations.

To avoid infeasible solutions when the total weight of selected items is higher than the capacity C, *penalties* may be assigned to them, or a *repair operator* [10] may be applied. We prefer the second approach because the application of penalties may significantly decrease the quality of genetic material in the current population.

The repair operator code for more general version of the knapsack problem, called the multi-knapsack problem, is shown in Sect. 3.

The genetic algorithm was tested using randomly generated instances with 50 items.

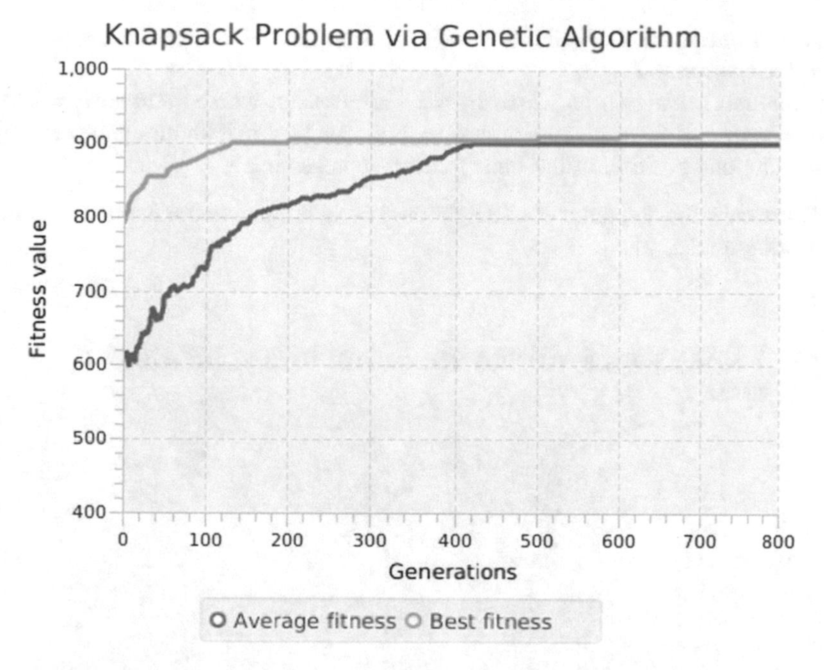

Fig. 1. Computation by genetic algorithm (instance with 50 items, roulette wheel selection, uniform crossover, mutation probability 5%, 6500 generations, the best solution of 10 runs)

Figure 1 shows the computation of a fitness function as depending on the sequence of generations of the algorithm. The upper characteristic shows the average value of the fitness function of all individuals in the population and the lower one shows the fitness function of the best individual in the population. It can be seen that the algorithm converges quite rapidly to a pseudo-optimal solution (no optimal solution is known for such a large instance).

The test results showed that one-point crossover gave worse results than two-point crossover, which, in turn, was even worse than a uniform crossover. Obviously, it would also be appropriate to monitor the width of the middle part of a chromosome in the two-point crossover because, when its width is very small, it brings very little change in comparison with the parental chromosomes and reduces the diversity of genetic material in the population, which increases the risk of small or almost no changes in the fitness function after only a small number of generations.

2.2 Simulated Annealing

The simulated annealing (SA) technique has been given much attention as a general-purpose way of solving difficult optimisation tasks. Its idea is based on simulating the cooling of a material in a heat bath - a process known as annealing. More details may be found in [9].

As in the previous section, we only mention the parameter settings:

- *initial temperature* T_0 = 10000,
- *final temperature* T_f = 1,
- *temperature reduction function* $\alpha(t) = t * decrement$, where *decrement* = 0.999,
- *neighbourhood* for a given solution $\mathbf{x}_0$ is formed by neighbours generated by the inverting one position in the binary string representing $\mathbf{x}_0$.

(This means that the number of neighbours of $\mathbf{x}_0$ is equal to the number of positions in this string.) (Fig. 2)

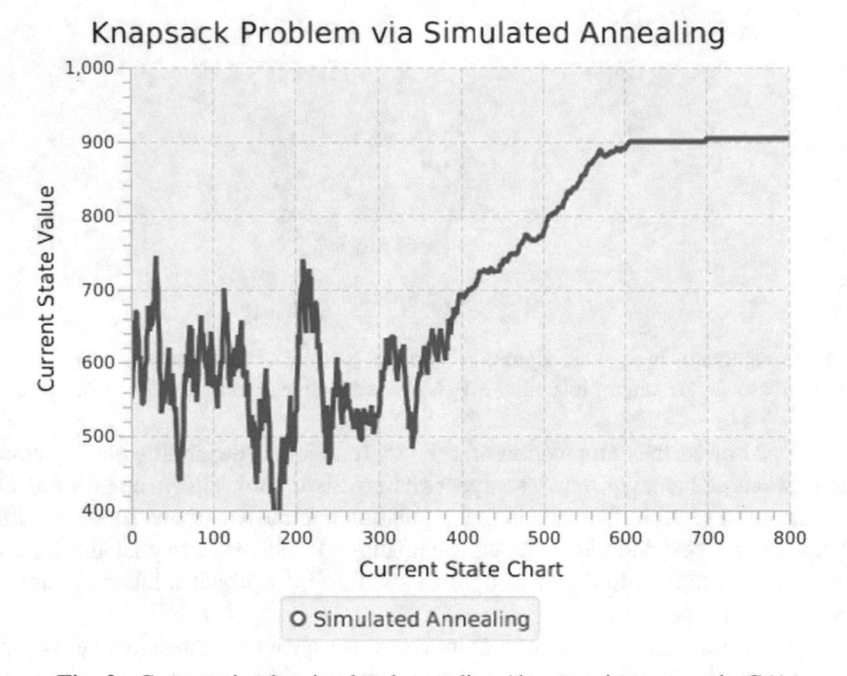

Fig. 2. Computation by simulated annealing (the same instance as in GA)

It is obvious that simulated annealing converges to the same, or a very close approximation of the optimal solution, which is consistent with the well-known "No free lunch theorem" [11]. Because the simulated annealing is a one-point method, only the dependence of the objective function for gradually updated points $\mathbf{x}_0$ (centres for generating neighbours) in the search space is plotted. The chart also shows that the likelihood of transfer to a worse neighbour is decreasing steadily as temperature decreases, which is also in line with the principle of the method.

As in the genetic algorithm, it is necessary to apply the repair operator for the selected solution in the neighbourhood, as mentioned in the previous section.

The computational time of GA for the tested instance was only 19 s on a computer with a processor frequency of 2.4 GHz and operating memory of 4 GB while SA takes more than 1 min. The reason is that, in each GA iteration, we generate only two children while SA creates the neighbourhood with 50 neighbours in each iteration.

An even worse situation would be for a hill-climbing algorithm, where each individual would have to be modified by the repair operator in order to determine the best neighbour. For simulated annealing, it is sufficient to apply the repair operator only to a randomly selected individual from the neighbourhood.

3 Multiple Constraints and Multiple Knapsack Problem

In a more complex case, we have more than one capacity constraint, e.g., not only the total weight but also the total volume. If the number of constraints is m, then the 0-1 knapsack problem changes to *multi-knapsack problem* (MKP) as follows:

$$\text{Maximise} \sum_{i=1}^{n} v_i x_i \tag{12}$$

subject to

$$\sum_{i=1}^{n} w_{ji} x_i \leq C_j,\ j = 1, \ldots, m \tag{13}$$

$$x_i \in \{0, 1\},\ i = 1, \ldots, n \tag{14}$$

For the solution of the task with multiple constraints, we must generalise the approaches mentioned above. The representations of chromosomes in GA and solutions in SA are the same.

To avoid infeasible solutions, a *repair operator* [10] is applied. Its pseudo-Pascal code for the MKP is shown as follows.

Let

$J := \{1, \text{K}\ , m\}$

$W_j := \sum_{i=1}^{n} w_{ji} x_i,\ \forall j \in J;$

for $i := n$ **downto** 1 **do**

if $(x_i = 1)$ **and** $(W_j > C_j;$ for any $j \in J)$

then begin $x_i := 0;$

$W_j := W_j - w_{ji},\ \forall j \in J$

end;

{ S is now a feasible solution to the MKP }

In practice, we can meet the requirements of not exceeding the separate capacities of several pieces of luggage, and the total capacity, too. For example, the airBaltic airlines specify luggage requirements as shown in Fig. 3.

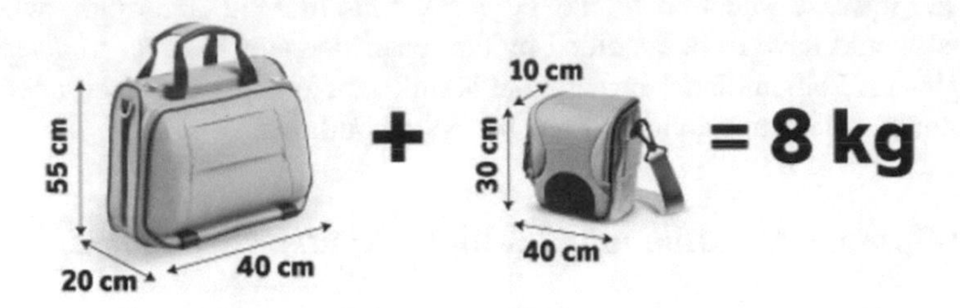

Fig. 3. airBaltic luggage requirements

Let us denote by w_{1i} the item volumes (assuming a size that fits the luggage), by w_{2i} the item weights, by C_1 and C_2, the volumes of knapsacks, and by C, the total weight of both knapsacks.

If a binary decision variable x_{ki} specifies whether or not item i is included in the knapsack k, then the Multiple Knapsack Problem for this case may be formally expressed as follows:

$$\text{Maximise} \sum_{k=1}^{2} \sum_{i=1}^{n} v_i x_{ki} \tag{15}$$

subject to

$$\sum_{i=1}^{n} w_{1i} x_{ki} \leq C_k, \; k = 1, 2 \tag{16}$$

$$\sum_{k=1}^{2} \sum_{i=1}^{n} w_{2i} x_{ki} \leq C \tag{17}$$

$$\sum_{k=1}^{2} x_{ki} \leq 1, \; i = 1, \ldots, n \tag{18}$$

$$x_{ki} \in \{0, 1\}, \; i = 1, \ldots, n, \; k = 1, 2 \tag{19}$$

Equation (18) means that each item may be included at most in one knapsack.

For solving the problem, many strategies for filling the knapsacks may be used such as starting with the bigger knapsack and then continuing with the smaller one, or putting items into knapsack alternately, completely randomly, etc. To use GA or SA, a straightforward modification can be used.

This model may, of course, be simply modified to a case with any number of constraints and knapsacks. If we have n items, m constraints for each knapsack,

r knapsacks and C_{kj} represents capacity constraint j of knapsack k, then the problem has the following model:

$$\text{Maximise} \sum_{k=1}^{r} \sum_{i=1}^{n} v_i x_{ki} \tag{20}$$

subject to

$$\sum_{i=1}^{n} w_{ji} x_{ki} \leq C_{kj}, \; k = 1, \ldots, r, \; j = 1, \ldots, m \tag{21}$$

$$\sum_{k=1}^{r} x_{ki} \leq 1, \; i = 1, \ldots, n \tag{22}$$

$$x_{ki} \in \{0, 1\}, \; i = 1, \ldots, n, \; k = 1, \ldots, r \tag{23}$$

It is evident that the extended versions of knapsack problems have only more constraints and need a modification of the repair operator, but the principle of the evaluation of objective functions is the same and the same is also the representation of the problems by binary strings.

4 Conclusion

In the previous paragraphs, several approaches to solving knapsack problem, were discussed. For small instances, deterministic methods such as the branch and bound method and, in a special case, the dynamic programming approach may also be used to obtain results better than in heuristic methods. On the contrary, the branch and bound method and dynamic-programming approach are not efficient for large instances because of the exponentially growing time in branch and bound method calculations and huge memory requirements for static arrays in the dynamic-programming approach.

In these cases, we choose heuristic techniques. The genetic algorithm and simulated annealing have been found to work efficiently. They provide good results in a reasonable amount of time for tens of items and capacity constraints.

In the future, we try to implement other heuristics and special multiple versions of the knapsack problem and compare their results.

References

1. Garey, M.R., Johnson, D.S.: Computers and Intractability: A Guide to the Theory of NP-Completeness. W.H. Freeman & Co., New York (1979)
2. Goldberg, D.E.: The Design of Innovation (Genetic Algorithms and Evolutionary Computation). Kluwer Academic Publishers, Dordrecht (2002)

3. Gulina, I., Matousek, R.: Efficient nearest neighbor searching in RRTs. In: Matousek, R. (ed.) Proceedings of 22nd International Conference on Soft Computing, MENDEL 2016 (2016)
4. Gutin, G., Punnen, A.P. (eds.): The Traveling Salesman Problem and Its Variations. Kluwer Academic Publishers, Dordrecht (2002)
5. Klapka, J., Dvořák, P., Popela, P.: Optimisation Methods. VUTIUM, Brno (2001). (in Czech)
6. Matousek, R., Popela, P.: Stochastic quadratic assignment problem: EV and EO reformulations solved by HC12. In: Proceedings of 20th International Conference on Soft Computing, MENDEL 2014. Brno University of Technology, VUT Press, Brno (2014)
7. Michalewicz, Z.: Genetic Algorithms + Data Structures = Evolution Programs, 3rd edn. Springer, Heidelberg (1996)
8. Michalewicz, Z., Fogel, D.B.: How to Solve It: Modern Heuristics. Springer, Berlin (2002)
9. Reeves, C.R.: Modern Heuristic Techniques for Combinatorial Problems. Blackwell Scientific Publications, Oxford (1993)
10. Šeda, M., Šeda, P.: A Minimisation of network covering services in a threshold distance. In: Matoušek, R. (ed.) Mendel 2015. Recent Advances in Soft Computing, vol. 378, pp. 159–169. Springer, Berlin (2015)
11. Wolpert, D.H., McReady, W.G.: No free lunch theorems for optimization. IEEE Trans. Evol. Comput. **1**, 67–82 (1997)
12. Zelinka, I., Snášel, V., Abraham, A. (eds.): Handbook of Optimization. From Classical to Modern Approach. Springer, Berlin (2013)

Waste Processing Facility Location Problem by Stochastic Programming: Models and Solutions

Pavel Popela[1(✉)], Dušan Hrabec[2], Jakub Kůdela[1], Radovan Šomplák[1], Martin Pavlas[1], Jan Roupec[1], and Jan Novotný[1]

[1] Faculty of Mechanical Engineering, Brno University of Technology, Technická 2, 616 69 Brno, Czech Republic
{popela,pavlas,roupec}@fme.vutbr.cz, jakub.kudela89@gmail.com, somplak@upei.fme.vutbr.cz, iannovot@gmail.com

[2] Faculty of Applied Informatics, Tomas Bata University, Nad Stráněmi 4511, 760 05 Zlín, Czech Republic
hrabec@fai.utb.cz

Abstract. The paper deals with the so-called waste processing facility location problem (FLP), which asks for establishing a set of operational waste processing units, optimal against the total expected cost. We minimize the waste management (WM) expenditure of the waste producers, which is derived from the related waste processing, transportation, and investment costs. We use a stochastic programming approach in recognition of the inherent uncertainties in this area. Two relevant models are presented and discussed in the paper. Initially, we extend the common transportation network flow model with on-and-off waste-processing capacities in selected nodes, representing the facility location. Subsequently, we model the randomly-varying production of waste by a scenario-based two-stage stochastic integer linear program. Finally, we employ selected pricing ideas from revenue management to model the behavior of the waste producers, who we assume to be environmentally friendly. The modeling ideas are illustrated on an example of limited size solved in GAMS. Computations on larger instances were realized with traditional and heuristic algorithms, implemented within MATLAB.

Keywords: Waste processing · Facility location problem Stochastic programming · Two decision stages · Uncertainty modeling Scenarios · Mathematical programming algorithms · Heuristics Genetic algorithms · GAMS · MATLAB · Pricing related ideas

1 Introduction

The growing concern for environment leads to integration of new solutions into traditional WM in practice. About 3 billion tonnes of waste are generated in the European Union countries yearly, see [2]. Moreover, due to the population

R. Matoušek (Ed.): MENDEL 2017, AISC 837, pp. 167–179, 2019.
https://doi.org/10.1007/978-3-319-97888-8_15

increase, migration of non EU inhabitants, and economic development in the EU countries, the amount of waste generated is rapidly increasing [3,6]. Therefore, municipal solid waste producers often face problems of insufficiency in available facility capacities to meet future waste disposal demand [8].

Municipal WM consists of various activities that can be clustered into four processing steps: waste generation, collection, transformation, and disposal [5]. This paper deals with the second stage: collection that also involves waste transportation to waste processing units. Hence, we concern on mathematical modeling and related decision support computations for the optimal WM including facility location planning in this step, see, e.g., [6] for an extensive review of WM modeling, and see also [19] for the facility location in the context of so called waste-to-energy plant planning. So, WM decision making problems belong to the class of optimization problems, which importance recently significantly increases in practice. Therefore, mathematical modeling of particular situations and its computational support can help to decision makers with control of the WM as well as to achieve cost savings [4].

Existing modeling and solution challenges are related to the fact that the studied problems often combine deterministic and stochastic parameters together with nonlinear terms and both continuous and discrete decision variables. Since many parameters in such WM system can be uncertain, straightforward applicability of deterministic mathematical programming methods can be doubtful [8]. Thus, to model the real world requirements in a suitable way, stochastic programming approach has been selected and applied in the model building process.

Among the above mentioned problems, we focus on a so called waste processing FLP that defines the task to choose the set of open and running waste processing units in the best way from the total expected cost point of view. Thus, the facility location decisions must be made when a logistics system is started from scratch i.e. when new products or services are launched or when existing product distributions or services are expanded [4]. Specifically, in this paper, we deal only with the waste producer preferences, and so, we minimize the related processing, transportation, and investment costs.

In this paper, the FLP is considered within the transportation network. In general, network design of transportation problems still belongs to interesting research topics in transportation planning [9,23]. Various approaches have been taken to solve network design problems, see [12,15] for a review of the network design problems and see [1] for a detailed review of solution techniques. See also [7] for our previous ideas and further references on a hybrid computational approach to network design problems where we deal mostly with switching on and off edges and arcs of the transportation network.

The next sections of the paper are organized as follows. Section 2 describes the developed FLP within waste transportation network design models. Two considered models are subsequently presented, described, and discussed. Firstly, a common transportation network flow is enriched with the on-off waste processing capacities in the chosen nodes to represent the facility location. Then, the randomly varying waste production is modeled by scenarios and two-stage stochastic integer linear program is obtained. As the second step, we suggest to model environmental friendly behavior of waste producers by the ideas inspired by utilization pricing mechanisms in operations research problems. Discussed modeling ideas are explained by an explanatory example in Sect. 2. Results of computations that were realized for various larger instances with utilization of both traditional and heuristical algorithms by using model and algorithm implementations in GAMS and MATLAB are commented in Sect. 3. Finally, Sect. 4 concludes the paper and outlines future research directions and suggests some new computational and modeling ideas for future development.

2 Models and Explanatory Examples

In this section, we develop the cost-minimizing stochastic mixed integer nonlinear program for the above mentioned problem in two steps. The introduced models use the following sets of indices, parameters and decision variables. The sets of indices are as follows:

I: set of transportation network related nodes representing places, $i \in I$,
J: set of transportation network related edges representing routes, $j \in J$,
S: set of included scenarios representing uncertainty, $s \in S$.

In this case, we can identify nodes with waste producers, transition places and waste processing units. In addition, we differ between existing processing units and those units that can be newly established. The edges model routes that may serve for transportation of waste. The structural information describing the network is completed with the following input parameters:

$a_{i,j}$: network description by node-edge incidence matrix,
$b^-_{i,s}$: available amount of produced waste in node i for scenario s,
b^+_i: available waste processing capacity in node i,
c_j: cost per transported unit of waste by edge j,
f_i: cost per processed unit of waste in node i,
g^-_i: cost per unprocessed waste left in node i,
g^+_i: cost per unit of unused capacity in node i,
h_i: cost per switched on processing unit in node i,
p_s: probability of achieving scenario s.

We further assume that waste producers considered in our model coordinate their decision steps and behave as one decision maker. So, among the model elements, the following decision variables are included:

$x_{j,s}$: waste transported by edge j for scenario s, bounded by $x_{U,j}$,
$y_{i,s}$: amount of waste processed in node i by scenario s,
$u^-_{i,s}$: amount of untransported waste from node i for scenario s,
$u^+_{i,s}$: amount of unused processing capacity in node i for scenario s,
$v^-_{i,s}$: amount of waste transported from node i for scenario s (negative),

$v_{i,s}^+$: amount of waste transported to node i for scenario s,
δ_i: indicator of switching on-off extra waste processing capacity in i.

The first model is a scenario-based two-stage mixed integer linear program that is described as follows:

$$\min \sum_{s \in S} p_s \left(\sum_{j \in J} c_j x_{j,s} + \sum_{i \in I} (f_i y_{i,s} + g_i^- u_{i,s}^- + g_i^+ u_{i,s}^+) \right) + \sum_{i \in I} h_i \delta_i \quad (1)$$

$$\text{s.t.} \quad \sum_{j \in J: a(i,j) > 0} a_{i,j} x_{j,s} = v_{i,s}^+, \quad \forall i \in I, s \in S, \quad (2)$$

$$\sum_{j \in J: a(i,j) < 0} a_{i,j} x_{j,s} = -v_{i,s}^-, \quad \forall i \in I, s \in S, \quad (3)$$

$$y_{i,s} + u_{i,s}^+ = b_i^+ \delta_i, \quad \forall i \in I, s \in S, \quad (4)$$

$$-b_{i,s}^- + v_{i,s}^+ = v_{i,s}^- + y_{i,s} + u_{i,s}^-, \quad \forall i \in I, s \in S, \quad (5)$$

$$x_{j,s}, y_{i,s}, u_{i,s}^-, u_{i,s}^+, v_{i,s}^-, v_{i,s}^+ \geq 0, \quad \forall i \in I,\ j \in J, s \in S, \quad (6)$$

$$x_{j,s} \leq x_{U,j}, \quad \forall j \in J, s \in S, \quad (7)$$

$$\delta_i \in \{0,1\}, \quad \forall i \in I. \quad (8)$$

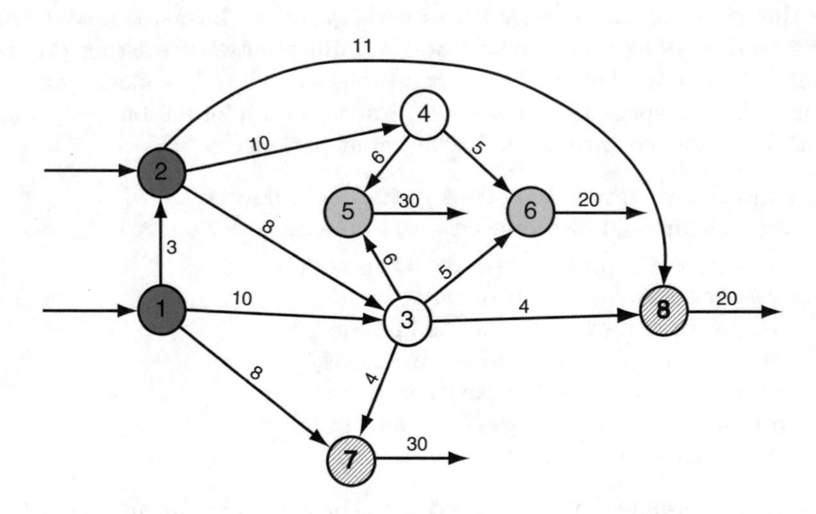

Fig. 1. Test network - visualization of simple input data

The objective function (1) minimizes the total cost that is a sum of scenario-related costs involving transportation costs, processing costs, penalizing costs for left waste, penalizing costs for unused capacity and investment costs following the investment decisions that must be the same for all scenarios. Equation (2) means that all flows entering node i are summarized to $v_{i,s}^+$. Similarly Eq. (3) says that all flows leaving node i are summarized to $v^-_{i,s}$. Equation (4) represents a constraint on the processed amount of waste that is bounded by processing unit capacity. This equation also allows to switch on new waste processing units. To make a difference between already built processing units and newly established ones the value of the first stage decision variables δ_i can be fixed. So, the value 0 is used for transition nodes and value 1 is utilized for the existing processing units. Equation (5) provides the balance constraint of inputs and outputs in node i. Finally, (6)–(8) specify domains of the decision variables. For the initial explanation, we have utilized the transportation network on Fig. 1. Such a simple example can be solved almost intuitively for one scenario case. Therefore, we list the output for the single scenario in the form of GAMS result file that also contains all input data:

```
Input data and results for data case: 01 - mid waste production
=================================================================================
Total optimal cost          zmin =          2650.00
Partial optimal cost         h*d =           600.00 investment of new units
Partial optimal cost p*(gM*uM) =              0.00 average for unprocessed waste
Partial optimal cost p*  (c*x) =            900.00 average transportation costs
Partial optimal cost p*  (f*y) =           1100.00 average for processing waste
Partial optimal cost p*(gP*uP) =             50.00 average for unused capacity
=================================================================================
S1 scenario optimal cost gM*uM =             0.00        p(S1) = 1.000000
   scenario optimal cost   c*x =           900.00
   scenario optimal cost   f*y =          1100.00
   scenario optimal cost gP*uP =            50.00
=================================================================================
nodes                    i |     N1     N2     N3     N4     N5     N6     N7     N8
investment costs         h |    0.0    0.0    0.0    0.0    0.0    0.0 1000.0  600.0
building unit            d |    0.0    0.0    0.0    0.0    1.0    1.0    0.0    1.0
product                h*d |    0.0    0.0    0.0    0.0    0.0    0.0    0.0  600.0
=================================================================================
Scenario S1 with probability p(S1) = 1.000000
---------------------------------------------------------------------------------
nodes                    i |     N1     N2     N3     N4     N5     N6     N7     N8
produced waste       bM>= |   35.0   30.0    0.0    0.0    0.0    0.0    0.0    0.0
---------------------------------------------------------------------------------
left waste              uM |    0.0    0.0    0.0    0.0    0.0    0.0    0.0    0.0
---------------------------------------------------------------------------------
related cost            gM |  100.0  100.0  100.0  100.0  100.0  100.0  100.0  100.0
product               gMuM |    0.0    0.0    0.0    0.0    0.0    0.0    0.0    0.0
---------------------------------------------------------------------------------
output flow             vM |   35.0   30.0   45.0    0.0    0.0    0.0    0.0    0.0
                       -vM |  -35.0  -30.0  -45.0    0.0    0.0    0.0    0.0    0.0
---------------------------------------------------------------------------------
    j   c*x     c      x | ax(N1) ax(N2) ax(N3) ax(N4) ax(N5) ax(N6) ax(N7) ax(N8)
E1a2    0.0   3.0    0.0 |    0.0    0.0    0.0    0.0    0.0    0.0    0.0    0.0
E1a3  350.0  10.0   35.0 |  -35.0    0.0   35.0    0.0    0.0    0.0    0.0    0.0
E1a7    0.0   8.0    0.0 |    0.0    0.0    0.0    0.0    0.0    0.0    0.0    0.0
E2a3   80.0   8.0   10.0 |    0.0  -10.0   10.0    0.0    0.0    0.0    0.0    0.0
E2a4    0.0  10.0    0.0 |    0.0    0.0    0.0    0.0    0.0    0.0    0.0    0.0
E2a8  220.0  11.0   20.0 |    0.0  -20.0    0.0    0.0    0.0    0.0    0.0   20.0
E3a5  150.0   6.0   25.0 |    0.0    0.0  -25.0    0.0   25.0    0.0    0.0    0.0
E3a6  100.0   5.0   20.0 |    0.0    0.0  -20.0    0.0    0.0   20.0    0.0    0.0
E3a7    0.0   4.0    0.0 |    0.0    0.0    0.0    0.0    0.0    0.0    0.0    0.0
E3a8    0.0   4.0    0.0 |    0.0    0.0    0.0    0.0    0.0    0.0    0.0    0.0
E4a5    0.0   6.0    0.0 |    0.0    0.0    0.0    0.0    0.0    0.0    0.0    0.0
E4a6    0.0   5.0    0.0 |    0.0    0.0    0.0    0.0    0.0    0.0    0.0    0.0
---------------------------------------------------------------------------------
c*x= 900.0  out:      vP |    0.0    0.0   45.0    0.0   25.0   20.0    0.0   20.0
              in:   -vM |  -35.0  -30.0  -45.0    0.0    0.0    0.0    0.0    0.0
---------------------------------------------------------------------------------
capacity              bP>= |    0.0    0.0    0.0    0.0   30.0   20.0   30.0   20.0
```

unused	uP	0.0	0.0	0.0	0.0	5.0	0.0	0.0	0.0
processed	y	0.0	0.0	0.0	0.0	25.0	20.0	0.0	20.0
related cost	f	0.0	0.0	0.0	0.0	20.0	20.0	10.0	10.0
product	fy	0.0	0.0	0.0	0.0	500.0	400.0	0.0	200.0
cost unused	gP	10.0	10.0	10.0	10.0	10.0	10.0	10.0	10.0
product	gPuP	0.0	0.0	0.0	0.0	50.0	0.0	0.0	0.0

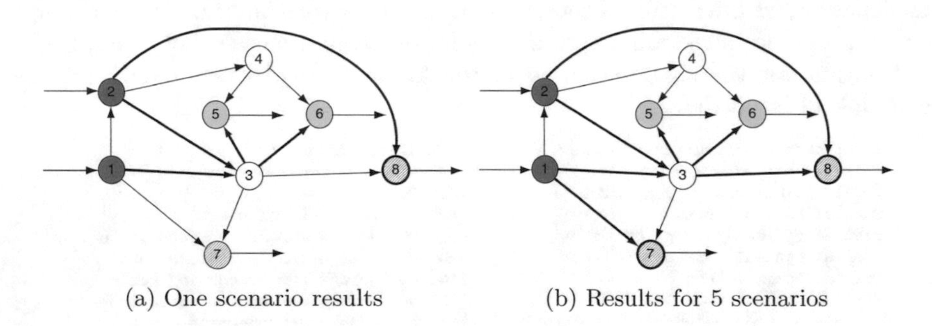

(a) One scenario results (b) Results for 5 scenarios

Fig. 2. Test network - visualization of results

Additionally, Fig. 2a shows the effect of one scenario that leads to the additional switching on available capacity in node 7 (see boldface circle) and extra routes (see boldface edges) used for waste transport. More scenarios taken into the account obviously lead to the increase of newly used processing units (see both nodes 7 and 8) and more routes used for transportation, see Fig. 2b.

To generalize our model, we introduce the pricing related ideas mentioned above in Sect. 1. Therefore, we assume that waste producers, who are trying to minimize their total cost can improve their behaviour and influence the amount of waste as the decision variable. Consequently, the prices may change. It is reasonable to assume the monopolistic type of behaviour from the set of waste processors and from the government who decide about the related prices. So, the second considered and generalized model is the following scenario-bases two-stage stochastic nonlinear mixed integer program:

$$\min \sum_{s\in S} p_s \left(\sum_{j\in J} c_j(x_{j,s})x_{j,s} + \sum_{i\in I} (f_i(y_{i,s})y_{i,s} + g_i^-(\bar{b}_i)u_{i,s}^- + g_i^+ u_{i,s}^+) \right) + \sum_{i\in I} h_i\delta_i \tag{9}$$

$$\text{s.t.} \quad \sum_{j\in J:a(i,j)>0} a_{i,j}x_{j,s} = v_{i,s}^+, \qquad \forall i\in I, s\in S, \tag{10}$$

$$\sum_{j\in J:a(i,j)<0} a_{i,j}x_{j,s} = -v_{i,s}^-, \qquad \forall i\in I, s\in S, \tag{11}$$

$$y_{i,s} + u^+_{i,s} = b^+_i \delta_i, \qquad \forall i \in I, s \in S, \tag{12}$$

$$-b^-_{i,s} + v^+_{i,s} = v^-_{i,s} + y_{i,s} + u^-_{i,s}, \quad \forall i \in I, s \in S, \tag{13}$$

$$\bar{b}_i + \varepsilon_{i,s} = b^-_{i,s}, \qquad \forall i \in I, s \in S, \tag{14}$$

$$x_{j,s}, y_{i,s}, u^-_{i,s}, u^+_{i,s}, v^-_{i,s}, v^+_{i,s} \geq 0, \qquad \forall i \in I, j \in J, s \in S, \tag{15}$$

$$x_{j,s} \leq x_{U,j}, \qquad \forall j \in J, s \in S, \tag{16}$$

$$\delta_i \in \{0,1\}, \qquad \forall i \in I, \tag{17}$$

$$b_{L,i} \leq \bar{b}_i \leq b_{U,i}, \; b^-_{i,s} \geq 0, \qquad \forall i \in I, s \in S. \tag{18}$$

In the second model (9)–(18), most of the constraints (see (10)–(13), (15)–(17) and compare it with (2)–(8)) remain the same, however, several important modifications have been included. The cost coefficients newly depend on decision variables (see the objective function (9)) and we have introduced functions $c_j(x_{j,s})$, $f_i(y_{i,s})$, and $g^-_i(\bar{b}_i)$ instead of coefficients c_j, f_i, and g^-_i respectively. We also assume that the decrease of the amount transported or processed will lead to the increase of the related unit cost specified by the price coordinating processing units. Similarly, we assume that the unit governmental penalty for the unprocessed waste will increase with decreasing production of the waste. See Fig. 3 for an example of $f_i(y_{i,s})$ function.

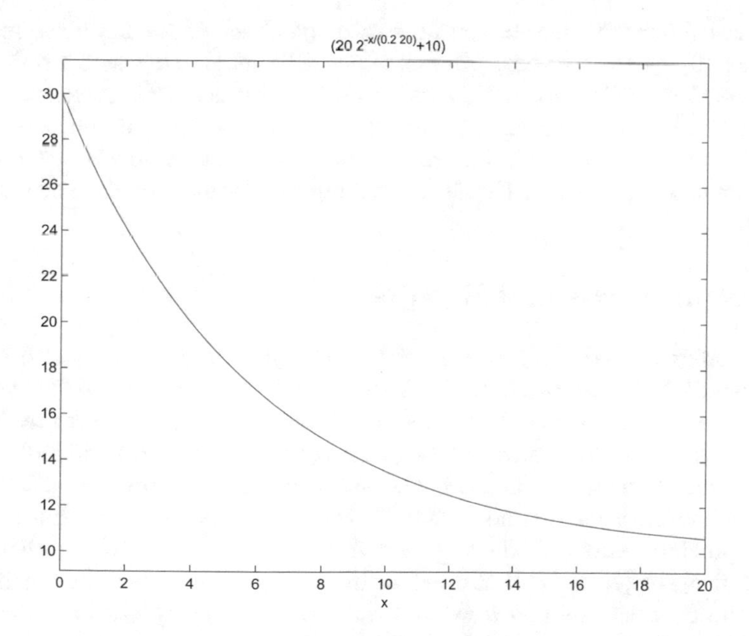

Fig. 3. Modeling pricing ideas

We suggest to notice that under the assumption of strict monotonicity of these functions, so traditional pricing related formulas can appear in the case that we decide to deal with inverse functions. However, the related interpretation derived from the viewpoint of the producers seems to be unrealistic for such case. Therefore, we have converted our original pricing ideas in the final ones that are included in the model. The decision of the waste producers about the amount of the waste delivered for the processing is denoted by $\bar{b}_i$ and changes only within the bounds $b_{L,i}$ and $\leq b_{U,i}$ are allowed, see (18). We expect random disturbances following this decision modeled by $\varepsilon_{i,s}$. Then, the $b^-_{i,s}$ is a dependent variable defined by (14).

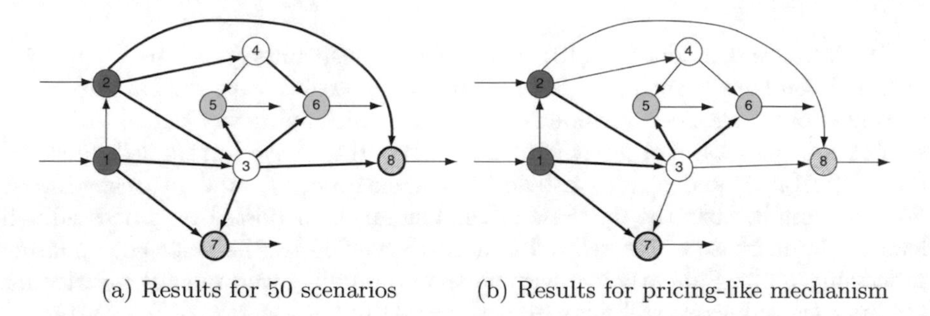

(a) Results for 50 scenarios (b) Results for pricing-like mechanism

Fig. 4. Test network - visualization of results

The last figures in this section illustrate the effect of pricing ideas included, see model (9)–(18). Allowing price changes will motivate waste producers to increase, e.g., recycling attempts and it may also reduce their total costs, waste produced and waste processed. Specifically, less processing units must be opened and less routes are used cf. Figure 4a involving solution for 50 scenarios for the first model (1)–(8) and Fig. 4b describing the results for the second model (9)–(18).

3 Computations and Results

We have programmed the abovementioned two models in GAMS and we have solved them by the use of BARON, MINOS and CPLEX solvers for small test instances obtaining acceptable results. The next computations were realized for larger instances of the model (9)–(18). However, the solution difficulties have appeared when the original GAMS code was applied as computations have led to increasing computational time needs. Therefore, heuristics have been discussed and the previous authors' ideas related to the suitable hybrid algorithm have been detailed, see [7]. Instead of previous implementations based on combination of the GAMS and C++ codes we have preferred the complete implementation in MATLAB. This implementation combines `fmincon` function with the genetic algorithm implementation to follow the algorithmic scheme:

1. Set up the instance of scenario-based two-stage mixed integer nonlinear program in MATLAB. Set up control parameters for the genetic algorithm implemented in MATLAB.

2. Create an initial population for GA instance. So, the initial values of 0–1 variables are generated and fixed to obtain a scenario-based (separable) nonlinear program.
3. Several runs of random generators are needed for a specified population size and number of considered scenarios. Repeatedly run `fmincon` procedure in MATLAB to obtain the set of scenario-related solutions. Each run solves the program for the fixed values of 0–1 variables.
4. The objective function values are computed, also for new individuals created by means of the genetic operators, initially in 2 and then in 3. Store the best results obtained from MATLAB (the optimal objective function values and optimal values of all variables for all scenarios) for comparisons.
5. Test the algorithm termination rules and stop in case of their satisfaction. Otherwise continue till the moment when the last scenario solution is obtained.
6. Generate input values for the GA from `fmincon` results, see step 4. Specifically, the objective function values for each member of population of the GA are obtained from results of the runs in 3.
7. Run GA to update the set of 0–1 variables (population), see, e.g., [13] for details. Return to step 3.

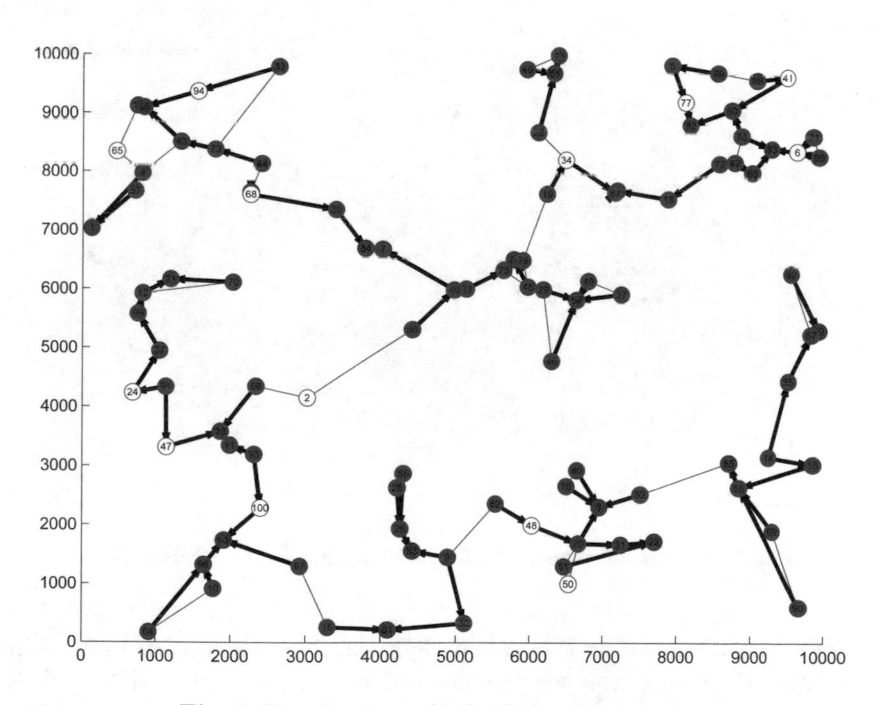

Fig. 5. Visualization of hybrid algorithm results

Table 1. Test results

Number of nodes	10			20		40		50
Number of scenarios	1	5	10	1	5	1	3	1
Computational time [s]	27	137	184	46	1070	288	3027	427
Number of nodes	12			24		42		55
Number of scenarios	1	5	10	1	5	1	3	1
Computational time [s]	32	151	193	62	1122	309	3227	493
Number of nodes	14			28		44		60
Number of scenarios	1	5	10	1	5	1	3	1
Computational time [s]	39	163	205	71	1197	327	3302	564
Number of nodes	16			32		46		65
Number of scenarios	1	5	10	1	5	1	3	1
Computational time [s]	45	182	226	83	1251	339	3411	617
Number of nodes	18			28		48		70
Number of scenarios	1	5	10	1	5	1	3	1
Computational time [s]	53	199	245	96	1307	378	3571	691

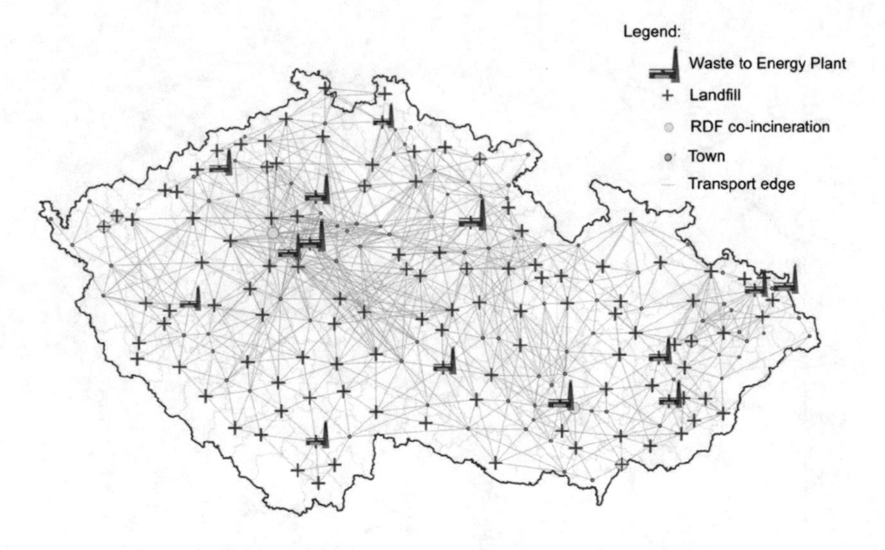

Fig. 6. Real-world transportation network for the Czech Republic

The results obtained by the hybrid algorithm implementation in MATLAB are illustrated for one instance of data and model (9)–(18) on Fig. 5. A special postprocessing procedure dynamically supporting the visualization of the obtained results have been implemented in MATLAB as well. Red nodes represent built waste processing units (e.g., incinerators), then the blue nodes identify waste producers and white nodes are transition nodes. Let us emphasize that for these test computations to simplify the data instance coding we do not assume any already built waste processing units. Then, the edges are different by the flow. The black edge denotes a non-zero flow while the blue edge identifies a zero flow. Similarly, the instances of the various size have been tested and the collected experience is contained in the Table 1. The average computational times show the expected increasing trends with the increase of number of nodes and number of scenarios.

4 Conclusions and Further Research

In the presented paper we have generalized a well known facility location problem to the specific problems of waste processing, see [19,22]. We have adopted the standpoint of the waste producers and we minimize the waste-management cost which they face, and which is derived from the related processing, transportation, and investment costs. We have built two stochastic programs starting from the transportation network flow model with on-and-off waste-processing capacities in selected nodes and randomly-varying waste production modelled by scenarios. Then, the pricing ideas from revenue management have been utilized to allow environmentally friendly behaviour of waste producers. For computational purposes a modified hybrid algorithm is implemented in MATLAB and obtained results are visualized.

Further research will lead to adaptation of the models and the algorithm to the case of real-world data and to the viewpoint of the waste processors, see Fig. 6. In general, similar mixed integer (either linear, bilinear or nonlinear) stochastic programs may appear in many application areas including design problems [10,20], control problems [21] or vehicle routing problems [16]. Moreover, a use of more advanced evolutionary algorithms seems to be necessary in further research. Therefore, we refer to genetic [13], differential evolution [17], particle swarm [14] and ant colony [18] optimization algorithms.

Acknowledgments. This work was supported by the Programme EEA and Norway Grants for funding via grant on Institutional cooperation project nr. NF-CZ07-ICP-4-345-2016 and by the specific research project "Modern Methods of Applied Mathematics for the Use in Technical Sciences", no. FSI-S-14-2290, id. code 25053. The authors gratefully acknowledge further support from the NETME CENTRE PLUS under the National Sustainability Programme I (Project LO1202) and support provided by Technology Agency of the Czech Republic within the research project No. TE02000236 "Waste-to-Energy (WtE) Competence Centre.

References

1. Babazadeh, A., Poorzahedy, H., Nikoosokhan, S.: Application of particle swarm optimization to transportation network design problem. J. King Saud Univ. - Sci. **23**, 293–300 (2011)
2. Blumenthal, K.: Generation and treatment of municipal waste. Technical report KS-SF-11-031 (2011)

3. Eiselt, H.A., Marianov, V.: Location modeling for municipal solid waste facilities. Comput. Oper. Res. **62**, 305–315 (2015)
4. Ghiani, G., Laporte, G., Musmanno, R.: Introduction to Logistic Systems Planning and Control. Wiley-Interscience Series in Systems and Optimization. Wiley, Chichester (2004)
5. Ghiani, G., et al.: Capacitated location of collection sites in an urban waste management system. Waste Manag. **32**, 1291–1296 (2012)
6. Ghiani, G., Lagana, D., Manni, E., Musmanno, R., Vigo, D.: Operations research in solid waste management: a survey of strategic and tactical issues. Comput. Oper. Res. **44**, 22–32 (2014)
7. Hrabec, D., Popela, P., Roupec, J., et al.: Hybrid algorithm for wait-and-see network design problem. In: 20th International Conference on Soft Computing MENDEL 2014, pp. 97–104. Brno University of Technology, VUT Press, Brno (2014)
8. Huang, G.H., Baetz, B.W., Patry, G.G., Terluk, V.: Capacity planning for an integrated waste management system under uncertainty: a North American case study. Waste Manag. Res. **15**, 523–546 (1997)
9. Kaya, O., Urek, B.: A mixed integer nonlinear programming model and heuristic solutions for location, inventory and pricing decisions in a closed loop supply chain. Comput. Oper. Res. **65**, 93–103 (2016)
10. Lániková, I., et al.: Optimized design of concrete structures considering environmental aspects. Adv. Struct. Eng. **17**(4), 495–511 (2014)
11. LeBlanc, L.J.: An algorithm for discrete network design problem. Transp. Sci. **9**, 183–199 (1975)
12. Magnanti, T.L., Wong, R.T.: Network design and transportation planning: models and algorithms. Transp. Sci. **18**, 1–55 (1984)
13. Matoušek, R., Žampachová, E.: Promising GAHC and HC12 algorithms in global optimization tasks. Optim. Methods Softw. **26**(3), 405–419 (2011)
14. Pluháček, M., Šenkeřík, R., Zelinka, I.: Particle swarm optimization algorithm driven by multichaotic number generator. Soft. Comput. **18**(4), 631–639 (2014)
15. Steenbrink, P.A.: Optimization of Transport Network. Wiley, New York (1974)
16. Stodola, P., Mazal, J., Podhorec, M., Litvaj, O.: Using the ant colony optimization algorithm for the capacitated vehicle routing problem. In: 16th International Conference on Mechatronics - Mechatronika (ME), pp. 503–510 (2014)
17. Šenkeřík, R., Pluháček, M., Davendra, D., Zelinka, I., Janoštík, J.: New adaptive approach for multi-chaotic differential evolution concept. In: Hybrid Artificial Intelligent Systems, pp. 234–243. Springer (2015)
18. Šoustek, P., Matoušek, R., Dvořák, J., Bednář, J.: Canadian traveller problem: a solution using ant colony optimization. In: 19th International Conference on Soft Computing MENDEL 2013, Brno, Czech Republic, pp. 439–444 (2013)
19. Šomplák, R., Pavlas, M., Kropác, J., Putna, O., Procházka, V.: Logistic model-based tool for policy-making towards sustainable waste management. Clean Technol. Environ. Policy **16**(7), 1275–1286 (2014)
20. Štěpánek, P., Lániková, I., Šimůnek, P., Girgle, F.: Probability based optimized design of concrete structures. In: Life-Cycle and Sustainability of Civil Infrastructure System, pp. 2345–2350. Taylor & Francis Group, London (2012)
21. Štětina, J., Klimeš, L., Mauder, T., Kavička, F.: Final-structure prediction of continuously cast billets. Mater. Tehnol. **46**(2), 155–160 (2012)

22. Yo, H., Solvang, W.D.: A general reverse logistics network design model for product reuse and recycling with environmental considerations. Int. J. Adv. Manuf. Technol. **87**, 1–19 (2016)
23. Zhao, J., Huang, L., Lee, D.-H., Peng, Q.: Improved approaches to the network design problem in regional hazardous waste management systems. Transp. Res. Part E **88**, 52–75 (2016)

Randomization of Individuals Selection in Differential Evolution

Roman Senkerik(✉), Michal Pluhacek, Adam Viktorin, Tomas Kadavy, and Zuzana Kominkova Oplatkova

Faculty of Applied Informatics, Tomas Bata University in Zlin, Nam T.G. Masaryka 5555, 760 01 Zlin, Czech Republic
{senkerik, pluhacek, aviktorin, kadavy, oplatkova}@fai.utb.cz

Abstract. This research deals with the hybridization of two computational intelligence fields, which are the chaos theory and evolutionary algorithms. Experiments are focused on the extensive investigation on the different randomization schemes for selection of individuals in differential evolution algorithm (DE).

This research is focused on the hypothesis whether the different distribution of different pseudo-random numbers or the similar distribution additionally enhanced with hidden complex chaotic dynamics providing the unique sequencing are more beneficial to the heuristic performance. This paper investigates the utilization of the two-dimensional discrete chaotic systems, which are Burgers and Lozi maps, as the chaotic pseudo-random number generators (CPRNGs) embedded into the DE. Through the utilization of either chaotic systems or equal identified pseudo-random number distribution, it is possible to entirely keep or remove the hidden complex chaotic dynamics from the generated pseudo random data series. This research utilizes set of 4 selected simple benchmark functions, and five different randomizations schemes; further results are compared against canonical DE.

Keywords: Differential evolution · Complex dynamics · Deterministic chaos Randomization · Burgers map · Lozi map

1 Introduction

This research deals with the mutual intersection of the two computational intelligence fields, which are the complex deterministic dynamics given by the selected chaotic systems driving the selection of indices in Differential Evolution (DE) algorithm and evolutionary computation techniques (ECT's). Currently, DE [1–4] is a well-known evolutionary computation technique for continuous optimization purposes solving many difficult and complex optimization problems. Some DE variants have been recently developed with the emphasis on adaptivity/self-adaptivity. DE has been modified and extended several times using new proposals of versions, and the performances of different DE variant instance algorithms have been widely studied and compared with other evolutionary algorithms. Over recent decades, DE has won most

R. Matoušek (Ed.): MENDEL 2017, AISC 837, pp. 180–191, 2019.
https://doi.org/10.1007/978-3-319-97888-8_16

of the evolutionary algorithm competitions in the leading scientific conferences [5–12], as well as being applied to several applications.

The importance of randomization within heuristics as a compensation of a limited amount of search moves is stated in the survey paper [2]. This idea has been carried out in subsequent studies describing different techniques to modify the randomization process [13, 14] and especially in [15], where the sampling of the points is tested from modified distribution. The importance and influence of randomization operations were also deeply experimentally tested in simple control parameter adjustment jDE strategy [16]. Together with this persistent development in such mainstream research topics, the basic concept of chaos driven DE have been introduced. Recent research in chaotic approach for heuristics uses various chaotic maps in the place of pseudo-random number generators (PRNG). The focus of this research is the direct embedding of chaotic dynamics in the form of chaos pseudo-random number generator (CPRNG) for the heuristic. The initial concept of embedding chaotic dynamics into the evolutionary/swarm algorithms is given in [17]. Also, the PSO (Particle Swarm Optimization) algorithm with elements of chaos was introduced as CPSO [18] followed by the introduction of chaos embedded PSO with inertia weigh strategy [19], and PSO with an ensemble of chaotic systems [20]. Recently the chaos driven heuristic concept has been utilized in ABC algorithm [20] and applications with DE [22].

The organization of this paper is following: Firstly, the motivation and novality for this research is proposed. The next sections are focused on the description of the concept of chaos driven DE, identification of chaotic series distribution and the experiment background. Results and conclusion follow afterward.

2 Motivation

This research is an extension and continuation of the previous successful initial experiment with the single/multi-chaos driven DE (ChaosDE), where the positive influence of hidden complex dynamics for the heuristic performance has been experimentally shown. This research is also a follow up to previous initial experiments with time continuous chaotic systems and different sampling rates used [23].

Nevertheless, the questions remain, as to why it works, why it may be beneficial to use the correlated chaotic time series for generating pseudo-random numbers driving the selection, mutation, crossover or other processes in particular heuristics.

The novelty of the research is given by the experiment investigating whether the chaos embedded heuristics concept belongs to the group of either "utilization of different PRNG with different distribution" or the unique chaos dynamics providing unique sequencing of pseudo-random numbers is the key of performance improvements. The last point was also inspired by recent advances in connection of complexity and heuristic [24] together with the research focused on the selection of indices in DE [8] where the indices (solutions) for mutation process were not selected randomly, but based on the complex behavior and neighborhood mechanisms.

To confirm or disprove the hypothesis above, a simple experiment was performed and presented here. Through the utilization of either chaotic systems or equal identified pseudo-random number distribution, it is possible to fully keep or remove the hidden complex chaotic dynamics from the generated pseudo random data series for obtaining the pseudo-random numbers for indices selection inside DE.

3 Chaotic Systems and Identification of CPRNGs Distributions

Following two well known and frequently utilized discrete dissipative chaotic maps were used as the CPRNGS for DE: Burgers (1), and Lozi map (2).

The Burgers mapping is a discretization of a pair of coupled differential equations to illustrate the relevance of the concept of bifurcation to the study of hydrodynamics flows. The Lozi map is a simple discrete two-dimensional chaotic map. With the typical settings as in Table 1, systems exhibit typical chaotic behavior [25].

Table 1. Definition of chaotic systems used as CPRNGs

Chaotic maps equations		Parameter settings
$X_{n+1} = aX_n - Y_n^2$ $Y_{n+1} = bY_n + X_n Y_n$	(1)	$a = 0.75$ and $b = 1.75$
$X_{n+1} = 1 - a\|X_n\| + bY_n$ $Y_{n+1} = X_n$	(2)	$a = 1.7$ and $b = 0.5$

For the comparisons of DE performance with indices selection driven either by CPRNG or identical PRNG distribution without chaotic dynamics, it was necessary to perform the CPRNGs distributions identification with 10000 samples and statistical distribution fit tests. *Statistica* and *Wolfram Mathematica* software were used for this task with following results (See also Figs. 1 and 2):

- Burgers map based CPRNG was identified as Beta distribution (α, β) with $\alpha = 0.63$ and $\beta = 3.54$.
- Lozi map based CPRNG was identified as Beta distribution (α, β) with $\alpha = 1.05$ and $\beta = 1.57$.

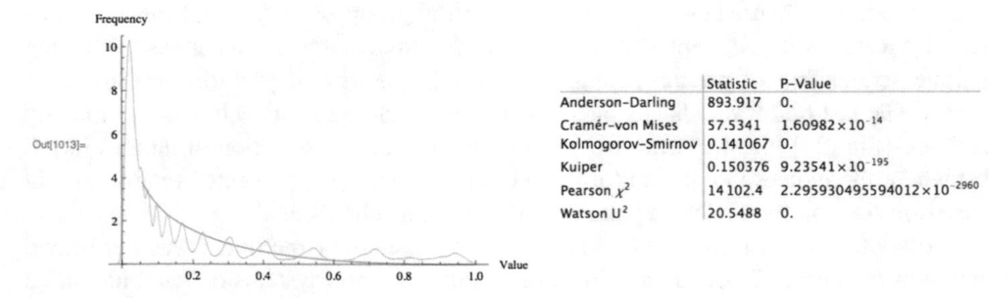

	Statistic	P-Value
Anderson-Darling	893.917	0.
Cramér-von Mises	57.5341	1.60982×10^{-14}
Kolmogorov-Smirnov	0.141067	0.
Kuiper	0.150376	9.23541×10^{-195}
Pearson χ^2	14102.4	$2.295930495594012 \times 10^{-2960}$
Watson U^2	20.5488	0.

Fig. 1. Identification (blue line) of Burgers map based CPRNG (green line – smooth histogram)

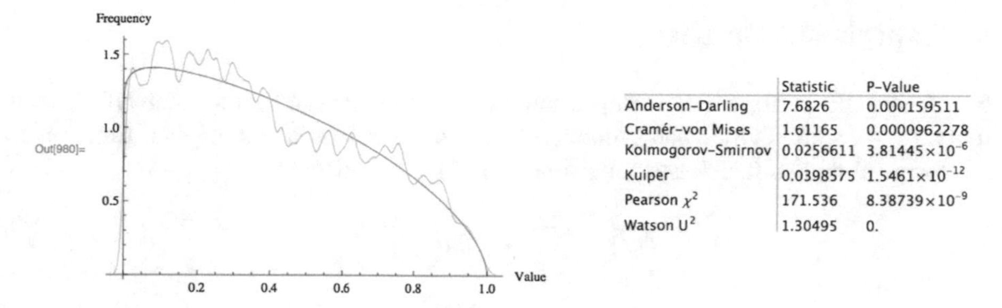

	Statistic	P-Value
Anderson-Darling	7.6826	0.000159511
Cramér-von Mises	1.61165	0.0000962278
Kolmogorov-Smirnov	0.0256611	3.81445×10^{-6}
Kuiper	0.0398575	1.5461×10^{-12}
Pearson χ^2	171.536	8.38739×10^{-9}
Watson U^2	1.30495	0.

Fig. 2. Identification (blue line) of Lozi map based CPRNG (green line – smooth histogram)

4 Differential Evolution

DE is a population-based optimization method that works on real-number-coded individuals [1]. DE is quite robust, fast, and efficient, with global optimization ability. It does not require the objective function to be differentiable, and it works well even with noisy and time-dependent objective functions. There are essentially five inputs to the heuristic. *Dim* is the size of the problem, *Gmax* is the maximum number of generations, *NP* is the total number of solutions, *F* is the scaling factor of the solution and *CR* is the factor for crossover. *F* and *CR* together make the internal tuning parameters for the heuristic. Due to a limited space and the aims of this paper, the detailed description of the well known canonical strategy of differential evolution algorithm basic principles is insignificant and hence omitted. Please refer to [1, 4] for the detailed description of the used *DE/Rand/1/Bin* strategy (both for ChaosDE and Canonical DE) as well as for the complete description of all other strategies.

5 The Concept of ChaosDE with Discrete Chaotic System as Driving CPRNG

The general idea of CPRNG is to replace the default PRNG with the chaotic system. As the chaotic system is a set of equations with a static start position, we created a random start position of the system, to have different start position for different experiments. Thus we are utilizing the typical feature of chaotic systems, which is extreme sensitivity to the initial conditions, popularly known as "butterfly effect". This random position is initialized with the default PRNG, as a one-off randomizer. Once the start position of the chaotic system has been obtained, the system generates the next sequence using its current position. Used approach is based on the following definition (3):

$$rndreal = \mathrm{mod}\left(\mathrm{abs}(rndChaos),\ 1.0\right) \tag{3}$$

6 Experiment Design

For ChaosDE performance comparison within this research, the Schwefel's test function (4), shifted Grienwang function (5), and shifted Ackley's original function in the form (6) and shifted Rastrigin's function (7) were selected.

$$f(x) = \sum_{i=1}^{\dim} -x_i \sin\left(\sqrt{|x_i|}\right) \tag{4}$$

Function minimum:

Position for E_n: $(x_1, x_2 \ldots x_n) = (420.969, 420.969, \ldots, 420.969)$
Value for E_n: $y = -418.983 \cdot dim$; Function interval: <−500, 500>.

$$f(x) = \sum_{i=1}^{\dim} \frac{(x_i - s_i)^2}{4000} - \prod_{i=1}^{\dim} \cos\left(\frac{x_i - s_i}{\sqrt{i}}\right) + 1 \tag{5}$$

Function minimum: Position for E_n: $(x_1, x_2 \ldots x_n) = \mathbf{s}$; Value for E_n: $y = 0$
Function interval: <−50, 50>.

$$f(x) = -20 \exp\left(-0.02 \sqrt{\frac{1}{D} \sum_{i=1}^{\dim} (x_i - s_i)^2}\right) - \exp\left(\frac{1}{D} \sum_{i=1}^{\dim} \cos 2\pi (x_i - s_i)\right) + 20 + \exp(1) \tag{6}$$

Function minimum: Position for E_n: $(x_1, x_2 \ldots x_n) = \mathbf{s}$; Value for E_n: $y = 0$
Function interval: <−30, 30>.

$$f(x) = 10 \dim \sum_{i=1}^{\dim} (x_i - s_i)^2 - 10 \cos(2\pi x_i - s_i) \tag{7}$$

Function minimum: Position for E_n: $(x_1, x_2 \ldots x_n) = \mathbf{s}$, Value for E_n: $y = -90000$ (dim 30)
Function interval: <−5.12, 5.12>.

Where s_i is a random number from the 90% range of function interval; $\mathbf{s}$ vector is randomly generated before each run of the optimization process.

The parameter settings for both canonical DE and ChaosDE were obtained based on numerous experiments and simulations (see Table 2). It was experimentally determined, that ChaosDE requires lower values of *Cr* parameter [26] for any used CPRNG. Canonical DE is using the recommended settings [4]. The maximum number of generations was fixed at 1500 generations. This allowed the possibility to analyze the progress of DE within a limited number of generations and cost function evaluations. Experiments were performed in the environment of *Wolfram Mathematica*; canonical DE, therefore, has used the built-in *Wolfram Mathematica* pseudo-random number generator *Wolfram Cellular Automata* representing traditional pseudorandom number generator in comparisons.

Table 2. Parameter set up for ChaosDE and Canonical DE

DE parameter	Value
Popsize	75
F (for ChaosDE)	0.4
CR (for ChaosDE)	0.4
F (for Canonical DE)	0.5
CR (for Canonical DE)	0.9
Dim	30

7 Results

Statistical results for the Cost Function (CF) values are shown in comprehensive Tables 3, 4, 5 and 6 for all 51 repeated runs of DE/ChaosDE. Overall four different benchmark functions and five randomization schemes were utilized.

The bold values within all Tables 3, 4, 5 and 6 depict the best-obtained results; italic values are considered to be similar. Since all experiments used different initialization, i.e., different initial population was generated within each run of Canonical or ChaosDE, statistical comparisons are based on the *Wilcoxon sum-rank test* with significance level 0.05; and performed for the pairs of ChaosDE with CPRNG and identified similar PRNG distribution. The graphical comparisons of the time evolution of average CF values for all 51 runs of five versions of DE/ChaosDE with different randomizations are depicted in Figs. 3, 4, 5 and 6. The notation in Tables and Figures is following: *Burgers/Lozi Map* represents the chaotic CPRNG, whereas *Burgers/Lozi Dist* represents identified distribution PRNG.

Table 3. Simple results statistics for the Canonical DE and ChaosDE; Schwefel's function

DE version	Avg CF	Median CF	Max CF	Min CF	StdDev	p-value
Canonical DE	−5493.26	−5339.34	−4944.96	−6628.4	**440.8144**	-
Burgers Dist	−10375.9	−10360	**−9245.86**	−11722.9	518.8032	0.0123
Burgers Map	**−10793.5**	**−11413.9**	−6787.51	−12328.1	1387.362	
Lozi Dist	−8709.74	−8530.74	−7814.36	−11042.5	661.7437	0.0002
Lozi Map	−9932.46	−9922.25	−8200.06	**−12530.9**	1043.777	

Table 4. Simple results statistics for the Canonical DE and ChaosDE; shifted Rastrigin's function

DE version	Avg CF	Median CF	Max CF	Min CF	StdDev	p-value
Canonical DE	−40188.49	−39264.95	−33629.64	−49994.34	3983.79	-
Burgers Dist	−81465.42	−82256.88	**−74959.67**	−85542.21	2521.65	0.02614
Burgers Map	**−82339.03**	**−83552.39**	−63945.49	**−87977.11**	5262.81	
Lozi Dist	−52641.81	−52722.69	−49731.38	−57271.93	**1851.70**	$7.1634.10^{-5}$
Lozi Map	−57054.66	−56667.65	−52235.56	−62969.12	2927.69	

Table 5. Simple results statistics for the Canonical DE and ChaosDE; shifted Ackley's function

DE version	Avg CF	Median CF	Max CF	Min CF	StdDev	p-value
Canonical DE	3.38E−09	2.68E−09	7.55E−09	9.48E−10	1.74E−09	-
Burgers Dist	4.333288	4.554203	7.464985	2.013873	1.328967	$2.7253.10^{-4}$
Burgers Map	1.43E−06	1.25E−12	1.775137	1.47E−14	0.391209	
Lozi Dist	1.64E−14	***1.47E−14***	3.6E−14	***7.55E−15***	5.26E−15	0.5738
Lozi Map	**1.54E−14**	***1.47E−14***	**2.89E−14**	***7.55E−15***	**4.11E−15**	

Table 6. Simple results statistics for the Canonical DE and ChaosDE; shifted Grienwang function

DE version	Avg CF	Median CF	Max CF	Min CF	StdDev	p-value
Canonical DE	***0***	***0***	***0***	***0***	***0***	-
Burgers Dist	0.525982	0.514041	0.998373	0.098319	0.26012	$1.6627.10^{-4}$
Burgers Map	6.89E−07	1.47E−09	0.15187	0	0.00323	
Lozi Dist	***0***	***0***	***0***	***0***	***0***	1
Lozi Map	***0***	***0***	***0***	***0***	***0***	

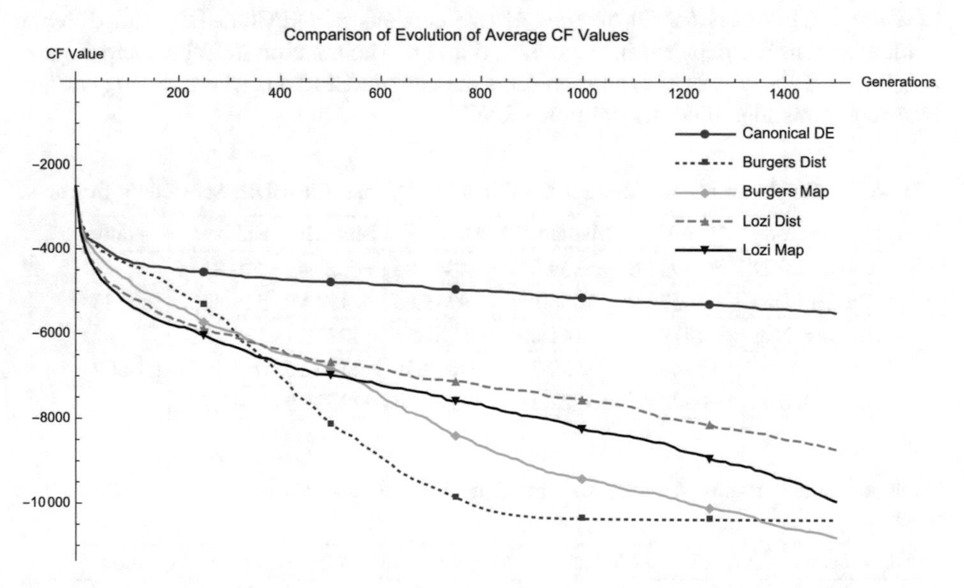

Fig. 3. Comparison of the time evolution of average CF values for the all 50 runs of Canonical DE, and four versions of ChaosDE with different randomization; Schwefel's function.

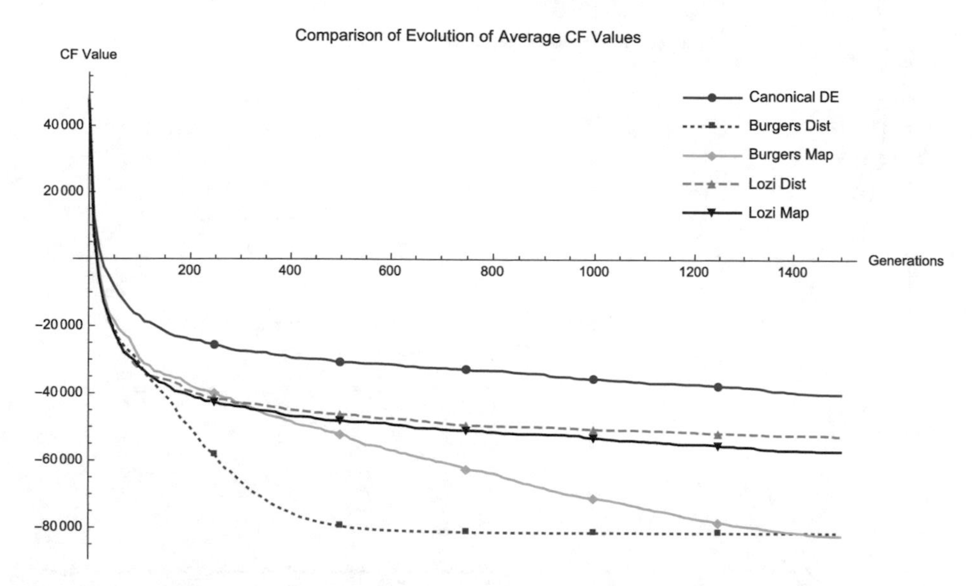

Fig. 4. Comparison of the time evolution of average CF values for the all 50 runs of Canonical DE, and four versions of ChaosDE with different randomizations; shifted Rastrigin's function.

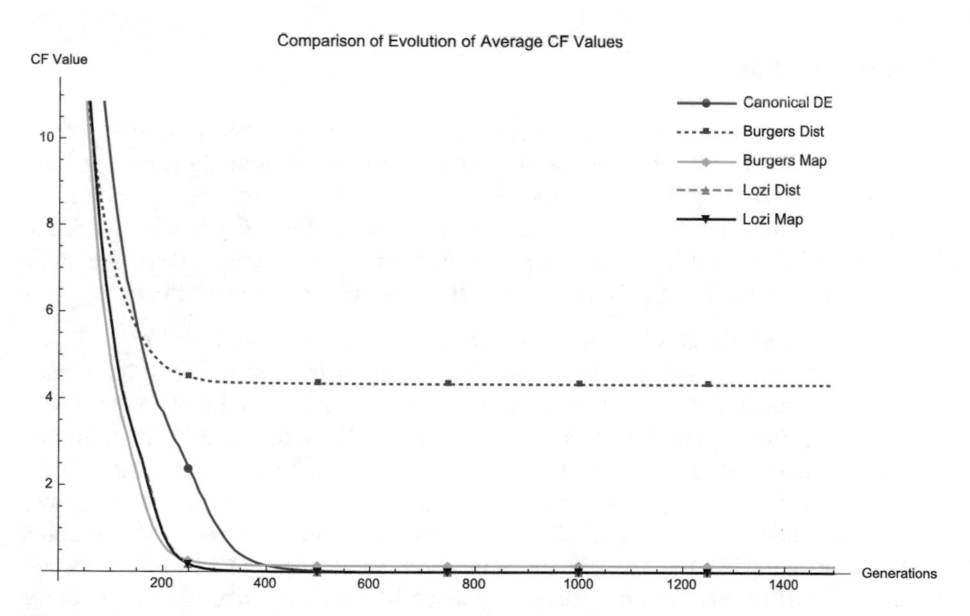

Fig. 5. Comparison of the time evolution of average CF values for the all 50 runs of Canonical DE, and four versions of ChaosDE with different randomization; shifted Ackley's function.

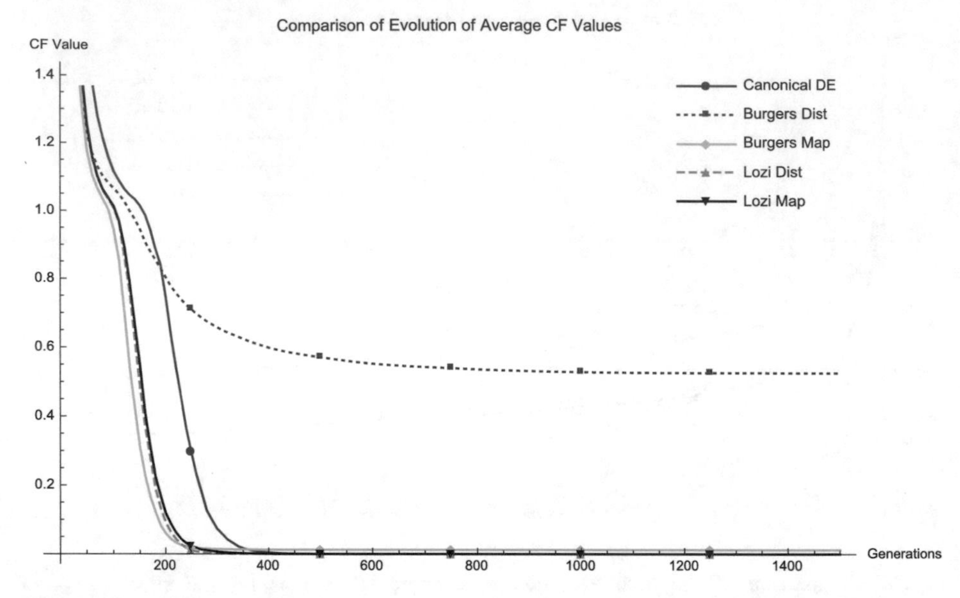

Fig. 6. Comparison of the time evolution of average CF values for the all 50 runs of Canonical DE, and four versions of ChaosDE with different randomizations; shifted Grienwang function.

8 Conclusions

The primary aim of this work is to experimentally investigate the utilization of the various discrete chaotic systems, as the chaotic pseudo-random number generator embedded into DE. Experiments are focused on the extended investigation, whether the different randomization and pseudo-random numbers distribution given by particular PRNG or hidden complex chaotic dynamics providing the unique sequencing is more beneficial to the heuristic performance. The findings can be summarized as:

- Obtained graphical comparisons and data in Tables 3, 4, 5 and 6 and Figs. 3, 4, 5 and 6 support the claim that chaos driven heuristic is more sensitive to the hidden chaotic dynamics driving the selection, mutation, crossover or other processes through CPRNG. The influence of different PRNG randomization (distribution) type is strengthened by the presence of chaotic dynamics and sequencing in the pseudo-random series given by the dynamics of discretized chaotic attractor/flow.
- It is clear that (selection of) the best CPRNGs are problem-dependent. By keeping the information about the chaotic dynamics driving the selection/mutation processes inside heuristic, its performance is significantly different: either better or worse against other compared versions.
- In the first two cases (Schwefel and shifted Rastrigin function – Tables 3 and 4), the performance of ChaosDE was significantly better in comparison with canonical DE. Furthermore, the effect of different PRNG distribution became even stronger with the chaotic dynamics kept inside CPRNG sequences. Lozi map based CPRNG has

given stable better performance than similar identified PRNG. Both Lozi map based PRNG/CPRNG have been outperformed by the utilization of Burgers map based PRNG/CPRNG. An interesting phenomenon has been revealed. The Burgers map based not-chaotic PRNG drives DE to the fast and robust progress towards function extreme (local) followed by premature population stagnation phase. Whereas Burgers map CPRNG with chaotic dynamics secured the continuous development of population towards a global best solution without stagnation.

- The third and the fourth case study (Tables 5 and 6) have given absolutely reversed character of results. Performance of Lozi based CPRNG/PRNG is comparable even with canonical DE (slightly better results for Lozi map CPRNG and Ackley function), whereas the Burgers map based randomization has given worse results. As aforementioned in the previous point, the premature stagnation for PRNG has also occurred here (more considerable), whereas the Burgers map based CPRNG with chaotic dynamics has driven the DE more or less towards the function extreme.
- Since the aim was to investigate the randomization/sequencing of indices selection inside DE, only the most straightforward canonical *DE/Rand/1/Bin* strategy has been utilized in this research. The parameter adjustment/strategy adaptation or ensembles techniques in jDE, EPSDE, SHADE may significantly interact with the dynamics of sequencing (selection) of indices driven by particular not-uniform PRNG/CPRNG.
- Sequencing of pseudo-random numbers and chaotic dynamics hidden inside pseudo-random series can be significantly changed by the selection of chaotic systems, thus to avoid the CF landscape dependency. The simplest way for changing the influence to the heuristic during the run is to swap currently used chaotic system for different one or to change the internal parameters of chaotic systems (Table 1).
- Furthermore many previous implementations of chaotic dynamics into the evolutionary/swarm-based algorithms (not-adaptive/adaptive/ensemble based) showed that it is advantageous since it can be easily implemented into any existing algorithm as a plug-in module.

Acknowledgements. This work was supported by Grant Agency of the Czech Republic - GACR P103/15/06700S, further by the financial support of research project NPU I No. MSMT-7778/2014 by the Ministry of Education of the Czech Republic and also by the European Regional Development Fund under the Project CEBIA-Tech No. CZ.1.05/2.1.00/03.0089, and by Internal Grant Agency of Tomas Bata University under the projects No. IGA/CEBIA-Tech/2017/004.

References

1. Price, K.V., Storn, R.M., Lampinen, J.A.: Differential Evolution: A Practical Approach to Global Optimization, ser. Natural Computing Series. Springer, Berlin (2005)
2. Neri, F., Tirronen, V.: Recent advances in differential evolution: a survey and experimental analysis. Artif. Intell. Rev. **33**(1–2), 61–106 (2010)

3. Das, S., Suganthan, P.N.: Differential evolution: a survey of the state-of-the-art. IEEE Trans. Evol. Comput. **15**(1), 4–31 (2011)
4. Das, S., Mullick, S.S., Suganthan, P.: Recent advances in differential evolution – an updated survey. Swarm Evol. Comput. **27**, 1–30 (2016)
5. Brest, J., Greiner, S., Bošković, B., Mernik, M., Zumer, V.: Self-adapting control parameters in differential evolution: a comparative study on numerical benchmark problems. IEEE Trans. Evol. Comput. **10**(6), 646–657 (2006)
6. Qin, A.K., Huang, V.L., Suganthan, P.N.: Differential evolution algorithm with strategy adaptation for global numerical optimization. IEEE Trans. Evol. Comput. **13**(2), 398–417 (2009)
7. Zhang, J., Sanderson, A.C.: JADE: adaptive differential evolution with optional external archive. IEEE Trans. Evol. Comput. **13**(5), 945–958 (2009)
8. Das, S., Abraham, A., Chakraborty, U., Konar, A.: Differential evolution using a neighborhood-based mutation operator. IEEE Trans. Evol. Comput. **13**(3), 526–553 (2009)
9. Mininno, E., Neri, F., Cupertino, F., Naso, D.: Compact differential evolution. IEEE Trans. Evol. Comput. **15**(1), 32–54 (2011)
10. Mallipeddi, R., Suganthan, P.N., Pan, Q.K., Tasgetiren, M.F.: Differential evolution algorithm with ensemble of parameters and mutation strategies. Appl. Soft Comput. **11**(2), 1679–1696 (2011)
11. Brest, J., Korosec, P., Silc, J., Zamuda, A., Boskovic, B., Maucec, M.S.: Differential evolution and differential ant-stigmergy on dynamic optimisation problems. Int. J. Syst. Sci. **44**(4), 663–679 (2013)
12. Tanabe, R., Fukunaga, A.S.: Improving the search performance of SHADE using linear population size reduction. In: 2014 IEEE Congress on Evolutionary Computation (CEC), pp. 1658–1665. IEEE (2014)
13. Weber, M., Neri, F., Tirronen, V.: A study on scale factor in distributed differential evolution. Inf. Sci. **181**(12), 2488–2511 (2011)
14. Neri, F., Iacca, G., Mininno, E.: Disturbed exploitation compact differential evolution for limited memory optimization problems. Inf. Sci. **181**(12), 2469–2487 (2011)
15. Iacca, G., Caraffini, F., Neri, F.: Compact differential evolution light: high performance despite limited memory requirement and modest computational overhead. J. Comput. Sci. Technol. **27**(5), 1056–1076 (2012)
16. Zamuda, A., Brest, J.: Self-adaptive control parameters' randomization frequency and propagations in differential evolution. Swarm Evol. Comput. **25**, 72–99 (2015)
17. Caponetto, R., Fortuna, L., Fazzino, S., Xibilia, M.G.: Chaotic sequences to improve the performance of evolutionary algorithms. IEEE Trans. Evol. Comput. **7**(3), 289–304 (2003)
18. dos Santos Coelho, L., Mariani, V.C.: A novel chaotic particle swarm optimization approach using Hénon map and implicit filtering local search for economic load dispatch. Chaos, Solitons Fractals **39**(2), 510–518 (2009)
19. Pluhacek, M., Senkerik, R., Davendra, D., Kominkova Oplatkova, Z., Zelinka, I.: On the behavior and performance of chaos driven PSO algorithm with inertia weight. Comput. Math Appl. **66**(2), 122–134 (2013)
20. Pluhacek, M., Senkerik, R., Davendra, D.: Chaos particle swarm optimization with Eensemble of chaotic systems. Swarm Evol. Comput. **25**, 29–35 (2015)
21. Metlicka, M., Davendra, D.: Chaos driven discrete artificial bee algorithm for location and assignment optimisation problems. Swarm Evol. Comput. **25**, 15–28 (2015)
22. dos Santos Coelho, L., Ayala, H.V.H., Mariani, V.C.: A self-adaptive chaotic differential evolution algorithm using gamma distribution for unconstrained global optimization. Appl. Math. Comput. **234**, 452–459 (2014)

23. Senkerik, R., Pluhacek, M., Zelinka, I., Davendra, D., Janostik, J.: Preliminary study on the randomization and sequencing for the chaos embedded heuristic. In: Abraham, A., Wegrzyn-Wolska, K., Hassanien, E.A., Snasel, V., Alimi, M.A. (eds.) Proceedings of the Second International Afro-European Conference for Industrial Advancement AECIA 2015, pp. 591–601. Springer International Publishing, Cham (2016)
24. Zelinka, I.: A survey on evolutionary algorithms dynamics and its complexity – mutual relations, past, present and future. Swarm Evol. Comput. **25**, 2–14 (2015)
25. Sprott, J.C.: Chaos and Time-Series Analysis. Oxford University Press, Oxford (2003)
26. Senkerik, R., Pluhacek, M., Kominkova Oplatkova, Z., Davendra, D.: On the parameter settings for the chaotic dynamics embedded differential evolution. In: 2015 IEEE Congress on Evolutionary Computation (CEC), 25–28 May 2015, pp. 1410–1417 (2015)

Grammatical Evolution for Classification into Multiple Classes

Jiří Lýsek[1(✉)] and Jiří Šťastný[1,2]

[1] Department of Informatics, Mendel University in Brno, Zemědělská 1, 613 00 Brno, Czech Republic
jiri.lysek@mendelu.cz

[2] Institute of Automation and Computer Science, Brno University of Technology, Technická 2, 616 00 Brno, Czech Republic

Abstract. In this contribution the authors deal with classification problems using an approach based on grammatical evolution. The named method is used to create short executable structures which are evolved to classify given input into multiple classes. Resulting structures are usable as computer programs for embedded devices with low computational resources. An universal formula for fitness value calculation of the evolved individual is introduced and an example of planar graphical objects classification in generated image dataset is presented. The presented approach is still applicable for general multi-class classification problems. The results of the proposed method are discussed and examined.

Keywords: Grammatical evolution · Multiclass classification
Fitness function · Object recognition · Learning

1 Introduction

Automatic classification is a task often solved in the field of computer science. Many methods have been introduced to perform this task robustly and quickly. This article shows an non-traditional approach to the classification problem based on grammatical evolution. A demonstration of proposed method is presented on a problem of classification of planar graphical objects into multiple classes. Presented method is suitable for a use-case where an embedded device with small computational power is used - this method can be used to generate a program for such device.

Grammatical evolution and generally all evolutionary methods received large attention in the last decade. The reason is the accessibility of very fast computers with their abilities of parallel processing of tasks and moreover large, scalable, clusters with enormous computing power which allow the evolutionary methods to show fully their potential. This was really not possible earlier, when evolutionary methods were used only to solve smaller optimisation tasks or as a demonstration of another possible approach to some problems.

R. Matoušek (Ed.): MENDEL 2017, AISC 837, pp. 192–207, 2019.
https://doi.org/10.1007/978-3-319-97888-8_17

Evolutionary methods are suitable to use with a large number of parallel processing units as the evaluation of all individuals in population is a computation easily performed in the parallel way. Some variations of discussed methods use multiple populations to extend the diversity of solutions.

Evolutionary methods are not the fastest ones compared to algorithms which are designed to solve one given problem [4]. On the other hand they provide a framework for solving very difficult problems which might not have any other problem-specific optimized solution developed yet [10,15].

The authors presented in their previous work [8] that the capability of evolutionary methods to train object classifiers is not negligible. They experienced problems with interpreting the output of a classifier because it returned only a single numerical value. In this contribution a way to output more than one value from recognition agent is introduced.

2 Survey of Classification Methods

The classification of objects is a very important task. The most complex living organisms perform this task hundreds of times every day. The classification is based on their senses such as smell, touch, hearing, visual perception and taste. Most of organisms are capable of learning new experience and extending the knowledge which can be used to classify new items encountered during their lifetime.

Unfortunately computers or robots which are used do not have this ability and if they are wanted to learn this ability, there must be created a system capable of acquiring information from sensors or other sources, learning knowledge from them and also capable of using the knowledge to classify objects or perform another requested operation [11]. Many learning algorithms were introduced during last decades to solve classification tasks using supervised and even unsupervised learning.

The authors are focused on the classification based on results of evolutionary process, namely in the field of computer vision, so a brief overview of statistical and artificial intelligence methods will be presented. Then the methods based on evolutionary methods with main focus on grammatical evolution will be described.

The classification problems can be divided into two categories: a binary classification and multiple class classification. The difference is that the binary classifier can be asked whether the presented input is or is not an object which the system is build to recognize. A multiple class classifier can also tell you more specific information about the presented object - it is capable of classifying the objects into one of many possible classes.

Binary classifiers can be chained to perform similarly as multiple class classifiers but due to linear chaining they tend to be slower.

The knowledge has to be stored into the classifier system before it can be used. This phase is called a training phase. After the training, when some measure of quality is fulfilled, a so called production phase begins. In this phase the classifier knowledge is usually not modified any more. There are many methods to train different types of classifiers [14].

2.1 Description of Classified Objects

To classify an object using a computer, one should first obtain a description of that object. This part of the classification process is very specific for every application. It depends on the system ability to sense surrounding area, computing power available and speed requirements. Usually we obtain a vector or a matrix of numbers which somehow describe the object to be classified. In some cases, a combination of description methods can be used [18].

Particularly for visual data there are two choices. The program can try to detect the object first using a binary classifier or feature extractor and obtain the description of the localised object. Another way is to present the image data directly on the classifier input, using a sliding window technique. This approach is suitable to use if the objects which are going to be recognized are roughly the same size. The advantage of this approach is that there is no need to develop a special algorithm for object extraction (localisation) and description. Other usage of the sliding window approach is the development of image filters and binary classifiers.

2.2 Traditional Statistical Methods for Classification

These methods generally use some statistical description of a given training set and divide the training set into clusters according to the extracted characteristics. In the production phase, the similarity between clustered training examples and input is measured and according to the result the classification is performed [14]. Any classification based on these methods is very powerful and methods like kNN or SVM are well known. The authors' focus is not aimed at these methods.

2.3 Artificial Intelligence Methods for Classification

Artificial intelligence methods are often used for classification nowadays. The most widely used tools for classification in the field of computer vision are various kinds of artificial neural networks. Mostly the feed forward multiple layer neural networks are used [16]. Lately the deep neural networks made of multiple layers of Boltzmann machines gained large attention.

The advantage of these methods over statistical methods is given by the combination of multiple research fields like neuroscience, evolutionary theories and of course statistics. This combination gives great flexibility and ability to solve hard problems. Usage of these methods is often justified by large scale of given task and time or computing power constraints. Most difficult task is the training phase of a classifier.

2.4 Classification Based on Evolutionary Methods

Evolutionary methods can be generally used in two roles in classification tasks. The first case is the usage of an evolutionary method as a tool to prepare, optimise [2] or cluster [12] the input data. Usually the usage of an evolutionary method in preprocessing step yields an increase in performance of any subsequent technique used. For example in [2] the evolutionary process is used to optimise PCA data for face recognition performed by neural network.

Another case of evolutionary methods usage is to build directly an agent which will be used for classification. This usually involves genetic programming techniques which are capable of storing knowledge into computer programs in form of a tree graph [5].

The authors' research is focused on the usage of grammatical evolution in computer vision. In this field, several innovative approaches which use grammatical evolution can be found.

Very interesting approach is presented in [13] where evolutionary process is used to form feature detectors similar to SIFT [6] or SURF [1]. These descriptors can be used to classify objects based on their similarity with descriptors stored in a database. But in this case, the evolutionary process does not lead to classification itself.

The idea of detection of place of interest can be transformed to direct detection of objects. This approach is used and extended in this article but the idea is not completely novel. Usage of short computer programs for planar object recognition is derived from the work [19] where various coins were used as unknown objects. In this work a computer program is created by grammatical evolution and is used on image input from a sliding window. The differences will be described in a separate section of this article.

Other researchers use similar techniques for detecting objects or features of a certain type in medical, aerial or radar images [7,17].

3 Proposed Method

In this section a proposed method for multiclass single-label classification and an example of grammatical evolution used for the classification of planar graphical objects by this method is presented. Authors' framework developed for evolutionary methods testing is introduced briefly and then the method for the classification itself is described. This method was used in parallel research [9].

3.1 Classifier Agent Architecture

Authors decided to test the approach on a small generated dataset of image data – this dataset will be introduced later. The main idea is to let the grammatical evolution create a moving window classifier, a computer program which is a result of the evolutionary process, similar to [8,19]. The difference is that the authors want to have multiple output values from generated structures. The amount of output values n is set to the value equal to the number of classes plus one. That particular additional output value is for the purpose of background detection. It is important to note that the grammatical evolution, which is a quiet slow process, is used only during the training phase. The result is a small and fast program which is used to analyse any input image by using the sliding window cut-outs as its input. Resulting program is suitable even for embedded devices with limited computational resources. The means used to achieve multiple-value output are described in the Sect. 3.3.

3.2 Classifier Agent Output Processing

The main problem with structures created by grammatical evolution in the past was that the produced programs had only one value on its output (see Eq. 1). The authors and also other researchers tried an approach where the output range of the classifier agent is divided into intervals – these intervals can be hard-coded [19] or adapted during the training process [8]. The first option gives a very hard constraint on the evolutionary process and the second option suffers from possible interval overlap as is demonstrated in Fig. 1.

$$f(\mathit{sample}) \rightarrow \mathit{single\ output\ value} \tag{1}$$

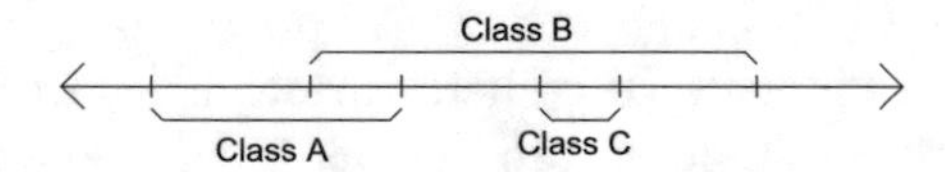

Fig. 1. Single value output of classifier – overlap of output intervals.

3.3 Semicolon and Register Node

To overcome the disadvantage of one output value described in Sect. 3.2 authors proposed the usage of a semicolon node, which can execute both its subtrees in sequence. A register is a place to write or read any numeric value. Thanks to combination of semicolon and register node the result can be composed of multiple output values from a single tree structured program. A multiple value output of a program can be used similarly to a vector output of a neural network. The example of a multiple value output shows the Fig. 2 and Eq. (2).

$$f(\mathit{sample}) \rightarrow (\mathit{class}\ A,\ \mathit{class}\ B, \mathit{class}\ C, \ldots) \tag{2}$$

The semicolon node is very beneficial because the structure of the program remains a tree but the behaviour of the program is sequential. The main reason to form tree structured programs is the implementation of crossover and mutation operators which can simply swap subtrees as is usual in grammatical evolution.

The function of the semicolon node is compared with the branching node in Fig. 3. The input value is set to 0 for this example. The bold line border represents graph nodes which are executed during the program run and dashed lines represent nodes which are not executed.

The semicolon node also creates a new ability to form a new individual just by combining two independent individuals into a new one by assigning them as arguments of a new semicolon node parent. This feature is implemented as an

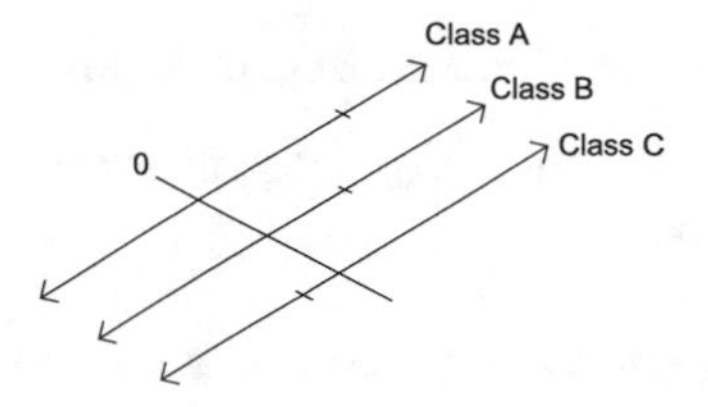

Fig. 2. Multiple value output of classifier – values can be combined into a vector.

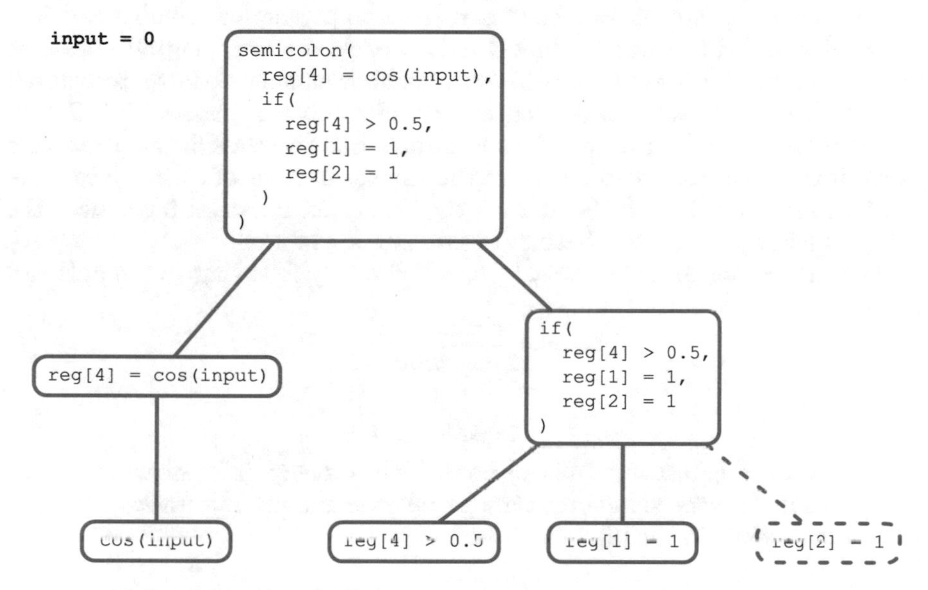

Fig. 3. The difference between semicolon and branching node.

optional behaviour of the crossover operator because two input individuals are needed. The probability of combining two individuals into one was set to 5%.

The registers are used to gather the program result. Usage of registers is also demonstrated in Fig. 3. Register values are used to read and write results of calculations. There are $2 \cdot n$ registers used in our implementation. The first half of registers represents the output vector and the additional registers can be used to store various computational results.

Output Processing. After the agent's response is acquired, all values are divided by their maximal absolute value. The euclidean distance between the normalized program output and each unit vector (versor) in n dimensional space, where n equals to the number of classes plus one, is calculated. Each unit vector is collinear with one of the n axes in that $n+1$ dimensional space – a basis of that space. The closest versor is selected and the position of value 1 of in that versor denotes the recognized object index.

3.4 Fitness Function for Classification Agents

There is proposed general formula (Eq. 3) for fitness value calculation during the process of classifier training.

$$f(member) = w_1 \cdot measureOfSuccess + w_2 \cdot \frac{recognizedCount}{classCount} \tag{3}$$

The resulting value lies inside of the interval $<0, 1>$. This formula generally adds a constraint, through its second part, that no population member can gain the maximum fitness value (1) unless it is capable of classifying all classes of training examples. The *recognizedCount* variable is number of classes recognized by a classifier. This value is divided by overall number of classes *classCount*. The constraint is not hardcoded into a grammar so the classifiers can steadily evolve from those which can recognize the smaller number of classes into more sophisticated ones. It is believed that the semicolon operator introduces the ability of mixing two whole classifier agents into a new one.

The weights for formula in Eq. (3) are defined by following two simple formulas:

$$w_1 = \frac{classCount}{classCount + 1} \tag{4}$$

$$w_2 = \frac{1}{classCount + 1} \tag{5}$$

The *measureOfSuccess* value can be defined in many different ways. It was decided to use a very simple measure of average success rate over all classes defined by formula in Eq. (6).

$$measureOfSuccess = \frac{\sum_{i=1}^{n} \frac{successReward_i}{maxPossibleReward_i}}{n} \tag{6}$$

where $n = classCount$. The resulting value of formula in Eq. (6) has to be from the interval $<0, 1>$.

During the training process, there was defined a successful detection of the object such that the classifier agent output is interpreted as correct inside an area with radius of the half of the sliding window size. The center of this area is the center of an object. The closer to the center, the higher value was added to that success-measuring variable of that particular class $successReward_i$ according to following Eq. (8).

For the whole image there were calculated values of $maxPossibleReward_i$ along with the $successReward_i$ values for each class i. Value $maxPossible Reward_i$ was calculated over all *points* in a circular area with defined radius around each object's center. Value $successReward_i$ was calculated only for points with a successful detection in that same area. The $distance_{point}$ value is a distance from a center point.

$$maxPossibleReward_i = \sum_{all\ points} \frac{1}{1 + distance_{point}} \tag{7}$$

$$successReward_i = \sum_{detected\ points} \frac{1}{1 + distance_{point}} \tag{8}$$

The sliding window classifier agent output is calculated on every fourth pixel (vertically and horizontally) to speed up the fitness value calculations during training.

3.5 Population Diversity Measure

During the evolutionary process, many local extremes are explored by the algorithm. Individuals which represent these local extremes receive a higher fitness value and they are more likely to be selected for reproduction. This can cause that the whole population shares a common form of resulting program and it will be very difficult for the mutation and crossover operators to form a new better solution. To prevent the algorithm of being stuck in one of such extremes, a good diversity measure of a population is needed [3].

Levenshtein string distance between all individuals in the population was measured. Only those were included into the new population which string distance was higher than certain selected threshold. Only individuals with similar fitness values were compared as the calculation of string distance is quite complex.

During experitmets was used a minimal Levenshtein distance value of 50 in the beginning with linear decrease to 10 in the end of the process. This decrease enables the process to explore a wide variety of program forms in the beginning (exploration phase) with possibility to focus on one certain form at the end (exploitation phase). These two values were set experimentally.

3.6 Invariance to Scale, Rotation and Side Flip

The scale invariance is lost due to the moving window architecture. Only suitable workaround is to change the size of the moving window or to resize the input image. The scale invariance is not always important because the size of observed objects and camera distance is given in many applications.

The system is invariant to object rotation and side flip thanks to the grammar terminals which are responsible for the extraction of statistical characteristics of planar areas. These terminals allow only the extraction of characteristics of circular areas so the rotation or side flip of the object is irrelevant.

4 Example of Usage

4.1 The Framework Used

For the authors' experiments there was created an environment in Java programming language where different algorithms based on evolutionary methods can be implemented, executed, tuned and measured. For each task a predefined structure of object classes is used. These classes implement individual, population and problem definition.

The authors' framework is capable of using multiple processor cores. One of the reason for selecting Java programming language was its ability to work well with multiple threads thanks to build-in synchronization services. The framework was written entirely from scratch with only minimum number of third party libraries.

An important advantage of this framework is that the grammatical evolution used not only to produce a string representation of an individual from the chromosome. During the translation process of a chromosome, an executable tree graph of functional nodes is built directly. This allows the authors to leave out the relatively slow parser which would normally be needed to translate the chromosome translation string result into an executable form. A parser is also implemented for the ability to execute evolved individuals separately but it is not used during the main algorithm execution.

4.2 Grammar

The grammar used for planar object detection is presented in the Table 1. The symbol n represents the number of registers.

Table 1. Grammar used for planar objects classification.

<table>
<tr><th>Non-terminal</th><th></th><th>Rewrite set</th></tr>
<tr><td><expr></td><td>::=</td><td><register_write> | semic(<expr>,<expr>) | if(<logic>,<expr>,<expr>)</td></tr>
<tr><td><register_write></td><td>::=</td><td>rWrite0(<math>) | ... | rWrite$n$ (<math></td></tr>
<tr><td><math></td><td>::=</td><td><number> | <math_fun>(<math>) | <matrix_stats>(<matrix>) | <register_read>() | <math_op>(<math>,<math>)</td></tr>
<tr><td><logic></td><td>::=</td><td>lgtn(<math>,<math>) | smtn(<math>,<math>)</td></tr>
<tr><td><matrix></td><td>::=</td><td>iIm | iSob | iTh | <matrix_op>(<matrix>,<matrix>) | <matrix_op_simple>(<matrix>,<math>) | <matrix_fun>(<matrix>)</td></tr>
<tr><td><number></td><td>::=</td><td>-20.0 | -19.5 | ... | 19.5 | 20.0</td></tr>
<tr><td><math_op></td><td>::=</td><td>add | sub | mul | pdiv</td></tr>
<tr><td><math_fun></td><td>::=</td><td>pow | sqrt | log | exp</td></tr>
<tr><td><matrix_op></td><td>::=</td><td>mAdd | mSub | mMul | mDiv</td></tr>
<tr><td><matrix_op_simple></td><td>::=</td><td>mAddC | mSubC | mMulC</td></tr>
<tr><td><matrix_fun></td><td>::=</td><td>mPow | mSqrt | mLog | mExp</td></tr>
<tr><td><matrix_stats></td><td>::=</td><td>mAvg0_3 | mAvg0_5 | mAvg0_10 | mAvg0_15 | mAvg0_20 | mAvg3_5 | mAvg5_10 | mAvg10_15 | mAvg15_20 | mSum0_3 | mSum0_5 | mSum0_10 | mSum0_15 | mSum0_20 | mSum3_5 | mSum5_10 | mSum10_15 | mSum15_20</td></tr>
<tr><td><register_read></td><td>::=</td><td>rRead0 | ... | rRead$n$</td></tr>
</table>

In Table 2, the grammar terminal symbols function is explained. The input and output data type is also listed.

Table 2. Grammar terminals explanation.

Terminal	Explanation	Input Output
semic(*arg1*, *arg2*)	Execute sequentially *arg1* and *arg2*	in: any out: none
if(*logic*, *arg1*, *arg2*)	If the *logic* expression evaluates to true, execute *arg1*, otherwise execute *arg2*	in: boolean, any, any out: none
lgtn(*arg1*, *arg2*), smtn(*arg1*, *arg2*)	Comparison function larger than and smaller than	in: number, number out: boolean
iIm	Input node – return grayscale image	in: none out: matrix
iSob	Input node – return image of edges detected by Sobel edge detector	in: none out: matrix
iTh	Input node – return thresholded image	in: none out: matrix
add(*arg1*, *arg2*), sub(*arg1*, *arg2*), mul(*arg1*, *arg2*), pdiv(*arg1*, *arg2*)	Basic mathematical operations: addition, subtraction, multiplication, division. Division by zero results into zero	in: number, number out: number
pow(*arg*), sqrt(*arg*), log(*arg*), exp(*arg*)	Basic mathematical functions. Square root of negative number results into 0. Logarithm of values smaller or equal to 0 results into 0	in: number out: number
mAdd(*arg1*, *arg2*), mSub(*arg1*, *arg2*), mMul(*arg1*, *arg2*), mDiv(*arg1*, *arg2*)	Basic mathematical operations on individual matrix cells. Division by zero results into zero	in: matrix, matrix out: matrix
mAddC(*arg1*, *arg2*), mSubC(*arg1*, *arg2*), mMulC(*arg1*, *arg2*)	Basic mathematical operations on individual matrix cells	in: matrix, number out: matrix
mPow(*arg*), mSqrt(*arg*), mLog(*arg*), mExp(*arg*)	Basic mathematical functions on matrix cells. Square root of negative number results into 0. Logarithm of values smaller or equal to 0 results into 0	in: matrix out: matrix
rWrite$n$(*arg*)	Store result of *arg* into register n	in: number out: none
rRead$n$()	Read value from register n	in: none out: number
mAvgm_n(*arg*)	Calculate average value of pixels in circular area between radii m and n from matrix *arg*	in: matrix out: number
mSumm_n(*arg*)	Calculate sum of pixel values in circular area between radii m and n from matrix *arg*	in: matrix out: number

4.3 Parameters of Grammatical Evolution

Evolutionary process has many parameters. There are presented those most important settings of the grammatical evolution process used in Table 3.

Table 3. Parameters of evolutionary process.

Parameter	Value	Explanation
Epochs	150	Number of iterations of whole algorithm
Population size	150	Number of members in population
Elitism	Enabled/10%	The best member of population in current epoch is always cloned into next epoch's population. This feature is enabled after certain number of iterations
Crossover rate	80%	The probability that two selected individuals will produce offsprings instead of moving themselves into new population without modification
Semicolon combination	5%	Two selected individuals are combined into a new one using the semicolon operator. The order of subtrees is selected randomly
Mutation rate	10%	The probability that the new individual will mutate after crossover
Selection type	Tournament/3	The selected tournament selection strategy selects three random members of the population and the best one is selected for reproduction
Chromosome length	80	Maximal length of the chromosome. If the grammar requires more genes to the translate chromosome into a program, this translation becomes invalid and such an individual is removed from the population
Register amount	12	The number of registers which the program has access to. First 6 are used as the program output

4.4 Input Image Preprocessing

There are two preprocessing steps. The first is a simple conversion of an image into the grayscale. This step is performed because it significantly decreases the amount of data to be processed. The resulting grayscale image is supplied as an output value of particular grammar terminal. The second step is the edge detection with Sobel operator and thresholding. This step is performed to supply an image of edges and a binary image to other grammar terminals.

4.5 Planar Objects

Here is presented the set of five classes of objects used for training and testing, similar objects were used previously in [16] with a neural network based classifier. The training images were randomly rotated. The testing image was one large image with many objects in random position present. The objects were not overlapping and the size was always the same. All used objects are displayed in Fig. 4.

An automated tool was created to generate images with training and testing objects in different positions. Along with the training and testing image a file with information about the center point position and classification for each object was supplied. Generated objects' center points were used during training and evaluation.

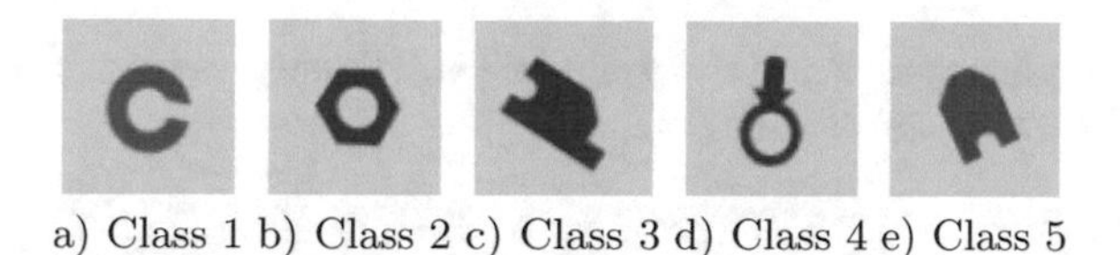

a) Class 1 b) Class 2 c) Class 3 d) Class 4 e) Class 5

Fig. 4. Testing objects.

The size of the sliding window was set to a value of 41×41 pixels in this test scenario. This size covers the area which any of the planar objects can occupy in any rotation. The value of the window size has to be an odd number so you can easily get the coordinate of a center pixel.

5 Results

Each run of grammatical evolution produces a bit different results. Two well performing classifier agents (see Figs. 5 and 6) were selected from independent runs and their performance is reported in this section. The development of the fitness value is also presented in Fig. 7 for both selected individuals.

The authors executed the algorithm 50 times with settings presented in Table 3. The fitness values of the best individuals were usually higher than 0.65.

```
semic(semic(rWrite4(mSum10_15(mDiv(iSob,mSub(iIm,iTh)))),rWrite2(log(
add(mSum5_10(iSob),mSum10_15(mDiv(iSob,mSub(iIm,iTh))))))),if(lgtn(5.
5,mSum5_10(iSob)),semic(rWrite5(pow(log(mSum5_10(mDiv(iIm,mSub(iIm,iT
h)))))),rWrite3(mSum10_15(mDiv(iSob,mDiv(iSob,mSub(iIm,iTh)))))),rWri
te1(mAvg15_20(iTh))))
```

Fig. 5. Program 1 – fitness value 0.80.

```
semic(if(lgtn(16.5,mSum5_10(iSob)),if(lgtn(5.5,mSum5_10(iSob)),semic(
if(lgtn(5.5,mSum5_10(iTh)),rWrite3(pdiv(mul(5.5,mSum15_20(iSob)),mAvg
0_20(iSob))),rWrite0(16.5)),rWrite5(10.5)),rWrite0(-7.5)),rWrite4(mAv
g0_20(iSob))),rWrite2(mAvg5_10(mExp(mDiv(mSubC(mSubC(iIm,mSum15_20(iS
ob)),mSum15_20(iSob)),iSob)))))
```

Fig. 6. Program 2 – fitness value 0.70.

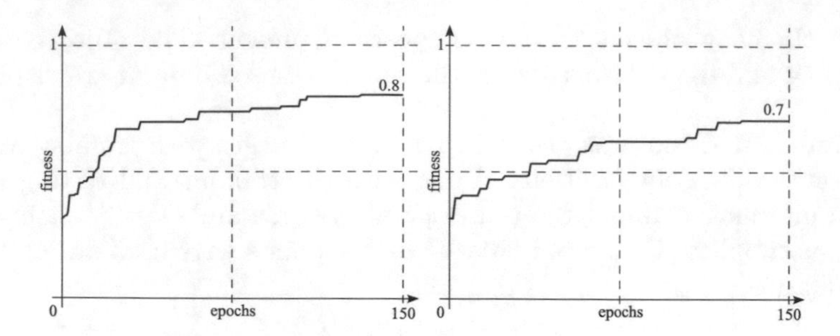

Fig. 7. Fitness value graphs for two runs of evolutionary process which resulted in program 1 (left) and program 2 (right).

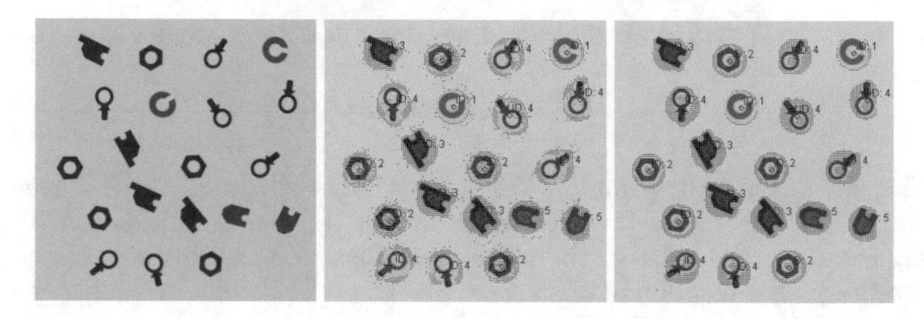

Fig. 8. Testing scene (left) and results of detections by program 1 (middle) and program 2 (right).

For testing purposes the step of the sliding window was set to the value of 2 pixels. On presented testing (Fig. 8 left) image (320 by 320 pixels) it means 19600 points had to be evaluated by the detector agent. The margins of the image are not processed (20 pixels on each side). Green dots represent correct classifications of objects. Correctly classified background is not highlighted. Red dots are errors (either objects or background). Blue dots are correct classifications in larger distance from the object centre.

Program 1 (in Fig. 5) with the fitness value of 0.8 detected correctly 15538 points out of 19600 (in Fig. 8 middle).

Program 2 (in Fig. 6) with the fitness value of 0.7 detected correctly 15569 points out of 19600 (in Fig. 8 right).

There are quite a lot of misclassification errors around the objects borders. These can be simply filtered out using thresholding to obtain the source image binary mask. Another way is to use two pass approach with a separate detector and classifier agent like in [8].

The first program, although it has a higher fitness value, performs a bit worse in the classification than the second one. The higher fitness value is caused by calculating the detections score according to the distance as defined in Eq. (8).

The result of the classification can be obtained by analysing the blobs of dots similarly to [8]. The blob is classified into a class with the most frequent occurrence. It is good to note that the red dots are often randomly assigned to different classes and the blob analysis can easily filter the right result.

6 Conclusion

A great advantage of grammatical evolution is not only the capability to find a classifier program but also that the program has an optimised structure. The grammatical evolution also provides a framework to solve many kinds of problems by its ability to produce any result based only on a supplied grammar. This is a big difference in comparison to the neural network which structure is given in the beginning and cannot change during the training process (the structure of neural network can be optimised but the training process has to start from scratch).

These advantages are, on the other hand, compensated by the time needed to find a good classifier program which is substantially longer than time needed to train a neural network. The speed of the grammatical evolution process can be increased by using more parallel computing cores. There are other options such as simplifying the grammar and limiting the maximal depth of the classifier tree program. But these modifications can affect negatively the result of the process. The grammar is crucial for the ability of the process to find a good classification agent and optimise it. By means of adding new grammar terminals or removing unnecessary ones there can be achieved substantial improvement in the final result.

The authors of the article demonstrated the proposed method on a recognition of planar graphical objects in a simulated technological scene. The final product of the learning process is a very fast moving window agent. This program is capable of detecting randomly rotated or flipped objects of the same scale in 2D images and can be used in some real application, for example a detection of objects in automatic sorting machines.

The execution of the resulting classification agent can be performed on parallel hardware by dividing the image area into smaller pieces.

Thanks to an independent process of chromosome translation through grammar into a string and an executable tree graph, even the grammar can be defined to produce programs in a selected target programming language. The authors used PHP language and Matlab environment often for agent testing. The final form of the evolved classifier is suitable to be executed in less sophisticated environment such as micro-controllers or PLCs.

In comparison to the authors' previous work [8] they achieved the defined goal: their recognition agents return multiple output values. The results of object recognition could be improved further by using two pass approach. Another possible usage of the proposed process is the construction of preprocessing filters which could extract image features for a subsequent classifier.

References

1. Bay, H., Ess, A., Tuytelaars, T.: Speeded-up robust features SURF. Comput. Vis. Image Underst. **110**(3), 346–359 (2008). ISSN 1077-3142
2. Boháč, M., Lýsek, J., Motyčka, A., Cepl, M.: Face recognition by means of new algorithms. In: 17th International Conference on Soft Computing, MENDEL 2011, pp. 324–329. Brno University of Technology, Czech Republic (2011). ISBN 978-80-214-4302-0
3. Burke, E.K., Gustafson, S., Kendall, G.: Diversity in genetic programming: an analysis of measures and correlation with fitness. IEEE Trans. Evol. Comput. **8**(1), 47–62 (2004). ISSN 1089-778X
4. Čížek, L., Šťastný, J.: Comparison of genetic algorithm and graph-based algorithm for the TSP. In: 19th International Conference on Soft Computing, MENDEL 2013, pp. 433–438. Brno University of Technology, Czech Republic (2013). ISSN 1803-381, ISBN 978-80-214-4755-4
5. Koza, J.R.: Genetic Programming: On the Programming of Computers by Means of Natural Selection. The MIT Press (1992). ISBN 0-262-11170-5
6. Lowe, G.D.: Distinctive image features from scale-invariant keypoints. Int. J. Comput. Vis. **60**(2), 91–110 (2004). ISSN 0920-5691
7. Liu, J., Tang, Y.Y., Cao, Y.C.: An evolutionary autonomous agents approach to image feature extraction. IEEE Trans. Evol. Comput. **1**(2), 141–158 (1997). ISSN 1089–778X
8. Lýsek, J., Šťastný, J., Motyčka, A.: Object recognition by means of evolved detector and classifier program. In: 18th International Conference of Soft Computing, MENDEL 2012, pp. 82–87. Brno University of Technology, Czech Republic (2012). ISSN 1803-3814, ISBN 978-80-214-4540-6
9. Lýsek, J., Šťastný, J.: Classification of economic data into multiple classes by means of evolutionary methods. In: Acta Universitatis Agriculturae et Silviculturae Mendelianae Brunensis, vol. 61, no. 7, pp. 2445–2449 (2013). ISSN 1211-8516
10. Marić, M.: An efficient genetic algorithm for solving the multi-level uncapacitated facility location problem. Comput. Inform. **29**(2), 183–201 (2010). ISSN 1335-9150
11. Munk, M., Kapusta, J., Švec, P., Turčáni, M.: Data advance preparation factors affecting results of sequence rule analysis in web log mining. E & M Ekonomie a Manag. **13**(4), 143–160 (2010). ISSN 1212–3609
12. Omran, M.G.H., Engelbrecht, A.P., Salman, A.: Differential evolution methods for unsupervised image classification. In: The 2005 IEEE Congress on Evolutionary Computation, Evolutionary Computation, vol. 2, pp. 966–973 (2005). ISBN 0-7803-9363-5
13. Perez, B.C., Olague, G.: Evolutionary learning of local descriptor operators for object. In: Proceedings of the 11th Annual Conference on Genetic and Evolutionary Computation, GECCO 2009, USA, pp. 1051–1058 (2009). ISBN 978-1-60558-325-9
14. Škorpil, V., Šťastný, J.: Comparison methods for object recognition. In: Proceedings of the 13th WSEAS International Conference on Systems, Rhodos, Greece, pp. 607–610 (2009). ISBN 978-960-474-097- 0
15. Stanimirović, Z.: A genetic algorithm approach for the capacitated single allocation P-Hub median problem. Comput. Inform. **29**(1), 117–132 (2010). ISSN 1335-9150
16. Šťastný, J., Škorpil, V.: Analysis of algorithms for radial basis function neural network. In: Personal Wireless Communications, no. 1, pp. 54–62. Springer (2007). ISSN 1571-5736, ISBN 978-0-387-74158-1

17. Tackett, W.A.: Genetic programming for feature discovery and image discrimination. In: Proceedings of the 5th International Conference on Genetic Algorithms, pp. 303–311. Morgan Kaufmann Publishers Inc., San Francisco (1993). ISBN 1-55860-299-2
18. Zhang, D., Lu, G.: Review of shape representation and description techniques. Pattern Recogn. **37**(1), 1–19 (2004). ISSN 1054-6618
19. Zhang, M., Ciesielski, V.B., Andreae, P.: A domain-independent window approach to multiclass object detection using genetic programming. EURASIP J. Appl. Sig. Process. **2003**, 841–859 (2003). ISSN 1687-6180

Neural Networks, Machine Learning, Self-organization, Fuzzy Systems, Advanced Statistic

A Spiking Model of Desert Ant Navigation Along a Habitual Route

Przemyslaw Nowak[1(✉)] and Terrence C. Stewart[2]

[1] Institute of Information Technology, Lodz University of Technology, Lodz, Poland
przemyslaw.nowak@p.lodz.pl

[2] Centre for Theoretical Neuroscience, University of Waterloo, Waterloo, ON, Canada
tcstewar@uwaterloo.ca

Abstract. A model producing behavior mimicking that of a homing desert ant while approaching the nest along a habitual route is presented. The model combines two strategies that interact with each other: local vector navigation and landmark guidance with an average landmark vector. As a multi-segment route with several waypoints is traversed, local vector navigation is mainly used when leaving a waypoint, landmark guidance is mostly used when approaching a waypoint, and a weighted interplay of the two is used in between waypoints. The model comprises a spiking neural network that is developed based on the principles of the Neural Engineering Framework. Its performance is demonstrated with a simulated robot in a virtual environment, which is shown to successfully navigate to the final waypoint in different scenes.

Keywords: Spiking neural networks · Neural Engineering Framework · Insect navigation

1 Introduction

Desert ants are solitary foragers, which exhibit amazing navigational skills [15]. When searching for food, they can go as far as 200 m away from their nest in relatively featureless terrain and then reliably return home [9]. Unlike other ants, however, these ants do not use pheromones to mark their trails because due to the high desert temperatures, those would evaporate too quickly. Consequently, despite their miniature brains, desert ants have developed a number of sophisticated mechanisms, including path integration and landmark guidance, to meet their navigational demands [9,15]. Moreover, they utilize different strategies depending on whether they are in familiar or unfamiliar terrain and whether the environment is mostly featureless or cluttered [14], which altogether has made them become model organisms for studying insect navigation [15].

Although much has been discovered about these mechanisms and strategies in recent years, the neural substrate is still largely unknown [1,9]. Here, we present a model using a spiking neural network (SNN) developed based on the principles of the Neural Engineering Framework (NEF), which produces behavior

R. Matoušek (Ed.): MENDEL 2017, AISC 837, pp. 211–222, 2019.
https://doi.org/10.1007/978-3-319-97888-8_18

similar to that of a homing ant approaching the nest along a habitual route. We demonstrate the performance of the model using a simulated robot in a virtual environment.

The rest of the paper is structured as follows: Sect. 2 provides an overview of desert ant navigation, Sect. 3 introduces the Neural Engineering Framework, the proposed model is described in Sect. 4 and preliminary results obtained from the model are presented in Sect. 5, and the final Sect. 6 is devoted to discussion and conclusions.

2 Desert Ant Navigation

Desert ants employ several strategies for navigation: path integration (also termed vector navigation), landmark guidance, and systematic search [15]. These strategies are utilized adaptively depending on the actual circumstances and interact with one another [15]. Path integration requires combining compass information, which is mainly provided by the polarization compass mechanism, with odometer information, which is mostly based on the pedometer mechanism (i.e., counting steps) and to a much lesser extent exploits optic flow [15]. Importantly, vector navigation can be broken down into global and local vector navigation: the global vector, obtained through integration of directions taken and distances covered since leaving the nest, maintains a continuously updated estimate of the direction and distance to the nest from the current location, whereas local vectors are associated with particular locations in the environment and store direction to the next waypoint on the route [6]. The next strategy, landmark guidance, involves "labeling" places using visual landmarks observed at these places from particular vantage points [14]. Those stored views can be later recalled to guide the ant to the corresponding places, using view-matching mechanisms that are retinotopically organized [14]. Several methods for such view-matching mechanisms have been proposed, the snapshot model being the canonical one, and its various variants such as the average landmark vector model later derived from it [9]. In the snapshot model, the current retinal image is compared with a stored one in terms of apparent sizes and bearings of individual landmarks, without the need for actual identification of the landmarks themselves, and the resulting differences collectively determine movement aimed at reducing the mismatch such that the two images coalesce. The average landmark vector (ALV) model is a parsimonious version, in which it is not individual landmarks that are compared, but average vectors defined by those landmarks all together. In other words, bearings to all landmarks in the view are lumped into a single two-dimensional ALV and only the ALVs of the current and stored views are compared [9]. Finally, systematic search strategy is employed when the ant has arrived close to the nest using its path integrator, but due to the inevitable cumulative errors in integration it is not exactly where it should be and thus begins to systematically search the vicinity of its current location to find the nest entrance. This strategy is mostly exploited in featureless terrain, where landmarks and view-matching mechanisms cannot be used for precise guidance to the nest [15].

When traveling over larger areas in featureless terrain, desert ants can rely only on the global vector navigation; on the other hand, in cluttered environments navigation using landmarks can override the global vector navigation, although path integration is continuously carried out [14]. Routes that are followed many times can become habitual; in such cases ants learn sequences of landmark scenes along the route and associate them with remembered heading directions (i.e., local vectors) [5,6], effectively partitioning the route into separate successive segments that can point in different directions [6]. Recognizing a particular scene triggers the associated local vector, which guides the ant to the visual catchment area of the next waypoint on the route [14]. An ant can store in its memory a number of routes and retrieve these memories correctly according to the current context [15]. However, this traditional approach to navigation along habitual routes has recently been challenged by a different, "view-familiarity" model, which postulates that the purpose of the movement aimed at aligning the current view with a stored one is not to ensure arrival at the corresponding waypoint, but rather to orient the ant in the correct direction for forward motion that will follow; thus, in the "view-familiarity" approach local vectors are actually eliminated [2].

To explain how navigation in desert ants is organized from the computational perspective, a number of models have been proposed, including those addressing homing in on a single goal using view-matching methods based on local landmarks [9], those concerning long-range navigation employing waypoints linked in a sequence and approached using view-matching methods [11], those concerning long-range navigation without explicit waypoints and using "view-familiarity" approach along with rotational scanning [1,2], and those dealing with long range navigation combining global and local vectors [7]; some of these models also implemented learning [1,2,11], utilized spiking neural networks [1], and were even applied to real robotic platforms [9,11].

3 Neural Engineering Framework

To create our SNN model, we used the Neural Engineering Framework [8]. This is a method for taking a high-level description of a desired algorithm and converting it into a set of spiking neurons that approximate that algorithm. In our case, we used standard Leaky Integrate-and-Fire neurons, connected with exponential synapses.

In the NEF, an algorithm is defined in terms of variables and functions on those variables. Variables are represented in a distributed manner by groups of neurons. For example, a group of 100 neurons may *encode* a 2-dimensional numerical value (such as the x and y coordinates of a local vector) in their firing pattern. Each neuron within this group has some particular preferred value for which it will fire the fastest. For the model discussed here, we choose these preferred stimuli (and thus the "tuning curve" for the neuron) randomly, but in future work it would be constrained based on neural data. The NEF also specifies the inverse operation, *decoding* variables, that is recovering their values from the observed firing patterns.

To implement a particular algorithm, the NEF forms connections between groups of neurons. Consequently, whenever two groups of neurons are connected, we define the numerical *function* that we want to have computed. For example, if one group of neurons represents x, y components of the velocity command and we have another group of neurons that we want to represent the steering angle θ, then we want to connect these two groups of neurons such that $\theta = tan^{-1}(y/x)$. The NEF provides a method for finding the synaptic connection weights between these two groups of neurons that best approximate this function. That is, if there are 100 neurons in the first group and 100 neurons in the second group, the NEF generates a 100×100 matrix such that, if the first group of neurons is stimulated with the pattern of activity for any x, y value, the second group will be driven to fire with the correct corresponding θ value.

Of course, these neurons will only *approximate* the desired function. In order to determine the actual behavior of the system, given this approximation (and given the variability caused by the spiking neuron activity itself), we need to run the neural system and observe its behavior. To do this, we use the software toolkit Nengo [3], which also includes a built-in implementation of the NEF methods.

Given this framework, everything in the model must be described as continuous real-valued variables and functions on those variables (The NEF also uses recurrent connections, which can approximate differential equations on those variables.) However, this means that it is not obvious how to implement discrete items in such a framework. For example, how can we have one group of neurons representing which waypoint the ant has reached most recently? For this, we take the approach of claiming that the neurons in this group represent some high-dimensional vector space (e.g., 32 dimensions), and we randomly choose points in that space to represent each waypoint. In other words, each waypoint is assigned a particular randomly chosen 32-dimensional vector. With this approach, symbol-like functions can be implemented with the NEF. For example, we can define a function where the input is the 32-dimensional vector representing which waypoint has been reached most recently, and the output is the next waypoint.

4 Model

Our SNN model is based on the classic approach that combines view-matching recognition of subsequent waypoints along a habitual route with expression of local vectors guiding to the visual catchment area of the successive waypoint. Rather than using the original snapshot model to implement the view-matching mechanism, we utilize the ALV model for this purpose. The SNN is coupled in a closed-loop manner to a simulated robot in a virtual environment and controls its movement. The robot model is based on the popular robot platform Pioneer P3-DX and is simulated using V-REP simulator [10].

In our model, only the critical parts of the whole system are implemented neurally, with the remaining components being computed numerically. In particular, we do not implement the neural circuitry of the ant compass and odometer

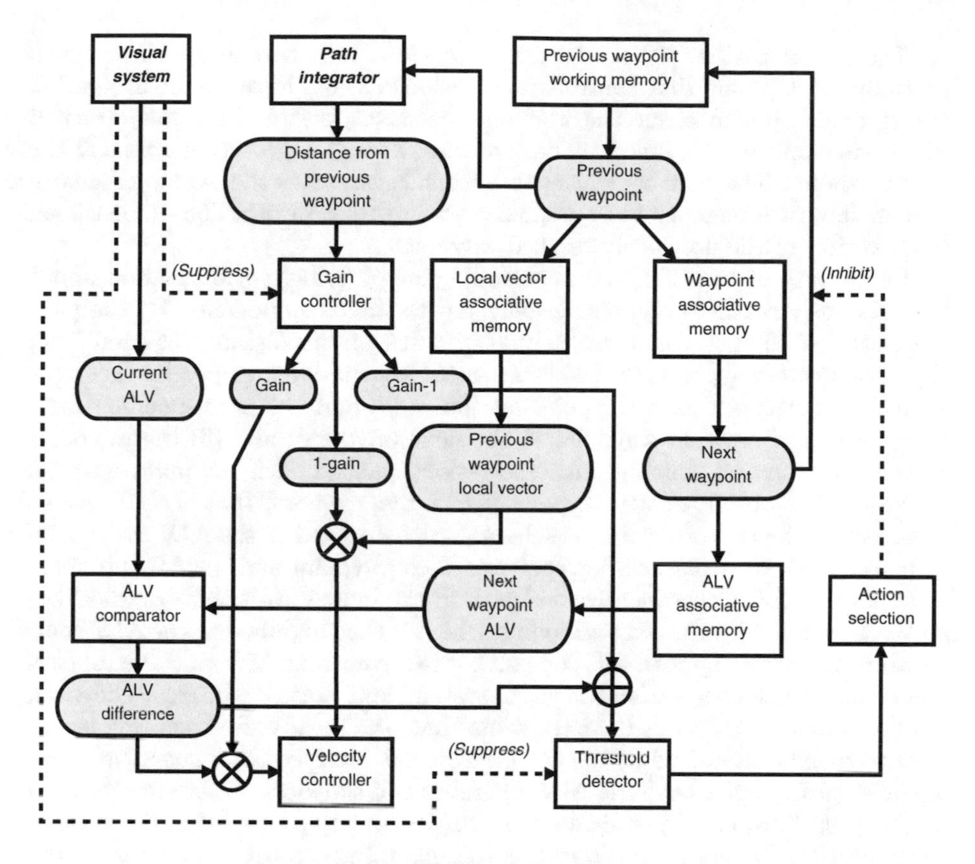

Fig. 1. Diagram of the model

mechanisms; instead, robot position and orientation are obtained directly from the simulator, while distance traveled from the previous waypoint is computed explicitly. Also, vision is not implemented, but substituted with direct calculations of ALVs based on the current robot pose and the position of landmarks within a predefined range. Finally, learning the habitual route is not considered; all the waypoints as well as the local vectors and ALVs associated with them are predetermined.

Figure 1 shows a diagram of the model. Rectangular boxes represent components of the system, while rounded boxes depict information that is computed by or transmitted between components. Bold italic font in rectangular boxes indicates that the corresponding components perform computations explicitly, to distinguish them from those that are implemented neurally.

There are two inputs to the system: vision and path integration. Vision obtains from the simulator positions of landmarks that are within the predefined range from the robot and calculates the currently perceived ALV in egocentric coordinates. Path integration obtains from the simulator robot current position and orientation and computes distance covered from the previously reached waypoint.

The neural part of the model can be divided into two subsystems, which partially overlap: one that controls robot velocity using local vector navigation and landmark guidance, and the other one that is responsible for keeping track of which waypoint has most recently been reached as well as detecting arrival at the next waypoint. The central element to both subsystems is the working memory that maintains a memory trace, represented in the form of a high-dimensional (HD) vector, of the previously reached waypoint.

In the case of the subsystem controlling robot velocity, information about the previously reached waypoint is delivered to three components: (1) the path integrator, so that it could properly keep track of the distance covered from the most recent waypoint; (2) the local vector associative memory, which stores associations between each waypoint on the route and the corresponding local vector leading from that waypoint to the next waypoint; and (3) the waypoint associative memory, which stores associations pairing each waypoint with its successor. The output of the waypoint associative memory, that is a HD vector representing the next waypoint, is subsequently conveyed to the ALV associative memory, which stores associations between each waypoint and the ALV observed at that waypoint. Consequently, this path in the model leads to retrieval of the ALV associated with the next waypoint, which is then supplied to the ALV comparator. The other input to this comparator is the current ALV that the robot is perceiving, which comes from the visual system, and thus the difference between the two ALVs is determined. At the same time, the local vector pointing to the next waypoint is obtained as the output from the local vector associative memory, and subsequently both the ALV difference and the local vector are combined with appropriate gain adjustments into a single velocity command in the velocity controller, which finally translates this command into motor commands for the robot wheels.

Gain adjustments applied to the ALV difference and to the local vector are a very important element in the model. Its purpose is to differentially weight contributions of the two signals depending on how far the robot has moved away from the previous waypoint and thus how close it has possibly come to the next one. Specifically, if the robot is near the previous waypoint, its velocity is mostly determined by the local vector associated with that previous waypoint and pointing to the next one; in turn, when the robot is far off from the previous waypoint and hence likely approaches the next one, it is the ALV difference that mainly dictates velocity. This design allows the robot to properly home in on the next waypoint while avoiding possibly detrimental confusion while going past a waypoint. Indeed, when the robot is leaving a waypoint that has just been reached in pursuit for the next one, landmarks belonging to that just reached waypoint are still visible and thus may confound the ALV difference, resulting in wrong velocity signals.

The other neural subsystem, which keeps track of which waypoint has most recently been reached and detects arrival at the next waypoint, shares some components with its counterpart controlling the robot velocity. Because detection of arrival at a waypoint is based on a threshold mechanism, the ALV difference determined by the ALV comparator is monitored by the threshold detector, which, when appropriate, triggers the action selection circuitry to update the working memory of the most recently reached waypoint. However, like in the case of velocity control, also in this process is the ALV difference subject to gain adjustment. The goal of this adjustment is to prevent another detrimental confusion that may arise when the ALV associated with the just-reached waypoint is very similar to that of the next waypoint. This confusion may occur because arrival at a waypoint makes the robot stop comparing the currently perceived ALV with that of the just-reached waypoint and start comparing it with that of the next waypoint, yet since at this moment the current ALV is almost identical to that of the just-reached waypoint, high similarity between the ALVs of the just-reached and of the next waypoints results in the current ALV being similar to that of the next waypoint as well. Consequently, the difference between the current ALV and the ALV of the next waypoint may happen to be below the threshold, which, if not adjusted, would in turn cause the robot to erroneously conclude that the next waypoint has also been reached, although the robot has even not left the previous one. To prevent such a misinterpretation, a negative gain is added to the ALV difference upon arrival at a waypoint, and then is progressively removed as the robot moves away from that waypoint until vanishing completely prior to approaching the next waypoint.

Because in vertebrates travel along a habitual route appears to be mediated by the basal ganglia [5], to implement the action selection circuitry we used the spiking model of the basal ganglia readily available in Nengo [12]. Obviously, the basal ganglia do not exist in ants, but since there is evidence suggesting that the central complex can be regarded as their homologue in arthropods [13], we adopted this approach as a first approximation. Consequently, the basal ganglia inhibit the connection between the waypoint associative memory and the working memory of the previous waypoint whenever the adjusted ALV difference is above the threshold, and only when this difference falls below the threshold does this connection become disinhibited. The result of disinhibition is that the output of the waypoint associative memory, that is the HD vector representing the next waypoint, which in this particular moment corresponds to the waypoint that has just been reached, overrides the current HD vector stored in the working memory of the previous waypoint, and thus appropriate update of the latter memory is achieved.

Finally, there are two additional suppression mechanisms employed in the model, which are activated when there are no landmarks in view. In such a case, the current ALV is undetermined, and therefore appropriate suppression of the gain controller cancels the contribution of the ALV difference to the velocity command, effectively making the robot steering depend only on the local vector, while similar suppression of the threshold detector prevents a possible false detection of arrival at the next waypoint.

5 Results

Figure 2 presents routes followed by the robot in two example scenes. Both routes consist of two segments, and therefore each scene contains three waypoints: the start one, which is in the upper right quadrant, the intermediate one, close to the middle, and the final one, close to the bottom. The start and the intermediate waypoints are depicted as orange circles, while the final waypoint is depicted as the red circle. The intermediate and final waypoints are surrounded by landmarks represented as green circles. Importantly, the robot does not know the locations of the waypoints; it only knows the ALVs associated with these waypoints and can perceive the landmarks. Moreover, at the beginning of simulations, it is not heading towards the intermediate waypoint, but is somewhat rotated to either side. The dark blue lines show the routes followed by the robot. As can be seen, the robot first goes to the intermediate waypoint, which requires an appropriate turn at the start location, and then turns accordingly and proceeds to the final waypoint. During the initial part of each segment of the routes, the robot is mainly driven by the corresponding local vector, while during the end part of each segment, it is mostly driven by the ALV difference. As a result, some minor turns appear in the second parts of the segments, which manifest corrections imposed by the ALV difference on the original directions taken at the beginning of the segments. Moreover, even if the robot does not home in precisely on the intermediate waypoint due to inherent noise in the SNN or imprecision of the ALV model, which can result in a direction slightly off from the optimal one during the beginning of the next segment, as is the case for the right scene, it can still reliably reach the final waypoint.

Some aspects of the model dynamics for the robot following the route in the right scene are presented in Fig. 3. The first plot shows distances to each waypoint over time. The second plot is a raster representing activity of a sample of 50 neurons in the working memory maintaining the trace of the previous waypoint. Evidently the pattern of activity changes when the robot reaches the intermediate waypoint. This is because the neuronal activity encodes the HD vector associated with the most recently reached waypoint, which changes upon arrival at the next waypoint. The value of this encoded HD vector can be decoded according to the NEF principles, and the similarity of the decoded value to the values of the actual HD vectors associated with each of the waypoints is visualized in the third plot, in which larger numbers correspond to higher similarity

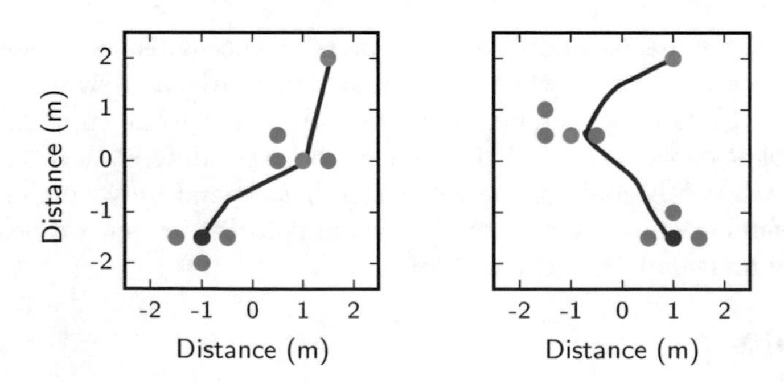

Fig. 2. Routes followed by the robot in two example scenes

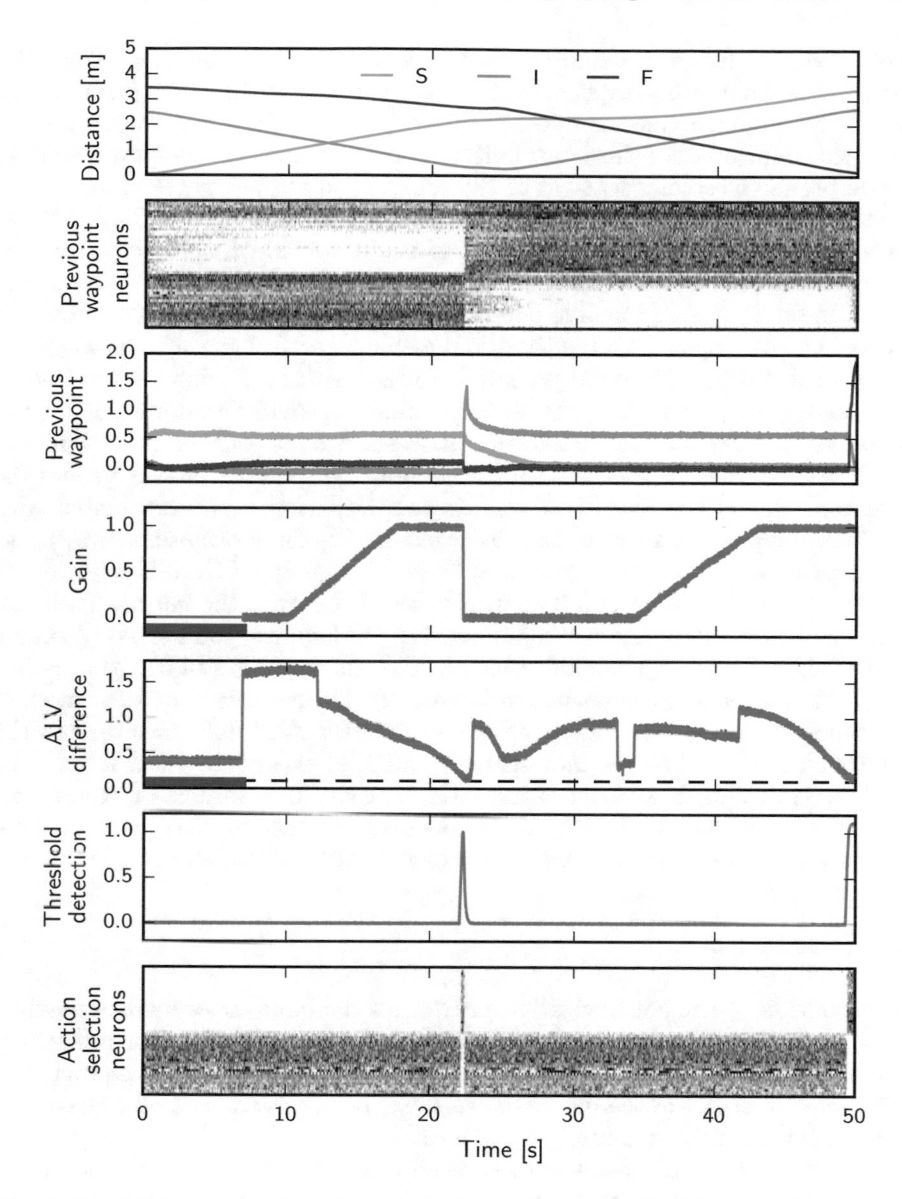

Fig. 3. Model dynamics when the robot travels along an example route (waypoints: S, start; I, intermediate; F, final)

(colors are the same as in the first plot). As expected, during the first segment of the route, the HD vector corresponding to the start waypoint is expressed, whereas during the second segment, the expressed HD vector corresponds to the intermediate waypoint. The similarity is not perfect, though, because the expressed HD vectors are noisy versions of the actual counterparts. The fourth plot shows modulation of the gain applied to the ALV difference. The gain, depicted as the blue line, equals 0 during the first part of the first segment of the route, then gradually increases to 1 in the second part of the segment, and maintains this value over the third part until the arrival at the intermediate waypoint, which begins the second segment of the route, resulting in the gain dropping to 0 and the whole cycle repeating. Moreover, whenever there are no landmarks in view, as in the case for the initial part of the route, which is indicated by the brown line at the bottom, the gain is forced to be 0 irrespective of its current value. In the fifth plot, the blue line represents the ALV difference, whereas the dashed line corresponds to the threshold at which arrival at the next waypoint is detected. During the initial part of the route with no landmarks in view, as indicated by the brown line, the ALV difference is small because the currently perceived ALV is a zero vector and the difference depends only on the magnitude of the ALV associated with the next waypoint. The ALV difference reaches the threshold around 23 s, and then increases rapidly as the ALV associated with the final waypoint, instead of that associated with the intermediate waypoint, starts to be used for its computation. Even though the ALV difference drops close to the threshold around 33 s, in the middle between the intermediate and final waypoints, there is no risk that arrival at the final waypoint could be falsely identified because the gain applied to the ALV difference is still 0 at that point. The sixth plot shows the output of the threshold detector: it is 0 for most of the time and displays a sudden peak only when the ALV difference reaches the threshold, that is when the intermediate and final waypoints are reached. The final, seventh plot, is a raster representing activity of a sample of 50 neurons in the action selection circuitry. As is evident, the pattern of activity of these neurons closely matches the time course of the output of the threshold detector.

6 Discussion

Our model of desert ant navigation provides a combination of ballistic navigation based on the local vector upon arrival at a waypoint, attractor navigation based on a simple view-matching mechanism when approaching a waypoint, and a weighted interplay of the two in between waypoints. Moreover, all crucial processes are implemented neurally using a SNN.

We have not implemented global vector navigation because experimental findings suggest that when an ant expresses a local vector, expression of the global vector is inhibited, which results in the global vector being ignored during navigation over a familiar, cluttered environment [6].

As a view-matching mechanism, we have adopted the ALV model, which, although parsimonious, has been successfully employed in several computational studies on desert ant navigation and even applied to real robots [9,11]. One problem with this model, however, is that it requires alignment of the perceived and stored ALVs to the common reference frame, such as the external compass reference, to allow their comparison, while experimental evidence suggests that in ants internal rotation of images can be excluded as can body rotation for this purpose [9]. A remedy would be to store a set of ALVs for a particular location, each corresponding to an image taken at a different orientation, which, in some circumstances, seems biologically plausible. In our model this problem does not exist because we further simplified the ALV model: ALVs associated with waypoints are stored only from the egocentric perspective of the approaching robot. This is because waypoints are maintained in a sequence and we assume that the local vector expressed at a previous waypoint should ensure approximately correct orientation of the robot while approaching the next waypoint, thus making rotations unnecessary.

Utilizing distance-dependent gain control applied to the ALV difference has made it possible to avoid the problem of perceptual aliasing, both when steering away from a previously reached waypoint and when detecting arrival at the next waypoint. It has also made our model consistent with experimental findings that in desert ants landmark memories are combined with vector information in a manner depending on how far along the route the ant is [4].

There are some limitations to our model as well. It suffers from the problems common to models based on thresholds [11], namely it is very sensitive to the actual choice of parameters, which so far had to be adjusted individually for each scene. Also, fine choice of neural time constants is required to achieve desired behavior.

In the future, the model could be extended in at least several ways: fixed thresholds could be replaced with some adaptive alternatives to make it more robust, more advanced view-matching mechanisms, such as the original snapshot model, could be adopted, and finally learning could be added.

References

1. Ardin, P., Peng, F., Mangan, M., Lagogiannis, K., Webb, B.: Using an insect mushroom body circuit to encode route memory in complex natural environments. PLoS Comput. Biol. **12**, e1004683 (2016). https://doi.org/10.1371/journal.pcbi.1004683
2. Baddeley, B., Graham, P., Husbands, P., Philippides, A.: A model of ant route navigation driven by scene familiarity. PLoS Comput. Biol. **8**, e1002336 (2012). https://doi.org/10.1371/journal.pcbi.1002336
3. Bekolay, T., Bergstra, J., Hunsberger, E., DeWolf, T., Stewart, T.C., Rasmussen, D., Choo, X., Voelker, A.R., Eliasmith, C.: Nengo: a Python tool for building large-scale functional brain models. Front. Neuroinform. **7**, 48 (2014). https://doi.org/10.3389/fninf.2013.00048
4. Bregy, P., Sommer, S., Wehner, R.: Nest-mark orientation versus vector navigation in desert ants. J. Exp. Biol. **211**, 1868–1873 (2008). https://doi.org/10.1242/jeb.018036
5. Collett, M.: How desert ants use a visual landmark for guidance along a habitual route. In: Proceedings of the National Academy of Sciences U.S.A., vol. 107, pp. 11638–11643 (2010). https://doi.org/10.1073/pnas.1001401107
6. Collett, M., Collett, T.S., Bisch, S., Wehner, R.: Local and global vectors in desert ant navigation. Nature **394**, 269–272 (1998)

7. Cruse, H., Wehner, R.: No need for a cognitive map: decentralized memory for insect navigation. PLoS Comput. Biol. **7**, e1002009 (2011). https://doi.org/10.1371/journal.pcbi.1002009
8. Eliasmith, C., Anderson, C.: Neural Engineering: Computation, Representation, and Dynamics in Neurobiological Systems. MIT Press, Cambridge (2003)
9. Lambrinos, D., Möller, R., Labhart, T., Pfeifer, R., Wehner, R.: A mobile robot employing insect strategies for navigation. Rob. Auton. Syst. **30**, 39–64 (2000). https://doi.org/10.1016/S0921-8890(99)00064-0
10. Rohmer, E., Singh, S.P.N., Freese, M.: V-REP: a versatile and scalable robot simulation framework. In: 2013 IEEE/RSJ International Conference on Intelligent Robots and Systems (IROS), pp. 1321–1326, Tokyo (2013). https://doi.org/10.1109/IROS.2013.6696520
11. Smith, L., Philippides, A., Graham, P., Baddeley, B., Husbands, P.: Linked local navigation for visual route guidance. Adapt. Behav. **15**, 257–271 (2007). https://doi.org/10.1177/1059712307082091
12. Stewart, T.C., Choo, X., Eliasmith, C.: Dynamic behaviour of a spiking model of action selection in the basal ganglia. In: Salvucci, D.D., Gunzelmann, G. (eds.) Proceedings of the 10th International Conference on Cognitive Modeling, pp. 235–240, Philadelphia (2010)
13. Strausfeld, N.J., Hirth, F.: Deep homology of arthropod central complex and vertebrate basal ganglia. Science **340**, 157–161 (2013). https://doi.org/10.1126/science.1231828
14. Wehner, R.: Desert ant navigation: how miniature brains solve complex tasks. J. Comp. Physiol. A. **189**, 579–588 (2003). https://doi.org/10.1007/s00359-003-0431-1
15. Wehner, R.: The architecture of the desert ant's navigational toolkit (Hymenoptera: Formicidae). Myrmecol. News **12**, 85–96 (2009)

The Application Perspective of Izhikevich Spiking Neural Model – The Initial Experimental Study

Adam Barton(✉), Eva Volna, and Martin Kotyrba

Department of Informatics and Computers, University of Ostrava, Ostrava, Czech Republic
{adam.barton, eva.volna, martin.kotyrba}@osu.cz

Abstract. In this paper we explore the Izhikevich spiking neuron model especially the synergy of the dimensionless model parameters and their implications to the spiking of the neuron itself. This spiking, principally the spike rate, is highly important from the application point of view. The understanding of the model is useful for better spiking network design, when the input neuronal stimulus is transferred to the spikes in order to produce faster network response. Whereas we can achieve the better neuronal response of the spiking network through utilization of the correct model parameters which impact to the neurons and the network neuronal dynamics significantly. The model parameters setup were described, demonstrated and spiking neuron model output and behaviour examined. The influence of the input current was also described in a given experimental study.

Keywords: Spiking neuron · Izhikevich model · Parameters impact Spiking · Spike rate · Experimental study

1 Introduction

The elementary units of the central nervous system are neurons, which are interconnected into a structure. The Spiking Neural Network (SNN) focuses only on the pulse neurons involved in the information process. The Spiking Neural Networks can be considered as the third generation of artificial neural networks which reduces abstraction of the neural simulation. The model itself works with the concept of time and the main idea of the SNN is that neurons in the network generate a pulse only if it reaches the critical value of the neuron membrane potential (threshold). Only then the pulse is emitted and further distributed to the network. Thus the neurons do not pulse in each cycle of impulse propagation.

A biological neuron can be generally divided into three parts: soma, dendrites and axon. Dendrites - the thin structures that arise from the cell body - receive signals from other neurons and transmit these signals to the core. Soma then performs a nonlinear processing. If the excitation exceeds the critical value, a pulse is generated and transmitted through an axon from one neuron to another neuron via dendrites connected to this particular axon. This connections are called synapses, the excitements spreads from the

R. Matoušek (Ed.): MENDEL 2017, AISC 837, pp. 223–232, 2019.
https://doi.org/10.1007/978-3-319-97888-8_19

presynaptic cells to the postsynaptic cells. One neuron is often associated with more than tens of thousands postsynaptic neurons and the axon may reach several centimetres.

Excitement itself consists of short electrical pulses - action potentials with an amplitude of ~ 100 mV and duration between 1 and 2 ms. An action potential is the basic unit of the transmission signal, and it is a brief moment when the cell membrane potential rapidly increases and then decreases. After the formation of the action potential, a refractory period occurs - a temporary negative shift value of the membrane potential following by the hyperpolarization (increasing potential difference). In this time window the negative membrane potential prevents further pulses. The minimum distance between two impulses defines an absolute refractory period which follows the relative refractory period. In this relative period is difficult to generate an action potential, a stronger than normal stimulus is needed to elicit an excitation. The form of the pulse is spreading throughout the axon unchanged. Sequences of these action potentials emitted by one neuron is called Spiking Train. The actual form of the action potential does not bear any information. So, the carrier of information is the number of pulses and the exact time among them.

Without the presence of any input signal, the neuron is relaxed in a constant membrane potential. The potential changes after the arrival of the pulse. However, the excited potential decays to the relaxed state. Thus the postsynaptic potential arrives as a response of the postsynaptic neuron to the presynaptic action potential. An actual generation of the action potential is typically not caused by only one of synaptic impulses, but as a result of a nonlinear interaction of the number of excitatory and inhibitory synaptic inputs (if we do not consider super-threshold input current).

2 Izhikevich Spiking Neuron Model

Whereas the simple integrate-and-fire spiking neuron model is not biologically accurate and it is not possible to create the pulse function of biological neurons, a more accurate spiking model was proposed in 2003 [1]. The model is able to produce a wide range of realistic behaviours by adding a recovery variable u to the quadratic integrate-and-fire model. The Izhikevich model is also accurate with the low computational cost (model require 13 FLOPs per simulation). Just as in other integrate-and-fire neuron models, the used terms in following equations are justified primarily due to the intention of spiking behaviour. The model is based on two differential equations:

$$dv/dt = 0.04v^2 + 5v + 140 - u + I, \tag{1}$$

$$du/dt = a(bv - u). \tag{2}$$

Variable v describes the trajectory of the neuron membrane potential, u represents the already mentioned recovery variable representing membrane recovery - activation of potassium currents and inactivation of sodium currents (negative feedback to the voltage) and variable I is the weighted synaptic current (incoming energy). Both parameters a and b are abstract parameters of the model.

When the voltage exceeds a dynamic threshold value between −70 and −50 mV, the neuron produces a spike with the +30 mV peak and variables v and u are modified, as follows:

$$v \leftarrow c, \tag{3}$$

$$u \leftarrow u + d. \tag{4}$$

By adjusting a, b, c, d parameters, we are able to mimic a range of biological neuronal responses. The parameter *a* represents the time scale of the recovery variable, *b* describes the sensitivity of the recovery variable to subthreshold fluctuations of the membrane potential variable, *c* represents the after-spike reset value of the membrane potential and *d* parameter defines the constant value added to the recovery variable after spikes (after-spike reset). The parameters of the important neuro-computational features of real neurons with a contribution to the temporal coding and the spike-timing information processing are shown in Table 1 [1].

Table 1. Different parameters of the known types of spiking neurons

Type of dynamic	A	B	C	D	I
Tonic Spiking	0.020	0.20	−65	6.00	14.00
Phasic spiking	0.020	0.25	−65	6.00	0.50
Tonic bursting	0.020	0.20	−50	2.00	15.00
Phasic bursting	0.020	0.25	−55	0.05	0.60
Bursting then spiking	0.020	0.20	−55	4.00	10.00
Spike frequency adaptation	0.010	0.20	−65	8.00	30.00
Class 1 excitability	0.020	−0.10	−55	6.00	0.00
Class 2 excitability	0.200	0.26	−65	0.00	0.00
Spike latency	0.020	0.20	−65	6.00	7.00
Subthreshold oscillations	0.050	0.26	−60	0.00	0.00
Frequency preference and resonance	0.100	0.26	−60	−1.00	0.00
Integration and coincidence detection	0.020	−0.10	−55	6.00	0.00
Rebound spike	0.030	0.25	−60	4.00	0.00
Rebound burst	0.030	0.25	−52	0.00	0.00
Threshold variability	0.030	0.25	−60	4.00	0.00
Bistability of resting and spiking states	1.000	1.50	−60	0.00	−65.00
Depolarizing after-potentials	1.000	0.20	−60	−21.00	0.00
Accommodation	0.020	1.00	−55	4.00	0.00
Inhibition-induced spiking	−0.020	−1.00	−60	8.00	80.00
Inhibition-induced bursting	−0.026	−1.00	−45	0.00	80.00

3 Model Parameters

From the application perspective, it is crucial to setup model constant parameters. These parameters directly affect the amount of energy propagated to the network. Since the Izhikevich spiking neuron model is able to simulate various biological neuronal responses satisfactorily, we will focus on the model dynamics (the pulsation itself).

3.1 Influence of the Input Current

To explain the impact of the input current to the model, we can simulate the membrane potential of the spiking neuron in the time window in ms scale and the model behaviour. For this case, we use the Tonic Spiking parameters setting specified in Table 1. Figure 1 demonstrates the evolution of variable v in mV scale for a various input current I and the corresponding Spike Trains. A larger amount of the energy injected to the model accelerates the spiking of the neuron. The initial faster pulsation is caused by the value of the membrane recovery variable u, see Fig. 2.

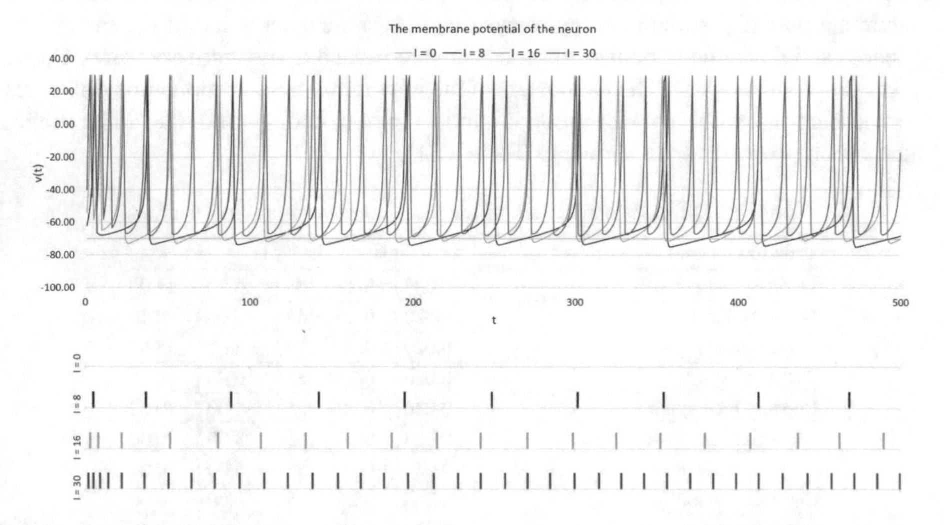

Fig. 1. At the top of the figure, there is the membrane potential of the spiking neuron in the time t in ms scale for several input current settings. The relevant spike trains are shown below the membrane potential chart. The pulse frequency increases with the larger amount of input current. The input current is constantly delivered to the model.

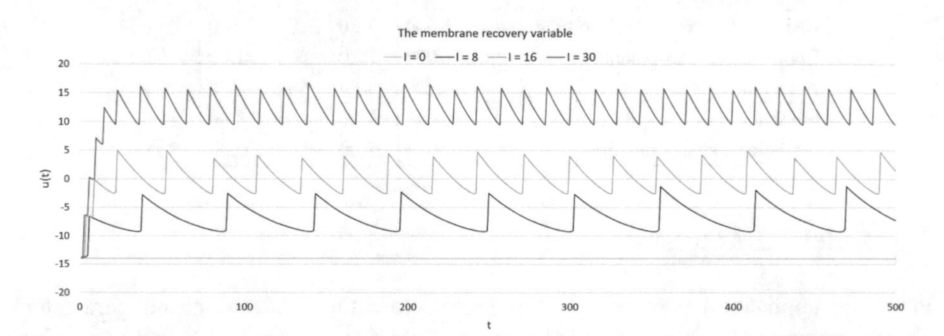

Fig. 2. The membrane recovery variable in time. We can clearly see the reason of the initial faster pulsing of the model since the variable u is step-increased by the value of parameter d. The smaller value caused the longer period of time to reach the pulsing stability.

3.2 Dimensionless Parameters of the Model

To describe the behaviour of the model with the various a, b, c, d parameters setting, the Tonic Spiking initial values are used and each parameter is described separately.

The parameter a defines the decay of the recovery variable (the time scale), see (2). Since the zero value is senseless, Fig. 3 shows only the meaningful setup. Unless we consider the case when the parameter a and d are both zero then the recovery variable u is neglected. The smaller values of this parameter result in the slower recovery of the neuron. The input current is fixed to the constant value I = 14 mV.

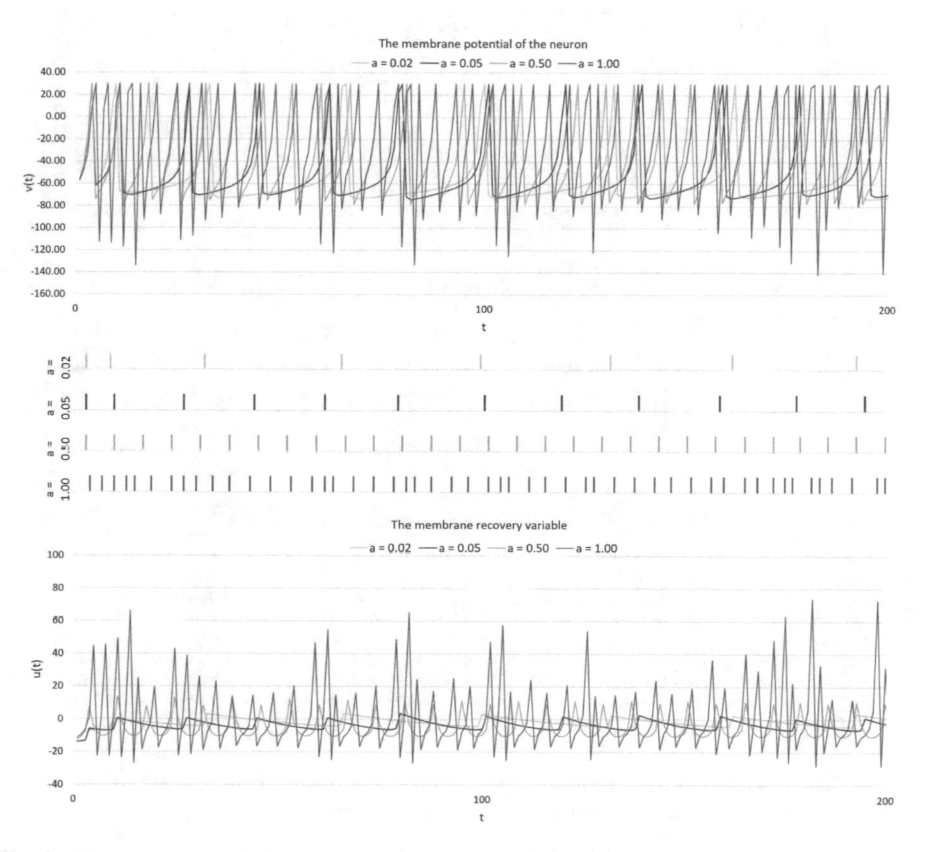

Fig. 3. The parameter a influence. The figure sequentially shows from the top to bottom: the membrane potential chart, spike trains below and the membrane recovery chart.

The parameter b defines the sensitivity of u to the subthreshold fluctuations of v, see (2). The greater values couple the membrane potential and the recovery variable more strongly resulting in possible subthreshold oscillations and low-threshold spiking dynamics. The initial setup of the parameters is based on the Tonic Spiking and the constant input current I = 30 mV. Note: the irregularity of the spiking for parameter b = 1.00, which is shown in Fig. 4.

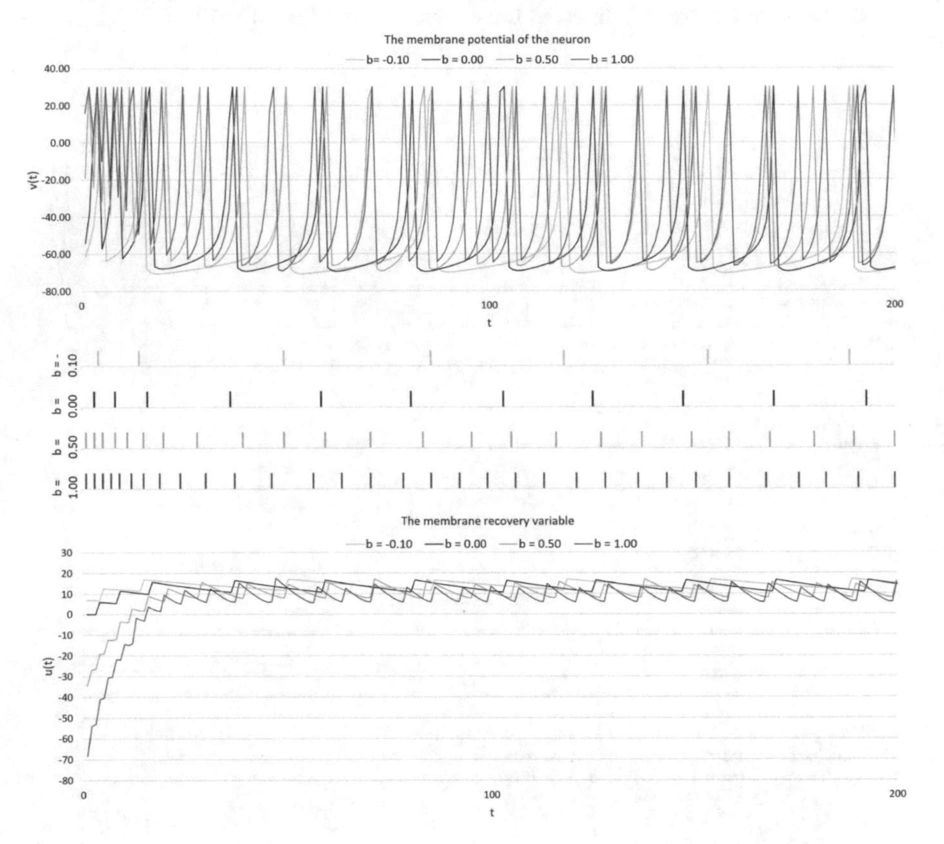

Fig. 4. The parameter b influence. The figure sequentially shows from the top to bottom: the membrane potential chart, spike trains below and the membrane recovery chart.

The parameter c describes the after-spike reset value of the membrane potential, see (3). The initial setup of the parameters is also based on the Tonic Spiking. The incoming energy is set to the constant value I = 14 mV (Fig. 5).

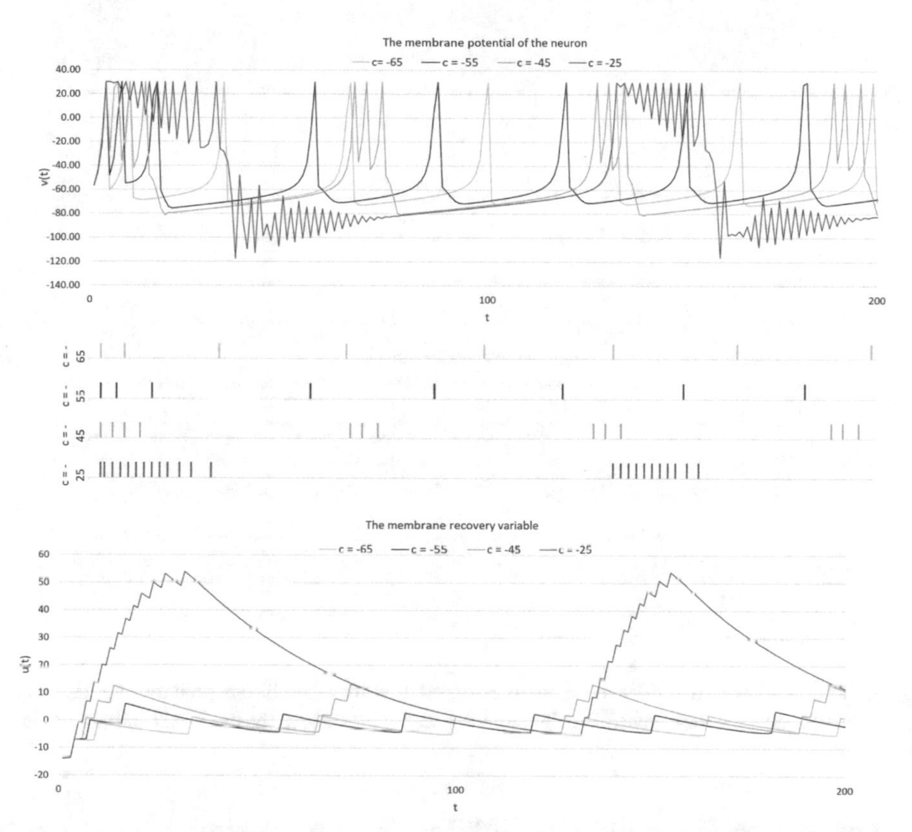

Fig. 5. The parameter c influence. From the top to bottom: membrane potential chart, spike trains, membrane recovery chart. Note that the parameter c significantly affects the pulsing of the neuron. The lower values can cause a burst spiking of the neuron.

The last parameter d describes the after-spike reset of the recovery variable, see (4). Tonic Spiking parameters setting is repeatedly used and the input current is set to constant value I = 14 mV (Fig. 6).

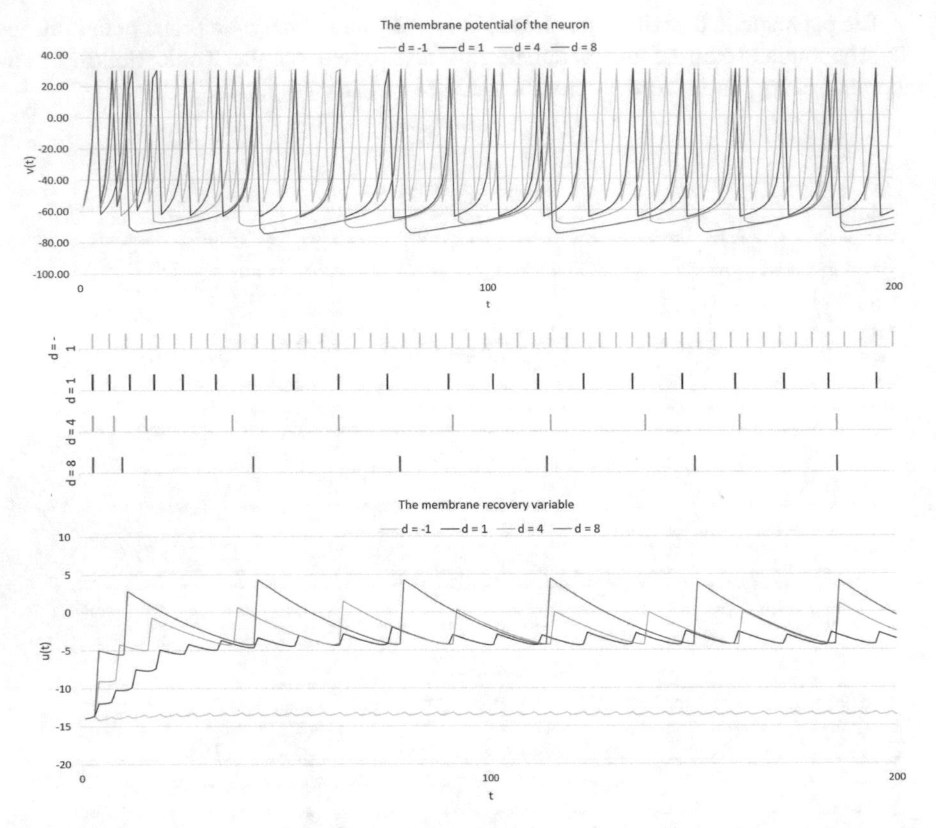

Fig. 6. The parameter d influence. From the top to bottom: membrane potential chart, spike trains, membrane recovery chart. The fast spiking can be reached by the small values of this parameter.

4 Application Possibilities of Spiking Neuron Model

The pattern recognition is related with automatic detection or classification of an object. The Izhikevich model finds applications in various fields, such as automated analysis of a medical image of blood contents, identification of humans from finger prints, handwriting recognition, etc.

The model suits for large scale cortical simulations due to its efficiency and biological accuracy. The model has been used in simulation of fully connected networks of cortical spiking neurons. In the feasibility study of using FPGAs for large scale simulations, there has been proposed a modularised processing element to evaluate a large number of Izhikevich spiking neurons in a pipelined manner to utilise a character recognition algorithm based on the Izhikevich model [2].

This spiking model of neuron has been also used in the networks to model human-like visual and auditory pathways [3]. The learning procedures has been designed to operate in an online evolvable adaptive way in the face and speech signal recognition in case of person authentication.

The network of Izhikevich neurons exhibited rhythms in the alpha and gamma frequency range, transient and sustained spike synchrony, spindle waves, sleep oscillations, and other temporal phenomena [1]. The network of interconnected neurons exhibits known types of cortical firing patterns, receptor kinetics, short-term plasticity and long-term spike-timing dependent plasticity (STDP) [4]. The Spike-timing dependent plasticity is a temporally asymmetric form of Hebbian learning induced by tight temporal correlations between the spikes of presynaptic and postsynaptic neurons and it is often used in combination with the spiking neuron models [5–8].

To capture the space-time-averaged behaviour of the network consisting of Izhikevich neurons, the Lumping Izhikevich neurons has been proposed. Thus the construction of a planar vector field that yields the firing rate of a bursting Izhikevich neuron can be read out, while leaving the sub-threshold behaviour intact [9].

5 Conclusion

Various model parameters setups were described and demonstrated. Since the Izhikevich model exhibits most of neural behaviours and does not require too much computational power, we looked at the model from an application point of view.

From this point of view, it is essential to examine the output of the spiking neuron and its behaviour. As shown in Fig. 1, the fundamental impact to the model is by an input current. Whereas the model transforms the current to the spikes, it is necessary to stress the importance of the input current to the neuron and the amount of this energy since the injected energy to the model directly affects the pulse rate.

Dimensionless parameters a, b, c, d significantly allow to model the spiking neuron output. The parameter a describes the time scale of the recovery variable. Smaller values of this parameter result in a slower recovery of the neuron, therefore a slower pulse rate. On the contrary, higher values result in faster pulsation and the time irregularity between the pulses. The parameter b also affects the pulse rate, as it describes the sensitivity of u to subthreshold fluctuations of v, see Fig. 4 and (2). The parameter c represents the after-spike reset value of v and fundamentally changing the pulsing of the neuron. Lower values causes a burst spiking of the neuron. By setting this parameter, we can influence the amount of pulses corresponding to a certain input. The parameter d represents the after-spike reset. Since the parameter determines the constant value added to the u after spikes, we can control the pulse rate. With knowledge of these parameters and their impacts on the output, the model becomes a powerful tool for the use of non-homogeneous neurons in the spiking network, especially in case of sensory input processing.

We will use the acquired knowledge about the behaviour of parameters in our future work, which will focus on the input sensor processing. Using appropriate parameters setting can inject the significant amount of energy (see Fig. 4 and burst spiking) to the network consist of Izhikevich neurons in order to produce faster network response. Consider the intended reactive behaviour on the stimulus. We can flood the network with energy by using the burst spiking setting for a certain sensor.

Designing a sophisticated parameters setting in the network can result in reaching a more complex and accurate function of the network instead of using only homogeneous spiking neurons in the network.

Acknowledgments. The research described here has been financially supported by University of Ostrava grant SGS07/PrF/2017. Any opinions, findings and conclusions or recommendations expressed in this material are those of the authors and do not reflect the views of the sponsors.

References

1. Izhikevich, E.M.: Simple model of spiking neurons. IEEE Trans. Neural Netw. **14**(6), 1569–1572 (2003)
2. Rice, K.L., et al.: FPGA implementation of Izhikevich spiking neural networks for character recognition. In: 2009 International Conference on Reconfigurable Computing and FPGAs, ReConFig 2009. IEEE (2009)
3. Wysoski, S.G., Benuskova, L., Kasabov, N.: Evolving spiking neural networks for audiovisual information processing. Neural Netw. **23**(7), 819–835 (2010)
4. Izhikevich, E.M., Gally, J.A., Edelman, G.M.: Spike-timing dynamics of neuronal groups. Cereb. Cortex **14**(8), 933–944 (2004)
5. Lee, P.R., et al.: Gene networks activated by specific patterns of action potentials in dorsal root ganglia neurons. Sci. Rep. **7**, 43765 (2017)
6. Rebola, N., Carta, M., Mulle, C.: Operation and plasticity of hippocampal CA3 circuits: implications for memory encoding. Nat. Rev. Neurosci. **18**(4), 208 (2017)
7. Asl, M.M., Valizadeh, A., Tass, P.A.: Dendritic and axonal propagation delays determine emergent structures of neuronal networks with plastic synapses. Sci. Rep. **7**, 39682 (2017)
8. Valtcheva, S., Venance, L.: Astrocytes gate Hebbian synaptic plasticity in the striatum. Nat. Commun. **7**, 13845 (2016)
9. Visser, S., Van Gils, S.A.: Lumping Izhikevich neurons. EPJ Nonlinear Biomed. Phys. **2**(1), 1–17 (2014)

Fuzzy Granular Calculations for the Sematic Web Using Some Mathematical Morphology Methods

Anna Bryniarska(✉)

Institute of Computer Science, Opole University of Technology,
ul. Proszkowska 76, 45-758 Opole, Poland
a.bryniarska@po.opole.pl

Abstract. In the Semantic Web, during searching information, we can get precise answer for our searching query, even if we have uncertain, vague or unclear information. This positive result of searching information depends on expert choices of acceptable fuzzy degrees for concepts and roles, and also depends on appropriate description of concepts and roles interpretations in the fuzzy sets algebra. In this paper is proposed such interpretation, by using methods of the mathematical morphology and granular computing. Moreover, in order to formulate this problem is used the fuzzy description logic and the postulates of searching information in the Semantic Web are widened.

Keywords: Semantic Web · Fuzzy disambiguation · Description logic FuzzyDL · Information retrieval logic · Fuzzy set algebra · Granulation Dilatation · Erosion

1 Introduction

The *Information Retrieval Systems* [2,8,9,14] are widely described in the literature. In this paper the theoretical apparatus of the *Information Retrieval Logic* [6,7] is widened. The information retrieval logic allows us to search knowledge in the semantic network, specifically in the Semantic Web, which is described by the *Fuzzy Description Logic* (*fuzzyDL*) [4,15,17,19,20].

In the paper [7] is presented a method of obtaining precise results for searched information, based on the *IRL* logic on the Web and also the postulates of this logic. Furthermore, it is shown that by the defuzzification process can be obtained some results, which are in *the confidence range*, accepted by experts. The confidence range is a set of acceptable fuzzy degrees of searching information. In this situation, even if we have uncertain, unclear or vague information, we can obtain exact answer about searched information. Then, the positive result of searching information not only depends on the appropriate choice of the acceptable fuzzy degrees by experts for concepts and roles, which describe searched information on the Web. It also depends on the appropriate description

R. Matoušek (Ed.): MENDEL 2017, AISC 837, pp. 233–244, 2019.
https://doi.org/10.1007/978-3-319-97888-8_20

of the concepts and roles interpretation in the fuzzy sets algebra. The problem of describing such interpretations is formulated in the paper [7]. Here, we use the theoretical apparatus from paper [7] to formulate this problem, and in order to solve it we propose to use mathematical morphology methods and granules computing. In this purpose we widen the postulates presented in [7].

2 An Accuracy of the Fuzzy Searching Result

Nowadays, *the information retrieval* in the Semantic Web usually means to search reliable source of this information. So far, information retrieval systems and semantic relations systems indicated only the semantically closest sources of the searched information. However, it is not always the case. Often, when we search information about some object, we get uncertain knowledge, described in a proper language – the ontology language. Even though, this uncertainty can lead to unequivocal determination of the information source for this object. In this way, we get the compatibility with the description of the object model (the thesaurus). Described situation is called *the fuzzy disambiguation paradox* of searching information.

Searching information in the Semantic Web is to find data copies which are:

- one-argument values of attributes – data representing knowledge about some features or types of objects,
- two-argument values of attributes – data representing knowledge about some properties of objects or relations between two objects.

In first case, data are called *concepts*, and in the second one they are called *roles*. To describe concepts and roles is used *the Description Logic* (*DL*) language [1]. The *DL* language describing concepts and roles can be extended to some formulas of the first order logic. In the extended language is created *a thesaurus*, which describes model concepts and roles. While, *the ontology* is a language which describes found concepts and roles, or searched data. If the interpretation of concepts and roles from the ontology, accordingly to experts knowledge and criteria, will result in the interpretation of concepts and roles of thesaurus, then this relationship is called *the residuum*.

Searching data on the Internet resources (in the set of the Web addresses) is an interpretation of these data which describe the degree of similarity of these data to the data from the thesaurus. This similarity degree is a number from the range [0,1] and is a measure of the searched data membership to the set of data available on the Internet. This degree also takes into account the semantic structure of these resources, described by the semantic network. Concepts and roles, from the ontology, are interpreted as fuzzy sets, in the space of the addresses on the Web and pairs of these addresses. In this way is made *the fuzzification* of the knowledge representation [5]. Furthermore, establishing the residuum is also *the defuzzification* of this knwoledge [5]. Then for a given query, a reliable-for-experts set of Web addresses representing this knowledge can be indicated.

Interpretation sets forming residuum will be further treated as an information search result.

During information search is used the following rule:

1. A question (a search query) is compared to the thesaurus;
2. If for all interpretations, the set of searched addresses is empty, then this question is compared to the ontology, so that the compared entry has the most similar meaning to the thesaurus;
3. Found, for this entry, the set of Web addresses represents knowledge which is identical or the most similar to the searched one.

This rule of *IR* is called *the residuum rule.* Based on this rule can be introduce *the information retrieval logic IRL.*

3 Postulates of the Fuzzy Morphology

In the *IRL* logic are used postulates. These postulates of the fuzzy morphology for searching information are:

P 1. There is a thesaurus which is a set of certain reference terms and formulas of the *IRL* language. Thesaurus terms represent knowledge in the same area as searched information and can be found in a text document (from the thesaurus).

P 2. The ontology includes all terms and formulas of the *IRL* language, which are semantically related to the searched information. The ontology includes the thesaurus. All thesaurus formulas are constructed of certain terms and assertions from the base set Tez. Likewise, all ontology formulas are constructed of certain terms and assertions belonging to the base set Ont. The degree of meaning similarity of ontology expressions to the thesaurus expressions is determined by an expert system based on the expert knowledge. In this system is defined the range of accepted similarity degrees V and also a set of interpretation of ontology expressions Fuz which establish the similarity degrees of these expressions to the thesaurus one.

P 3. The space of information retrieval is a group of addresses of knowledge resources on the Web, semantically related to the terms and formulas representing the searched information. Knowledge resources are text documents available at these addresses.

P 4. Finding information is to search in the text document (terms from the Semantic Web created by the ontology), which semantic structure is the most similar to the structure of the thesaurus document [6]. If both documents contain expressions that are equally used by agents in the communication process, then these expressions represent the same knowledge. Furthermore information retrieval is searching for the text document that represents the same knowledge or most similar knowledge to the one from the thesaurus. Therefore, *the residuum rule* of information retrieval is applied.

P 5. The information retrieval of the intersection of concepts represents collective knowledge of these concepts. *The complementary descriptions* φ, ϕ are denoted as $\varphi \& \phi$ and represent the collective knowledge represented by the data set of these descriptions. For any description φ exists such a dual description φ^d, which complements φ to description $\varphi \& \varphi^d$, and represents the same knowledge as some description from the thesaurus.

P 6. During searching information, if $x, y \in [0,1][0,1]$ are the semantic similarity degrees of some instances of descriptions φ, ϕ, to the thesaurus descriptions, then the degree of such similarity $\varphi \& \phi$ is a number $x \mathbf{~t~} y$, where operation $\mathbf{t} : [0,1] \times [0,1] \rightarrow [0,1]$ is some t-norm. If the semantic similarity degree to the thesaurus description φ^d, we denoted as x^d, then $x^d = 1 - x$. Then, the operation $\mathbf{s} : [0,1] \times [0,1] \rightarrow [0,1]$ described by the formula: $x \mathbf{~s~} y = 1 - (1-x) \mathbf{~t~} (1-y)$ is some s-norm.

Similar situation is when x, y are the semantic similarity degrees of the formulas φ, ϕ to some concept instance or thesaurus formulas. Then, $x = 1, y = 1$, $x \mathbf{~t~} y = 1, x^d = 0, y^d = 0$. If the description φ is not similar to any description from the thesaurus, then $x = x^d = 0$.

Normally, the t-norm has properties as follow (for all $x, y, z, x_0, y_0 \in [0,1]$):

$$x \mathbf{~t~} y = y \mathbf{~t~} x \tag{1}$$

$$(x \mathbf{~t~} y) \mathbf{~t~} z = x \mathbf{~t~} (y \mathbf{~t~} z) \tag{2}$$

$$x \leqslant x_0 \text{ and } y \leqslant y_0 \text{ implies } x \mathbf{~t~} y \leqslant x_0 \mathbf{~t~} y_0 \tag{3}$$

$$1 \mathbf{~t~} x = x; 0 \mathbf{~t~} x = 0 \tag{4}$$

Additional, the property of the t-norm, accordingly to the **P6.** postulate, is the dual operation $x^d = 1 - x$.

Every t-norm describes an unique implication $\rightarrow$ (the residuum), which determine the measurement of the formulas implication similarity (for all $x, y, z \in [0,1]$):

$$z \leqslant (x \rightarrow y) \text{ iff } x \mathbf{~t~} z \leqslant y \tag{5}$$

or

$$(x \rightarrow y) = \sup\{z \in [0,1] : x \mathbf{~t~} z \leqslant y\}. \tag{6}$$

The implication of these formulas is semantically similar to the thesaurus formulas, if the similarity degree of the implication predecessor is the closest to its successor. Furthermore, this means that in found text document is the formula φ, which implies a certain formula ϕ from this document, or represents the searched information. When ϕ represents the searched information and is not contained in this document, then the formula φ is supplemented with the formula θ representing the experts knowledge, so that the complementary formulas $\varphi \& \phi$ have the same similarity degree as search formula ϕ (the degree equal 1). Thus, in the case of imprecise implications predecessor φ, the successor is a sharp expression with the semantic similarity degree equal 1.

P 7. The similarity degree of the conjunction $\varphi \wedge \phi$ to the thesaurus formulas is defined as (for all $x, y \in [0,1]$):

$$x \boxtimes y := x \ \mathbf{t} \ (x \to y) \tag{7}$$

where x, y are the similarity degrees of formulas φ, ϕ to the thesaurus ones.

P 8. The similarity degree of the alternative $\varphi \vee \phi$ to the thesaurus formulas is defined as (for all $x, y \in [0,1]$):

$$x \boxplus y := ((x \to y) \to y) \boxtimes ((y \to x) \to x) \tag{8}$$

where $x, y \in [0,1]$ are the similarity degrees of formulas φ, ϕ to the thesaurus ones.

P 9. The algebra $AL = \langle L, \boxtimes, \boxplus, \ \mathbf{t}, \to, 0, 1\rangle$ is *a regular residuated lattice* (or the *AL-algebra*) accordingly to definition from [12]. The operations $\boxtimes, \boxplus, \mathbf{t}, \to$ are described accordingly to postulates **P6** – **P8**. The algebra $\langle L, \boxtimes, \boxplus, 0, 1\rangle$ is a complete lattice with the largest element 1 and the least element 0. While $\langle L, \mathbf{t}, 1\rangle$ is a commutative semigroup with the unit element 1, i.e. $\mathbf{t}$ is commutative, associative, and $1 \ \mathbf{t} \ x = x$ for all x.

In the AL-algebra can be defined operations of *the completeness* $'$ and *the equivalence* $\leftrightarrow$:

$$x' := (x \to 0) \tag{9}$$

$$x \leftrightarrow y := (x \to y) \boxtimes (y \to x) \tag{10}$$

and also *the dual* operation ^{d} described in the postulate **P6**.

With such determined postulates, can be specified *IRL* logic in the Semantic Web.

4 The IRL Logic Language

The semantic network, or the Semantic Web, can be identified with an ordered, indexed graph [13]. In the semantic network the nodes of this graph are the individuals, the one-node relation is called a concept and the two-node relation is called a role.

The descriptions of individuals $(t_1, t_2, ...)$, concepts $(C_1, C_2, ...)$ and roles $(R_1, R_2, ...)$ are *the terminology.* When the connection between objects is represented by the semantic network, then it is called *the assertion.* The assertion that 't_1 is C_1', we written as '$t_1 : C_1$', while the role R_1 between objects t_1 and t_2, we written as '$(t_1, t_2) : R_1$'.

In the context of the semantic network research [4,11,15,17,18] the knowledge representation can be described in *the attributive language* (AL) of the description logic [1,10]. Then knowledge is represented by: concepts *TBox*, roles *RBox* and assertions *ABox*. The semantic network can be widened by the edges which connect concepts and roles with each other and describe the relationship between them. These relationships are *the axioms.*

Furthermore, in the semantic network can be implemented the fuzziness of all elements of this network, ie. concepts, roles, assertions and axioms. Then, *the Fuzzy Description Logic* (*fuzzyDL*) [19,20] is used in order to describe elements of the semantic network and assign a fuzzy degree to knowledge described by these elements. We introduce the syntax fo the AL language for the *Information Retrieval Logic*, analogically to the *fuzzyDL* language.

Syntax of TBox.

The following names are included to the set of concepts and roles names:

- $\top$ (*Top*) – the universal concept or role,
- $\bot$ (*Bottom*) – the empty concept or role.

Let C, D be the names of the concepts, R be the name of a role, and m be *the modifier*. Then *complex* concepts are:

- $\neg C$ – *the concept negation*; it means all instances of concepts which are not an instance of the concept C;
- C^d, R^d – *the dual concept* for C, *the dual role* for R;
- $C \& D$ – *the completion of concepts* C and D (see **P5**);
- $C \sqcap D$ – *the intersection of concepts* C and D; it means all instances of both concepts C and D;
- $C \sqcup D$ – *the union of concepts* C and D; it means all instances either of the concept C or the concept D;
- $\exists R.C$ – *the existential quantification*; it means all instances of the concept C which are in role R with at least once occurrence of the concept C;
- $\forall R.C$ – *the universal quantification*; it means all occurrence of the concept C which is in role R with some occurrence of the concept C;
- $m(C)$ – *the modification* m *of the concept* C; it means the concept C which is modified by the word m. For example m can occur as a word: very, more, the most, high, higher or the highest.

Concepts which are not complex are called *atomic*.

Syntax of ABox.

For any concepts instances t_1, t_2, the concept name C and the role name R, *the assertions* are "$t_1 : C$", "$(t_1, t_2) : R$". We read them: t_1 is an instance of the concept C, the pair (t_1, t_2) is an instance of the role R.

Syntax of axioms TBox.

For any concepts names C, D and any number $\alpha \in [0, 1]$, *the axioms* are: $C \sqsubseteq D$, $C = D$, $\langle C \sqsubseteq D, \alpha \rangle$, $\langle C = D, \alpha \rangle$.

Syntax of axioms RBox.

For any roles R_1, R_2 and any number $\alpha \in [0, 1]$, *the axioms* are: $R_1 \sqsubseteq R_2$, $R_1 = R_2$, $\langle R_1 \sqsubseteq R_2, \alpha \rangle$, $\langle R_1 = R_2, \alpha \rangle$.

Syntax of formula.

Any assertion or axiom is a formula. For any formulas φ, ϕ, a variable x and a number $\alpha \in [0, 1]$, formulas are: $\neg\varphi$, $\langle \varphi, \alpha \rangle$, $\forall_x \varphi$, $\exists_x \varphi$, $\varphi \Rightarrow \phi$, $\varphi \wedge \phi$, $\varphi \vee \phi$, $\varphi \Leftrightarrow \phi$, φ^d.

5 Information Retrieval Logic in the Semantic Web

The Information Retrieval Logic is a theoretical foundation of the system for searching knowledge in the Semantic Web. Here, we present interpretation of knowledge fuzzification for the *IRL* logic and also the semantic of all elements in the Semantic Web.

5.1 Fuzzification in the Semantic Web

The expressions of the *IRL* logic are interpreted in the regular residuated lattice $AL = \langle L, \boxtimes, \boxplus, \mathbf{t}, \rightarrow, 0, 1\rangle$ and in the chosen, ordered algebra of fuzzy sets $\boldsymbol{F} = \langle F, \&^F, \wedge^F, \vee^F, \neg^F, c^F, e^F, 0^F, 1^F, M, (\cdot)^d, F_0\rangle$. This algebra contains of a fuzzy set family F and listed bellow operations and functions:

- $\&F$ is an completion of fuzzy sets;
- $\wedge^F$ is an intersection of fuzzy sets;
- $\vee^F$ is a sum;
- $\neg^F$ is a complement operation;
- c^F is a function called the degree of containment of fuzzy sets;
- e^F is a function called the degree of equality of fuzzy sets;
- 0^F is a fuzzy set which is a function with only one numeric value 0 (the empty set);
- 1^F is a fuzzy set which is a function with only one numeric value 1;
- M is a set of one-argument operations called *the modification functions*; $(\cdot)^d$ is an one-argument modification function of fuzzy sets with a dual value (the postulate **P6**);
- $F_0 \subseteq F$ is a family of fuzzy sets such that when $\mu \in F_0$, then exists $Y \subseteq X \cup X \times X$ and for $y \in Y, \mu(y) = 1$, otherwise $\mu(y) = 0$.

The sum and intersection operations of fuzzy sets are described by the appropriate t-norms and s-norms, respectively.

These operations are defined in *the algebra* $AL = \langle L, \boxtimes, \boxplus, \mathbf{t}, \rightarrow, 0, 1\rangle$ as follows. For any fuzzy sets $\mu_A, \mu_B \in F$ and $x \in [0, 1]$:

$$(\mu_A \&^F \mu_B)(x) = \mu_A(x) \ \mathbf{t} \ \mu_B(x) \tag{11}$$

$$(\mu_A \wedge^F \mu_B)(x) = \mu_A(x) \boxtimes \mu_B(x) \tag{12}$$

$$(\mu_A \vee^F \mu_B)(x) = \mu_A(x) \boxplus \mu_B(x) \tag{13}$$

$$c^F(\mu_A, \mu_B)(x) = \mu_A(x) \rightarrow \mu_B(x) \tag{14}$$

$$e^F(\mu_A, \mu_B)(x) = \mu_A(x) \leftrightarrow \mu_B(x) \tag{15}$$

$$(\neg^F \mu_A)(x) = \mu_A(x) \rightarrow 0 \tag{16}$$

$$(\mu_A^d)(x) = 1 - \mu_A(x) \tag{17}$$

Let X is a set of all objects (data copies), which are part of the Semantic Web and $X \times X$ is a set of all ordered pairs of the set X. For further interpretation, we assumed that the interpretation $I = (\boldsymbol{F}^I)$ satisfied conditions:

F 1. For the concept instances t assigns certain values $t^I \in X$ and for the pair of instances (t_1^I, t_2^I) assigns pairs $(t_1^I, t_2^I) \in X \times X$.

F 2. For the concept name C assigns fuzzy set $C^I : X \cup X \times X \rightarrow [0,1]$, that for any $x, y \in X, C^I(x,y) = C^I((x,y)) = C^I(y)$.

F 3. For the role name R assigns a fuzzy set $R^I : X \cup X \times X \rightarrow [0,1]$, which equals 0 for arguments from X,

F 4. For the modifier m assigns a function $m^I : [0,1] \rightarrow [0,1]$, where $m^I \in M$,

F 5. For formulas φ, including assertions and axioms, assigns some number $\varphi^I \in [0,1]$.

5.2 Semantic of IRL Logic

We describe the semantic of the Information Retrieval Logic for concepts, assertions, and axioms. The semantic of formulas is not presented, because in this paper it is not further used (see [7]).

Semantic of Concepts Tbox. For any $x \in X$, concept names C, D, the role name R and the modifier m:

$$\top^I(x) = 1 \tag{18}$$

$$\bot^I(x) = 0 \tag{19}$$

$$(C^d)^I(x) = (C^d(x))^d \tag{20}$$

$$(R^d)^I(x) = (R^d(x))^d \tag{21}$$

$$(\neg C)^I(x) = (\neg^F C^I)(x) \tag{22}$$

$$(C \& D)^I(x) = (C^I \&^F D^I)(x) \tag{23}$$

$$(C \sqcap D)^I(x) = (C^I \wedge^F D^I)(x) \tag{24}$$

$$(C \sqcup D)^I(x) = (C^I \vee^F D^I)(x) \tag{25}$$

$$(\exists R.C)^I(x) = \sup_{y \in X} (R^I \wedge^F C^I)(x,y) \tag{26}$$

$$(\forall R.C)^I(x) = \inf_{y \in X} (\neg^F R^I \vee^F C^I)(x,y) \tag{27}$$

$$(m(C))^I(x) = m^I(C^I(x)) \tag{28}$$

Semantic of Assertions ABox. For any instance t of the concept C and any instances t_1, t_2 of the role R:

$$(t : C)^I = C^I(t^I) \tag{29}$$

$$((t_1, t_2) : R)^I = R^I(t_1^I, t_2^I) \tag{30}$$

Semantic of Axioms. For any concept names C, D and roles R_1, R_2:

$$(C \sqsubseteq D)^I = c^F(C^I, D^I) \tag{31}$$

$$(R_1 \sqsubseteq R_2)^I = c^F(R_1^I, R_2^I) \tag{32}$$

$$(C = D)^I = e^F(C^I, D^I) \tag{33}$$

$$(R_1 = R_2)^I = e^F(R_1^I, R_2^I) \tag{34}$$

6 Granulation in the IRL Logic

The information retrieval system *IRS* is used to search reliable for experts subsets $X \cup X \times X$, where X is some set of available Internet resources from a specific knowledge field. These subsets allow us to reliably interpret expressions of the *IRL* logic. The *IRL* expressions are built of atomic concepts and roles. In order to search information, similarly to the statistic, we use *the confidence range* V. The most important is that all experts, based on the confidence range, accept the set of membership degrees of objects to the fuzzy set. In this expert system we use a database $K = \langle Tez, Fuz, V, Ont\rangle$, where:

- Tez is a set of used, atomic concepts and roles from the thesaurus,
- Ont is a set of used, atomic concepts and roles from the ontology (accordingly to the postulate **P2**),
- Fuz is a set of possible to use interpretation of fuzzification, accordingly to postulates **P1** – **P9**, conditions **F1** – **F5** and formulas (18)–(33),
- V is a function, which for any concept C or role R from the set $Tez \cup Ont$ assigns *the fuzzy confidence ranges* $V(C) = [\alpha_C, 1]$ or $V(R) = [\alpha_R, 1]$, accepted by all experts.

We want to satisfy the confidence ranges for the most possible *pattern Web addresses*, established by experts. In order to more adequate describe the set Fuz, in the $\boldsymbol{F}$ algebra are established some subsets F, called *the algebra granules* F for database K (short: *granules*).

The granules are the values of a function $^{Gr} : Tez \cup Ont \to \wp(F)$ called *a granulation for database* $K = \langle Tez, Fuz, V, Ont\rangle$, if for any concepts C and roles R from the set $Tez \cup Ont$ these formulas are satisfied:

Gr 1. $\bot^{Gr} = \{0^F\}, \top^{Gr} = \{1^F\}$;

Gr 2. $C^{Gr} = \{\mu_C \in F : \exists(I \in Fuz)\mu_C = C^I\}$;

Gr 3. $R^{Gr} = \{\mu_R \in F : \exists(I \in Fuz)\mu_R = R^I\}$.

To mentioned granules contain fuzzy sets, which values satisfy the confidence range, for some pattern Web addresses. We can perform two operations on granules: (1) in order to increase the number of pattern addresses, which satisfy the confidence degree, we might use the operation $\&^F$ or (2) leave such complementation of these fuzzy sets, which do not give new fuzzy sets. These operations are the morphological operations *dilatation* and *erosion*, respectively.

7 Fuzzy Granules Morphology

The operation of dilatation and erosion of the algebra granules F can be described analogically in the vector space like in papers [3,16,21].

Firstly, for the database $K = \langle Tez, Fuz, V, Ont\rangle$, the confidence range $V = [\alpha, 1]$ and any granule G, we join a new granule $B(G)$, called *a structuring element* of a granule G. This granule is described as follow:

$$B(G) = \{\mu_g : \text{ for } g \in G \text{ and any } y \in X \cup X \times X, \text{ if } \alpha \leq g(y),$$
$$\text{then } \mu_g(y) = 0, \text{ otherwise } \mu_g(y) = g(y)\} \cup \{0^F\} \quad (35)$$

Intuitively, the structuring element of a granule G points to the Web addresses, in which the fuzzy degrees should be increased by the dual fuzzy set from the set $B(G)$.

As the next postulate of fuzzy morphology in the information retrieval on the Web, we assume:

P 10. $B(G) \subseteq F$.

Further, we need some definitions (for any $A, B \in \wp(F), \alpha \in [0, 1]$):

Definition 1. *A set $B^d = \{\mu : \mu^d \in B\}$ is called the dual set for a set B. If B is a granule, then B^d is also a granule.*

Definition 2. *A set $A_c = \{\mu \in F : \mu = a\&^F c, a \in A\}$ is called the fuzzy translation of a set A in operation $\&^F$ to a fuzzy set c. If A is a granule, then A_c is also a granule.*

Definition 3. *Let $Y_A(\alpha) = \{x \in X \cup X \times X :$ for any $a \in A, \alpha \leq a(x)\}$.*

A set $A_\alpha = \{\mu_a :$ for $a \in A, \mu_a$ is a fuzzy set such that for any $y \in Y_A(\alpha), \mu_a(x) = a(x)$, otherwise $\mu_a(x) = 0\}$, is called a set A *cut off* to the confidence range $V = [\alpha, 1]$. If A is a granule with the confidence range $V = [\alpha, 1]$, then A_α is also a granule.

To use this definition in the information retrieval, we need to accept a postulate:

P 11. $A_\alpha \subseteq F$.

Definition 4

$$A \oplus B = \cup\{(A)_b : b \in B^d\} \quad (36)$$
$$A \ominus B = \cap\{(A)_b : b \in B^d\} \quad (37)$$

Definition 5. *If A is a granule and B is its structuring element, then granules are:*

- *a set $A \oplus B = \cup\{(A)_b : b \in B^d\}$,called the dilatation A,*

- *a set* $A \ominus B = \cap\{(A)_b : b \in B^d\}$, *called the erosion* A,
- *a set* $A \bullet B = (A \oplus B) \ominus B$, *called the closing of a set* A,
- *a set* $A \circ B = (A \ominus B) \oplus B$, *called the opening of a set* A.

Based on above definitions and because for any granule $A, 1^F \in B^d = (B(A))^d$ and $\mu\&^F 1^F = \mu$, we have the following facts:

$$y \in Y_A(\alpha) \Rightarrow \forall \mu \in B^d(\mu(y) = 1) \tag{38}$$

$$A_\alpha \ominus B = A_\alpha \tag{39}$$

$$A_\alpha \subseteq A \Rightarrow A_\alpha \subseteq A \ominus B \tag{40}$$

$$A_\alpha \subseteq A \Rightarrow A_\alpha = (A \ominus B)_\alpha \tag{41}$$

$$A \ominus B \subseteq A \subseteq A \oplus B \tag{42}$$

$$Y_{(A \ominus B)}(\alpha) \subseteq Y_A(\alpha) \subseteq Y_{(A \oplus B)}(\alpha) \tag{43}$$

We also accept a postulate:

P 12. For a granule A, if $V = [\alpha, 1], x \in X \cup X \times X, a \in A, \alpha > a(x)$, then $\alpha \leq (a\&^F a^d)(x)$.

It means that in the set $A \oplus B$ appears fuzzy set $a\&^F a^d$, which satisfy the confidence range in such Web addresses or their pairs, that did not satisfy the confidence range for the granule A.

8 Conclusion

Presented theoretical apparatus allows us to precise information retrieval in the Semantic Web, in the fuzzy environment. The erosion operation is used to get possible fields of interpretation of concepts and roles, while the dilatation operation – possible fields of interpretation ie. granules which interpretation is in the confidence range. In granules established by the morphology operations, are fuzzy sets, which for as biggest as possible numbers of Web addresses, satisfy the fuzzy confidence range. In other words, we establish such granules C^{Gr} and R^{Gr}, for which sets $Y_A(\alpha)$ are as biggest as possible. Such analysis of using Semantic Web allows us to accurately correct the base set of interpretations Fuz. Then postulates **P10** – **P12** are satisfied. Moreover, accordingly to the expert criteria and knowledge, searched data from ontology are compatible, with some degree, to the data from the thesaurus. It means that the information retrieval logic [7] can be successfully used.

References

1. Baader, F., Calvanese, D., McGuinness, D.L., Nardi, D., Patel-Schneider, P.F. (eds.): The Description Logic Handbook: Theory, Implementation and Applications. Cambridge University Press, Cambridge (2003)
2. Baeza-Yates, R., Ribeiro-Neto, B.: Modern Information Retrieval. Addison Wesley, Boston (1999)

3. Bloch, I.: Mathematical morphology. In: Aiello, M., Pratt-Hartmann, I., van Benthem, J. (eds.) Handbook of Spatial Logics, pp. 857-944. Springer (2007)
4. Bobillo F., Straccia, U.: FuzzyDL: an expressive fuzzy description logic reasoner. In: IEEE World Congress on Computational Intelligence, Hong Kong, pp. 923–930 (2008)
5. Bryniarska, A.: The algorithm of knowledge defuzzification in semantic network. In: Grzech, A., Borzemski, L., Świątek, J., Wilimowska, Z. (eds.) Information Systems Architecture and Technology, Networks Design and Analysis, pp. 23–32. OW Politechniki Wrocawskiej, Poland (2012)
6. Bryniarska, A.: An information retrieval agent in web resources. In: IADIS International Conference Intelligent Systems and Agents 2013 Proceedings, Prague, Czech Republic, pp. 121–125 (2013)
7. Bryniarska, A.: The paradox of the fuzzy disambiguation in the information retrieval. (IJARAI) Int. J. Adv. Res. Artif. Intell. **2**(9), 55–58 (2013)
8. Ceglarek, R., Rutkowski, W.: Automated acquisition of semantic knowledge to improve efficiency of information retrieval systems. In: Proceedings of the Business Information Systems 10th International Conference, BIS 2006. LNCS, pp. 329–341. Springer (2006)
9. Ceglarek, R., Haniewicz, K., Rutkowski, W.: Semantic Compression for Specialized Information Retrieval Systems. In: Advances in Intelligent Information and Database Systems, Studies in Computational Intelligence, vol. 283, pp. 111–121. Springer (2010)
10. Cerami, M., Esteva, F., Bou, F.: Decidability of a description logic over infinite-valued product logic. In: Proceedings of the 12th International Conference on the Principles of Knowledge Representation and Reasoning, pp. 203–213 (2010)
11. Galantucci, L.M., Percoco, G., Spina, R.: Assembly and disassembly by using fuzzy logic and genetic algorithms. Int. J. Adv. Robot. Syst. **1**(2), 67–74 (2004)
12. Háje, P.: Metamathematics of Fuzzy Logic. Trends in Logic, Studia Logica Library, vol. 4. Kluwer Academic Publishers, Dordrecht (1998)
13. Kowalski, R.A.: Logic for Problem Solving. North Holland (1979)
14. Manning, Ch.D., Raghavan, P., Schütze, H.: Introduction to Information Retrieval. Cambridge University Press, Cambridge (2008)
15. Pan, J.Z., Stamou, G., Stoilos, G., Thomas, E.: Expressive querying over fuzzy DL-lite ontologies. In: Scalable Querying Services over Fuzzy Ontologies, 17th International World-Wide-Web Conference, Beijin (2008)
16. Serra, J.: Image Analysis and Mathematical Morphology. Academic Press, Cambridge (1982)
17. Simou, N., Stoilos, G., Tzouvaras, V., Stamou, G., Kollias, S.: Storing and querying fuzzy knowledge in the semantic web. In: Proceedings of 4th International Workshop on Uncertainty Reasoning for the Semantic Web, Karlsruhe, Germany (2008)
18. Simou, N., Mailis, T., Stoilos, G., Stamou, S.: Optimization techniques for fuzzy description logics. In: Proceedings of 23rd International Workshop on Description Logics, vol. 573, pp. 244–254. CEUR-WS, Waterloo (2010)
19. Straccia, U.: A fuzzy description logic. In: Proceedings of the 15th National Conference on Artificial Intelligence, Madison, USA, pp. 594–599 (1998)
20. Straccia, U.: Reasoning with fuzzy description logics. J. Artif. Intell. Res. **14**, 137–166 (2001)
21. Wu, Q.H., Lu, Z., Ji, T.: Protective Relaying of Power Systems Using Mathematical Morphology. Springer, Dordrecht (2009)

Exploring the Map Equation: Community Structure Detection in Unweighted, Undirected Networks

Rodica Ioana Lung(✉), Mihai-Alexandru Suciu, and Noémi Gaskó

Center for the Study of Complexity, Babeş-Bolyai University, Cluj-Napoca, Romania
rodica.lung@econ.ubbcluj.ro
http://csc.centre.ubbcluj.ro

Abstract. The map equation is one of the most efficient methods of evaluating a possible community structure of a network, computed by using Shanon's entropy and probabilities that a random walker would exit a community or would wonder inside it. Infomap, the method that optimizes the map equation to reveal community structures, is one of the most efficient computational approach to this problem when dealing with weighted, directed or hierarchical networks. However, for some unweighted and undirected networks, Infomap fails completely to uncover any structure. In this paper we propose an alternate way of computing probabilities used by the map equation by adding information about 3-cliques corresponding to links in order to enhance the behavior of Infomap in unweighted networks. Numerical experiments performed on standard benchmarks show the potential of the approach.

Keywords: Community structure · Social networks
Unweighted graphs · Infomap

1 Introduction

The community structure detection problem in complex networks is one of the most challenging problems in network analysis [2]. From the very outset the concept of community has not yet been formally defined, as all existing attempts can be contradicted by counter-examples. An explanation for this is that a community, intuitively described as a group of nodes that are more connected to each other than to the rest of the network can take many forms in different types of networks. Thus, finding the community structure of a network - similar to an optimization problem - cannot be solved by one universal search algorithm suited to all types of network structures.

However, this drawback also makes the problem suitable to be approached by computational intelligence tools, as their adaptability makes them appropriate to tackle this challenge. Thus, a plethora of methods that deal with this problem in various way have emerged. Out of them, the performance of Infomap [10] is

R. Matoušek (Ed.): MENDEL 2017, AISC 837, pp. 245–253, 2019.
https://doi.org/10.1007/978-3-319-97888-8_21

outstanding: there is no method that reports results better than it on *all* networks when testing 'standard' community structure benchmarks, for weighted, directed or hierarchical networks. However, when considering unweighted graphs with less clear community structure its performance drops significantly, indicating a need for improvement.

Infomap optimizes a map equation using a greedy search algorithm, followed by a heath-bath simulated annealing algorithm. The map equation is based on probabilities computed by simulating a random walk in the network. In this paper we propose a modified version of Infomap that uses different probabilities to compute the map equation to deal with unweighted networks. Numerical experiments are presented in order to assess differences between the two methods.

2 Infomap and Infomap$_t$

The intuition behind Infomap is that using a two level Huffman coding [1] - one for naming communities and a second for naming nodes within communities - yields the shortest code length for the *correct* community structure. The names of the communities are irrelevant, only the minimum possible length is considered to construct the map equation, described in what follows.

2.1 The Map Equation

The map equation assigns to the partition M of n nodes with m modules the number $L(M)$ representing the average description length when considering the two-level Huffman coding and Shanon's source coding theorem [11]:

$$L(M) = q_{\curvearrowright} H(\mathcal{Q}) + \sum_{i=1}^{m} p^i_{\circlearrowright} H(\mathcal{P}^i) \tag{1}$$

with the step probability that the random walker switches modules

$$q_{\curvearrowright} = \sum_{i=1}^{m} q_{i\curvearrowright}$$

and

$$H(\mathcal{Q}) = \sum_{i=1}^{m} \frac{q_{i\curvearrowright}}{\sum_{j=1}^{m} q_{j\curvearrowright}} \log\left(\frac{q_{i\curvearrowright}}{\sum_{j=1}^{m} q_{j\curvearrowright}}\right). \tag{2}$$

The movement within module i is estimated using

$$p^i_{\circlearrowright} = q_{i\curvearrowright} + \sum_{node \in i} p_{node} \tag{3}$$

and

$$\begin{aligned} H(\mathcal{P}^i) = & \frac{q_{i\curvearrowright}}{q_{i\curvearrowright} + \sum_{\beta \in i} p_\beta} \log\left(\frac{q_{i\curvearrowright}}{q_{i\curvearrowright} + \sum_{\beta \in i} p_\beta}\right) + \\ & + \sum_{\alpha \in i} \frac{p_\alpha}{q_{i\curvearrowright} + \sum_{\beta \in i} p_\beta} \log\left(\frac{p_\alpha}{q_{i\curvearrowright} + \sum_{\beta \in i} p_\beta}\right). \end{aligned} \tag{4}$$

$H(\cdot)$ is actually the entropy estimating the minimum length of the Huffman encoding of the module names $\mathcal{Q}$ at one level and nodes within module $\mathcal{P}^i$ at the second level, based on the fact that the Huffman coding provides the minimum code length for given data. The fact that the actual code is irrelevant for the optimization of $L(M)$ comes from using Shanon's entropy as a lower limit for the length of the average Huffman code length as the minimum length that can be used to store this information.

2.2 Probabilities

The intuition used by Infomap is that the minimum length illustrates the structure of the network grouping nodes that have a higher probability to 'keep' a random walker traveling among them. Within Infomap this probability is computed based on the distribution of node's visit frequencies, which is derived from the node degrees. For weighted networks the weighted degree is used.

Thus, probability $q_{i\curvearrowright}$ that a random walker leaves module i is computed as

$$q_{i\curvearrowright} = \sum_{\alpha \in i} \sum_{\beta \notin i} p_\alpha w_{\alpha\beta} \tag{5}$$

with $w_{\alpha\beta}$ representing the weight of the link from α to β. When no weights are used, $w_{\alpha\beta}$ is 1. In (3), the probability p_{node} (and p_α, p_β in subsequent formulas) is the ergodic node visit frequency in the random walk, directly proportional to the (weighted) degree of a node. The node flow is computed as the weighted degree of the node divided by the double sum of all network weights. The link flow is computed as its weight divided by the same sum.

In this form Infomap exploits the link's weights as the map equation is entirely based on probabilities p_α. In the absence of weights, p_α are computed using the weight 1 for each link. For a network with nodes having the same degree these probabilities will be equal. If the community structure is defined such that the number of links a node has in its community is equal or less than the number of links it has outside its community, it is impossible for the random walker to identify exit nodes (as they have all the same probability) and thus to find the correct structure.

2.3 Infomap$_t$

Probabilities p_α are used both to derive the probability of a move within a module and to exit a module. To enhance the search in unweighted networks we can replace the ergodic node frequencies when considering an infinite walk - which is how Infomap computes these probabilities - with corresponding frequencies when considering also the number of triangles (3-cliques) between nodes, based on the premise that if there are more triangles between two nodes the probability that a random walker moves back and forth between the two nodes - thus considering them in the same community - is higher.

For each link in the network the number of triangles between corresponding nodes is computed. This task can be performed beforehand and the number of triangles can be used as weights for the unweighted network. To preserve the structure, this number is added to the default value of 1 for each link. Thus the weight of a link $w_{\alpha,\beta}$ will be computed as $w_{\alpha,\beta} = 1 + T(\alpha, \beta)$, where $T(\alpha, \beta)$ represents the number of triangles between α and β, i.e. the number of nodes adjacent to both α and β. If two nodes in the network are not connected, neither the number of triangles is computed.

Thus, the node flow will be proportional to the total number of triangles the node has with its neighbors and the link flow to the number of triangles the corresponding nodes share (considering also the existence of the links). By introducing this feature in Infomap, a new variant is obtained, which we call Infomap$_t$. The same map equation is optimized, with probabilities derived from the combined number of links and number of triangles between nodes.

3 Numerical Experiments

In this section we evaluate the potential of using triangle-based probabilities within Infomap$_t$ by comparing numerical results obtained with Infomap on a set of synthetic benchmarks and real-world networks with known community structure. We used the code available at http://www.mapequation.org/code.html. Both variants are run using default options as set in the code.

Benchmarks. Standard benchmark sets for evaluating community structure detection algorithms include the Girvan-Newman (GN) and LFR sets [3,5]:

- The GN set consists of networks of 128 nodes having equal degrees of 16, divided in 4 communities of 32 nodes each. The community structure is characterized by the z_{out} parameter that counts the number of links a node has outside its community. We generated[1] 8 sets of 30 networks each, corresponding to $z_{out} = \{1, \ldots, 8\}$. The most challenging set is the one with $z_{out} = 8$, as each node has equal number of links to nodes in its community and outside.
- The LFR set is generated (See footnote 1) by assigning node degrees using a power law distribution in order to create a more realistic setting. Communities are characterized by the mixing parameter μ computed as the number of links a node has outside its community to the total node degree. We used 3 LFR sets: one with 128 nodes and μ values from 0.1 to 0.5 and two with 1000 nodes, one having small communities (S) and one with large communities (B) with $\mu \in \{0.1, \ldots, 0.6\}$. Each set contains 30 networks, generated with parameters from [7].

Four real-world networks with known community structure are used: the bottle-nose *dolphin* network [9], the *football* network [3], the Zachary *karate* club network [12], and the *books* about US politics network.[2]

[1] Using the source code available at https://sites.google.com/site/andrealancichinetti/software, last accessed May, 2015.

[2] http://www.orgnet.com, last accessed 9/3/2015.

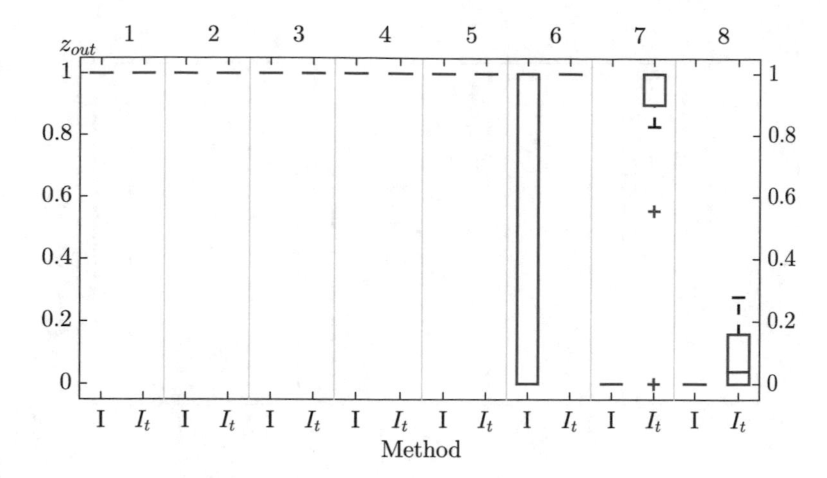

Fig. 1. Boxplots representing NMI values obtained by the two methods on the GN sets. 'I' denotes Infomap and I_t denotes Infomap$_t$.

Performance Evaluation. To evaluate the performance of the two methods we use the normalized mutual indicator (NMI) [6] to compare their output to what is known to be the correct structure of the network. For each synthetic 30 networks set NMI values are compared using the Wilcoxon sign-rank test to assess statistical significance. The null hypothesis that there is no difference in median is rejected if the p value is smaller than the confidence threshold of 0.05. For each real world network we perform 30 independent runs and compare results using the rank-sum test in a similar manner.

3.1 Results and Discussion

Results are represented by using boxplots of NMI values obtained for the 30 networks in each synthetic set and for the 30 runs for each real-world set. Box-plots illustrate the interquartile range, median values and outliers for a distribution, offering a comprehensive view over results obtained in multiple runs.

GN set. The most simple networks in the GN set from the community structure point of view are those having $z_{out} \leq 5$. Both Infomap and Infomap$_t$ found the correct cover for all networks, reporting results with NMI values of 1 (Fig. 1). Starting with $z_{out} = 6$ Infomap does not manage to find the correct cover, with median NMI values of 0. In fact, for $z_{out} = 7$ and 8 in all cases the NMI value is 0, indicating this specific behavior of Infomap of not being able to cope with some unweighted networks. However, Infomap$_t$ finds the correct communities for all networks having $z_{out} = 6$, reports a median value of 1 for $z_{out} = 7$ and positive values for $z_{out} = 8$. Differences between results obtained for these z_{out} values are significant indicating a definite improvement when using the triangle based probabilities within Infomap$_t$ compared to Infomap.

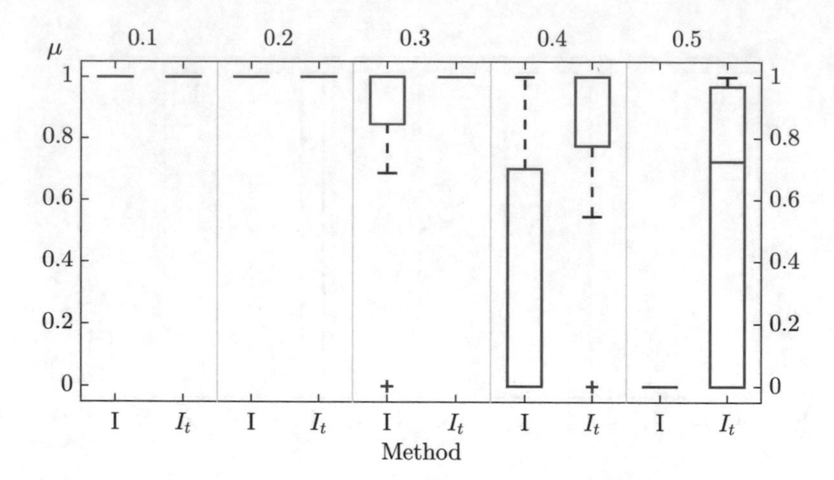

Fig. 2. Boxplots representing NMI values obtained by the two methods on the LFR 128 nodes set. 'I' denotes Infomap and I_t denotes Infomap$_t$.

Other methods in literature report NMI values close to 0.9 for $z_{out} = 6$, close to 0.7 for $z_{out} = 7$ and to 0.25 for $z_{out} = 8$ [7]. Methods based on the evolutionary computation paradigm may report better results: for $z_{out} = 7$ average NMI values of 0.98 by NoisyEO [8] and 0.89 by both Oslom [7] and Meme-Net [4]; for $z_{out} = 8$ average NMI of 0.65 by NoisyEO [8] and of 0.31 by Meme-Net [4]. The main drawback of these methods - except Oslom which is a statistical method - is related to their computational complexity and scalability. However, the two are main strengths of Infomap.

LFR sets, 128 nodes. The LFR sets with 128 nodes are more challenging than the GN sets as the distribution of nodes in communities is not uniform.

However, when comparing the two methods (Fig. 2) the situation seems similar to the one before: for $\mu = 0.1, 0.2$ both methods find the correct structure for all networks. For $\mu = 0.3$, both report median values of 1, but the difference in results is significant as the mean NMI reported by Infomap for this set is 0.86.

For $\mu = 0.4$ and 0.5 differences are obvious: Infomap reports median values of 0 while Infomap$_t$ reports a median of 1 for $\mu = 0.4$ and of 0.72 for $\mu = 0.5$. The Wicoxon sign-rank test confirms that the difference is significant. Other methods do not usually report values for this set.

LFR sets, 1000 nodes. Results obtained for the sets with 1000 nodes are displayed in Figs. 3 and 4. For the S (small communities, Fig. 3) sets with mixing probabilities up to $\mu = 0.4$, and for the B (big communities, Fig. 4) sets with $\mu \leq 0.3$ both methods find the correct structure for all networks in the sets. Small differences appear for the S sets with $\mu = 0.5$ and 0.6, but they are not significant from a statistical point of view. For the B sets with $\mu = 0.5$ and 0.6 differences are significant: for $\mu = 0.6$ Infomap$_t$ clearly reports better results

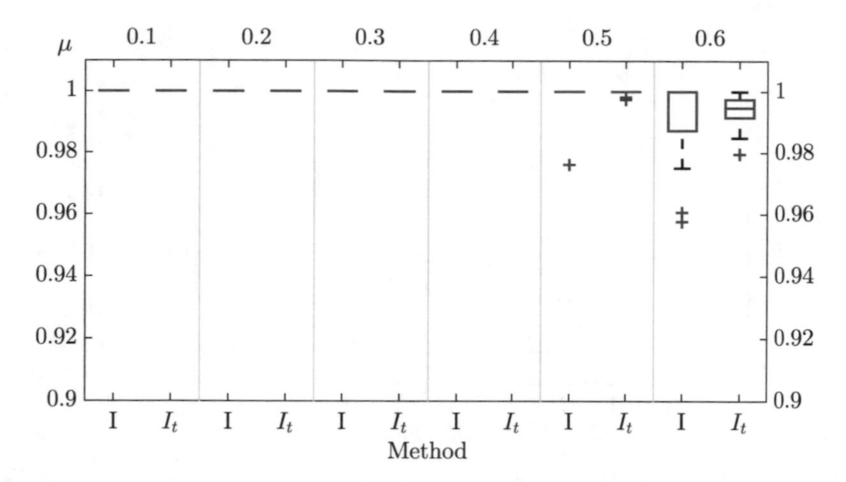

Fig. 3. Boxplots representing NMI values obtained by the two methods on the LFR 1000 S nodes sets. 'I' denotes Infomap and I_t denotes Infomap$_t$.

than Infomap. When considering corresponding weighted benchmarks, Infomap reports NMIs equal to 1 for these sets.

Real-World Networks. Real world networks with a given structure are more challenging as they do not exhibit properties such as those build in the synthetic benchmarks, e.g. constant mixing parameter or node degree. Moreover, sometimes the given structure considered does not reflect directly the network but takes into account supplementary information that may or may not be embedded in the network structure. For example, the *books* network is constructed by using books purchasing data: books are nodes; a link is added if two books have been purchased together [13]. Nodes are manually labeled as 'liberal', 'neutral', or 'conservative'. It may be that some - for example - 'neutral' customers willing to be informed purchase together all three types of books thus inducing a structure of links that will not reflect the political views expressed in the books. Infomap and Infomap$_t$ converged to the same (yet different) cover in all 30 runs. The NMI reported by Infomap was 0.30 while Infomap$_t$ found a cover with NMI 0.26. NoisyEO reports a NMI of 0.5 for this network.

For the other real-world networks results were as follows: for the *dolphins* network the NMI values were 0.46 and 0.30 by Infomap and $Infmap_t$, respectively; for the *football* network 0.79 and 0.80; and for the *karate* 0.54 and 0.46. These results show that the triangles may not always represent a valid option; but when we are not aware about the real structure, both methods could be used to study the network and results compared in order to extract the most reliable information about the possible structure.

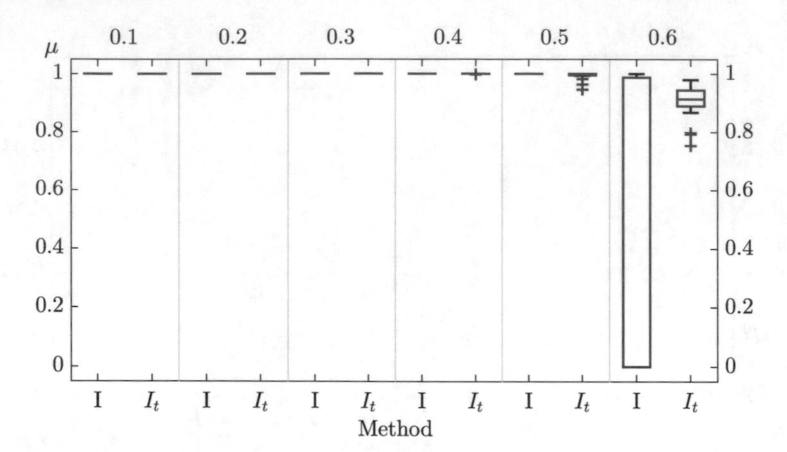

Fig. 4. Boxplots representing NMI values obtained by the two methods on the LFR 1000 B nodes sets. 'I' denotes Infomap and I_t denotes Infomap$_t$.

4 Conclusions

Finding the community structure of networks has been the focus of many recent studies. Given the many forms networks and communities may take, we cannot expect one method to tackle all possible situations. But there are methods that perform best in most situations, and one of these is Infomap, except for unweighted networks. The maximum flow paradigm based on a random walker does not reflect the network structure even if other methods manage to do so. In this paper we propose an improvement for Infomap by substituting the probabilities derived from the random walk with probabilities proportional to the number of triangles two nodes share. Numerical experiments show that this approach has potential in spite of the fact that it does not always yield better results. Future work may consist in exploring other means of quantifying probabilities that the random walker leaves a community in order to better explore the map equation.

Acknowledgments. This work was supported by a grant of the Romanian National Authority for Scientific Research and Innovation, CNCS - UEFISCDI, project number PN-II-RU-TE-2014-4-2332.

References

1. Cormen, T.H., Stein, C., Rivest, R.L., Leiserson, C.E.: Introduction to Algorithms, 2nd edn. McGraw-Hill Higher Education, New York City (2001)
2. Fortunato, S.: Community detection in graphs. Phys. Rep. **486**, 75–174 (2010). http://www.sciencedirect.com/science/article/pii/S0370157309002841
3. Girvan, M., Newman, M.E.J.: Community structure in social and biological networks. Proc. Nat. Acad. Sci. **99**(12), 7821–7826 (2002). https://doi.org/10.1073/pnas.122653799

4. Gong, M., Ma, L., Zhang, Q., Jiao, L.: Community detection in networks by using multiobjective evolutionary algorithm with decomposition. Phys. A: Stat. Mech. Appl. **391**(15), 4050–4060 (2012). http://www.sciencedirect.com/science/article/pii/S0378437112002579
5. Lancichinetti, A., Fortunato, S.: Benchmarks for testing community detection algorithms on directed and weighted graphs with overlapping communities. Phys. Rev. E **80**, 016118 (2009). https://doi.org/10.1103/PhysRevE.80.016118
6. Lancichinetti, A., Fortunato, S., Kertész, J.: Detecting the overlapping and hierarchical community structure in complex networks. New J. Phys. **11**(3), 033015 (2009)
7. Lancichinetti, A., Radicchi, F., Ramasco, J.J., Fortunato, S.: Finding statistically significant communities in networks. PloS One **6**(4), e18961 (2011)
8. Lung, R., Suciu, M., Gasko, N.: Noisy extremal optimization. Soft Comput. 1–18 (2015). https://doi.org/10.1007/s00500-015-1858-3
9. Lusseau, D., Schneider, K., Boisseau, O., Haase, P., Slooten, E., Dawson, S.: The bottlenose dolphin community of doubtful sound features a large proportion of long-lasting associations. Behav. Ecol. Sociobiol. **54**(4), 396–405 (2003). https://doi.org/10.1007/s00265-003-0651-y
10. Rosvall, M., Bergstrom, C.T.: Maps of random walks on complex networks reveal community structure. Proc. Nat. Acad. Sci. **105**(4), 1118–1123 (2008). http://www.pnas.org/content/105/4/1118.abstract
11. Shannon, C.E.: A mathematical theory of communication. Bell Syst. Tech. J. **27**(3), 379–423 (1948). https://doi.org/10.1002/j.1538-7305.1948.tb01338.x
12. Zachary, W.W.: An information flow model for conflict and fission in small groups. J. Anthropol. Res. **33**(4), 452–473 (1977)
13. Zhang, Z.Y., Sun, K.D., Wang, S.Q.: Enhanced community structure detection in complex networks with partial background information. Sci. Rep. **3**, 3241 (2013). http://dx.doi.org/10.1038/srep03241

One of the Possible Ways to Find Partial Dependencies in Multidimensional Data

Libor Žák[1(✉)], David Vališ[2], and Lucie Žáková[3]

[1] University of Technology in Brno, Technická 2, 616 69 Brno, Czech Republic
zak.l@fme.vutbr.cz

[2] University of Defence in Brno, Kounicova 65, 662 10 Brno, Czech Republic
david.valis@unob.cz

[3] Jan Amos Komensky University,
Roháčova 63, 130 00 Prague, Czech Republic
zakova.lucie@ujak.cz

Abstract. The article deals with one of the possible ways of determining the impact of only one (or selected group) input variable on the output one. When measuring in a real process, the output variable is affected by multiple input variables. The linear dependence of the output variable on the input ones can be determined by the correlation coefficient (multiple correlation coefficient). The influence of only one input variable (or group) can be expressed using the partial correlation coefficient, which is often different from the correlation coefficient. The aim of the paper is to find a way to modify the measured data to match the correlation coefficient of the modified data to the partial correlation coefficient for the original data. This is illustrated by the amount of engine oil soot in the number of hours of operation and the number of days from oil change.

Keywords: Partial regression · Regression analysis · Field data

1 Introduction

Let us suppose we have a process in which the output variable depends on multiple input variables. Data describing the process was obtained from real traffic and therefore the input variables are random. This means that we do not have output values depending on inputs, when one (or more) input variable changes and the others are constant.

For a planned experiment – with simulated inputs – the entire space of possible inputs is appropriately covered by measuring, and the independence (linear independence) of the input variables is usually met. However, for some real processes, we cannot measure output dependence on inputs as in the planned experiment. The aim is to find also in these processes a dependence of output quantity only on selected input variables provided that the other input variables are constant.

An example of such a process is the production of soot in engine oil, when the carbon black content is based on two input variables, time from exchange (Calendar time [day]) and number of hours of exchange (Operating time [Mh]). It is assumed that as the number of operating time increases, the amount of soot in oil increases, while the

R. Matoušek (Ed.): MENDEL 2017, AISC 837, pp. 254–261, 2019.
https://doi.org/10.1007/978-3-319-97888-8_22

number of soot decreases due to the additive substances in the oil with increasing number of days from replacement and engine inactivity. When analyzing the soot content, the influence of both input variables is combined. We can ask, how would depend the amount of soot on the calendar time only (engine did not run) or on operating time only (the engine is still running – of course, there is a 24Mh = 1 day dependence, but this dependence is not significant).

We will try to modify the output variables (soot) as a result of the influence of other input variables by examining the output variable, depending on only one input.

2 A Brief Description of the Procedure

To eliminate the influence of other input variables a regression analysis of the whole process is used. The input variables are divided into two groups – the variables whose influence we examine and the ones whose effect we want to eliminate. We will perform a complex regression analysis focusing on these two groups. If the regression analysis shows that the influence of these two groups can be separated, then by subtracting the influence of the selected input variables (which we want to make constants) we obtain the dependency of the modified output variable only on pre-selected input variables. This is a similar procedure to calculating the partial correlation coefficient.

Let us suppose that $\boldsymbol{X} = (X_1, \cdots, X_k)$ represents independent variables and a dependent variable is represented by a random variable Y, regression function:

$$Y = \varphi(\boldsymbol{X}, \boldsymbol{\beta}), \tag{1}$$

where $\boldsymbol{\beta} = (\beta_1, \ldots, \beta_m)$ is vector of regression coefficients.

Let us denote $\boldsymbol{X} = (\boldsymbol{X}^1, \boldsymbol{X}^2)$, where $\boldsymbol{X}^1$ are the input variables whose influence on the output variable Y is examined and $\boldsymbol{X}^2$ are the input variables whose effect we want to eliminate. The aim is to find dependence:

$$Z = \psi(\boldsymbol{X}^1, \boldsymbol{\beta}^1) \tag{2}$$

We are getting ready for realization

$$y = \psi(\boldsymbol{x}^1, \boldsymbol{\beta}^1) = E(Y|\boldsymbol{X}^1 = \boldsymbol{x}^1, \boldsymbol{X}^2 = \widehat{\boldsymbol{x}}^2), \tag{3}$$

where $\widehat{\boldsymbol{x}}^2$ is the value vector for the constants of the variables $\boldsymbol{X}^2$. These values are chosen to apply:

$$E(Y) = E\big(\psi(\boldsymbol{X}^1, \boldsymbol{\beta}^1)\big). \tag{4}$$

Separating the influence of these groups means that the regression model $Y = \phi(\boldsymbol{X}, \boldsymbol{\beta})$ can be expressed in the form of:

$$Y = \varphi_1(\boldsymbol{X^1}, \boldsymbol{\beta^1}) * \phi_2(\boldsymbol{X^2}, \boldsymbol{\beta^2}), \tag{5}$$

where * is appropriated binary operation.

In looking for a regression model, we limit ourselves to a linear regression model. Then $Y = \sum_{j=1}^{m} \beta_j f_j(\boldsymbol{X})$ and the separation is considered in the form:

$$Y = \sum_{j=1}^{p} \beta_j^1 g_j^1(\mathbf{X}^1) + \sum_{j=1}^{q} \beta_j^2 g_j^2(\boldsymbol{X}^2). \tag{6}$$

By removing the effect $\boldsymbol{X}^2$ we get the modified output variable Z:

$$Z = Y - \sum_{j=1}^{q} \beta_j^2 h_j(\boldsymbol{X}^2) \tag{7}$$

and we are looking for the final model in form:

$$Z = \sum_{j=1}^{r} \lambda_j h_j(\mathbf{X}^1), \tag{8}$$

where $\boldsymbol{\lambda} = (\lambda_1, \ldots, \lambda_r)$ is vector of regression coefficients.

3 Assessment of Oil Field Data

3.1 Oil Field Data

The procedure of separation is shown on searching for the dependence of the soot formation (in [%]) in the engine oil (output variable) on the number of calendar time (in [day]) and the number of operating time (in [Mh]) (input variables). For clarity, we will indicate in the formulas the calendar time - day and operating time - Mh. Data was obtained from a real process – 443 measurements are available.

The correlation between the day and Mh inputs is 0.407558, there is the strong dependence on the input variables (the more days, the higher the number of Mh). Thus the measured values of output data (soot) are dependent on both inputs. In this case, if we examine the dependence of soot on only one input variable so that the second input variable is neglected, we get distorted values. This illustrates the comparison of partial correlation coefficients (see Table 1) with coefficients of linear regression function found (see Figs. 2 and 3).

For the regression model, we chose the regression line (the regression line is sufficient for the description (see [1–3, 11]) (Fig. 1).

Table 1. Correlation and partial correlation between variables

correlation	Mh	soot
day	0.40755821	-0.0796209
Mh		0.5070207

partial correlation	Mh	soot
day	0.52133265	-0.3636912
Mh		0.59264278

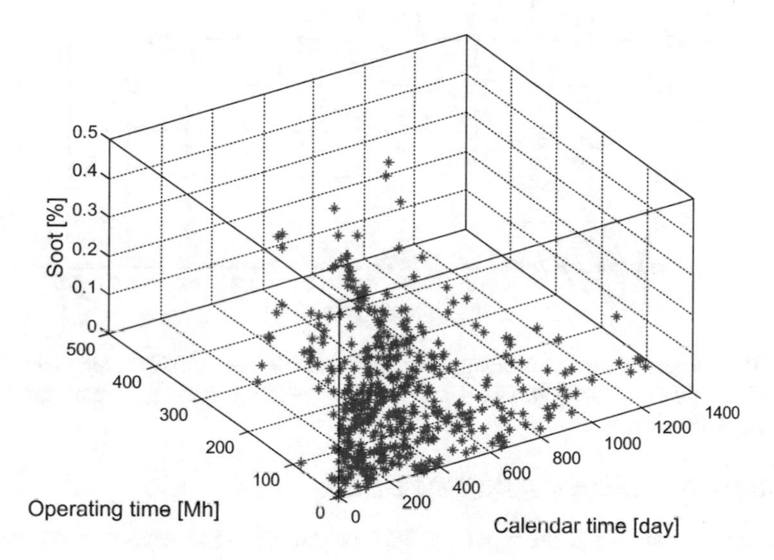

Fig. 1. Dependence of soot on calendar time [day] and operating time [Mh].

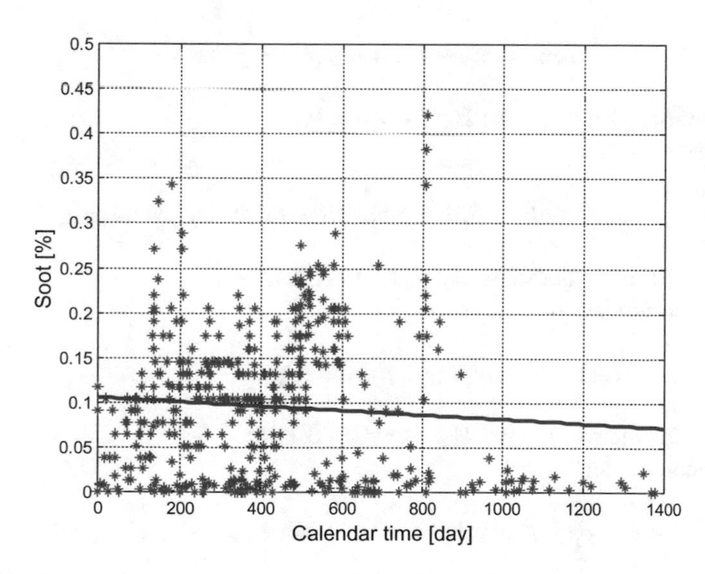

Fig. 2. Dependence of soot on calendar time [day], correlation: −0.0792, regression function: soot = 0.1056876 − 2.38662E-05day, determination coefficient: 0.0059, p-value for the whole model test: 0.0476.

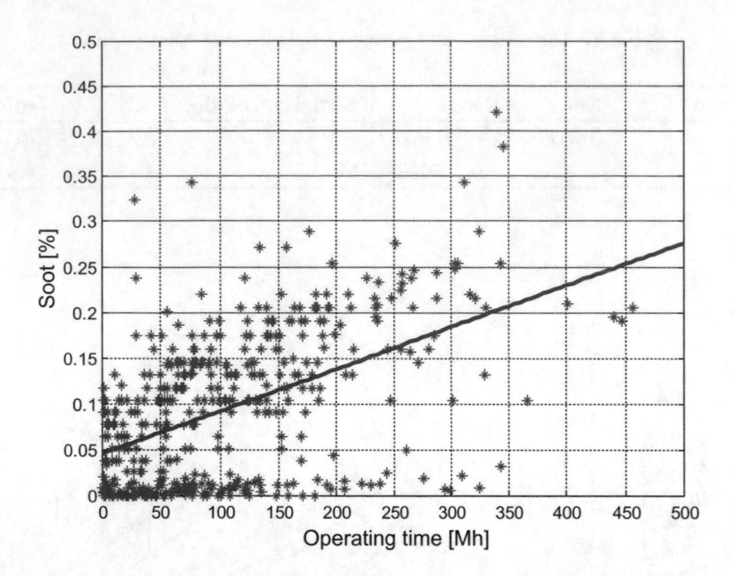

Fig. 3. Dependence of soot on operating time [Mh], correlation: 0.507021, regression function: soot = 0.0461660 + 0.000459698Mh, determination coefficient: 0.2567, *p*-value for the whole model test: 2.55182E-30.

3.2 Regression Analysis of Oil Field Data

Regression analysis – can the inputs of the day and Mh be separated in the regression model? The following linear regression models were tested:

Regression model:

$$\text{soot} \; = \beta_0 + \beta_1 \text{day} + \beta_2 \text{Mh} + \beta_3 \text{dayMh}, \tag{9}$$

p- value for the hypothesis: H_0: $\beta_3 = 0$ is 0.740.

Regression model:

$$\text{soot} \; = \beta_0 + \beta_1 \text{day} + \beta_2 \text{day}^2 + \beta_3 \text{Mh} + \beta_4 \text{dayMh}, \tag{10}$$

p- value for the hypothesis: H_0: $\beta_4 = 0$ is 0.542.

Regression model:

$$\text{soot} = \beta_0 + \beta_1 \text{day} + \beta_2 \text{Mh} + \beta_3 \text{Mh}^2 + \beta_4 \text{dayMh}, \tag{11}$$

p- value for the hypothesis: H_0: $\beta_4 = 0$ is 0.199.

Regression model:

$$\text{soot} = \beta_0 + \beta_1 \text{day} + \beta_2 \text{day}^2 + \beta_3 \text{Mh} + \beta_4 \text{Mh}^2 + \beta_5 \text{dayMh}, \tag{12}$$

p- value for the hypothesis: H_0: $\beta_5 = 0$ is 0.078.

Regression model:

$$\text{soot} = \beta_0 + \beta_1\text{day} + \beta_2\text{Mh} + \beta_3\text{Mh}^2 + \beta_4\text{dayMh}, \tag{13}$$

p- value for the hypothesis: H_0: $\beta_4 = 0$ is 0.054.
Regression model:

$$\text{soot} = \beta_0 + \beta_1\text{day}^2 + \beta_2\text{Mh} + \beta_3\text{dayMh}, \tag{14}$$

p- value for the hypothesis: H_0: $\beta_3 = 0$ is 0.527.
Zero hypotheses: H_0: $\beta_j = 0$ for other parameters are rejected for all models.

3.3 Modified Oil Field Data

Based on the regression analysis, the inputs of the day and Mh can be separated. Based on previous investigations (see [1–6, 11]), we restrict to a regression plane:

soot = β_0 + $\beta_1$day +β_2Mh and we get:

$$\text{soot} = 0.076197205 - 0.000102898\text{day} + 0.000586548\text{Mh}. \tag{15}$$

This relationship is used to convert the amount of soot in dependence only on days:

$$\text{soot}_{\text{day}} = \text{soot} - 0.000586548\text{Mh} \tag{16}$$

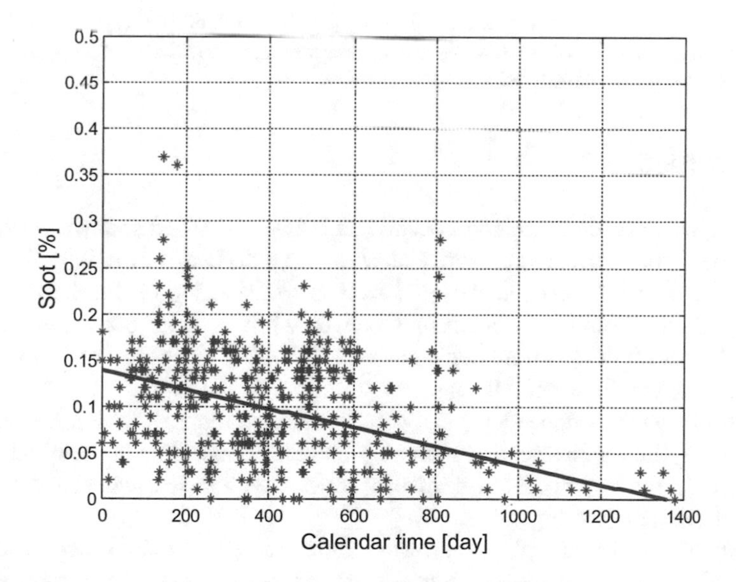

Fig. 4. Dependence of soot on calendar time [day], correlation: −0.393124063, regression function: $\text{soot}_{\text{day}} = 0.13922203 - 0.000102898\text{day}$, determination coefficient: 0.15455, p-value for the whole model test: 7.94426E-18.

and only on Mh:

$$soot_{\text{Mh}} = \text{soot} + 0.000102898\text{day}. \quad (17)$$

The values are adjusted to the extent that: $E(Y) = E\left(\psi(\boldsymbol{X^1}, \boldsymbol{\beta^1})\right)$. (see Figs. 3 and 4).

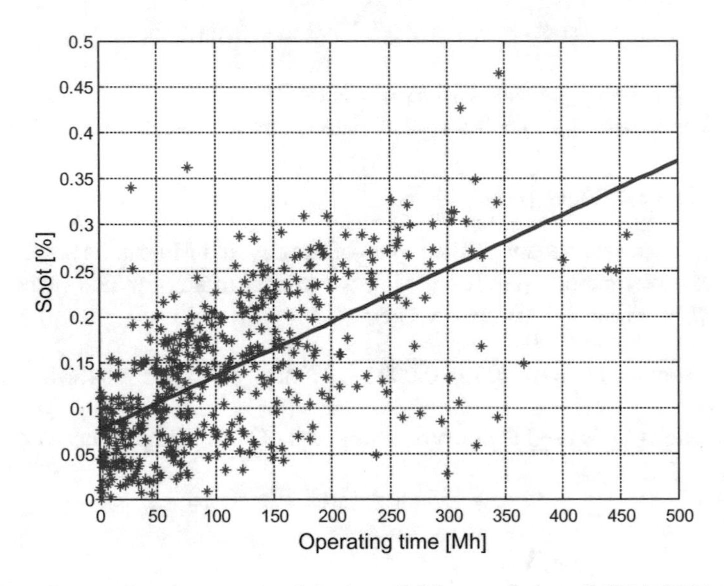

Fig. 5. Dependence of soot on operating time [Mh], correlation: 0.627412386, regression function: $\text{soot}_{\text{Mh}} = 0.076197 + 0.00058655\text{Mh}$, determination coefficient: 0.39365, *p*-value for the whole model test: 7.0502E-50.

4 Conclusion

For the original data, the correlation coefficient between day and soot was −0.0796209, whereas the partial correlation coefficient was −0.3636912. After editing the data, the correlation coefficient between day and soot is −0.393124063. Furthermore, in the original data, the correlation coefficient between Mh and the soot was 0.5070207, whereas the partial correlation coefficient was: 0.59264278. After correction, the correlation coefficient between Mh and the soot is 0.627412386. Thus, the data after modification resembles dependence only on the selected input variables. Because this is a real process, the requirement of normality should be met. For the original data (see Figs. 2 and 3), the hypothesis of normality residual is rejected, whereas the modified data is not rejected (see Figs. 4 and 5).

For simplicity, the simplest regression functions (line, plane) were used in the article. A similar procedure would repeat even for more complex regression functions. The line and plane were selected with respect to dates when these regression functions describe the data with sufficient precision. It is advisable to consult the expert on the measured process.

Acknowledgements. Outputs of this project LO1202 were created with financial support from the Ministry of Education, Youth and Sports under the "National Sustainability Programme I".

References

1. Žák, L., Vališ, D.: Comparison of Regression and Fuzzy Estimates on Field Oil Data, MENDEL 2015, pp. 83–89. Brno University of Technology, VUT Press, Brno (2015)
2. Vališ, D., Žák, L., Pokora, O.: Engine residual technical life estimation based on tribo data. Eksploatacja i niezawodnosc-Maint. Reliab. **2014**(2), 203–210 (2014)
3. Vališ, D., Žák, L., Pokora, O.: Contribution to systém failure occurence prediction and to system remaining useful life estimation based on oil field data. J. Risk Reliab. **2014**(1), 33–42 (2014)
4. Abrahamsen, E.B., Asche, F., Milazzo, M.F.: An evaluation of the effects on safety of using safety standards in major hazard industries. Saf. Sci. **59**, 173–178 (2013)
5. Woch, M., Kurdelski, M., Matyjewski, M.: Reliability at the checkpoints of an aircraft supporting structure. Eksploatacja i niezawodnosc-Maint. Reliab. **17**(3), 457–462 (2015)
6. Vališ, D., Žák, L., Vintr, Z., Hasilová, K.: Mathematical analysis of soot particles in oil used as system state indicator. In: IEEE International Conference on Industrial Engineering and Engineering Management, IEEM 2016. IEEE, Bali (2016). ISBN 978-1-5090-3664-6
7. Mouralová, K., Matoušek, R., Kovář, J., Mach, J., Klakurková, L., Bednář, J.: Analyzing the surface layer after WEDM depending on the parameters of a machine for the 16MnCr5 steel. Measur., J. Int. Measur. Confed. (IMEKO) **2016**(94), 771–779 (2016). ISSN 0263-2241
8. Mouralová, K., Bednář, J., Kovář, J., Mach, J.: Evaluation of MRR after WEDM depending on the resulting surface. Manuf. Technol. **2**, 396–401 (2016)
9. Woch, M.: Reliability analysis of the pzl-130 orlik tc-ii aircraft structural component under real operating conditions. Eksploatacja i niezawodnosc-Maint. Reliab. **19**(2), 287–295 (2017)
10. Hasilova, K.: Iterative method for bandwidth selection in kernel discriminant analysis. In: Talasova, J., Stoklasa, J., Talasek, T. (eds.) 32nd International Conference on Mathematical Methods in Economics (MME), Olomouc, Czech republic, Mathematical methods in economics (MME 2014), pp. 263–268 (2014)
11. Valis, D., Zak, L., Vintr, Z., Hasilova, K.: Mathematical analysis of soot particles in oil used as system state indicator. In: International Conference on Industrial Engineering and Engineering Management (IEEM), Bali, pp. 486–490 (2016)
12. Valis, D., Hasilova, K., Leuchter, J.: Modelling of influence of various operational conditions on li-ion battery capability. In: IEEE International Conference on Industrial Engineering and Engineering Management (IEEM), Bali, pp. 536–540 (2016)

Emergent Phenomena in Complex Systems

Jiri Bila(✉)

Institute of Instrumentation and Control Engineering,
Czech Technical University in Prague,
Technicka 4, 166 07 Prague 6, Czech Republic
bila@vc.cvut.cz

Abstract. The paper turns attention to opinion of Robert Laughlin (Nobel Prize laureate) saying that “Emergence is an organization principle”. There are recapitulated our recent results in the field of emergence phenomena in Complex systems and especially the detection of emergent situation indicating violations of structural invariants. The paper moves in the border between physics and cybernetics with rich references to special fields of mathematics. A lot of concepts and approaches are original ones and open the space for novel ways of the basic research.

Keywords: Emergent phenomena · Structural invariants
Detection of emergent situation · Violation of structural invariants

1 Introduction

This paper has been inspired by the book of Robert B. Laughlin “A Different Universe (Reinventing Physics from the bottom down.)” [1]. Laughlin understands and introduces “emergence as an organization principle” and from this point of view explores a wide field of physics. He does not write directly “a principle of organization of matter” – may be because, he got Nobel Prize for quantum liquids. This principle appears (according to Laughlin) not only in creation of particles but also in formation of sounds, tunes (and as a consequence in formation of speech) and in many other areas. In the context of our recent works [6–9] is pleasant to see that Laughlin introduces some emergent situations that we described and detected by means of violation of structural invariants (e.g., violation of symmetries, supersymmetries or structural invariants in hydrodynamics (Re)) in Table 1 in paper [9] and then in works [6–8].

The factor which was for Laughlin very important and may be changed slightly his point of view was the “collective behavior” and the condition of many elements in observed system. Collective behavior has already been pointed out in works of Haken [4] and in many works exploring synchronization phenomena of large groups of living elements (fish in ocean [17], birds in long way travels, spontaneous light impulses of fire flies). The same conditions of many elements we find nowadays in papers with topics of Genetic Algorithms, Genetic Programming and Evolution Computing [23, 24, 26] or in papers with multiagent problems [5].

R. Matoušek (Ed.): MENDEL 2017, AISC 837, pp. 262–270, 2019.
https://doi.org/10.1007/978-3-319-97888-8_23

In this paper we want to discuss the statements of R. Laughlin in a wider context of disciplines and to compare his conclusions with our result in detecting emergent situations [6–9].

2 Related Works

Papers introduced in this Section are not the best works of their authors. However – because I want to attract an interest in the described field of problems I introduce works that are available and easy for study.

Robert Laughlin approaches the complex systems as systems which are able to change their structure (and execute emergence phenomena) from inside the system not explicitly excited by external actions. (Though these external influences (regarding their intensity) is impossible to neglect.) The same approach we see in works [3, 4, 6–10, 17, 18]. On the other side stays approach of authors [5, 27] that understand the emergence phenomena dependent on the external effects and influences. This second approach brings problems if it is understood as an implication that emergent situations are possible to model and simulate and that the defense against negative emergent situations is in thousands of simulation experiments. (Let us imagine five hundred airplanes above an airport. If we accept that this complex system contains a sufficient portion of chaos then any number of simulations experiments (preformed before) does not bring success and it is impossible to use it for management of airplanes in real time.) Literature sources [19, 20] point out one approach of computational chemistry that is near to understanding complex systems with some temporary lines of research [11, 12].

There exists a large number of sources referring artificial intelligence in connection with complex systems and especially with emergence phenomena in creative activities and problem solving. From this field we name only one source [18] in which are another references pointed out research line managed by John Gero.

3 Emergence as an Organization Principle

Some properties of "Complex system" seem to be stable nowadays. Here there are:

- Many elements in mutual interactions,
- Multidimensionality,
- Quasistability in state changes,
- Nonlinear characteristics,
- Self-organizing processes,
- Emergent behavior,
- Motions in the border of chaos,
- Non stochastic future,
- Inclination to network and multiagent organizations.

Similarly the concept "Emergent situation" has a certain conceptual background [6–8]:

Emergent Situations of Type A – Weak Emergent Situations. The causes of these situations and their output forms (outputs, shapes) are known. They can be recognized and their appearance can be predicted. Examples of processes and systems that generate such situations are: the Belousov-Zhabotinski reaction; environments for initiating solitons; the oregonator; the brusselator. They all belong to the field of Synergetics.

Emergent Situations Class B. The causes of these situations are not known, but their output forms are known. Such situations have the following properties:

(b1) The situation appears suddenly without an explicit association with situations of the previous relevant "time-space" context of the system. (The reason of it may be in insufficient evidence of possible previous situations (cognitive reason) or the system changes its structure "from inside". In most cases we assume a mixture of both the variants.)
(b2) The situation appears as a discrete concept meaningful in the mind of the observer, e.g., a behavior (of a group of termites), an object (a photograph), a shape (e.g., a design of a sculpture) or a property (super conductivity).
(b3) The global reason for the appearance of this type of situation is a violation of the system structure (not a violation of the system function). In other words, the situation is induced by a jump in the system structure (the author of [17] speaks about saltatory changes).
(b4) The detailed reasons and the internal causes of the appearance of the situation are not known (and therefore is impossible to propose a complete prediction model). (However "the shape" of the emergent situation is known - i.e., it is known how it looks like.)
(b5) The appearance of a situation of this type can be detected. Situations belonging to this class include: a change in behavior strategy in a swarm colony; the appearance of floods; the appearance of rough waves; traffic jams.

Emergent Situations of Class C. Neither the causes nor the output forms of these situations are known. Such situations have the same properties as the situations from class (B), with the exception of item (b4), which has the following content:

(c4) No model of a situation of this kind is available before it first operates. (It is not known how the situation looks like.) Situations that belong to this class include: potential instabilities in ecosystems; the appearance of artifacts in nano-structures; discovery situations in Problem Solving; a violation of supersymmetries in quantum mechanics.

For some cases of EMSs of type (B), and especially of type (C), the model of EMSs is unavailable (b4, c4). In these cases, an investigation is made of the structure of the environment in which an emergent situation appearance is "anticipated", and the theory of the violation of structural invariants is applied.

We may see that characteristics of emergent situations could gradually fulfill Laughlin image about an organization principle. In work [9] was introduced the table with appearance of emergent situations in various applications with their detection by means of structural invariants. Though the results of this work have not been cited by anyone (as I know) the table will be hard to avoid for someone who will deal with emergent situations in complex systems.

The fact that is good known by Laughlin is an emergence of new particles as a consequence of violation of symmetry (as an structural invariant). Laughlin did not tell explicitly where and in which is violated the symmetry (but we know it from quantum mechanics). As for the violation of SUSY (SuperSymmetry) we anticipate gravitation particles (gravitons) [3, 13–15], and as for the violation of more essential symmetries we can name axions [16]. Before we reconstruct table from work [9] we recapitulate what Laughlin understood as emergences:

- Generation of new particles as a consequence of violation of symmetry (especially in quantum mechanics).
- Appearance of discrete (better is to say decomposable) components in continuous environments (e.g., sounds and tunes in gases, liquids and solid materials and bodies). (In this sense to this group belongs also the rings and vortices of Pavel Osmera [25].)
- Emergence of novel properties (e.g., superconductivity, Debye temperature).

To this collection belongs a long file of emergent phenomena, emergent situations and emergent components that are not directly related to Laughlin:

- Emergent situations in networks of macro world (e.g., floods, traffic jams).
- Emergence of novel solutions and technological designs (e.g., gas mantle of Carl F. Auer von Welsbach).

(Some other will be referred in Table 1 and in the legend to this Table.)

There are also some small interesting confluences in Laughlin's statements. For example – the whole book [1] is a presentation of emergent phenomena, emergent situations and self organizing processes in various fields of physics (from Newton laws, through acoustics, superconductivity till quantum mechanics). On the other hand he told (in work [2]) word for word "… in environments where appear self organization and emergences do not hold laws of quantum mechanics". (Yes. It is selected from the dialog with a newsman from Der Spiegel however they were words of Nobel Laureate.) Some beginning student of physics could be internally excited by firing question "What is valid from nowadays physics?".

What is not in concordance with Laughlin's image of emergence principle are nano technological structures that are formed by self organization or by self assembly [28]. (There appears a natural question "What is more powerful in structuring of matter: self organization and self assembly (as natural procedures of Nature) or emergence (as an organization principle?".) Very interesting is Laughlin's "meeting" with artificial intelligence. Though we form this discipline it is worthwhile to read one of many comments to Laughlin's book: "A Different Universe" proposes that consciousness and intelligence are emergent properties. Building artificial intelligence based on conventional computer technology is therefore a fools errand. Computers are machines that are designed and programmed to do specific things. No matter how big or fast the computer is, it will only do what it is programmed to do. If consciousness is indeed an emergent property, it would also, by definition, be insensitive to the underlying physics from which it emerged. So even if carbon-based lifeforms were not present in the universe, life and even intelligence could (in principle) emerge from a completely different underlying physical process, perhaps in a self-organized plasma. The possibility of intelligence without carbon-based life should give proponents of a questionable "finely-tuned universe" based on the silly "anthropic principle" second thoughts.

4 Emergent Situations and Structural Invariants

Legend for Table 1:

Table 1. DESs (DUXs) … detection of emergent situations (detection of unexpected situations).

Field of application →	Networks		Monitoring		Cycles		Creativity		Special physical systems		Super gravity D = 11	
Reason of emergence ↓	IND	SI	IND	SI	IND	SI	IND	SI	IND	SI	IND	SI
Formation of new dimension	I1 I4	(MB, M)			I5	ALG TRS	I3 I5	DE IAT			I1 I5	DFM
Appearance of new properties			I1 I2 I3	HD RUL	I5	ALG TRS			I1	SYM S		
Violation of some structural or physical law									I1 I4	SYMS	I1 I5	Dual of FM
Restructuring from inside	I1 I2 I3	(MB, M)	I1 I2 I3	HD RUL	I2 I4	ALG TRS	I3 I7	DE IAT	I1 I4	SYMS	I1 I4	Dual of FM

Application or the area of application: **Networks** (hydrological and transport networks) [6, 7], **Monitoring** … DESs in monitoring of ecosystems or bioengineering systems [27], **CYCLES** … DESs in processes with characteristic cycles (machines, biorhythms, circadian cycles, ECG signals) [8], **Creativity** … DESs in creative processes [18], **SPECPH SYSTEMS** … DESs in special physical and chemical systems [1, 10, 28], **Super gravity** D = 11 … DESs in systems inspired by Kaluza-Klein mechanism in D = 11 supergravity, [13–15].

Reasons of Emergent situations: **Formation of new dimension** … dimension is considered as an element of some basis (matroid basis, vector space basis), **Appearance of new properties** … appearance of new relevant variables, **Violation of some structural or physical law** … e.g., violation of transitivity in an equivalence relation, **Restructuring from inside** … in cases where restructuring form inside of the system prevails external influences.

INDicators of emergent situations: **I1** … very large or very small quantities of standard variables, **I2** … unexpected sequences and coincidences of actions in behavior, **I3** … non resolvable situations, **I4** … unexpected configurations of I/O quantities, **I5** … appearance of novel solution – artifact.

Structural invariants:

- **(MB, M)** … **M**atroid **B**ases, **M**atroid,
- **(DM, RUL)** … **D**ulmage-**M**endelsohn Decomposition, Set of evaluated **RUL**es,
- **(HD, RUL)** … **H**asse **D**iagram, Set of evaluated **RUL**es,
- **(DE, IAT)** … **D**egree of **E**mergence, **I**nterpretation **AT**tractors,
- **(ALG TRS)** … **ALG**ebra of **TR**ansformation**S**,
- **(SYMS)** … physical and geometrical **SYM**metrie**S**,
- **(DFM)** … **D**ual of **F**ano **M**atroid.

There have been introduced model examples of environments where these invariants as detectors of emergent situations appeared. In the recent years where investigated at first invariants (**MB, MC, EZ**), lately denoted as (**MB, M**) [6–8, 29] and (**ALG TRS**), [8, 10]. In both cases there have been used theoretical background from Matroid theory [30] and the Ramsey theory of graph [21, 22].

Table 1 is actualized table from work [9] (without references to older cases).

5 Computer Chemistry (CCH) and Problem Solving (PS)

In this section will be investigated similarities and differences in motivations and formal means of Computer Chemistry and Problem Solving. CCH is one of disciplines which uses consequences of emergent phenomena and that works with some formal means as PS. From many sources we cite here the approach of prof. V. Kvasnicka and his colleagues [19] that began to use theory of sets, theory of graphs, algebra and powerful numerical methods for modeling of molecules. They developed alternative to application of methods of quantum chemistry (at first to operations with Schrödinger equation) – that indicated some numerical problems (at that time). It was not only computations of geometrical positions of atoms but also computations of values of some other important variables as electronic charge density, bind structure, optical susceptibility and others. (Today are used special program packages corrected by exacting experiments [11, 12].)

Let us investigate in more details the description of complex systems in disciplines introduced in the title of this Section.

Very important for complex system description are two factors: level of the description and the basic group (compartment) of complex system elements. Here does not exist any rule how to do it. However the both are formed with the goal to visualize the main features of the considered complex system. For our disciplines we have:

- CCH … molecules (e.g. in [19, 20]) or cells of crystals (e.g. in [11]),
- PS … knowledge and situations (especially initial and final situations, boundary situations) (e.g. in [18]).

Emergent phenomena in complex systems depend more than in cases of classical systems on cognitive dispositions of the observer. At first we have to say what we consider as “cognitive” and “not cognitive”. Simply we can say that all what is consciously registered by our mind is cognitive. It means that all what we observe and in further we process is cognitive. It means that does not exist any “objective reality”.

Exists only model of reality. There are many models of reality simply ordered according to “a distance” of “reality”. Models that are near to “reality” explain how the processes of “reality” function. Models more distanced of reality introduce how the processes of reality can be represented, signed, computed. For the needs of this paper we will consider only two levels of models: a level very near to reality (denoted as NAT) and the level more distanced of “reality” (denoted here as SYMB). The description of process that we want to indicate in complex system compartment has in NAT and SYMB the following form:

$$\mathrm{NAT} : \mathrm{S1} \rightarrow (\mathrm{S1} \oplus \mathrm{s}) \rightarrow \text{Chaotic phase} \rightarrow \mathrm{SOP} \rightarrow \mathrm{EP} \rightarrow \mathrm{S2}, \tag{1}$$

S1, S2 are compartments of complex systems. S1 is extended by a sub compartment “s” and goes through a chaotic phase, phase of self organization and phase of emergent phenomenon into S2. Symbol “⊕” has no specific significance and depends on a real case of method application. SOP symbolizes “Self Organizing Process” and EP is “Emergent Phenomenon”.

$$\mathrm{SYMB} : \mathrm{SM1} \rightarrow (\mathrm{SM1} \otimes \mathrm{sm}) \rightarrow \mathrm{SM2}, \tag{2}$$

where SM1, SM2 are sign models representing S1 and S2, and sm is a sign sub model that extends SM1.

In our investigated fields we find:

- CCH:

$$\mathrm{SYMB}_{\mathrm{CCH}} : \mathrm{G_e} \rightarrow (\mathrm{G_e} \otimes \mathrm{G_R}) \rightarrow \mathrm{G_P}, \tag{3}$$

  where G_e is a graph of so called educts, G_R is a graph of chemical reaction and G_P is a graph of a product of the reaction.
- PS:

$$\mathrm{SYMB}_{\mathrm{PS}} : \mathrm{K_i} \rightarrow (\mathrm{K_i} \otimes \mathrm{K_a}) \rightarrow \mathrm{K_f}, \tag{4}$$

  where K_i is a structure of initial knowledge, K_a is a structure of additional knowledge and K_f is a structure of final knowledge.

What is similar in both the discussed disciplines: The proper processes go below the NAT level.

What is different: The operations with graphs in CCH model the syntheses of new matters and mainly their products, not emergences. Though, according to Laughlin, chemistry is a natural space for self organizing processes and emergent phenomena in CCH they are completely below NAT. On the other hand – in PS are described conditions for appearance of emergent situation. If we consider expression (4) - after addition of K_a appears self organization and emergence that lead to K_f.

6 Conclusion

In the proposed paper were introduced concepts of complex systems and emergent situations. There was actualized the table od structural invariants originally introduced in [9] and there was extended the Laughlin's thesis "Emergence is an organization principle" [1] into fields outside the physics. The paper corresponds to Laughlin's interest in complex systems and emergences (that was motivated by a final phase of reductionism) and contributes to a starting phase of age of emergences and uncertainties.

Acknowledgment. This research has been supported by means of SGS17/P2301/OHK2-015. This support is very gratefully acknowledged.

References

1. Laughlin, R.B.: A Different Universe (Reinventing Physics from the Bottom Down). Basic Books, New York (2006)
2. Grolle, J., Schmundt, H., Laughlin, R.B.: Der Urknall ist nur Marketing. Der Spiegel (1) (2008)
3. Nills, H.P.: Supersymmetry, supergravity and particles physics. Phys. Rep. (Rev. Sect. Phys. Lett.) **110**(1–2), 1–162 (1984)
4. Haken, H., Wagner, M.: Cooperative Phenomena. Springer, Heidelberg (1973)
5. Navarro, I., Matia, F.: A survey of collective movement of mobile robots. Int. J. Adv. Rob. Syst. **10**, 1–9 (2013)
6. Bila, J.: Processing of emergent phenomena in complex systems. Int. J. Enhanc. Res. Sci. Technol. Eng. **3**(7), 1–17 (2014)
7. Bila, J.: Emergent phenomena in natural complex systems. In: Synaiei, A., et al. (eds.) Proceedings of Interdisciplinary Symposium of Complex Systems, ISCS 2014. Emergence, Computation and Complex Systems, vol. 8, pp. 89–100. Springer, Heidelberg (2014)
8. Bila, J., Mironovova, M., Rodríguez, R., Jura, J.: Detection of emergent situations in complex systems by violations of structural invariants on algebras of transformations. Int. J. Enhanc. Res. Sci. Technol. Eng. **4**(9), 38–46 (2015)
9. Bila, J.: The detection of emergent situations by structural invariants. In: Matousek, R. (ed.) Proceedings of 17th International Conference on Soft Computing, MENDEL 2011, MENDEL, vol. 17, pp. 534–539. Brno University of Technology, VUT Press, Brno (2011)
10. Bila, J.: The syntheses of technological materials as emergences in complex systems. In: Matousek, R. (ed.) Proceedings of 20th International Conference on Soft Computing, MENDEL 2014, pp. 305–310. Brno University of Technology, VUT Press, Brno (2014)
11. Reshak, A.H., Khan, S.A., Kamarudin, H., Bila, J.: NaAuS chicken-wire-like semiconductor: electronic structure and optical properties. J. Alloy. Compd. **582**, 6–11 (2014)
12. Reshak, A.H., Alahmed, Z.A., Bila, J., et al.: Exploration of the electronic structure of monoclinic α-$Eu_2(MoO4)_3$: DFT-based study and X-ray photoelectron spectroscopy. J. Phys. Chem. C **120**(19), 10559–10568 (2016)
13. Horava, P., Witten, E.: Nucl. Phys. B **460**, 506–524 (1996)
14. Nieto, J.A.: Matroid theory and supergravity. Rev. Mexicana de Física **44**(4), 358–361 (1998)

15. Toppan, F.: Extended supersymmetries in one dimension. Acta Polytech. **48**(2), 58–74 (2008)
16. Hagmann, C., Van Bibber, K., Roseneberg, L.J.: Axions and other very light bosons. J. Phys. **G33**(1), 431–432 (2006)
17. Reid, R.G.B.: An Emergence Theory, in Biological Emergences. Evolution by Natural Experiment, pp. 361–400. Massachusetts Institute of Technology, Massachusetts (2007)
18. Kryssanov, V.V., Tamaki, H., Kitamura, S.: Understanding design fundamentals: how synthesis and analysis drive creativity, resulting in emergence. Artif. Intell. Eng. **15**, 329–342 (2001)
19. Kvasnicka, V., Kratochvil, M., Koca, J.: Mathematical Chemistry and Computational Solutions of Syntheses. Academia, Prague (1987). (in Czech)
20. Jensen, F.: Introduction to Computational Chemistry. Wiley, New York (1999)
21. Ramsey, F.P.: On a problem of formal logic. Proc. Lond. Math. Soc. **30**, 264–286 (1930)
22. Nesetril, J., Rödl, V.: A structural generalization of Ramsey theorem. Bull. Am. Math. Soc. **83**, 127–128 (1977)
23. Matoušek, R.: GAHC: hybrid genetic algorithm. In: Advances in Computational Algorithms and Data Analysis. Lecture Notes in Electrical Engineering, vol. 14, pp. 549–562 (2009)
24. Brandejsky, T.: The use of local models optimized by genetic programming in biomedical-signal analysis. In: Handbook of Optimization from Classical to Morden Approach, pp. 697–716. Springer, Heidelberg (2012)
25. Osmera, P.: Vortex-fractal-ring structure of electron. In: Matousek, R. (ed.) Proceedings of 14th International Conference on Soft Computing, MENDEL 2008. MENDEL, pp. 115–120. Brno University of Technology, VUT Press, Brno (2008)
26. Kotyrba, M., Volna, E., Bujok, P.: Unconventional modelling of complex system via cellular automata and differential evolution. Swarm Evol. Comput. **25**, 52–62 (2015)
27. Bila, J., Pokorny, J., Jura, J., Bukovsky, I.: Qualitative modeling and monitoring of selected ecosystem functions. Ecol. Model. **222**, 3640–3650 (2011)
28. Fojtík, A., et al.: NANO, a fascinating phenomenon of nowadays. COMTES FHT, Prague (2014). (in Czech)
29. Bila, J.: Detection of emergent situations in complex systems by structural invariant (BM, M). In: Submitted for the 22nd International Conference on Soft Computing, MENDEL (2017)
30. Oxley, J.G.: Matroid Theory. Oxford Science Publications, Oxford (2001). Reprinted Edition

Utilization of Machine Learning in Vibrodiagnostics

Daniel Zuth(✉) and Tomáš Marada

Faculty of Mechanical Engineering, Institute of Automation and Computer Science, Brno University of Technology, Brno, Czech Republic
{zuth,marada}@fme.vutbr.cz

Abstract. The article deals with possibilities of use machine learning in vibrodiagnostics to determine a fault type of the rotary machine. Sample data are simulated according to the expected vibration velocity waveform signal at a specific fault. Then the data are pre-processed and reduced for using Matlab Classification Learner which creates a model for identifying faults in the new data samples. The model is finally tested on a new sample data. The article serves to verify the possibility of this method for later use on a real machine. In this phase is tested data preprocessing and a suitable classification method.

Keywords: Vibrodiagnostics · Neuron network
Classification learner · Machine learning · Matlab · PCA
Classification method · SVN · Static unbalance · Parallel Misalignment

1 Introduction

The article deals with the use of machine learning in vibrodiagnostics. The aim is to create a signal generation model that simulates a specific failure of the rotating machines and another aim is to select appropriate classification methods. Basic faults are selected for which the expected vibration waveform is evident and their simulated signals are tested in the Matlab environment in the Classification learner application where different classification methods are tested.

For learning, a 1,000 sample signals will be generated, with a random sequence and a distribution of the individual faults. The learned model will be tested on new 10,000 sample signals. The result will be an overview of the success of the predictions of a particular failure of the rotation machines. The selected classification method and created model will be implemented in the embedded machines in the future and will be tested on real data. In this article, the time domain data only is used against the publication [2,3] where frequency analysis was used.

2 Selected Faults of Rotating Machines

Three faults that have a known signal in the frequency domain are selected for initial testing. They are "*Static Unbalance*" (F1), "*Parallel Misalignment*" (F2)

R. Matoušek (Ed.): MENDEL 2017, AISC 837, pp. 271–278, 2019.
https://doi.org/10.1007/978-3-319-97888-8_24

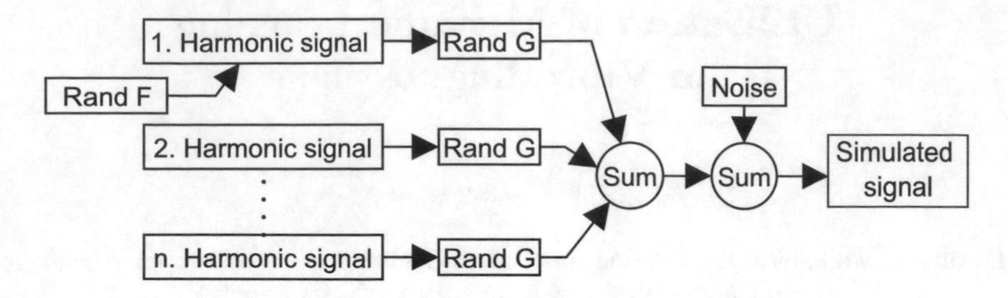

Fig. 1. Signal simulation diagram

and "*Mechanical Looseness - Loose base*" (F3). A faultless machine (F0) is also simulated for comparison.

F0 - Faultless Machine - In this case, only a moderately increased frequency of the first harmonic and the noise signal is added. Schematic representation of a rotary machine and the expected FFT signal is in Fig. 2 (F0). Generated simulated signal (Time domain and FFT) for this case is in Fig. 3 (F0).

F1 - Static Unbalance - This fault manifests itself by a significant increase in the first harmonic frequency and by a slight increase of the second one. The phase of signal on bearing stands is the same, unlike from the dynamic unbalance. Schematic representation of a rotary machine fault is in Fig. 2 (F1) and the expected FFT signal according to [1,4,5] is in Fig. 2 (F1). Generated simulated (Time domain and FFT) signal for this case is in Fig. 3 (F1).

F2 - Parallel Misalignment - This fault manifests by a strong vibration with a phase shift, usually dominates the second harmonic frequency. Schematic representation of a rotary machine fault is in Fig. 2 (F2) and the expected FFT signal according to [1,4,5] is in Fig. 2 (F2). Generated simulated signal (Time domain and FFT) for this case is in Fig. 3 (F2).

F3 - Mechanical Looseness - Loose Base - This fault manifests itself by a significant increase of the second harmonic frequency and the presence of subharmonics frequencies. Schematic representation of a rotary machine fault is in Fig. 2 (F3) and the expected FFT signal according to [1,4,5] is in Fig. 2 (F3). Generated simulated (Time domain and FFT) signal for this case is in Fig. 3 (F3).

3 Fault Simulation

The final waveform for the machine learning is formed by mixing harmonic signals. To get closer to the real one, the signal is affected by a noise signal (80% of the normalized first harmonic amplitude of machines in faultless condition). Harmonic signals have randomly shifted frequency from −10% to +10% ("RAND F" in Fig. 1) and amplitude of each component from −10% to 10% ("RAND G" in Fig. 1). This is schematically shown in Fig. 1. For the purposes of the machine learning is generated 1000 samples with defects, which are randomly chosen from random sequences.

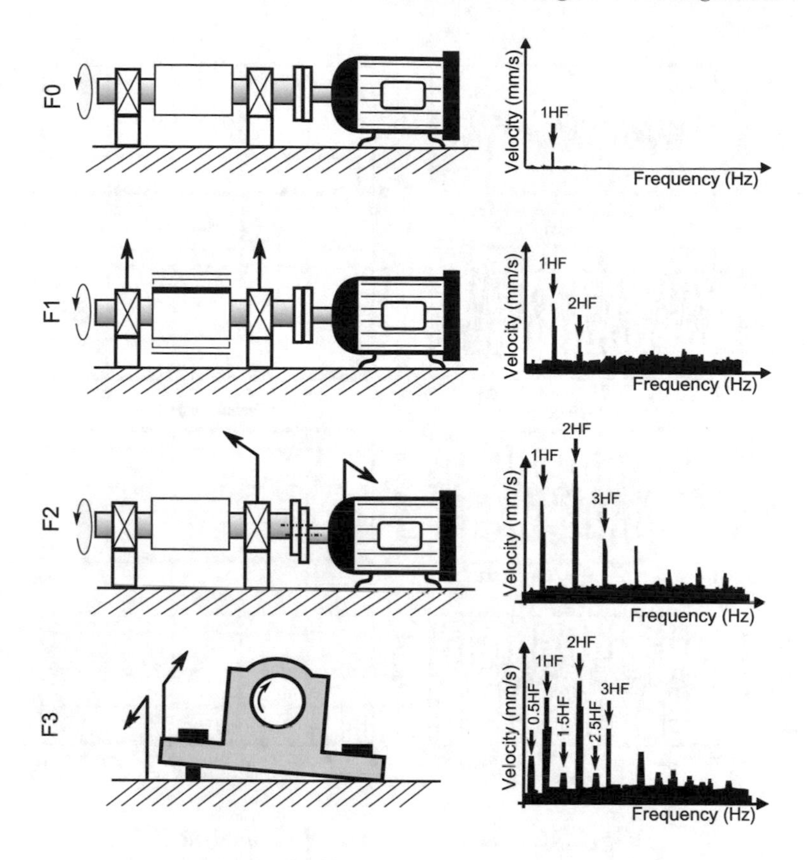

Fig. 2. Fault and expected FFT signal

4 Verification of Simulated Signal

In our laboratory it is possible to simulate the fault *F1- Static unbalance* on vibrodiagnostics models with the instrument Viditech 2500CV [8]. Figure 4 (F0) is the final frequency analysis of faultless machine running and Fig. 4 (F1) is the frequency analysis of machine with fault F1 with unbalance weighing 5 g. The instrument 2500CV displays only the frequency analysis deprived of noise thanks to the filtering and averaging. From the measurement, it is obvious a rough 300% increase in the first harmonic frequency and a slight increase in the second harmonic frequency. Machine revolutions speeds were set at about 1350 rpm. The same speeds were set in fault simulations, so the first harmonic frequency is $1350/60 = 22.4$ Hz.

5 Data Processing

Classification of faults is performed according to the measured sample time domain data (not frequency domain) as can be measured via an AD converter

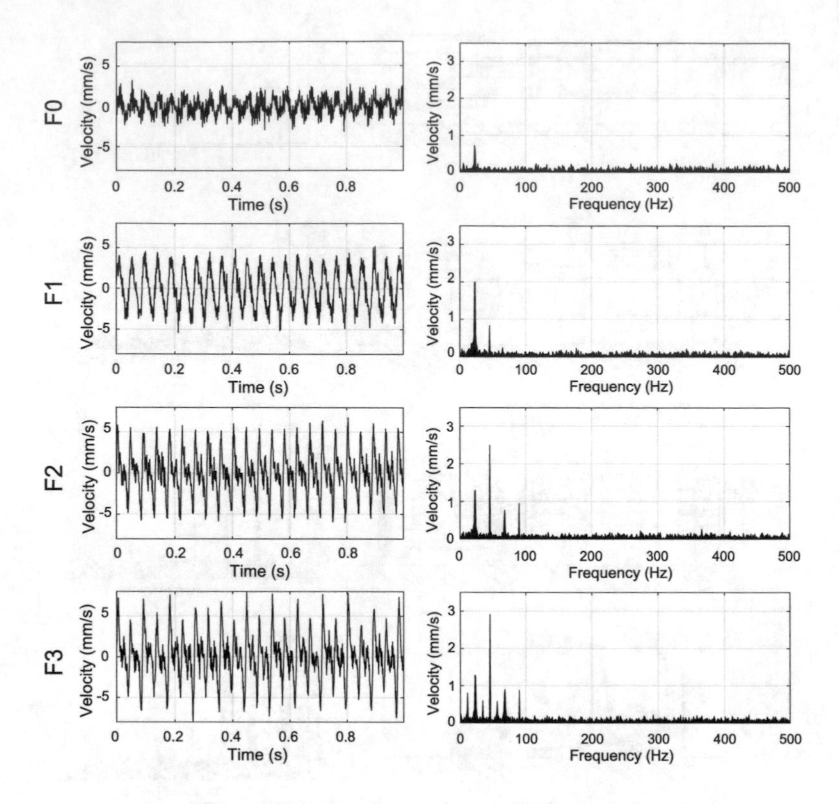

Fig. 3. Simulated signal with FFT analysis

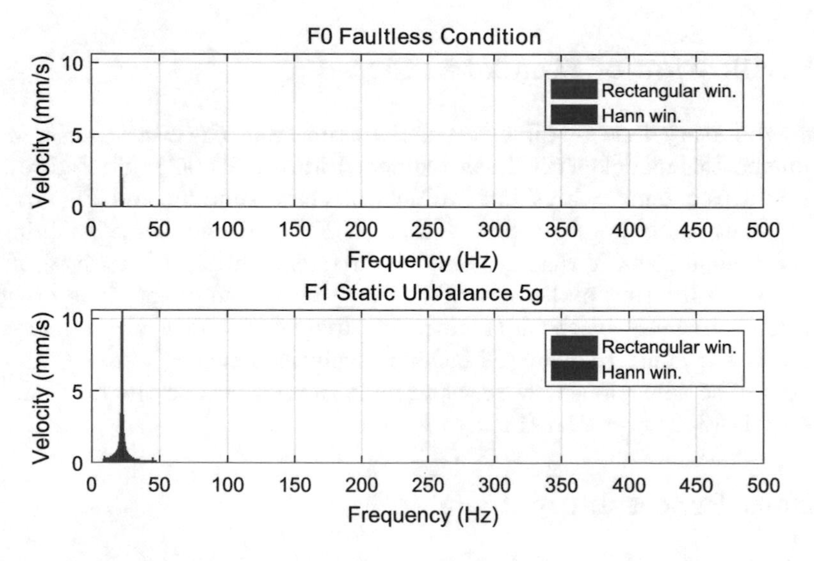

Fig. 4. Measured signal from CV2500 for F0 and F1

with a sampling frequency 1000 sps and number of samples 1000. This signal is processed using three functions that are RMS, STD and PCA. Selecting functions will be a subject for further testing, the goal will be to select a function with the least demands on a computing power but with good properties for the classification of faults. The pre-processed data are used to classify faults using Classification Learner (Fig. 5).

RMS Function - The RMS value is commonly used in vibrodiagnostics to determine machine condition for exemple ISO 10816 (Mechanical Vibration of Machines with Operating speed from 10 to 200 rev/s) [6]. In this case, the `rms()` Matlab function is used.

STD Function - Standard deviation. In this case, the `std()` Matlab function is used.

PCA Function - Principal component analysis of raw data. In this case, the `pca()` Matlab function is used.

6 The Choice of Classification Methods

Classification Learner can test a sample data for different classification methods and choose the most successful one to generate a classification model. All used methods are compared in Table 1. The best result is "Linear SVM" model and the success prediction is presented via confusion matrix in Fig. 6. The same results were also "Coarse Gaussian SVM" method, but the final classification model use "Linear SVM".

6.1 Testing the Classification Model

The final model is tested on a sample of new data, when 10000 samples are generated for each fault, and then the result is classified according to a previously created classification model. The success rate faults prediction is in Fig. 7.

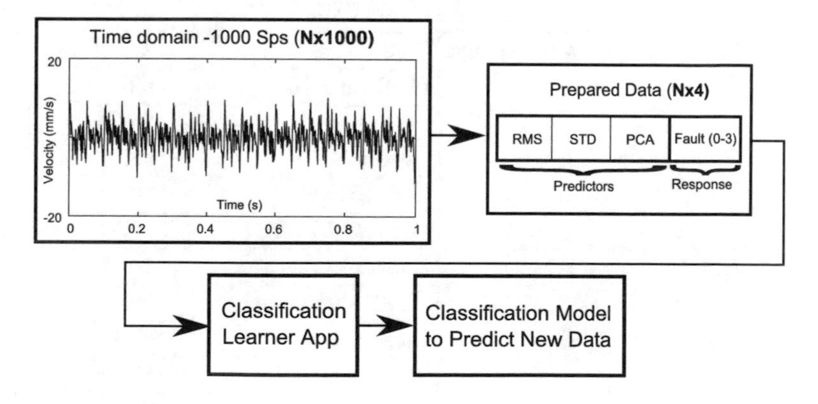

Fig. 5. Data processing for classification learner app

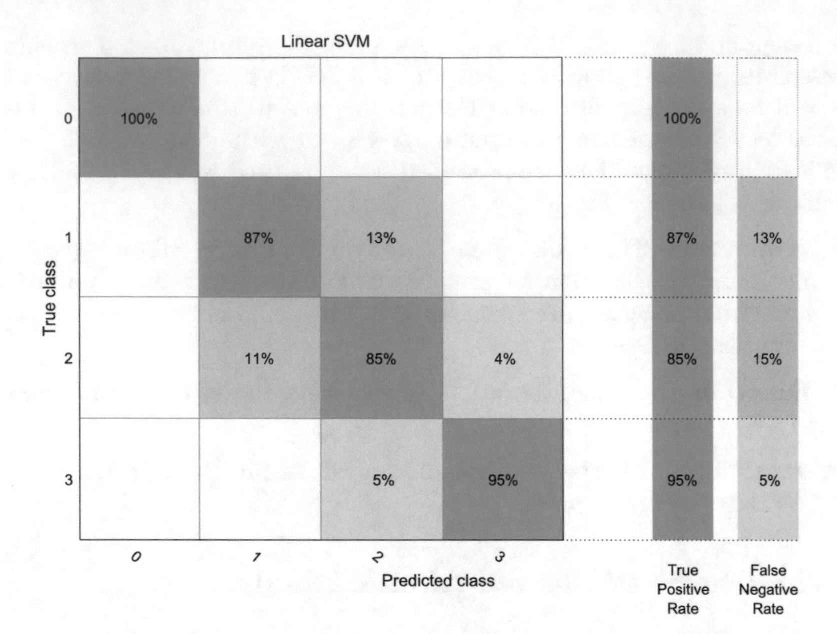

Fig. 6. Result of the best model from classification leraner

Table 1. Compare classification method from classification learner

Decision trees:	Complex tree	87.9%
	Medium tree	89.8%
	Simple tree	90.9%
Support vector machines:	Linear SVM	**91.6%**
	Fine Gaussian SVM	90.6%
	Medium Gaussian SVM	91.2%
	Coarse Gaussian SVM	**91.6%**
	Quadratic SVM	91.5%
	Cubic SVM	91.2%
Nearest neighbor classifiers:	Fine KNN	87.9%
	Medium KNN	90.4%
	Coarse KNN	86.0%
	Cosine KNN	68.9%
	Cubic KNN	90.6%
	Weighted KNN	89.4%
Ensemble classifiers:	Boosted trees	89.8%
	RUSBoost trees	90.4%
	Bagged trees	88.9%
	Subspace KNN	78.1%
	Subspace discriminant	91.3%

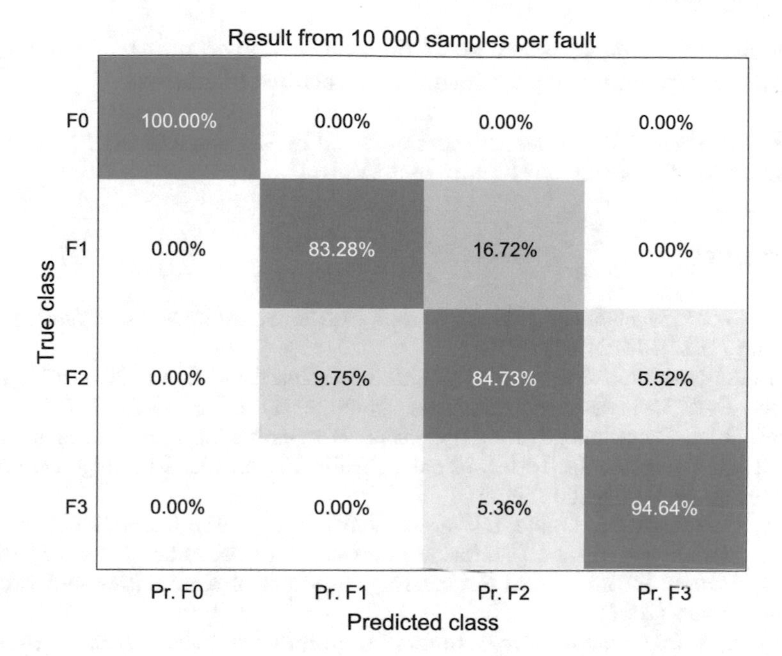

Fig. 7. The result of the model on the new test data

Even in this case the signal is formed by mixing harmonic signals and the signal is affected by a noise signal (80% of the normalized first harmonic amplitude of machines in faultless condition) and harmonic signals have randomly shifted frequency (from −10% to +10%) and amplitude of each component (from −10% to 10%).

7 Conclusion

This article deals with the possibilities of using an machine learning to classify faults of the rotary machine. The principle is based on a simulation of expected waveforms of individual faults and generate a sufficient amount of sample for machine learning and then obtain a classification model.

The simulation contains some factors on purpose as they may occur on a real machine. They are: an added noise signal, a random change of amplitudes of the harmonic signals and a shifty revolution frequency. The success of the classification rate in practice depends on realistic simulations and also depends on data preprocessing functions before the machine learning, and these factors will be tested in the future. Specific methods are compared in Classification Learner app and the results of each method for a specific package samples is in Table 1.

The results summarized in Fig. 7 indicate a successful application of this method where the classification of the faultless state is uniquely identified (100%)

and classifying the faults is not lower than 83%. The accurate determination of fault type is very important for future maintenance planning.

Acknowledgment. This research was supported by the grant of BUT IGA No. FSI-S-14-2533: "Applied Computer Science and Control".

References

1. Broch, J.T.: Mechanical Vibration and Shock Measurements. Brüel & Kjaer, Naerum (1984). ISBN 8787355361
2. Ypma, A.: Learning Methods for Machine Vibration Analysis and Health Monitoring Theses. Delft University of Technology, Delft (2001). ISBN 90-9015310-1
3. Maluf, D.A., Daneshmend, L.: Application of machine learning for machine monitoring and diagnosis. In: 10th International Florida Artificial Intelligence Research Symposium, pp. 232–236 (1997)
4. SKF: Vibrodiagnostic Guide, DIF s.r.o., San Diego (1994). CM5003-CZ
5. Zuth, D.: Analýza vibrací (Vibration analysis). Theses. Brno University of Technology, Faculty of Mechanical Engineering, Institute of Automation and Computer Science, Brno (2004)
6. Zuth, D., Vdoleček, F.: Měření vibrací ve vibrodiagnostice. Automa: časopis pro automatizační techniku. FCC Public, Praha (1994). ISSN 12109592
7. Vdoleček, F.: Terminology in branch of measurement uncertainties [Terminologie v oboru nejistot měření]. Akustika **16**(1), 40–42 (2012). ISSN 1801-9064
8. ViDiTech :: 2500CV. http://www.viditech.cz/index.php/home-cs/products/online-monitory/2000cv/. Accessed 20 Apr 2017

Deep Analytics Based on Triathlon Athletes' Blogs and News

Iztok Fister Jr.[1(✉)], Dušan Fister[2], Samo Rauter[3], Uroš Mlakar[1], Janez Brest[1], and Iztok Fister[1]

[1] Faculty of Electrical Engineering and Computer Science, University of Maribor, Smetanova 17, 2000 Maribor, Slovenia
iztok.fister1@um.si
[2] Faculty of Mechanical Engineering, University of Maribor, Smetanova 17, 2000 Maribor, Slovenia
[3] Faculty of Sport, University of Ljubljana, Gortanova 22, 1000 Ljubljana, Slovenia

Abstract. Studying the lifestyle of various groups of athletes has been a very interesting research direction of many social sport scientists. Following the behavior of these athletes' groups might reveal how they work, yet function in the real-world. Triathlon is basically depicted as one of the hardest sports in the world (especially long-distance triathlons). Hence, studying this group of people can have a very positive influence on designing new perspectives and theories about their lifestyle. Additionally, the discovered information also helps in designing modern systems for planning sport training sessions. In this paper, we apply deep analytic methods for discovering knowledge from triathlon athletes' blogs and news posted on their websites. Practical results reveal that triathlon remains in the forefront of the athletes' minds through the whole year.

Keywords: Artificial Sport Trainer · Data mining · Data science
Lifestyle · Triathletes · Websites

1 Introduction

Triathlon is regarded as one of the more popular sports today. It consists of three different sports disciplines, i.e., swimming, cycling and running, that are conducted consecutively one after the other. Because of competing in three different sport disciplines, triathlon is also called a multi-sport. It requires a holistic approach for planning sport training from the sports trainer's standpoint and demands a so-called healthy active lifestyle from the athlete's standpoint [8]. Due to the complexity of the triathlon, almost all muscles of the athlete's body need to be captured during the sports training process.

On the other hand, there are various distances available for triathlon athletes. Beside the Ironman triathlon [5] that is recognized as the king of all triathlons or "holy grail" for triathlon athletes[1], there also exist, among others, Ultra

[1] http://www.ironman-slovenia.com/content/view/229/1/.

R. Matoušek (Ed.): MENDEL 2017, AISC 837, pp. 279–289, 2019.
https://doi.org/10.1007/978-3-319-97888-8_25

triathlon, Half ironman triathlon, Olympic triathlon and Sprint triathlon. The Ironman triathlon was born in Hawaii and consists of three marathon distances in each sports discipline, i.e., 3.8 km of swimming, 180 km of biking and 42.2 Km of running, Ultra triathlons multiply these distances by factors of 2x for Double, 3x for Triple, 4x for Quadruple, 5x for Quintuple and 10x for Deca Ultra triathlon, while Half ironman, Olympic and Sprint triathlons divide the distances of the Ironman's disciplines by a half, quarter (almost exactly) and fifth (almost exactly), respectively.

Triathlons have attracted a really high number of athletes in recent years. For the majority of people, a sport event is more than just the passive or active spending of leisure time [4]. Moreover, the amount of those athletes that classify themselves as "serious" participants in a triathlon [10] has increased from year to year. Consequently, their participation in the past events, and experience acquired thereby [6], affects an athlete's active lifestyle. Many of them see their participation in sport activities either as a way of asserting themselves, a social event, or a reason for physical activity.

In any case, "serious" participants put the selected sport activity at the forefront of their life. They often develop a sense of belonging to a particular circle of like-minded, physically active people, where everything turns around their preferred sport [7]. Two factors are crucial for distinguishing a professional from amateur athletes. Actually, the professional athletes are recognized due to their immense strenuous endeavors in their sports disciplines that grow into some kind of challenge for them, where they not only compete, but also have faith in themselves. In order to achieve their competition goals, the athletes are also ready to suffer [12]. Thus, it holds that, the more strenuous and difficult the challenge, the stronger the expected pride and satisfaction of the athletes.

Interestingly, the border between the professional and amateur athletes in triathlon have been becoming thinner every day, especially due to the fact that training and racing in this sport demand from all participants to be organized and fully focused on achieving their goals [2]. Actually, triathlon becomes an active lifestyle for the triathlon athletes, regardless of whether they are professionals or amateurs. As a matter of fact, both kinds of athletes need to reconcile their life activities, like family, training and even leisure.

Studying the lifestyle of triathlon athletes becomes really an interesting research topic [11]. A lot of research has been produced in this domain during recent years. However, the lifestyle of triathlon athletes is treated in a slightly different way in this contribution study. Using modern data science approaches, we analyzed triathlon athletes' blogs and news found on their personal websites. Indeed, we are focused on the deep analytic methods, where data from more than 150 personal websites of triathlon athletes were analyzed (some athletes have only blogs). Our goal is to extract features which are in the forefront of the athletes' minds in particular month of the year.

The structure of the remainder of the paper is as follows. Section 2 deals with background information underlying the deep analytic methods. In Sect. 3, outlines of analyzed data are described, as well as illustrating the proposed method. The experiments and results are the subjects of Sect. 4. The paper concludes with Sect. 5, where the directions are also outlined for the further work.

2 Background

Knowledge about particular athletes' groups in sport is very important. On the one hand, it helps psychologists, sociologists, and sport scientists to understand how they work, behave, coordinate liabilities, and so on. On the other hand, it can also be very helpful for computer scientists in building the more efficient intelligent systems for decision-making in place of the real sports trainers. One of such systems is also the Artificial Sport Trainer proposed by Fister et al. in [3]. At the moment, the Artificial Sport Trainer operates on data that were produced basically by the following two sources:

- **Sport activity files:** Are produced by sport trackers or other wearable mobile devices. From these files, several parameters of sports training sessions can be extracted: Duration, distance, heart rate measures, calories, etc.
- **Athlete's feedback:** Is usually obtained from particular athletes in the sense of questionnaires.

Normally, social sport scientists have not relied on the data extracted from blogs residing on websites until now. On the other hand, data from websites seems to be also a very important source of discovered information that can be used by deep analytic methods. Deep analytics enable organizations to learn about entitles occurring in big data, such as people, places, things, locations, organizations and events, and use the derived information in various decision-making processes [9]. Typically, entities can consist of actions, behaviors, locations, activities and attributes. Nowadays, social networks represent one of the biggest origins for producing the big data. In line with this, Facebook and Twitter produce more than 200 million text based messages per day as generated by 100s of millions of users. The big data analytics are composed of a complex layered technology running on multiple infrastructures. Indeed, big data vary in volume, variety and velocity of processing. The primary output of the deep analytics are discovered patterns and models that can be applied to knowledge discovery (e.g., prediction analysis) and visualization. In our pioneering work, we propose a novel method for analyzing websites of particular athletes' groups in long-distance triathlons based on deep analytic methods. The method, together with used materials, is presented in the next section in detail.

3 Materials and Methods

The purpose of this section is twofold. On the one hand, data sources that were applied in the study are illustrated, while, on the other hand, the proposed method for deep analytics is described in detail. This method is intended for discovering those entities that most highlight the active lifestyle of triathlon athletes in the duration of one year.

3.1 Outline of Data

Data have been collected on various blogs and news residing on personal websites of triathlon athletes that were found through search engines. Thus, data of amateur and professional athletes were collected, although data from the latter group of athletes prevail. Indeed, the professional athletes live from their own publicity and, consequently, they need to have a good digital identity. Therefore, some professionals have great and attractive websites, where much valuable information can be found. For instance, the more exciting websites that are also included in this study are disposed by the following athletes: Skye Moench[2], Adam Kacper[3], Cody Beals[4], etc. However, the most important data from these websites were blog posts and news reports.

3.2 Proposed Method for Deep Analytics on Triathlon Athlete Feeds

The proposed method for deep analytics on triathlon athlete blogs and news consists of the following steps:

- Defining objectives
- Obtaining data
- Preprocessing
- Visualization of results.

In the remainder of the paper, the mentioned steps are presented in detail.

Defining Objectives. The main characteristics of triathlon athletes refer to their specific lifestyle, where everything is subordinated to the triathlon. Consequently, athletes' thoughts, beliefs and acts are reflected in blogs and news (also feeds) that are published on their websites. On the other hand, the subjects on these modern communication medias depend on the history of events that have occurred during the triathlon season. For instance, the final triathlon competition takes place traditionally in Hawaii each year. As a result, this fact is observed in increasing feeds that refer to this competition and those islands.

In line with this, the objective for the proposed deep analytic method is to collect and analyze the new feeds posted by the observed triathlon athletes. The analysis is focused on discovering the new feeds, where the frequencies of words detected in the corresponding documents are counted (Fig. 1). Actually, these frequencies highlight those areas in the athlete's active lifestyle that prevail in a specific year season. When these word frequencies are treated historically, relations between words can be indicated. Indeed, the word represents an entity in our deep analysis.

[2] http://skyemoench.com/.
[3] http://www.kacperadam.pl/en/.
[4] http://www.codybeals.com/.

Fig. 1. The figure presents a word cloud generated using English feeds that are saved in a dataset. The word cloud is an image made of words that together resemble a cloudy shape. The size of a word shows how important it is. The bigger word the is, the more frequently it is presented in feeds. In fact, the figure perfectly outlines the active lifestyle of the triathlon athlete. The most frequently used words refer to concepts describing the triathlon sport, as the 'race', 'time', 'run', 'bike', 'swim', 'day', 'course'. The 'race' is the most important concept in an active lifestyle of the triathlon athletes that demands setting the specific race goals. These goals are, typically, posted at the beginning of the triathlon season and demand adapting the sports training plan according to the athlete's current form. On the other hand, concepts such as 'swim', 'bike' and 'run' refer to sports training sessions for the specific triathlon disciplines. Indeed, the remaining concepts in the Figure only complete a whole picture of the athlete's lifestyle. This means that the triathlon athlete really sets triathlon sports activities in the forefront of his life.

Obtaining Data. The new blogs and news data are downloaded from the websites through Rich Site Summary (RSS) feeds using web scraping. The RSS feed enables users to access updates to online content. In our study, the following elements of RSS feed were the object of our interest: Description, published date, link and content. Thus, the last element is assigned with value "EMPTY" in the case that it is not available (due to various RSS feed generators).

Web scraping is used for extracting data from websites. Web scraper software may access the World Wide Web directly using the Hypertext Transfer Protocol, or through a web browser. Our "web scraper" is an automated process implemented as a bot that checks for new data at least once per day. The new data is gathered from the website and saved directly into a dataset of the JavaScript Object Notation (JSON) file format for later retrieval or analysis. The scraper is developed by using the Python package **feedparser**[5] and launching the corresponding methods.

[5] https://pypi.python.org/pypi/feedparser.

Interestingly, we can also indicate by using the web scraper that some triathlon athletes are very active in updating the matter on their websites, while the others, unfortunately, do not really like this. Additionally, our scraper is also capable of fetching non-English feeds. Although the postprocessing of these feeds is not implemented at the moment, they are reserved in datasets for the future work.

Preprocessing. This step is really important in the sense of processing raw data. After processing datasets in JSON format, firstly we removed all HTML tags using beautifulsoup package[6]. After this step, we divided feeds according to the months of their publishing. This means that we divided feeds into 12 groups, as presented in Table 1.

Table 1. Distribution of feeds per months

Month	Number of feeds
January	140
February	135
March	161
April	137
May	190
June	135
July	155
August	194
September	158
October	153
November	169
December	113

Totally, 1,840 feeds have been considered in the study, since we selected only feeds written in the English language. Detection of the language has been performed by Python package langdetect[7]. After this step, a pure text was extracted from the RSS element description and archived in different dataset groups on a monthly basis. When feeds for all months were collected, we started with calculating word frequencies. The NLTK package [1] was used for that task. In the NLTK process, a text is tokenized, then short words and numbers are removed (length <3), all words are written in lower case and, at the end, stop words are removed. After this, frequencies can been calculated for each month.

[6] https://pypi.python.org/pypi/beautifulsoup4.

[7] https://pypi.python.org/pypi/langdetect.

Presentation of Results. The last step is intended for presenting the results. Since humans are visual beings, it is more convenient for them to represent results in a graphical way. For our visualization we used Graphviz[8]. The results of the proposed method are visualized using a social network visualization technique, where a social network is defined as a directed graph with nodes (entities) and edges (relations between nodes) denoting a direction between nodes.

4 Results and Discussion

The purpose of our experimental work was to show that the lifestyle of a typical triathlon athlete really turns around the magic word "triathlon". This word represents the central concept that crucially affects the athlete's thoughts, beliefs and actions. The effects of this focusing are reflected into blogs and news posted on websites by the observed athletes. In this study, data of the athletes were used as proposed in Sect. 3.1.

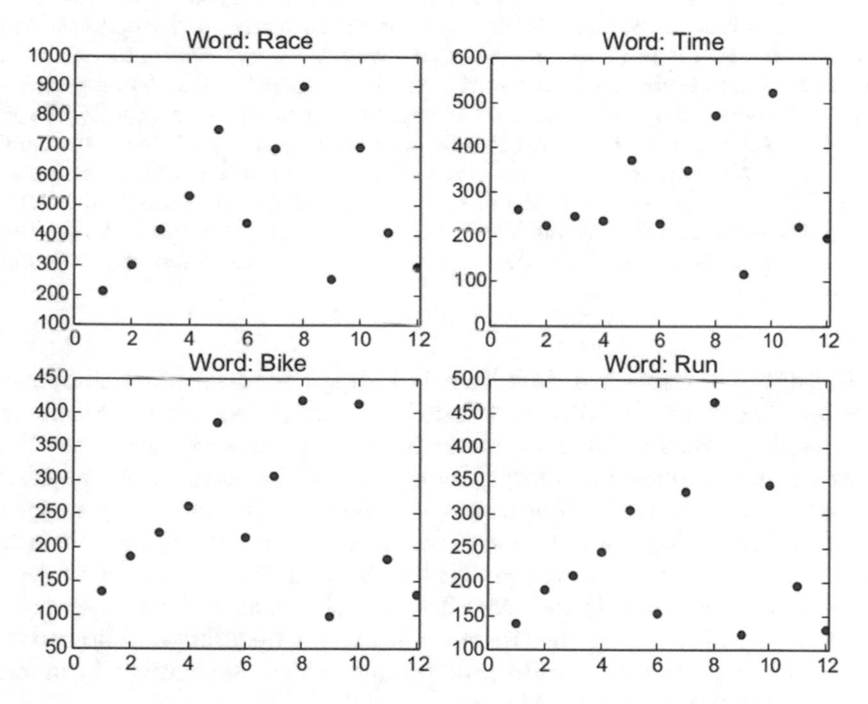

Fig. 2. Frequency of the four most frequently occurring words in feeds according to their appearance throughout the months - frequencies of the words "race", "bike" and "run" increase at the start of the first half of the year, while the word "time" remains relatively constant. In the second half of the year, the frequencies appear disorderly at first sight, although they correlate in all four diagrams.

[8] http://www.graphviz.org/.

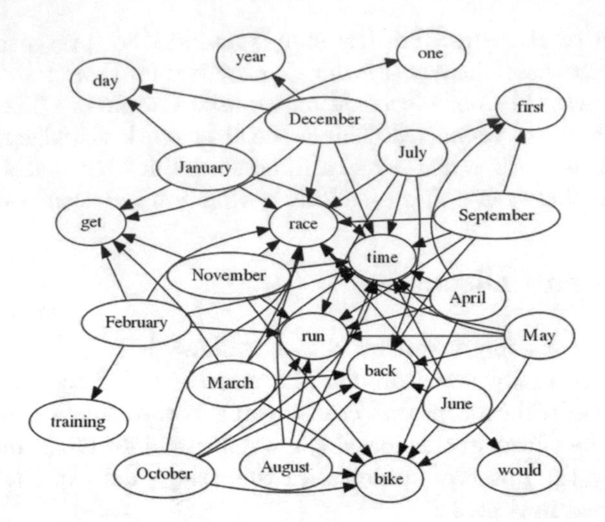

Fig. 3. The results of the coarse-grained network analysis by $K = 5$ - These deep analytics show that, in January, the observed triathlon athletes usually make the sports training plans for the forthcoming year. The entity "January" relates to words "get" and "race" that indicate questions like: "To which race to go?", "How to get permission for participation in the race?" and "How to organize the athlete's schedule?". The word "training" stands out in February, which indicates the beginning of the sports training cycle. The words "running" and "bike" are in the forefront of the athletes' minds from March to June. From July to September, triathlon athletes are focused especially on racing, as can be indicated by the higher frequency of the word "race". At the end of the year, especially in December, they think about events that happened in racing in the outgoing year.

Usually, only the last 10 new feeds (i.e., blogs or news) are maintained in websites. Therefore, the selected websites are monitored by our web scraper and the new feeds are collected automatically into the server. From the feeds, a raw text was extracted and incorporated into word clouds, where frequencies of words which occurred in documents are counted. The word frequencies are then visualized using a social network visualization technique, where relations between months and words are presented as directed edges in graphs. Moreover, a weight is also attached to each edge denoting the number of word occurrences in specific months. The frequencies of the four most frequent word occurring in feeds of observed triathlon athletes according to their appearance in different months of year are illustrated in Fig. 2.

Relations between events, types of sports training sessions and people can be indicated according to a specific year season from the social network. Actually, the time season is referred to by months representing the source node for edge, while the drain node is specified by a word. Actually, the number of edges incident to drain nodes can be limited in the social network by using the parameter K. Indeed, the parameter K determines the graph density. In line with this, when the value of the parameter K is low, the fine-grained analysis of a network

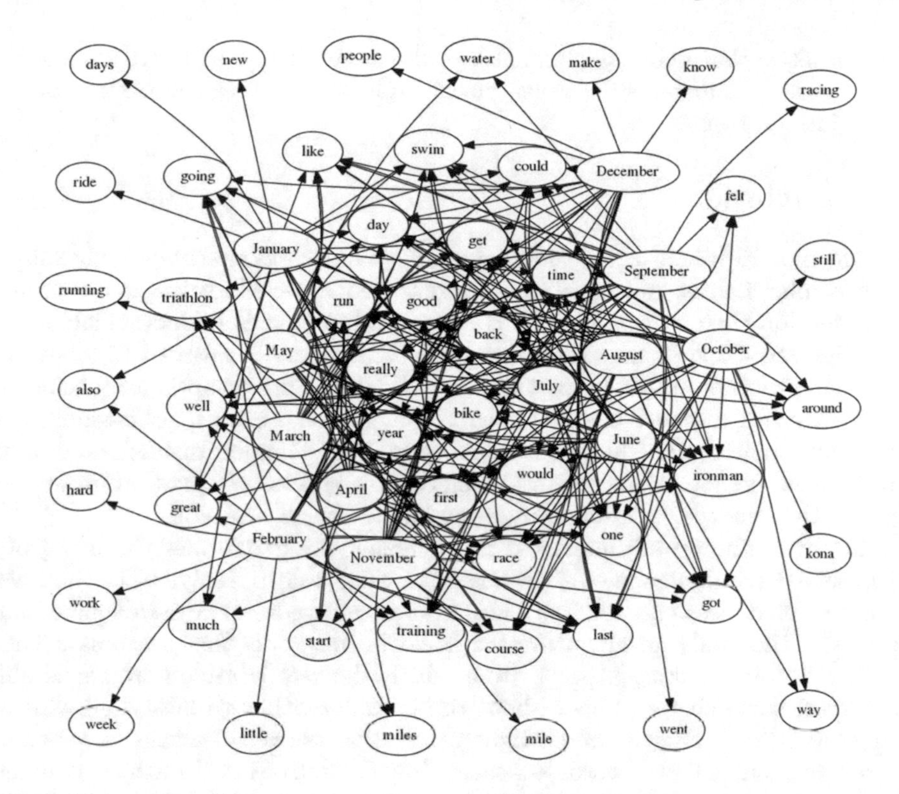

Fig. 4. The results of the fine-grained network analysis K = 25 - It is demonstrated that, in January, triathlon athletes usually think about which cycling sports training sessions to perform in the forthcoming year. In February, they are focused on hard work, as indicated by the words "hard" and "work". The words "well" and "great" indicate that they are satisfied with the performed sports training in the past. In May, they are focused on running as indicated by the word "running". June and July are devoted to thinking about cycling and competing in Ironmans. In the forefront of athletes/minds in August and September, racing prevails due to the higher occurrences of the word "race". The word "Kona" refers to the famous Ironman on Hawaii that represents the official end of the Ironman competition season. In December, social events come to the forefront of the triathlon athletes' minds as can be indicated by the words "know" and "people".

can be performed, and vice versa, when K is high, the network is capable of coarse-grained analysis.

Two analyses of the social network were performed in the study:

- Coarse-grained network analysis, where the five most frequently occurring words in feeds of the observed triathlon athletes were taken into consideration (Fig. 3),
- Fine-grained network analysis, where the 25 most frequently occurring words in feeds of the observed triathlon athletes were taken into consideration (Fig. 4).

Let us notice that detailed explanations about obtained results are assigned to each figure as comments. However, these explanations were contributed by the real triathlon trainer.

5 Conclusion

Yet the old truth holds, that different groups of athletes share the same values, have similar habits, are capable of similar thoughts. Moreover, some groups even develop their own jargon for communication that is understandable only for a limited circle of the group members. This paper is devoted to analyzing one such group of athletes, i.e., triathlon athletes. This group is interesting for deep analytics, because there is a very thin border between professional and amateur athletes' thoughts, beliefs and acts. The common denominator of all group members represents the total focus on the sports training that grows into their active lifestyle, where nothing is as important as triathlon.

In our study, we are interested in analyzing the active lifestyle of a group of observed triathlon athletes. In line with this, a deep analytics method was proposed that uses the updated blogs and news (feeds) as a data source. The results of the analytics are presented in social networks that serve as a basis for the decision-making process, from which the real triathlon trainer is able to extract some characteristics about the triathlon athlete's lifestyle. Based on our study, the trainer observed tight correlation between months of year and increased occurrences of words appearing in the forefront of the athletes' minds. These words reflect faithfully athletes' thoughts, belief and acts that are characteristic for the particular year season. For instance, the triathlon racing is carried on in cycles, i.e., at the beginning of the year, athletes start with planning their sports training sessions, in the middle of the year, training sessions of the specific triathlon disciplines that need to be improved and triathlon competitions are in forefront, while, at the end of the year, the final Ironman in Hawaii and evaluating the past season are the most frequently used subjects from the posted feeds. The results of the proposed method show that it is possible to determine the triathlon active lifestyle by using deep analytic methods from feeds posted in triathlon athletes' websites. Obviously, the mentioned methods are also suitable for application in the other sports disciplines. Beside the blogs and news, Facebook feeds could also serve as a data source for the future. Finally, this method could be applied in other domains as well.

References

1. Bird, S., Klein, E., Loper, E.: Natural Language Processing with Python: Analyzing Text with the Natural Language Toolkit. O'Reilly Media, Inc., Sebastopol (2009)
2. Bridel, W.F.: Finish whatever it takes considering pain and pleasure in the Ironman Triathlon: a socio-cultural analysis. Ph.D thesis. Queens University (2010)
3. Fister Jr., I., Ljubič, K., Suganthan, P.N., Perc, M., Fister, I.: Computational intelligence in sports: challenges and opportunities within a new research domain. Appl. Math. Comput. **262**, 178–186 (2015)

4. Green, B.C., Jones, I.: Serious leisure, social identity and sport tourism. Sport in Soc. **8**(2), 164–181 (2005)
5. Knechtle, B., Nikolaidis, P.T., Rosemann, T., Rüst, C.A.: Der Ironman-Triathlon. Praxis (16618157) **105**(13), 761–773 (2016)
6. Rauter, S.: Mass sports events as a way of life (differences between the participants in a cycling and a running event). Kinesiol. Slov. **20**(1), 5 (2014)
7. Richard, S., Jones, I.: The great suburban Everest: an insiders perspective on experiences at the 2007 Flora London Marathon. J. Sport Tour. **13**(1), 61–77 (2008)
8. Shipway, R., Holloway, I.: Running free: embracing a healthy lifestyle through distance running. Perspect. Public Health **130**(6), 270–276 (2010)
9. Sokol, L., Chan, S.: Context-based analytics in a big data world: better decisions. IBM Redbooks (2013)
10. Stebbins, R.A.: Serious Leisure: A Perspective for our Time, vol. 95. Transaction Publishers, Piscataway (2007)
11. Wicker, P., Hallmann, K., Prinz, J., Weimar, D.: Who takes part in triathlon? An application of lifestyle segmentation to triathlon participants. Int. J. Sport Manage. Mark. **12**(1–2), 1–24 (2012)
12. Willig, C.: A phenomenological investigation of the experience of taking part in extreme sports'. J. Health Psychol. **13**(5), 690–702 (2008)

Weighted Multilateration in Volumetry of CNC Machine Tools

Barbora Navrátilová(✉)

Faculty of Mechanical Engineering, Institute of Mathematics,
Brno University of Technology, Technicka 2896/2, 61669 Brno, Czech Republic
115877@vutbr.cz

Abstract. Our research study describes utilization of a weighted multilateration principle in the multi axis machine error modeling. Multilateration plays on of the most important role in the identification of the exact position of the point in the machine tool work space. Using laser tracer, the measured data were obtained in spherical coordinates. The multilateration attempts to improve data accuracy using only a radial component. Each of measurements has a different quality of the laser-TRACER signal. According to this information, three ways of the weight assignment are presented. The weight of each value of the measurement distance was given as a ration of each signal quality to: the best signal quality that was obtained during the measurement; the sum of all signal quality for each position of the measuring device; the norm of vector of total signal quality for each position of the measuring device. All of our results were compared with the data that were obtained by the original software TRAC-CAL.

Keywords: Accuracy of measurement · Multilateration
Weighted multilateration · Machine tool · LaserTRACER

1 Principle of Multilateration

With increasing demands on the accuracy and the efficiency of machine tools (MT), we still need better means of determination deviation in the work space. Generally, sources of possible errors in the work space can be classified into quasi-static errors and dynamic errors. In the case quasi-static errors, we talk about 60–70% of the total error of the MT [1]. This is one of the main inspiration for many research papers e.g. [2–8]. There are two possible attitudes for description of geometric errors: relative to particular position of the axis or relative to the work space. Regarding geometric errors relative to the work space, we talk about volumetric accuracy of the MT, which is given by volumetric deviation (ve). Generally, ve is a combination of all geometric errors and is defined as the difference between the pairs of nominal points introduced by the numerical control and the actual points achieved by the MT (ref). When the deviations are characterized, their impact can be compensated. The first step is the determination of the exact

R. Matoušek (Ed.): MENDEL 2017, AISC 837, pp. 290–298, 2019.
https://doi.org/10.1007/978-3-319-97888-8_26

position in the work space. The influence of heat sources is neglected in the next section (these issues can be found [9,10]).

The measured data were obtained in the spherical coordinates by using a portable measurement system laserTRACER (LT). However, we need to eliminate the possible sources of inaccuracy in the measurement. In the case of the laser interferometer, there are many different disturbing factors including environmental errors, geometric errors or instrument errors. In order to enhance precision of the measurements, there is an importance of monitoring environmental conditions [11], which alters the refractive index of measuring beam (air pressure, ambient temperature, and humidity).

The maximization of accuracy of the measuring instruments means a reduction of noise effects caused by the inaccuracy of the interferometer or angle encoders. For each point in the spherical coordinates there are two possible angle errors (azimuth, polar) and the error of radial component, see Fig. 1.

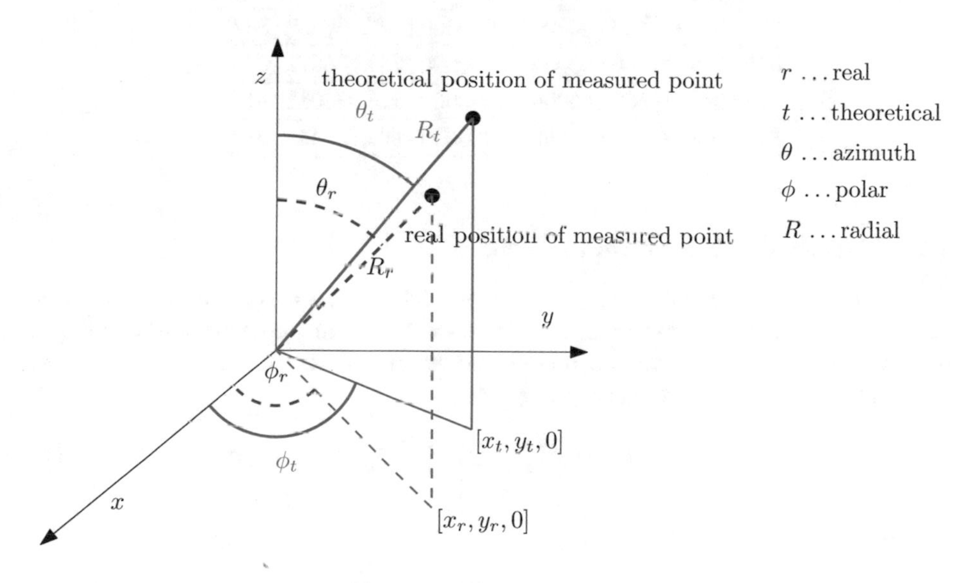

Fig. 1. Measurement noise

The aim of multilateration is to reduce the measurement uncertainty, therefore only the radial component of each point is used. In this case, it is necessary to use more that one measurement apparatus. The accuracy of determination of the point position also depends on the number of LTs and their location [12].

We have to use three or four LTs. The accurate position of the point is given by the intersection of three or four spheres. Each of the spheres is defined by radial component of each LTk and the center of spheres is the position of each LT. In case of four different LT (four different coordinates system), there is a need to transform of these coordinates to one central coordinates system (selfcalibration) [13].

2 Problem Formulation

In our case, only one LT in four different positions for each measurement was used. During the measurement, the enviromental conditions were monitoring. The distance from LT was measured for each point $[x_i, y_i, z_i]$ of the work space. Table 1 reflects the example of output of measurement for one of four LT positions. The data creates input for other calculation.

Table 1. LaserTRACER data – input data of position 4

X [mm]	Y [mm]	Z [mm]	L [mm]	Q [%]
420.000000	500.000000	0.000000	462.742071	84.00
470.000000	500.000000	0.000000	424.331010	85.00
520.000000	500.000000	0.000000	388.424453	86.00
570.000000	500.000000	0.000000	355.736539	84.00
570.000000	450.000000	0.000000	333.076671	80.00
570.000000	400.000000	0.000000	316.343011	81.00
570.000000	350.000000	0.000000	306.438445	83.00

2.1 Multilateration Without Weight

First, we determined the exact position of each LT. In case, that all the measurements are equally important, input parameters are coordinates of all the points introduced by numerical model and measured distances between LT on each position and the measured point [14]:

$$\min_x \|F(x)\|_2^2 = \min_x \sum_i F_i^2(x), \quad (1)$$

where

$$F_i = \sum_{k=1}^{4} \left(\sqrt{(x_i - LT_x^k)^2 + (y_i - LT_y^k)^2 + (z_i - LT_z^k)^2} - (LT_D^k - Pk_i) \right)^2, \quad (2)$$

where index k $(k = 1, \ldots, 4)$ indicates the position of LT, point $[LT_x, LT_y, LT_z]$ is the position of LT, index D denotes the dead distance. The object of minimization is the LTs position.

Then we applied the same process to determine the exact position of point in the work space. The object of minimization is the exact position of the points.

2.2 Multilateration with Weight

Futher, we consider that the determination of a final position dependent on quality of the signal. The vector $\mathbf{q}k$ $(k = 1, \cdots, 4)$ is the vector of value of quality of TRACER signal of LT^k, it is clear that $qk_i > 0$. Weight is given to each measured distance of each LT^{k}

$$w_i = \frac{qk_i}{\max\limits_i \bigcup\limits_k qk_i} \tag{3}$$

then it is clear in this case of weighting that $\forall i \; w_i \in \langle 0, 1 \rangle$. Another possibility

$$w_i = \frac{qk_i}{\sum\limits_{i=1}^{n} qk_i} \tag{4}$$

then it is clear for this case of weighting for each position of LT that $\sum\limits_i w_i = 1$. Another way

$$w_i = \frac{qk_i}{\sqrt{\sum\limits_{i=1}^{n} (qk_i)^2}} \tag{5}$$

then it is clear for this case of weighting for each position of LT that $\|w\| = 1$. To apply this weight to our problem, the function F_i has a following form

$$F_i = \sum_{k=1}^{4} N^k w_i, \tag{6}$$

where

$$N^k = \left(\sqrt{(x_i - LT_x^k)^2 + (y_i - LT_y^k)^2 + (z_i - LT_z^k)^2} - (LT_D^k - Pk_i)\right)^2 \tag{7}$$

When we used Eq. (3), **w** was denoted **w1** (simillary for Eqs. (4) or (5) it was used the expression **w2** or **w3**).

3 Implementation

Previous multilateration formulations were simulated in MALTAB. The trust-region-reflective least squares algorithm was used to provide better stability. We denote a distance vector between LT on the first position and the measured point **P1** (similarly expression **P2**–**P4** for LT on another position). **X**, **Y**, **Z** are vectors of x, y and z coordinates of all points given by a numerical model.

In the first step – we determined of LT positions, the default-input data **X**, **Y**, **Z** and **P1**–**P4**. All the possibilities were used for the calculation $[LT_x, LT_y, LT_z]$: without weight, with weight **w1** (weight given by Eq. (3)), **w2**, **w3**. The obtained results were compared with TRAC_CAL Table 1.

TRAC_CAL data were given to two decimal places. All results of determination of LT position are very similar. These data were used in the second step.

In the second step – object of minimization is the vector composed of **X**, **Y**, **Z**, the default-input data are LT positions and **P1**–**P4**.

Table 2. The comparison of results between data obtained by our algorithm and TRAC_CAL^{TM} data.

Position	Weight	x coordinate	y coordinate	z coordinate
LT1	Without	−113.507274	130.501108	−479.639119
	w1	−113.508087	130.501531	−479.638851
	w2	−113.507710	130.501464	−479.639013
	w3	−113.508080	130.501569	−479.638866
	TRAC_CAL	−113.51	130.5	−479.64
LT2	Without	−114.570762	303.531734	−479.570818
	w1	−114.570956	303.532639	−479.570233
	w2	−114.571365	303.532421	−479.570716
	w3	−114.570986	303.532664	−479.570271
	TRAC_CAL	−114.57	305.53	−479.56
LT3	Without	790.266920	126.768899	−350.205633
	w1	790.269402	126.769190	−350.206325
	w2	790.268415	126.769129	−350.206051
	w3	790.269381	126.769173	−350.206345
	TRAC_CAL	790.27	126.77	−350.21
LT4	Without	802.635818	308.951119	−350.195037
	w1	802.639266	308.952214	−350.195628
	w2	802.638248	308.951829	−350.195450
	w3	802.639264	308.952217	−350.195646
	TRAC_CAL	802.65	308.95	−350.19

Following figures illustrate the influence of selected weighting on the final position error. For better illustration, each of axis error is shown separately. Figure 2 shows x component of volumetric error for each measured point 1–88. There are four information (four different shades of blue) about each point. First information is error in x axis, when we used our algorithm without the weight. Another shades show the values of this error, when different weights were used. The effect of weight is evident. The influence of **w1** and **w2** on the error seems similar.

The sample of exact values of x error is in the Table 3. The similarity between using **w1** or **w2** is evident.

The value of error is less than original value without weight for all of the considered case of weight. The situation **w2** is the closest situation to the value without the weight.

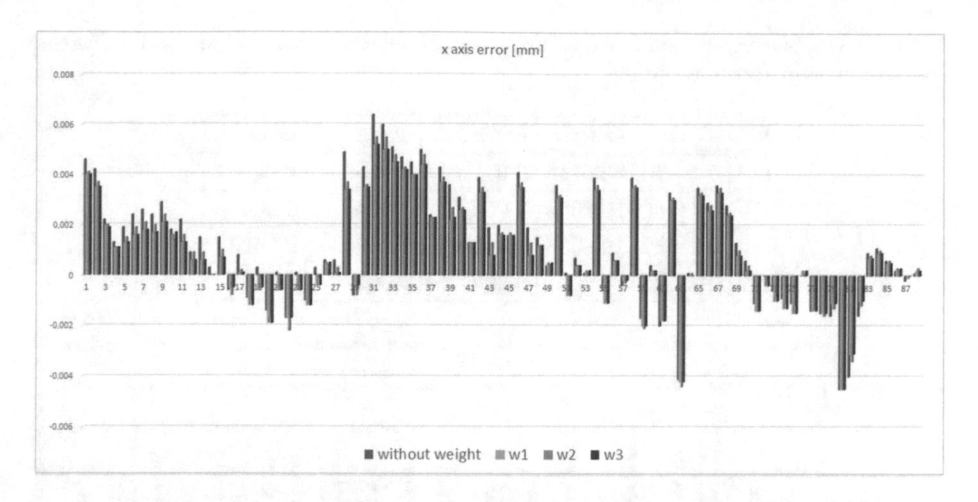

Fig. 2. x axis errors [mm] for measured points in the work space

Table 3. x-axis compression between the measurement of error without weight (error) and with weight (error_w) [mm]

X	Y	Z	Error	Error_w1	Error_w2	Error_w3
170	0	−400	0.0046	0.004	0.0041	0.004
170	0	−350	0.0042	0.0035	0.0037	0.0035
170	0	−300	0.0022	0.002	0.002	0.0019
170	0	−250	0.0013	0.0011	0.0011	0.0011

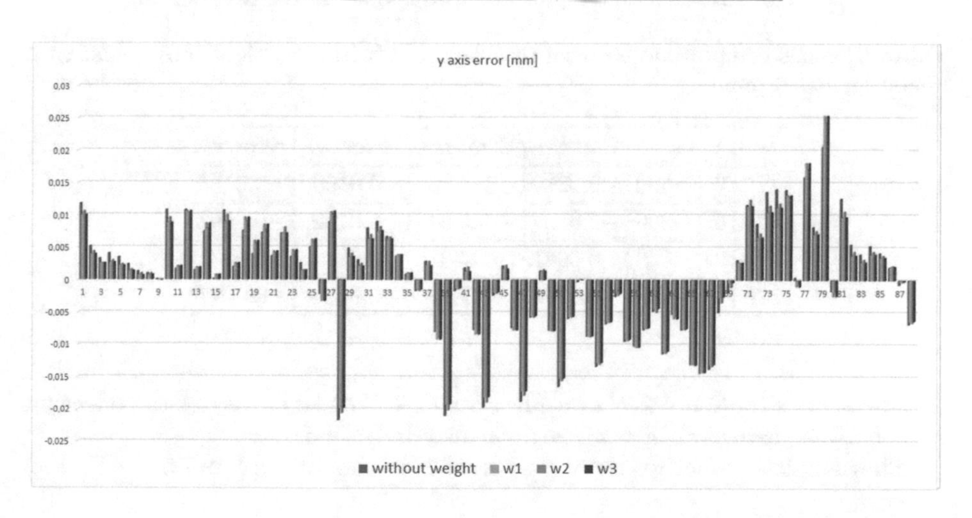

Fig. 3. y axis errors [mm] for measured points in the work space

Table 4. y-axis compression between the measurement of error without weight (error) and with weight (error_w) [mm]

X	Y	Z	Error	Error_w1	Error_w2	error_w3
170	0	−400	0.0118	0.01	0.0106	0.01
170	0	−350	0.0052	0.004	0.0044	0.004
170	0	−300	0.0033	0.0026	0.0026	0.0026
170	0	−250	0.0041	0.0027	0.003	0.0027

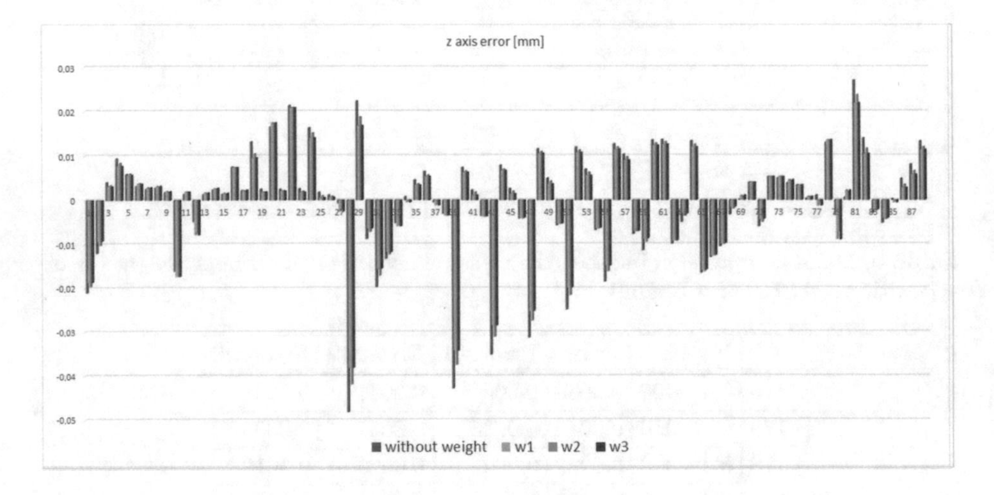

Fig. 4. z axis errors [mm] for measured points in the work space

Table 5. z-axis compression between result of error without weight (error) and with weight (error_w) [mm]

X	Y	Z	Error	Error_w1	Error_w2	error_w3
170	0	−400	−0.021	−0.0186	−0.0196	−0.0186
170	0	−350	−0.012	−0.0093	−0.0102	−0.0092
170	0	−300	0.0041	0.0031	0.0034	0.0031
170	0	−250	0.0094	0.0078	0.0084	0.0077

The same situation reflects on Figs. 3 and 4. The similarity of the effect on error between using **w1** or **w2** is significantly in both cases.

The sample of exact values of $y(z)$ error is in the Tables 4 and 5.

4 Conclusion

The presented work is focused on determination of position of points in the work space by using the multilateration technique. Two approaches are presented - with and without weighing. In the first case, we considered the influence of the quality change of the laserTRACER signal during the measurements. According to this information, there were three ways presented how to assign the weight. Weights were given for each measured distance. The developed algorithm operates in two steps: determination of the exact position of LT and determination of the exact position of the points in the work space. In the first step, we compared both of approaches with the data obtained by the original software TRAC_CAL. Based on the results presented in Table 2 which compares TRAC_CAL^{TM} data with data enumerated by our algorithm, we can assume correctness of our method. The difference between the corresponding position obtained by TRAC_CAL and our considerations are negligible. In the second step, we compared our results with different ways of weighting only. The effect of weights **w1** and **w3** on final error is very similar.

Our future work will be focused on incorporation of the temperature change influence to model precision. In this situation, we assume using multiple–criteria analysis.

Acknowledgement. This work was supported by the project No. FAST/FSI-J-17-4753.

References

1. Ramesh, R., Mannan, M.A., Poo, A.N.: Error compensation in machine tools - a review: part I: geometric, cutting-force induced and fixture-dependent errors. Int. J. Mach. Tools Manuf. **40**(9), 1235–1256 (2000). https://doi.org/10.1016/S0890-6955(00)00009-2
2. Hrdina, J., Vašík, P.: Dual numbers approach in multiaxis machines error modeling. J. Appl. Math. **2014**, 1–6 (2014). https://doi.org/10.1155/2014/261759
3. Hrdina, J., Vašík, P., Matoušek, R.: Special orthogonal matrices over dual numbers and their applications. Mendel J. ser. **2015**(6), 121–126 (2015)
4. Hrdina, J., Vašík, P., Holub, M.: Dual numbers arithmentic in multiaxis machine error modeling. MM Sci. J. **2017**(1), 1769–1772 (2017)
5. Holub, M., Hrdina, J., Vašík, P., Vetíška, J.: Three-axes error modeling based on second order dual numbers. J. Math. Ind. **5**(2), 1–11 (2015). https://doi.org/10.1186/s13362-015-0016-y
6. Holub, M., Blecha, P., Bradac, F., Kana, R.: Volumetric compensation of three-axis vertical machining centre. MM Sci. J. **1**, 677–681 (2015). https://doi.org/10.17973/MMSJ.2015-10-201534
7. Holub, M., Knobloch, J.: Geometric accuracy of CNC machine tools. In: Proceedings of the 16th International Conference on Mechatronics, Mechatronika 2014, pp. 260–265 (2014). https://doi.org/10.1109/MECHATRONIKA.2014.7018268
8. Holub, M., Michalíček, M., Vetíška, J., Marek, J.: Prediction of machining accuracy for vertical lathes. In: Proceedings of the 10th International Conference on Mechatronics 2013, Mechatronics 2013: Recent Technological and Scientific Advances, pp. 41–48 (2013). https://doi.org/10.1007/978-3-319-02294-9-6
9. Mayr, J., Jedrzejewski, J., Uhlmann, E., Donmez, M.A., Knapp, W., Härtig, F., Wendt, K., Moriwaki, T., Shore, P., Schmitt, R., Brecher, Ch., Würz, T., Wegener, K.: Thermal issues in machine tools. CIRP Ann. - Manuf. Technol. **61**(2), 771–791 (2012). https://doi.org/10.1016/j.cirp.2012.05.008

10. Gomez-Acedo, E., Olarra, A., Zubieta, M., Kortaberria, G., Ariznabarreta, E., López de Lacalle, L.N.: Method for measuring thermal distortion in large machine tools by means of laser multilateration. Int. J. Adv. Manuf. Technol. **80**(1), 523–534 (2015). https://doi.org/10.1007/s00170-015-7000-y
11. Teoh, P.L., Shirinzadeh, B., Foong, Ch.W., Alici, G.: The measurement uncertainties in the laser interferometry-based sensing and tracking technique. Measurement **32**(2), 135–150 (2002). https://doi.org/10.1016/S0263-2241(02)00006-4
12. Takatsuji, T., Goto, M., Kirita, A., Kurosawa, T., Tanimura, Y.: The relationship between the measurement error and the arrangement of laser trackers in laser trilateration. Measure. Sci. Technol. **11**(5), 477–483 (2000)
13. Aguado, S., Santolaria, J., Samper, D., Aguilar, J.J.: Study of self-calibration and multilateration in machine tool volumetric verification for laser tracker error reduction. Proc. Inst. Mech. Eng. Part B: J. Eng. Manuf. **228**(7), 659–672 (2014). https://doi.org/10.1177/0954405413511074
14. Navratilova, B., Hrdina, J.: Multilateration in volumetry: case study on demonstrator MCV 754QUICK. Mendel J. Ser. **2016**(1), 295–300 (2016)

Numerical Solving Stiff Control Problems for Delay Differential Equations

Zdeněk Šmarda(✉)

Department of Mathematics, Faculty of Electrical Engineering and Communication, Brno University of Technology, Technická 8, 616 00 Brno, Czech Republic
smarda@feec.vutbr.cz

Abstract. In the paper we present a numerical method for solving stiff control problems for delay differential equations based on the method of steps and the differential transformation method (DTM). Approximation of the solution is given either in the form of truncated power series or in the closed solution form. An application on a two-dimensional stiff delay systems with single delay and multiple delays with a parameter is shown the high accuracy and efficiency of the proposed method.

Keywords: Stiff problem · Method of steps
Differential transformation method · Initial value problems

1 Introduction

Stiff problems arise in many areas such as chemical kinetics, nuclear reactor theory, control theory, biochemistry, climatology, electronics, fluid dynamics, etc. (see, for example [1–3]). Solving stiff problems numerically by a given method with assigned tolerance, a step size is restricted by stability requirements rather than by the accuracy demands. This behaviour is usually observed in problems that have some components that decay much more rapidly than other components. In general, there is no universally accepted definition of stiffness. Among others stiffness with DDEs models can be defined by phenomena such as:

- Strong contractivity of neighbouring solutions.
- Multiple time scales (fast transient phases).
- The fact that explicit numerical integrators are not able to reproduce a correct approximation of the solution in an efficient way.

Another definition, stiff differential equation is an equation for which certain numerical methods for solving the equation are numerically unstable, unless the step size is taken to be extremely small. In other words, the step size is restricted by stability and not accuracy considerations (see [4]). In [5], stiff equations are defined to be such equations where implicit methods perform tremendously better than explicit ones. Practically, stiff equation solvers are based on implicit methods and not on explicit methods.

R. Matoušek (Ed.): MENDEL 2017, AISC 837, pp. 299–310, 2019.
https://doi.org/10.1007/978-3-319-97888-8_27

Numerical methods of solving stiff delay differential equations (DDEs) are in the forefront of study for several decades. El-Safty and Hussien [6] obtained the numerical solution of stiff DDEs in the form of Chebyshev series. Guglielmi and Hairer [7] applied the Radau IIA methods for solving stiff DDEs. Huang et al. [8] investigated the error behaviour of general linear methods for stiff DDEs. Bellen and Zennaro [9] dealt with the numerical treatment of DDEs (including stiff problems) in different fields of science and engineering. Zhu and Xiao [10] applied the parallel two-step ROW-methods for solving stiff DDEs and analyzed the stability behaviours of these methods. Asadi et al. [11] presented the new form of the homotopy perturbation method (NHPM) for solving stiff DDEs.

In the paper we present a numerical method for solving stiff systems of delay differential equations based on the method of steps and the differential transformation method (DTM) for the following Cauchy problem

$$\epsilon \mathbf{x}'(t) = \mathbf{A}\mathbf{x}(t) + \sum_{i=1}^{n} \mathbf{B}_i \mathbf{x}(t-h_i) + \mathbf{u}(t), \tag{1}$$

where $\mathbf{x}'(t) = (x_1'(t), x_2'(t), \ldots, x_n'(t))^T$, $\mathbf{x}(t) = (x_1(t), x_2(t), \ldots, x_n(t))^T$ is the state vector, $\mathbf{A}, \mathbf{B}_i$, $i = 1, 2, \ldots, n$ are constant $n \times n$ matrices, $\mathbf{u}(t) = (u_1(t), u_2(t), \ldots, u_n(t))^T$ is the vector of control functions, $\epsilon > 0$ is a parameter, h_i, $i = 1, \ldots, n$ are positive constants.

Let $h^* = \max\{h_1, h_2, \ldots, h_n\}$. Initial vector function $\mathbf{\Phi}(t) = (\phi_1(t), \phi_2(t), \ldots, \phi_n(t))^T$ needs to be assigned for system (1) on the interval $[-h^*, 0]$ and

$$\mathbf{x}(0) = \mathbf{\Phi}(0) = \mathbf{x_0}, \tag{2}$$

where $\mathbf{x_0}$ is a constant vector.

2 Differential Transformation

The differential transformation of a real function $x(t)$ at a point $t_0 \in \mathbb{R}$ is

$$\mathcal{D}\{x(t)\}[t_0] = \{X(k)\}_{k=0}^{\infty},$$

where the kth component $X(k)$ of the differential transformation of the function $x(t)$ at t_0 is defined as

$$X(k) = \frac{1}{k!}\left[\frac{d^k x(t)}{dt^k}\right]_{t=t_0}. \tag{3}$$

The inverse differential transformation of $\{X(k)\}_{k=0}^{\infty}$ is defined as follows:

$$x(t) = \mathcal{D}^{-1}\Big\{\{X(k)\}_{k=0}^{\infty}\Big\}[t_0] = \sum_{k=0}^{\infty} X(k)(t-t_0)^k. \tag{4}$$

In applications, the function $x(t)$ is expressed by a finite series

$$x(t) = \sum_{k=0}^{N} X(k)(t-t_0)^k. \tag{5}$$

The following well-known formulas for $t_0 = 0$ can be easily derived from definitions (3), (4):

Lemma 1. *Assume that $F(k)$, $H(k)$ and $X(k)$ are differential transformations of functions $f(t)$, $h(t)$ and $x(t)$, respectively. Then:*

If $f(t) = t^n$, then $F(k) = \delta(k-n)$, where δ is the Kronecker delta.

If $f(t) = e^{\lambda t}$, then $F(k) = \dfrac{\lambda^k}{k!}$.

If $f(t) = \dfrac{d^n x(t)}{dt^n}$, then $F(k) = \dfrac{(k+n)!}{k!} X(k+n)$.

If $f(t) = x(t)h(t)$, then $F(k) = \sum_{l=0}^{k} X(l)H(k-l)$.

Lemma 2. *Assume that $F(k)$, $G(k)$ are differential transformations of functions $f(t)$, $g(t)$, where $a > 0$ is a real constant. If*

$$f(t) = g(t-a), \text{ then } F(k) = \sum_{i=k}^{N} (-1)^{i-k} \binom{i}{k} a^{i-k} G(i), \; N \to \infty. \tag{6}$$

Proof. Using the binomial formula we have

$$\begin{aligned}
f(t) &= \sum_{k=0}^{\infty} G(k)(t-t_0-a)^k = \sum_{k=0}^{\infty} G(k) \sum_{i=0}^{k} (-1)^{k-i} \binom{k}{i} (t-t_0)^i a^{k-i} \\
&= \sum_{k=0}^{\infty} \sum_{i=0}^{k} (-1)^{k-i} \binom{k}{i} (t-t_0)^i a^{k-i} G(k) \\
&= \sum_{i=0}^{\infty} \sum_{k=i}^{\infty} (t-t_0)^i (-1)^{k-i} \binom{k}{i} a^{k-i} G(k) \\
&= \sum_{i=0}^{\infty} (t-t_0)^i \sum_{k=i}^{\infty} (-1)^{k-i} \binom{k}{i} a^{k-i} G(k) \\
&= \sum_{k=0}^{\infty} (t-t_0)^k \sum_{i=k}^{\infty} (-1)^{i-k} \binom{i}{k} a^{i-k} G(i).
\end{aligned}$$

Hence, from (4) we get

$$F(k) = \sum_{i=k}^{N} (-1)^{i-k} \binom{i}{k} a^{i-k} G(i).$$

The proof is complete. □

Using Lemmas 1 and 2, the differential transformation formula for function $f(t) = \frac{d^n}{dt^n} g(t-a)$ can be deduced.

Lemma 3 (in [12]). *Assume that* $F(k)$, $G(k)$ *are differential transformations of functions* $f(t)$, $g(t)$, $a > 0$. *If* $f(t) = \dfrac{d^n}{dt^n} g(t-a)$, *then*

$$F(k) = \frac{(k+n)!}{k!} \sum_{i=k+n}^{N} (-1)^{i-k-n} \binom{i}{k+n} a^{i-k-n} G(i), N \to \infty. \tag{7}$$

Using Lemmas 2 and 3, any differential transformation of a product of functions with delayed arguments and derivatives of that functions can be proved. However, such formulas are complicated and not easily applicable for solving functional differential equations with multiple constant delays (see for example [12–14]).

3 Main Results

At first, we apply the method of steps (see [15]) to system (1).

Let $\hat{h} = \min\{h_1, h_2, \dots, h_n\}$. To find a solution on interval $(0, \hat{h}]$ we substitute the initial state $\mathbf{\Phi}(t)$ in all places where state vector with delays appears. Then system (1) changes to a system of differential equations without delays in the state

$$\epsilon \mathbf{x}_1'(t) = \mathbf{A}\mathbf{x}_1(t) + \sum_{i=1}^{n} \mathbf{B}_i \mathbf{\Phi}(t-h_i) + \mathbf{u}(t) \tag{8}$$

subject to state vector at time $t = 0$

$$\mathbf{x}_1(0) = \mathbf{\Phi}(0) = (\phi_1(0), \phi_2(0), \dots, \phi_n(0)).$$

Applying DTM to system (8), we get the following system of recurrence relations

$$\epsilon(k+1)\mathbf{X}_1(k+1) = \mathbf{A}\mathbf{X}_1(k) + \sum_{i=1}^{n} \mathbf{B}_i \mathbf{F}_i(k) + \mathbf{U}(k), \tag{9}$$

where $\mathbf{X}_1(k)$, $\mathbf{F}_i(k)$, $\mathbf{U}(k)$ are differential transformation of vectors $\mathbf{x}_1(t)$, $\mathbf{\Phi}(t-h_i)$, $\mathbf{u}(t)$, $i = 1, 2, \dots, n$. The differential transformation of the state vector at time $t = 0$ is defined as

$$\mathbf{X}_1(k) = \frac{1}{k!} \mathbf{x}_1^{(k)}(0).$$

Using transformed initial conditions and then inverse transformation rule, we obtain approximate solution of Eq. (1) in the form of the finite Taylor series

$$\begin{aligned} \mathbf{x}_1(t) &= (x_{11}(t), x_{12}(t), \dots, x_{1n}(t)) \\ &= \left(\sum_{k=0}^{N} X_{11}(k) t^k, \sum_{k=0}^{N} X_{12}(k) t^k, \dots, \sum_{k=0}^{N} X_{1n}(k) t^k \right) \end{aligned}$$

on the interval $(0, \hat{h}]$. Then we continue with solving system (1), generally on interval $(0, T]$, $T > 0$, as follows:
If $T \in (0, \hat{h}]$ then $\mathbf{x}_1(t)$ is a desired approximate solution of (1).

If not, we continue to find a solution in the interval $[\hat{h}, \hat{h} + 2\hat{h}]$. Then we solve the system

$$\epsilon \mathbf{x}_2'(t) = \mathbf{A}\mathbf{x}_2(t) + \sum_{i=1}^{n} \mathbf{B}_i \mathbf{x}_1(t - h_i) + \mathbf{u}(t) \tag{10}$$

with the initial function

$$\mathbf{\Phi}(t) = \mathbf{x}_1(t), \ t \in [0, \hat{h}]$$

and subject to the state vector at time $t = \hat{h}$

$$\mathbf{x}_2(\hat{h}) = \mathbf{x}_1(\hat{h}).$$

Applying DTM to Eq. (10), we obtain the approximate solution of (10) in the form

$$\begin{aligned}\mathbf{x}_2(t) &= (x_{21}(t), x_{22}(t), \dots, x_{2n}(t)) \\ &= \left(\sum_{k=0}^{N} X_{21}(k)(t - \hat{h})^k, \sum_{k=0}^{N} X_{22}(k)(t - \hat{h})^k, \dots, \sum_{k=0}^{N} X_{2n}(k)(t - \hat{h})^k \right).\end{aligned}$$

In general, using the above mentioned method, we get the approximate solution of (1)

$$\begin{aligned}\mathbf{x}_j(t) &= (x_{j1}(t), x_{j2}(t), \dots, x_{jn}(t)) \\ &= \left(\sum_{k=0}^{N} X_{j1}(k)(t - (j-1)\hat{h})^k, \sum_{k=0}^{N} X_{j2}(k)(t - (j-1)\hat{h})^k, \right. \\ &\quad \left. \dots, \sum_{k=0}^{N} X_{jn}(k)(t - (j-1)\hat{h})^k \right)\end{aligned}$$

as long as for some j, $T \in ((j-1)\hat{h}, (j)\hat{h}]$. Then we get the solution of (1) in the form

$$\mathbf{x}(t) = \begin{cases} \mathbf{x}_1(t), \, t \in (0, \hat{h}], \\ \mathbf{x}_2(t), \, t \in (\hat{h}, 2\hat{h}], \\ \vdots \\ \mathbf{x}_j(t), \, t \in ((j-1)\hat{h}, T]. \end{cases}$$

We demonstrate potentiality of our approach on the following examples:

Example 1. Consider the stiff delay system with single delay (see [10,16])

$$\begin{bmatrix} x_1'(t) \\ x_2'(t) \end{bmatrix} = \begin{bmatrix} -1001 & -125 \\ 8 & 0 \end{bmatrix} \begin{bmatrix} x_1(t) \\ x_2(t) \end{bmatrix} + \begin{bmatrix} 1000\mathrm{e}^{-0.1} & 125\mathrm{e}^{-0.2} \\ -8\mathrm{e}^{-0.1} & -2\mathrm{e}^{-0.2} \end{bmatrix} \begin{bmatrix} x_1(t-0.1) \\ x_2(t-0.1) \end{bmatrix} + \begin{bmatrix} 1126 - 1000\mathrm{e}^{-0.1} - 125\mathrm{e}^{-0.2} \\ -8 + 8\mathrm{e}^{-0.1} + 2\mathrm{e}^{-0.2} \end{bmatrix} \tag{11}$$

for $t \in (0, 0.1]$ with $\mathbf{\Phi}(t) = [1 + \mathrm{e}^{-t}, 1 + \mathrm{e}^{-2t}]^T$ on $[-0.1, 0]$. The exact solution of system (11) is

$$x_1(t) = 1 + \mathrm{e}^{-t},$$
$$x_2(t) = 1 + \mathrm{e}^{-2t}.$$

Applying the method of steps to system (11), we obtain only the system of ordinary equations

$$x_{11}'(t) = -1000x_{11}(t) - 125x_{12}(t) + 1000\mathrm{e}^{-0.1}(1 + \mathrm{e}^{-(t-0.1)}) + 125\mathrm{e}^{-0.2}(1 + \mathrm{e}^{-2(t-0.1)}) + 1126 - 1000\mathrm{e}^{-0.1} - 125\mathrm{e}^{-0.2}, \tag{12}$$

$$x_{12}'(t) = 8x_{11}(t) - 8\mathrm{e}^{-0.1}(1 + \mathrm{e}^{-(t-0.1)}) - 2\mathrm{e}^{-0.2}(1 + \mathrm{e}^{-2(t-0.1)}) - 8 + 8\mathrm{e}^{-0.1} + 2\mathrm{e}^{-0.2}. \tag{13}$$

From (12) we have

$$x_{11}'(t) = -1001x_{11}(t) - 125x_{12}(t) + 1000\mathrm{e}^{-t} + 125\mathrm{e}^{-2t} + 1126,$$
$$x_{12}'(t) = 8x_{11}(t) - 8\mathrm{e}^{-t} - 2\mathrm{e}^{-2t} - 8. \tag{14}$$

Taking the differential transform method to system (14), we obtain very simple recurrence relations

$$X_{11}(k+1) = \frac{1}{k+1}\left[-10001X_{11}(k) - 125X_{12}(k) + 1000\frac{(-1)^k}{k!} + 125\frac{(-2)^k}{k!} + 1126\delta(k)\right],$$

$$X_{12}(k+1) = \frac{1}{k+1}\left[8X_{11}(k) - 8\frac{(-1)^k}{k!} - 2\frac{(-2)^k}{k!} - 8\delta(k)\right], \tag{15}$$

$k = 0, 1, 2, \ldots$. From initial conditions, we have $X_{11}(0) = 2,\ X_{12}(0) = 2$. Solving recurrence system (15) for $k \geq 1$, we get

$$X_{11}(1) = -1,\ X_{11}(2) = \frac{1}{2!},\ X_{11}(3) = -\frac{1}{3!},\ X_{11}(4) = \frac{1}{4!}, \ldots, X_{11}(k) = \frac{(-1)^k}{k!},$$

$$X_{12}(1) = -2,\ X_{12}(2) = 2 = \frac{2^2}{2!},\ X_{12}(3) - \frac{4}{3} = \frac{(-2)^3}{3!},\ X_{12}(4) = \frac{2}{3} = \frac{2^4}{4!}, \ldots,$$

$$X_{12}(k) = (-1)^k \frac{2^k}{k!}.$$

Hence the approximate solution of system (11) has the form

$$\begin{aligned} x_{11}(t) &= X_{11}(0) + X_{11}(1)t + X_{11}(2)t^2 + X_{11}(3)t^3 + \cdots + X_{11}(k)t^k \\ &= 1 + \left(1 - t + \frac{t^2}{2!} - \frac{t^3}{3!} + \cdots + \frac{(-1)^k}{k!}t^k\right) \\ &= 1 + \sum_{k=0}^{N}(-1)^k \frac{t^k}{k!}, \\ x_{12}(t) &= X_{12}(0) + X_{12}(1)t + X_{12}(2)t^2 + X_{12}(3)t^3 + \cdots + X_{12}(k)t^k \\ &= 1 + \left(1 - 2t + \frac{(-2)^2}{2!}t^2 - \frac{(-2)^3}{3!}t^3 + \cdots + \frac{(-2)^k}{k!}t^k\right) \\ &= 1 + \sum_{k=0}^{N}(-1)^k \frac{(2t)^k}{k!}. \end{aligned}$$

It is obvious that for $N \to \infty$ we get the exact solution $\mathbf{x}_1(t) = (1+e^{-t}, 1+e^{-2t})^T$ of system (11) on the interval $[0, 0.1]$. Now we will find a solution $\mathbf{x}_2(t)$ on the interval $(0.1, 0.2]$. with the initial function

$$\mathbf{\Phi}_1(t) = \mathbf{x}_1(t) = (1 + e^{-t}, 1 + e^{-2t})^T$$

for $t \in [0, 0.1]$ and the initial condition

$$\mathbf{x}_2(0.1) = (1 + e^{-0.1}, 1 + e^{-0.2})^T.$$

Applying the method of steps to system (11), we get for $t \in (0.1, 0.2]$ the following system

$$\begin{aligned} x_{21}'(t) &= -1001x_{21}(t) - 125x_{22}(t) + 1000e^{-0.1}e^{-(t-0.1)} \\ &\quad + 125e^{-0.2}e^{-2(t-0.1)} + 1126, \end{aligned} \tag{16}$$

$$x_{22}'(t) = 8x_{21}(t) - 8e^{-0.1}e^{-(t-0.1)} - 2e^{-0.2}e^{-2(t-0.2)} - 8. \tag{17}$$

Then transformed system (16) has the form

$$\begin{aligned} X_{21}(k+1) &= \frac{1}{k+1}\left[-10001X_{21}(k) - 125X_{22}(k) + 1000e^{-0.1}\frac{(-1)^k}{k!}\right. \\ &\quad \left. + 125e^{-0.2}\frac{(-2)^k}{k!} + 1126\delta(k)\right], \end{aligned} \tag{18}$$

$$X_{22}(k+1) = \frac{1}{k+1}\left[8X_{21}(k) - 8e^{-0.1}\frac{(-1)^k}{k!} - 2e^{-0.2}\frac{(-2)^k}{k!} - 8\delta(k)\right], \tag{19}$$

$k = 0, 1, 2, \ldots$. From initial conditions, we have

$$\begin{aligned} X_{21}(0) &= 1 + e^{-0.1}, \\ X_{22}(0) &= 1 + e^{-0.2}. \end{aligned}$$

Solving recurrence system (18) for $k \geq 1$, we get

$$X_{21}(1) = -\mathrm{e}^{-0.1},\ X_{21}(2) = \mathrm{e}^{-0.1}\frac{1}{2!},\ X_{21}(3) = -\mathrm{e}^{-0.1}\frac{1}{3!},$$
$$X_{21}(4) = \mathrm{e}^{-0.1}\frac{1}{4!},\ldots, X_{21}(k) = \mathrm{e}^{-0.1}\frac{(-1)^k}{k!},$$
$$X_{22}(1) = -2\mathrm{e}^{-0.2},\ X_{22}(2) = 2\mathrm{e}^{-0.2} = \mathrm{e}^{-0.2}\frac{2^2}{2!},\ X_{22}(3) = -\mathrm{e}^{-0.2}\frac{4}{3}$$
$$= \mathrm{e}^{-0.2}\frac{(-2)^3}{3!}, X_{22}(4) = \mathrm{e}^{-0.2}\frac{2}{3} = \mathrm{e}^{-0.2}\frac{2^4}{4!},\ldots, X_{22}(k)$$
$$= \mathrm{e}^{-0.2}(-1)^k\frac{2^k}{k!}.$$

Hence the approximate solution of system (11) has the form

$$\begin{aligned}
x_{21}(t) &= X_{21}(0) + X_{21}(1)(t-0.1) + X_{21}(2)(t-0.1)^2 + X_{21}(3)(t-0.1)^3 \\
&+ \cdots + X_{21}(k)(t-0.1)^k \\
&= 1 + \mathrm{e}^{-0.1}\left(1 - (t-0.1) + \frac{(t-0.1)^2}{2!} - \frac{(t-0.1)^3}{3!} + \ldots \right. \\
&\left. + \frac{(-1)^k}{k!}(t-0.1)^k\right) = 1 + \mathrm{e}^{-0.1}\sum_{k=0}^{N}(-1)^k\frac{(t-0.1)^k}{k!}, \\
x_{22}(t) &= X_{22}(0) + X_{22}(1)(t-0.1) + X_{22}(2)(t-0.1)^2 + X_{22}(3)(t-0.1)^3 \\
&+ \cdots + X_{22}(k)(t-0.1)^k \\
&= 1 + \mathrm{e}^{-0.2}\left((1 - 2(t-0.1) + \frac{(-2)^2}{2!}(t-0.1)^2 - \frac{(-2)^3}{3!}(t-0.1)^3 \right. \\
&\left. + \cdots + \frac{(-2)^k}{k!}(t-0.1)^k\right) = 1 + \mathrm{e}^{-0.2}\sum_{k=0}^{N}(-1)^k\frac{(2(t-0.1))^k}{k!}.
\end{aligned}$$

Then for $N \to \infty$ we get the exact solution $\mathbf{x}_2(t) = (1+\mathrm{e}^{-t}, 1+\mathrm{e}^{-2t})^T$ of system (11) on the interval $(0.1, 0.2]$. In general, for $t \in ((j-1)0.1, (j)0.1]$, we obtain a solution of system (11) in the form

$$\begin{aligned}
x_{j1}(t) &= X_{j1}(0) + X_{j1}(1)(t-0.1(j-1)) + X_{j1}(2)(t-0.1(j-1))^2 \\
&+ X_{j1}(3)(t-0.1(j-1))^3 + \cdots + X_{j1}(k)(t-0.1(j-1))^k \\
&= 1 + \mathrm{e}^{-0.1(j-1)}\left(1 - (t-0.1(j-1)) + \frac{(t-0.1(j-1))^2}{2!} \right. \\
&\left. - \frac{(t-0.1(j-1))^3}{3!} + \cdots + \frac{(-1)^k}{k!}(t-0.1(j-1))^k\right) \\
&= 1 + \mathrm{e}^{-0.1(j-1)}\sum_{k=0}^{N}(-1)^k\frac{(t-0.1(j-1))^k}{k!}, \\
x_{j2}(t) &= X_{j2}(0) + X_{j2}(1)(t-0.1(j-1)) + X_{j2}(2)(t-0.1(j-1))^2
\end{aligned}$$

$$
\begin{aligned}
&+ X_{j2}(3)(t-0.1(j-1))^3 + \cdots + X_{j2}(k)(t-0.1(j-1))^k \\
&= 1 + \mathrm{e}^{-0.2(j-1)}\left(1 - 2(t-0.1(j-1)) + \frac{(-2)^2}{2!}(t-0.1(j-1))^2\right. \\
&\left. - \frac{(-2)^3}{3!}(t-0.1(j-1))^3 + \cdots + \frac{(-2)^k}{k!}(t-0.1(j-1))^k\right) \\
&= 1 + \mathrm{e}^{-0.2(j-1)}\sum_{k=0}^{N}(-1)^k\frac{(2(t-0.1(j-1)))^k}{k!},
\end{aligned}
$$

$j = 1, 2, \ldots$. It is obvious that the series

$$
\mathrm{e}^{-0.1(j-1)}\sum_{k=0}^{N}(-1)^k\frac{(t-0.1(j-1))^k}{k!}
$$

converges for $N \to \infty$ to functions e^{-t} on interval $t \in ((j-1)0.1, (j)0.1]$, $j = 1, 2, \ldots$ and the series

$$
\mathrm{e}^{-0.2(j-1)}\sum_{k=0}^{N}(-1)^k\frac{(2(t-0.1(j-1)))^k}{k!}
$$

converges for $N \to \infty$ to functions e^{-t} on interval $t \in ((j-1)0.1, (j)0.1]$, $j = 1, 2, \ldots$.

Thus the exact solution of system (11) is

$$
\begin{aligned}
x_1(t) &= 1 + \mathrm{e}^{-t}, \\
x_2(t) &= 1 + \mathrm{e}^{-2t}.
\end{aligned}
$$

for $t > 0$.

Remark 1. The same problem was solved by Zhu and Xiao [10] and Pushpam and Dhayabaran [16] using the parallel two-step ROW-methods and the single term Walsh series method, respectively. But in both cases, the authors obtained only an approximate solution.

Example 2. Consider the stiff delay system with multiply delays (see [6])

$$
\begin{aligned}
\begin{bmatrix}\epsilon x_1'(t)\\ \epsilon x_2'(t)\end{bmatrix} &= \begin{bmatrix}-0.5 & 0\\ 0 & -1\end{bmatrix}\begin{bmatrix}x_1(t)\\ x_2(t)\end{bmatrix} + \begin{bmatrix}0 & 0\\ -0.5 & 0\end{bmatrix}\begin{bmatrix}x_1(t-0.5)\\ x_2(t-0.5)\end{bmatrix} \\
&+ \begin{bmatrix}0 & -0.5\\ 0 & 0\end{bmatrix}\begin{bmatrix}x_1(t-1)\\ x_2(t-1)\end{bmatrix} + \begin{bmatrix}0.5\mathrm{e}^{-(t-1)/\epsilon}\\ 0.5\mathrm{e}^{-(t-0.5)/2\epsilon}\end{bmatrix}
\end{aligned} \tag{20}
$$

for $t \in (0, 0.5]$ with $\mathbf{\Phi}(t) = [\mathrm{e}^{-t/2\epsilon}, \mathrm{e}^{-t/\epsilon}]^T$ on $[-1, 0]$. The exact solution of system (11) is

$$
\begin{aligned}
x_1(t) &= \mathrm{e}^{-t/2\epsilon}, \\
x_2(t) &= \mathrm{e}^{-t/\epsilon}.
\end{aligned}
$$

Applying the method of steps to system (20), we obtain the system of ordinary equations

$$\begin{aligned} \epsilon x_1'(t) &= -0.5x_1(t) - 0.5\mathrm{e}^{-(t-1)/\epsilon} + 0.5\mathrm{e}^{-(t-1)/\epsilon}, \\ \epsilon x_2'(t) &= -x_2(t) - 0.5\mathrm{e}^{-(t-0.5)/2\epsilon} + 0.5\mathrm{e}^{-(t-0.5)/2\epsilon}. \end{aligned} \tag{21}$$

Taking the differential transform method to system (21), we obtain trivial recurrence relations

$$\begin{aligned} X_1(k+1) &= \frac{1}{\epsilon(k+1)}(-0.5X_1(k)), \\ X_2(k+1) &= \frac{1}{\epsilon(k+1)}(-X_2(k)), \end{aligned} \tag{22}$$

$k = 0, 1, 2, \ldots$. From initial conditions, we have $X_1(0) = 1,\ X_2(0) = 1$. Solving recurrence system (22) for $k \geq 1$, we get

$$\begin{aligned} X_1(1) &= \frac{-1}{2\epsilon},\ X_1(2) = \frac{1}{8\epsilon^2},\ X_1(3) = -\frac{1}{48\epsilon^3},\ X_1(4) = \frac{1}{384\epsilon^4}, \ldots, \\ X_2(1) &= -\frac{1}{\epsilon},\ X_2(2) = \frac{1}{2\epsilon^2},\ X_2(3) = -\frac{1}{6\epsilon^3},\ X_2(4) = \frac{1}{24\epsilon^4}, \ldots. \end{aligned} \tag{23}$$

The coefficients (23) can be written as follows:

$$\begin{aligned} X_1(1) &= -\frac{1}{2\epsilon},\ X_1(2) = \frac{1}{2!}\frac{1}{(2\epsilon)^2},\ X_1(3) = -\frac{1}{3!}\frac{1}{(2\epsilon)^3},\ X_1(4) \\ &= \frac{1}{4!}\frac{1}{(2\epsilon)^4}, \ldots, X_1(k) = (-1)^k \frac{1}{k!}\frac{1}{(2\epsilon)^k}, \\ X_2(1) &= -\frac{1}{\epsilon},\ X_2(2) = \frac{1}{2!}\frac{1}{\epsilon^2},\ X_2(3) = -\frac{1}{3!}\frac{1}{\epsilon^3},\ X_2(4) \\ &= \frac{1}{4!}\frac{1}{\epsilon^4}, \ldots, X_2(k) = (-1)^k \frac{1}{k!}\frac{1}{\epsilon^k}. \end{aligned}$$

Then the approximate solution of system (20) has the form

$$\begin{aligned} x_1(t) &= X_1(0) + X_1(1)t + X_1(2)t^2 + X_1(3)t^3 + \cdots + X_1(k)t^k \\ &= 1 - \frac{t}{2\epsilon} + \frac{1}{2!}\frac{t^2}{(2\epsilon)^2} - \frac{1}{3!}\frac{t^3}{(2\epsilon)^3} \\ &+ \frac{1}{4!}\frac{t^4}{(2\epsilon)^4} + \cdots + (-1)^k \frac{1}{k!}\frac{t^k}{(2\epsilon)^k} \\ &= \sum_{k=0}^{N} \frac{(-1)^k}{k!}\left(\frac{t}{2\epsilon}\right)^k, \\ x_2(t) &= X_2(0) + X_2(1)t + X_2(2)t^2 + X_2(3)t^3 + \cdots + X_2(k)t^k \\ &= 1 - \frac{t}{\epsilon} + \frac{1}{2!}\frac{t^2}{\epsilon^2} - \frac{1}{3!}\frac{t^3}{\epsilon^3} \\ &+ \frac{1}{4!}\frac{t^4}{\epsilon^4} + \cdots + (-1)^k \frac{1}{k!}\frac{t^k}{\epsilon^k} = \sum_{k=0}^{N} \frac{(-1)^k}{k!}\left(\frac{t}{\epsilon}\right)^k. \end{aligned}$$

The same procedure as in Example 1, for $N \to \infty$, we obtain the exact solution of system (20) in the form $x_1(t) = \mathrm{e}^{-t/2\epsilon}$, $x_2(t) = \mathrm{e}^{-t/\epsilon}$.

Remark 2. El-Safty and Hussien [6] solved system (20) using Chebyshev series method but they obtained only an approximate solution.

4 Conclusion

In the paper we propose an efficient semi-analytical technique based on the combination of the method of steps and the differential transformation method. The main advantage of the presented method is to provide the user an analytical approximation of the solution, in many cases, an exact solution, in a rapidly convergent sequence with elegantly computed terms. Finally, a subject of further investigation is to develop the presented technique for stiff systems with proportional delays

$$\epsilon \mathbf{x}'(t) = \mathbf{A}\mathbf{x}(t) + \sum_{i=1}^{n} \mathbf{B}_i \mathbf{x}(q_i t) + \mathbf{u}(t),$$

where $0 < q_i < 1,\ i = (1, 2, \ldots, n)$ and for stiff systems with time varying delays

$$\epsilon \mathbf{x}'(t) = \mathbf{A}\mathbf{x}(t) + \sum_{i=1}^{n} \mathbf{B}_i \mathbf{x}(t - \tau_i(t)) + \mathbf{u}(t),$$

where $\tau_i(t) \geq \tau_{i0} > 0,\ t > 0,\ i = 1, 2, \ldots, n$ are real functions.

Acknowledgement. The author was supported by the Czech Science Foundation under the project 16-08549S and by the Grant FEKT-S-17-4225 of Faculty of Electrical Engineering and Communication, Brno University of Technology.

References

1. Bujurke, N.M., Salimath, C.S., Shiralashetti, S.C.: Numerical solution of stiff systems from nonlinear dynamics using single-term Haar wavelet series. Nonlinear Dyn. **51**, 595–605 (2008). https://doi.org/10.1007/s11071-007-9248-8
2. Tian, Y.C., Zhang, T.: Wavelet-based collocation method for stiff systems in process engineering. J. Math. Chem. **44**, 501–513 (2008). https://doi.org/10.1007/s10910-007-9324-9
3. Bocharov, G.A., Marchuk, G.I., Romanyukha, A.A.: Numerical solution by LMMs of a stiff delay-differential system modelling an immune response. Numer. Math. **73**, 131–148 (1996). https://doi.org/10.1007/s002110050188
4. Baker, C.T.H., Paul, C.A.H.: Issues in the numerical solution of evolutionary delay differential equations. J. Adv. Comput. Math. **3**, 171–196 (1995). https://doi.org/10.1007/BF02988625
5. Kirlinger, G.: Linear multistep methods applied to stiff initial value problems. J. Math. Comput. Model. **40**, 1181–1192 (2004). https://doi.org/10.1016/j.mcm.2005.01.012

6. El-Safty, A., Hussien, M.A.: Chebyschev solution for stiff delay differential equations. Int. J. Comput. Math. **68**, 323–335 (1998). https://doi.org/10.1080/00207169808804699
7. Guglielmi, N., Hairer, E.: Implementing Radau IIA methods for stiff delay differential equations. Computing **67**, 1–12 (2001). https://doi.org/10.1007/s006070170013
8. Huang, C., Chen, G., Li, S., Fu, H.: D-convergence of general linear methods for stiff delay differential equations. Comput. Math. Appl. **41**, 627–639 (2001). https://doi.org/10.1016/S0898-1221(00)00306-0
9. Bellen, A., Zennaro, M.: Numerical Solution of Delay Differential Equations. Oxford University Press, Oxford (2003). https://doi.org/10.1093/acprof:oso/9780198506546.001.0001
10. Zhu, Q., Xiao, A.: Parallel two-step ROW-methods for stiff delay differential equations. Appl. Numer. Math. **59**, 1768–1778 (2009). https://doi.org/10.1016/j.apnum.2009.01.005
11. Asadi, M.A., Salehi, F., Mohyud-Din, S.T., Hosseini, M.M.: Modified homotopy perturbation method for stiff delay differential equations (DDEs). Int. J. Phys. Sci. **7**(7), 1025–1034 (2012). https://doi.org/10.5897/IJPS11.1670
12. Arikoglu, A., Ozkol, I.: Solution of differential-difference equations by using differential transform method. Appl. Math. Comput. **181**, 153–162 (2006). https://doi.org/10.1016/j.amc.2006.01.022
13. Karakoc, F., Bereketoglu, H.: Solutions of delay differential equations by using differential transform method. Int. J. Comput. Math. **86**, 914–923 (2009). https://doi.org/10.1080/00207160701750575
14. Mohammed, G.J., Fadhel, F.S.: Extend differential transform methods for solving differential equations with multiple delay. Ibn Al-Haitham J. Pure Appl. Sci. **24**(3), 1–5 (2011)
15. Kolmanovskii, V., Myshkis, A.: Introduction to the Theory and Applications of Functional Differential Equations. Kluwer, Dordrecht (1999). ISBN 978-94-017-1965-0
16. Pushpam, A.E.K., Dhayabaran, D.P.: Applicability of STWS technique in solving linear system of stiff delay differential equations with constant delays. Recent Res. Sci. Technol. **3**(10), 63–68 (2011)

Dynamic Generalized Berge-Zhukovskii Equilibrium

Noémi Gaskó(✉), Mihai Suciu, and Rodica Ioana Lung

Babeş-Bolyai University, Cluj-Napoca, Romania
gaskonomi@cs.ubbcluj.ro
http://csc.centre.ubbcluj.ro

Abstract. The Generalized Berge-Zhukovskii equilibrium extends the Berge-Zhukovskii equilibrium problem by introducing constraints over the set of strategy profiles. The new equilibrium is computed in a dynamic environment by using an evolutionary dynamic equilibrium tracking algorithm. Numerical experiments for the generalized Cournot duopoly illustrate the capability of the approach.

Keywords: Generalized Berge-Zhukovskii equilibrium
Dynamic games

1 Introduction

While, as central concept in Game Theory, the Nash equilibrium [10] has been widely studied and numerous algorithms for its computation or approximation have been proposed, other types of equilibria did not benefit from the same kind of attention from the computing community, one of the reasons being their intractable character. A computational approach that is efficient not only in computing but also in tracking the set of Berge-Zhukovskii equilibria in a dynamic setting is presented in this paper.

Dynamic games, introduced for the first time in [12], are important tools that try to capture dynamic aspects related to real game situations.

There are several definitions and characterizations of dynamic games: in [8] five important classes of dynamical games are mentioned: Stackelberg games, standard repeated games, stochastic games, difference and differential games, and evolutionary games. In [11] dynamic games are described by using the following characteristics:

- several epochs are taken into account;
- players receive different payoffs each epoch;
- the overall payoff of a player is the sum/integral of its payoffs over all the epochs of time;
- the received payoff of each player in one epoch depends on the state of the system at that time;

R. Matoušek (Ed.): MENDEL 2017, AISC 837, pp. 311–319, 2019.
https://doi.org/10.1007/978-3-319-97888-8_28

- the state of the system changes in time; changes may depend on the actions of the players in previous epochs;
- the change may be described by a difference or differential equation.

By this characterization only differential and difference games can be considered as dynamic games. Several solution concepts based on best-response maximization exist for dynamic games. The most common ones are the Open-loop Nash equilibrium (OLNE) and the Markov-perfect (or feedback) Nash equilibrium.

A different approach to discrete-time games is to consider each state of the system as an independent game to be solved by approximating an equilibrium for each state. We propose a method for tracking the generalized Berge-Zhukovskii equilibrium of a discrete-time dynamic game by using a Differential Evolution (DE) that has been adapted to compute generalized Berge-Zhukovskii equilibria and also to deal with dynamic environments.

The paper is organized as follows: Sect. 2 introduces the generalized Berge-Zhukovskii equilibrium. Section 3 describes the dynamic version of the Cournot game. In Sect. 4 the dynamic generalized equilibrium tracking algorithm for generalized Berge-Zhukovskii equilibrium detection is presented. Numerical experiments are described in Sect. 5. Section 6 concludes the paper.

2 Generalized Berge - Zhukovskii Equilibrium

Berge-Zhukovskii equilibrium is an alternate solution concept for non-cooperative games with applications in trust games (for example centipede game, prisoners dilemma, etc.), where it ensures the highest possible payoff for each player (as opposed to Nash equilibrium for these games).

The Generalized Nash equilibrium problem (GNEP) [1,5] is a generalization of the classical equilibrium problem [10] that better models real world situations (for example energy markets [2]) by taking into account restrictions among players strategies.

In this paper the Generalized Berge-Zhukovskii equilibrium problem (GBZEP), which is an extension of the simple Berge-Zhukovskii equilibrium detection problem is proposed. In this case the strategies of players are not independent and constraints are used to model that.

Formally a finite strategic non-cooperative game can be described as a system $G = ((N, S_i, u_i), i = 1, \ldots, n)$, where:

- N represents the set of players, n is the number of players;
- for all players $i \in N$, S_i is the set of actions available, and $S = S_1 \times S_2 \times \ldots \times S_n$ is the set of all possible situations of the game.
  $s = (s_1, s_2, \ldots s_n) \in S$ is a strategy (or strategy profile) of the game;
- for each player $i \in N$, $u_i : S \rightarrow R$ represents the payoff function of player i.

In a generalized non-cooperative game the strategy space is modelled in the following way: for all player $i \in N$ the **common strategy set** S is formed by the decisions $s = (s_1, \ldots, s_n)$ where $s_i \in K_i(s)$, $K_i : S \to \mathcal{P}(S_i)$, where $\mathcal{P}(S_i)$ represents the power set of S_i.

A strategy s^* is a Berge-Zhukovskii (BZ) equilibrium, if the payoff of each player i does not decrease considering any deviation of the other $N - \{i\}$ players. Formally we have the following definition:

Definition 1. *A strategy profile $s^* \in S$ is a Berge-Zhukovskii equilibrium if the inequality*

$$u_i(s^*) \geq u_i(s_i^*, s_{N-i})$$

holds for each player $i = 1, \ldots, n$, and $s_{N-i} \in S_{N-i}$.

The notation (s_i^*, s_{N-i}) represents the situations in which all other players apart from $i(N - \{i\})$ change their strategies from s^* to s: $(s_i^*, s_{N-i}) = (s_1, s_2, \ldots, s_i^*, \ldots, s_n)$.

The Generalized Berge-Zhukovskii can be described:

Definition 2. *A strategy profile $s^* \in S$ is a generalized Berge-Zhukovskii equilibrium if the inequality*

$$u_i(s^*) \geq u_i(s_i^*, s_{N-i})$$

holds for each player $i = 1, \ldots, n$, and $s_{N-i} \in \mathcal{S}$.

Generative relations [9] can be used to characterize a certain equilibrium. Furthermore, they may be used within optimization heuristics for fitness assignment purposes in order to guide their search to the desired equilibrium type.

The first generative relation was introduced for the Nash equilibrium detection in [9]. A generative relation for the detection of Berge-Zhukovskii equilibrium is introduced in [6].

Consider two strategy profiles s and q from S. Denote by $b(s, q)$ the number of players who lose by remaining to the initial strategy s, while all other players are allowed to play the corresponding strategies from q and at least one player switches from s to q.

We may express $b(s, q)$ as:

$$b(s, q) = card\{i \in N, u_i(s) < u_i(s_i, q_{N-i})\},$$

where $card\{M\}$ denotes the cardinality of the set M.

Definition 3. *Let $s, q \in S$. We say the strategy s is better than strategy q (s Berge-dominates q) with respect to Berge-Zhukovskii equilibrium, and we write $s \prec_B q$, if and only if the inequality $b(s, q) < b(q, s)$ holds.*

Definition 4. *The strategy profile $s^* \in S$ is a Berge-Zhukovskii non-dominated strategy (BNE), if and only if there is no strategy $s \in S, s \neq s^*$ such that s dominates s^* with respect to $\prec_B$ i.e. $s \prec_B s^*$.*

The relation $\prec_B$ is a the generative relation of the Berge-Zhukovskii equilibrium, i.e. the set of the non-dominated strategies with respect to the relation $\prec_B$ equals the set of Berge-Zhukovskii equilibria. The exact same relation will be used in the case of the generalized Berge Zhukovskii equilibrium by taking into account only strategy profiles from the common strategy set $\mathcal{S}$.

3 Dynamic Generalized Cournot Game

A dynamic game is a mathematical model of the interactions between players, who control a dynamical system [7,11].

Let us consider the Cournot oligopoly model [3], where n firms produce a quantity of s_i products, $i = 1, \dots, n$.

Let $P(Q)$ be the market clearing price, where $Q = \sum_{i=1}^{n} s_i$.

$$P(Q) = \begin{cases} a - Q, & \text{if } Q \leq a; \\ 0, & \text{if } Q > a. \end{cases}$$

Each firm has the common cost function $C(s_i)$, $C_i(s_i) = cs_i$.

The payoff for the firm i is its profit, that can be described with the following equation: $u_i(s) = s_i P(Q) - C_i(s_i)$, where $s = (s_1, s_2, \dots, s_n)$. The payoff for the firm i can be summarized:

$$u_i(s) = \begin{cases} s_i(a - Q - c), & \text{if } \sum_{i=1}^{n} s_i \leq a; \\ -cs_i, & \text{if } \sum_{i=1}^{n} s_i > a. \end{cases}$$

If we consider the following constraints related to the quantity of goods to be produced, we obtain a generalized Cournot oligopoly model (the strategies of the players are interconnected):

$$\sum_{i=1}^{n} \alpha_i \cdot s_i > \alpha, \alpha \in [0, 15], \alpha_i \in [0, 1], s_i \in [l_1, l_2], i = 1, \dots, n. \tag{1}$$

Dynamic feature. By considering that the coefficients α_i, $i = 1, \dots, n$ vary in time, a dynamic generalized Cournot oligopoly model is obtained. A simple model is that can be used is $\alpha_{i+1} \in U(0, 1)$ where $U(0, 1)$ represents a random uniform number between 0 and 1. Any other variation can be considered.

4 Dynamic Generalized-Equilibrium Tracking Algorithms

In order to compute the GBZEs in a constrained dynamic environment we use an Evolutionary Algorithm for Equilibria Detection (EAED) extended with a mechanism for tracking and adapting to environment changes.

An EAED is an evolutionary algorithm that uses a selection for survival operator that guides the search using the generative relation described in Sect. 2 (an offspring replaces its parent only if it is better than it in *Berge-Zhukovskii* sense).

The proposed algorithm, the Dynamic Generalized Equilibrium Tracking Differential Evolution (*DGET-DE*) is based on Differential Evolution [13] and it is used in [15] for Nash equilibria detection and in [14] for generalized Nash equilibrium detection. The following steps are used for dealing and adapting to changes in the dynamic environment [15]:

1. When a change in the game is detected (the change is detected using a sentinel) the magnitude of change m is estimated:
$$m = \frac{|constraint_{Old} - constraint_{New}|}{\max(constraint_{Old}, constraint_{New})} \tag{2}$$
where $constraint_{Old}$ and $Constraint_{New}$ represent the old and new constraint violation values for the sentinel.
2. Increase population diversity uniform mutation, with $p_m \in [p_{min}, p_{max}]$, where p_{min} and p_{max} are parameters of the algorithm, is applied to each individual. The probability of mutation is computed using the formula: $p_m = p_{min} + m \cdot (p_{max} - p_{min})$.
3. Apply uniform mutation with probability p_m and mutation step $N(0, \sigma)$ where σ is linearly correlated to the estimated magnitude of the change m: $\sigma_m = \sigma_{min} + m \cdot (\sigma_{max} - \sigma_{min})$.
4. The value of crossover probability C_r is decreased (in order to promote exploration) and linearly increases at each generation until it reaches a predefined threshold. The value of of the scaling factor F is increased, and linearly decreases at each generation until it reaches a certain threshold.

The general description of the *DGET-DE* is presented in Algorithm 1. The algorithm is compared with a simple DE version, called Adaptive Equilibrium Differential Evolution Algorithm (*AE-DE*), that only uses step 4 of the adaptation mechanism presented here.

Algorithm 1. Dynamic Generalized Equilibrium Tracking Differential Evolution Algorithm for BZ detection (*DGET-DE-BZ*)

```
Randomly generate initial population of game situations;
repeat
   Create offspring by mutation and recombination (DE/rand/1/bin);
   Evaluate offspring (compute payoff functions for all players);
   if the offspring is better (in Berge-Zhukovskii sense) than parent then
      Replace the parent by offspring in the next generation;
   end if
   if change detected then
      Apply uniform mutation with the adapted value of p_m and step N(0, σ) according to Steps 1–3;
   end if
   Apply adapted Differential evolution; (using Step 4. for varying F);
until termination condition is met;
```

The constraint handling is based on the rules presented in [4]:

1. If both individuals aren't feasible then the one that violates restriction less is chosen;
2. If one individual is feasible and the other is not then the feasible one chosen;

3. When both individuals are feasible then the one that Berge-Zhukovskii dominates the other is chosen. If these vectors are indifferent (no individual is dominated) then keep the parent.

An individual is deemed feasible if it satisfies the constraints defined by (1).

5 Numerical Experiments

The constrained (generalized) dynamic Cournot duopoly is considered for the numerical experiments. The Berge-Zhukovskii Generational Distance (BZ-GD) indicator and inverse Berge-Zhukovskii Generational Distance indicator (BZ-IGD), similar to the multiobjective GD and IGD [16], are computed in order to measure the quality of the results. These performance indicators use the Eucledian distance to measure the closeness between the approximation set and the true BZ equilibria.

The BZ-GD indicator measures the average distance of the approximation set A to the nearest point from the theoretical front:

$$\text{BZ-GD} = \frac{1}{|A|} \sum_{u \in A} d(u, p),$$

where $d(u, p)$ represents the minimum Euclidean distance between each point from the approximation set A and the reference set.

The BZ-IGD indicator computes the average distance between the theoretic Berge-Zhukovskii equilibria (denoted by B) and the nearest solution in the solution set found (A):

$$\text{BZ-IGD} = \frac{1}{|B|} \sum_{u \in B} d(u, v),$$

where $d(u, v)$ represents the Euclidean distance between each point from B and the nearest point from A.

Parameters used for the Cournot duopoly are as follows: $a = 24$, $c = 9$, $s_i \in [1, 15], n = 2$. Every value which satisfies the equation $\alpha_1 \cdot s_1 + \alpha_2 \cdot s_2 = \alpha$ is a BZ strategy profile of the game. Figures 1 and 2 present the obtained payoffs for two different cases.

We consider 20 independent runs for each algorithm with each run consisting of 50 epochs. A population of $N = 150$ individuals is used. For the algorithms initial parameter values are: $C_r = 0.8$ and $F = 0.2$. After 200 generations (1 epoch) the constraints of the game change by generating new α_i values following a uniform distribution.

Figure 3 presents boxplots of the average and best BZ-GD indicator computed over 20 independent runs. Results for the AE-DE and DGET-DE are similar. Figure 4 indicates the inverse BZ-IGD indicator. Table 1 presents average, standard deviation, minimum and maximum value of the BZ-GD and BZ-IGD for each run.

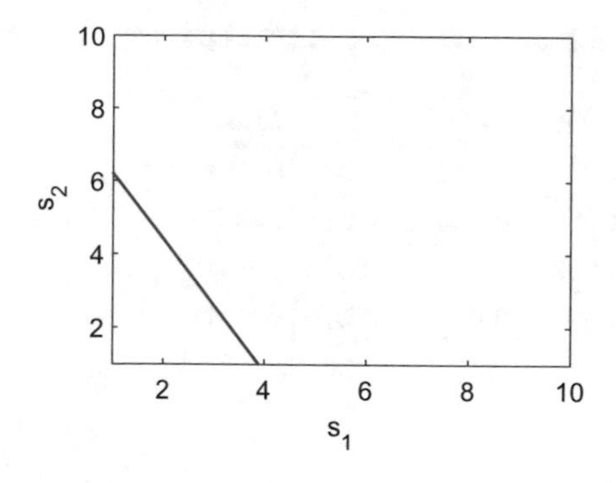

Fig. 1. GBZE for $\alpha_1 = 1.8, \alpha_2 = 1$, $\alpha = 8$ for one epoch

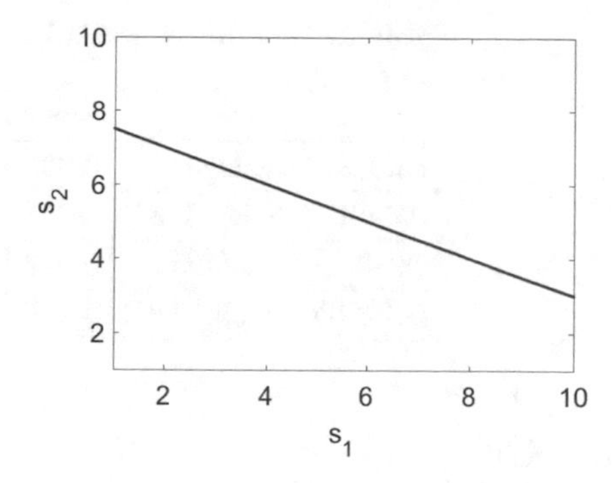

Fig. 2. GBZE for $\alpha_1 = 0.5, \alpha_2 = 1$, $\alpha = 8$ for the next epoch

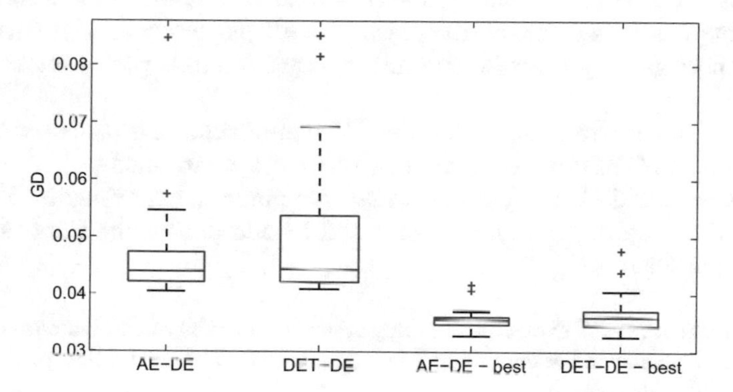

Fig. 3. BZ-GD boxplots: mean and best results for AE-DE and DGET-DE

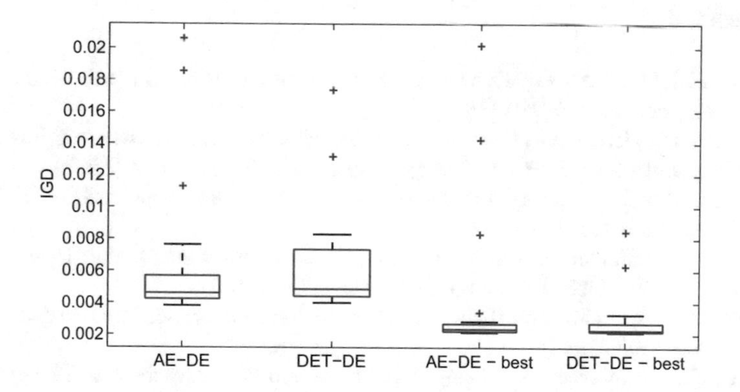

Fig. 4. BZ-IGD boxplots: mean and best results for AE-DE and DGET-DE

A Wilcoxon rank sum test was employed to compare the two methods showing no significant differences between the BZ-GD indicators of the two methods; another Wilcoxon test confirms however a significant difference between BZ-GD and BZ-IGD indicators.

Table 1. Descriptive statistics of obtained BZ-GD and BZ-IGD

	Mean	St. Dev	Median	Min	Max
BZ-GD (AE-DE)	0.037	0.008	0.034	0.034	0.069
BZ-GD (DGET-DE)	0.039	0.039	0.035	0.033	0.060
BZ-IGD (AE-DE)	0.017	0.0009	0.017	0.015	0.018
BZ-IGD (DGET-DE)	0.017	0.0008	0.016	0.015	0.018

6 Conclusions

The generalized Berge-Zhukovskii equilibrium problem is introduced and studied in a dynamic environment. As Berge-Zhukovskii equilibrium is significant in social dilemmas (ensures a better payoff for all players than the Nash equilibrium for these games), the new generalized equilibrium is promising in economic games.

Using a generative relation for the BZ equilibrium the proposed method is able to find the GBZE of the game in a dynamic environment.

The generalized Cournot game is used for numeric experiments. The performance of the proposed method is evaluated by computing the Berge-Zhukovskii Generational Distance.

Acknowledgments. This work was supported by a grant of the Romanian National Authority for Scientific Research and Innovation, CNCS - UEFISCDI, project number PN-II-RU-TE-2014-4-2560.

References

1. Arrow, K.J., Debreu, G.: Existence of an equilibrium for a competitive economy. Econometrica **22**, 265–290 (1954)
2. Cardell, J.B., Hitt, C.C., Hogan, W.W.: Market power and strategic interaction in electricity networks. Resour. Energy Econ. **19**(1–2), 109–137 (1997)
3. Cournot, A.: Recherches sur les Principes Mathematique de la Theorie des Richesses. Hachette, Paris (1838)
4. Deb, K.: An efficient constraint handling method for genetic algorithms. Comput. Methods Appl. Mech. Eng. **186**(2–4), 311–338 (2000)
5. Debreu, G.: A social equilibrium existence theorem. Proc. Natl. Acad. Sci. USA **38**(10), 886–893 (1952)
6. Gaskó, N., Dumitrescu, D., Lung, R.I.: Evolutionary detection of Berge and Nash equilibria. In: Pelta, D., Krasnogor, N., Dumitrescu, D., Chira, C., Lung, R. (eds.) Nature Inspired Cooperative Strategies for Optimization (NICSO 2011). Studies in Computational Intelligence, vol. 387, pp. 149–158. Springer, Heidelberg (2012)
7. Haurie, A., Krawczyk, J.: An introduction to dynamic games (2001)
8. Lasaulce, S., Tembine, H.: Game Theory and Learning for Wireless Networks: Fundamentals and Applications, 1st edn. Academic Press, Cambridge (2011)
9. Lung, R.I., Dumitrescu, D.: Computing Nash equilibria by means of evolutionary computation. Int. J. Comput. Commun. Control **3**, 364–368 (2008)

10. Nash, J.: Non-cooperative games. Ann. Math. **54**(2), 286–295 (1951)
11. Van Long, N.: A Survey of Dynamic Games in Economics: Volume 1 of World Scientific Books. World Scientific Publishing Co. Pte. Ltd. (2010)
12. Roos, C.F.: A mathematical theory of competition. Am. J. Math. **47**(3), 163–175 (1925)
13. Storn, R., Price, K.: Differential evolution - a simple and efficient adaptive scheme for global optimization over continuous spaces. J. Glob. Optim. **11**, 341–359 (1997)
14. Suciu, M., Gaskó, N., Lung, R.I., Dumitrescu, D.: Nash equilibria detection for discrete-time generalized cournot dynamic oligopolies. In: Terrazas, G., Otero, F., Masegosa, A. (eds.) Nature Inspired Cooperative Strategies for Optimization (NICSO 2013). Studies in Computational Intelligence, vol. 512, pp. 343–354. Springer, Heidelberg (2014)
15. Suciu, M., Lung, R.I., Gaskó, N., Dumitrescu, D.: Differential evolution for discrete-time large dynamic games (2013)
16. Van Veldhuizen, D.A., Lamont, G.B.: On measuring multiobjective evolutionary algorithm performance. In: Proceedings of the 2000 Congress on Evolutionary Computation, vol. 1, pp. 204–211 (2000)

Fuzzy Logic Control Application for the Risk Quantification of Projects for Automation

Olga Davidova and Branislav Lacko(✉)

Brno University of Technology, Technická 2, 616 69 Brno, Czech Republic
{davidova.o,lacko}@fme.vutbr.cz

Abstract. This article describes the classical approach to risk quantification. This is followed by recommendations of fuzzy sets for advanced risk quantification in the automation project. Different models for fuzzification and defuzzification are presented and the optimum model variants are found with the help of the MATLAB program system.

Keywords: Project · Risk control · Risk quantification · Fuzzy sets
Fuzzy control · Defuzzification

1 Introduction

The article draws attention to the danger posed by bad quality risk quantification. The bad results of quantification can lead to an incorrect measure for reducing risks and distorting the results of overall risk analysis. The article presents a clear reminder of the options of the possible expression of risk values and method of determining these values. Special fuzzy approach to risk quantification is recommended and analysed.

2 Classical Value Expression in Risk Quantification

Risk, risk probability and risk impact values can be expressed:

In financial terms as the amount calculated from the risk probability p and scope of its impact d with subsequent calculation for i-th risk $hr_i = p_i \times d_i$ [As the probability value is a dimensionless figure the risk is enumerated in units of the currency in which the risk impact was enumerated] (see for example the recommendations mentioned in publication [1]).

By specification of a numerical value on a scale, for example in the case of the scoring method with a risk map, see Table 1, as p = <1;10> and d = <1; 10> with subsequent score calculation replacing the risk value s = <1; 100> or by another scale and calculation [2].

Verbally as an assessment[1] using tables (for example for the probability value as: high, medium and low probability, for the impact value as: big, medium and small impact). The tables include: a matrix for the risk value class allocation, a probability class table, an impact class table, and the risk value class, for example by the RIPRAN method [3]).

[1] This method is mainly used in cases of quantification of risks of social systems, the "soft systems".

R. Matoušek (Ed.): MENDEL 2017, AISC 837, pp. 320–326, 2019.
https://doi.org/10.1007/978-3-319-97888-8_29

Each of these approaches has its advantages and disadvantages. Therefore the risk analysis team must agree about the approach to be used and select the most appropriate one subject to the circumstances of the situation, unless the method of value expression is directly defined by the methodology used.

3 Traditional Methods Quantification

Table 1 shows a brief survey of basic advantages and disadvantages of the individual methods of value expression in connection with risk quantification. Please note that project teams of companies and institutions or other teams performing risk analysis do not discuss these advantages and disadvantages and hardly notice the differences between the methods. They simply use the established method.

A shared property of all three value expression methods is their acuity in distinguishing between cases. That is an advantage and simplification on the one hand, but on the other this simplified value rendering may pose an issue.

The very acuity of distinction between individual cases and situations of risk analysis in complex development projects of industrial process automation has caused that proposals have emerged for application of the fuzzy sets apparatus [4–6, 9, 10]. Regarding the relative novelty of the idea and lack of readiness of the current project teams use of this theory in practical project risk management is still scarce.

4 Fuzzy Sets Application for Risk Quantification

The Fuzzy sets were introduced by Lofti Zadeh in 1965. What's the difference between a crisp set and a fuzzy set? Let U be the universe and its elements be denoted as x. The crisp set A of U is defined as a function $f_A(x)$, and called the characteristic function of A. For any

Table 1. Advantages and disadvantages of different methods of value expression in risk quantification

Method of value expression principle	Advantage(s)	Disadvantage(s)
Direct value expression	Directly reflects the risk value definition Natural meaning of numerical values	In many cases direct specification of the probability and/or impact value is difficult
By means of defined values on a defined numerical scale	Scales with different numbers of values can be chosen Risk value is easy to define and allows for risk value comparisons	The meanings of the individual values on the scale must be specified
By means of selected verbal expressions	Verbal expressions can vary. Verbal expressions are easy to understand (often intuitively)	The meanings of the individual verbal expressions must be defined Verbal expression of the risk value on the basis of verbal expression of probability and impact must be determined

element x of universe U, the characteristic function $f_A(x)$ is equal to 1 if x is an element of set A, and is equal to 0 if x is not an element of A [7].

The fuzzy set A of universe U is defined as function $\mu_A(x)$, and called the membership function of set A. $\mu_A(x)$: $U \rightarrow [0, 1]$, where $\mu_A(x) = 1$ if x is totally in A; $\mu_A(x) = 0$ if x is not in A; $0 < \mu_A(x) < 1$ if x is partly in A. For any element x of universe U, the membership function $\mu_A(x)$ equals the degree to which x is an element of set A. This degree, a value between 0 and 1, represents the degree of membership, and is called the membership value, of an element x in set A [7]. The membership functions allow us to graphically represent the fuzzy sets. The simplest membership functions are formed using straight lines. In this paper the *triangular* membership function is used. The triangular function (Fig. 1) is defined by a lower limit $\boldsymbol{a}$, an upper limit $\boldsymbol{c}$, and a value $\boldsymbol{b}$, where $\boldsymbol{a} < \boldsymbol{b} < \boldsymbol{c}$.

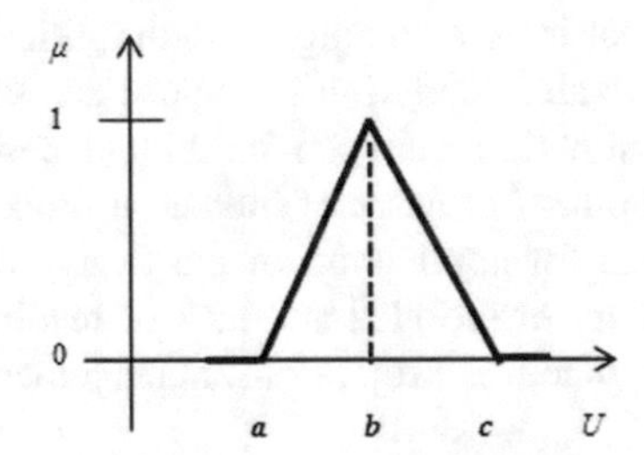

Fig. 1. Triangular membership function

Fuzzy sets work with linguistic variables. Linguistic variables are described by the linguistic variable name, the set of verbal values of the linguistic variable (the set of terms), the universe where the individual terms are defined, and the function mapping the verbal values to the universe values. We chose two input linguistic variables for the start, the probability and the project impact, and an output linguistic variable, the risk value defined by the total project budget. In all cases we used three triangular membership functions details of which are shown in Table 2.

Table 2. Survey of membership functions

Variable name	Term name	Membership function values			Scope of universe U
		a	b	c	
Probability	Low probability LP	0	0	40	0–100
	Medium probability MP	20	50	80	[%]
	High probability HP	60	100	100	
Impact on the project	Small impact SI	0	0	$160 * 10^3$	$0–50 * 10^4$
	Medium impact MI	$400 * 10^3$	$250 * 10^3$	$460 * 10^3$	[CZK]
	Big impact BI	$340 * 10^3$	$500 * 10^3$	$500 * 10^3$	
Value of risk	Small risk SR	0	0	40	0–100
	Medium risk MR	15	50	85	[%]
	High risk HR	60	100	100	

The fuzzy model was created with the help of the Fuzzy Logic Toolbox of the Matlab software environment [8] with its FIS editor. Table 3 shows the rule basis used for the interference mechanism. The interference mechanism utilises Mamdani implication and aggregation. The maximum aggregated membership function is hereinafter identified as μ_{Ag}.

Table 3. Rules

Rules	SI	MI	BI
LP	SR	SR	MR
MP	SR	MR	HR
HP	MR	HR	HR

To obtain the acute risk value from the output fussy set there is the defuzzification process. In our case more methods were selected to allow for comparisons. In this paper the following defuzzification methods are used:

Centroid (Center of Area-COA) defuzzification (Fig. 2) returns the centre of the area under the curve of the membership function Figure.

$$COA = \frac{\int \mu_{Ag}(u)udu}{\int \mu_{Ag}(u)du} \quad (1)$$

Bisector (Bisector of Area-BOA) defuzzification (Fig. 2) uses a vertical line that divides the area under the curve into two equal areas.

$$BOA = \int_{\alpha}^{BOA} \mu_{Ag}(u)du = \int_{BOA}^{\beta} \mu_{Ag}(u)du \quad (2)$$

Mean of Maximum (MOM) defuzzification method (Fig. 3) uses the average value of the aggregated membership function outputs.

Smallest of Maximum (SOM) defuzzification method (Fig. 3) uses the minimum value of the aggregated the membership function outputs.

Table 4 shows different values of the input membership functions, i.e. risk prob-

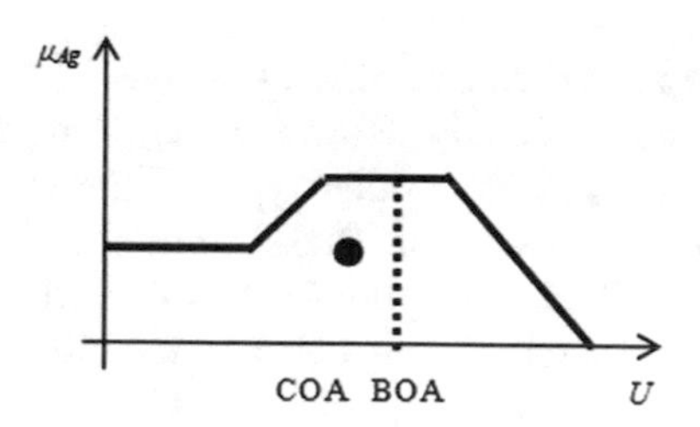

Fig. 2. Deffuzification COA, BOA

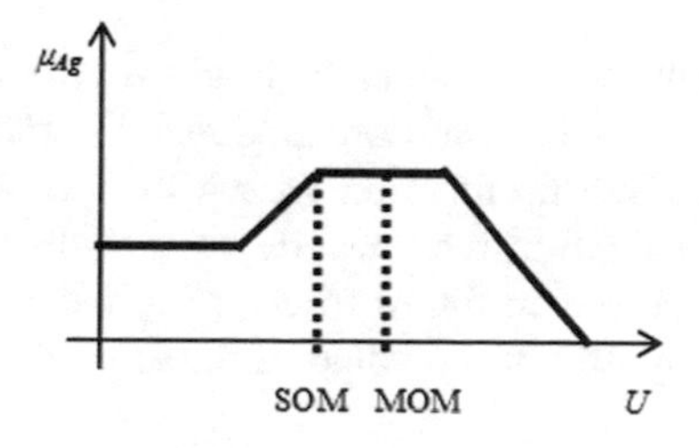

Fig. 3. Deffuzification SOM, MOM

ability and its project impact, and 4 variants of the output risk value subject to the defuzzification method used.

Figures 4, 5, 6 and 7 show surfaces for different variants of the defuzzification

Table 4. Input and output value comparisons

Probability	Impact	Value of risk			
		COA	BOA	MOM	SOM
12.4	$25*10^3$	14	13	6	0
12.4	$111*10^3$	16.6	16	13	0
12.4	$261*10^3$	14	13	6	0
12.4	$370*10^3$	32.6	26	11	0
12.4	$475*10^3$	50	50	50	40
30.7	$25*10^3$	16.4	16	12.5	0
30.7	**$111*10^3$**	**40.3**	**41**	**50**	**27**
30.7	$261*10^3$	42.7	44	50	28
30.7	$370*10^3$	48.7	49	50	28
30.7	$475*10^3$	64	68	87.5	75
51.8	$25*10^3$	13.3	12	3	0
51.8	$111*10^3$	40.3	41	50	27
51.8	$261*10^3$	50	50	50	48
51.8	$370*10^3$	55.4	54	50	30
51.8	$475*10^3$	86.7	88	97	94
71.1	$25*10^3$	39.2	39	14	0
71.1	$111*10^3$	49.4	49	37	0
71.1	$261*10^3$	59.7	60	50	26
71.1	$370*10^3$	59.7	60	50	26
71.1	$475*10^3$	83	83	86	72
91.3	$25*10^3$	50	50	50	43
91.3	$111*10^3$	61.4	62	87	74
91.3	$261*10^3$	86.4	88	96	92
91.3	$370*10^3$	84.1	85	89	78
91.3	$475*10^3$	86.4	88	96	92

methods. The values highlighted in the table with red colour were used here.

This contribution modified the deffuzification methods. However the triangular membership function type with three terms was used. Next other types of the membership function may be used with different term numbers. These models will be processed in future research by the RIPRAN method. The probability value will be composed of the value of threat probability and the relevant scenario probability.

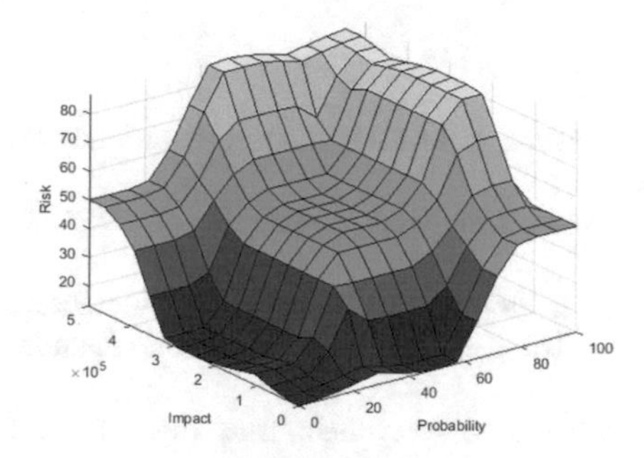

Fig. 4. Surface – deffuzification COA

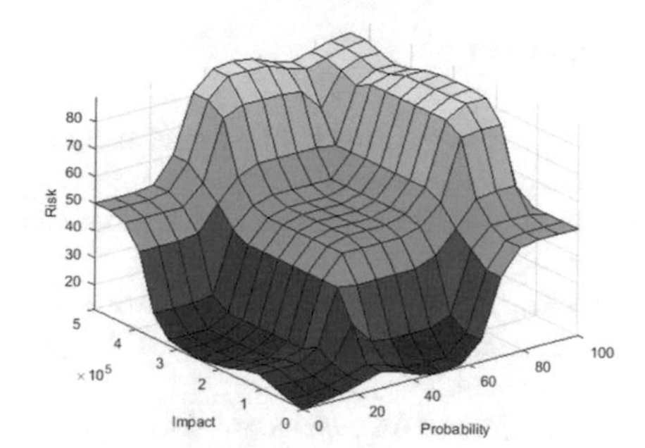

Fig. 5. Surface – deffuzification BOA

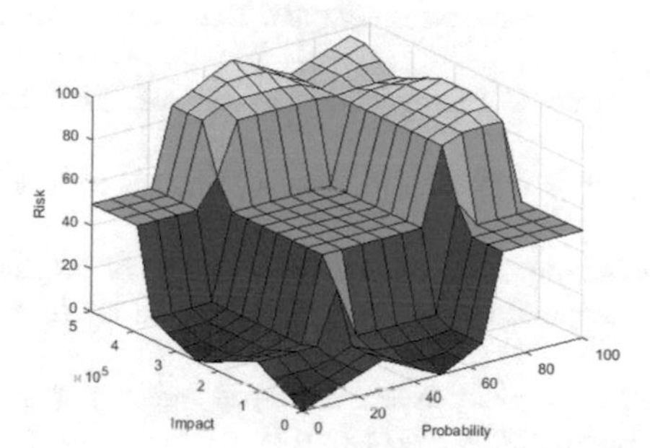

Fig. 6. Surface – deffuzification MOM

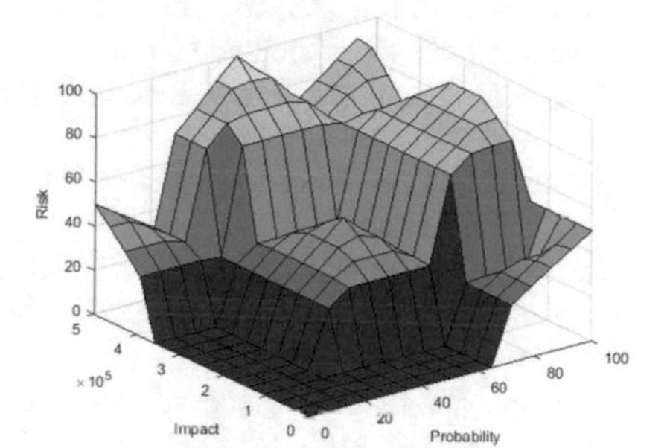

Fig. 7. Surface – deffuzification SOM

5 Conclusion

The purpose of the present research is finding suitable models for fuzzification and defuzzification of project risks in their quantification. Nevertheless clearly will structured projects are able to use one of the classical methods of risk value expression in risk quantification. On the other hand complex situations of soft projects are better off with fuzzy sets. Fuzzy sets better model non-acute values of quantified risks. In addition these values may subsequently be used for defining rules of risk management process control through predefined procedures of fuzzy control. The obtained rules can then be used for implementation of an expert system supporting risk management, which is planned as the next research stage.

Acknowledgments. Supported by grant BUT IGA No.: FSI-S-17-4785 Engineering application of artificial intelligence methods.

References

1. A Guide to the Project Management Body of Knowledge (PMBOK Guide)–Fourth edition. Project Management Institute, Pennsylvania (2008)
2. EN 60812: 2006 Procedure for failure mode and effects analysis (FMEA). CENELEC, Brussels (2006)
3. Lacko, B.: The Risk Analysis of Soft Computing Projects. In: Proceedings International Conference on Soft Computing – ICSC 2004, pp. 163–169. European Polytechnic Institute, Kunovice (2004)
4. Doskočil, R.: An evaluation of total project risk based on fuzzy logic. Bus. Theor. Pract. **1** (17), 23–31 (2016)
5. Dikmen, I., Birgonul, M.T., Han, S.: Using fuzzy risk assessment to rate cost overrun risk international construction projects. Int. J. Project Manag. **25**, 494–505 (2007)
6. Shang, K., Hossen, Z.: Applying Fuzzy Logic to Risk Assessment and Decision-Making. Canadian Institute of Actuaries, Canada (2013)
7. Klir, G.J., Yuan, B.: Fuzzy Sets and Fuzzy Logic. Theory and Applications. Prentice Hall PRT, New Jersey (1995)
8. Gulley, N., Roger Jang, J.-S.: Fuzzy Logic Toolbox for Use with Matlab. The MathWorks Inc., Berkeley (1995)
9. Doskočil, R., Doubravský, K.: Qualitative evaluation of knowledge based model of project time- cost as decision making support. Econ. Comput. Econ. Cybern. Stud. Res. **1**(51), 263–280 (2017)
10. Doskočil, R.: Evaluating the creditworthiness of a client in the insurance industry using adaptive neuro-fuzzy inference system. Eng. Econ. **1**(28), 15–24 (2017)
11. Brožová, H., Bartoška, J., Šubrt, T.: Fuzzy approach to risk appetite in project management. In: Proceedings of the 32nd International Conference on Mathematical Methods in Economics, pp. 61–66. Palacky University, Olomouc (2014)

Author Index

R. Matoušek (Ed.): MENDEL 2017, AISC 837, pp. 327–328, 2019.
https://doi.org/10.1007/978-3-319-97888-8

Lecture Notes in Networks and Systems 1040

The series “Lecture Notes in Networks and Systems” publishes the latest developments in Networks and Systems—quickly, informally and with high quality. Original research reported in proceedings and post-proceedings represents the core of LNNS.

Volumes published in LNNS embrace all aspects and subfields of, as well as new challenges in, Networks and Systems.

The series contains proceedings and edited volumes in systems and networks, spanning the areas of Cyber-Physical Systems, Autonomous Systems, Sensor Networks, Control Systems, Energy Systems, Automotive Systems, Biological Systems, Vehicular Networking and Connected Vehicles, Aerospace Systems, Automation, Manufacturing, Smart Grids, Nonlinear Systems, Power Systems, Robotics, Social Systems, Economic Systems and other. Of particular value to both the contributors and the readership are the short publication timeframe and the world-wide distribution and exposure which enable both a wide and rapid dissemination of research output.

The series covers the theory, applications, and perspectives on the state of the art and future developments relevant to systems and networks, decision making, control, complex processes and related areas, as embedded in the fields of interdisciplinary and applied sciences, engineering, computer science, physics, economics, social, and life sciences, as well as the paradigms and methodologies behind them.

Indexed by SCOPUS, INSPEC, WTI Frankfurt eG, zbMATH, SCImago.

All books published in the series are submitted for consideration in Web of Science.

For proposals from Asia please contact Aninda Bose (aninda.bose@springer.com).

Marcelo Zambrano Vizuete ·
Miguel Botto-Tobar · Sonia Casillas ·
Carina Gonzalez · Carlos Sánchez ·
Gabriel Gomes · Benjamin Durakovic
Editors

Innovation and Research – Smart Technologies & Systems

Proceedings of the CI3 2023, Volume 1

Springer

Editors
Marcelo Zambrano Vizuete
Instituto Tecnológico Superior Rumiñah
Sangolquí, Ecuador

Miguel Botto-Tobar
Eindhoven University of Technology
Eindhoven, The Netherlands

Sonia Casillas
universidad de salamanca
Salamanca, Spain

Carina Gonzalez
Universidad de La Laguna
Tenerife, Spain

Carlos Sánchez
Universidad de Zaragoza
Zaragoza, Spain

Gabriel Gomes
Universidad Estatal de Campinas
Campinas, Brazil

Benjamin Durakovic
International University of Sarajevo
Sarajevo, Bosnia and Herzegovina

ISSN 2367-3370 ISSN 2367-3389 (electronic)
Lecture Notes in Networks and Systems
ISBN 978-3-031-63433-8 ISBN 978-3-031-63434-5 (eBook)
https://doi.org/10.1007/978-3-031-63434-5

This Springer imprint is published by the registered company Springer Nature Switzerland AG
The registered company address is: Gewerbestrasse 11, 6330 Cham, Switzerland

Preface

The 4th edition of the International Research and Innovation Congress – Smart Technologies and Systems, CI3 2023, took place from August 30 to September 1, 2023, at the facilities of the Instituto Tecnológico Universitario Rumiñahui, located in the city of SangolquÍ, Pichincha, Ecuador.

The conference was organized by the Red de Investigación, Innovación y Transferencia de Tecnología RIT2, made up of the most relevant university institutes in Ecuador, among which are ITCA, BOLIVARIANO, ARGOS, VIDA NUEVA, ESPÍRITU SANTO, SUDAMERICANO, ISMAC, SAN ISIDRO, ARTES GRÁFICAS, ORIENTE, HUMANE, SUCRE, CENTRAL TÉCNICO, POLICÍA NACIONAL and RUMIÑAHUI.

Additionally, the event is sponsored by the Secretaría de Educación Superior, Ciencia, Tecnología e Innovación SENESCYT, Labortorio de Comunicación Visual de la Universidad Estatal de Campinas—Brazil, Universidad Ana G. Méndez—Puerto Rico, Centro de Investigaciones Psicopedagógicas y Sociológicas—Cuba, Instituto Superior de Diseño de la Universidad de La Habana—Cuba, GDEON and the Corporación Ecuatoriana para el Desarrollo de la Investigación y la Academia—CEDIA.

The main objective of CI3 2023 is to generate a space for dissemination and collaboration, where academia, industry and government can share their ideas, experiences and results of their projects and research.

"Research as a pillar of higher education and business improvement" is the motto of the Conference and suggests how research, innovation and academia must coincide with the productive sector to leverage social and economic development.

CI3 2023 had 145 papers submitted, of which 52 were accepted for publication and presentation. To guarantee the quality of the publications, the event has a staff of more than 70 experts, from different countries such as Spain, Argentina, Chile, Mexico, Peru, Brazil, Ecuador, among others, who carry out an exhaustive review of each proposal sent.

Likewise, during the event a series of keynote conferences were held, given by both national and international experts, allowing attendees to get in touch with the latest trends and technological advances around the world. Keynote speakers included: Ph.D. Carina González, University of La Laguna, Spain; Ph.D. Gabriel Gómez, State University of Campinas, Brazil; Ph.D. Carlos Sanchez, University of Zaragoza, Spain; Ph.D. Juan Minango, Instituto Tecnológico Superior Rumiñahui; Dr. Iván Cherrez, Universidad Espíritu Santo, Ecuador; and Ph.D. Nela Paustizaca, Escuela Politécnica del Litoral, Ecuador.

The content of this proceeding is related to the following topics:

- Smart Cities
- Innovation and Development
- Applied Technologies
- Economics and Management
- ICT for Educations.

Organization

General Chairs

Marcelo Zambrano Vizuete	Instituto Tecnológico Universitario Rumiñahui, Ecuador

Organizing Committee

Marcelo Zambrano Vizuete	Instituto Tecnológico Universitario Rumiñahui, Ecuador
Miguel Botto-Tobar	Eindhoven University of Technology, The Netherlands
Sonia Casillas	Universidad de Salamanca, Spain
Carina Gonzales	Universidad de La Laguna, Spain
Carlos Sánchez	Universidad de Zaragoza, Spain
Gabriel Gomes	Universidad Estatal de Campinas, Brazil
Benjamin Durakovic	International University of Sarajevo, Bosnia and Herzegovina

Steering Committee

Marcelo Zambrano	Instituto Tecnológico Superior Rumiñahui, Ecuador
Miguel Botto	Eindhoven University of Technology, The Netherlands
Karla Ayala	Instituto Tecnológico Superior Rumiñahui, Ecuador
Xavier Duque	Instituto Tecnológico Superior Rumiñahui, Ecuador
Ernesto Huerta	Instituto Tecnológico Superior Rumiñahui, Ecuador
Carmita Suárez	Instituto Tecnológico Superior Rumiñahui, Ecuador
Maritza Salazar	Instituto Superior Tecnológico Espíritu Santo, Ecuador
Wilfrido Robalino	Instituto Superior Tecnológico Vida Nueva, Ecuador

Roberto Tolozano	Instituto Superior Tecnológico Bolivariano, Ecuador
Carlos Pérez	Instituto Tecnológico Sudamericano, Ecuador
Alicia Soto	Instituto Superior Tecnológico ITCA, Ecuador
Lorena Ávila	Instituto Superior Tecnológico Policía Nacional, Ecuador
Denis Calvache	Instituto Superior Tecnológico ISMAC, Ecuador
Efraín Paredes	Instituto Superior Tecnológico IGAD, Ecuador
Paola Vasconez	Instituto Superior Tecnológico Oriente, Ecuador
Sandra El Khori	Instituto Superior Tecnológico San Isidro, Ecuador
Santiago Illescas	Instituto Superior Tecnológico Sucre, Ecuador
Jorge Calderón	Instituto Superior Tecnológico Argos, Ecuador
Trotsky Corella	Instituto Superior Tecnológico Central Técnico, Ecuador
Javier García	Instituto Superior Tecnológico Humane, Ecuador
Sandra Salazar	Instituto Superior Tecnológico American College, Ecuador
David Flores	Instituto Superior Tecnológico Cordillera, Ecuador
Marcelo Aguilera	Instituto Superior Tecnológico del Azuay, Ecuador

Program Committee

Carina González	Universidad de La Laguna, Spain
Carlos Sánchez	Universidad de Zaragoza, Spain
Sergey Balandin	Universidad de Helsinki, Finland
Silva Ganesh Malla	SKR Engineering College, India
Alberto Ochoa	Universidad Autónoma Ciudad de Juárez, Mexico
Antonio Orizaga	Universidad de Guadalajara, Mexico
Rocío Maciel	Universidad de Guadalajara, Mexico
Víctor Larios	Universidad de Guadalajara, Mexico
Yorberth Montes	Universidad de Zulia, Venezuela
Juan Minango	Universidad Estatal de Campinas, Brazil
Francisco Pérez	Universidad Politécnica de Valencia, Spain
Víctor Garrido	Universidad Politécnica de Valencia, Spain
Alberto García	Universidad Politécnica de Valencia, Spain
Javier Hingant	Universidad Politécnica de Valencia, Spain
Javier Prado	Universidad Técnica Federico Santa María, Chile
Marco Heredia	Universidad Politécnica de Madrid, Spain

Diego Paredes	Universidad de Zaragoza, Spain
Angela Cadena	Universitat de Valencia, Spain
Sonia Casillas	Universidad de Salamanca, Spain
Marco Cabezas	Universidad de Salamanca, Spain
Francesc Wilhelmi	Centre Tecnològic de Telecomunicacions de Catalunya, Spain
Santiago Vidal	Universidad Nacional Del Centro de la Provincia de Buenos Aires, Argentina
Adailton Antônio Galiza Nunes	Universidad Estatal de Campinas, Brazil
Leandro Bezerra De Lima	Universidad Federal de Mato Grosso Do Sul, Brazil
Andrea Carolina Flores	Embraer Defensa y Seguridad, Brazil
Elmer Levano Huamacto	Universidad Federal Do Parana, Brazil
Ari Lazzarotti	Universidad Federal de Goiás, Brazil
Ellen Synthia Fernandes de Oliveira	Universidad Federal de Goiás, Brazil
Fernanda Cruvinel Pimentel	Universidad Federal de Goiás, Brazil
Yuzo Lanoi	Universidad Estatal de Campinas, Brazil
Gabriel Gómez	Universidad Estatal de Campinas, Brazil
Pablo Minango	Universidad Estatal de Campinas, Brazil
Benjamin Durakovic	International University of Sarajevo, Bosnia and Herzegovina
Jhonny Barrera	Universidad Nacional de La Plata, Argentina
Daniel Ripalda	Universidad Nacional de La Plata, Argentina
Alonso Estrada Cuzcano	Universidad Nacional Mayor de San Marcos, Peru
Marcelo Zambrano	Instituto Tecnológico Superior Rumiñahui, Ecuador
Ana Zambrano	Escuela Politécnica Nacional, Ecuador
Angel Jaramillo	Universidad de las Américas, Ecuador
Ángela Cadena	Universidad de Guayaquil, Ecuador
Oscar Leon	Universidad de Guayaquil, Ecuador
Wladimir Paredes	Consejo de Aseguramiento de la Calidad de la Educación Superior, Ecuador
Miguel Angel Zúñiga	Universidad de Cuenca, Ecuador
Alex Santamaría	Universidad Laica Eloy Alfaro, Ecuador
Fabián Sáenz	Universidad de las Fuerzas Armadas ESPE, Ecuador
Darwin Aguilar	Universidad de las Fuerzas Armadas ESPE, Ecuador
Sergio Montes	Universidad de Las Fuerzas Armadas Espe, Ecuador
Jhenny Cayambe	Pontificia Universidad Católica del Ecuador Sede Ibarra, Ecuador

RIT2

Contents

Innovation and Development

Smart Cities

IoT Applied to Improve Production Controls in the Ecuadorian Floriculture Sector

Claudio Arcos[1(✉)], Pablo Calvache[2], and Ricardo Calderón[3]

[1] San Francisco Global Foundation and ISMAC Technological Institute, Quito, Ecuador
claudioarcos@sanfranciscoglobal.org
[2] ISMAC Technological Institute, Quito, Ecuador
[3] World Roses Center Farm, Tabacundo, Ecuador

Abstract. For the development of this study, the internet of things IoT concept was used, in the context of which a network of sensors was deployed within a rose crop that is planted in an area of 30,000 m^2, which are covered under greenhouses. The crop is located at 2,900 m above sea level and, for a year and a half, humidity data, dew point data, temperature data and vapor pressure deficit VPD data were collected with the use of the sensor network. in such a way that a database of more than 49 million data was obtained that was used to contrast and analyze the data collected in the same period related to diseases and pests that can occur in rose production. The results show that there are levels of correlation that allow determining mechanisms for measuring environmental conditions in production to improve efficiency and effectiveness in the decision-making process for better crop management.

Keywords: IoT · Digital transformation · Precision agriculture

1 Introduction

Between the years 2018 and 2022, Ecuador exported 20.46% of a total of USD$15,850′580,000.00 of the fresh cut roses product, thus becoming the second most exporting country of this product during that period, below the Netherlands and above Columbia [1].

In this context, rose crops in Ecuador must implement control technologies in productive management to meet local and international demands in various areas. The phytosanitary field is one of the areas whose management, adequate or not, positively or negatively affects access to international markets, compliance with local controls, quality improvement and waste reduction.

The fact is that damage to the rose plant can occur in different ways. The main damage is caused by pests such as thrips and mites or diseases such as botrytis, powdery mildew and downy.

In this sense, the concept of precision agriculture is transcendental to implement mechanisms that favor the control and management of potential effects on plants. In this context, it is important to first understand what IoT is as a mechanism to promote precision agriculture in rose production, and secondly what are the main diseases and pests that affect the cultivation of roses.

M. Z. Vizuete et al. (Eds.): CI3 2023, LNNS 1040, pp. 3–17, 2024.
https://doi.org/10.1007/978-3-031-63434-5_1

1.1 IoT as a Mechanism to Promote Precision Agriculture in Rose Production

At present, the domain of agriculture results from the application of methods that, by using technological instruments, allow the improvement of traditional cultivation methods in the areas that are possible [2].

The fact is that knowledge-intensive solutions such as the use of sensors, drones, automated irrigation systems, real-time monitoring, timed disinfection, etc., make it possible to enhance the organization's resource and capacity management and favor productivity, efficiency, control of pests and diseases, quality of the final product and competitive access to markets, mainly.

For this reason, farmers are constantly looking for mechanisms to manage the risks of climate change. In this context, the Internet of Things, known as IoT, is a technological evolution that provides access to data, information and services of the activities in which it is applied, through intelligent networks that deliver data in real time [3].

The IoT applied to agriculture can improve several processes, because with better information it is possible to deploy actions to sustain the quality of the product during the production process and thus satisfy customer requirements [4].

In the flower sector, with the application of sensors within the IoT concept, it is possible to know what is happening with the production processes in terms of their incidence environment. In other words, in order to maintain optimal growth and development conditions for rose plants inside a greenhouse, it is necessary to control and manage, among other things, the temperature, humidity, vapor pressure deficit and the dew point within the crop environment [5].

The use of IoT technology can simplify this management by obtaining data from said measurements in real time and, with the analysis of these data, favor decision-making for the good of the production process, as well as the management and control of pests and diseases.

The diffusion of these technologies for the improvement of competitiveness in the flower sector of Ecuador is essential. This is because it has been shown that the presence of innovation activities depends on the execution of various skills within the organization, and its innovative approach is stimulated by mastering a portfolio of technological skills from which they are carried out. R&D activities and, if digital transformation and absorption are not promoted, innovation will not happen [6].

1.2 Diseases and Pests that Affect the Cultivation of Roses

Mainly: the development of cultural practices, adequate air circulation, correct protection of the growing area and irrigation, help prevent and control diseases and pests [7], This, plus the productive capacities of the peasants, which include their traditional and ancestral knowledge on sowing and harvesting mechanisms and practices, have contributed to the development of the flower sector in Ecuador [8].

Additionally, there are organic and chemical treatments to manage pests and diseases, mainly pests such as thrips and mites or diseases such as botrytis, powdery mildew and downy [9–11].

As for pests, thrips are small insects that feed on the petals and cause cosmetic damage to the flower. For their part, the mites that affect roses are sap suckers, therefore, they feed on the sap of the leaves, causing them to turn yellow and dry.

Among the most common diseases are: botrytis, which is a fungal disease that generates brown spots mainly on the petals. Powdery mildew is also a fungal disease that consists of the appearance of a white coating mainly on the leaves and stems, the same that is similar to dust on the leaves, buds and stems. For its part, downy is another fungal disease characterized by the appearance of reddish or orange pustules on the underside of the leaves [12].

The different affectations can cause decomposition of the flower, premature wilting, weakness of the plant, disorder of the growth cycle, deformations, premature fall of the leaves, affectation of the photosynthesis process; all of this decreases the quality of the flowers, reduces their life time in the vase, and therefore decreases the competitiveness of the final product, with the consequent loss of the export market.

The control of environmental conditions inside the greenhouse facilitates decision-making regarding agricultural activities as well as the application of fungicides to protect the health of plants. The fact is that both pests and diseases are the result of poor production management in the face of changing environmental conditions that favor the appearance of pests and diseases, mainly temperature, humidity, vapor pressure deficit and the dew point.

2 Materials and Methods

2.1 Database and Variables

For the development of the study, five sensors were installed on June 25, 2021, the same ones that were distributed in 30,000 m^2 of a rose farm located in Tabacundo - Ecuador, at 2,900 m above sea level.

The sensors used are dataloggers that record and store the data in their internal memory, and transmit it in real time through the internet or it can be downloaded via Bluetooth. For the present study, the data was obtained by the transmission via the internet to a PC.

The characteristics of the sensors are:

- Wireless access to data.
- The recorded data can be downloaded but cannot be edited.
- It has user-configurable alerts.
- Unlimited number of users that can monitor the same sensor.
- Connection to a portal to monitor in real time through a mobile application that allows the storage and download of data in CSV format.
- Measures temperature ranges from −40 °C to +60 °C.
- Measures relative humidity ranges from 0% to 100% (prolonged exposure to >80% can create a lag of up to +3% in readings, this effect gradually reverses after returning to < 80% conditions).
- Temperature accuracy (0 °C to 60 °C): +−0.3 °C typical, +−0.5 °C maximum.
- Temperature Accuracy (full range): +−0.7 °C typical, +−1.2 °C maximum.
- Humidity Accuracy (@25 °C, 20% to 80%): +−3% typical, +−4.5% maximum.
- Humidity Accuracy (@25 °C, 0% to 100%): +−4.5% typical, +−7.5% maximum.

- Signal range: 100 m in line of sight. It is greater with the use of an amplifying antenna that increases the coverage radius to connect the sensor signal to a router.
- Battery: CR2477 × 1.3 V (duration of one year approximately).
- Dimensions: 40 × 40 × 16.5 mm.
- Weight: 40 gr.
- Compatibility: iOS 8+, Android 4.3+, Bluetooth 4.0+

Each sensor obtains data on temperature, humidity, vapor pressure deficit, and dew point. Sensors are IoT technology that record, store, and transmit data over the Internet in real time to one or more terminals with Android or iOS systems. The network architecture is shown below (see Fig. 1).

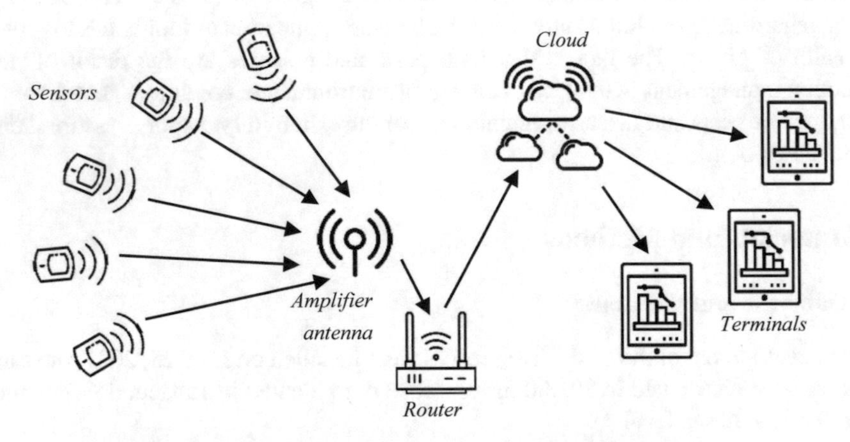

Fig. 1. Network architecture. Source: direct investigation.

Once the sensors were distributed, an amplifying antenna was fixed in the center of the five sensors at a distance of approximately 100 m from each sensor to the antenna. The antenna was connected to a router with internet access via a network cable. The router receives and transmits the data in real time to any terminal that has installed access to the information transmitted by the five sensors.

The data is permanently displayed in the sensors' own software, which allows all the data to be downloaded in CSV format. The databases constitute structured repositories with observations managed by storage systems and software packages of the sensors. In this sense, to obtain the data under study, observations collected from June 25, 2021 at 12:40 p.m. to January 19, 2023 at 5:40 p.m. were considered.

The database consists of minute-by-minute observations of four readings: temperature, relative humidity, vapor pressure deficit (VPD) and dew point, within the crop environment, obtaining the following observations and data:

- Block 1 database: 825,421 observations and 9,905,064 data
- Block 2 database: 825,407 observations and 9,904,884 data
- Block 3 database: 825,411 observations and 9,904,932 data
- Block 4 database: 825,411 observations and 9,904,932 data

- Block 5 database: 825,414 observations and 9,904,968 data

In order to compare the measurements of the five sensors with the state of the plants, the postharvest database of the company participating in the study was used. Said database stores daily information on harvest about production for exporting and discarded production. Discarded production is the number of stems discarded by pests and/or diseases caused in the rose, which go through the composting process because they do not comply with quality parameters for export. Within the discarded production, the different causes analyzed in this study are recorded, such as: in pests: mites and thrips; in diseases: botrytis, downy and powdery mildew. In the same way, the period of time considered for the analysis of the post-harvest database is from June 25, 2021 to January 19, 2023, obtaining the following observations and data:

- Postharvest database: 289,270 observations and 5,785,400 data

Since in Ecuador there are critical points in the productive systems, such as the precariousness of governance and management capacity [13], it is essential to consider that the determining factors of innovation in organizations include the adoption of digital transformation tools, understanding this action as the acquisition or absorption of innovation.

Therefore, demonstrating the application of a system based on IoT to improve production from the implementation of innovation solutions stimulates the mastery of technological skills as determinants of innovation [6].

In this sense, it is understood that the IoT applied acts as a catalyst for innovation, for which the question for this research is: how to improve efficiency in the Ecuadorian floriculture sector using IoT?

3 Results

Prior to the statistical analysis of the databases, these were refined and organized, disaggregating the date and time data as individual variables, to facilitate the grouping of data in intervals by hours, days, weeks, quarters, and years; and the days June 25, 2021 and January 19, 2023 were discarded as they did not have the total number of observations on those days.

To treat the database, a very important criterion for the study was incorporated, such as solar luminosity, so that the observations were grouped considering the sunrise from 6:00 a.m., and the sunset at 6:00 p.m., as well as the highest point (zenith) 12:00 pm. Therefore, a weighted average of four indicators per day was obtained considering the daily luminosity and a simple average was applied for grouping the days per week.

In reference to the postharvest database, the data was broken down by dates in format (mm/dd/yy), to facilitate their grouping by days and weeks. The production data were disaggregated to obtain the discarded production due to pests and diseases, and the data from each of the five sensors were grouped to obtain the discarded production and its causes. Sundays and holidays were not considered for not having postharvest records.

For the analysis, the databases were grouped and the observations were structured in years divided into weeks with the readings of the sensors and the causes of national production obtained from the postharvest record.

Once the data of the sensor variables and the causes of discarded production have been organized by years and weeks, the normality test is calculated. The normality test is performed because it is necessary to examine whether the data follows a normal distribution or not. For the normality test we use the Kolmogórov-Smirnov test, which is performed at a 95% confidence level. According to the result of this test, the statistical model should be selected, either for parametric or non-parametric data, and the following hypotheses are proposed:

H1: The data set does not follow the normal distribution
H0: The data set follows the normal distribution

If the result of the Kolmogórov-Smirnov test is less than 0.05, then the H0 hypothesis is rejected, that is, the data set does not follow a normal distribution.

The database was processed with the SPSS software, and the results show that the pest and disease variables obtain significance values less than 0.05 (see Table 1, 2, 3, 4 and 5), therefore the null hypothesis H0 is rejected and the alternative hypothesis H1 is accepted. Consequently, a statistical model for non-parametric data was then applied, it was Spearman Rho correlation coefficient, that is a nonparametric test used when the data does not follow a normal distribution.

Table 1. Normality test with data from sensor #1. Source: direct investigation.

Kolmogórov-Smirnov[a]			
	Stat	df	Sig
Humidity	,079	83	,200*
Dew point	,113	83	,011
Temperature	,058	83	,200*
VPD	,110	83	,015
Botrytis	,305	83	<,001
Powdery mildew	,306	83	<,001
Downy mildew	,308	83	<,001
Thrips	,191	83	<,001
Mites	,197	83	<,001

* This is a lower limit of the true meaning.
a. Lilliefors significance correction.

Table 2. Normality test with data from sensor #2. Source: direct investigation.

Kolmogórov-Smirnov[a]			
	Stat	df	Sig
Humidity	,064	83	,200*
Dew point	,124	83	,003
Temperature	,063	83	,200*
VPD	,096	83	,056
Botrytis	,271	83	<,001
Powdery mildew	,308	83	<,001
Downy mildew	,353	83	<,001
Thrips	,259	83	<,001
Mites	,220	83	<,001

* This is a lower limit of the true meaning.
a. Lilliefors significance correction.

Table 3. Normality test with data from sensor #3. Source: direct investigation.

Kolmogórov-Smirnov a			
	Stat	df	Sig
Humidity	,068	83	,200*
Dew point	,118	83	,006
Temperature	,063	83	,200*
VPD	,111	83	,013
Botrytis	,323	83	<,001
Powdery mildew	,302	83	<,001
Downy mildew	,273	83	<,001
Thrips	,283	83	<,001
Mites	,213	83	<,001

* This is a lower limit of the true meaning.
a. Lilliefors significance correction.

Table 4. Normality test with data from sensor #4. Source: direct investigation.

Kolmogórov-Smirnov a			
	Stat	df	Sig
Humidity	,122	51	,057
Dew point	,139	51	,016
Temperature	,209	51	<,001
VPD	,132	51	,026
Botrytis	,280	51	<,001
Powdery mildew	,260	51	<,001
Downy mildew	,402	51	<,001
Thrips	,346	51	<,001
Mites	,183	51	<,001

* This is a lower limit of the true meaning.
a. Lilliefors significance correction.

Table 5. Normality test with data from sensor sensor #5. Source: direct investigation.

Kolmogórov-Smirnov a			
	Stat	df	Sig
Humidity	,085	82	,200*
Dew point	,117	82	,008
Temperature	,092	82	,085
VPD	,102	82	,033
Botrytis	,291	82	<,001
Powdery mildew	,305	82	<,001
Downy mildew	,299	82	<,001
Thrips	,309	82	<,001
Mites	,204	82	<,001

* This is a lower limit of the true meaning.
a. Lilliefors significance correction.

Once the Spearman correlation coefficient statistical method was applied to the database, the respective correlation coefficient levels were obtained. The interpretation of Spearman's correlation coefficient ranges between -1 and + 1, indicating negative or positive associations respectively; 0 zero means that there is no correlation. The results can be seen in Table 6, 7, 8, 9 and 10.

Table 6. Correlation of pests and diseases with data from sensor #1. Source: direct investigation.

	Humidity				
	Botrytis	Powdery mildew	Downy mildew	Thrips	Mites
Rho =	0,122	0,204	0,041	−0,054	0,165
	Dew point				
	Botrytis	Powdery mildew	Downy mildew	Thrips	Mites
Rho =	0,081	0,245	−0,021	−0,09	0,097
	Temperature				
	Botrytis	Powdery mildew	Downy mildew	Thrips	Mites
Rho =	−0,003	0,072	−0,085	−0,034	−0,108
	VPD				
	Botrytis	Powdery mildew	Downy mildew	Thrips	Mites
Rho =	−0,086	−0,078	−0,057	−0,064	−0,15

Table 7. Correlation of pests and diseases with data from sensor #2. Source: direct investigation.

	Humidity				
	Botrytis	Powdery mildew	Downy mildew	Thrips	Mites
Rho =	0,533	0,093	−0,028	−0,434	0,382
	Dew point				
	Botrytis	Powdery mildew	Downy mildew	Thrips	Mites
Rho =	0,664	0,3	−0,144	−0,45	0,554
	Temperature				
	Botrytis	Powdery mildew	Downy mildew	Thrips	Mites
Rho =	0,251	0,317	−0,174	−0,038	0,267
	VPD				
	Botrytis	Powdery mildew	Downy mildew	Thrips	Mites
Rho =	−0,284	0,107	−0,024	0,318	−0,165

Table 8. Correlation of pests and diseases with data from sensor #3. Source: direct investigation.

	Humidity				
	Botrytis	Powdery mildew	Downy mildew	Thrips	Mites
Rho =	0,573	0,442	−0,014	−0,236	0,377
	Dew point				
	Botrytis	Powdery mildew	Downy mildew	Thrips	Mites
Rho =	0,698	0,62	−0,187	−0,128	0,477
	Temperature				
	Botrytis	Powdery mildew	Downy mildew	Thrips	Mites
Rho =	0,335	0,391	−0,296	0,118	0,292
	VPD				
	Botrytis	Powdery mildew	Downy mildew	Thrips	Mites
Rho =	−0,222	0,09	−0,123	0,229	−0,066

Table 9. Correlation of pests and diseases with data from sensor #4. Source: direct investigation.

	Humidity				
	Botrytis	Powdery mildew	Downy mildew	Thrips	Mites
Rho =	0,452	−0,009	−0,248	−0,142	0,22
	Dew point				
	Botrytis	Powdery mildew	Downy mildew	Thrips	Mites
Rho =	0,689	0,295	−0,233	−0,03	0,384
	Temperature				
	Botrytis	Powdery mildew	Downy mildew	Thrips	Mites
Rho =	0,455	0,393	−0,018	0,168	0,236
	VPD				
	Botrytis	Powdery mildew	Downy mildew	Thrips	Mites
Rho =	−0,071	0,18	0,181	0,178	−0,02

Table 10. Correlation of pests and diseases with data from sensor #4. Source: direct investigation.

	Humidity				
	Botrytis	Powdery mildew	Downy mildew	Thrips	Mites
Rho =	0,524	0,299	0,106	−0,298	0,405
	Dew point				
	Botrytis	Powdery mildew	Downy mildew	Thrips	Mites
Rho =	0,713	0,373	−0,081	−0,079	0,425
	Temperature				
	Botrytis	Powdery mildew	Downy mildew	Thrips	Mites
Rho =	0,449	0,227	−0,276	0,234	0,144
	VPD				
	Botrytis	Powdery mildew	Downy mildew	Thrips	Mites
Rho =	−0,116	−0,054	−0,219	0,348	−0,195

According to the statistical analysis of the data and its correlation, the following can be noted:

Humidity. It is verified that there is a positive correlation between the relative humidity variable with the presence of botrytis. The relationship is directly proportional with values of Rho min = 0.122 in sensor #1 and Rho max = 0.573 in sensor #3.

It is verified that there is a positive correlation between the relative humidity variable with the presence of powdery mildew. The relationship is directly proportional with values of Rho min = 0.093 in sensor #2 and Rho max = 0.442 in sensor #3. In this relationship, sensor #4 is the only one with a negative correlation with Rho = −0.009, so the possibility of an investigation by rose variety is opened considering the types of roses existing in the analysis radius measured by this sensor.

There is not enough significance to prove a correlation between the variable relative humidity with downy. The values shown are negative Rho max = −0.248 in sensor #4 and positive Rho max = 0.106 in sensor #5.

It is verified that there is a negative correlation between the relative humidity variable with the presence of thrips. The relationship is inversely proportional with values of Rho min = −0.054 in sensor #1 and Rho max = −0.434 in sensor #2.

It is verified that there is a positive correlation between the relative humidity variable with the presence of mites. The relationship is directly proportional with values of Rho min = 0.165 in sensor #1 and Rho max = 0.405 in sensor #5.

Dew Point. It is verified that there is a positive correlation between the dew point variable with the presence of botrytis. The relationship is directly proportional with values of Rho min = 0.081 in sensor #1 and Rho max = 0.713 in sensor #5.

It is verified that there is a positive correlation between the dew point variable with the presence of powdery mildew. The relationship is directly proportional with values of Rho min = 0.245 in sensor #1 and Rho max = 0.620 in sensor #3.

It is verified that there is a negative correlation between the variable dew point with the presence of downy. The relationship is inversely proportional with values of Rho min = −0.021 in sensor #1 and Rho max = −0.233 in sensor #4.

It is verified that there is a negative correlation between the dew point variable with the presence of thrips. The relationship is inversely proportional with values of Rho min = −0.030 in sensor #4 and Rho max = −0.450 in sensor #2.

It is verified that there is a positive correlation between the dew point variable with the presence of mites. The relationship is directly proportional with values of Rho min = 0.097 in sensor #1 and Rho max = 0.554 in sensor #2.

Temperature. It is verified that there is a positive correlation between the temperature variable with the presence of botrytis. The relationship is directly proportional with values of Rho min = 0.251 in sensor #2 and Rho max = 0.449 in sensor #5. In this relationship, sensor #1 is the only one with a negative correlation with Rho = −0.003, so the possibility of an investigation by rose variety is opened considering the types of roses existing in the analysis radius measured by this sensor.

It is verified that there is a positive correlation between the temperature variable with the presence of powdery mildew. The relationship is directly proportional with values of Rho min = 0.072 in sensor #1 and Rho max = 0.391 in sensor #3.

It is verified that there is a negative correlation between the variable temperature with the presence of downy. The relationship is inversely proportional with values of Rho min = −0.085 in sensor #1 and Rho max = −0.296 in sensor #3.

There is not enough significance to prove a correlation between the variable temperature with thrips. The values shown are negative Rho max = −0.038 in sensor #2 and positive Rho max = 0.234 in sensor #5.

It is verified that there is a positive correlation between the temperature variable with the presence of mites. The relationship is directly proportional with values of Rho min = 0.236 in sensor #4 and Rho max = 0.292 in sensor #3. In this relationship, sensor #1 is the only one with a negative correlation with Rho = −0.108, so the possibility of an investigation by rose variety is opened considering the types of roses existing in the analysis radius measured by this sensor.

VPD. It is verified that there is a negative correlation between the VPD variable with the presence of botrytis. The relationship is inversely proportional with values of Rho min = −0.086 in sensor #1 and Rho max = −0.284 in sensor #2.

There is not enough significance to prove a correlation between the variable VPD with powdery mildew. The values shown are negative Rho max = −0.078 in sensor #1 and positive Rho max = 0.078 in sensor #2.

There is not enough significance to prove a correlation between the variable VPD with downy. The values shown are negative Rho max = −0.219 in sensor #5 and positive Rho max = 0.181 in sensor #4.

It is verified that there is a positive correlation between the VPD variable with the presence of thrips. The relationship is directly proportional with values of Rho min = 0.178 in sensor #4 and Rho max = 0.348 in sensor #5. In this relationship, sensor #1 is the only one with a negative correlation with Rho = −0.064, so the possibility of an investigation by rose variety is opened considering the types of roses existing in the analysis radius measured by this sensor.

It is verified that there is a negative correlation between the VPD variable with the presence of mites. The relationship is inversely proportional with values of Rho min = −0.020 in sensor #4 and Rho max = −0.195 in sensor #5.

4 Discussion

The use of IoT technology helps to obtain data in real time so a farm is able to take decisions also in real time looking to improve the production process, as well as the management and control of pests and diseases.

Controlling affectations that cause damage of the flower can help to get a better competitive level with the consequent increase of the possibilities for exporting.

As a result of the present study, the control of environmental conditions inside the greenhouse can help decision-making regarding agricultural activities as well as the application of fungicides to protect the health of plants. In this sense, this research proposes a data sheet that combines both, pests and diseases, to show when the farm should be alert about the changing environmental conditions that favor the appearance of pests and diseases, considering temperature, humidity, vapor pressure deficit and the dew point, to deploy actions to prevent the damage of the roses.

Table 11 shows a set of indicators that consider the conditions of temperature, humidity, dew point and VPD, which when combined, resulted in the 20% of highest presence and incidence of diseases and pests. These indicators represent the possibility of improving production control through the IoT applied to the production of roses, and mean a guide for daily work on a farm that decides to implement digital transformation tools such as IoT and data analytics.

Table 11. Data sheet of estimated conditions for the prevention of pests and diseases in the production of roses. Source: direct investigation.

Estimated conditions for the incidence of botrytis				
Range	*Humidity*	*Dew point*	*Temperature*	*VPD*
MIN:	49.20	6.95	16.57	0.54
MAX:	77.50	14.77	22.02	1.66
Estimated conditions for the incidence of powdery mildew				
Range	*Humidity*	*Dew point*	*Temperature*	*VPD*
MIN:	57.72	9.80	16.03	0.59
MAX:	77.50	12.89	23.60	1.95
Estimated conditions for the incidence of downy mildew				

(*continued*)

Table 11. (*continued*)

Estimated conditions for the incidence of botrytis				
Range	*Humidity*	*Dew point*	*Temperature*	*VPD*
Range	*Humidity*	*Dew point*	*Temperature*	*VPD*
MIN:	55.57	7.11	13.82	0.36
MAX:	79.40	12.95	20.16	1.39
Estimated conditions for the incidence of thrips				
Range	*Humidity*	*Dew point*	*Temperature*	*VPD*
MIN:	48.90	8.36	16.56	0.66
MAX:	71.69	12.83	21.30	1.86
Estimated conditions for the incidence of mites				
Range	*Humidity*	*Dew point*	*Temperature*	*VPD*
MIN:	49.20	6.95	14.73	0.45
MAX:	79.10	14.02	21.27	1.71

5 Conclusions

In all cases it was evidenced that the incidence of pests and diseases has different levels of correlation with the environmental conditions inside the greenhouses. This information is useful to establish a data sheet with a set of indicators that allow the producer to know when it must be carried out preventive and even corrective actions due to the effect of knowledge about environmental conditions in real time.

The present study demonstrates the importance of the IoT applied to control production in the floriculture production sector, however, this study does not cover the economic impact and administrative management due to the fact of handling data on environmental conditions that could affect the crop. Nevertheless, the data sheet of estimated conditions for the prevention of pests and diseases in the production of roses is an important advance for this agricultural sector, the same one that should be measured in a new study.

The possibility of deepening the current study and developing decision-making methods based on statistical information built from the IoT applied to the flower sector remains open.

References

1. Trade Map: Trade Map, 21 05 2023. [En línea]. www.trademap.org
2. Tamoghna, O., Sudip, M., Narendra, S.R.: Wireless sensor networks for agriculture: The state-of-the-art in practice and future challenges. Comp. Electron. Agric. **118**, 66–84 (2015)
3. Want, R., Schilit, B., Jenson, S.: Enabling the Internet of Things. Computer **48**, 28–35 (2015)
4. Alagarsamy, G., Gowthaman, N.: Intelligent IoT (IIoT)-based food supply monitoring system with alert messages in transmission and distribution environment. In: de Technology, Business, Innovation, and Entrepreneurship in Industry 4.0, Gewerbestrasse, EAI/Springer Innovations in Communication and Computing, pp. 353–368 (2023)

5. Grower Talks: Grower Talks, 01 11 2018. [En línea]. www.growertalks.com
6. Arcos, C., Padilla, A.: Model of technological competencies as determinants of innovation: a comparative intersectoral study in Ecuador. In: de Trends in Artificial Intelligence and Computer Engineering. ICAETT 2022. Lecture Notes in Networks and Systems, vol. 619. Springer, Cham, pp. 561–574 (2023)
7. Gostinchar, J.: Cultivo del rosal. Cofisa, Madrid (1954)
8. Arcos, C.: Obstacles to innovative activities: case of the floriculture sector in Ecuador. In: de Technology, Business, Innovation, and Entrepreneurship in Industry 4.0, Gewerbestrasse. EAI/Springer Innovations in Communication and Computing, pp. 67–94 (2023)
9. Calvache, Á.: Elaboración de un manual técnico-práctico del cultivo de rosas para exportación. Researchgate (2010)
10. Messelink, G., Kruidhof, M.: Advances in pest and disease management in greenhouse cultivation. In: de Achieving sustainable greenhouse cultivation, London, Burleigh Dodds Science Publishing, p. 46 (2019)
11. Yong, A.: El cultivo del rosal y su propagación. Cultivos Tropicales **25**(2), 53–67 (2004)
12. Daughtrey, M., Buitenhuis, R.: Integrated pest and disease management in greenhouse ornamentals. In: de Integrated Pest and Disease Management in Greenhouse Crops, Berna, Plant Pathology in the 21st Century book series (ICPP, vol. 9), pp. 625–679 (2020)
13. Heredia, M., Falconí, V., Silva, J.H., Amores, K., Endara, C., Ausay, K.F.: Technological innovation for the sustainability of knowledge and natural resources: case of the Choco Andino biosphere reserve,» de innovation and research: a driving force for socio-econo-technological development. In: Botto-Tobar, M., Zambrano Vizuete, M., Díaz Cadena, A. (eds.). Springer, Cham, pp. 464–476 (2021)

Transesterification of Waste Oils for Sustainable Transport in the Tourism Sector

Edwin Chiliquinga[1](✉), Dennis Ugeño[1], Edwin Machay[1], and Viviana Silva[2]

[1] Instituto Tecnológico Universitario Vida Nueva, Quito, Ecuador
omar.chiliquinga@istvidanueva.edu.ec

[2] Instituto Tecnológico Universitario Central Técnico, Quito, Ecuador

Abstract. This research aims to analyse the properties of biodiesel obtained from residual oils that have been collected in sites dedicated to community tourism, taking into consideration that at present, Ecuador does not have solid strategies to recycle and rationally reuse the volume of residual oils generated in the preparation of food in different sectors. In the tourism field, the generation of residual oils is not an isolated case, residual oils, produced in food preparation, are generally discarded; However, this work developed a strategy for obtaining biodiesel from waste oils, defining the reagents necessary for the development of the chemical reaction called transesterification, from which biodiesel was obtained that will be used as biofuel and on the other hand the glycerine which will be used as a degreaser, the necessary reagents were fixed from the bibliographic investigation, and then proceeded to characterise the oil used as raw material, for later carry out the experimental phase to analyse the physical and chemical properties at laboratory level, finally used in an internal combustion engine diesel obtaining an average opacity level of 16%, with these values proceeds to validate the application of the product obtained in agricultural machinery commonly used in tourism community.

Keywords: Transesterification · Biodiesel · Sustainability · Pollution · Opacity

1 Introduction

Currently tourism in Ecuador is among the four main sources of non-oil economic income in the country, in general the flow of national and foreign tourists to different tourist destinations report a considerable income of foreign currency in establishments dedicated to the service of food and beverages, many of these places offer great gastronomic variety that are mostly prepared with palm oil, soybean, soybean, colsa, palm kernel, olive, sunflower, among others; This undoubtedly entails an environmental problem that must be solved urgently, since these establishments do not have good environmental practices or resources for the management of residual oils produced in cooking, today it has caused great environmental and public health problems by polluting rivers and lands [1].

For years it has been known that oils should not be reused in the kitchen, at least not very often, since the high temperatures used for food preparation by frying can release

M. Z. Vizuete et al. (Eds.): CI3 2023, LNNS 1040, pp. 18–29, 2024.
https://doi.org/10.1007/978-3-031-63434-5_2

water vapour that when mixed that when reacting with triglycerides cause the formation of fatty acids, monoglycerides, diglycerides, glycerol and even some carcinogenic substances such as benzopyrene, an element also found in tobacco smoke. Most common diseases caused by eating trans fats are related to colon problems, blood vessel problems, and possibly stomach cancer [2].

Cooking by frying is one of the most important techniques in food preparation, this technique was used since 1600 BC; however, after preparing food with this technique, some people continue to use the oils indiscriminately either because of working conditions or because of the cost of the oil, in other cases, the residual oil is poured directly into the soil, rivers, oceans, etc., affecting the local ecosystem. This situation has led to the search for alternatives worldwide through the development of biofuels as a solution to this problem [1].

The main objective of the development of biofuels is to reduce the rate of environmental pollution left by the consumption of petroleum-based fuels. Currently biodiesel can replace diesel completely or can be used as an additive for internal combustion engines in the form of mixtures containing 20% and 50% biodiesel (B 20, B 50) thus offering improvement of the properties and characteristics of diesel [3].

The purpose of this research is the manufacture of an industrial equipment used to obtain biodiesel through the transformation of vegetable waste oils, through the transesterification process that combines the oil with methanol and sodium hydroxide (NaOH), which allows the production of a compound that can be used directly in the burner, electricity generator or in a diesel combustion engine, also as a residual product of this reaction glycerol is obtained, commonly called glycerine, it can be used in other industries such as pharmaceuticals for the manufacture of cleaning products, degreasers, etc. [4].

The evaluation of the biodiesel obtained focuses on the development of physical, chemical tests and opacity measurement, thus determining the feasibility of using this biofuel as an advantage from the energy point of view and recommended from the environmental point of view, since it has a lower emission of harmful gases, especially carbon dioxide (CO2), which is the main cause of the greenhouse effect; In this way, the production and use of alternative energies, friendly to the environment and in turn progressively reduce the use of fossil fuels, renewed interest in the efficient management of energy and economic resources that allow a globalised growth in the production and economy of the country in different social sectors [5].

2 Biodiesel

The ASTM (American Society for Testing and Materials) defines biodiesel as "the long-chain monoalkyl ester of fatty acids derived from renewable resources, such as vegetable oils or animal fats, for use in diesel engines". It is presented in a liquid state and is obtained from renewable resources such as vegetable oils of soybean, rapeseed/canola, sunflower, palm and others, as well as animal fats, through a process called Transesterification [2].

Transesterification consists of mixing vegetable oil or fats with an alcohol (usually methanol) and an alkali (caustic soda). After a resting time, by decantation, the biodiesel is separated from its by-product glycerol. This process requires a reactor that is the heart of every biodiesel plant. Despite different experiences made since the 1920s in the production of biodiesel, it achieved real notoriety in the early years of the twenty-first century caused by the increase in international oil prices [13].

According to the research of Tejada Tovar [6], the different possibilities of obtaining biodiesel from fatty waste generated in the livestock farm: chicken, pork, and beef fat, for this the variables molar alcohol/animal fat ratio and amount of catalyst are analysed, “since they are the ones that most affect the quality of biodiesel, when evaluating their characteristics and yields of the transesterification reaction.

In addition, Tejada Tovar [6] mentions that the yield in the process of extracting fat from chicken waste material is 70.5%, and pork fat 90%, so they are proposed as viable alternatives from the technical point of view, allowing the valorisation of these organic wastes and alleviating the pollution generated in this type of industries. From the characterization of chicken and pork fat, it was established that both raw materials are low acidity, which guarantees a high degree of transesterification, taking into account that it had a yield of 96% from chicken fat and 91.2% from pork fat.

2.1 Raw Materials for the Production of Biodiesel

The most used raw material for the manufacture of biodiesel should be one that contains a high triglyceride index such as: sunflower oil, rapeseed, soybean, used frying oil, tallow, etc. Table 1 shows the main feedstocks used to make biodiesel.

Table 1. Raw material for the production of biodiesel

Conventional oils	Alternative vegetable oils	Other Sources
Sunflower	Brassica carinata	Genetically modified seed oil
Colsa	Cynara curdunculus	Animal fats (beef and buffalo tallow)
Coconut	Camelina sativa	Microalgae oils
Soybean	Cambre abyssinica	Microbial production oils
Palm	Pogianus	Frying oils

Each type of feedstock used for biodiesel production is described below.

Conventional vegetable oils: The most common raw materials used are oilseed oils such as sunflower oil, canola oil (Europe), soybean oil (USA) and coconut oil (Philippines); and oily fruits such as palm oil used in Malaysia and oil in Indonesia [7].

Alternative vegetable oils: These are better adapted to the conditions of the countries where they are developed and are more beneficial for energy crops. It highlights, for example, camelina, camellia and jatropha oil, in Spain, the crops most suitable to the conditions of the country are Brassica carinata and Cynara cur-dunculus [8].

GMO vegetable oil: The main difference between oil and fat is the fatty acid content. Oils with a high proportion of unsaturated fatty acids, such as sunflower oil or camelina oil, can improve the performance of biodiesel at low temperatures, but reduce its oxidative stability, resulting in a high iodine index. Therefore, highly unsaturated oils that have been genetically modified to reduce the establishment ratio, such as sunflower oil with high oil content, can be considered as feedstock for biodiesel production [9].

Used frying oils: It is an alternative for the production of biodiesel with better prospects, since it is the most economical raw material for the production of this fuel. The use of used oil promotes good treatment and waste disposal [3].

Animal fats: Animal fats, especially cow tallow, can also be used for the production of biodiesel [10].

2.2 Use of Biodiesel in Internal Combustion Engines

The use of biodiesel has several environmental benefits, compared to diesel, it reduces the pollution load caused by the engines. Reported biodiesel test results where a 20% reduction in sulphur compound emissions, a 10% reduction in carbon monoxide (CO), a 14% reduction in hydrocarbons (HC), and a 26% reduction in particulate matter for diesel fuel were observed. As for nitrogen oxide (NOx) emissions, there is no single standard, as some studies reported an increase in NOx emissions, while others reported a decrease (Piloto Rodríguez, 2010). Adding biodiesel to diesel has several advantages [16]. Biodiesel has a higher cetane number. Biodiesel improves lubrication, which prolongs the life of engine components. Biodiesel also has a higher flash point than diesel. Although the flash point does not directly affect combustion, it makes biodiesel safer for storage and transportation [9].

2.3 Transesterification

Vegetable oils usually contain free fatty acids, phospholipids, sterols, water and other impurities. Therefore, the oil cannot be used directly as fuel [20]. To overcome these problems, vegetable oils undergo a slight chemical modification called transesterification. As a result of this modification, a cleaner and more environmentally safe fuel (biodiesel) was obtained, whose main components are mono alkyl esters of long-chain fatty acids [11].

Transesterification, or alcoholysis, named for the use of short-chain alcohols such as methanol or ethanol, is the replacement of an alcohol group in an ester provided by triglycerides of animal or vegetable fats by another alcohol in a process similar to hydrolysis. Use alcohol instead of water. This process is widely used to reduce the high viscosity of triglycerides. Figure 1 shows the general reaction for transesterification [14].

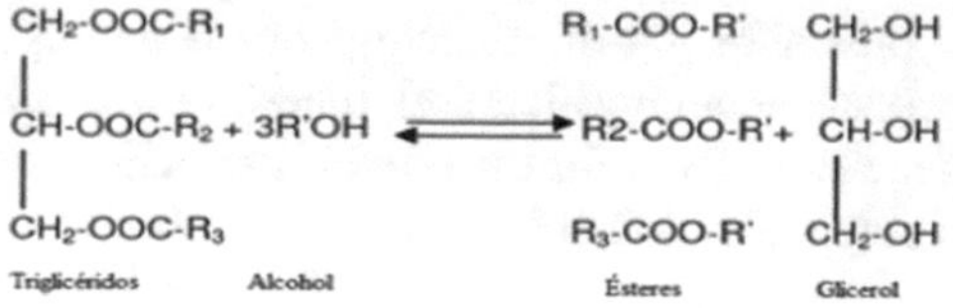

CH_2-OOC-R_1 | R_1-COO-R' | CH_2-OH
CH-OOC-R_2 + 3R'OH ⇌ R2-COO-R'+ CH-OH
CH_2-OOC-R_3 | R_3-COO-R' | CH_2-OH
Triglicéridos Alcohol Ésteres Glicerol

Fig. 1. General equation of the transesterification reaction [11]

For transesterification methanol or ethanol can be used. If methanol is used the process is called methanolysis, and if ethanol is used it is called ethanolysis, as shown in Fig. 2.

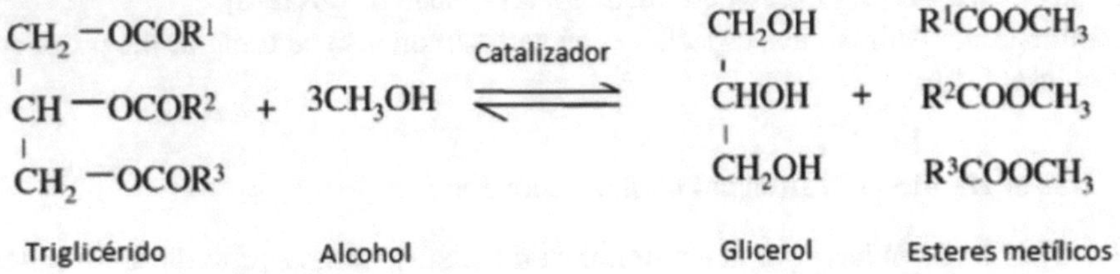

$$\begin{array}{l} CH_2-OCOR^1 \\ | \\ CH-OCOR^2 \\ | \\ CH_2-OCOR^3 \end{array} + 3CH_3OH \xrightleftharpoons{\text{Catalizador}} \begin{array}{l} CH_2OH \\ | \\ CHOH \\ | \\ CH_2OH \end{array} + \begin{array}{l} R^1COOCH_3 \\ R^2COOCH_3 \\ R^3COOCH_3 \end{array}$$

Triglicérido Alcohol Glicerol Esteres metílicos

Fig. 2. General equation for triglyceride ethanolysis [11]

Both processes can be catalysed by acids or bases. The most commonly used bases are sodium hydroxide, potassium hydroxide and carbonates. Commonly used acid catalysts are: sulfuric acid, sulfonic acid, hydrochloric acid [15]. After transesterification of triglycerides, the product is a mixture of esters, glycerol, alcohol, catalyst and triglycerides, diglycerides and monoglycerides, which are then separated. A layer of glycerine was deposited at the bottom of the reaction vessel. The formation of diglycerides and monoglycerides is an intermediate step in the process. Figure 3 shows the mechanism of the transesterification reaction [11].

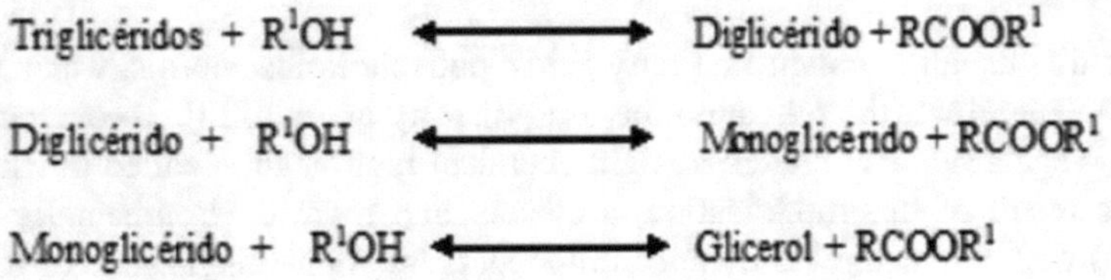

$$\text{Triglicéridos} + R^1OH \longleftrightarrow \text{Diglicérido} + RCOOR^1$$

$$\text{Diglicérido} + R^1OH \longleftrightarrow \text{Monoglicérido} + RCOOR^1$$

$$\text{Monoglicérido} + R^1OH \longleftrightarrow \text{Glicerol} + RCOOR^1$$

Fig. 3. Reaction Mechanisms [11]

3 Methodology

To obtain biodiesel it is important to analyse the density and viscosity of animal, automotive and hydraulic oils, allowing the selection of the appropriate oil to apply the transesterification process in obtaining biodiesel, therefore, density measurements of the different types of oils are carried out for the previous selection of the most appropriate raw material and thus obtain the biofuel (Fig. 4).

Fig. 4. Density measurement of vegetable and animal oils

With the data obtained, a table is generated showing the relative density of each of the oils (Table 2).

Table 2. Relative density of oils

Type of oil	Relative density
Vegetable Oils	0,90
Animal Oils	0,93
Automotive Oil	0,95
Hydraulic oil	0,98

At this point the Ford viscosity cup is used, which helps to quickly and accurately determine viscosity in liquids." However, you should keep in mind that it is only possible with dense liquids, since this method is applied to determine other mechanical variables, such as the average speed which will allow the selection of the pipe necessary for the construction of the biodiesel distillation system [17] (Fig. 5).

Fig. 5. Viscosity Measurement by FORD Cup Method # 4

Table 3 shows the values of the times obtained from each of the raw materials with the Ford cup, using these data, we proceed to calculate the viscosities as a function of the time of each of the oils.

Table 3. Time obtained from Ford Cup #4

Type of oil	Average time(s)
Vegetable oils	10,40
Animal oils	11,00
Automotive oil	11,80

Next, the mathematical model for calculating the viscosity of oils is presented.

$$Vp = |2{,}74 \times t - 100 \times t| \tag{1}$$

where: Vp = Viscosity (mPa.s) and t = Time (s). Then we proceed to replace the average time in the equation previously raised to obtain the viscosity of the different types of oils considered as raw material.

Calculation of the viscosity of vegetable base oil. Time of 10,40 s.

$$Vp = |2{,}74 \times t - 100 \times t|$$

$$Vp = |2,74(10{,}40) - 100(10{,}40)|$$

$$Vp = 1011{,}50 mPa.s$$

Calculation of the viscosity of animal base oil for time of 11,00 s.

$$Vp = |2{,}74 \times t - 100 \times t|$$

$$Vp = |2{,}74(11{,}00) - 100(11{,}00)|$$

$$Vp = 1069{,}86 mPa.s$$

Calculation of automotive oil viscosity for time of 11,00 s.

$$Vp = |2{,}74 \times t - 100 \times t|$$

$$Vp = |2{,}74(11{,}80) - 100(11{,}80)|$$

$$Vp = 1147{,}66 mPa.s$$

Calculation of the viscosity of hydraulic systems for time of 11,00 s.

$$Vp = |2{,}74 \times t - 100 \times t|$$

$$Vp = |2{,}74(12{,}20) - 100(12{,}20)|$$

$$Vp = 1186{,}57 mPa.s$$

Through the measured and calculated values, the following table is made which indicates the physical and chemical properties of each of the raw material analysed to select the most optimal substance that serves the transesterification process.

Table 4. Properties of oils.

Type of oil	Density kg/m^3	Viscosity mPa s	Temperature °C
Vegetable oils	900	1011,50	100 a 180
Animal oils	930	1069.86	180 a 200
Automotive oil	950	1147.66	200 a 250
Hydraulic oil	980	1186,57	250 a 300

The results obtained presented in Table 4, show the density of each of the oils, taking into account that higher density oils need a higher temperature (200 °C to 300 °C) for decomposition and lower density oils decompose at a lower temperature (100 °C to 180 °C), reaching the conclusion that vegetable and animal oils are suitable for the process of obtaining biodiesel using the transesterification process [18].

3.1 Property Analysis

After obtaining the biodiesel, the properties of the resulting product are extracted through experiments and calculations to compare with some national and international specifications that validate whether the biodiesel obtained is suitable for consumption (Table 5).

Table 5. Density of diesel and biodiesel at 15 °C.

Fuel	ISO 12185 kg/m^3	NTE INEN 1498 kg/m^3	ASTM D6751 kg/m^3	Measured Value kg/m^3
Commercial diesel	820–845	850	–	–
Commercial biodiesel	–	–	860 – 900	–
Biodiesel obtained	–	–	–	880

Another property to be evaluated was the kinematic viscosity, for which it was necessary to heat the biodiesel obtained since the NTE INEN 1498 standard specifies the measurement of this parameter at 37.8 °C [19] (Table 6).

Table 6. Kinematic viscosity of diesel and biodiesel cSt

Fuel	NTE INEN 1498 cSt	ASTM D975 cSt	Measured Value cSt
Diesel 1	1,3–3	–	–
Diesel 2	2,5–6	–	–
Premium diesel	2,5–6	–	–
Commercial biodiesel	–	1,9 – 6	–
Obtained biodiesel	–	–	4

The average kinematic viscosity value obtained from tests allows the product obtained to be validated, since it is of the national and international parameters of commercial diesel from Ecuador and international biodiesel.

The pH mediation was developed with the aim of identifying if the substance obtained is acidic or not, this would allow to validate the application of the biofuel in an internal combustion engine, if the result is acidic it could cause severe damage to the metal parts of the engine (Table 7).

Table 7. pH index of commercial biodiesel and biodiesel obtained.

Fuel	EN 14214 pH	pH Measured value
commercial biodiesel	6–8,4	6
obtained biodiesel	–	6–7

The measured pH value of biodiesel allows to define as a fuel suitable for internal combustion engines.

The determination of density, kinematic viscosity and pH index support the application of biodiesel for the operation of an internal combustion engine.

Below, the opacity results obtained when running a diesel internal combustion engine with the biofuel produced are detailed, the results will be compared with the NTE INEN 017 standards (ENVIRONMENTAL MANAGEMENT. AIR. MOTOR VEHICLES. DETERMINATION OF THE OPACITY OF EXHAUST EMISSIONS OF DIESEL ENGINES BY STATIC TEST FREE ACCELERATION METHOD) (Table 8).

Table 8. Results obtained from opacity (%) [12].

Temp: (85–90) °C	Model and year: Motor Great Wall 2016			Fuel type: Biodiesel	
	Acceleration 1 (%)	Acceleration 2 (%)	Acceleration 3 (%)	Acceleration 4 (%)	Acceleration 5 (%)
Test 1	15	17	14	15	15
Test 2	17	17	15	18	18
Test 3	17	18	18	15	15
Opacity result					
Test 1: 15%					
Test 2: 17%					
Test 3: 16%					

Note. To obtain more realistic measurements, three tests were performed

The average result of the opacity of the engine using biodiesel as an alternative fuel is 16%, which guarantees that the engine produces a low level of particulate matter (PM) according to the regulations is within the ranges.

4 Conclusions

With the studies carried out in the research it was possible to determine the most appropriate process such as transesterification to obtain biodiesel from waste oils produced by animal fats that are often used in the cooking of food in the tourism sectors.

Through mathematical expressions I can identify the parameters and modifiable variables within the transesterification process to select the raw material that allows to obtain diesel 2 also known as biodiesel.

The values obtained in the calculations will help the design of the prototype for the biodiesel distillation machine. By means of a thermodynamic process that will decompose the oil molecules for the transesterification process and the obtaining of the product.

The research carried out allowed the generation of biodiesel from residual oils with physical and chemical properties that meet national and international specifications which guarantee the optimal operation of the engine, avoiding damage due to overheating, oxidation, corrosion or lubrication.

The community must have knowledge of waste management so, in the first instance, it is necessary to guide citizens on the initial classification of the same and then, on activities such as reuse, recovery, recycling; breaking the traditional scheme of "collection - final disposal" reaching the conclusion that it is necessary to generate a program of collection of residual oils in the community to promote sustainable tourism.

References

1. de León Benítez, J.B., Abreu Rodríguez, L., Matiauda, M., Miño Valdés, J.E.: Diseño de una planta de obtención de Biodiesel a partir de un residual de la industria azucarera (cachaza, Centro Azúcar, vol. 43, no. 1, pp. 1–9 (2019)
2. García-Díaz, M., Gandón-Hernández, J., Maqueira-Tamayo, Y.: Estudio de la obtención de biodiesel a partir de aceite comestible usado, Tecnología Química, vol. 33, no. 2, pp. 162–169 (2019)
3. Marquínez, A.N.M., et al.: Obtención de biodiesel a partir de aceite de coco (Cocos nucifera L.), Revista de Iniciación Científica, vol. 6, no. 1, pp. 9–14 (2020)
4. Pardal, A.C.D.V.: Obtención de biodiesel por transesterificación de aceites vegetales: nuevos métodos de sínteses, 2012, Accedido: 21 de junio de 2022. [En línea]. Disponible en: https://repositorio.ipbeja.pt/handle/20.500.12207/738
5. Diseño de un reactor de transesterificación para la obtención de biodiesel a partir de aceites vegetales. https://riunet.upv.es/handle/10251/73470 (accedido 21 de junio de 2022)
6. Tejada Tovar, C., Tejeda Benítez, L., Villabona Ortiz, Á., Monroy Rodríguez, L.: Obtención de biodiesel a partir de diferentes tipos de grasa residual de origen animal, Luna Azul, no. 36, pp. 10–25 (2019)
7. Tejeda Benitez, L.: Obtención de biodiesel a partir de diferentes tipos de grasa residual de origen animal, Extracting biodiesel from different types of animal origin residual fat, 21 de septiembre de 2022. http://www.scielo.org.co/scielo.php?script=sci_arttext&pid=S1909-24742013000100002 (accedido 21 de junio de 2022)
8. Oliva-Montes, J., Flores-Rodríguez, J., López-Medina, R., Santos-Camacho, J., Contreras, J., Vaca-Mier, M.: Producción de biodiesel a partir de grasa animal utilizando catálisis heterogénea, vol. 2, no. 5, p. 10
9. Garrido, S.M.: Tecnología, territorio y sociedad. Producción de biodiesel a partir de aceites usados, Íconos, vol. 0, no. 37, p. 75 (2013). https://doi.org/10.17141/iconos.37.2010.422
10. Valderrama Negrón, A., Jacinto Hernández, C., Ponce García, S., Manrique Pollera, L.: Uso del diseño factorial en los ensayos de liberación controlada del ácido 1,3 indolacético cargado en matrices de quitosano, Revista de la Sociedad Química del Perú, vol. 83, no. 3, pp. 354–365 (2017)
11. Esterificación y transesterificación de aceites residuales para obtener biodiesel. http://www.scielo.org.co/scielo.php?script=sci_arttext&pid=S1909-24742015000100003 (accedido 21 de junio de 2022)
12. INEN: RTE INEN 017 Control de emisiones contaminantes de fuentes móviles terrestres. Quito (2016)
13. Rožić, M., Babac, M.T., Radić Stojanović, D., Božić, M.: Trans-esterification reactions in the synthesis of bioactive compounds. Molecules **26**(12), 3503 (2021)

14. Zhu, S., et al.: Bio-diesel production and environmental impacts. Renew. Sustain. Energy Rev. **123**, 109758 (2020)
15. Pousa, J.R., Ballesta Claver, J., Piqueras Céspedes, J., Rodríguez Avi, J., Viola, E.: Biodiesel production: raw material and extraction methods of lipids from vegetable oils and animal fats. Grasas Aceites **70**(3), e318 (2019)
16. Phan, A.N., Phan, T.M.: Biodiesel production from non-edible plant oils. In Biodiesel-A feasible alternative in present scenario, pp. 75–98. IntechOpen (2018)
17. Kryvda, V., Pisárčik, M., Matúš, M., Lazor, M.: Energy properties of biodiesel fuels. In: MATEC Web of Conferences, vol. 157, p. 02020. EDP Sciences (2018)
18. Silitonga, A.S., Atabani, A.E., Mahlia, T.M.I.: A review on pro-spect of Jatropha curcas for biodiesel in Indonesia. Energy Convers. Manage. **181**, 197–215 (2019)
19. Li, G., Xue, J., Han, B.: Emission characteristics of biodiesel–diesel blends: a review. Renew. Sustain. Energy Rev. **118**, 109512 (2020)
20. Freedman, B., Butterfield, R.O., Pryde, E.H.: Transesterification kinetics of soybean oil. J. Am. Oil Chem. Soc. **63**(10), 1375–1380 (1986)

LoRaWAN Applied to WSN as Support for Sustainable Agriculture in Rural Environments. Case Study: Pedro Moncayo Canton, Pichincha Province

Evelyn Bonilla-Fonte, Edgar Maya-Olalla, Alejandra Pinto-Erazo, Ana Umaquinga-Criollo, Daniel Jaramillo-Vinueza, and Luis Suárez-Zambrano(✉)

Universidad Técnica del Norte, Avenida 17 de Julio 5-21 y General María Córdoba, 100105 Ibarra, Ecuador
{eamaya,lesuarez}@utn.edu.ec

Abstract. This work describes the development of a Long-Range Low-Power Wireless Sensor Network (LPWAN) implemented in a rural environment to monitor environmental variables that influence the growth of grass crops. The sensor node collects and transmits data on soil moisture, UV radiation, temperature, and rainfall in a cultivation area covering 1000 m^2. The data collected by the sensor node is sent to the central node or gateway via LoRa communication, where it is processed and transmitted to a cloud platform called ThingSpeak using the MQTT protocol. The data is displayed in the form of graphs. The analysis, design, implementation, and testing of the wireless link are carried out, and the data is sent to an IoT platform where it is collected, visualized, and analyzed in real-time using MQTT as the central communication protocol. Functional tests are performed, including path loss, receiver power, signal-to-noise ratio, receiver sensitivity, link budget, Fresnel zone, azimuth angle, and elevation angle. Finally, a stable scenario for data transmission and reception is determined, which can be applied to an optimization system for crop cultivation in rural agricultural areas.

Keywords: LoRaWAN · MQTT · propagation loss · scattering factor

1 Introduction

Tocachi, a rural parish in the province of Pichincha, located in the Pedro Moncayo canton, is internationally recognized for its cultivation and production of roses. However, livestock farming is also important in this region. In the area, there is a dairy association made up of small-scale farmers who raise cattle for this purpose. These animals require quantities of green forage for their feed, making pasture cultivation essential and cost-effective. It is common to find large tracts of land with cattle. The farmers are a fundamental part of the economic growth in the area, both in milk and meat production [1].

M. Z. Vizuete et al. (Eds.): CI3 2023, LNNS 1040, pp. 30–45, 2024.
https://doi.org/10.1007/978-3-031-63434-5_3

The duration and performance of grasses depend on factors such as crop planning, fertilization, proper cutting, and especially irrigation [2]. Plants in their early stages are highly sensitive to lack or excess moisture, so it is crucial to maintain constant moisture for the crops to achieve optimal growth and production.

The local farmers have traditionally implemented sprinkler irrigation systems, but the water distribution is not adequate, resulting in wastage as they cannot determine which part of the land needs irrigation and how much water should be provided. Additionally, climatic factors such as temperature, rainfall, and solar radiation significantly affect the production and quality of the Grass [3]. It is important for the person in charge of the cultivation to consider these parameters for proper pasture management.

In the study area, the average annual precipitation varies from 400 to 1300 mm, with a dry period between July and October, resulting in a water deficit of 0 to 330 mm [4]. The daily growth rate of grass in summer ranges from 5–20 kg/MS/ha/day, while in winter it increases to levels of 60–80 kg/MS/ha/day [5]. Therefore, it is important to apply irrigation to increase grass production. Improving the efficiency of the sprinkler irrigation system will allow farmers to reduce unnecessary water consumption.

In [6], the design of a Wireless Sensor Network (WSN) for the control and monitoring of a drip irrigation system in a strawberry plantation is presented. The primary objective is to collect data from multiple sensor nodes distributed throughout the crop, which are then transmitted to a central node and stored in a local database. Zigbee technology is employed for communication between the sensor nodes and the central node. Additionally, the system uses reference values for soil moisture and temperature to activate or deactivate the solenoid valve and a fan, respectively.

In [7], the design of an intelligent irrigation system based on fuzzy logic for vegetable crops at the La Pradera farm of the Technical University of the North is presented. The main goal of this research is to develop a drip irrigation system based on fuzzy logic, which includes a WSN defined in the IEEE 802.14 protocol. Sensor data is wirelessly transmitted to a control station using Zigbee technology. In this station, the data is analyzed, and the opening time of the solenoid valve for crop irrigation is determined.

In [8], a system for irrigation control using machine learning techniques is designed for alfalfa crops at the La Pradera farm of the Technical University of the North. The collected data comes from a sensor network developed in [9] and is stored in a local database. Wireless communication is carried out using LoRaWAN modules.

Considering the aforementioned works, the present work proposes the design of a long-range and low-power wireless sensor network (LPWAN) for a grass crop in a rural environment. The system collects data from variables through a WSN located in the cultivation area, which covers an area of 1000 m^2. The analysis, design, implementation, and testing of the prototype are carried out. Subsequently, the data is sent to an IoT platform where it is collected, visualized, and analyzed in real time using the MQTT protocol as the central point of communication [10]. Path loss, receiver power, signal-to-noise ratio, receiver sensitivity, link budget, Fresnel zone, azimuth angle, and elevation angle are analyzed.

2 Methods

A methodology was followed to design and develop the proposed system, considering the requirements, analysis, design, and testing. The WSN LoRa wireless sensor network was designed with its corresponding design phases:

2.1 Hardware and Software Selection

The selection of components is based on the developer's requirements and the system architecture, considering coverage, compatibility, compact size, cost, and availability in the national market. For the node responsible for sensing environmental variables, the Arduino UNO board based on the ATmega328 microchip is chosen, as it has the necessary number of analog/digital pins to establish wireless communication with the central node. For wireless communication between the sensor node and the central node, as well as between the central node and the actuator node, the Dragino LoRa Shield based on the Sx1276 chip, and the Raspberry Pi 3 board are selected. These components allow for coverage of distances greater than 2 km. The chosen platform for registering the data sent by the sensors is ThingSpeak, as it provides suitable visualization of the collected data. The selected sensors are capacitive soil moisture sensor, LM35 temperature sensor, GUVA-S12SD UV radiation sensor and MH-RD YL-83 water or rain sensor. Figure 1 shows the general architecture of the WSN nodes with LoRa technology.

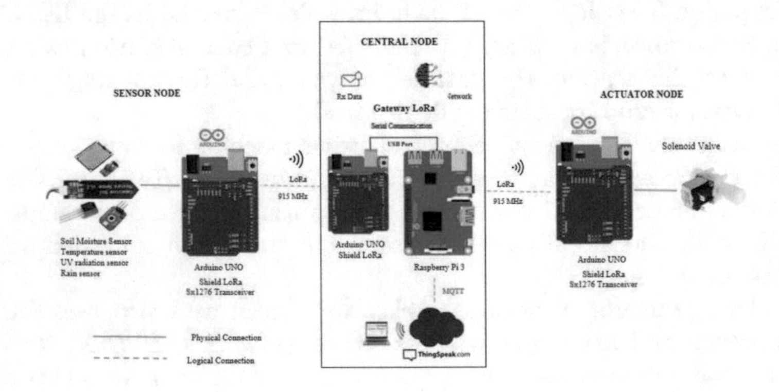

Fig. 1. General System Architecture.

2.2 Analysis of the Current Situation

There is access to a grass crop owned by one of the dairy association members, which covers a total area of 3400 m^2. For the development of this project, a sample of 1000 m^2 will be used, as shown in Fig. 2, the distribution of spaces between the plot, water source, and control center is shown.

Fig. 2. The blue area is the water source, the orange area is the control center.

2.3 Design of the System

In this phase, a star-type communication is established using LoRa technology be- tween the sensor and actuator nodes towards the gateway located in the central node. The central node receives sensor data through the MQTT protocol and visualizes it on an IoT platform.

2.4 Power Supply

Lithium batteries of 3.7 V and 6000 mAH are used to power the sensor node and actuator node. Additionally, a solar panel is available to recharge the batteries. On the other hand, the gateway is powered through a connection to the electrical grid using a 110 VAC to VDC power adapter.

2.5 Wireless Communication Design

The wireless communication is designed using a simulated radio link in Radio Mobile, with a frequency of 915 MHz, within the LoRa technology operating range of 902 to 928 MHz. Figure 4 shows the network diagram with the parameters of the nodes and the distances between them and the gateway (Fig. 3).

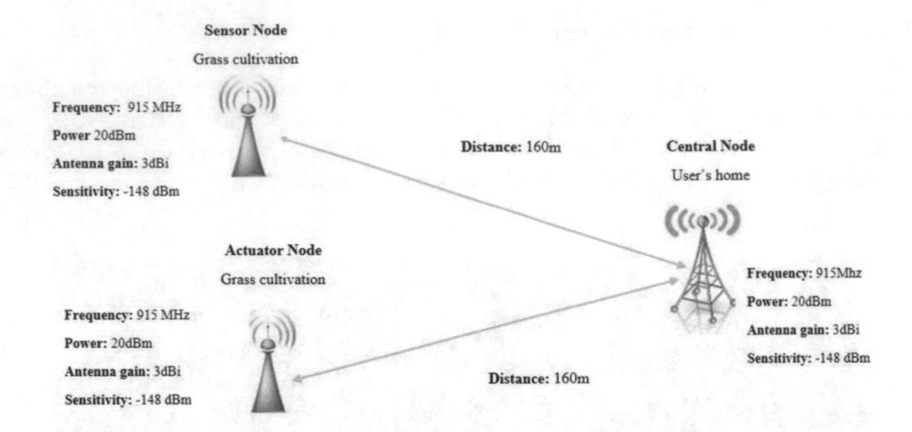

Fig. 3. Network diagram of the system.

Table 1. Coordinates of each node that forms the network.

Nodes	(Latitude)	(Longitude)	(Latitude)	(Longitude)
Gateway	0° 2′35.21″ N	78°17′35.03″ O	0.0431	−78.2930
Sensor Node	0° 2′31.65″ N	78°17′38.81″ O	0.0421	−78.2941
cre Actuator Node	0° 2′31.58″ N	78°17′38.77″ O	0.0421	−78.2941

Google Earth allows setting the latitude and longitude of the nodes and gateway, as specified in Table 1.

The communication yielded the following results: path loss between the gateway and the sensor node in free space of 75.75 dB, receiver power of 25 dBm, signal-tonoise ratio for LoRa communication between −20 dB and +10 dB, receiver sensitivity of −125 dBm, link margin of 150 dBm, link budget of 145 dBm, Fresnel zone of 3.62 m, azimuth angle of 47.72°, and elevation angle of 1.611°, see Fig. 4. Also, the heights of the antennas are varied to achieve a completely clear Fresnel zone, and the different simulated scenarios are detailed in Table 2.

Table 2. Propagation Characteristics

	Antenna height Sensor-Central	Worst Fresnel	Free Space Loss (dB)	Obstruction (dB)	Propagation loss (dB)	Worst (dB) reception (dB)
1	0.5–6	0.6 F1	75.8	0.5	81.3	91.7
2	1–6	0.8 F1	75.8	−0.5	80.3	92.7
3	1.5–6	1.0 F1	75.7	−0.8	80.0	93.0
4	3–6	1.2 F1	75.7	0.0	80.8	92.2
5	4.5–6	1.6 F1	75.7	0.2	81.0	92.0

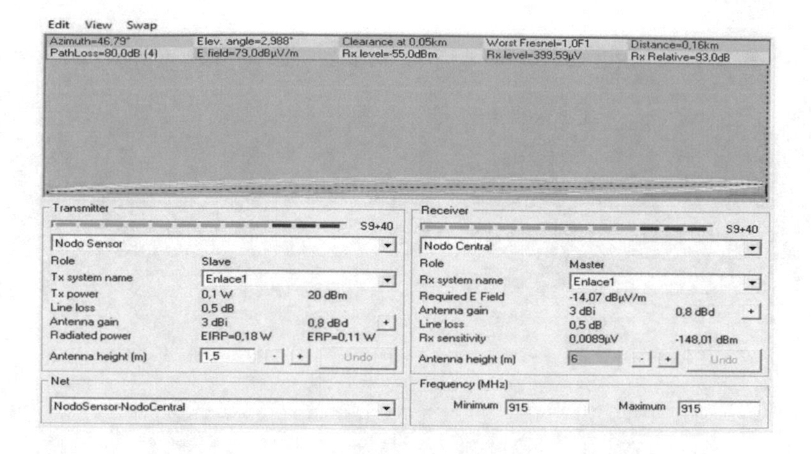

Fig. 4. Radio Mobile simulation - LoRa link between the sensor and gateway nodes.

In the link profile, free space losses remain unchanged with the antenna heights, while total propagation losses decrease as the antenna heights vary. A minimum Fresnel clearance of 0.6F1 is achieved, ensuring a smooth signal, as it should always be greater than or equal to 60%. The worst reception varies from 91.7 to 93 dB, indicating a safety margin with respect to the receiver sensitivity. The higher this value, the greater the assurance that the signal reaches the receiver. Figure 5 shows the results of the link between the sensor node and the central node. Also, the line of sight between the central node and the actuator node with the different simulated scenarios is observed in Table 3.

Table 3. Propagation Characteristics

	Antenna height Central-Actuator	Worst Fresnel	Free Space Loss (dB)	Obstruction (dB)	Propagation loss (dB)	Worst (dB) reception (dB)
1	6-0.5	0.6 F1	75.8	0.3	81.2	91.9
2	6-1	0.9 F1	75.8	−0.6	80.2	92.8
3	6-1.5	1.1 F1	75.8	−0.8	80.1	92.9
4	6-3	1.3 F1	75.8	0.7	81.6	91.4
5	6-4.5	1.6 F1	75.8	−0.2	80.7	92.4

The worst Fresnel value is 0.6 F1, the free space losses are approximately 76 dB, and the obstructions range from −0.8 dB to 0.7 dB. The total propagation losses reach 81.6 dB, and the worst reception approaches 93 dB. Based on the data obtained from the different scenarios, it can be concluded that all radio links are acceptable, as the first Fresnel zone exceeds 60%. However, for this project, Scenario 3 is implemented, with an antenna height of 1.5 m for the sensor and actuator nodes, and a height of 6 m for the central node.

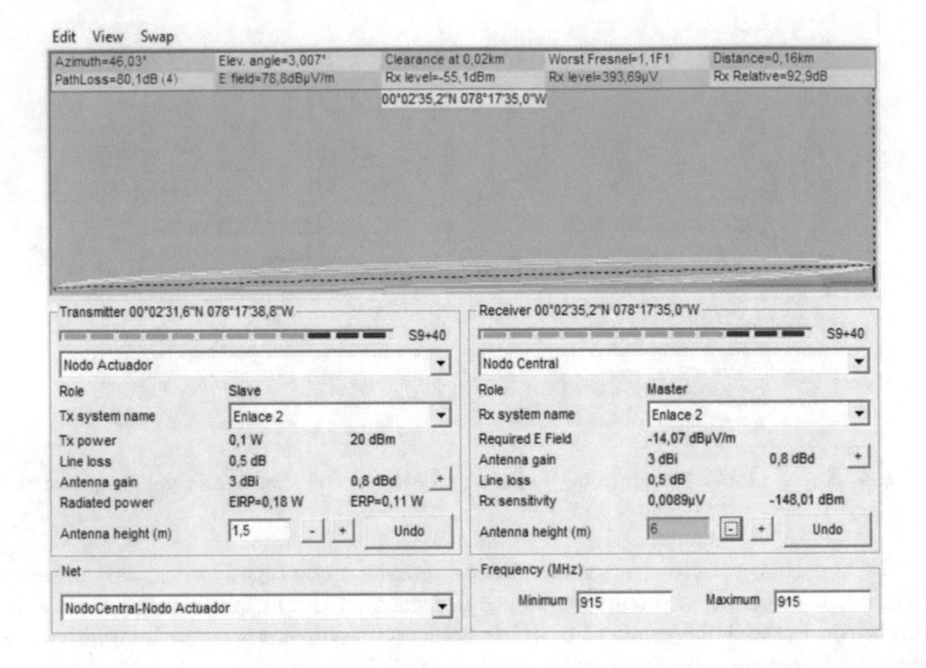

Fig. 5. Link between central node and actuator node.

2.6 Software Encoding of the System

The sensor and central nodes are encoded with the same configuration parameters to enable communication. The sensor node is responsible for reading the environmental variables, while the central node configures the gateway to send the data to the cloud using the MQTT protocol.

In Fig. 6 shows the flowchart of the sensor and gateway nodes. The sensor node waits to receive a request from the gateway and then sends the data obtained from the sensors. In the central node, the data sent by the sensor node is read. Finally, the sensor data is retrieved and sent to the ThingSpeak cloud server using the MQTT protocol.

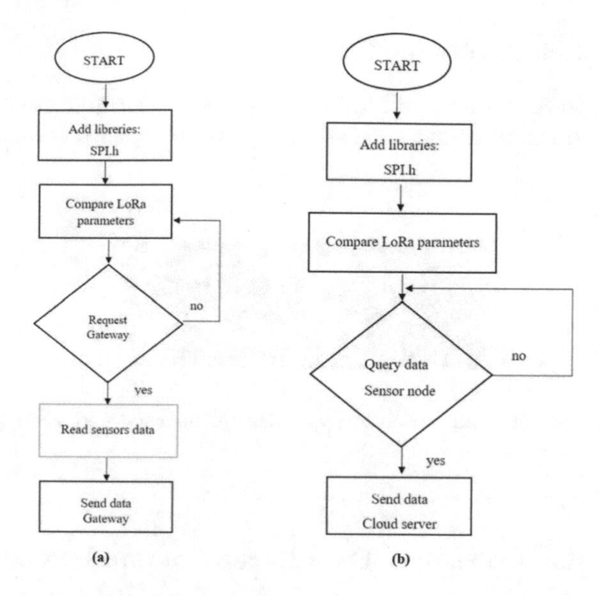

Fig. 6. Flowchart of the sensor node and central node.

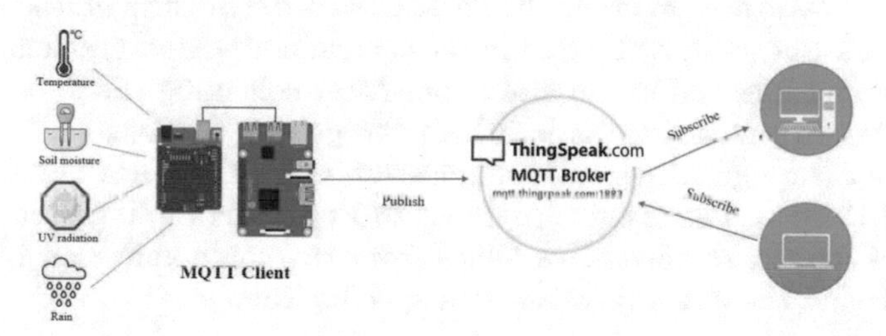

Fig. 7. Publish/Subscribe MQTT Architecture.

The system utilizes a basic star topology with the MQTT protocol, where all clients connect directly to the central point, known as the broker, which acts as a server. The system architecture is shown in Fig. 7. The Raspberry Pi3 board configures MQTT using the paho.mqtt.publish library. The corresponding Channel ID and Write API Key for the created channel are entered. Additionally, the MQTT host address is stored in a variable, and the topic is created for publishing the data. Finally, the temperature, humidity, UV radiation, and rain values are assigned to the fields created in the platform, allowing the updated values of each field to be observed in the channel.

3 Results and Discussion

The following aspects are verified: the electrical test of the sensor node, data collection, data reception, data presentation and wireless communication tests.

3.1 Sensor Node Electrical Test

The successful powering on of all system devices and proper voltage supply are verified. It is also ensured that the solar panel is charging the battery, as shown in Fig. 8.

Fig. 8. Sensor Node powered by a lithium battery and solar panel.

3.2 Information Collection, Data Reception and Presentation

The sensor node measures environmental variables such as soil moisture, UV radiation, temperature, and rainfall. These data are sent through the LoRa wireless network to the gateway located in the central node. Measurements are taken both in the morning and in the afternoon. After collecting the data, the sensor node sends it to the central node, where the gateway is located. This process is carried out through wireless communication using the LoRa module at a frequency of 915 MHz. The results are shown in Fig. 9. Figure 10 graphically shows the data of the environmental variables from the communication between the sensor node and the gateway, taken at a specific time.

```
Serial Monitor ×
Message (Ctrl+Enter to send mess
1 UV=0 H=284 T=17 ll=1007
1 UV=0 H=284 T=16 ll=1007
1 UV=0 H=285 T=16 ll=1007
1 UV=0 H=284 T=16 ll=1007
1 UV=0 H=284 T=16 ll=1006
1 UV=0 H=284 T=16 ll=1007
1 UV=0 H=284 T=16 ll=1006
```

```
Serial Monitor ×
Message (Ctrl+Enter to send mes
UV=1 H=312 T=13 ll=1005
UV=1 H=312 T=13 ll=1005
UV=1 H=312 T=13 ll=1005
UV=1 H=312 T=13 ll=1005
UV=1 H=312 T=13 ll=1004
UV=2 H=312 T=30 ll=1005
UV=2 H=312 T=27 ll=1005
UV=2 H=312 T=31 ll=1005
```

Fig. 9. Sending and receiving data from the central node-gateway.

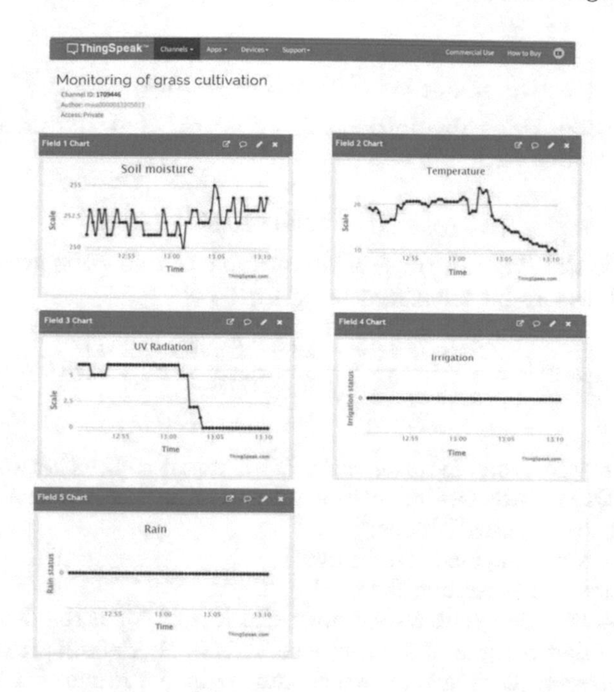

Fig. 10. Data Presentation in ThingSpeak.

3.3 Wireless Communication Testing

The wireless communication tests with LoRa technology proved to be adequate between the sensor node and the gateway at a maximum distance of 205 m without any data loss.

3.4 Airtime of Transmission

The airtime of a LoRa packet is calculated, which is the time it takes for a signal to be transmitted from a sensor node to be received by the gateway (receiver) node. This calculation is based on the spreading factor (SF), coding rate (CR), and signal bandwidth (BW). The airtime can be determined using the Eq. 1 [11].

$$ToA = T_{\text{preamble}} + T_{\text{payload}} \tag{1}$$

Where the preamble time is calculated using Eq. 2, where n_preamble is the preamble length, which according to the Semtech datasheet is programmed as 8 for the transceivers SX1276/77/78/79.

$$T_{\text{preamble}} = (n_{\text{preamble}} + 4.25) \cdot T_{\text{symbol}} \tag{2}$$

The symbol time is calculated using Eq. 3, where SF is the spreading fac-tor and BW is the bandwidth.

$$T_{\text{symbol}} = 2^{\frac{SF}{BW}} = 2^{\frac{7}{125}} = 1024 \tag{3}$$

By substituting this value into Eq. 2, we obtain that the preamble time is 12.54 ms.

$$T_{\text{preamble}} = (8 + 4.25) \cdot 1024 = 12.54 \tag{4}$$

To calculate the Payload time, it is necessary to know the symbol length in the payload. This can be calculated using Eq. 5.

$$N_{\text{payload}} = 8 + \frac{8PL - 4SF + 28 + 16CRC - 20IH}{4(SF - 2DE)}(CR + 4) \tag{5}$$

where:

- IH: Header, when enabled, the value IH is equal to 0; otherwise, it is 1.
- DE: Low Data Rate Optimize, the value DE is 1 when enabled, and the value DE is 0 for the other case.
- PL: Indicates the payload size in bytes.
- SF: Indicates the spreading factor.
- CRC: Indicates the cyclic redundancy check used for error detection. When enabled, it takes a value of 1; otherwise, it is 0. By default, it is disabled.
- CR: Indicates the coding rate, which can be in the range of 1 to 4 (1 corresponding to 4/5, 4 corresponding to 4/8).

By replacing this value in Eq. 5, we obtain a value of 58 symbols in the payload.

$$N_{\text{payload}} = 8 + \max\left(\left\lceil \frac{8(35) - 4(7) + 28 + 16(0) - 20(0)}{4(7 - 2(0))} \right\rceil \cdot (1 + 4), 0\right) = 58 \text{ symbols}$$

Once this value is obtained, we can calculate the duration of the payload using Eq. 7.

$$T_{\text{payload}} = N_{\text{payload}} \cdot T_{\text{symbol}} \tag{6}$$

Replacing the values, we obtain a payload time of 59.392 ms.

$$T_{\text{payload}} = 58 \cdot 1.024 = 59.39 \text{ ms} \tag{7}$$

Finally, we obtain the value of the airtime by replacing it in Eq. 1.

$$ToA = T_{\text{preamble}} + T_{\text{payload}} = 12.54 + 59.39 = 71.93\,\text{ms} \tag{8}$$

Figure 11 shows that the airtime is 71.94 ms, which is consistent with the calculations performed. This value can be verified using an online calculator for LoRa packets from Semtech modems, where the transmission parameters are entered, and the duration performance is obtained, including symbol time, airtime, preamble duration, and equivalent bit rate.

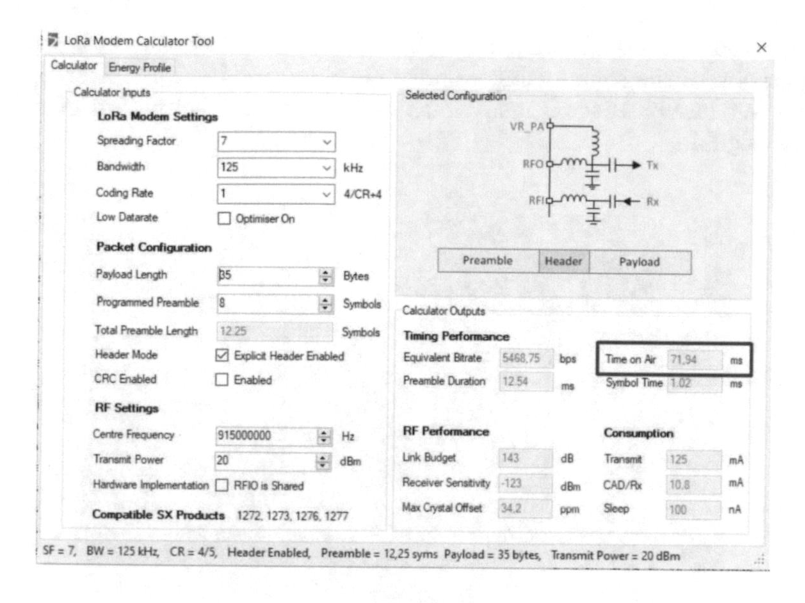

Fig. 11. LoRa Calculator for Airtime Calculation.

3.5 Spectral Analysis of LoRa Transmission

The Keysight N9322C spectrum analyzer is used to record and playback the captured signal traces. The spectrogram is used to visualize the spectrum of the signal in a LoRa transmission with different spreading factors (SF7 to SF12) in the LoRa nodes. The results obtained from the spectrum analyzer are presented in the Fig. 12.

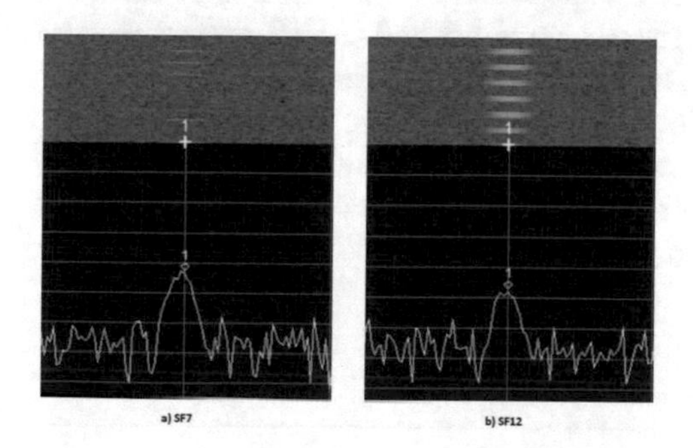

Fig. 12. Spectrum of the signal when varying the spreading factor from SF7 to SF12.

Each spectrum shows a colored horizontal line representing different signal amplitudes. In Fig. 13, the spectrum of a signal in MATLAB can be observed using a spreading factor SF7, with a packet airtime of 71.94 ms. Due to the short duration of this airtime, the signal is very faint and shorter compared to the spectrum of the signal using a spreading factor SF12. The latter signal has a longer airtime of 1646.59 ms., which allows for the distinction of a longer-duration signal.

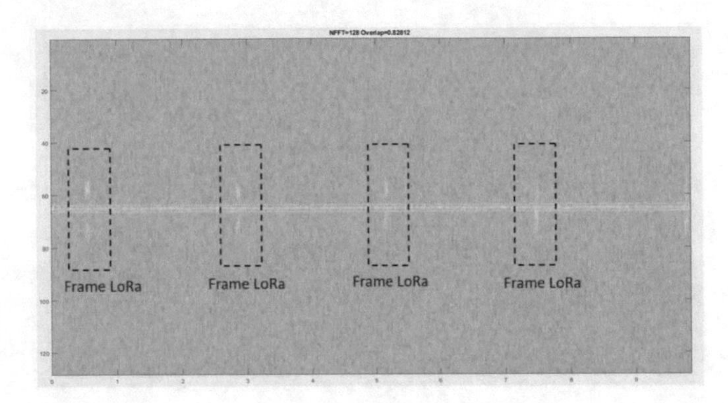

Fig. 13. Transmitted packets from the sensor node to the central node at 915 MHz.

3.6 MQTT Protocol Analysis

Once the TCP three-way handshake is completed, the client can connect to the MQTT broker to publish and subscribe to messages. The client sends a CONNECT packet to the broker to establish the connection. Then, access is granted, and a CONNECT ACK packet is responded, which contains return codes indicating the transmission status (success or error). Once the connection is accepted, messages can be published (PUBLISH) from the client to the broker or vice versa. In this field, the Topic is specified, which identifies the ThingSpeak channel ID. The last captured message is the DISCONNECT, indicating the closure of the connection sent from the client to the broker or vice versa. In this case, the client only publishes sensor data, so only the mentioned MQTT messages are captured. See Fig. 14.

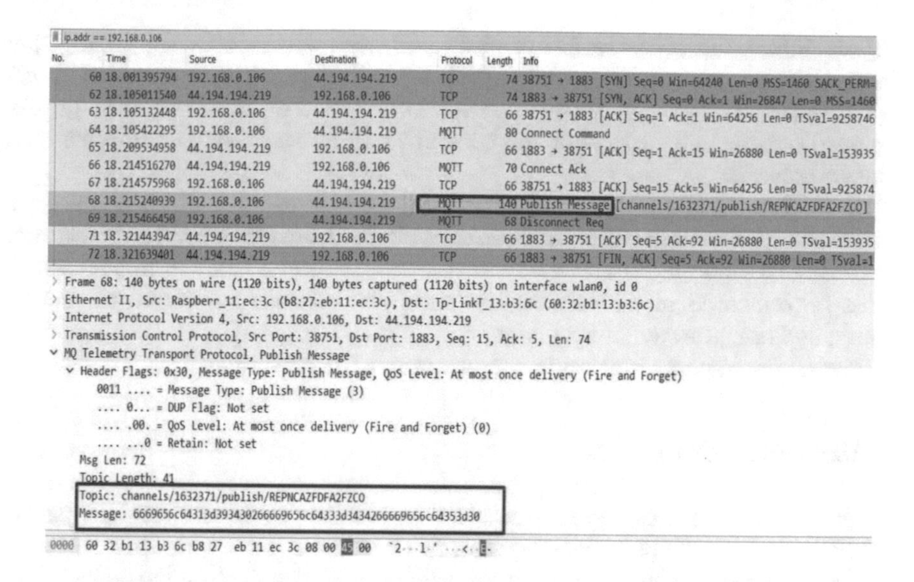

Fig. 14. MQTT Messages: Connect, Connect ACK, Publish, Disconnect.

4 Conclusions

- Appropriate devices were used for the sensor node, actuator, and central node, meeting thc system requirements to establish a stable communication and perform data transmission and reception.
- An electrical test was conducted to verify that the power board supplies power to the sensor nodes, actuator, and gateway for approximately 24 h. If necessary, the battery can be recharged using the micro-USB port of the TP5046 charging module with 5 V.
- The transmission of environmental variables data from the sensor node to the central node was successfully carried out, complying with the parameters established in the radio link simulation to ensure an obstruction free transmission.
- For communication using LoRa technology, the following parameters were determined: spreading factor (SF7), bandwidth of 125 KHz, error coding rate of 4/5, and transmission power of 20 dBm, resulting in an approximate airtime of 70 ms.
- The data of the environmental variables sent by the sensor node were published via the MQTT protocol on the ThingSpeak platform, where they are represented through graphs with an update frequency of every 15 s.
- A LoRa WSN wireless sensor network was designed with four components (humidity, temperature, UV radiation, and rainfall sensors), enabling the monitoring of environmental variables. Historical records and graphs were generated based on the measurements from the sensor node, which will serve as background information for future studies.

- To facilitate the presentation of clear and accurate results, the ThingView software was used. It is a mobile application that provides access to ThingSpeak channels through a user-friendly interface. The data can also be displayed on the LCD screen of the central node in case the user doesn't have internet access.
- The implementation of a wireless sensor network using LoRa technology is versatile, as there are various solutions available from different manufacturers, catering to different budgets, objectives, and ranges. Although in this case it was implemented in a plot with less than one kilometer between the sensor node and the gateway, LoRa technology offers coverage of over two kilometers, allowing for implementation in larger areas without difficulties.

5 Recommendations

- The gateway requires a stable internet connection to ensure efficient packet routing and enable data visualization in the cloud.
- It is recommended to perform an initial charge of the lithium batteries of the sensor and actuator nodes through the micro-USB port.
- It is important to ensure a clear line of sight between the sensor node and the gateway, as obstacles can weaken the transmitted signal. This may require adjustments in the network design during development.
- It is recommended to adjust the airtime of the transmission to optimize performance and minimize propagation losses, maximizing the power used.

6 Future Work

This phase and its preliminary results lay the foundation for the implementation of crop-oriented systems in rural environments.

References

1. Castro, M., Espin, F.: Análisis de los factores que determinan la sostenibilidad y sustentabilidad de la economía social y solidaria para la crianza y comercialización de vacuno en pie y faenado en los cantones Quito, Cayambe y Pedro Moncayo. Tesis **1**, 17 (2016)
2. Chacón, P.: Cultivo de pastos. SN Power Proyecto Cheves **1**, 40 (2015)
3. Red IAPD. Libro de actas del I Congreso en Investigación en Agricultura para el Desarrollo (2011)
4. Hidalgo, K., et al.: Plan de Desarrollo y Ordenamiento Territorial-Pedro Moncayo 2018-2025 (2018)
5. Leon, R., Bonifaz, N., Gutierrez, F.: Pastos y Forrajes del Ecuador (2018)
6. Jhomaira, B.: Red WSN para el control y monitoreo de un sistema de riego por goteo de una plantación de fresas en la granja experimental Yuyucocha - UTN. Universidad Técnica del Norte (2014)

7. Leandro, S.: Diseño de un sistema de riego inteligente para cultivos de hortaliza basado en Fuzzy Logic en la granja La Pradera de la Universidad Técnica del Norte. Universidad Técnica del Norte (2019)
8. Dario, C.: Diseño de un sistema para el control de riego mediante técnicas de aprendizaje automático aplicadas a la agricultura de precisión en la granja La Pradera de la Universidad Técnica del Norte. Universidad Técnica del Norte (2020)
9. Dominguez, A.: Diseño de una red de sensores inalámbricos LPWAN para el monitoreo de cultivos y materia orgánica en la granja experimental La Pradera de la Universidad Técnica del Norte," Universidad Técnica del Norte (2020)
10. Nettikadan, D., Raj, S.: Smart community monitoring system using thingspeak IoT plaform. Int. J. Appl. Eng. Res. **13**(October), 13402–13408 (2018)
11. Semtech. Semtech SX1276 (2020)

Artificial Intelligence, Towards the Diffusion of Cultural Tourism in the D.M. Quito

Geovanny Javier Cujano Guachi(✉) and Victoria Estefanía Segovia Mejia

Instituto Tecnológico Vida Nueva, Quito 170126, Ecuador
geovanny.cujano@istvidanueva.edu.ec

Abstract. The need to generate a digital participatory experience approach requires the adaptation of tourism to the automation of processes, these induce the digitalization and revolution of the immersive experience of tourists, through virtual reality, management of big data and artificial intelligence (AI).). It is therefore essential to analyze the AI to process opinions and interests of the visitors and determine the effectiveness of the interaction of the AI with the tourist. The use of methodologies and models such as safety and security, lean and agile, georeferenced information systems, seek to articulate tourism products to the digital environment. The study applies an AI model with social interaction in cultural visiting sites, the results demonstrate the feasibility of analog conversion to a digital environment, presenting schemes and necessary means for the adoption and adaptation of cultural tourism to the digital world.

Keywords: Cultural tourism · Artificial intelligence · Digital transformation

1 Introduction

The concepts of artificial intelligence (AI) are present in the data analysis process, mainly from those businesses that massify and universalize products or services. [1] considers that a corporate strategy used is the use of an information system, tourism faces the need to generate an experience design focused on the user in addition to combining the participatory experience. That is why the research seeks to formulate the analysis of the AI to process the opinions and interests of the visitors, determine the active interaction of the AI with the tourist, to achieve the purpose the research focuses on parameters and methodology of [2] who develops the term safety and security in tourism, and the Lean and Agile methodology. The results will allow to establish the changing preferences of tourists, to apply it to the AI so that the algorithm understands the preference qualities. All this embodied in a web page that collects the data and presents results. The main function is to digitize the tourism sector, providing both establishments and tourists with information and preference profiles, while ensuring the data and well-being of the visitor in matters of health and integrity. So that the research question is answered: Is it necessary to generate a user-focused experience design that can combine the participatory experience through tourist contact with AI, to determine preferences? [2] When the tourist experience enters a virtual world, [1] The application of Virtual

M. Z. Vizuete et al. (Eds.): CI3 2023, LNNS 1040, pp. 46–58, 2024.
https://doi.org/10.1007/978-3-031-63434-5_4

Reality (VR) in conjunction with Artificial Intelligence (AI) represents a revolution in the sector such as those led by the arrival of the first websites dedicated to tourism or applications for mobile devices with services tourist.

1.1 Technological Disruption

[3] artificial intelligence is in an embryonic state. Despite the condition, 3 stages are developed that the process of adopting technological innovation must follow, focused on the implementation of artificial intelligence.

- Progressive process innovation
- Innovation in new products and services (incremental)
- Disruptive technology innovation [4]

The interpretation of artificial intelligence, through processes, takes the following scenarios.

- Replacement: ratio of readaptation of human tasks by technology
- Increase: increase of attributes of the machines
- Symbiosis: mutual interaction of each of the individuals [4]

When engaging the service sector, it is necessary to categorize [5] Artificial Narrow Intelligence, (ANI), the cognitive capacity in relation to common sense is not very objective, inferior to human capacity [6] Artificial General Intelligence (AGI), objective ability to solve broad problems. In this sense, projects such as receptionist robots in hotels in Japan are being developed, or human symbiotic-empathic translator machines, such as assistants with artificial intelligence (Google, Alexa…).

Digital Tourism Sector. The trace of tourists on the internet, based on their products and search for destinations, through mobile devices, payments, consumption of both suppliers and opinions, rating of platforms and services as well as social networks.

By combining artificial intelligence and tourism, data is obtained on behaviors, information, needs, heterogeneous nature. Data that can be potentiated once the type of user is defined, offering successful tourist products and solutions, and at the same time personalized [7] Technological generation has an impact on the tourism industry Perfection and the search for process perfection drive digitalization [8].

The contribution of artificial intelligence to tourism has been framed in the ease of communication with the tourist, in this way both; Machines like robots carry artificial intelligence and perform some service, [9] the interaction of elements in the digital composition of a service involves the ability of AI, in continuous learning this allows perfection in customer preferences based on the service.

The AI applications used in tourism are:

Internet Of Things [10]. It is the interrelation between elements with AI principles, it interconnects, monitors, manages and identifies preferences in a product or service.

Virtual assistants and chatbots: [11] blend hybrid skills integrating human qualities that bring AI closer to learning tasks including natural language processing.

Blockchain: time chain information blocks in digital transaction actions

Tourism as an interdisciplinary study [12] tourism is a phenomenon that encompasses open systems that make it an interdisciplinary study, it is from this point the opening of tourism to work with ICTs and therefore with the digitization process. The structuring of the tourist market classifies the tourist activity as follows [13].

From this foundation, the definition based on the digitization of tourism, openness to the application of the safety and security methodology, this type of environment generates an analysis of the content to be presented based on the priorities of the tourist and their profile.

Tourism in cultural places of visit In this category are considered the tourist activities carried out to Museums, monuments, historical sites [14]. The fundamental purpose is to know the history, art, and culture found in a destination. From this perspective, it can be defined that:

Museums monopolize the distribution and syncretism of a culture [15], Museums monopolize the distribution and syncretism of a culture, it is considered as a center that presents culture and preservation through the collection of cultural manifestations that influence lifestyle [8].

Within the category of historical monument, it focuses on places with particular historical background determined by a structure that mixes urbanity and identity [8].

Historical ensembles: A historical ensemble or [10] historic center involve aspects that represent social development [16] the preservation of this style of resource allows to preserve the identity and cultural knowledge of a locality [18] the following stages should be considered:

- Control of load capacity, through controls of crowded areas
- Use of AI, to reduce intermediation
- Unique digital identification of the traveler (Safety and security)

The benefits of the implementation of AI, in the cultural sites of visit, allows to save costs of productivity, interpret the quality of the service and innovation.

To revolutionize the tourist experience through the use of ICTs, 3 dimensions must be considered to follow [3]:

- Digitization of content: this process will make it possible to visit the destination from a controlled environment, in addition to providing a universal panorama due to the availability of dialects
- Technological scope: platforms and means of reaching Artificial Intelligence
- Marketing: focuses on the lines of promotion that maintain social relations.
- Security: through public-private cooperation they must regulate the sector to be digitized [3].

2 Methods

2.1 Desing

The study methodology focused on the review of the literature, this with the purpose of obtaining quantitative and qualitative aspects through primary studies, with the objective of understanding the information on a particular topic, for its subsequent analysis, comparison and description of results [19].

For the review and orientation of the research, the quantitative study was proposed, based on the statistics of technical aspects of artificial intelligence, in cultural sites visited, for which models of digital transformation of services are used based on of cultural policies [13].

The focus of the research is of a simple descriptive nature, due to the intention of detailing the phenomenon of artificial intelligence in a cultural interpretation center, be it a museum, historic center or monument.

As a phenomenon for the study, the environment of the museums within the MD Quito is considered, which have their expression in painting, sculpture or architecture, combining the elements corresponding to cultural tourism [20] For the implementation of artificial intelligence, the approach of the [21] an agile museum, which applies in the context of a cultural center the deployment of virtual assistants with artificial intelligence for relative interaction with the visitor.

The methodology, used in the model of [22] It covers the articulation of artificial intelligence and a museum: this method called Lean and Agile, covers the design of a digital, multimedia experience between the tourist and the object.

The theoretical axis of the research is based on models proposed in various investigations.

- Geographic information systems: articulation of businesses focused on tourism, forms of digitization of georeferenced tourism products and services.
- Virtual reality/big data and artificial intelligence: Immersive virtual reality application format
- Internet of things: Management of virtual environments in areas of universal access to tourism
- Cultural and arts digital organizational development model: attributes to the organization the hierarchical pyramid, organization and work forces, for hybrid work, human and artificial intelligence
- Agile Museum: Applies in the context of a cultural center the deployment of virtual assistants with artificial intelligence for relative interaction with the visitor.
- Safety and security: Tourist card or passport model with tourist profiles and preferences, which protects integrity.
- Virtual assistants/chartboots and Blockchain: Applications for tourism with artificial intelligence
- Lean and Agile: Implementation of digitization in museums focused on the visitor experience.

The exposed models articulate the process of implementation of artificial intelligence in cultural visiting sites.

2.2 Population and Sample

For the purpose of the population, all the tourist companies involved in the physical and digital market are incurred, through the SITURIN and MINTUR records, the establishments considered in the study are. Food and beverages, Cafeterias, Restaurants, Intermediation Operation, Tourist transport and museums, in addition to the number of tourists who visit the D.M Quito, the total result is 4542. The population is classified as indefinite due to lack of updated data in the records of government entities. Once the total amount has been determined with the statistical calculation of an infinite sample, the total of 202 tourists who receive the product is determined, this will allow defining their preferences before the study created, and 123 tourist establishments, which will be surveyed randomly.

2.3 Instrument and Data Collection

For the data collection of tourists, a questionnaire used in the investigation is adapted and taken: exploration of social benefits for tourism Performing arts and industrialization in culture - tourism integration based on deep learning technology and artificial intelligence. The questionnaire is applied in dimensions of a social nature, taking cultural tourism (performing art and industrialization) and artificial intelligence as the central axis. It uses 11 items, to which a 5-point Likert scale is adapted. The questionnaire is divided by indices (G1, G2, G3).

G1 Index: From the perspective of (kim & others, 2022), it is necessary to analyze the social benefit in the city in relation to cultural tourism, use of resources, that is, how cultural tourism contributes to improving the living conditions through the inclusion of social economic relationship.

- Cultural heritage: articulates cultural identity and collective memory, safeguards the values of social heritage through the transferability of values
- Employment: Virtual Reality (VR) and Artificial Intelligence (AI) to tourism will mean a new revolution in the sector such as those led by the cheaper means of transport or the arrival of the first websites dedicated to tourism or the appearance of applications for mobile devices with tourist services, therefore, better living and employment conditions
- Quality of life: Data that can be potentiated once the type of user is defined, offering successful tourism products and solutions, and at the same time personalized.
- Infrastructure: the infrastructure of services interrelates the disruptive innovation of services

G2 Index: The importance of environmental care is part of large cities and their culture, [17] the cities carry the cultural heritage of the aboriginal peoples and these are promoted with cultural tourism, the higher the level of conservation, the greater the probability of maintaining and sustaining a representative tourist resource.

- Ecological environment: focuses on concepts of sustainability through the preservation of a legacy for future generations, uses the potential of tourism to conserve natural and cultural areas.
- Cultural protection: it constitutes a fundamental anthropological element for social sustainability.
- Environmental protection: focuses on the cultural component from the proactivity of individuals for the conservation of ecological zones.

G3 Index [18] the importance of cultural tourism in relevant social benefit in the economic aspect, which allows covering the implementation of digital systems.

- Economy: The use of economic systems through articulated information allows the digitization of processes focused on social and tourism benefits.
- Profitability: forms a visible finite chain by studying social enactment factors
- Utilities: the benefits of technologies are considered based on connectivity
- Return: tourism is a phenomenon that encompasses open systems that make it an interdisciplinary study, it is from this point the opening of tourism to work with ICTs

Dimensions: For the structure of internal coherence, the development of dimensions is incurred as:

- Cultural tourism: In the publication of ICOMOS in the Charter of Cultural Tourism (1976) it has the function of promulgating the historical and artistic social relationship.
- Environmental cultural heritage includes digitization actions in the conservation of customs and traditions of tourist spaces.
- Artificial Intelligence Artificial Narrow Intelligence (ANI), and Artificial General Intelligence (AGI) is one that can only provide solutions to one or a spectrum of cognitive and common-sense abilities that are inferior to human capacity.

Items for Validity Analysis

Table 1. Questionnaire adapted from the publication [18] exploration of social benefits for tourism Performing arts and industrialization in culture - tourism integration based on deep learning technology and artificial intelligence.

Variable	Item
cultural tourism	
TC1	Building artificial intelligence systems can link local cultural resources and the developing city
TC2	Cultural tourism generates jobs for the local population
TC3	The economy grows and the quality of life improves with cultural tourism
TC4	The infrastructure focused on cultural tourism, such as transportation, provision of basic services, improves considerably
Environmental cultural heritage	
HCA1	The ecological environment linked to cultural tourism as a heritage of the peoples is adequately treated
HCA2	The cultural environmental environments of the city are preserved and valued

(continued)

Table 1. *(continued)*

Variable	Item
HCA3	Local residents feel the benefit of protecting and conserving a cultural natural asset
Artificial intelligence	
IA1	The implementation of artificial intelligence can attract foreign tourists to promote local development and related industries
IA2	The implementation of artificial intelligence is capable of generating a net benefit of profitability to a cultural tourism project
IA3	The generation of artificial intelligence projects can cover its operating costs with the dissemination of cultural tourism
IA4	The inclusion of companies related to cultural tourism to artificial intelligence programs may reflect the ability to use capital to generate income

The data adapted from the questionnaire were analyzed through the validity of constructs and dimensions (Table 1).

Reliability analysis of the construction of the questionnaire: According to, [26] This process makes it possible to define the validity and precision of an instrument before its application, considering this aspect for the execution, 10% of the sample was taken so that a vision and perspective of the formulation of the questions can be obtained, contrasted with the previous results. The results reflect the 0.824 level of reliability, in Cronbach's Alpha.

Once the data collection instruments have been applied, the analysis is carried out based on the following antecedents:

- To determine the valid variables applied in the questionnaire addressed to tourists and the instrument focused on tourism companies, the construction of the tourist passport was defined through the analysis of communalities of each one.
- Structure of constructs of and dimensions of the variables adopted.

Analysis and Data Processing

[19], determines that the validity of constructs is fundamental to determine the data and the design that will be used to evaluate them, so that the applied instruments are within the variables and results presented, which is why in the processing of cases in SPSS the data was established in two parameters, one for the population of 202 tourists and 123 tourist companies.

The approach that was given to the processing of cases, to the companies related to the tourism field, who contribute their criteria to the development of the investigation based on this new form of content dissemination.

Analysis of construct strengths and information relationship in the instrument applied to tourists. Once the data is processed, the relationship of the two instruments is considered, determining that there is a relationship between the data and these will allow as a result to know the aspects with the greatest impact to be reflected in the development of the virtual passport (Tables 2 and 3).

Table 2. Analysis of the intelligence system construct

Construct	Number of indicators	Reliability	Cronbach's alpha
TC	6	0.96	0.955
HCA	3	0.86	0.770
IA	5	0.90	0.869

Table 3. Result of the analysis of the tourist passport construct

Construct	Number of indicators	Reliability	Cronbach's alpha
IAT	5	0.90	0.871
AV	4	0.97	0.967
CTU	5	0.93	0.911

Finally, the proportions of communalities obtained were analyzed so that the questions and results of the proposed items can be related (Table 4).

Table 4. Analysis of the tourist intelligence system

Construct items	C	CF
Building artificial intelligence systems can link local cultural resources and the developing city	.821	.507
Cultural tourism generates jobs for the local population	.596	.618
The economy grows and the quality of life improves with cultural tourism	.552	.363
The infrastructure focused on cultural tourism, such as transportation, provision of basic services, improves considerably	.804	.657
The ecological environment linked to cultural tourism as a heritage of the peoples is adequately treated	.751	.409
The cultural environmental environments of the city are preserved and valued	.552	.423
Local residents feel the benefit of protecting and conserving a cultural natural asset	.708	.485
The implementation of artificial intelligence can attract foreign tourists to promote local development and related industries	.547	.524

(*continued*)

Table 4. (*continued*)

Construct items	C	CF
The generation of artificial intelligence projects can cover its operating costs with the dissemination of cultural tourism	.678	.536
La generación de proyectos de inteligencia artificial puede cubrir sus costes operativos con la difusión del turismo cultural	.607	.364
The inclusion of companies related to cultural tourism to artificial intelligence programs may reflect the ability to use capital to generate income	.529	.494

Table 5. Analysis of the tourist intelligence system

Items of the tourist passport construct	C	CF
Do you believe in the reinvention of tourism through artificial intelligence?	.670	.534
Is artificial intelligence part of tourism competitiveness?	.799	.539
Do you think artificial intelligence favors the tourism industry?	.636	.559
What kind of tourists do you receive in your travel agency?	.560	.553
Can a personalized service be provided through the virtual passport?	.722	.417
With the creation of a virtual passport, what types of services would it include?	.716	.554
Can the needs of tourists be identified with the virtual passport?	.754	.427
What would you like to see inside the virtual passport?	.708	.478

Result of the communities found tourist passport variable (Table 5)

Analysis of communalities of the virtual passport variable

Once the data has been analyzed, the means that the development of the project will cover can be defined, especially the presentation and common data that will be placed in the tourist passport such as:

- Active tourism companies
- Contribution of tourism companies to the sustainable development of the truism
- Acts to protect and conserve a cultural natural asset.

3 Discussion

Tourism products maintain a level of degradation due to physical contact with the tourist. Artificial intelligence, and digitization means allow to preserve elements of the cultural heritage, for future generations [4] Immersive virtual reality allows you to explore beyond time and space and even beyond other barriers such as language. From the results obtained, it can be defined that the main function of AI is to digitize the tourism sector, providing both establishments and tourists with information and preference profiles, while ensuring data and well-being. of the visitor in matters of health and integrity. Thus, it is considered necessary to generate a user-focused experience design that can combine the participatory experience through tourist contact with AI, to determine preferences.

To achieve this purpose, it is essential to consider that immersive VR can improve historical and geographical knowledge and the sensory approach allows you to enjoy a different tourist experience in a specific destination. To this same sense joins [10], who defends the context of the use of artificial intelligence as a strength of tourism as a means of sustainability of resources and tourist attractions, becoming a tool in favor of cultural destinations (Fig. 1).

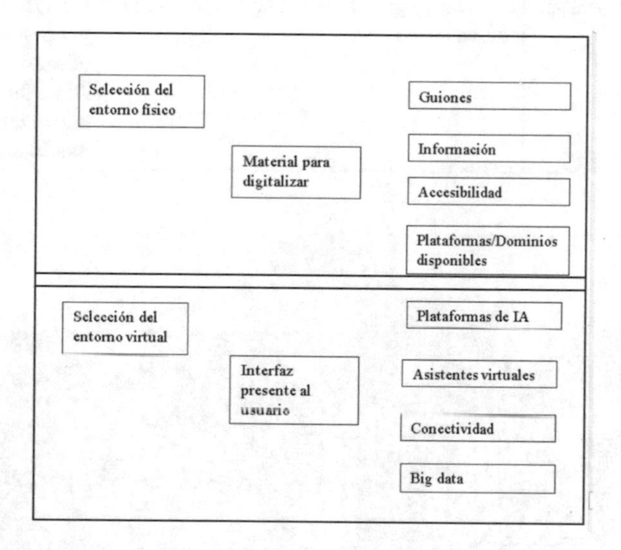

Fig. 1. Initial outline of the AI application

Schemes focused on the analysis of [13] The approaches of artificial intelligence in the tourist experience are determined in three perspectives (Table 6).

The models approach takes the analysis and transcription of [13]. By adopting the Lean and Agile model. In this way, the results are articulated in the provision of the page in which the tourist passport is hosted, as an initiative for the dissemination and virtualization of the experience through companies dedicated to tourist activity (Fig. 2).

Table 6. Models applied to cultural centers through AI

Design	*Model and methodology*	*Approach*
User-focused experience design	Hybrid experience model that qualifies human-computer interaction	AI must process the opinions and interests of visitors
Participatory experience design	Qualitative experience model (HISTOQUAL/SERVPERF/SERVQUAL)	Analyze the interaction of AI with the tourist
Agile experience design	Lean and Agile model, adapted to AI in a tourist space	Analyzes the changing goals and preferences of tourists, to apply it to AI so that the algorithm understands the qualities of preference

Fig. 2. The page is located at the address: https://pasaporteuio2023.mystrikingly.com/

4 Conclusion

The feasible means for the application of artificial intelligence are available to tourism, Models such as Safety and security, guarantee the integrity of the tourist in addition to determining the profile and preferences, shortening the intermediation process, the qualities of the Lean and Agile Model, focus on the transformation of a physical environment into a universal digital space within the reach of tourists. Artificial intelligence is the future of cultural tourism, the physical means are degrading, but the elements stored in the big data allow us to obtain a retelling of the lived history. Artificial intelligence, from the results obtained, allows to disseminate visitor preferences, in addition to keeping updated records of visits and even generating habits and frequency of visits, the models considered analyze social, economic and even political aspects of visitors.

The proposed methodological approach determines that, in the tourism field, especially in the approach to consumer experiences through artificial intelligence, the methodology of the hybrid experience model qualifies the human-computer interaction, the qualitative model of experiences (HISTOQUAL/SERVPERF/SERVQUAL) makes it possible to define satisfaction through the expectations and perception of the tourist and the Lean and Agile model, adapted to AI, adapts a service to a digital environment. The results, embodied in the development of the page, allow us to identify the articulated qualities of the interaction necessary for the development of a tourist adaptation system through the digitization of needs, taking the use of artificial intelligence as a strength of tourism. as a means of sustainability of natural and cultural resources and tourist attractions.

Work presented is an adaptation to artificial intelligence, thus increasing efficiency and tourist quality, giving a technological turn to tourist services, through the analysis of outbound tourists and tourism companies that collaborate actively with the product obtained. Project follow-up is vital, to correct alternative methodologies to the application, from the perspective of the study, the Safety and security and Lean and Agile methodology, allow defining the preferences for product selection, in future research, These methodologies could focus the tourism sector on the development of digitization.

References

1. Sastre, D., Martín, I., Martín, L.: La experiencia 5G en el turismo: el caso de Segovia, Ciudad Patrimonio de la Humanidad. Rev. Ibérica Sistemas Tecnol. Inf. **E24**, 336–348 (2019)
2. de Oliveira, D.: Tarjeta turística safety and security: el pasaporte para turistas y destinos seguros. Cuadernos Turismo **46**, 489–504 (2020)
3. de Oliveira, D.P.: Safety and security tourist card: the passport for tourists and safe destinations. Cuadernos Turismo **46**, 653–654 (2020)
4. González, D.: Modelos de gestión de museos con Inteligencia Artificial. Editorial Universitat Politècnica de València, pp. 685–697 (2022)
5. Almeida, M.: Robots, inteligencia artificial y realidad virtual: una aproximación en el sector del turismo. Cuadernos Turismo **44**, 13–26 (2019)
6. Barrera, C., González, J., Cáceres, G.: Toma de decisiones en el sector turismo mediante el uso de Sistemas de Información Geográfica e inteligencia de negocios. Rev. Científica **38**(2), 160–173 (2020)
7. Car, T.: Internet of things (IoT) in tourism and hospitality: opportunities and challenge. Tour. South East Eur. **5**, 163–175 (2019)
8. Chión, S., Charles, V.: Analítica de datos para la modelación estructural. Pearson Educación, Lima (2016)
9. Hegley, D., Tongen, M., David, A.: The Agile museum. In: MW2016: Museums and the web (2016). https://mw2016.museumsandtheweb.com/paper/the-agile-museum/
10. Lulia, D.: Aplicaciones de la inteligencia artificial en el sector turístico (2020). http://repositori.uji.es/xmlui/bitstream/handle/10234/194538/TFG_2020_Diana%20Iulia%20Iosivan.pdf?sequence=1
11. Melançon, M.: The USPTO's sisyphean plan: Increasing manpower will not match artificial intelligence's inventive capabilities. Texas Law Rev. **96**(4), 873–889 (2018)
12. Rodriguez, J.: Conocimiento y aprendizaje en las grandes cadenas hoteleras españolas en Iberoamérica: internet como herramienta de aprendizaje organizativo. Cuadernos Turismo **21**, 135–137 (2008)
13. Sabiescu, A.: A critical reflection on three paradigms in museum experience design. In: Vermeeren, A., Calvi, L., Sabiescu, A. (eds.) Museum Experience Design. Springer Series on Cultural Computing, pp. 259–276. Springer, Cham (2018). https://doi.org/10.1007/978-3-319-58550-5_13

14. Canavire, B.: Inteligencia artificial, cultura y educación: una plataforma latinoamericana de podcast para resguardar el patrimonio cultural. Rev. Científica Ciencias Soc. **21**, 59–71 (2023)
15. Cortes, S.S.: Evolución del concepto de servicescape: revisión sistemática de literatura 1995–2017. Rev. Espacios **39**(04) (2018)
16. Determinación de indicadores pertinentes para utilizar dentro de un sistema de gestión de información turística para el departamento de Casanare. Turismo Soc. XX **20**(10), 191–209 (2018)
17. Hui-Min, Y.W.: Design of eco-tourism service quality evaluation model based on artificial neural network. In: 2020 IEEE International Conference on Industrial Application of Artificial Intelligence (IAAI), pp. 119–124. IEEE (2020)
18. Zhang, R.: Exploración de beneficios sociales para el turismo Artes escénicas Industrialización en la integración cultura-turismo basada en tecnología de aprendizaje profundo e inteligencia artificial. Front. Psychol. **12**, 1–11 (2021)
19. Arévalo, D.S.G.: La gestión de departamentos y destinos turísticos de Colombia según el Índice de Competitividad Turística Regional (ICTRC). Rev. Int. Turismo Empresa Territorio **4**(2), 158–177 (2020)
20. Manterola, C., Astudillo, P., Arias, E., Claros, N.: Revisiones sistemáticas de la literatura. Qué se debe saber acerca de ellas. Cirugia Española **91**(3), 149–155 (2013)
21. Olmos, L., Garcia, R.: Estructura del Mercado turístico. Paraninfo, España (2016)
22. Panosso, A., Lohmann, G.: Teoria del Turismo. Trillas, Colombia (2012)

Real-Time Data Acquisition Based on IoT for Monitoring Autonomous Photovoltaic Systems

Fernando Jacome(✉), Henry Osorio, Luis Daniel Andagoya-Alba, and Edison Paredes

Instituto Superior Universitario Rumiñahui, Sangolquí, Ecuador
fernando.jacome@ister.edu.ec

Abstract. Technological development has made photovoltaic systems more accessible for applications in domestic self-generation microsystems. In this context, new monitoring techniques are necessary to visualize the operating status of photovoltaic systems components. The main objective of this work is to monitor the operation status of a photovoltaic system through remote measurements in real-time using IoT technology. The study used IoT technology for the remote acquisition of principal elements of a photovoltaic system data to show them in time series through a graphical interface. The study resulted in a real-time remote monitoring system that allows the observation of photovoltaic system variables. The study concludes that through IoT technology, complex real-time monitoring systems of photovoltaic system operating variables can be configured to store them in a database that allows the visualization and analysis of historical data behavior of photovoltaic systems.

Keywords: Photovoltaic System · Remote Monitoring · IoT · Electrical Variables

1 Introduction

Technological development has made photovoltaic systems more accessible for domestic self-generation microsystems. In this context, new challenges are presented for network operators and for end users who implement these systems, for self-consumption and for distributed generation. These new challenges not only have to do with the installations and their impacts but also with the monitoring and management systems of this type of equipment. In this context, new monitoring techniques are needed to visualize the state of the components of the photovoltaic systems [1].

The current technological advances in the field of monitoring techniques present a constant evolution, going from being manual to being carried out through automatic processes using advanced devices and complex processing procedures, both for the acquisition and analysis of data. In [2], a review of the development of some data monitoring techniques for the diagnosis of the state of photovoltaic panels is carried out.

M. Z. Vizuete et al. (Eds.): CI3 2023, LNNS 1040, pp. 59–70, 2024.
https://doi.org/10.1007/978-3-031-63434-5_5

In this study, a general classification is made into three groups. Manual methods that are applied to small systems, these techniques include visual inspection, reflectometry methods, and ground capacitance measurements. Semi-automatic methods, in this group are thermal cameras, infrared or electroluminescent images for locating faults in a large-scale system. The automatic methods used in a general way with algorithms based on modern techniques of analysis and data treatment are also proposed.

One of the ways used in automatic methods is the development of monitoring systems based on IoT applications, these have become a viable alternative in remote monitoring systems development that could be very useful in places with difficult access and where is difficult to have the necessary infrastructure to carry out some type of manual or semi-automatic monitoring [3, 4]. IoT-based monitoring systems are mainly made up of sensors, Arduino brand prototype development boards and Raspberry Pi microcomputers which, depending on the configuration, obtain data from a photovoltaic system that are displayed on display screens and/or mobile applications or also stored. in a physical database or in a data cloud. Initially, these methodologies were developed for monitoring the variables of the photovoltaic panels, but later they have been used for other applications such as estimates of the operating state of the system in general [3, 5–8].

In other studies, new methodologies are developed to improve the processes of acquisition, presentation and treatment of data. In [9], a data acquisition methodology based on IoT technology is presented. This study develops a data acquisition scheme with three levels, in the first one the current, voltage, temperature and solar irradiance data are taken through sensors. At the next level, these data are sent through the internet to finally present them in a third phase on a web platform or stored in a database. The device is made with Arduino technology both for the data acquisition from the sensors and for the transmission of them for them to be presented in real-time. In this same context, the study [10], presents a methodology where the use of a SCADA system is proposed as a means of monitoring variables. This methodology also uses sensors, an Arduino device to receive the variables and Raspberry Pi to manage and send the data.

IoT-based methodologies have also been developed as automatic techniques for monitoring photovoltaic operating variables through remote data acquisition. In general, these methodologies use sensors, data acquisition devices such as Arduino and data processing through a Rasberry Pi, however, the last one is not always used, but a remote server can also be used for the analysis of the data obtained. The variables that are taken through sensors are voltages, currents, temperature and irradiance at certain points of a photovoltaic system, this can be variable depending on the study to be carried out [11–14].

The main objective of this work is to monitor a photovoltaic system operation state through remote measurements in real-time using IoT technology. For this, the process has been divided into four stages, data acquisition, data processing, sending the data to a cloud database and displaying them. The main contribution of the project is related to the unification of the data acquisition process, treatment and presentation of them in online visualization platforms, which makes this model a remote and real-time monitoring system, in addition to show of a backup of the monitored variables stored in time series data, all of this, based on IoT technology. These systems are suitable to be applied in autonomous photovoltaic systems, especially in those with difficult access and where the necessary infrastructure to establish conventional communication systems is not available.

2 Methodology

The proposed methodology consists of four stages as indicated in Fig. 1. The first consists of the variables measurement through the sensors installed in certain points of the photovoltaic system, the second consists of the acquisition of sensor data through an Arduino device and the sending to a cloud through a Rasberry Pi. The third consists of the treatment of the data obtained in a Rasberry Pi microcomputer and finally a fourth stage where the data from the sensors are sent through the Internet to a display device which can be a mobile or a server. Each of the stages is detailed below in Fig. 1.

For filtering of data obtained from the sensors, the Arduino MeanFilter library was used, which implements a moving average filter. This filter stores a certain number of samples to define the window used by the filter to later obtain the average. A window

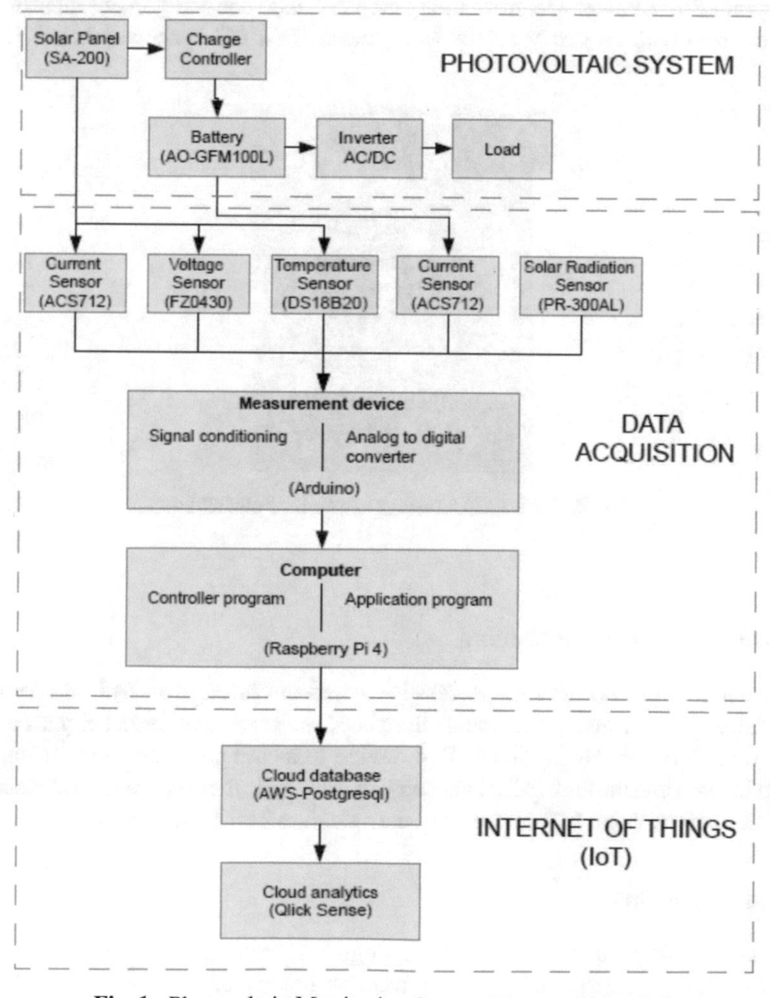

Fig. 1. Photovoltaic Monitoring System Diagram Using IoT.

of 10 samples was used. The moving average filter is widely used in signal acquisition systems and does not have a high computational cost to be implemented on a prototyping board such as Arduino.

2.1 Variable Measurement Process.

The measurement process is carried out through sensors located at specific points of the photovoltaic system. The variables taken are Voltages, Current, Irradiance, Temperature in the photovoltaic panel and Current in the battery. ACS712 Current [15], FZ0430 Voltage [16], DS18B20 Temperature [17] and Radiation PR-300AL [18] sensors are used.

To minimize the measurement error, the values from sensor of the solar irradiance variable were contrasted with a commercial solar irradiance meter such as the TES 132 DataLogging Solar Power Meter. In the same way, Current and voltage measurements were contrasted with Proskit MT-3109 Multimeter. TES 132 is showed in the Fig. 2.

Fig. 2. TES 132 Data-logging Solar Power Meter.

2.2 Data Acquisition and Sending

The data acquisition and filtering process is carried out through an Arduino device that is responsible for converting the analog signals of the sensors to digital signals and then sending them through the Raspberry Pi 4 device to a storage cloud and a visualization platform in real-time through IoT technology. In this stage, there is a remote measurement process through sensors, IoT devices and data storage in time series.

2.3 Data Processing

Data processing for visualization is done through IoT processes directly on the Rasberry Pi, which acts as a central node that receives, processes, and sends the data. The data collected from the sensors and monitoring devices is processed on the Raspberry Pi using specific algorithms and models designed to identify normal operation patterns and detect possible parameters whit any error that could affect the complete measurement process.

The monitored variables are Radiation, Solar Panel Voltage, Solar Panel Current, Solar Panel Temperature and Battery Current.

2.4 Results Visualization

Data processing for visualization is done through IoT processes; in the case of storage, it was done through an open-source relational database AWS-PostgreSQL. This database management system is known for its open source, reliability, scalability, and ability to manage a large volume of data. Choosing AWS-PostgreSQL ensures secure and efficient storage of collected data in the system.

For data visualization, the Qlik Sense application has been used, which is a tool that allows the interactive dashboards and graphics creation to analyze and visualize data effectively, with Qlik Sense.

The flow diagram of the proposed methodology is presented in Fig. 2.

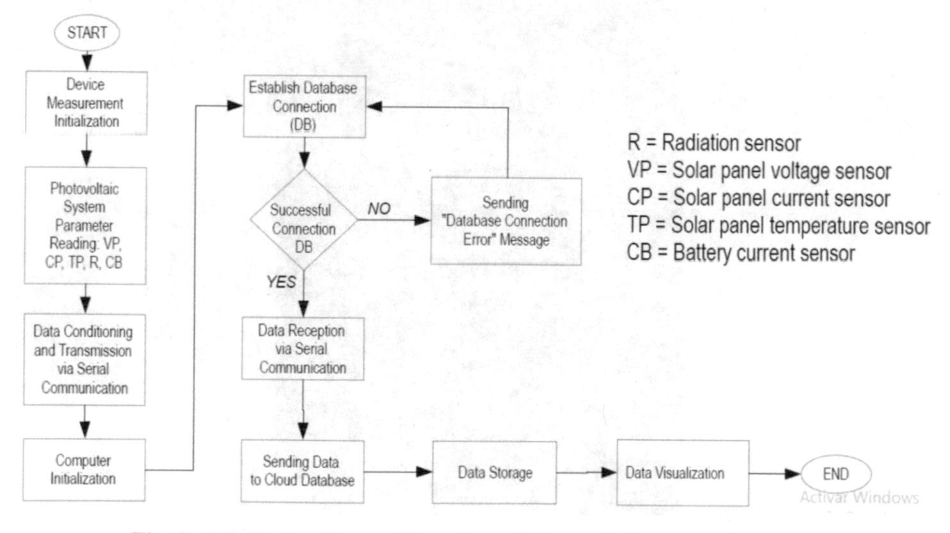

Fig. 3. Monitoring System Flowchart of photovoltaic system using IoT.

3 Results

The present work is based on the data acquisition system development for a photovoltaic system characteristic variables observation. The project is based on IoT technology for data acquisition from a photovoltaic system.

The methodology has been applied in a real test photovoltaic system located in the Instituto Superior Tecnológico Rumiñahui building located in the Pichincha province.

The test system consists of the next elements:

Photovoltaic module: 200 W photovoltaic panel, 21 V of open circuit voltage and 12.82 A of short circuit current.
Battery module: Gel battery, capacity of 100 Ah at 12 V.
Charge regulator: Solar Charge Controller PWM, nominal voltage of 12–24 V and maximum current of 20A.
Voltage Inverter: 1KW DC/AC inverter, nominal voltage 12 VDC/110 VAC.
Current sensors: ACS712.
Loads: 4 LED lights of 9 W.
Raspberry Pi: Model 4 with 8 GB of RAM and 32 GB of SSD.
Arduino: UNO R3 Model.
Voltage Sensor: FZ0430.
Temperature Sensor: DS18B20.
Radiation Sensor: PR-300AL.

The complete test of the Photovoltaic System is shown in Fig. 4.

Fig. 4. Test Photovoltaic System.

The application results of this methodology are detailed below:

The first process of the present study is focused on the remote acquisition of data through IoT technology use. This methodology allowed the collection of relevant information for the analysis of the test photovoltaic system state. These data are related to voltage, current, temperature in the solar panel and solar irradiation that it receives. For this, the system was configured to monitor, acquire and record the data every 5 min during an initial period of 3 months. This sampling frequency made it possible to obtain a large amount of data in a significant time interval, which is essential to perform a representative analysis and understand the behavior of the system in real-time. The results obtained were validated by real measurements with the respective equipment intended for these purposes.

The collection and storage of this data have two main purposes. To observe clearly the values of the photovoltaic system operating variables in real time to verify its operation, which allows observing the efficiency of the system and determining the need to carry out some type of maintenance in it.

The data obtained from the IoT system during the week of June 11 to June 15, 2023, are presented below. These data are of great importance because they will provide detailed information about the behavior of the system under specific conditions and will be useful as a guide for the analysis and estimation of the photovoltaic system operation state.

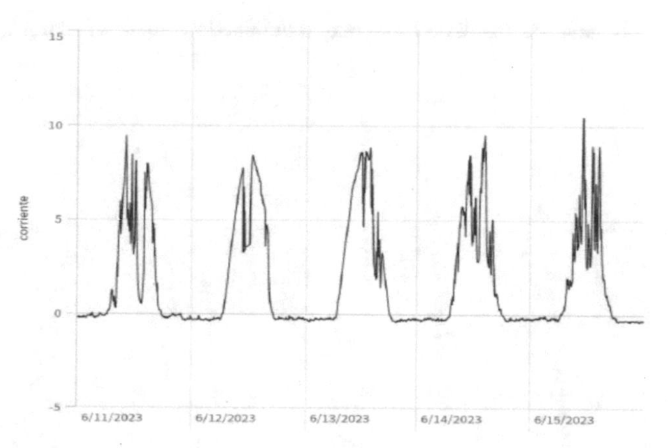

Fig. 5. Current at Photovoltaic Panel output obtained by the Monitoring System.

The graphic of Fig. 5 presents the current data obtained at the photovoltaic panel output of the test system, it clearly shows the variation of the current generated by the photovoltaic panel, which allows us to identify patterns, trends, and potential irregularities in panel performance. The fluctuations in the generated current values are influenced by solar radiation and ambient temperature. During the hours of maximum solar radiation, it is expected that the generated current reaches its highest values, while, in periods of less solar radiation, the current may decrease.

The graphic of Fig. 6 represents the voltage values recorded by the monitoring and measurement device for five days. This graphic provides detailed information on how the voltage varies in time, which helps us to identify patterns, trends and possible problems in the operation of the panel through the voltage values recorded in time series. The voltage variation in a day is directly influenced by solar radiation and ambient temperature. During the hours of greatest solar radiation, it is supposed that the voltage reaches its maximum values, while, at times of less solar radiation, the voltage can decrease to its minimum values.

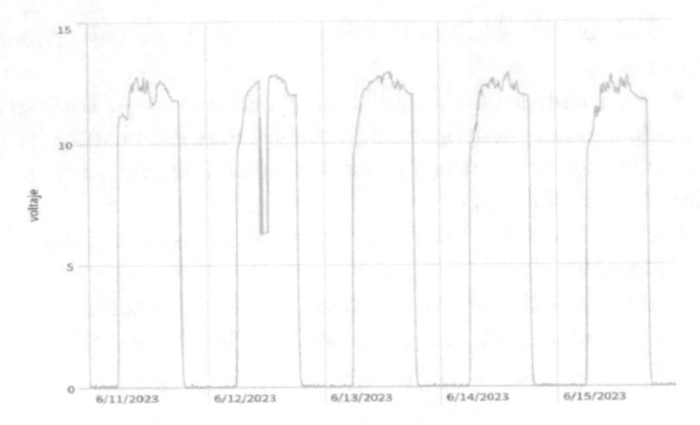

Fig. 6. Voltage at Photovoltaic Panel output obtained by the Monitoring System.

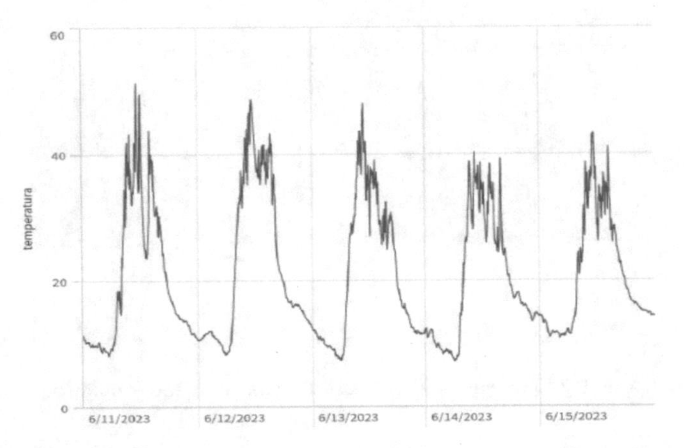

Fig. 7. Temperature on Photovoltaic Panel obtained by the Monitoring System.

The graphic in Fig. 7 provides detailed information about the temperature fluctuations produced on the photovoltaic panel throughout the day. The efficiency of the photovoltaic panel is influenced by its temperature, an increase in temperature can decrease the efficiency of solar cells and reduce energy production. Therefore, it is important to monitor this parameter to assess system performance and take corrective actions on time.

The graphic in Fig. 8 provides detailed information about the incident solar radiation on the photovoltaic panel and its variation during the day, which allows us to understand how solar radiation affects the energy production of the photovoltaic panel. The graphic shows how solar radiation varies throughout the day, with radiation peaking during the hours of greatest solar intensity, generally at noon. The graphic could also give radiation information based on seasonal patterns and differences in solar radiation between cloudy and clear days.

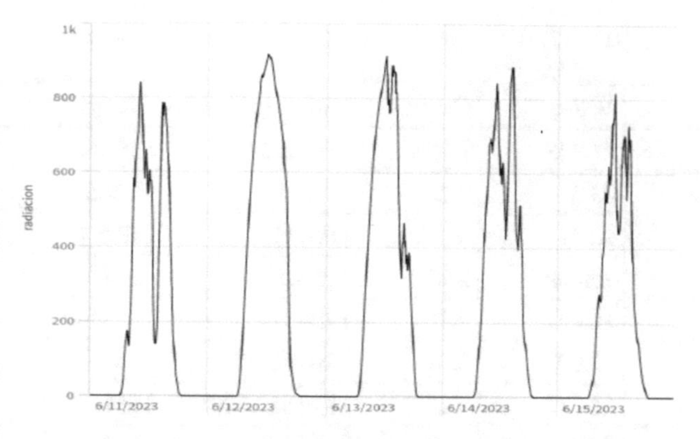

Fig. 8. Radiation on Photovoltaic Panel obtained by the Monitoring System.

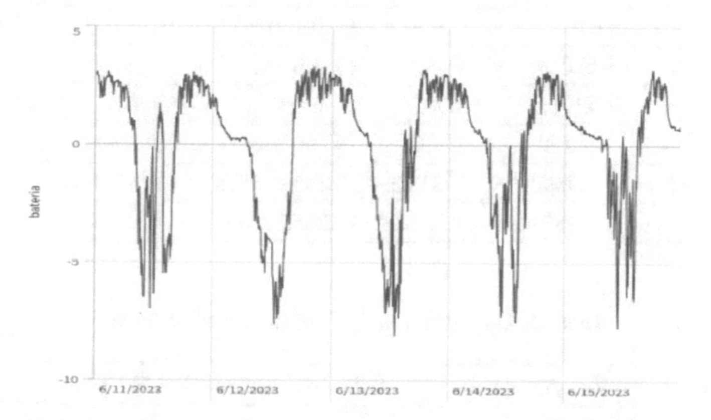

Fig. 9. Current from the battery obtained by the Monitoring System.

The graphic of Fig. 9 shows the current measurements in the battery throughout a week, in it, we can see positive values that represent an energy contribution to the load and negative values that correspond to the battery charge with the energy from the photovoltaic panel. With these values, it is possible to have patterns of battery behavior that could give information about the battery state operation.

Table 1 shows a descriptive statistic applied to data acquired from the monitoring system. In this table, solar irradiance has a maximum of 1100 W/m^2 with standard deviation of 414.19 W/m2 that represents solar irradiation of one day. Instead, module temperature has a maximum temperature of 56.75 °C with a standard deviation of 11.59 °C and a minimum of 5.88 °C.

Table 2 shows a correlations analysis of measurement variables. In the table, Solar irradiance and PV temperature have a high correlation with PV power generation.

Compared to the method presented in [3] and [4] which use the ThingSpeak Platform, the data of the present work is stored in a cloud database for further analysis and visualization in different time frames. In this way, you can see how the measured variables behave over days, weeks or months.

Table 1. Descriptive Statistics of measured variables

	Voltage	Current	Solar Irradiance	Temperature	Power > 0
Mean	11.228	3.208	594.529	25.960	36.936
Standard Error	0.031	0.026	3.741	0.105	0.305
Mode	9.877	0.481	0.000	12.500	0.107
Median	11.402	2.270	658.540	24.000	26.262
First Quartile	10.037	0.744	154.086	15.937	8.385
Third Quartile	12.081	5.071	1024.525	34.687	58.189
Variance	11.591	8.127	171553.990	134.537	1140.081
Standard Deviation	3.405	2.851	414.191	11.599	33.765
Kurtosis	4.680	−0.508	−1.589	−0.839	−0.226
Skewness	0.080	0.833	−0.180	0.437	0.902
Range	23.056	11.399	1100.351	50.875	149.866
Minimum	0.123	0.005	−0.001	5.875	0.100
Maximum	23.179	11.404	1100.350	56.750	149.966
Sum	137598.389	39318.393	7285953.798	318143.546	452651.517
Count	12255.000	12255.000	12255.000	12255.000	12255.000

Table 2. Correlation analysis of measured variables

Correlations	Voltage	Current	Solar Irradiance	PV Temperature	Power
Voltage	1.0000				
Current	0.0941	1.0000			
Solar Irradiance	0.4486	0.8024	1.0000		
PV Temperature	0.4313	0.8250	0.9095	1.0000	
Power	0.1828	0.9876	0.8127	0.8431	1.0000

As mentioned in [2], the most widely used technology is WIFI, which shows good performance for distances of up to 100 m. Regarding security, the device with WIFI connection and Internet access is the Raspberry PI that has a Linux-based operating system installed that is less prone to security flaws, although updates to this operating system must be made regularly.

4 Conclusions

IoT technology is an effective and promising solution in the field of remote monitoring photovoltaic systems variables, the main advantage of these systems is the ability to present real values of the monitored variables and facilitates the visualization of the monitored system operating status.

The study demonstrated the feasibility of using IoT technology for real-time remote monitoring of photovoltaic systems variables. The results obtained allowed us to observe the variability of the characteristic variables of monitored systems by time series analysis techniques.

The adequate treatment of the data acquired through IoT technology, especially through time series analysis techniques, provides important information for decision-making based on data and the implementation of corrective actions in a timely manner in the case to detect any anomaly in the data obtained by the monitoring system.

The remote monitoring system based on IoT technology can be the initial point for the development of new advances in the field of remote monitoring. The system stores the results obtained in time series in a database, which makes it possible to identify behavior patterns and therefore the possibility of observing an anomaly that could be due to a fault at some point in the photovoltaic system.

References

1. Masson, G., Bosch, E., Kaizuka, I., Jäger-Waldau, A., Donoso, J.: Snapshot of global PV markets 2022 task 1 strategic PV analysis and outreach PVPS (2022)
2. Mellit, A., Kalogirou, S.: Artificial intelligence and internet of things to improve efficacy of diagnosis and remote sensing of solar photovoltaic systems: challenges, recommendations and future directions. Renew. Sustain. Energy Rev. **143**, 110889 (2021). https://doi.org/10.1016/j.rser.2021.110889
3. Priharti, W., Rosmawati, A.F.K., Wibawa, I.P.D.: IoT based photovoltaic monitoring system application. J. Phys. Conf. Ser. **1367**(1), 012069 (2019). https://doi.org/10.1088/1742-6596/1367/1/012069
4. Muthusamy, S.: Fault detection and monitoring of solar PV panels using internet of things. Int. J. Ind. Eng. **2**, 146–149 (2018)
5. Tellawar, M.P., Chamat, N.: An IOT based smart solar photovoltaic remote monitoring system. Int. J. Eng. Res. **8**(09) (2019)
6. Gopal, M., Prakash, T.C., Ramakrishna, N.V., Yadav, B.P.: IoT based solar power monitoring system. In: IOP Conference Series: Materials Science and Engineering, vol. 981, no. 3, p. 032037 (2020). https://doi.org/10.1088/1757-899X/981/3/032037
7. Rerhrhaye, F., et al.: IoT-based data logger for solar systems applications. In: ITM Web Conference, vol. 46, p. 01003 (2022). https://doi.org/10.1051/itmconf/20224601003
8. Vujović, I., Koprivica, M., Đurišić, Ž.: Monitoring and management solution for distributed PV systems based on cloud and IoT technologies. In: 2022 30th Telecommunications Forum (TELFOR), pp. 1–4 (2022). https://doi.org/10.1109/TELFOR56187.2022.9983768
9. Sasikala, G.: Arduino based smart solar photovoltaic remote monitoring system. Malays. J. Sci. 58–62 (2022). https://doi.org/10.22452/mjs.vol41no3.8
10. Aghenta, L.O., Iqbal, M.T.: Development of an IoT based open source SCADA system for PV system monitoring. In: 2019 IEEE Canadian Conference of Electrical and Computer Engineering (CCECE), pp. 1–4 (2019). https://doi.org/10.1109/CCECE.2019.8861827

11. Ahmad, S., Shakir, M., Ahmad, B., Aslam, Z., Qaiser, M., Aslam, S.: Development of an IoT based DAQ system for PV monitoring and management. In: 2022 International Conference on Engineering and Emerging Technologies (ICEET), pp. 1–6 (2022). https://doi.org/10.1109/ICEET56468.2022.10007393
12. Hojabri, M., Kellerhals, S., Upadhyay, G., Bowler, B.: IoT-based PV array fault detection and classification using embedded supervised learning methods. Energies **15**(6), Art. No. 6 (2022). https://doi.org/10.3390/en15062097
13. Mellit, A., Hamied, A., Lughi, V., Pavan, A.M.: A low-cost monitoring and fault detection system for stand-alone photovoltaic systems using IoT technique. In: Zamboni, W., Petrone, G. (eds.) ELECTRIMACS 2019. LNEE, vol. 615, pp. 349–358. Springer, Cham (2020). https://doi.org/10.1007/978-3-030-37161-6_26
14. Hamied, A., Boubidi, A., Rouibah, N., Chine, W., Mellit, A.: IoT-based smart photovoltaic arrays for remote sensing and fault identification. In: Hatti, M. (ed.) Smart Energy Empowerment in Smart and Resilient Cities. LNNS, vol. 102, pp. 478–486. Springer, Cham (2020). https://doi.org/10.1007/978-3-030-37207-1_51
15. Current Sensor ACS712 (AC or DC) 30A. Future Electronics Egypt. https://store.fut-electronics.com/products/current-sensor-acs712-ac-or-dc-30a. Accessed 19 Mar 2023
16. Módulo Sensor Medidor de Voltaje Breakout FZ0430. Carrod. https://www.carrod.mx/products/modulo-sensor-fz0430-medidor-de-voltaje-breakout. Accessed 19 Mar 2023
17. J. A. K. Electronics. DS18B20+PAR Maxim Integrated - Analog Output Temperature SensorsJAK Electronics. https://www.jakelectronics.com/productdetail/maximintegrated-ds18b20par-6131397 Accessed 19 Mar 2023
18. Intelligent Detection of Solar Radiation Energy. https://rainbowsensor.en.made-in-china.com/product/uwrtdLzYggRo/China-Intelligent-Detection-of-Solar-Radiation-Energy.html Accessed 19 Mar 2023

Georeferenced Maintenance Management of Public Lighting Systems Using IoT Devices

Angel Toapanta(✉), Daniela Juiña, Byron Silva, and Consuelo Chasi

Instituto de Investigación Geológico y Energético IIGE, Quito, Ecuador
angel.toapanta@geoenergia.gob.ec

Abstract. Public lighting plays an essential role in our communities; it not only provides visibility during the nighttime hours, but also has a significant impact on safety, quality of life, economic activity, and social cohesion in urban areas. Its importance stems from its ability to create secure, attractive, and functional environments for all citizens. Quito's city reports of failures or breakdowns in luminaires or lamp pole are typically obtained from users who detect the issues and the time it takes to resolve the reported problem is high, making the lighting service inefficient for pedestrians and drivers. In the present work, a public lighting maintenance management system has been developed to optimize maintenance tasks for luminaires and lamp pole. This system employs technologies like the Internet of Things (IoT), data-driven planning, and automated tracking to improve efficiency, costs reduction, and maintain reliable and safe public illumination. The proposed system includes an IoT device responsible for real-time data collection about the state of luminaires, lamp pole, energy consumption, operation, and other relevant parameters. This is possible due to integrated sensors that detect faults, changes in lighting intensity, and other anomalies. The data collected by the IoT device is transmitted to a broker and stored in a NoSQL database. Subsequently, this information is analyzed using the proposed software, enabling pattern and trend identification, set thresholds and alerts to detect issues like inactive lights, changes in energy consumption, or performance deterioration, planning maintenance, task tracking, and route optimization.

Keywords: Street Lighting · IoT · Maintenance · Management · LoRa · MQTT · NOSQL

1 Introduction

About 2900 TWh of electrical energy is consumed annually for indoor and outdoor lighting; considering the expansion of cities and companies worldwide, in the next 20 years, there will be an approximate increase of 50% in current demand levels [1].

We use lighting daily in all the spaces where we live and work. Artificial lighting systems are a tool that greatly impacts productivity and the quality of life of living beings. Environments with an adequate lighting system provide safety and allow road and pedestrian mobility [2].

M. Z. Vizuete et al. (Eds.): CI3 2023, LNNS 1040, pp. 71–82, 2024.
https://doi.org/10.1007/978-3-031-63434-5_6

The public lighting system is one of the sources of the greatest energy consumption and also requires good management and investment to maintain and control energy consumption [3]. In Ecuador, public lighting systems are in charge of the electric companies around the country [4].

When thinking about new lighting installations, must be a strategy integrated to concentrate services such as internet, lighting, and fiber optic connections for building within the same infrastructure; this is how, in recent years, there has been interest from tell managers in the field of public lighting systems to integrate intelligent devices through a network with internet with which it is possible to: monitor, control, and even keep historical records of the inventory of street lighting luminaires [5], as well as the status of all its components, thus allowing to reduce the time of attention to failures to maintain the quality indexes of the street lighting service [2].

To repair a luminaire, it is crucial to know the technical specifications such as the type of technology, auxiliary equipment for its operation, power, voltage, current, the height at which the luminaire is installed, and many other details that facilitate the maintenance of this and thus reduce operating costs for this type of corrective maintenance. When there is a record of damage to the components or the installed luminaire itself, can perform failure analysis, can be planned preventive and corrective maintenance that may require. This action reduces operating costs and the acquisition of consumables, accessories, and spare parts.

1.1 Previous Works

In the literature, there are different proposals for the design and implementation of monitoring and control of public lighting systems to reduce energy consumption and maintenance costs.

Chunguo Jing et al. in their work show a wireless sensor network system consisting of a sensor node installed on each pole to detect and control the luminaires; the remote terminal unit that serves as a transmission station using the GPRS network between the control center and the sensor nodes; and the control center that monitors and controls all the luminaires in real time [6].

Lavric et al. presents a monitoring and control system based on a WSN network that allows remote control of street lighting luminaires. When the system detects a vehicle, the light intensity of the luminaires is increased to a predetermined level so as not to affect traffic safety and reduced in the opposite case; furthermore, employing current sensors, allows the identification of any possible malfunction of the luminaires [7].

Also, one of the most popular and versatile schemes proposed is the Zigbee network-based scheme; several authors use them effectively in their models; Yusoff et al. suggest a system for street lighting control based on Zigbee. The wireless Zigbee network monitors street lighting from a base station. In addition, it incorporates an automatic mode of operation that uses light sensors to automatically turn on the luminaires when the daylight intensity drops below a certain level [8].

On the other hand, Kurman et al. proposes a system that uses ZigBee-based wireless devices and also combines with sensors to control and guarantee the lighting parameters of the system. They have designed and implemented a monitoring system based on Internet technology integrated with street lighting. The mode adopted in this monitoring system is the Browser/Server mode, which uses the technologies of an embedded web server (standard web browser over the Internet) [9].

Other authors present schemes for the control and monitoring of public lighting systems, focusing mainly on providing facilities and information for the maintenance of these systems; Bezbradica et al. present a system for measuring the light intensity and GPS positioning of public lighting luminaires, creating a database that stores more than just the coordinates. This intelligent system consists of a central control and database which allows the management, programming, and analysis of the lighting system and coordinates information from traffic and weather sensors; communication between the software and the switch is established via Ethernet, fiber optics, and GPRS. The switch cabinet allows data collection from street lighting sensors for every 50 luminaires [10].

Additionally, Kumar et al. propose a system for collecting and analyzing information about street lighting infrastructure. They develop an automobile-mounted sensor platform that allows for collecting and recording data from luminaires during night-time tours. This platform enables mapping street lighting levels, identifying luminaires, estimating their heights, and geo-tagging them. It highlights an image recognition algorithm to identify luminaires from the video data collected by the sensor platform and its subsequent use in estimating heights. With this, they develop a semi-live virtual three-dimensional street lighting model at an urban scale [9].

2 Methodology

The proposed block diagram for the geo-referenced maintenance management system for street lighting using IoT devices is shown in Fig. 1 shows the proposed block diagram for the geo-referenced maintenance management system for public lighting luminaires using IoT devices [11]; The system consists of three stages:

1. IoT Public Light Device: This is responsible for data collection and transmission. It replaces traditional photocells and will be installed on each luminaire.
2. Maintenance Management System: This is responsible for analysis, monitoring, and maintenance management. It is hosted on the free tier of Amazon Web Services (AWS), which offers flexibility, scalability, security, and a variety of cloud services [12] cal interface through which the end user can interact with the system [13].

The block diagram in Fig. 2 shows the components of the IoT field devices, which are constituted by a VAC/VDC converter that will provide power to all the electronic components of the device, a microcontroller with built-in memory for processing and treatment of signals from the sensors, a wireless communication module to transmit the information from each luminaire to the router reaching stage 2, a GPS module to provide the location of the luminaire, a current sensor to determine the operating status of the luminaire, an ambient light sensor to provide information for switching the luminaire on or off, an accelerometer to detect events such as impacts and changes in orientation on the lamp pole, a power relay to activate the output of the luminaire's power supply circuit [14].

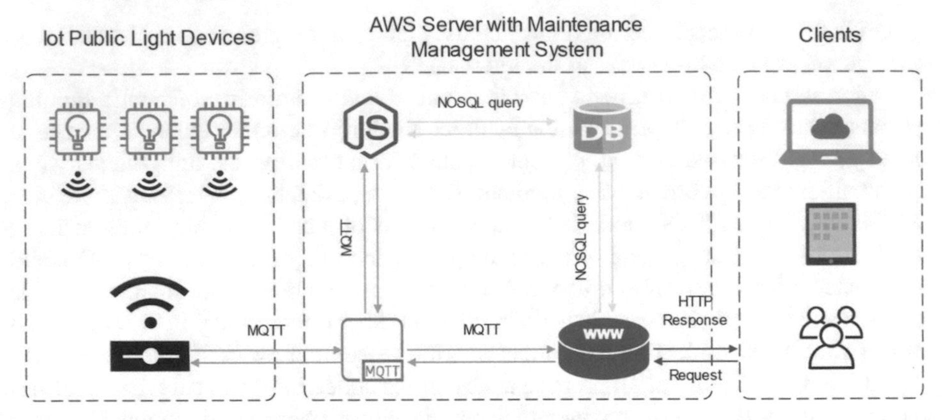

Fig. 1. Proposed system block diagram

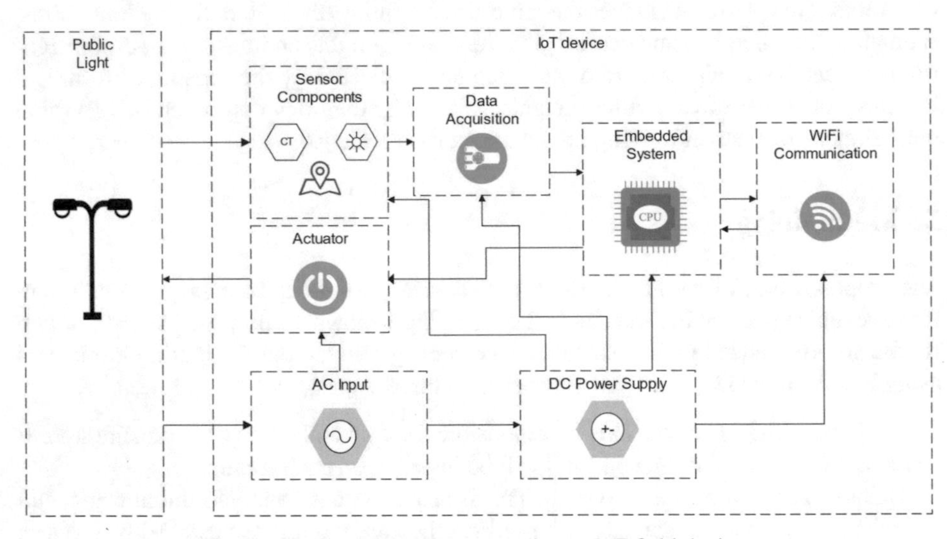

Fig. 2. Block diagram corresponding to IoT field devices.

2.1 Sensors of the IoT Field Device

Current Sensor TA12–200. El sensor de corriente TA12–200 permite adquirir los datos de corriente que circula a través de la luminaria para establecer su estado de funcionamiento. Las principales características de este sensor son:

Transformation ratio: 2000:1
Input current: 0-5A
Non-linearity: $\leq 0.2\%$
Phase shift: $\leq 5°$

BH1750 Light Sensor. The BH1750 photodetector is based on a photodiode, a trans-impedance amplifier, and a filter. Its single-chip construction reduces noise sources as a result of the integration of the amplifier and the photodiode; the sensor has a spectral response capability close to the response of the human eye in the 400 nm and 700 nm range, and the amplifier output signal is digitized by a 16-bit ADC [15].

NEO-6M GPS Positioning Sensor. The NEO-6M GPS is a 50-channel device with a time-to-first-correction (TTFF) of less than 1 s; its design and technology suppress sources of interference and mitigate the effects of multiple trajectories. In addition, it has a dedicated acquisition engine, with 2 million correlators, capable of performing massively parallel searches in time/frequency space, allowing it to find satellites instantly. The information is extracted from the minimum recommended position data of the $GPRMC sentence NMEA 0183 standard packet protocol Quetec: L80 GPS Protocol Specification (2014) [16].

MPU-6050 Accelerometer. The MPU-6050 accelerometer is a six-axis sensor that combines accelerometer and gyroscope on a single chip. It uses I2C communication and offers accurate measurement with configurable sensitivity. It has digital filters to improve measurement quality and can detect events such as impacts and changes in orientation. Low power and compact size [17].

2.2 Acquisition of Data

Each IoT device has a conditioning stage for the signals of each analog sensor for which an ADC converter of 4 analog channels with a resolution of 16 bits with a sampling rate of up to 860sps and an I2C serial data bus is used.

In the case of the current sensor, a precision resistor of 800 ohms recommended by the manufacturer is used to convert the current signal to a voltage signal of 2V, and this signal enters the ADC for further processing.

The light, gps and accelerometer sensors have embedded conditioning and transmit data through I2C serial data buses.

2.3 Embedded System

IoT applications generally use embedded SoC (System-On-Chip) controllers and Open Source Hardware, as is the case of Arduino and ESP32; this type of Hardware does not usually carry an operating system, and its programming is based on C and C++ [18]. This component performs the logic (algorithms) related to acquisition, conversion, quantification, and co-communication. The proposed IoT field device is developed on a 240 MHz Tensilica LX6 dual-core + 1 ULP, 600 DMIPS ESP32 SoC, including a 19 dB maximum output power SX1276/SX1278 LoRa chip. The ESP32 SoC is dual-core so that separate tasks can be assigned to the different cores, which is a great advantage as it alleviates CPU usage and allows having more than one task running at the same time, each on a different core [19, 20].

2.4 Communication

The IoT device based on the ESP32 SoC has several wireless communication channels such as WiFi 802.11 b/g/n, Bluetooth V4.2, and Node-To-Node Lo-RaWAN with a frequency of 915 MHz in the ISM band without a license due to geographical location and corresponding regulations. This allows it to connect to existing wireless networks deployed in smart cities. The MQTT protocol is used for messaging transport, which is built to provide orderly, lossless, and bidirectional connections. MQTT is widely used because it is lightweight, open, and designed for low bandwidth, high latency networks. MQTT comprises three members: publisher, broker, and subscriber; the broker manages messages between publishers and subscribers without needing to identify a publisher and consumer based on physical aspects. MQTT allows data acquisition in a simple and simplified way [21, 22].

For sending messages, the IoT device will process all the information and send it to the MQTT broker hosted in the cloud. The QoS QoS for MQTT is defined and controlled by the sender, i.e., each IoT device can have a different policy; for our case, and to alleviate the bandwidth, we will select a QoS = 0.

The software developed for our management system allows the registration of unique topics for each IoT device in a structured hierarchy, and the MQTT broker only processes the data of each device previously registered.

2.5 Power Supply

The IoT device is designed to be powered with a voltage between 100V to 240V, which will depend on the nominal voltage of the luminaire so that it works with the same power supply already available in lighting networks. This alternating voltage is converted into 5V DC voltage to operate all electronic components.

2.6 Description of Software

The management system allows to visualize the data in real time or the historical values of all the parameters configured in each IoT device. The data is published to the MQTT broker, this information is requested by the software for real-time visualization or by the node JS server, which in turn is responsible for storing the data in the NoSQL database and its subsequent visualization in the software as historical data. The software was developed using the Laravel framework in its version 10 in the backend and VueJs in the frontend and its implementation has been done based on design patterns so that developers can understand it in order to ensure that the software can be updated.

This software allows:

- Analyze data to identify patterns and trends.
- Establish thresholds and alerts to detect problems such as luminaires going out, changes in energy consumption or deterioration in performance.
- Maintenance planning based on the analyzed data.
- Automatically generate preventive and corrective maintenance schedules. This includes scheduling work orders using location and geolocation data so that technical personnel perform their tasks efficiently and reduce mobilization times.
- Maintain historical data of the parameters sent by each IoT device, as well as maintenance and work orders performed.
- Report generation.

The software has five modules:

1. User authentication
   a. Login
   b. Email verification
   c. Password reset
   d. Two-factor authentication
2. User administration and permissions
   a. Register new users
   b. Edit user information
   c. Create roles and permissions
   d. Associate users with roles and permissions
3. Inventory Management
   a. Create, view, and edit technical specifications for luminaires, poles, and fixtures
4. Installation and maintenance manager
   a. Register facilities
   b. Maintenance planning
   c. Alarms for luminaire and pole failures.
   d. Automatic elaboration of work orders
   e. Maintenance record
   f. Equipment change log
5. Reports

Module 1 user authentication and module 2 user administration and permissions are related to program security for user authentication and user authorizations. Also, module 1 allows the creation and edition of information for all users, while Module 2 establishes user authorizations through roles and permissions so that each based role has defined access to each part of the software.

Module 3, inventory management, is in charge of inventory management, i.e., it allows storing the technical specifications of luminaires, auxiliary equipment, and poles, as well as their availability.

Module 4, installation and maintenance manager, allows the automatic creation of work orders, which are created based on the parameters established in the maintenance planning, whether preventive or corrective, for luminaires and poles.

If a luminaire or the pole on which it is installed presents failures, the system automatically displays alerts and notifications so that system users are informed. The application identifies the number of failures of nearby luminaires and sets it as a decision parameter. In case the first parameter is not met, a time limit is set since the loss has occurred.

This information is analyzed in order, to deploy maintenance personnel automatically, optimizing the response time by the claimant; in addition, the application has a connection to Google Maps so that the person follows the route optimally chosen to perform the maintenance.

If the failure is in the pole structure, the application instantly generates the work order since these failures are structural and can cause accidents to pedestrians and vehicles.

All this can be done thanks to the field device proposed in this document, which provides the necessary information to know the status of the luminaires in real time.

Module 5 allows for generating dynamic reports through the use of filters.

3 Results and Discussion

The software has been deployed on Amazon's free Amazon EC2 server with 99.99% service availability, which allows up to 750 h per month at the free level (Fig. 3).

Fig. 3. Public Lighting Maintenance Management System

The IoT devices proposed in this document are installed on 10 street lighting poles with an average installation distance of 30 m (Fig. 4).

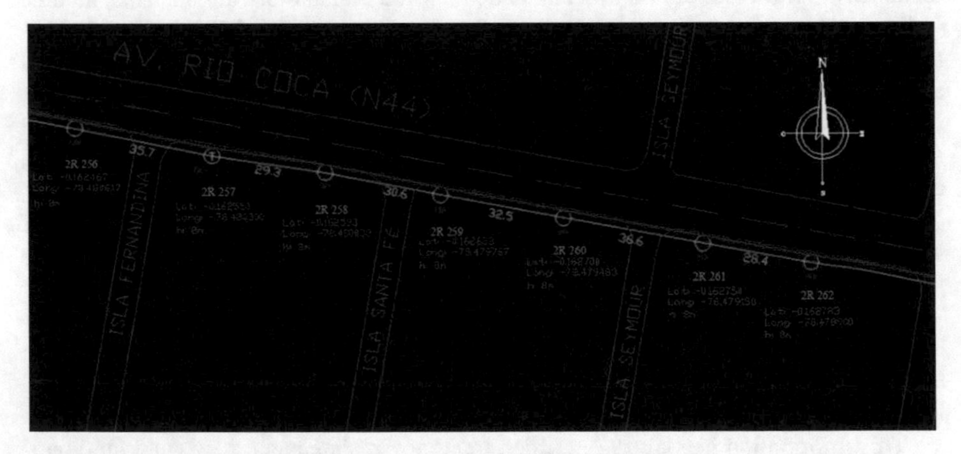

Fig. 4. Layout of public lighting poles

The first test performed is to the IoT device by console so that the parameters of the luminaire in which it is installed can be seen. The message received (topic) has the following format: Syste Name/Company/Gateway Code/Circuit Identification/Luminaire Code/option. Example: PLM/EEQ/2R-GW1/2R/258/data(field1, field2,…, fieldn). The data of the luminaire and pole states come from field 5, a defined code depending on their state (Table 1).

Table 1. Pole and luminaire status codes

Code	Description
M0100	Ok
M0101	ON
M0102	OFF
E0300	Fault: powered, but does not turn on
E0301	Fault: circuit without energy
E0302	Fault: pole broke down
E0303	Fault: lost of communication

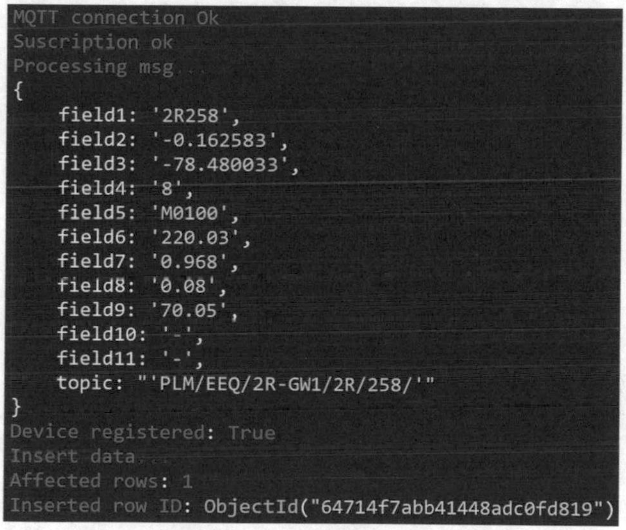

```
MQTT connection Ok
Suscription ok
Processing msg...
{
    field1: '2R258',
    field2: '-0.162583',
    field3: '-78.480033',
    field4: '8',
    field5: 'M0100',
    field6: '220.03',
    field7: '0.968',
    field8: '0.08',
    field9: '70.05',
    field10: '-',
    field11: '-',
    topic: "'PLM/EEQ/2R-GW1/2R/258/'"
}
Device registered: True
Insert data...
Affected rows: 1
Inserted row ID: ObjectId("64714f7abb41448adc0fd819")
```

Fig. 5. IoT device test run

Once the IoT devices have been configured and registered in the management software, we proceed to test the operation of the proposed Maintenance Management System by simulating failures in different IoT nodes (Figs. 5, 6 and 7).

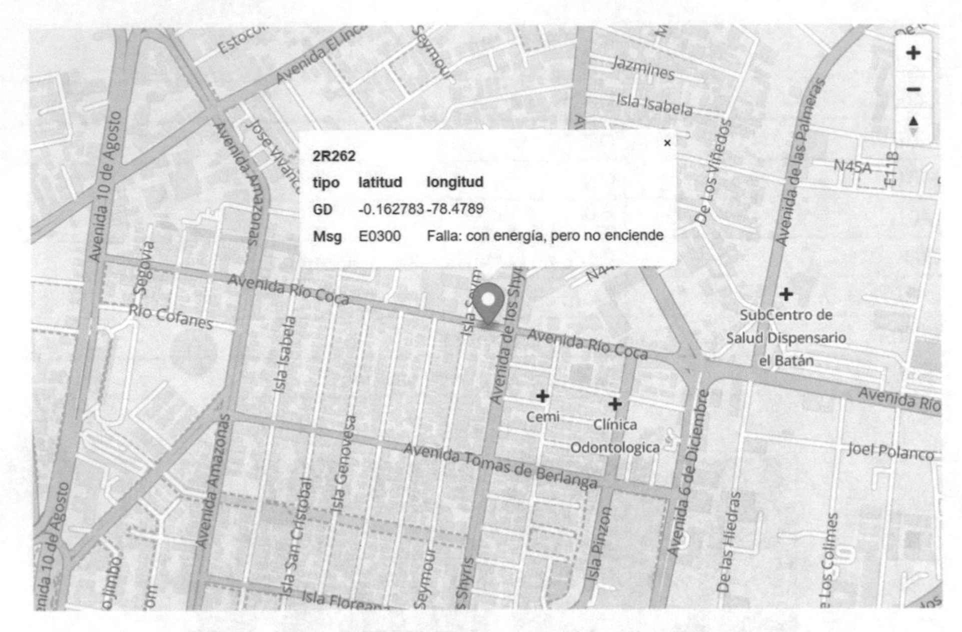

Fig. 6. Message E0300, Fault: powered, but does not turn on

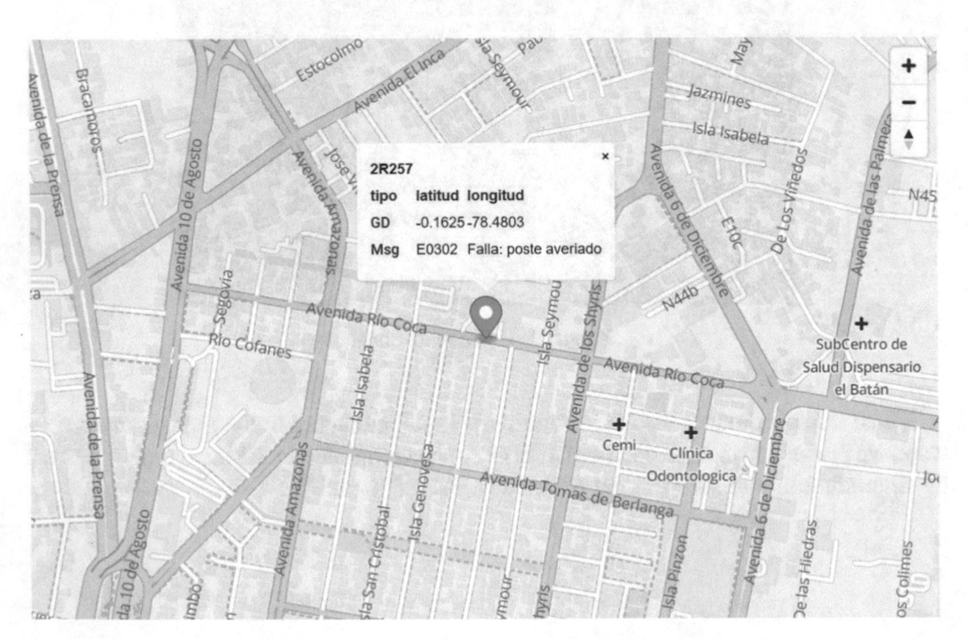

Fig. 7. Message E0302, Fault: pole broken down

4 Conclusion

The management of georeferenced maintenance of public lighting systems through the "internet of Things" IoT arises from the need to have energy savings to regulate, schedule, and manage the use of lighting efficiently.

The damage report in the city of Quito is currently obtained from the user who detects the failure or breakdown of a luminaire; the time it takes to solve the reported problem is high. Therefore, the lighting service becomes inefficient for passers-by and drivers. However, it is now within reach of managers to implement a management system that allows monitoring and storing information so that after the corresponding analysis, preventive maintenance routes can be planned and created, effectively managing the inventory of luminaires and increasing the quality index of the Public Lighting service.

The information collected can also be used to differentiate the levels of illumination required by each lighting system, whether for road or pedestrian traffic, guaranteeing, depending on vehicle traffic, adapting the lighting to the context of traffic and time, providing a feeling of greater security combined with aesthetically pleasing designs.

The initiative to implement these georeferenced management systems is a stepping stone to achieving an integrated system that will allow in the near future to control an intelligent city that can obtain data with motion sensors, which in addition to controlling lighting control traffic through traffic lights, which acquire images and videos of the conflict zones in the city to create safe spaces for citizens.

All these changes will allow us to evolve the behavior of the operation and management of public lighting, lighting when it is needed, where it is needed, reducing the light pollution that is caused, and even conserving the environment.

References

1. PNUMA, FMAM, U4E: Aceleración de la adopción mundial de la iluminación energéticamente eficiente (2016)
2. Szalai, A., Szabó, T., Horváth, P., Timár, A., Poppe, A.: Smart SSL: application of IoT/CPS design platforms in LED-based street-lighting luminaires. In: 2016 IEEE Lighting Conference of the Visegrad Countries (Lumen V4), pp. 1–6 (2016)
3. Ageed, Z.S., et al.: A state of art survey for intelligent energy monitoring systems. Asian J. Res. Comput. Sci. **8**, 46–61 (2021). https://doi.org/10.9734/ajrcos/2021/v8i130192
4. Conelec: Regulación Nro. CONELEC 005/14 Prestación del Servicio de Alumbrado Público General (2014)
5. Pachamanov, A., Pavlov, D., Kassev, K., Nikolova, K.: Individual control of street luminaries by speed of change of natural light in the evening and by calendars in the morning. In: 2021 Sixth Junior Conference on Lighting (Lighting), pp. 1–4 (2021)
6. Jing, C., Shu, D., Gu, D.: Design of streetlight monitoring and control system based on wireless sensor networks (2007)
7. Lavric, A., Valentin, P., Finis, I., Simion, D.: The design and implementation of an energy efficient street lighting monitoring and control system. Przeglad Elektrotechniczny. **88**, 312–316 (2012)
8. Yusoff, Y.M., Rosli, R., Karnaluddin, M.U., Samad, M.: Towards smart street lighting system in Malaysia. In: 2013 IEEE Symposium on Wireless Technology & Applications (ISWTA), pp. 301–305 (2013)

9. Kumar, S., Deshpande, A., Ho, S., Ku, J., Sarma, S.: Urban street lighting infrastructure monitoring using a mobile sensor platform. IEEE Sens. J. **16**, 1 (2016). https://doi.org/10.1109/JSEN.2016.2552249
10. Bezbradica, M., Trpovski, Ž.: Advanced street lighting maintenance using GPS, light intensity measuring and incremental cost-effectiveness ratio. In: 2014 International Conference on High Performance Computing & Simulation (HPCS), pp. 670–675 (2014)
11. Gupta, A.K., Johari, R.: IOT based electrical device surveillance and control system. In: 2019 4th International Conference on Internet of Things: Smart Innovation and Usages (IoT-SIU), pp. 1–5 (2019)
12. Domínguez-Bolaño, T., Campos, O., Barral, V., Escudero, C.J., García-Naya, J.A.: An overview of IoT architectures, technologies, and existing open-source projects. Internet Things. **20**, 100626 (2022). https://doi.org/10.1016/j.iot.2022.100626
13. Toapanta, R., Chafla, J., Toapanta, A.: Physical variables monitoring to contribute to landslide mitigation with IoT-based systems. In: Botto-Tobar, M., S. Gómez, O., Rosero Miranda, R., Díaz Cadena, A. (eds.) Advances in Emerging Trends and Technologies, ICAETT 2020, Advances in Intelligent Systems and Computing, vol. 1302, pp. 58–71. Springer, Cham (2021). https://doi.org/10.1007/978-3-030-63665-4_5
14. Sanchez-Sutil, F., Cano-Ortega, A.: Development and implementation of a PQ analyser to monitoring public lighting installations with a LoRa wireless system. Internet Things. **22**, 100711 (2023). https://doi.org/10.1016/j.iot.2023.100711
15. Araguillin, R., Toapanta, A., Juiña, D., Silva, B. (2022). Design and characterization of a wireless illuminance meter with IoT-based systems for smart lighting applications. In: Botto-Tobar, M., Zambrano Vizuete, M., Diaz Cadena, A., Vizuete, A.Z. (eds.) Latest Advances in Electrical Engineering, and Electronics, LNEE, vol. 933, pp. 129–140. Springer, Cham (2022). https://doi.org/10.1007/978-3-031-08942-8_10
16. Quectel: L80 GPS protocol specification (2014)
17. Chokshi, R.: MPU-6000 and MPU-6050 register map and descriptions revision 4.0 (2012)
18. Mohamed, K.S.: The era of internet of things: towards a smart world (2019)
19. Prade, L., Moraes, J., de Albuquerque, E., Rosário, D., Both, C.B.: Multi-radio and multi-hop LoRa communication architecture for large scale IoT deployment. Comput. Electr. Eng. **102**, 108242 (2022). https://doi.org/10.1016/j.compeleceng.2022.108242
20. Leonardi, L., Lo Bello, L., Patti, G.: MRT-LoRa: a multi-hop real-time communication protocol for industrial IoT applications over LoRa networks. Comput. Commun. **199**, 72–86 (2023). https://doi.org/10.1016/j.comcom.2022.12.013
21. Alahmadi, H., Bouabdallah, F., Al-Dubai, A.: A novel time-slotted LoRa MAC protocol for scalable IoT networks. Futur. Gener. Comput. Syst. **134**, 287–302 (2022). https://doi.org/10.1016/j.future.2022.04.003
22. Nakamura, K., et al.: A LoRa-based protocol for connecting IoT edge computing nodes to provide small-data-based services. Digit. Commun. Netw. **8**, 257–266 (2022). https://doi.org/10.1016/j.dcan.2021.08.007

Degradation of Synthetic Oils: Physicochemical Viscosity Tests

José Vicente Manopanta-Aigaje(✉), Fausto Neptali Oyasa-Sepa, Oswaldo Leonel Caiza-Caiza, and Marcela Liliana Herrera-Mueses

Instituto Universitario ISMAC (ISMAC University Institute), Quito, Ecuador
vmanopanta@tecnologicoismac.edu.ec

Abstract. The oils used in compression-ignition engines involving the running time of the automobiles under certain mileages within the preventive and corrective maintenance, were established as the object of this study, being analyzed the 15w 40 CI synthetic base oil to validate the degradation level of a lubricant in compression-ignition engines. The sample was taken in containers with capacity of 120 ml. International standards were considered the procedures: kinematic viscosity testing, ASTM D 445 Standard; total alkalinity and acidity value, ASTM D 4739/ASTM D 664 Standards; infrared spectrum in used oils, NVE 751 Standard: oxidation, water, sulfation, nitration, fuel, soot; metal content by spectrometry, wear metals, contamination metals and additive metals, ASTM D 6595 Standard; oil spot test. For this study, literary analysis was used based on a comprehensive review of scientific papers, academic impact projects and reports that will make the readers have a clear view of the physical and mechanical characteristics of the lubricant. In the tribology analysis in engine oils, establishing the wear due to the movement of internal motor components, the degradation of a lubricant should be lower to maintain a better performance in a combustion engine, thus establishing a physicochemical test to determine at different mileages its durability with respect to friction and wear in a given sample of the lubricant.

Keywords: Lubrication · Degradation · Viscosity · Tests · Oxidation · Standard

1 Introduction

The lubricant used for engines contains various chemical compounds such as heavy metals (e.g., chromium, cadmium, arsenic, lead, among others), polycyclic aromatic hydrocarbons, benzene and sometimes chlorinated solvents, PCBs, etc. [1]. Some researchers have analyzed the combustion and emission of recycled engine oil-diesel blends. The recycled oil was prepared in two stages. First, waste engine lubricant was treated and blend with diesel in different proportions. Properties such as flash point, kinematic viscosity, calorific value, cetane number, cloud point, pour point and density were measured according to (ASTM) Standards. The blends were fed into a diesel engine and the tests indicate an increase in brake thermal efficiency, gas temperature compared to that of diesel. A decrease in brake specific fuel consumption and NOx emissions in the exhaust

M. Z. Vizuete et al. (Eds.): CI3 2023, LNNS 1040, pp. 83–98, 2024.
https://doi.org/10.1007/978-3-031-63434-5_7

gas combustion phase is determined [2]. In the local market, the oil used for category N3 engines belongs to the class called semi-synthetic, which depending on the parameters and quality can extend its change period, depending on the package of additives added, which can be antioxidant, detergent, diluent or anti-corrosive [3]. The analysis and monitoring of engine oil provide greater reliability about the real condition of the engine and prevents unexpected corrective maintenance. In diesel and gasoline internal combustion engines, where fuel is burned, lubrication is extremely difficult due to the additional and more demanding phenomena that must be faced: high temperatures, combustion products and residues that can contaminate the lubricant, high stresses, among others [4]. Using the standards of the Ecuadorian standardization service (INEN, D445, ASTM D5185, D2896, E2412), we will verify the correct change interval of the semi-synthetic oil used in the wear, viscosity, degradation and total base number tests [5]. Sampling is the most critical aspect of oil analysis. If a representative sample is not obtained, all subsequent oil analysis efforts will be nullified. The main objectives for obtaining a representative sample are: 1. To maximize information density. To obtain as much information as possible per milliliter of oil. 2. To minimize information distortion. The concentration of information should be represented in tables of results, and graphs at different mileage where the metals present in the lubricant are visualized. It is important that the sample is not contaminated during the analysis procedure in the tests. Limits in oil analysis are sometimes referred to as alarms, which are devices created to assist in the interpretation of reports [6]. During the analysis and verification of the physicochemical degradation tests of an engine lubricating oil by means of results obtained from the sampling used in compression-ignition engines, kinematic viscosity and infrared analysis were determined establishing parameters by mileage in oxidation, nitration and sulfation.

2 Theoretical Framework

2.1 Lubricating Oils

Lubricating oils are constituted by a base, which provides the primary lubrication characteristics; the base can be mineral, synthetic or vegetable. Viscosity is a factor that is affected by temperature. It is important to consider the operating temperatures to which the oil will be subjected [30]. Engine lubricants are a combination of paraffinic, naphthenic and aromatic hydrocarbons obtained by distillation of crude oil (mineral oils) or by synthesis from petrochemical products (synthetic oils). The variation in the proportion of the different types of hydrocarbons in the mixture determines the physical and chemical characteristics of the oils. A high proportion of paraffinic hydrocarbons gives the oil a higher resistance to oxidation, while a high content of aromatic hydrocarbons favors thermal stability [31].

2.2 Types of Lubricating Oils

Mineral Oil. Base oil or lubricant base is one of the products derived from the distillation of crude oil. During petroleum refining, lubricant bases are produced, which must strictly comply with the viscosity range that characterizes them [32]. Mineral oils are derived from the distillation of petroleum and, therefore, their origin is 100% natural. Mineral base oils are made up of three types of compounds: paraffinic, naphthalenic and aromatic, the first ones being the those found in greater proportion (60 to 70%), because they have the best lubricating properties, but there are naphthalenic and aromatic compounds that provide properties that paraffins do not have, such as good behavior at low temperatures and solvent power, among others [17].

Semi-synthetic Oil. They are the result of mixing or combining minerals and synthetics. No more than 30% of synthetic compounds and the remaining 30% of mineral. Thanks to this combination, excellent advantages of both are obtained, since they are more economical than synthetic compounds.

The superior characteristics of a semi-synthetic oil are:

- Higher viscosity index (withstands extreme temperatures better).
- It extends oil change intervals, because it withstands oxidation better than synthetic oils.
- Semi-synthetic lubricant improves lubrication with the help of another synthetic lubricant.
- Semi-synthetic oil is more environmentally friendly.
- Certain benefits provided by synthetic oils are obtained without having to invest in the oil [18].

Synthetic Oil. It is a highly refined lubricant to prevent premature engine wear, which has outstanding flow characteristics at low temperatures. As a result, the components are lubricated avoiding frictional wear, being very effective in cold starting where a significant amount of wear can occur in the moving parts of the engine [28].

Viscosity stability. A higher viscosity index is synonymous with a more stable viscosity over a wider temperature range produced in the engine.

Higher thermal and oxidation stability.- The higher thermal and oxidation stability of synthetic lubricants results in a lower increase in viscosity over time with respect to engine operation, thus prolonging the respective change of lubricating oil by working hours or mileage [29].

2.3 Use of Lubricating Oils

The lubricating oils inside the internal combustion engine of a diesel vehicle have an exclusive function with the care of the internal mobile mechanisms inside the automobile. This is how the fishing industry in the last years has prioritized to use an ideal lubricant for its diesel engines since the high cost of repairs in the short term has been a disadvantage in the last years is. For that reason, it should be clarified that the corrective maintenance of an engine is inevitable in the short or long term, but in the automotive industry it is important that a lubricating oil with excellent characteristics prolongs the useful life of the engine for a longer period of time. The study of the viscosity index at different temperatures is essential to analyze it by means of parameters with respect to its additives [9].

2.4 Synthetic Oil for Diesel Engines

A synthetic lubricant has base oils that are highly refined in a laboratory more than those used in usual mineral engine oils, obtaining better protection and performance features in the moving and fixed parts of the engine. Synthetic lubricants are manufactured with more advanced refining processes and therefore have a special chemical treatment and have a higher purity and quality as opposed to a mineral oil. This not only removes more impurities from the crude oil, but also allows the individual molecules in the lubricant to be modified to match the demands of today's automobiles [29]. When the engine is started, the mineral lubricant takes considerable time to circulate throughout the engine system, resulting in frictional wear between moving engine parts. On the other hand, synthetic oil begins to circulate quickly, protecting every moving part within the engine. Synthetic oils can also significantly increase fuel economy. During the warm-up period of a normal truck run, mineral oils have the disadvantage of being thicker and circulating slowly throughout the engine's internal mechanism, causing wear, making the engine more fuel-intensive and less efficient. In contrast, synthetic lubricants start working faster and the engine reaches its maximum operating efficiency effectively prolonging its service life [28].

3 Method

The research paper is presented with the literature review methodology, whose purpose is to consult several authors to discuss conclusions and results. Literature review means to discover, consult and obtain references, and other materials useful for research, from which the inquiry, plus the collection of important and necessary information to pose the research problem (Hernandez et al., 2014). On the one hand, it is of the documentary type since researchers perform a second-hand information search when looking for and selecting information that is already documented: recorded, compiled and classified.

This research was conducted in a review of several articles, books, theses, projects, scientific journals and verifiable sources that ensure the credibility of the concepts and analyses presented.

3.1 Viscosity

As an example of viscosity in a diesel road transport vehicle, a quantitative and qualitative analysis of the lubricant is considered necessary, the same that complies with the operating conditions suitable for a diesel engine, with respect to its durability under progressive operating conditions. Laboratory tests in relation to the study of the lubricant is a trend of relevant importance since a baseline that represents an interpretation of the engine wear can be determined through statistical limits and if the lubricant meets the ideal characteristics through the frequency of use of the same engine, so that, through a chemical data report, it can be established by means of indicators and color codes whether the oil has a favorable engine characterization [8].

Taking as a reference a heavy equipment, it is deduced that there is one of the most relevant advantages since the lubricant allows to establish a wide durability and reduces the corrective maintenance postponing it to a longer time, therefore, the study of the ideal lubricant is reflected in the preventive maintenances having to be carried out in the established period after the pertinent oil change, and an accompaniment of the viscosity study in a specialized chemical laboratory establishing parameters of temperature, mileage, and elements that produce the oxidation phase in the moving and fixed parts of the engine, being the ideal indicator for warning tests, and establishing limit results according to the hours of operation if it is construction equipment such as backhoes, caterpillars and mechanical shovels [9] (Table 1).

Table 1. Lubricants according to the equipment in operation [9]

Engine Manufacturer			
Oil Analysis	Caterpillar	Cummins	Detroit Diesel
	All models	All models	All models
Iron	100 ppm.	84 ppm.	150 ppm.
Copper	45 ppm.	20 ppm.	90 ppm.
Lead	100 ppm.	100 ppm.	–
Aluminum	15 ppm.	15 ppm.	–
Chromium	15 ppm.	15 ppm.	–
Spectroscopy	20 ppm.	20 ppm.	–
Sodium	40 ppm.	20 ppm.	50 ppm.
Boron	20 ppm.	25 ppm.	20 ppm.
Silicon	10 ppm.	15 ppm.	None specified
Viscosity	+20% to −10% of SAE nominal grade	±1 SAE grade or 4 cSt of new oil (Visc. at 100 °C)	+40% to −15% of the nominal value (Visc. at 40 °C)
Water	0.25% max.	0.2% max.	0.3% max.
TBN	1.0 mg KOH/g min. Estimate	2.0 mg KOH/g min. or half the new oil or equivalent to TAN	1.0 mg KOH/g min. Estimate
Fuel dilution	5% max.	5% max.	2.5% max.
Coolant dilution	0.1% max.	0.1% max.	0.1% max.
Ferrography	In exceptions	In exceptions	In exceptions

One of the quick test alternatives to determine the degradation and wear of a diesel lubricating oil can be achieved by means of an oil spot test, establishing comparative image parameters to determine the continuous wear with respect to the oil. This basic procedure comes from an analysis with a suitable diagnostic tool to determine the scope of ideal lubrication. These images determine the wear that this oil has through color spectra. This verification can be determined as a quick test to observe the oil condition after a certain time of use. Additionally, CNG engines show a higher demand due to their high thermal and mechanical demand, exposing a high degradation after a certain time of use [10] (Table 2).

Table 2. Database of Maxter Multigrade CI-4 SAE 15W40 oil [10]

Name	TBN (mgKOH)	Viscosity AT 100 °C (cSt)	Soot FT-IR (abs/1 mm)	DI	CI	WD
M01-A	9.06	14.07	0.74	94.5168	2.3854	13.0794
M02-A	9.08	12.35	0.63	90.6937	2.3664	22.0229
M03-A	9.88	13.42	0.85	94.7259	2.3766	12.5346
M04-A	6.25	13.34	0.58	93.0169	2.3832	16.6424
M05-A	9.63	14.01	0.63	88.9408	2.3585	26.0832
M06-A	8.81	13.68	0.84	95.3611	2.3826	11.0525
M07-A	6.94	13.33	0.91	93.1826	2.3687	16.1481
M08-A	8.68	13.91	1.01	95.7843	2.4104	10.1615

3.2 Engine Category

Machinery manufactured by Caterpillar shown in a data sheet according to its engine characteristics is considered [13] (Table 3).

Table 3. Main characteristics of the Caterpillar 2014 excavator engine [13]

Engine model	Cat•C6.6 ACERT™
Lubrication:	Circulating pressure
Gross power: SAE J1995	111 kW
Piston diameter	105 mm
Stroke	127 mm
Displacement	6.6
Pistons:	Inline-six
Compression Ratio:	16.2: 1

3.3 Measurement of Wear Elements in the Lubricant

For the collection of the samples, initially the cleaning of the oil intake port that is right on the oil filter was performed as in Fig. 1, then the sample was taken in a transparent plastic container of 120 ml that has an airtight lid. The label has information on the customer, unit, unit code, date of sampling, type of oil, oil hours, unit hours, so that the recommendations given regarding the condition of the oil are as accurate as possible.

In addition, the volume to be collected in each sample is 100 ml of oil and an oil sampling kit was used for this purpose. It was established to take the oil sample at intervals of 250 working hours, which corresponds to the preventive maintenance time suggested by the manufacturer [11].

Fig. 1. Oil sampling points in the engine [12]

4 Analysis of Results

4.1 SAE 15W-40 Oil Properties

Nomenclature of SAE 15W-40 oil, represented by the letter W, meaning Winter, referring to the degree of cold viscosity in temperature, the second numbering represents the lubricant viscosity in hot temperature. The higher the percentage of viscosity, the greater the protection to the mechanical parts. It is also emphasized that an excess of viscosity can cause internal friction and decrease the performance of the vehicle, which means a deterioration of the engine. In reference to the characteristics of the manufacturer of (2020) 15W40 Bardahl oil, it is understood that it is a multigrade oil, designed for the lubrication of 4-stroke turbocharged and naturally aspirated diesel engines. Therefore, these vehicles require a specific lubrication according to the type of engine. As an outstanding point about the 15W40 oil, it fulfills the function of protecting the engine from wear and corrosion to which it is exposed according to the activity it regularly performs, thus allowing the lubricant to adapt to the temperature diversity according to the climate to which the engine is exposed [27].

4.2 Infrared Oil Analysis

Over time, diesel engine oil degrades due to exposure to high temperatures and contact with combustion products. Infrared spectroscopy can detect the presence of degradation products, such as acids, resins and varnishes, which can affect the viscosity and lubricating properties of the oil. Early detection of oil degradation allows timely changes to be made and potential engine downtime to be avoided. Infrared analysis of diesel engine oil is a technique used to determine the chemical composition and quality of diesel engine lubricant by using infrared spectroscopy. This technique makes it possible to identify and quantify the different molecular components present in the oil, as well as to detect the presence of contaminants or degradation. Among those are:

- Measurement of elements in oil: Cu, Fe and Cr.
- Measurement of elements in oil: Al, Pb and Sn.
- Measurement of elements in oil: Si, Na and K.
- Measurement of elements in oil: Mo and Ni.

The measurement of elements such as aluminum Cu, Fe and Cr, Al, Pb and Sn, Si, Na and K, Mo and Ni in diesel engine oil is commonly performed through Optical Emission Spectroscopy (ICP-OES) or Atomic Absorption Spectroscopy (AA) analysis techniques. These techniques experimentally obtain the manifestation of metals in the engine lubricant and are widely used in the analysis of oils and fluids. Figure 2 shows the evolution of the acidity and alkaline reserve depletion measurements in mg KOH/g of the samples analyzed for diesel engines, respectively. It is indicated in the first instance that the intersection of the trend curves of the TAN and TBN measurements show the oil change intervals for each of the lubricating oils. As we can see, type II oil shows a clear reduction of the established period by approximately 50% (15,000 km).

Reaching the oil change intervals established by the manufacturer (30,000 km) using this type could pose serious risks of engine failure. The better performance of type I compared to type II can be attributed to its additive packages and probably to its lubricant base. It should be noted that type I has higher levels of TBN than type II, thus neutralizing the high levels of acidification of the oil at the end of the oil change interval. Diesel engines undoubtedly show a lower demand from the point of view of oil acidity, being not too much affected by this variable. Their alkaline reserve reaches without any problem the periods that have been established for these engines and can reach higher levels if we extrapolate the data obtained [26] (Table 4).

Table 4. Comparative Table

Characteristics	Type I	Type II	Type III
SAE Grade	10W40	15W40	15W40
Density at 15 °C (kg/m^3)	865	865	881
Viscosity at 40 °C (cSt)	91.8	112.0	108.0
Viscosity at 100 °C (cSt)	14.3	14.5	14.5
Viscosity Index	160	125 min.	130 min.
T.B.N. (mg KOH/g)	13.2	7.0	10
Aminic additives (Abs cm^{-1}/0.1 mm)	17,991*	12,978*	1,275*

(continued)

Table 4. (*continued*)

Characteristics	Type I	Type II	Type III
Antiwear additives (Abs cm^{-1}/0.1 mm)	8,048*	10,903*	12,950*
Flash point, open cup (°C)	>220	215	215 min.
Pour point (°C)	<−33	−27	−27 max.
Specifications	IVECO 18-1809	API CF-4	ACEA E7/E5, API CI – 4/CH-4/SL

Table 4: Main characteristics of the lubricating oils
(*) These results correspond to measurements carried out in the laboratory using the FT-IR technique.

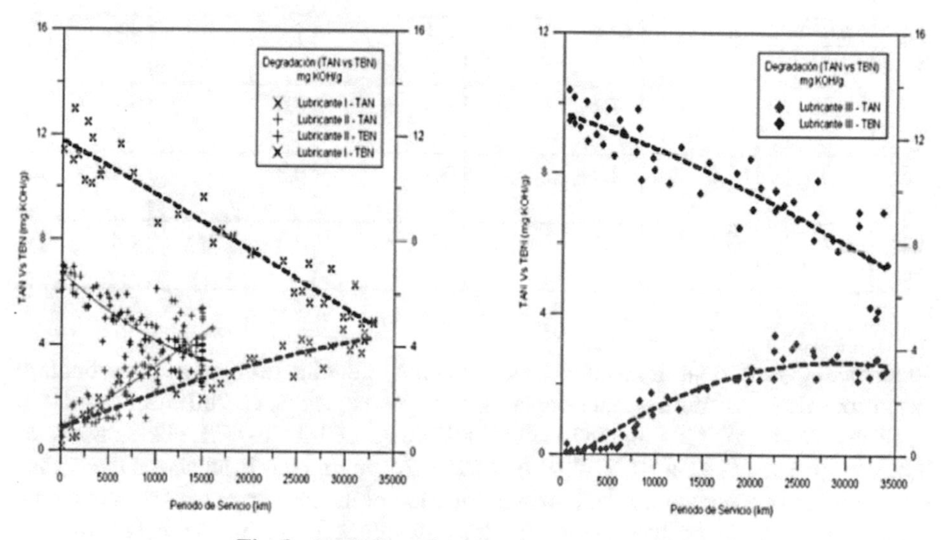

Fig. 2. Evolution of acidity measurements [26].

5 Discussion

5.1 Interpretation of the Wear of New 15W40 Oil

It can be seen in Table 5 that the proportion of polymer additives was similar in almost all the lubricants in the tests, and they are between 7% by weight and 8% by weight. Some irregularities were found for engine lubricants C and H, where the amount of polymer additives was above or below this range, proportionally. With reference to the absolute values of the polymeric additives, they can be somewhat overestimated due to the mechanism of size-exclusion chromatography [19].

Kinematic viscosity is relatively unaffected in all long-range automotive lubricant tests, gasoline cars identify a minor increase, diesel cars a minor decrease. Although the changes are not relevant, at least in this mileage range, it is suggested to involve these

Table 5. Chromatographic segmentation of SAE 15W- 40 lubricants [19]

	Polymer additives in oil (wt%)	Composition of base oil + low MW additives (wt%)				
		Saturates	Monoaromatics	Diaromatics	Polyaromatics	Polar compounds
Oil A	8.0	72.5	17.7	2.7	1.6	5.5
Oil B	8.2	71.4	15.2	2.7	2.9	7.8
Oil C	10.4	69.6	16.9	3.3	3.2	6.9
Oil D	8.1	67.3	18.3	4.2	3.4	6.8
Oil E	7.2	83.7	10.9	1.7	1.6	2.1
Oil F	7.4	84.7	7.2	1.4	0.6	6.1
Oil G	8.0	93.6	2.9	0.5	0.3	2.8
Oil H	6.4	73.1	16.4	3.3	2.3	4.8

fundamentally different trends in viscosity change in engine development, particularly when considering modern extended operating ranges exceeding 20,000 km covered [20].

The 15W-40 API CI-4 and M7ADS V 15W-40 API CI-4 CH-4/SL oils show similar lubrication. The M7ADS III 15W-40 API CF-4/SG engine lubricant shows the highest surface wear, in other words, the lowest lubrication of the new engine lubricants seen in the tests. Correlation analysis of the experimental values found that the fuel content that penetrated the lubricants correlates negatively with viscosity ($R = -0.87$). Low water contamination in the engine oil does not cause a revealing negative effect on lubrication. A significant correlation was confirmed between the oxidation, nitration and sulfation products of chemical degradation of the lubricants used in the tests ($R \geq 0.90$). These degradation products improve lubrication due to their polarity, i.e., they caused better lubrication of worn lubricants compared to new engine oils [21].

In the case of tests of used 15W-40 engine oils, those are shown in samples no. 2-2, 2-3, 2-4 (see Fig. 3 and Table 6) with 12,335 km, 25,888 km and 25,900 km covered. The concentration of the ZDDP additive was inspected at the level of 74%, 67% and 59%, i.e., it did not exceed the limit amount of 30%. Based on the identification of the fuel in the oil in relation to the viscosity of the new engine lubricant [21].

The change in KV at 100 °C was less than 15.96%. The oxidative-onset temperature (OOT) of the lubricants decreased gradually with the working mileage. All OOT values of the worn lubricants are at 210 °C effectively. A general test indicated that the used engine

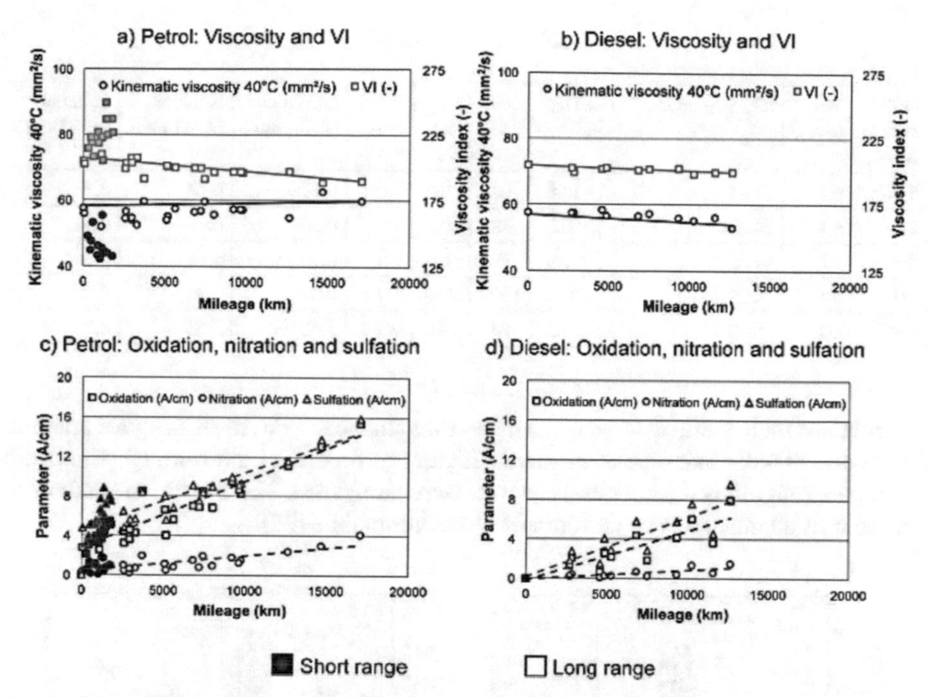

Fig. 3. Kinematic viscosity and viscosity index (VI) as a function of mileage. a) Gasoline cars, b) diesel cars. Oxidation, nitration and sulfation according to mileage. c) Gasoline cars, d) diesel cars. The lines shown only facilitate the appreciation of the results with respect to the tests carried out [20].

Table 6. Engine oil test assessment [21]

Oil sample No.	WS (mm^2)	$KV_{100°C}$ ($mm^2.S^{-1}$)	ZDDP (%)	Soot (%T)	Fuel (wt. %)	Water (wt. %)	Oxidation (A/0.1 mm)	Nitration (A/0.1 mm)	Sulfation (A/0.1 mm)
1-1	9.0	14.62	100.0	100	0.0	0.0	0.00	0.00	0.00
1-2	7.9	12.15	69.3	72	4.4	0.0	0.17	0.08	0.16
1-3	8.3	9.23	69.3	72	20.0	0.0	0.17	0.08	0.16
1-4	8.8	7.47	69.3	72	30.0	0.0	0.17	0.08	0.16
1-5	6.3	14.75	63.4	54	2.7	0.2	0.05	0.08	0.14
2-1	5.9	14.29	100.0	100	0.0	0.0	0.00	0.00	0.00
2-2	7.6	12.94	74.0	81	17.0	0.0	0.21	0.12	0.19
2-3	6.3	13.30	67.0	78	0.0	0.1	0.07	0.10	0.14
2-4	7.9	15.48	59.0	79	7.0	0.0	0.07	0.09	0.17
3-1	5.8	14.14	100.0	100	0.0	0.0	0.00	0.00	0.00
3-2	5.9	14.07	11.5	93	0.0	0.0	0.15	0.23	0.26

(*continued*)

Table 6. (*continued*)

Oil sample No.	WS (mm^2)	$KV_{100°C}$ ($mm^2.S^{-1}$)	ZDDP (%)	Soot (%T)	Fuel (wt. %)	Water (wt. %)	Oxidation (A/0.1 mm)	Nitration (A/0.1 mm)	Sulfation (A/0.1 mm)
3-3	6.1	14.07	13.8	93	0.0	0.0	0.14	0.22	0.26
3-4	6.3	14.68	5.0	88	0.0	0.0	0.26	0.40	0.38
3-5	5.8	14.65	6.0	88	0.5	0.0	0.26	0.40	0.38
4-1	3.5	14.37	100.0	100	0.0	0.0	0.00	0.00	0.00
4-2	2.1	15.58	23.2	75	0.0	0.2	0.42	0.62	0.57

oils retained their useful detergent and dispersant characteristics in an adequate amount. The four-ball wear scar diameters and coefficient of friction of the worn lubricants did not increase significantly after the road tests were completed. These tests are a reference for the next oil change to be performed on the automobile [22].

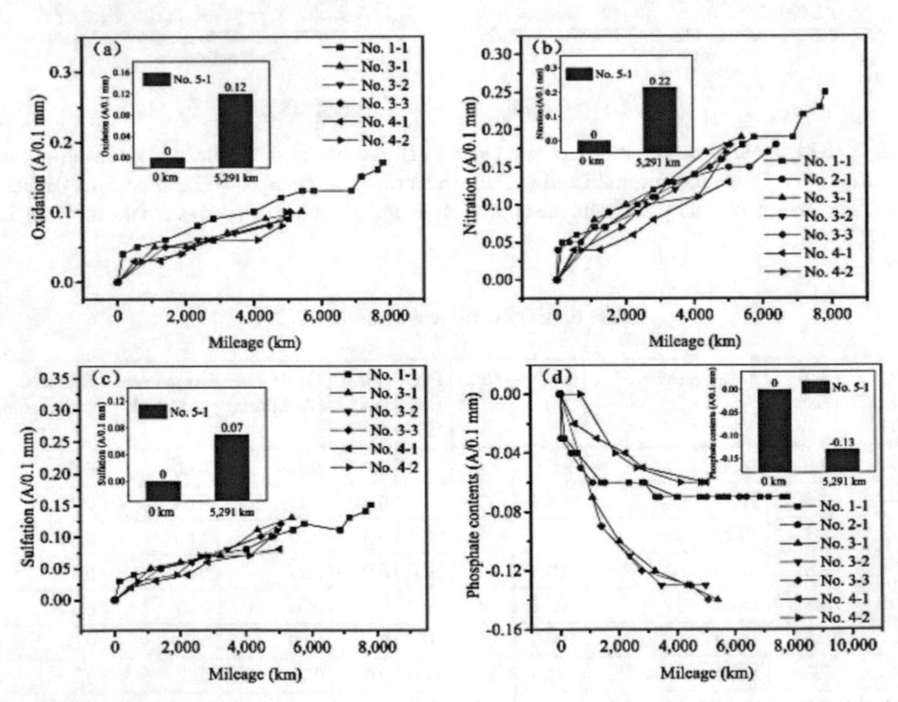

Fig. 4. Existing component test parameters of engine lubricant: (a) oxidation value, (b) nitration value, (c) sulfation value and (d) phosphate content [22].

The results of the tested lubricant components tend to be assessed by means of Fig. 4. As Figs. 4(a)–4(c) show, all oxidation, nitration and sulfation values of the tested engine lubricants increased with the operating mileage, and phosphate content decreased with an amplification of the tests (Fig. 4(d)). The oxidation, nitration and sulfation values increased rapidly, while the phosphate content declined rapidly during the primary test period. This was caused by the mix of residual lubricant and dissolution of residues of the previous lubricant. The phosphate content changed during the parallel tests with the same car model (1-1 and 2-1), the repeated test with mineral oil (3-1, 3-2 and 3-3) and the repeated test with synthetic lubricant (4-1 and 4-2). These had excellent prolongation, indicating that the consumption of antiwear components over the course of the road tests was well repeated. The experimental car 5-1 was brand new and was started with 5,291 km. The lubricant component change characteristics of car 5-1 were almost identical to those of the other road tests. The total oxidation, nitration and sulfation values of the worn lubricants were below the warning limit (1.0 A/0.1 mm) reported in Ref. [23]. Briefly, it can be stated that the authors focused on the study of the most common form of degradation of engine lubricating oil. Oxidation occurs under mild conditions by the gradual weakening of antioxidants. Concurrently with the oxidation products, nitrates are formed. Nitration originates in the crankcase due to the occurrence of combustion gases from the explosion time in the engine. Due to the mixing of oil with air and combustion gases at high temperatures, ideal conditions are created in the crankcase for oxidation and nitration. On the same principle [21, 22], the formation of sulfates is argued, in experimental words, the formation of SOx, reactions of combustion gases with lubricant. Soot also contributes to increase the viscosity of lubricants in a diesel engine. Soot generated in the engine can cause consistent sludge, high oil viscosity or gelling of the lubricant itself [24]. Contamination of these compounds in the lubricant with soot did not cause a significant increase in viscosity due to dispersants that dispersed them in isolation [25].

5.2 Mathematical Models - Tribology "Wear Coefficient"

The normalization method represents an important contribution to the use of techniques in wear evaluation and failure diagnosis, since it makes the analysis independent of engine type and size, lubricant capacity and operating conditions. The normalized concentration for a metallic element in the oil sample Cn is obtained by multiplying the concentration of the wear element by coefficients that adjust this concentration according to the ratios of size (Kt), volume of lubricant in the system (Kv) and chemical composition of the engine components (Km) [14].

The correction of the concentrations measured by the spectrometer with the objective of considering the effect of particle loss in the sample taken was performed through the constant velocity model for systems with leakage and additives, developed by Espinoza [15], and which is summarized in the following equations:

$$Cm(t) = \frac{P}{Z * Vo}\left(C_{mo} - \frac{P}{Z * V_o}\right) * e^{Z.t} \tag{1}$$

where:

$$Z = \frac{Qa}{Vo} \tag{2}$$

With Eq. (1), (P) is determined. In this investigation, the effect of the filter was not considered since the tests were carried out with the engine without oil filter. The corrected concentrations, which represent the number of particles that would exist in the crankcase if there were no leaks, additives or filters at time (t), are calculated through Eq. (3) and (4).

If (t0) is the time of oil change, so it is equal to zero, the above equation changes to this:

$$Cc = Co + \frac{P}{V_o}(t_1 - t_0) \tag{3}$$

The normalization model showed below was developed by Espinoza [15, 16] in 990. The normalized concentration for a metallic element (i) in the oil sample Cn(i) is obtained by multiplying the concentration of the wear element measured in the oil by coefficients that adjust this concentration according to the ratios of size (Kt), volume of lubricant in the system (Kv) and chemical composition of the components of tribology (Km(i)) [15].

$$Cc = Co + \frac{P}{V_o}t \tag{4}$$

$$\mathrm{Cn(i) = Cd(i)\,Kt\,Kv\,Km(i)} \tag{5}$$

$$\mathrm{Kt = Z/Zn} * 1/\lambda^2 \tag{6}$$

$$\mathrm{Kv = V/Vn} \tag{7}$$

$$\mathrm{Km(i) = Yn(i)/Y(i)} \tag{8}$$

In Eq. 6, Z is the number of cylinders, λ is the similarity ratio between the analyzed engine, V is the volume of lubricant in the system. The subscript n refers to the normalized or reference engine, as expressed in Eq. 9.

$$\mathrm{D/D_n} = \lambda \tag{9}$$

6 Conclusion

In the use of lubricants for vehicles, a diversity of analyses has been reported, where it is shown that lubricants, based on the components of which they are a part, can produce wear in the engine, so it is feasible to use a type of lubricant that minimizes wear through proactive maintenance.

Lubricating oils start from a general identification of category with respect to their initial manufacture, whereby mineral, semi-synthetic and synthetic oils are denoted, showing diverse physicochemical analyses in combustion engines depending on the additive that are included in the manufacture.

The diverse qualitative bibliometric analyses by description of physicochemical test methods generalize a study of kinematic viscosity considering the current viscosity trends present in synthetic oils. Additionally, the quantitative results focus on infrared analyses to determine the quantity and chemical composition of the degradation agents by concentration of metallic elements.

The analyses show that if an oil was contaminated, its hydrodynamic lubrication will lose its lubrication, which will cause severe wear in the engine, directly affecting the rings, cylinders, oil pump and in the latter producing foam, which will cause it to break, inducing cavitation. Another cause can be the content of dirt or soot in the lubricant, damaging the crankshaft, camshaft, valves and other parts. Thus, due to the above-mentioned causes, it is highly important that the oil is in perfect condition to promote the good performance of the vehicle.

In a lubricant, the presence of oxidation is observed, which is the main cause of degradation of engine oils and in general of any organic compound, therefore, synthetic oils must have agents and additives that produce a better lubricant that acts effectively in hard conditions and high working temperatures inside the engine to better preserve the moving parts of the engine.

References

1. Hamawand, I., Yusaf, T., Rafat, S.: Recycling of waste engine oils using a new washing agent. Energies **6**(2), 1023–1049 (2013)
2. Prabakaran, B., Zachariah, Z.: Production of fuel from waste engine oil and study of performance and emission characteristics in a diesel engine. Int. J. Chem Tech Res. **9**(5), 474–480 (2016)
3. Caiza, S.D.S., Martínez, S.A.B., Terán, J.C.R.: Análisis Físico-Químico de la vida útil en aceites sintéticos. Dominio de las Ciencias **8**(3), 49 (2022)
4. Saldivia, F.: Aplicación de mantenimiento predictivo. Caso Estudio: analisis de aceite usado en un motorde combustion interna. In: XI Latin American and Caribbean Conference for Engineering and Technology (LACCEI'2013), Mexico (2013)
5. Saldivia, F.: Comportamiento de las propiedades físico-químicas de un aceite lubricante usado en un motor ce combustion interna (2013)
6. Troyer, D., Fitch, J.: Oil analysis basics en español (2004). Gómez García, C. -(2012)
7. Rondón, P.P., Pereira, A.G., Meneces, M.D.L.Á.R., Kruskova, E.R., Águila, M.V.G. (2015)
8. Manjon Patiño, Á.A., Villadiego del Villar, J.D., Salazar Salcedo, S. (2022)
9. Apaza Pastor, J.H.: Prueba de desempeño de motor con el aceite Chevron en la flota de camiones CAT 785D-Mina Shougang Hierro Perú para el cliente San Martín Contratistas Generales-San Juan de Marcona-Ica (2022)
10. Arellano, G., Helguero, M.: Implantación de Análisis de Aceite en Motores de Combustión Interna de Ciclo Diesel. Revista ESPOL 1–6 (2022)
11. Buchelli-Carpio, L., García-Granizo, V.: Detección temprana de fallas en motores de combustión interna a diésel mediante la técnica de análisis de aceite. Revista Ciencia UNEMI **8**(15), 84–85 (2015)
12. Cummins. The Complete Guide to Fluid Analysis (2012)
13. CATEPILLAR: Cat SOS Services. Understanding Your Results (2007)
14. Espinoza, H.: La normalización de las concentraciones de particulasde desgaste en aceite aplicada a motores diésel, vol. 8, no. 2, pp. 1–6. Instituto de Investigación y Desarrollo Anzoátegui, Universidad de Oriente, AVANCES Investigación en Ingeniería (2011)
15. Espinoza, H.: Diagnostico de fallos en motores de encendido por compresión de automoción. Tesis doctoral, Universidad Politécnic a de Valencia, España (1990)

16. Espinoza, H.: Diagnóstico de motores de encendido por compresión mediante análisis de aceite. Universidad Politécnica de Valencia, SPUPV-90-1018, España (1990)
17. García, A.: Diseño, selección y producción de nuevos biolubricantes. Tesis Doctoral Universitat Ramon Llull IQS-Bioenginyeria (2011)
18. Pérez, D.S., et al.: El ciclo de evolución de los lubricantes. ContactoS No. 97, 62 (2015)
19. Cerny, J., Strnad, Z., Sebor, G.: Composition and oxidation stability of SAE 15W–40 engine oils. Tribol. Int. **34**(2), 127–134 (2001). https://doi.org/10.1016/s0301-679x(00)00150-x
20. Agocs, A., et al.: Comprehensive assessment of oil degradation patterns in petrol and diesel engines observed in a field test with passenger cars – conventional oil analysis and fuel dilution. Tribol. Int. **161**, 107079 (2021). Cited 16 times
21. Sejkorová, M., Hurtová, I., Jilek, P., Novák, M., Voltr, O.: Estudio del efecto de la degradación fisicoquímica y la contaminación de los aceites de motor en su lubricidad. Revestimientos **11**(1), 60 (2021)
22. Wei, L., Duan, H., Jin, Y., et al.: Degradación del aceite de motor durante las pruebas de ciclismo urbano. Fricción **9**(5), 1002–1011 (2021)
23. Kral Jr, J., Konecny, B., Kral, J., Madac, K., Fedorko, G, Molnar, V. (2014)
24. Ziółkowska, M.: Wpływ sadzy na proces żelowania oleju silnikowego. Nafta-Gaz **3**, 178–184 (2019)
25. Wolak, A., Zając, G., Fijorek, K., Janocha, P., Matwijczuk, A.: Investigación experimental de los rangos de parámetros de viscosidad: estudio de caso de aceites de motor en el grado de viscosidad seleccionado. Energías **13**, 3152 (2020)
26. Macian-Martinez, V., Tormos-Martinez, B., Gomez, Y., Bermudez-Tamarit, V. (2013)
27. Bardahl (2020). https://www.bardahl.com.mx/ventajas-del-aceite-15w-40/
28. Robles, M., David, K.: Re-refinación de Aceites Lubricantes Usados Mediante Procesos Físico-Químicos (2016)
29. E. M. Corporation: ¿Qué es un lubricante sintético? (2015)
30. Corella Jiménez, F.D.: Caracterización de Aceites por sus Propiedades Ópticas (2016)
31. Ziolo, L.F.B., Gómez, J.R., Prieto-Cadavid, S., Gallo, S.C.: Biorremediación de suelos contaminados con aceites usados de motor. Revista CINTEX **20**, 69–96 (2015)
32. Dubois, R.A.: Introducción a la refinación del petróleo, Eudeba (2005)

CONNECTED Project. Connecting to the Disconnected

Marcelo Zambrano-Vizuete[1(✉)], Juan Minango-Negrete[1], Segundo Toapanta-Toapanta[1], Edgar Maya-Olalla[2], Patricia Nuñez-Panta[3], and Eddye Lino-Matamoros[3]

[1] Instituto Tecnológico Superior Rumiñahui, Av. Atahualpa 1701 and 8 de febrero, 171103 Sangolquí, Ecuador
marcelo.zambrano@ister.edu.ec

[2] Universidad Técnica del Norte, Av. 17 de julio 5-21, 100105 Ibarra, Ecuador

[3] Instituto Espíritu Santo, Av. Juan Tanca Marengo and Av. Las Aguas, 090510 Guayaquil, Ecuador

Abstract. The pandemic caused by COVID19 had a disruptive effect throughout the planet, bringing to light the social, economic, and technological gaps that exist in current societies around the world, including such important aspects as connectivity (internet service) and education. Within the educational field, the closure of schools and colleges, as well as the lack of internet service in rural areas, prevented millions of children from continuing with their formal education. The CONNECTED project uses cutting-edge telematics tools and technologies to develop a communications system that allows delivering internet service to rural areas lacking this service; This, in turn, makes it possible for children and young people residing in these areas to continue with their formal education through tele-education processes.

Keywords: Rural Communications · Teleeducation · School Dropout

1 Introduction

During 2020 and part of 2021, COVID19 caused a third of the world's population to be subjected to confinement and strong mobility restrictions [1]. The first case of this virus (SARS-CoV-2.67 - Covid19) was identified in December 2019, in the city of Wuhan, capital of Hubei province, in the People's Republic of China. The World Health Organization (WHO) recognized it as a pandemic on March 11, 2020; By date, more than 118,000 cases had already been reported in 114 countries, and 4,291 people had lost their lives [2]. As of January 16, 2022, more than 350 million confirmed cases of this disease have been reported, with more than 5 million deaths, in more than 250 countries [3].

In addition to the human losses mentioned above, the measures adopted to combat the virus brought with them a drastic reduction in socioeconomic activities such as employment, tourism and education. One of the most important problems has been

M. Z. Vizuete et al. (Eds.): CI3 2023, LNNS 1040, pp. 99–109, 2024.
https://doi.org/10.1007/978-3-031-63434-5_8

reflected in the education sector: in more than 138 countries, schools and colleges were closed, affecting more than 1.37 billion children and young people who were unable to continue with their basic and secondary academic training [3, 4]. Under these conditions, connectivity and internet service became critical since they made it possible to compensate for the lack of physical contact and distancing using telematic applications and virtuality [5].

In the case of Ecuador, nearly one million children and young people were unable to continue with their formal education due to the difficulties in carrying out effective tele-education processes, mainly due to connectivity problems and the lack of internet service [6, 7]. Approximately, during the 2020–2021 period, more than 150,000 dropouts were registered in the basic and secondary education system of Ecuador [8].

According to figures from the National Institute of Statistics and Censuses (INEC), by April 2021, only 50% of Ecuadorian households, in the urban sector, had internet access and, for the rural sector, the figures worsen, only 40% of households have access to this service. This technological gap has a direct impact on the existing economic and social gaps between the urban and rural sectors [9–11].

To solve the problems, the CONNECTED project is developed, using cutting-edge tools and technologies that make it possible to provide connectivity and internet service to rural areas not covered by conventional providers of this service (dark areas).

It is important to emphasize that the objective of the project has a strictly social nature: to enable children and young people residing in rural areas without internet service to have a formal basic and intermediate level education. This, in turn, enable to improve the quality of life of citizens residing in these areas, reducing the existing socioeconomic gaps between this rural and urban sector.

This document is organized as follows: first, an introduction is made regarding the topics related to the project; second, the methodology used during its development and implementation is described; Third, the tests and results obtained are presented and, finally, fourth, the main conclusions resulting from the project are presented.

2 Methodology

This project is part of applied, qualitative and quantitative research. Applied, since its objective is the practical implementation (prototype) of a technical and technological solution that provides a solution to a real problem. Qualitative, the ultimate goal of the project is to provide connectivity and internet service, yes or no, to rural areas not covered by conventional Internet providers. Quantitative, the quality of Internet service is measured through quantitative variables such as transmission speed and bandwidth.

In general, the methodology used in this project is experimental, that is, after the design and sizing of the solution, the implementation and functionality tests (trial and error) will be carried out. If this is the case, the necessary corrective measures will be made to re-run tests and verify the functionality of the system.

The project is divided into four phases or stages:

1. Literature review and determination of the state of the art. It allowed us to learn about the advances, challenges and current state of the Internet service and tele-education processes in rural areas, as well as the successful models and implementations that

have been developed around this topic. According to this study, the problem was established, determining that in Ecuador, there is a significant lack of this service in rural areas.

2. Design and sizing of the solution. First, the generic architecture of the system was designed, which establishes the requirements, conditions and general structure of the system so that it can provide a solution to the identified problem. Second, the Target Geographic Area that will benefit from the development of this project was determined. Third, the specific architecture was designed taking into consideration the particular characteristics of the geographical area taken as the objective and adapting the generic architecture to these particularities. Finally, the characteristics and quantity of the equipment, devices and electrical and communications elements required for the implementation were determined.
3. Implementation. A prototype of the communications system is implemented in accordance with the area chosen as the objective, the specific architecture and the sizing of the equipment previously carried out.
4. Functionality tests. Since the main objective of the project is to allow the effective execution of tele-education processes within the Target Area, functionality tests were carried out using computer applications such as Microsoft Teams [12] and Zoom [13]. Measurements of bandwidth and data transmission speeds were taken both in the communications backbone and in each of the end users.

2.1 Architecture

The generic architecture of communications system is shown in Fig. 1 and consists of the following elements:

1. Donor, institution that donates a portion of its internet bandwidth for the development of the project.
2. Target Geographic Area, beneficiary area of the project and in which the donated bandwidth will be used.
3. Communications backbone, a set of point-to-point radio links that provides connectivity between the Donor and the Target Geographic Area. It is important to consider that, for licensing reasons, radio links must operate in free bands such as 2.4 GHz or 5 GHz. At these frequencies, the existence of line of sight between the transmitters and receivers of each of them becomes essential. Links deployed for communications backbone can be made up of one or several radio frequency links (hops), everything will depend on the distance between the donor and the Target Geographic Area, as well as the physical obstacles that prevent the line of sight between transmitters and receivers.
4. Base Station, point multipoint telecommunications equipment that allows wireless coverage of the Target Geographic Area. The base station must have connectivity with the last node of the communications backbone installed in the Target Geographic Area.
5. User terminals, telecommunications equipment, installed in the end user locations and that enable communication with the Base Station.
6. Access points, telecommunications equipment that receives the Internet signal from user terminals and distributes it to user locations through an IEEE 802.11 network.

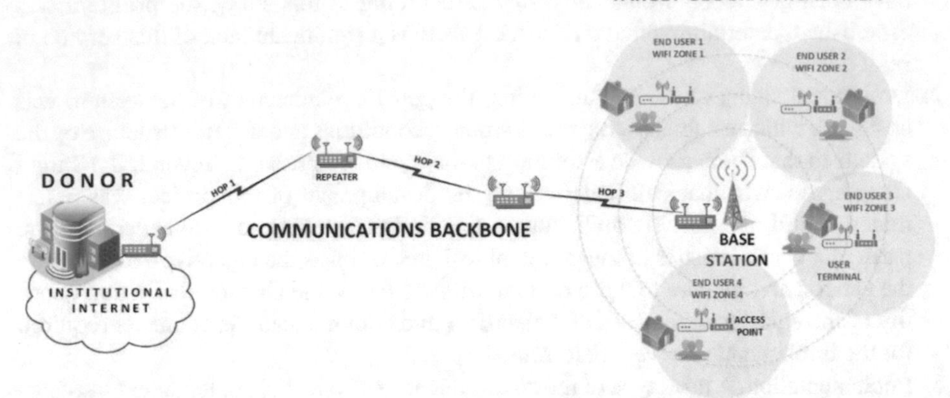

Fig. 1. Generic CONNECTED project architecture.

The San Antonio neighborhood (Ecuador country, Pichincha Province, Rumiñahui canton) was chosen as the Target Geographic Area; a small town located on the slopes of the Pasochoa volcano, at an approximate distance of 15 km from the Rumiñahui Higher Technological Institute (ISTER), which serves as a donor for this project. It was identified that in this area there was a group of approximately 15 people, including children and young people, who had not been able to continue with their formal education during the COVID19 pandemic due to lack of internet service and the impossibility of carrying out teleeducation processes. Figure 2 shows the geographic location of the San Antonio neighborhood and the line of sight study that demonstrates the feasibility of reaching the Target Geographic Area through a single radio link.

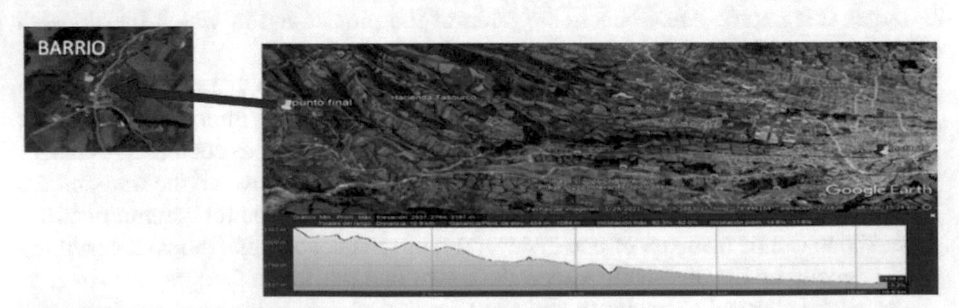

Fig. 2. Line of sight study for the Target Geographic Area: Barrio San Antonio/Pichincha/Rumiñahui.

Once the Target Geographic Area was established, the generic architecture was adapted to the characteristics of that area (geographic, accessibility, electric energy, etc.). Figure 3 shows the specific architecture of the communications system for the San Antonio neighborhood, taking ISTER as a donor, with a bandwidth of 100 Mbps.

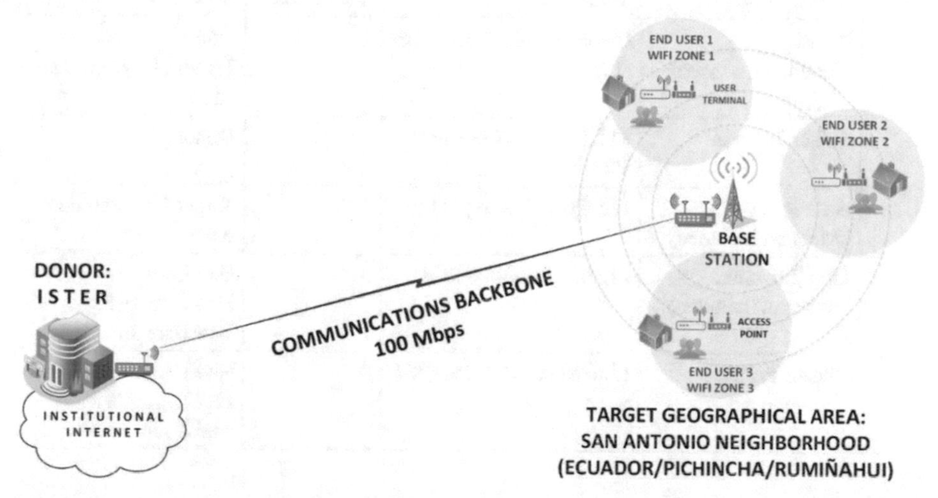

Fig. 3. Specific architecture of the CONNECTED project for the San Antonio neighborhood (Ecuador/Pichincha/Rumiñahui).

2.2 Sizing, Implementation, and Functionality Testing of the System

For the implementation of the system, LigoWave brand equipment [14] was chosen, given its operating frequency characteristics, protection against dust and liquids, traffic management, high tolerance to noise and interference, among others. Although LigoWave equipment can operate under the IEEE 802.11 standard [15, 16], many of the aforementioned features are related to its proprietary layer 2 protocol called Ipoll3 [17, 18].

Ipoll3 makes an improvement to the HCF (Hybrid Coordinate Function) medium access technique, changing the typical control carried out sequentially by the base station, for an intelligent polling carried out individually by each of the access points (APs) subscribed to that station base. Each AP sends a polling frame to the base station and must receive a response before starting to send data. Before polling other APs, the base station must receive data from the AP in question. APs that require less airtime are registered to a low activity or inactivity list, while stations that generate more traffic are registered to an active list. This allows the network to self-regulate so that each wireless link gets the maximum bandwidth with the lowest possible latency.

The equipment used is described in Table 1 below.

Table 1. Equipment used for the implementation of the prototype of the CONNECTED project.

No.	Description	Model	Quantities	Ubication
1	Radio (Antenna included)	LigoPTP 5-23 RapidFire [19]	2	Donor Target Geographical Area
2	Switch layer 2	TPLINK TL-WR840N V2 [20]	1	Donor
3	Base Station (Antenna included)	LigoDLB 5-90n [21]	1	Target Geographical Area
4	User Terminal (Antenna included)	LigoDLB 5-20ac [22]	3	End User site 1 End User site 2 End User site 3
5	Access Point	LigoWave NFT 2ac [23]	3	End User site 1 End User site 2 End User site 3

In addition to the equipment described above, an instance of the LigoWave network management software, called WNMS [24], was implemented. Among the functionalities of this software, can highlight the collection of inventory data, generation of alarms, statistical reports, remote configuration of equipment, among others. In the case of the CONNECTED project, WNMS allowed continuous and real-time monitoring of the radio links and the wireless communications network deployed in the Target Geographic Area.

A prototype of the communications system was implemented, for which ISTER serves as a donor, providing 100 Mbps of its institutional bandwidth for this project. All equipment operates at a frequency of 5 GHz. Figure 4 shows a diagram of the installed equipment.

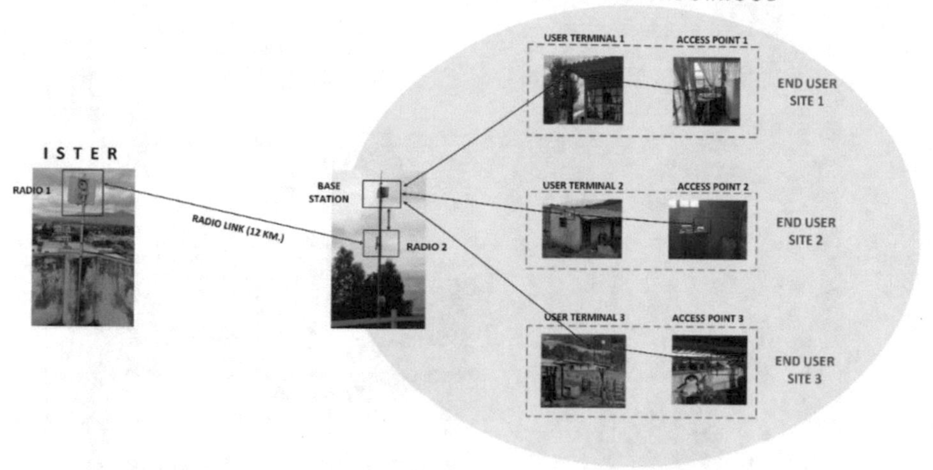

Fig. 4. Prototype of the communications system for the CONNECTED project, San Antonio neighborhood (Ecuador/Pichincha/Rumiñahui).

3 Results and Functionality Test

To verify the functionality of the system, multiple audio and video tests were carried out with 9 recurring users, 4 of them used Microsoft Teams and the remaining 5 using ZOOM. 7 of the end users used their smartphone during the tests and 2 used their laptops. The results obtained were satisfactory in 100% of the cases. No packet losses or significant delays were detected during test communications.

Additionally, 4 transmitters and 4 receivers were chosen at random to carry out text file transfer tests (.docx), each with an approximate size of 10 MB. As in the previous case, the tests were carried out successfully.

Although LigoWave equipment allows it, quality of service (QoS) was not implemented, with the objective of verifying the response of the Ipoll3 protocol to network traffic. It is important to highlight the stability of LigoWave equipment, maintaining minimal variations in transmission speeds and delays in transmissions.

Figure 5 shows the results of the network parameter measurements carried out through the LigoWave WNMS software, both at the donor site and at the Target Geographical Area.

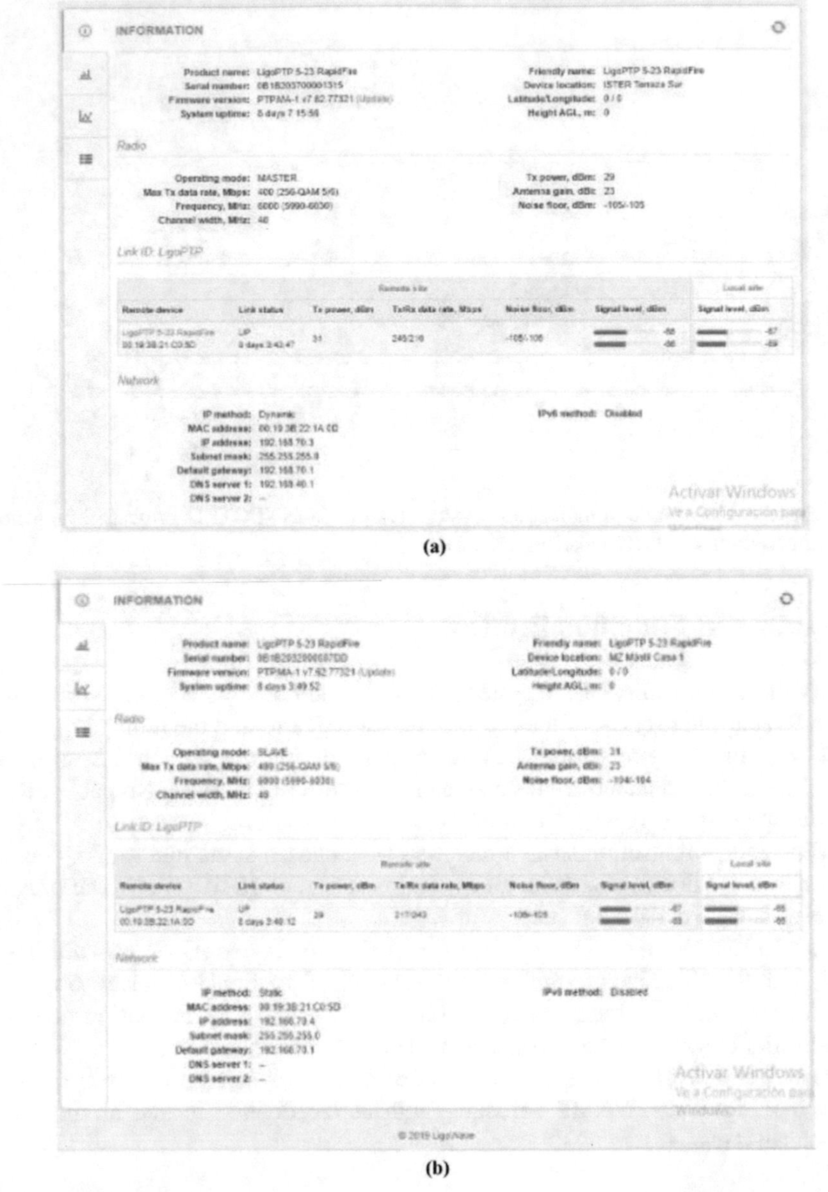

(b)

Fig. 5. **(a)** Network parameters LigoWave 5-23 RapidFire – ISTER. **(b)** Network parameters LigoWave 5-23 RapidFire – Target Geographical Area. **(c)** Network parameters LigoWave 5-90 Base Station – Target Geographical Area. **(d)** Network parameters LigoWave 5-20ac User Terminal – End User site 3.

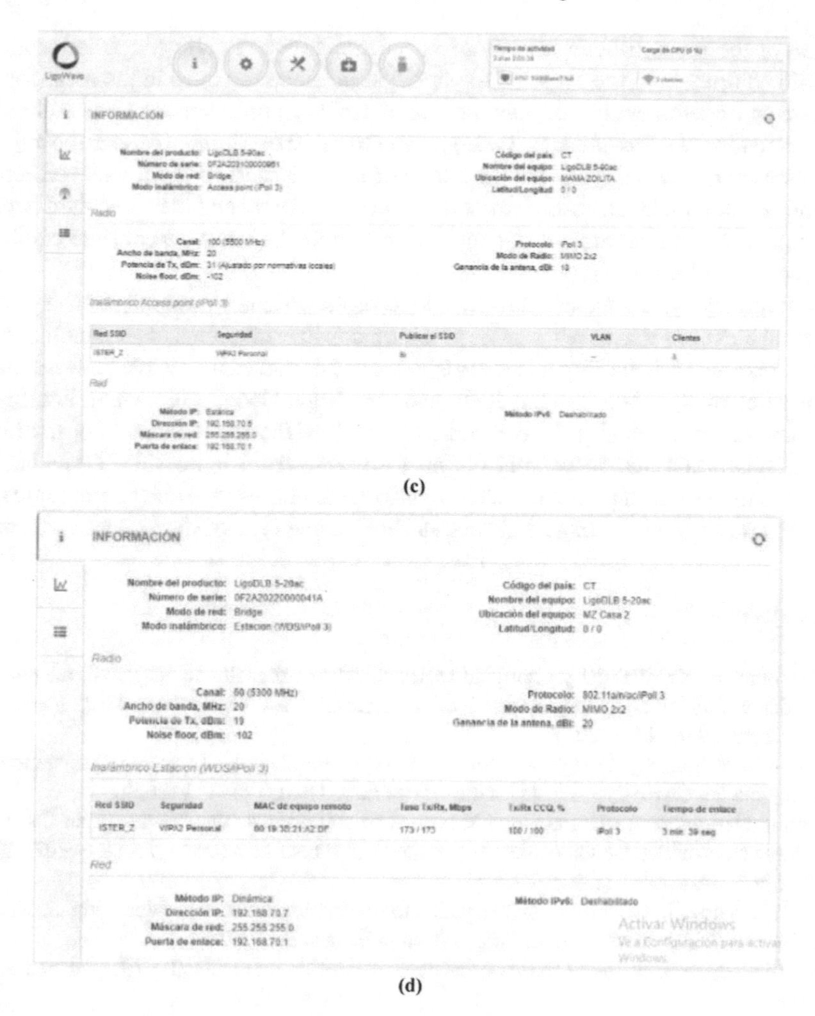

(c)

(d)

Fig. 5. (*continued*)

4 Conclusions

The CONNECTED project has a strictly social character, that is, it does not seek economic but social profitability. In the case of this first prototype of the system, the *Instituto Tecnológico Superior Rumiñahui* (ISTER) functions as a donor of the internet service, assigning the project a bandwidth of 100 Mbps. All functionality tests were successful, corroborating the feasibility of the system for the development of Teleeducation effective processes.

It is important to remember that the CONNECTED project is developed jointly by the ISTER and the *Instituto Espíritu Santo* (TES). In addition, it has the support of the *Corporación Ecuatoriana para el Desarrollo de la Investigación y la Academia* (CEDIA).

One of the most difficult activities to accomplish was establishing the location of the radio equipment in the Target Geographic area, since the trees in the area (forested area) are an impediment to reaching the line of sight required between transmitters and receivers to form the link. At first, it was planned to make two jumps to reach the objective point, however, after carrying out the site visit and the field study, it was verified that with the installation of a 6-m-high mast it was possible to reach the line of sight between the radios with a single jump, this, considering that the Ligowave RapidFire equipment has a range of 30 km.

As future work, it is expected to be able to replicate this communications system in other rural locations that have the same problem, considering, if necessary, multiple hops or radio links to reach the Target Geographic Area (this makes it possible to overcome the lack of a line of sight between the donor and the Target Geographic Area). For this, the use of layer 2 switches must be considered in each of the implemented hops, allowing signal regeneration and preventing the bit error rate from increasing. Likewise, it is planned to implement a portable system solution that allows providing communications and/or internet service to areas that have suffered some type of disaster and are cut off.

References

1. Dalponte, B.: COVID-19 y seguridad regional. Macro-seguritización global, adaptaciones locales y dilemas de la cooperación latinoamericana. URVIO. Revista Latinoamericana de Estudios de Seguridad (2021)
2. Maguiña Vargas, C., Gastelo Acosta, R., Tequen Bernilla, A.: El nuevo Coronavirus y la pandemia del Covid-19. Revista Médica Herediana **31**(2), 125–131 (2020)
3. United Nations News Home Page, Más de 156 millones de estudiantes están fuera de la escuela en América Latina debido al coronavirus. https://news.un.org/es/story/2020/03/1471822,2020. Accessed 25 Sept 2023
4. Sanz, I., Sainz, J.: Organización de Estados Iberoamericanos para la Educación, la Ciencia y la Cultura (OEI), Efectos del Coronavirus en la Educación (2020)
5. Bonilla-Guachamin, J.A.: Las dos caras de la educación en el COVID-19. Ciencia América (2020)
6. El Pais. Ecuador: la educación online desde casa es imposible e injusta. Planeta Futuro (2020)
7. El Universo. Un millón de estudiantes sin acceso a educación virtual durante la emergencia sanitaria (2020)
8. Primicias. La pandemia empujó a 150.0000 estudiantes hacia la deserción escolar (2020)
9. Instituto Nacional de Estadísticas y Censos. Indicadores de tecnologías de la información y comunicación. Gobierno en cifras (2021)
10. CEPAL. Pandemia provoca aumento en los niveles de pobreza sin precedentes en las últimas décadas e impacta fuertemente en la desigualdad y el empleo (2021)
11. Reliefweb. El choque COVID-19 en la pobreza, desigualdad y clases sociales en el Ecuador: Una mirada a los hogares con niñas, niños y adolescentes. OCHA Services (2020)
12. Microsoft, Microsoft Teams (2023). https://www.microsoft.com/es/microsoft-teams/group-chat-software. Accessed 16 June 2023
13. Zoom Home Page (2023). https://zoom.us/es. Accessed 16 June 2023
14. LigoWave Home Page. https://www.ligowave.com/es. Accessed 16 June 2023
15. Bellalta, B.: Delay Analysis of IEEE 802.11 be multi-link operation under finite load. IEEE Wireless Commun. Lett. **12**(4), 595–599 (2023)

16. Zubow, A., Memedi, A., Dressler, F.: Towards hybrid electronic-mechanical beamforming for IEEE 802.11 ad. In: 18th Wireless On-Demand Network Systems and Services Conference (WONS). IEEE (2023)
17. LigoWave Ipoll Protocol Home Page. https://www.ligowave.com/wiki/faq/apc-ipoll-protocol/. Accessed 16 June 2023
18. LigoWave, iPoll™- PTMP Protocol, Technical Paper (2016)
19. LigoWave Rapidfire Home Page. https://www.bhphotovideo.com/c/product/1411609-REG/ligowave_ptp_rapidfire_5_23_5ghz_ptp_mimo_700mbps.html?ul=S&gclid=CjwKCAjwyNSoBhA9EiwA5aYlb9w-ytkL6xqdcVAc-TDWSxMw5E3cYO41sD3I_selnEc3_nAyU4mmWxoCX-cQAvD_BwE. Accessed 21 Aug 2023
20. TPlink Wireless Router TL-WR840 Home Page. https://www.tp-link.com/ec/home-networking/wifi-router/tl-wr840n/. Accessed 21 Aug 2023
21. LigoWave LigoDBL-590n Home Page. https://www.ligowave.com/products/ligodlb-5-90n. Accessed 21 June 2023
22. LigoWave LigoDBL-520ac Home Page. https://www.ligowave.com/es/products-2/ligodlb-5-20-ac-3. Accessed 21 June 2023
23. LigoWave NFT-2AC Home Page. https://www.ligowave.com/es/products-2/nft-2ac-3. Accessed 21 June 2023
24. LigoWave Network Monitoring Software Home Page. https://www.ligowave.com/network-monitoring-software. Accessed 21 June 2023

Performance Analysis of Coherent and Non-coherent Detection Techniques in Chirp Spread Spectrum for Internet of Things Applications

Juan Minango[1(✉)], Marcelo Zambrano[1], Moisés Toapanta[1], Patricia Nuñez[2], and Eddye Lino[2]

[1] Instituto Tecnológico Superior Universitario Rumiñahui, Av. Atahualpa 1701 and 8 de Febrero, 171103 Sangolquí, Ecuador
juancarlos.minango@ister.edu.ec

[2] Instituto Espirítu Santo, Av. Juan Tanca Marengo and Av. Las Aguas, 090510 Guayaquil, Ecuador

Abstract. This paper presents an analysis of Chirp Spread Spectrum (CSS) for Internet of Things (IoT) applications, focusing on the performance of coherent and non-coherent detectors in various channel environments. The study evaluates scenarios with additive white Gaussian noise (AWGN), frequency-selective channels, and Rayleigh fading channels, considering a system with 10 receiving antennas. The results demonstrate that the coherent detector outperforms the non-coherent detector in terms of performance. However, the non-coherent detector offers the advantage of lower complexity. To further improve the performance of non-coherent detection in CSS-based IoT systems, the need for exploring new techniques is emphasized. Future research should aim to bridge the performance gap between coherent and non-coherent detection, considering adaptive signal processing algorithms, advanced filtering techniques, or hybrid detection schemes. By addressing the challenges associated with non-coherent detection, CSS can become a reliable and efficient modulation scheme for low signal-to-noise ratio conditions in IoT applications. This research contributes to the advancement of CSS in IoT, enabling seamless connectivity and data exchange in diverse IoT scenarios.

Keywords: Chirp Spread Spectrum (CSS) · Internet of Things (IoT) · Coherent detection · Non-coherent detection · Performance analysis

1 Introduction

In recent years, the Internet of Things (IoT) has emerged as a transformative technology, connecting an unprecedented number of devices and enabling a wide range of applications that enhance efficiency, convenience, and productivity in

M. Z. Vizuete et al. (Eds.): CI3 2023, LNNS 1040, pp. 110–122, 2024.
https://doi.org/10.1007/978-3-031-63434-5_9

various domains. With the explosive growth of IoT deployments, there is a pressing need for robust and efficient communication protocols that can handle the unique challenges posed by IoT environments. Chirp Spread Spectrum (CSS) has emerged as a promising modulation technique for overcoming some of these challenges, providing a reliable and versatile solution for IoT communication [1].

The concept of Chirp Spread Spectrum dates back to the early 1940s when it was initially used in military applications such as radar systems. However, it is only in recent years that CSS has gained significant attention as a viable modulation technique for IoT applications. Unlike traditional modulation schemes like frequency shift keying (FSK) or amplitude shift keying (ASK), CSS employs a unique approach that allows for increased range, improved robustness, and enhanced coexistence with other wireless systems [2].

CSS utilizes the principles of spread spectrum communication, which involves spreading the signal energy across a wide frequency band. This spreading not only provides inherent resistance to interference but also enables multiple devices to transmit simultaneously without causing significant mutual interference [2,3]. By employing linear frequency modulation, known as chirping, CSS achieves enhanced spectral efficiency and resilience against fading and multipath effects.

The Internet of Things, often referred to as the "network of smart devices," encompasses a vast ecosystem of interconnected sensors, actuators, and embedded systems, all working together to gather and exchange data. From smart homes and industrial automation to healthcare monitoring and environmental sensing, the IoT promises to revolutionize how we interact with technology and the world around us [4]. However, the massive number of IoT devices, their varied communication requirements, and the presence of other wireless systems demand innovative solutions to ensure reliable and efficient data transmission.

This is where CSS shines. By leveraging its unique characteristics, CSS addresses several key challenges encountered in IoT deployments. Its inherent resistance to interference makes it particularly suitable for crowded frequency bands, where numerous devices coexist, such as in urban environments or industrial settings. The low power requirements of CSS enable energy-efficient operation, making it well-suited for battery-powered IoT devices that are often deployed in remote or inaccessible locations [5].

Moreover, CSS enables long-range communication, allowing IoT devices to transmit data reliably over extended distances. This capability is especially valuable in applications such as environmental monitoring, precision agriculture, and asset tracking, where seamless connectivity across large areas is essential.

The purpose of this paper is to explore the applications and advantages of Chirp Spread Spectrum in the Internet of Things domain, with a specific focus on the comparison between coherent and non-coherent detection in CSS. We will investigate the performance of these detection schemes in two different environments: one with only noise and another with additive white Gaussian noise (AWGN) and Rayleigh fading.

The structure of the rest of the paper is as follows. Section 2 provides an in-depth explanation of Chirp Spread Spectrum modulation, including its principles, signal processing techniques, and advantages over traditional modulation schemes. In Sect. 3, the concept of coherent and non-coherent detection in CSS is discussed focusing on their benefits and limitations. Section 4 analyzes the performance comparison in terms of bit error rate (BER) between coherent and non-coherent detectors in the presence of additive white Gaussian noise (AWGN) and Rayleigh fading. Finally, the conclusions are presented in Sect. 5.

2 Chirp Spread Spectrum Concepts

From a mathematical standpoint, Chirp Spread Spectrum (CSS) is a modulation technique that utilizes linear frequency modulation to spread the signal energy across a wide frequency band [5]. This spreading technique provides inherent resistance to interference and enables multiple devices to transmit simultaneously without significant mutual interference [6]. In this introduction, we will explore the mathematical concepts underlying CSS and its advantages in the context of the Internet of Things (IoT).

CSS modulation is based on the principle of linear frequency modulation, also known as chirping. It involves sweeping the frequency of the transmitted signal linearly over time, resulting in a chirp waveform. Mathematically, the chirp waveform can be described as:

$$s(t) = \cos\left(2\pi\left(f_c t + \frac{K}{2}t^2\right)\right) \tag{1}$$

where $s(t)$ represents the chirp signal as a function of time t, f_c is the carrier frequency, and K represents the chirp rate. The chirp rate, defined as the change in frequency over time, determines the spread of the signal across the frequency band. A larger chirp rate results in a wider spread, providing increased resistance to interference.

CSS offers several advantages in IoT applications. Firstly, its resistance to interference allows for reliable communication in crowded frequency bands where multiple devices coexist [7]. By spreading the signal energy over a wide frequency band, CSS reduces the impact of narrowband interference sources, enabling robust communication in challenging environments. Additionally, CSS enables long-range communication, making it ideal for IoT deployments requiring connectivity over extended distances. The ability to transmit data reliably over larger distances enhances the coverage and scalability of IoT networks.

Furthermore, CSS exhibits low power requirements, making it energy-efficient and well-suited for battery-powered IoT devices. The efficient utilization of power resources ensures prolonged battery life, enabling long-term operation and reducing maintenance needs. This characteristic is particularly valuable for IoT deployments in remote or inaccessible locations, where frequent battery replacement or recharging may not be feasible.

Firstly, it is important to note that Chirp Spread Spectrum (CSS) is closely related to orthogonal Frequency Shift Keying (FSK) modulation, as both techniques employ orthogonal modulation schemes in different domains. While FSK modulation operates in the frequency domain, CSS operates in the time domain [8].

In FSK modulation, the transmitted signal is modulated by switching between different discrete frequencies, with each frequency representing a specific symbol. This modulation scheme is widely used in various communication systems, including wireless networks and digital audio broadcasting. FSK enables efficient transmission of digital data by using a set of orthogonal frequency carriers, ensuring minimal interference between symbols [3].

On the other hand, CSS takes a different approach by modulating the transmitted signal using linear frequency modulation, also known as chirping, in the time domain. This linear frequency sweep creates a chirp waveform, which spreads the signal energy across a wide frequency band. The spreading of the signal in CSS provides inherent resistance to interference and allows for the simultaneous transmission of multiple devices without significant mutual interference.

Despite the differences in the modulation domains, CSS and FSK share similarities in terms of their use of orthogonal modulation. Both techniques aim to minimize interference and improve the reliability of data transmission. While FSK achieves orthogonality in the frequency domain through the use of orthogonal frequency carriers, CSS achieves orthogonality in the time domain through the spreading of the signal energy [9].

Understanding the relationship between CSS and FSK can provide valuable insights into the principles and benefits of CSS modulation. By leveraging the orthogonal modulation characteristics, CSS enables robust and efficient communication in IoT applications, overcoming challenges such as interference, multipath fading, and coexistence with other wireless systems.

The complex base-band of the basic chirp $x_0(t)$, which has a frequency that linearly increases from $\frac{-BW}{2}$ at $t = \frac{-T}{2}$ to $\frac{BW}{2}$ at $t = \frac{T}{2}$ is given by:

$$x_0(t) = \exp\left\{j\pi\frac{BW}{T}t^2\right\}p_T(t), \quad \forall t \in R \tag{2}$$

where BW denotes the bandwidth of the signal, and T the time to transmit a symbol, then $\log_2(BW.T) = SF$ is the spreading factor which gives the number of bits carried by a single symbol that can take $M = BW \times T = 2^{SF}$ values and $p_T(t) = 1$ for all $t \in [-T/2, T/2[$, and $p_T(t) = 0$ otherwise.

To modulate a symbol $m \in M$, the initial frequency of $x_0(t)$ is shifted by $m\frac{BW}{M} = \frac{m}{T}$ Hz and the chirp must be wrapped between $[-BW/2; BW/2]$, that we have:

$$x_m(t) = \exp\left\{j2\pi(\frac{BW}{2T} + \frac{m}{T})t\right\}\exp\{-j2\pi lBWt\}\,p_T(t), \tag{3}$$

$$\forall t \in Rl \in Z \ \text{ s.t.} \frac{BW}{T}t + \frac{m}{T} \in [lBW - \frac{BW}{2}; lBW + \frac{BW}{2}].$$

Thus, the modulated signal consists of multiple symbols sent sequentially, with a time spacing of T seconds:

$$s(t) = \sum_{n=0}^{L/SF} x_{\mathbf{m}_n}(t - nT), \forall t \in R, \quad (4)$$

where $\mathrm{m}_n \in M$ is the n-th element of $\mathbf{m}$.

To obtain the discrete-time of $s(t)$, we must sample at the rate $F_{samp} = \frac{1}{T_samp} = BW$. Using the relation $BWT = M$, we have $T = MT_{samp}$. Thus, the discrete-time expression of $s(t)$ is:

$$s[k] = s(kT_{samp}) = \sum_{n=0}^{L/SF} x_{\mathbf{m}_n}(kT_{samp} - nT) = \sum_{n=0}^{L/SF} x_{\mathbf{m}_n}[k - nM] \quad \forall k \in Z. \quad (5)$$

Therefore, the discrete-time expression of a modulated chirp is:

$$\begin{aligned} x_m[k] &= x_m(kT_{samp}) \\ &= \exp\left\{ j2\pi \left(\frac{BW}{2T} kT_{samp} + \frac{m}{T} \right) kT_{samp} \right\} \\ &\times \exp\left\{ -j2\pi lBW kT_{samp} \right\} \\ &\times p_T(kT_{samp}), \end{aligned} \quad (6)$$

$$\forall k, l \in Z \text{ s.t.} \frac{BW}{kT_{samp}} + \frac{m}{T} \in [lBW - \frac{BW}{2}; lBW + \frac{BW}{2}].$$

where $p_T(kT_{samp}) = p_{\frac{T}{T_{samp}}}(k) = p_M[k]$ and $\exp\{-j2\pi lBWkT_{samp}\} = \exp\{-j2\pi lk\} = 1 \forall l, k \in Z$. Thus, the expression of $x_m(t)$ is simpler in the discrete-time domain:

$$x_m[k] = \exp\left\{ j2\pi (\frac{B}{2T} kT_{samp} + \frac{m}{T}) kT_{samp} \right\} p_M[k] = \exp\left\{ j\pi \frac{k^2}{M} \right\} \exp\left\{ j2\pi \frac{m}{M} k \right\} p_M[k]. \quad (7)$$

$$x_m[k] = \exp\left\{ j2\pi (\frac{B}{2T} kT_{samp} + \frac{m}{T}) kT_{samp} \right\} p_M[k] = \exp\left\{ j\pi \frac{k^2}{M} \right\} \exp\left\{ j2\pi \frac{m}{M} k \right\} p_M[k]. \quad (8)$$

Incorporating (8) in (5), we obtain:

$$s[k] = \sum_{n=0}^{L/SF} p_M[k - nM] \exp\left\{ j\pi \frac{(k - nM)^2}{M} \right\} \exp\left\{ j2\pi \frac{\mathbf{m}_n}{M} (k - nM) \right\} \quad (9)$$

$$s[k] = \underbrace{\exp\left\{ j\pi \frac{k^2}{M} \right\}}_{Chirp} \underbrace{\sum_{n=0}^{L/SF} p_M[k - nM] \exp\left\{ j2\pi \frac{\mathbf{m}_n}{M} k \right\}}_{M-FSK} \quad (10)$$

The Eq. 10 shows that CSS signals are basically chirped versions of M-arry Frequency Shift Keying (M-FSK) signals [6].

3 Coherent and Non-coherent Detector in CSS

Considering an AWGN channel, we have that:

$$y[k] = s[k] + z[k] \quad \forall k \in Z, \tag{11}$$

where $z[k]$ is a zero-mean, complex-circular white Gaussian noise. Later, we proceed with dechirping operation:

$$y_d[k] = y[k]\exp\left\{-j\pi\frac{k^2}{M}\right\} \quad \forall k \in Z, \tag{12}$$

where $y_d[k]$ is the dechirped observed signal. This process does not change the statistical properties of the noise, thus, the system involving $y_d[k]$ instead of $y[k]$ is equivalent to $M - FSK$ over AWGN.

3.1 Coherent Detector

Coherent detectors play a crucial role in Chirp Spread Spectrum (CSS) by extracting the transmitted information from the received CSS signal. These detectors rely on maintaining phase and frequency coherence between the transmitted and received signals, enabling precise demodulation and recovery of the original data [1].

In CSS, coherent detection involves aligning the carrier frequencies of the received signal with a replica of the transmitted chirp signal. This alignment is achieved through carrier synchronization, ensuring that the local replica of the transmitted chirp signal is accurately matched in frequency and phase with the received signal. By multiplying the received signal with the synchronized replica, the coherent detector can extract the information embedded in the CSS signal [1].

Coherent detection is essential for separating the desired signal from other interference sources, as it takes advantage of the alignment of the carrier frequencies. This alignment allows for precise demodulation, resulting in accurate recovery of the original data. It is particularly useful in environments with low signal-to-noise ratio or when dealing with interfering signals that are closely located in the frequency spectrum.

Overall, coherent detectors in CSS enable reliable and accurate demodulation, making them an important component in the successful implementation of CSS-based communication systems.

Over perfect AWGN channels, it allows for maximum likelihood detection (MLD). Its use is mandatory for demodulating derivative of CSS that use the phase to encode symbols [16]. Thus, a MLD strategy for M-CSS in the presence of AWGN is to find the symbol m whose associated signal $\exp\left\{2\pi\frac{m}{M}k\right\}$ correlates the best with the observed signal $y_d[k]$. This is given by:

$$\hat{\mathbf{m}}_n = \text{argmax } Real\left\{\sum_{k=-\infty}^{\infty} p_M[k]\exp\left\{-2\pi\frac{m}{M}k\right\} y_d[k+nM]\right\}. \tag{13}$$

From the above expression, we observe the term in the real part is the M point Discrete Fourier Transform (DFT), denoted by $F_M\{.\}$ of the received signal:

$$\hat{\mathbf{m}}_n = \text{argmax } Real\left\{F_M\left\{y_d[k+nM]\right\}[m]\right\}. \quad (14)$$

Expanding (16), we have:

$$\begin{aligned} F_M\left\{y_d[k-nM]\right\}[m'] &= F_M\left\{s[k-nM]\right\}[m'] + F_M\left\{z[k]\right\}[m'] \\ &= M\delta_{n-n'}\delta_{m_n-m'_n} + F_M\left\{z[k]\right\}[m'], \end{aligned} \quad (15)$$

where $\delta_n = 1$ if $n = 0$ and zero otherwise. As we can see, the correlator output corresponding to the symbol that was sent has its amplitude augmented by M, as opposed to the other correlators. This means that this modulation, along with coherent demodulation gets more robust to noise as M increases.

3.2 Non-coherent Detector

In Chirp Spread Spectrum (CSS), a non-coherent detector is an alternative to the coherent detection scheme. Unlike coherent detection, which relies on maintaining phase and frequency coherence between the transmitted and received signals, non-coherent detection does not require such synchronization [5].

In non-coherent detection, the received CSS signal is processed without aligning the carrier frequencies or maintaining phase coherence. Instead, the detector focuses on extracting the modulating information from the received signal using techniques that do not rely on precise carrier synchronization [6].

One common non-coherent detection technique used in CSS is energy detection. This technique involves measuring the energy level of the received signal over a certain integration period. By comparing the energy level to a predetermined threshold, the detector makes a decision on the presence or absence of the transmitted signal.

Since non-coherent detection does not require precise synchronization, it offers several advantages in certain scenarios. Non-coherent detectors are more robust in environments where carrier frequency offsets or phase variations occur, as they are not affected by these impairments. They also eliminate the need for complex carrier recovery circuits, simplifying the receiver design and reducing implementation complexity.

However, non-coherent detection typically has lower performance compared to coherent detection, particularly in terms of signal detection sensitivity and demodulation accuracy. Non-coherent detectors may suffer from higher detection error rates and reduced data throughput, especially in low signal-to-noise ratio (SNR) conditions. Additionally, non-coherent detection may have limitations in scenarios where high-precision demodulation or decoding is required [4].

Nonetheless, non-coherent detection is suitable for applications where carrier synchronization is challenging or not feasible, or when the emphasis is on simplicity and robustness rather than achieving the highest possible performance. It offers an alternative approach for CSS systems in certain environments where precise synchronization is difficult to achieve or maintain.

Non-coherent detector replaces the real part operator by the magnitude operator in (16):

$$\hat{\mathbf{m}}_n = \text{argmax} \left| F_M \left\{ y_d[k+nM] \right\} [m] \right|. \tag{16}$$

3.3 Quadrature Demodulation

In sub-section, we have included the description about quadrature demodulation for comparison purposes which is a technique used in communication systems to extract information from a modulated signal. It is commonly employed in various modulation schemes, including amplitude modulation (AM), frequency modulation (FM), and phase modulation (PM).

The term "quadrature" refers to the use of two separate signal paths, known as the in-phase (I) and quadrature (Q) channels, which are 90° out of phase with each other. The modulated signal is split into these two channels, allowing for simultaneous processing of both the in-phase and quadrature components.

The quadrature demodulation process typically involves two main steps: mixing and low-pass filtering. In the mixing stage, the received signal is multiplied by two local oscillators, one in-phase and one quadrature, which are synchronized with the carrier frequency of the modulated signal. This mixing process results in the separation of the modulated signal into its in-phase and quadrature components.

After mixing, the resulting signals are passed through low-pass filters to remove high-frequency components and noise. These filters are designed to allow only the baseband information, typically the amplitude, frequency, or phase modulation, to pass through. The filtered in-phase and quadrature signals can then be further processed or demodulated to recover the original modulating information.

Quadrature demodulation offers several advantages in communication systems. By using the I/Q channels, it enables the receiver to capture both the amplitude and phase characteristics of the modulated signal simultaneously. This allows for accurate demodulation and recovery of the original information, even in the presence of noise and interference.

Additionally, quadrature demodulation is compatible with various modulation schemes, making it a versatile technique in modern communication systems. It is widely used in applications such as wireless communication, software-defined radio (SDR), and digital broadcasting.

Quadrature demodulation can be used in M-CSS signal, once the M-CSS signal has passed by the dechirping process, we have a M-FSK signal which can be seen as frequency modulated M-arry Amplitude Shift Keying (ASK) modulations. Thus, we have two-stage demodulation process. The first stage performs quadrature detection of the frequency variation in $y_d[k]$, and the second performs M-ASK demodulation on this frequency variation.

Quadrature demodulation computes the product of $y_d[k]$ with a unit delayed and conjugated version of itself. In order to determine its frequency variation. Thus, we have:

$$a[k] = \arg \left\{ y_d[k] y_d^*[k-1] \right\}. \tag{17}$$

Once the frequency is the derivative of the phase, we have that quadrature demodulator corresponding to the digital derivative of the phase of $y_d[k]$. After quadrature detection, the signal is analogous to an M-ASK modulated signal:

$$a[k] = \frac{2\pi}{M} \sum_{n=0}^{L/SF} p_M[k - nM]\mathbf{m}_n + z'[k], \tag{18}$$

where $z'[k]$ has no guarantee to be Gaussian nor white. Later, a hard decision by comparing its amplitude to $M - 1$ thresholds can be taken in order to obtain the transmitted symbols $\mathbf{m}$.

4 Performance Results Comparisons

In this section, we analyze and compare the performance of CSS in different physical environments, such as AWGN, frequency-selective and Rayleight channels. We evaluate the performance metrics, including signal-to-noise ratio (SNR) and bit error rate (BER).

4.1 AWGN Channel

Figure 1 compares the three demodulators described in Sect. 3 for a CSS with SF = 9. We note that quadrature demodulation has the worst performance, which can however be improved by oversampling the received signal with a factor of 2. However, quadrature demodulation continues to present an E_b/N_0 difference as high as 23 dB for a BER of 10^{-4} in comparison with coherent demodulation.

4.2 Frequency-Selective Channel

In a frequency-selective channel in order to evaluate the performance of coherent and non-coherent demodulation, we have considered a static frequency-selective channel with impulse response $h[k] = 0.407\delta_{k-1} + 0.815\delta_k + 0.407\delta_{k+1}$ with $\delta_k = 1$ if $k = 0$ and $\delta_k = 0$ otherwise. Note that this channel is non-invertible. The received signal is given by:

$$y[k] = \sum_{l=-\infty}^{\infty} h[l]s[k - l] + z[k] \tag{19}$$

From Fig. 2 we note that CSS performs better than FSK on this channel due to the fact that it is better at spreading the information in frequency.

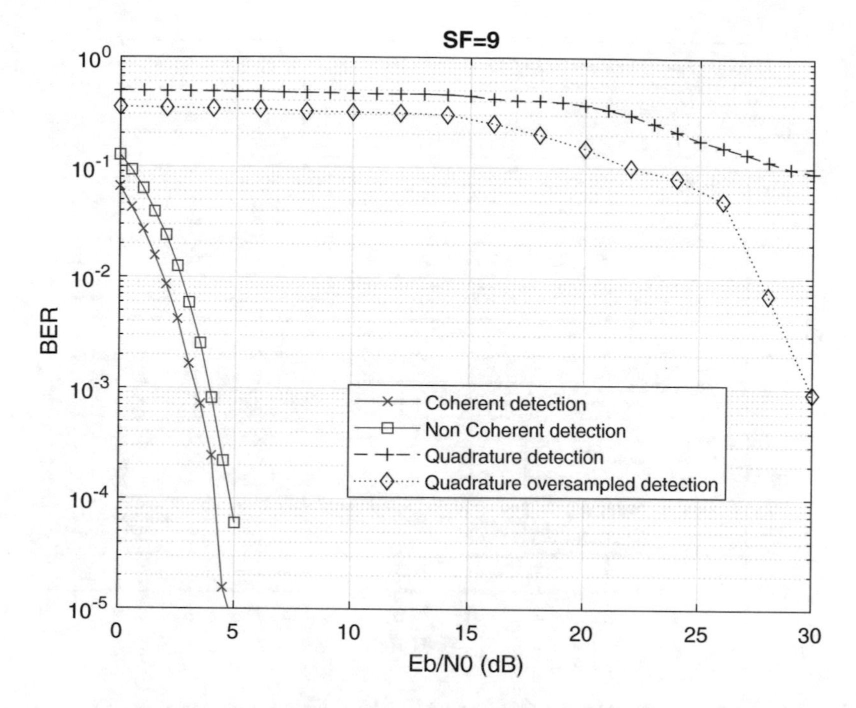

Fig. 1. Comparison between different detectors for CSS with SF = 9 in an AWGN channel

4.3 Rayleigh Channel

In this analysis, a slow and flat fading channel is assumed. Thus, the received signal is given by:

$$y[k] = h[k]s[k] + z[k], \tag{20}$$

where $h[k] = \alpha[k]\exp\{j\theta\}$ with $\alpha[k]$ and θ being attenuation factor and uniformly distributed phase variations.

Figure 3 shows the comparison between coherent and non-coherent CSS modulation with SF = 6 in a Rayleigh channel. As it is evident coherent detection presents a better performance of about 0.5 dB in comparison with non-coherent detection. For the theoretical non-coherent detection BER curve the 'beading' tool from Matlab was used. However, this same tool for the case of the theoretical coherent detection BER curve is only defined for $M = 2$.

Figure 4 compares the performance of a CSS modulation with SF = 6 employing $N_r = 1$ and $N_r = 10$ receive antennas for coherent and non-coherent detection. We can observe better performance when we use more than 1 receive antenna. For this case, we note that the performance difference between coherent and non-coherent detection increase as N_r increase. Thus, for $N_r = 10$ the performance difference is about 4 dB for a BER$= 10^{-3}$.

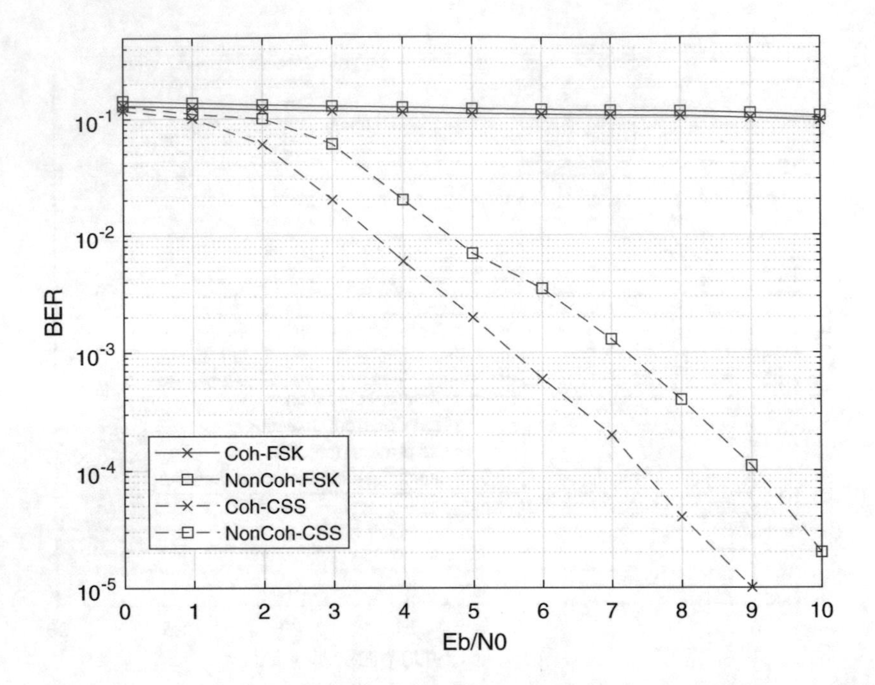

Fig. 2. Comparison between CSS and FSK in a Frequency-selective Channel.

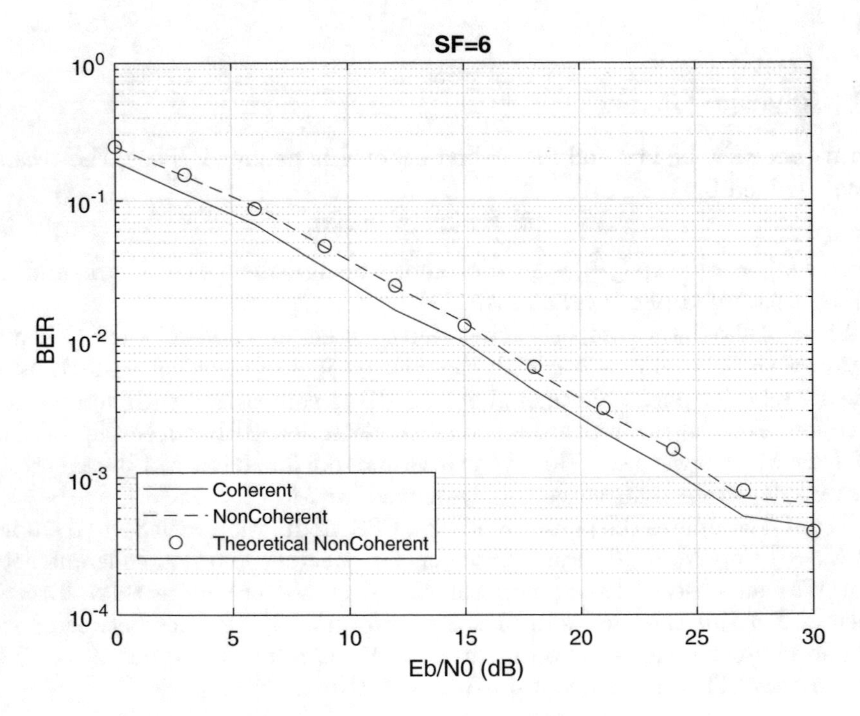

Fig. 3. CSS with coherent and non-coherent detection in a Rayleigh Channel.

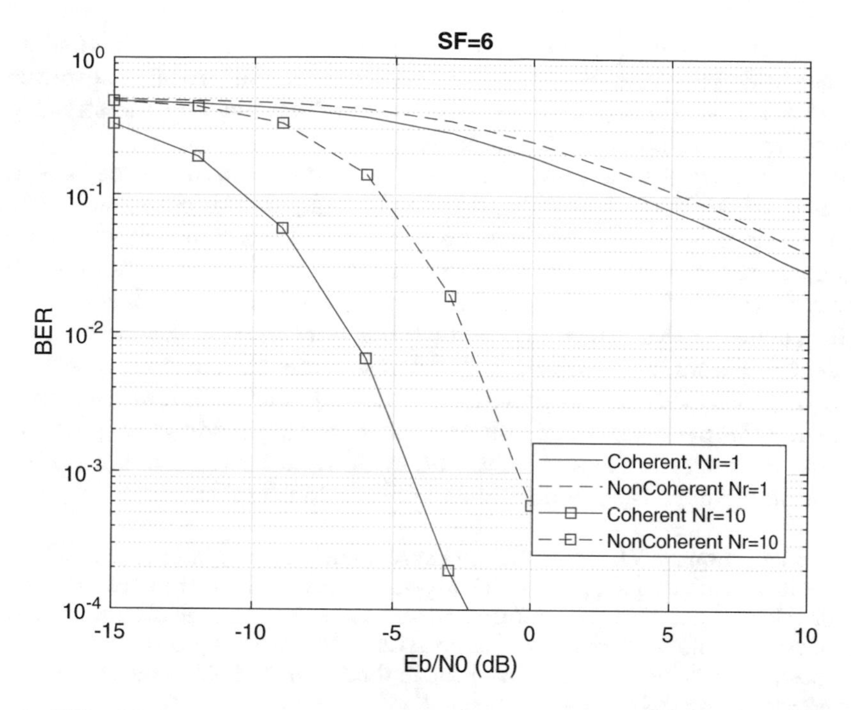

Fig. 4. CSS with coherent and non-coherent detection and diversity $N_r = 10$ receive antennas in a Rayleigh Channel.

5 Conclusion

In conclusion, the utilization of Chirp Spread Spectrum (CSS) in Internet of Things (IoT) applications has been presented and discussed in this paper. The performance of both coherent and non-coherent detectors was thoroughly analyzed in various channel environments, including scenarios with additive white Gaussian noise (AWGN), frequency-selective channels, and Rayleigh fading channels.

The results demonstrate that the coherent detector exhibits superior performance compared to the non-coherent detector in the studied system with 10 receiving antennas. The coherent detection scheme, with its ability to maintain phase and frequency coherence between the transmitted and received signals, proves to be more effective in extracting the desired information from the CSS signal.

However, it is worth noting that the non-coherent detector offers the advantage of lower complexity due to its independence from precise carrier synchronization. This makes it an attractive option for resource-constrained IoT devices or scenarios where maintaining synchronization may be challenging.

Nevertheless, to further enhance the performance of non-coherent detection in CSS-based IoT systems, it is essential to explore and evaluate new techniques. Future research should focus on developing innovative methods that can mitigate the performance gap between coherent and non-coherent detection. These advancements may include adaptive signal processing algorithms, advanced filtering techniques, or hybrid detection schemes that combine the strengths of both coherent and non-coherent approaches.

By addressing the challenges associated with non-coherent detection, such as increased bit error rates or reduced data throughput, the potential of CSS in low signal-to-noise ratio conditions can be fully realized. This will contribute to the advancement of reliable and efficient IoT communication systems, enabling seamless connectivity and data exchange in diverse IoT applications.

In summary, this study highlights the significance of coherent and non-coherent detectors in CSS-based IoT systems. While the coherent detector currently demonstrates superior performance, the exploration of new techniques for improving the performance of non-coherent detection holds great promise. By bridging the performance gap, we can leverage the advantages of both detection schemes and unlock the full potential of CSS in IoT applications.

Acknowledgments. We would like to express our sincere gratitude to CEDIA for their generous financial support, which played a pivotal role in the successful realization of this project. Their commitment to our vision has been instrumental, and we appreciate the confidence they have shown in our mission. This project would not have been possible without the invaluable contribution from CEDIA, and we extend our deepest thanks for their unwavering support.

References

1. Tselikis, G., Askoxylakis, I.G.: Chirp spread spectrum for IoT applications: challenges and opportunities. IEEE Commun. Mag. **55**(11), 147–153 (2017)
2. Zheng, Y., Dong, X., Chen, L.: Chirp spread spectrum in Internet of Things: a comprehensive survey. IEEE Internet Things J. **8**(8), 6604–6617 (2021)
3. Ali, S., Ismail, M., Anpalagan, A.: Chirp spread spectrum modulation for energy-constrained IoT systems. IEEE Access **8**, 214312–214327 (2020)
4. Shah, D., Garg, S.: Chirp spread spectrum-based IoT localization technique for smart cities. In: Proceedings of the International Conference on IoT in Social, Mobile, Analytics and Cloud, pp. 143–148 (2019)
5. Han, Y., Han, C.: Chirp spread spectrum communication scheme for IoT applications in wireless sensor networks. Sensors **21**(6), 2129 (2021)
6. Ahmad, N., Rahman, A.F., Isa, N.A.M.: Chirp spread spectrum modulation for secure and reliable IoT communications. In: Proceedings of the International Conference on IoT in Social, Mobile, Analytics and Cloud, pp. 223–228 (2019)
7. Suh, J., Oh, C., Hong, D.: Chirp spread spectrum-based communication for cooperative Internet of Things. In: Proceedings of the International Conference on Information Science and Applications, pp. 1–4 (2019)
8. Haq, R.U., Baig, I., Khan, M.A.U.: Chirp spread spectrum modulation for reliable communication in IoT. In: Proceedings of the International Conference on Advanced Communication Technologies and Networking, pp. 8–13 (2021)
9. Gupta, S., Nath, B., Pandey, V.: Performance evaluation of chirp spread spectrum modulation for IoT applications. In: Proceedings of the International Conference on Inventive Computation Technologies, pp. 384–389 (2021)

Exploring the Use of Blockchain for Academic Certificates: Development, Testing, and Deployment

Juan Minango(✉), Marcelo Zambrano, and Cesar Minaya

Instituto Tecnológico Superior Universitario Rumiñahui, Av. Atahualpa 1701 and 8 de Febrero, 171103 Sangolquí, Ecuador
juancarlos.minango@ister.edu.ec

Abstract. This study presents the development and deployment of a smart contract for issuing academic certificates on the Sepolia network. The smart contract, built using the Solidity language and the ERC721 standard, enables secure and decentralized certification, ensuring the integrity and immutability of academic records. The deployment process was facilitated by the use of the Alchemy development tool, allowing for seamless interaction with the blockchain network. The successful deployment was confirmed through the acquisition of the contract address and the verification of transaction details on Etherscan. Unit tests were conducted to ensure the functionality and reliability of the smart contract, validating its ability to mint and verify academic certificates. The results demonstrated the potential of blockchain technology in transforming traditional certification systems by providing transparency, tamper-proof records, and enhanced security. This study contributes to the growing field of blockchain applications in the academic sector, highlighting the importance of decentralized solutions for efficient and trustworthy certification processes.

Keywords: Smart contract · Academic certificates · Blockchain · deployment · Alchemy · ERC721 · Decentralized certification

1 Introduction

Blockchain technology has emerged as a groundbreaking innovation with far-reaching implications across industries. Its decentralized and transparent nature has the potential to disrupt traditional systems, establishing new standards of trust, security, and efficiency [1,2]. By providing a decentralized ledger that records transactions and information across a network of computers, blockchain eliminates the need for intermediaries, reduces the risk of fraud, and enhances transparency [3].

One of the key advantages of blockchain is its ability to establish trust in a trustless environment. Through cryptographic techniques and consensus

M. Z. Vizuete et al. (Eds.): CI3 2023, LNNS 1040, pp. 123–137, 2024.
https://doi.org/10.1007/978-3-031-63434-5_10

mechanisms, blockchain ensures the integrity and immutability of data, making it resistant to manipulation and tampering. This feature has the potential to revolutionize industries such as finance, supply chain management, healthcare, the academic sector, and many more, where trust and transparency are paramount [4].

In finance, blockchain technology enables secure and efficient peer-to-peer transactions without the need for traditional intermediaries such as banks [5]. It provides a decentralized and tamper-proof record of transactions, enhancing security and reducing costs associated with traditional payment systems. Additionally, blockchain-based smart contracts automate and enforce the terms of agreements, streamlining processes and minimizing the risk of disputes [5,6].

In supply chain management, blockchain facilitates end-to-end traceability, enabling businesses and consumers to track the journey of products from origin to destination. This transparency helps in verifying the authenticity and quality of goods, ensuring compliance with regulations, and combating issues such as counterfeiting and supply chain fraud. By providing an immutable record of transactions and product information, blockchain enhances trust among participants and simplifies complex supply chain processes [7].

In the healthcare industry, blockchain has the potential to revolutionize data sharing and patient records management. With blockchain, patients can have control over their medical records, granting access to healthcare providers as needed. This decentralized approach improves the security and privacy of sensitive medical information while enabling interoperability and efficient sharing of data among healthcare providers, leading to better patient outcomes and more effective research [8].

Furthermore, blockchain technology has demonstrated adaptability across different sectors, including energy, voting systems, intellectual property, and more. Its decentralized and transparent nature can enhance the efficiency, security, and accountability of various processes. By utilizing blockchain, organizations can streamline operations, reduce costs, mitigate risks, and establish new business models that were previously not feasible [9].

Besides, the academic sector can also benefit significantly from the implementation of blockchain technology. One of the key applications is in the issuance and verification of academic certificates. By leveraging blockchain, educational institutions can create tamper-proof digital certificates as non-fungible tokens (NFTs). These NFT-based certificates provide a secure and decentralized record of achievements, enabling easy verification by employers and other educational institutions. This eliminates the risks associated with counterfeit certificates and simplifies the verification process, enhancing trust and reliability in academic credentials [10].

Thus, in this paper, we present the development of a smart contract[1] for the emission of academic certificates by Instituto Tecnológico Universitario Rumiñahui (ISTER). The smart contract was deployed on a blockchain net-

[1] We have included the link of the repository of the Academic Certificate smart contract [https://github.com/jminangoIster/corn_leaves_diseases] for reproducibility.

work, allowing us to interact with it using the Metamask wallet. By leveraging the capabilities of blockchain technology and the Ethereum network, we aim to change the certification process and enhance the transparency, security, and accessibility of academic credentials issued by ISTER.

The remainder of this paper is organized as follows. Section 2 provides an introduction to the Ethereum network, specifically the Sepolia network, which was utilized for deploying and testing our smart contracts. In Sect. 3, we provide a concise overview of the development tools and network connection, which served as the foundation for the implementation and interaction with our smart contracts. Section 4 delves into the details of the smart contract developed specifically for issuing academic certificates, highlighting its key functionalities and mechanisms. The deployment results of the smart contract are presented in Sect. 5, showcasing the outcomes and performance metrics. Finally, in Sect. 6, we draw conclusions based on our findings and discuss the implications and potential future directions for further research in this domain.

2 Ethereum and the Sepolia Network

The Ethereum network has gained significant attention as a decentralized, open-source blockchain platform that enables the development and deployment of smart contracts. With its native cryptocurrency, Ether (ETH), Ethereum provides a robust and versatile environment for building decentralized applications (dApps) and executing programmable transactions [4,11].

In this section, we introduce the Sepolia network, a specific instance of the Ethereum network that we utilized for deploying and testing our smart contract for issuing academic certificates. Sepolia is a test network specifically designed for development and experimentation purposes, providing developers with a sandbox-like environment to deploy and interact with smart contracts before deploying them on the Ethereum mainnet [12].

Sepolia offers several advantages for blockchain tests. Firstly, it allows us to test the functionality and performance of our academic certificate smart contract without incurring any real-world costs or risks associated with the deployment on the mainnet. This enables us to iterate and refine our implementation before taking it live [13].

Secondly, Sepolia provides a network with a similar architecture and consensus mechanism to the Ethereum mainnet, allowing us to simulate real-world conditions and gauge the scalability and efficiency of our smart contract deployment [10]. By using Sepolia, we can evaluate the network's transaction processing capabilities, gas costs, and potential bottlenecks that may affect the issuance and verification of academic certificates.

2.1 Faucet in the Sepolia Network

A faucet is a service specifically designed for test networks like Sepolia. Its purpose is to distribute test ETH to developers, allowing them to cover transaction costs and simulate real-world scenarios without incurring any expenses. Test ETH obtained from the faucet serves as a valuable resource for deploying and interacting with smart contracts in a risk-free environment [14,15].

To obtain test ETH through the Sepolia faucet, we must follow these steps:

1. Visit the Sepolia testnet faucet website or a compatible faucet service designed for the Sepolia network.
2. Provide the Ethereum address associated with the wallet that will be used for interacting with the academic certificate smart contract on Sepolia.
3. Submit a request to receive test ETH by clicking on the "Request" or "Get Test ETH" button.
4. The faucet service will process the request and verify the provided Ethereum address.
5. Upon successful verification, the faucet will transfer a predetermined amount of test ETH to the provided Ethereum address.
6. By obtaining test ETH through the faucet, we can seamlessly engage with the academic certificate smart contract on Sepolia without the need for real funds. This enables thorough testing, debugging, and validation of the smart contract's functionality before deploying it on the live Ethereum network.

3 Development Tools and Network Connection

3.1 Hardhat: A Development Framework

In the development of the academic certificate smart contract, the Hardhat framework was utilized. Hardhat is a popular development environment for Ethereum smart contracts, providing a robust set of tools and utilities that facilitate the entire development lifecycle. It offers features such as smart contract compilation, deployment, testing, and debugging.

Hardhat simplifies the development process by providing a command-line interface (CLI) for executing tasks related to smart contracts. It supports the Solidity programming language and integrates seamlessly with various Ethereum networks, including the Sepolia network. With Hardhat, developers can efficiently write, compile, and deploy their smart contracts, making it an ideal choice for this research.

3.2 Alchemy: Connecting to the Sepolia Network

To interact with the Sepolia network and deploy the academic certificate smart contract, the Alchemy platform was used. Alchemy acts as an infrastructure provider, offering developers easy access to Ethereum networks and providing powerful APIs for blockchain connectivity.

Alchemy simplifies the process of establishing a connection with the Sepolia network by providing a reliable and user-friendly interface. It offers advanced features such as transaction monitoring, debugging tools, and real-time network analytics. By leveraging Alchemy, developers can seamlessly interact with the Sepolia network, deploy smart contracts, and monitor their transactions.

Figure 1 showcases the Ister Academic Certificates decentralized application (DApp) running on the Alchemy platform. The DApp interface displays the network connection details, including the Sepolia network, which enables seamless interaction with the academic certificate smart contract. Users can mint and verify academic certificates securely through the user-friendly interface provided by the DApp. The integration of Alchemy ensures reliable and efficient network connectivity, enhancing the overall experience of issuing and verifying academic certificates on the blockchain.

Fig. 1. Ister Academic Certificates DApp on Alchemy.

3.3 Metamask: Interacting with the Academic Certificate and Faucet

Metamask is a widely used browser extension that serves as a digital wallet and allows users to interact with Ethereum-based applications. In the context of the academic certificate smart contract, Metamask plays a crucial role in facilitating user interactions and accessing the Sepolia network.

By installing the Metamask extension in our web browser, users can create Ethereum wallets and securely store their cryptographic keys. Metamask also provides an intuitive user interface for interacting with the academic certificate smart contract, enabling users to mint and verify certificates.

Metamask was configured to connect to the Sepolia network by adding the appropriate network details, including the network name, chain ID, RPC URL, and optionally the faucet's address. Once connected, users can seamlessly interact with the smart contract deployed on the Sepolia network, minting and verifying academic certificates.

Figure 2 shows the Metamask browser extension interface with an account named "Ister" connected to the Sepolia network. The account balance of 0.5 SepoliaETH indicates that the account has successfully received faucet funds. The integration of Metamask provides a user-friendly and convenient experience for managing blockchain transactions and interacting with the academic certificate system.

By combining Hardhat for development, Alchemy for network connection, and Metamask for user interactions, the academic certificate smart contract can be efficiently developed, deployed, and interacted with on the Sepolia network. This combination of tools and technologies ensures a seamless and user-friendly experience for all stakeholders involved.

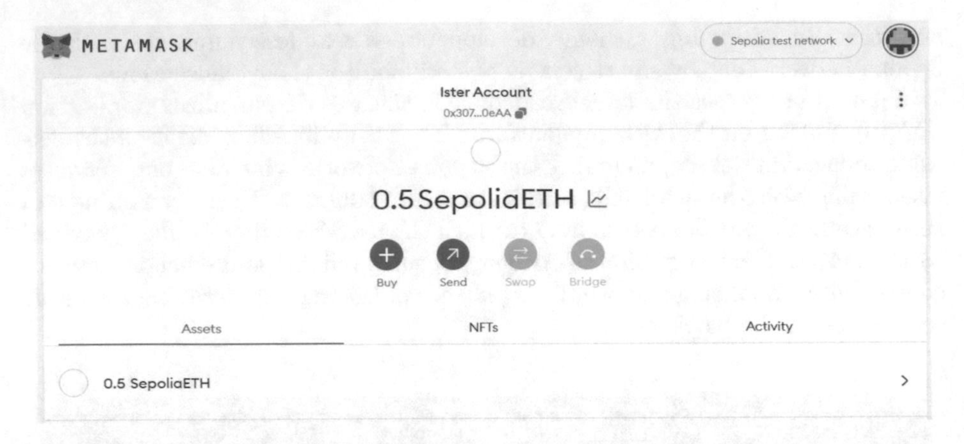

Fig. 2. Metamask Browser Interface with Ister Account on Sepolia Network.

4 Academic Certificate Smart Contract: Functionality and Mechanisms

In this section, we provide an in-depth explanation of the smart contract developed for issuing academic certificates. The smart contract, named "Academic-Certificate", is built on the Ethereum blockchain using the Solidity programming language.

The contract inherits from the ERC721 standard which is shown in Listing 1.1, which is a widely adopted standard for non-fungible tokens (NFTs) on Ethereum. Unlike fungible tokens, which are interchangeable and identical to each other (like cryptocurrencies), non-fungible tokens represent unique assets that can have distinct characteristics and properties. Each ERC721 token is assigned a unique identifier, or token ID, which distinguishes it from other tokens in the same contract. Thus, our contract ensures that each academic certificate issued is unique and can be easily transferred and verified.

Listing 1.1. AcademicCertificate contract

```
1 contract AcademicCertificate is ERC721 {
2 }
```

The structure of an academic certificate is defined within the contract using a struct. It includes fields such as recipient name, date of issue, issuer, course, grade, and duration. These fields capture the essential information associated with each certificate and it is presented in Listing 1.2.

Listing 1.2. AcademicCertificate contract

```
1 struct Certificate {
2         string recipientName;
3         uint256 dateOfIssue;
4         string issuer;
```

```
5         string course;
6         string grade;
7         string duration;
8     }
```

To mint a new certificate NFT, the contract includes a function named "mintCertificate". This function requires that the caller is the designated "Ister Account", which ensures that only authorized entities can emit academic certificates. Upon minting, a new certificate ID is generated, and the certificate data is stored in a mapping. See Listing 1.3.

Listing 1.3. AcademicCertificate contract

```
1  function mintCertificate(
2          string memory _recipientName,
3          uint256 _dateOfIssue,
4          string memory _issuer,
5          string memory _course,
6          string memory _grade,
7          string memory _duration
8      ) public {
9          require(_isterAccount == msg.sender, ''Just Ister
               Account can emit Academic Certificates'');
10         uint256 certificateId = _totalCertificates + 1;
11         certificates[certificateId] = Certificate(
12             _recipientName,
13             _dateOfIssue,
14             _issuer,
15             _course,
16             _grade,
17             _duration
18         );
19         _mint(msg.sender, certificateId);
20         _totalCertificates++;
21         emit NewCertificate(
22             certificateId,
23             _recipientName,
24             _dateOfIssue,
25             _issuer,
26             _course,
27             _grade,
28             _duration
29         );
30     }
```

The contract also includes a function named "verifyOwnership" which allows anyone to verify their ownership of a certificate NFT by providing its ID. This function returns a boolean value indicating whether the caller is the owner of the specified certificate as is shown in Listing 1.4.

Listing 1.4. AcademicCertificate contract

```
1 function verifyOwnership(uint256 _certificateId) public
      view returns (bool) {
2         return ownerOf(_certificateId) == msg.sender;
3     }
```

Additionally, the contract emits an event named "NewCertificate" whenever a new certificate is minted. This event includes the certificate ID, recipient name, date of issue, issuer, course, grade, and duration. See Listing 1.5.

Listing 1.5. AcademicCertificate contract

```
1 event NewCertificate(
2         uint256 indexed certificateId,
3         string recipientName,
4         uint256 dateOfIssuem,
5         string issuer,
6         string course,
7         string grade,
8         string duration
9     );
```

5 Results and Performance Metrics

5.1 Unitary Tests

In the repository accompanying this article, a comprehensive set of test cases has been developed to validate the functionality and performance of the AcademicCertificate smart contract. These tests utilize the Mocha and Chai testing frameworks, along with the Hardhat development environment.

The tests cover various aspects of the contract, including the deployment process, the minting of new certificate NFTs, and the verification of ownership. By running these tests, it is possible to ensure that the smart contract behaves as expected and meets the desired requirements.

The first test case, "should mint a new certificate NFT," verifies that the contract successfully mints a new certificate with the specified details and assigns ownership to the correct address. It ensures that the certificate data is correctly stored and retrievable.

The second test case, "should verify ownership of a certificate NFT," ensures that the contract correctly verifies the ownership of a certificate NFT. It validates that only the rightful owner can claim ownership and that unauthorized users are unable to do so.

These test cases provide a solid foundation for ensuring the robustness and reliability of the AcademicCertificate contract. They enable developers and users to have confidence in the contract's functionality and its ability to issue and verify academic certificates securely and accurately.

Figure 3 shows the successful execution of the tests using the command 'npx hardhat test'. The tests verify the functionality of the AcademicCertificate smart contract, ensuring that it properly mints a new certificate NFT and verifies the ownership of the certificate. The tests were executed using the Hardhat development environment, which provides a robust and reliable framework for testing Ethereum smart contracts."

```
PS C:\ws-ethereum\educationalCertificates> npx hardhat test


  AcademicCertificate
    Deployment
      ✓ should mint a new certificate NFT (924ms)
      ✓ should verify ownership of a certificate NFT (46ms)


  2 passing (977ms)
```

Fig. 3. Successful execution of tests using 'npx hardhat test'. The tests verify the functionality of the AcademicCertificate smart contract, ensuring proper certificate issuance and ownership verification.

5.2 Deployment Results and Verification on Alchemy and Etherscan

To deploy the AcademicCertificate smart contract, we execute the command 'npx hardhat run scripts/deploy.ts –network sepolia' which will run the deployment script using the sepolia network.

Upon successful deployment, we received a message indicating that the "Academic Certificate" contract has been deployed. The contract address was displayed as: "Contract deployed at address: 0x4e86B73205B95c6537B307e778158 01b2ee7F7e4" as it is shown in Fig. 4.

```
PS C:\ws-ethereum\educationalCertificates> npx hardhat run scripts/deploy.ts --network sepolia
Academic Certificate Successfully Deployed!
Contract deployed at address: 0x4e86B73205B95c6537B307e77815801b2ee7F7e4
```

Fig. 4. Successful Deployment of the Academic Certificate Smart Contract.

The contract address shown in Fig. 4 represents the unique identifier assigned to the deployed Academic Certificate smart contract on the blockchain network. This address serves as a reference or identifier to interact with the specific instance of the smart contract.

In simpler terms, the contract address can be thought of as the location of the smart contract on the blockchain. It allows users to interact with the deployed contract, such as minting new certificate NFTs or verifying ownership of existing certificates.

Alchemy Dashboard Results. In this subsubsection, the results from the successful deployment of the Academic Certificate smart contract on the Alchemy dashboard are presented and discussed.

Figure 5 shows the Alchemy Dashboard, displaying important deployment statistics of the Academic Certificate smart contract. The dashboard reveals key metrics such as the median response time of 22 ms, indicating the efficiency and responsiveness of the deployed contract. Additionally, the success rate of 98% highlights the robustness and reliability of the deployment process. These results demonstrate the successful deployment and smooth operation of the Academic Certificate smart contract on the Ethereum network, ensuring seamless issuance and verification of academic certificates.

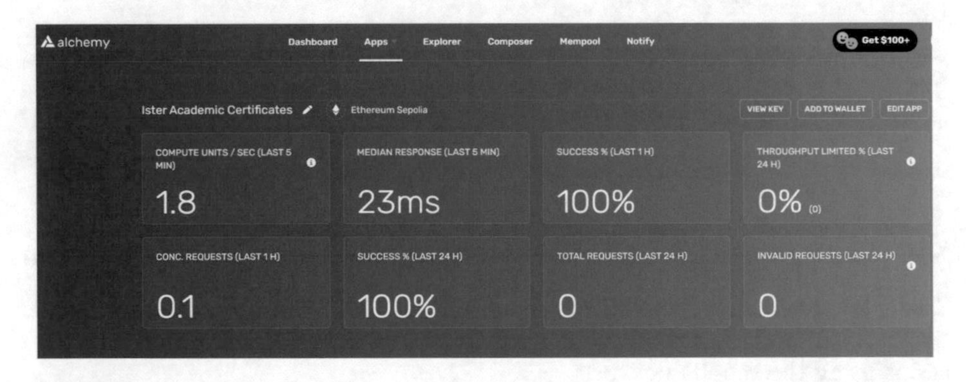

Fig. 5. Alchemy Dashboard showing deployment statistics of the Academic Certificate smart contract.

Figure 6 showcases the Alchemy Explorer, providing detailed information about the Ister Account and various Ethereum network parameters. The public key associated with the Ister Account is displayed, ensuring secure and authenticated interactions with the blockchain. Additionally, key network parameters such as eth_chainId, eth_getTransactionReceipt, eth_blockNumber, and eth_getTransactionCount are visible, providing insights into the current state of the Ethereum network. These parameters play a crucial role in facilitating transactions, verifying blocks, and ensuring the integrity of the blockchain. Thus, the Alchemy Explorer serves as a valuable tool for monitoring and analyzing blockchain activity, enhancing transparency and reliability in the deployment and utilization of the Academic Certificate smart contract.

Figure 7 provides essential information about individual transactions related to our academic certification smart contract. This information includes:

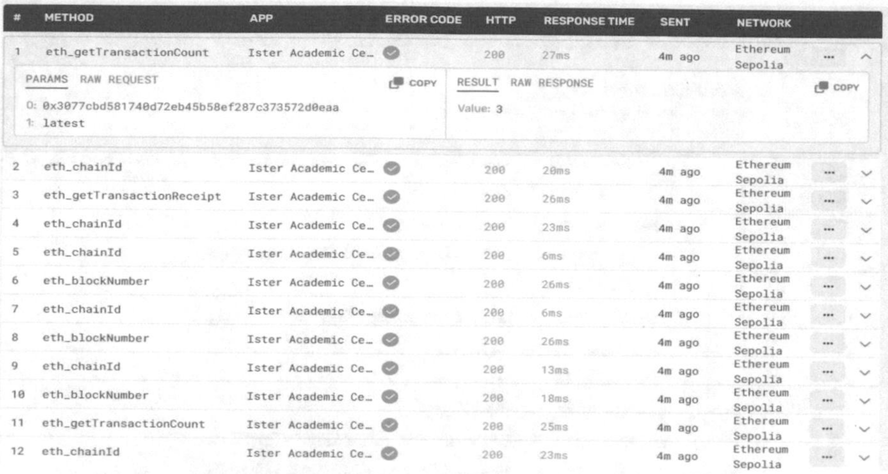

#	METHOD	APP	ERROR CODE	HTTP	RESPONSE TIME	SENT	NETWORK
1	eth_getTransactionCount	Ister Academic Ce…		200	27ms	4m ago	Ethereum Sepolia
	PARAMS RAW REQUEST COPY 0: 0x3077cbd581740d72eb45b58ef287c373572d0eaa 1: latest			RESULT RAW RESPONSE Value: 3			COPY
2	eth_chainId	Ister Academic Ce…		200	20ms	4m ago	Ethereum Sepolia
3	eth_getTransactionReceipt	Ister Academic Ce…		200	26ms	4m ago	Ethereum Sepolia
4	eth_chainId	Ister Academic Ce…		200	23ms	4m ago	Ethereum Sepolia
5	eth_chainId	Ister Academic Ce…		200	6ms	4m ago	Ethereum Sepolia
6	eth_blockNumber	Ister Academic Ce…		200	26ms	4m ago	Ethereum Sepolia
7	eth_chainId	Ister Academic Ce…		200	6ms	4m ago	Ethereum Sepolia
8	eth_blockNumber	Ister Academic Ce…		200	26ms	4m ago	Ethereum Sepolia
9	eth_chainId	Ister Academic Ce…		200	13ms	4m ago	Ethereum Sepolia
10	eth_blockNumber	Ister Academic Ce…		200	18ms	4m ago	Ethereum Sepolia
11	eth_getTransactionCount	Ister Academic Ce…		200	25ms	4m ago	Ethereum Sepolia
12	eth_chainId	Ister Academic Ce…		200	23ms	4m ago	Ethereum Sepolia

Fig. 6. Alchemy Explorer displaying information related to the Ister Account and Ethereum network parameters.

- Transaction Hash: This is a unique identifier for each transaction. It serves as a reference point to track and verify the transaction's details and status on the blockchain.
- Network: Indicates the specific network on which the transaction is taking place. This helps ensure that the transaction is being executed on the intended network.
- Block Number: The block number in which the transaction is included. It helps determine the position of the transaction in the blockchain's chronological order.
- Gas: Gas represents the computational cost required to execute the transaction. It is a fundamental concept in Ethereum and ensures that participants are fairly incentivized for their computational work.
- From Address: The address of the sender who initiated the transaction. In the context of our academic certification smart contract, this could be the address of the academic institution or the authorized issuer.
- To Address: The address of the recipient or the destination of the transaction. In the case of academic certifications, this would typically be the address of the individual receiving the certificate.
- Transaction Type: Indicates the type of transaction being executed. In this case, it is labeled as EIO-1559, which refers to a specific type of transaction related to the Ethereum Improvement Proposal (EIP) 1559. EIP-1559 aims to improve transaction fee predictability and efficiency on the Ethereum network.

Etherscan. Etherscan is a widely used blockchain explorer and analytics platform for the Ethereum network. It provides users with a comprehensive view

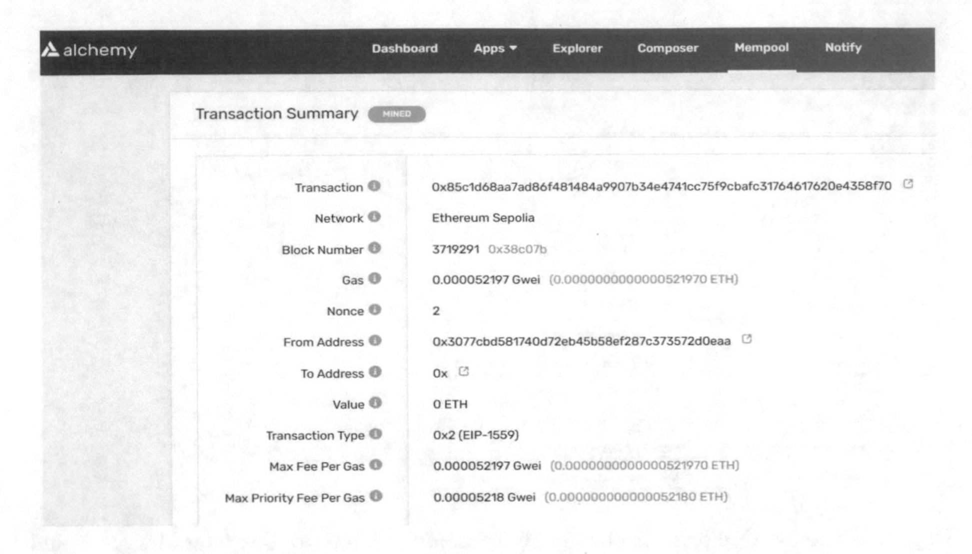

Fig. 7. Transaction Summary in Alchemy Mempool.

of the Ethereum blockchain, allowing them to explore and verify transactions, smart contracts, addresses, and other on-chain activities. Etherscan offers various features, including the ability to search for specific transactions, view account balances, track token transfers, and monitor contract interactions. It also provides detailed information about block confirmations, gas fees, and network statistics. Overall, Etherscan is a valuable tool for individuals and businesses to gain insights and transparency into the Ethereum blockchain ecosystem.

Figure 8 shows the transaction details on Etherscan for the deployment of the Academic Certificate smart contract. It includes important information such as the contract address and transaction hash.

The contract address represents the unique identifier of the deployed smart contract on the Ethereum blockchain. It serves as the entry point for interacting with the contract and accessing its functionalities.

The transaction hash is a unique identifier for the transaction that deployed the smart contract. It can be used to track and verify the transaction's status and details, including the sender, recipient, gas used, and block confirmation.

Etherscan provides a comprehensive view of these transaction details, enabling users to validate and audit smart contract deployments. It also offers additional information such as the transaction timestamp, gas price, and cumulative gas used. This data helps ensure transparency and accountability in the deployment process.

Finally, Fig. 9 displays the transaction details on Etherscan for the successful deployment of the Academic Certificate smart contract on the Sepolia network. The information presented in this figure is similar to the transaction summary shown in the Alchemy dashboard.

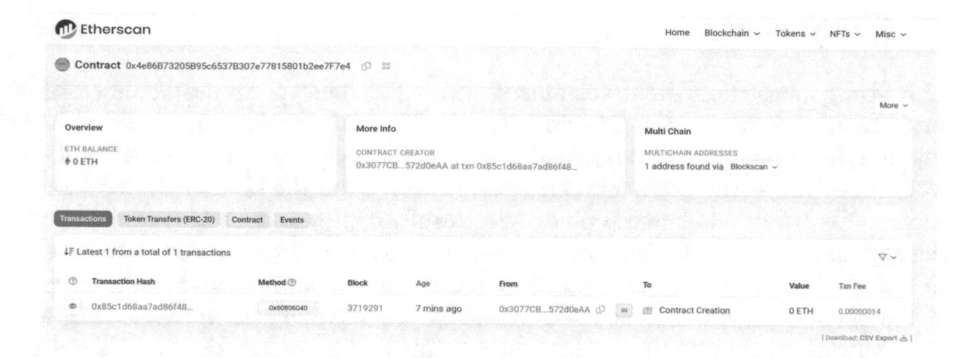

Fig. 8. Etherscan Transaction Details for Academic Certificate Deployment.

The transaction details include the transaction hash, which uniquely identifies this specific transaction on the blockchain. Additionally, it provides essential information such as the network (Sepolia), block number, gas used, and the sender and recipient addresses. These details confirm that the Academic Certificate smart contract was deployed on the Sepolia network and validate its successful execution.

The inclusion of this figure reinforces the successful deployment of the smart contract and highlights the transparency and public accessibility of transaction details on the Ethereum blockchain through platforms like Etherscan.

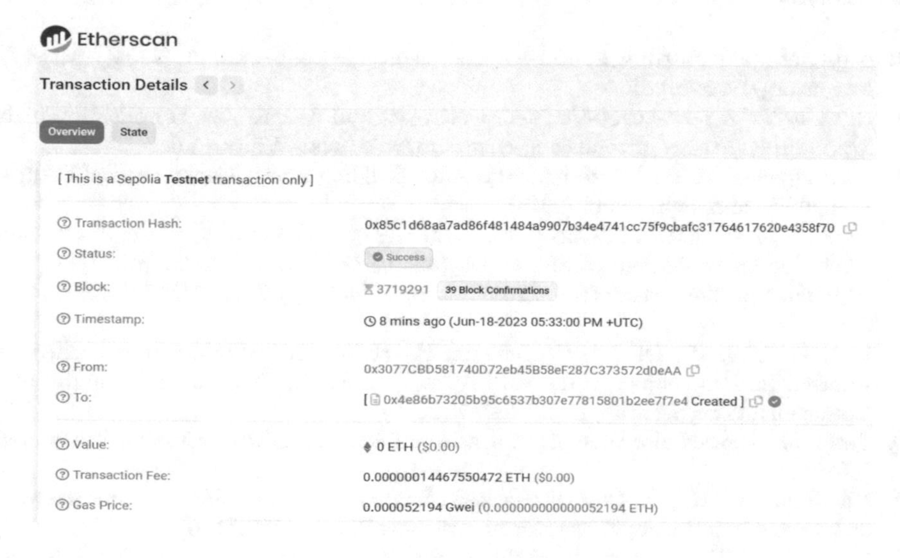

Fig. 9. Etherscan Transaction Details for Academic Certificate Deployment.

6 Conclusion

This study focused on the development and deployment of an academic certificate smart contract on the Sepolia network. Through rigorous unit testing, we ensured the reliability and functionality of the smart contract, validating its ability to issue and verify academic certificates securely on the blockchain. Utilizing Alchemy as our development tool, we successfully deployed the smart contract on the Sepolia network. The deployment process was smooth, and we observed positive outcomes, including the acquisition of the contract address and confirmation of successful deployment. The use of Alchemy provided us with valuable insights into network performance metrics, such as median response time and success percentage, further reinforcing the efficiency and effectiveness of the deployed smart contract. Furthermore, our exploration of Etherscan allowed us to delve deeper into the transaction details, confirming the transparency and accessibility of blockchain data. This demonstration solidifies the successful deployment of the academic certificate smart contract on the Sepolia network. Overall, this study showcases the potential of blockchain technology, specifically smart contracts, in revolutionizing academic certification systems. By leveraging the capabilities of the Sepolia network, Alchemy, and Etherscan, we have achieved a robust and transparent solution for issuing and verifying academic certificates. Moving forward, further research and development in this area can unlock even more opportunities for secure and decentralized certification systems.

References

1. Nakamoto, S.: Bitcoin: a peer-to-peer electronic cash system (2008). https://bitcoin.org/bitcoin.pdf
2. Buterin, V.: A next-generation smart contract and decentralized application platform (2014). https://github.com/ethereum/wiki/wiki/White-Paper
3. Antonopoulos, A.M.: Mastering Ethereum: Building Smart Contracts and DApps. O'Reilly Media, Sebastopol (2018)
4. Atzei, N., Bartoletti, M., Cimoli, T.: A survey of attacks on ethereum smart contracts (SoK). In: Maffei, M., Ryan, M. (eds.) POST 2017. LNCS, vol. 10204, pp. 164–186. Springer, Heidelberg (2017). https://doi.org/10.1007/978-3-662-54455-6_8
5. Luu, L., Chu, H., Olickel, H., Saxena, P., Hobor, A.: Making smart contracts smarter. In: Proceedings of the 2016 ACM SIGSAC Conference on Computer and Communications Security, pp. 254–269 (2016)
6. Swan, M.: Blockchain: Blueprint for a New Economy. O'Reilly Media, Sebastopol (2015)
7. Yli-Huumo, J., Ko, D., Choi, S., Park, S., Smolander, K.: Where is current research on blockchain technology? a systematic review. PLoS ONE **11**(10), e0163477 (2016)
8. Li, Q., Xu, L.D., Zhao, S.: Blockchain-based smart homes: a survey. IEEE Access **5**, 22927–22942 (2017)
9. Sivaraman, V., Maesa, D.D.: Blockchain applications and use cases: a systematic review. Futur. Gener. Comput. Syst. **97**, 512–529 (2019)

10. Koens, T., Harmsze, F.: Blockchain in education: an overview of opportunities and challenges. In: Proceedings of the 2017 IEEE Frontiers in Education Conference (FIE), pp. 1–9 (2017)
11. Choo, K.K.R., Dhillon, G., Sun, H.: Security of blockchain: a systematic review. Digital Commun. Netw. **4**(2), 118–137 (2018)
12. Tödling, F., Quirchmayr, G., Seufert, M.: Blockchain for academic certificates: what are the requirements? In: Proceedings of the 2nd International Workshop on Cryptocurrencies and Blockchain Technology (CBT), pp. 105–116 (2019)
13. Nascimento, A.F., Silva, A., Moreira, E., Peixoto, G.: An overview of blockchain technology: architecture, consensus, and future trends. J. Inf. Syst. Eng. Manag. **4**(3), 17 (2019)
14. Xu, M., Chen, Y., Yu, S.: A blockchain-based framework for secure and efficient certificate verification. In: Proceedings of the 2019 IEEE International Conference on Communications (ICC), pp. 1–6 (2019)
15. Bhargava, B., Garg, S., Kapoor, A.: Blockchain technology for secure storage of academic certificates. In: Proceedings of the 2019 IEEE International Conference on Smart Computing and Electronic Enterprise (ICSCEE) (2019)

Designing a Water Level Measurement System to Monitor the Flow of Water from the Primary Catchment Source of Tulcán City Using LoRa Communication Technology

Paul Ramírez Ortega[1], Edgar Maya-Olalla[1(✉)], Hernán M. Domínguez-Limaico[1], Marcelo Zambrano[1,2], Carlos Vásquez Ayala[1], and Marco Gordillo Pasquel[1]

[1] Universidad Técnica del Norte, Avenida 17 de Julio 5-21 y General María Córdoba, 100105 Ibarra, Ecuador
eamaya@utn.edu.ec

[2] Instituto Superior Tecnológico Rumiñahui, Avenida Atahualpa 1701 y 8 de Febrero, 171103 Sangolquí, Ecuador

Abstract. This study focuses on utilizing LoRa wireless communication technology to create a sensor network that measures the water level of a river. The collected data can help assess the current state of the river and detect any changes in its level. Alerts can be sent via email if any changes are detected. The system consists of various components. Firstly, sensor nodes are utilized that employ HC-SR04 sensors for level calculation. Secondly, the ESP32 microcontroller is connected to the LoRa RFM95 module for wireless transmission using LoRa. Finally, the microcontroller handles the processing tasks. The central node acts as a receiver for the data gathered by the sensor nodes. It consists of an Arduino Uno compatible with the LoRa RFM95 module. The central node is connected to a Raspberry PI 4 for processing and then sending the data using the MQTT protocol to the server. The information can then be viewed on the ThingSpeak platform. Tests have been conducted to verify the established communication between the nodes and the Gateway. The results have shown that the LoRa communication parameters are accurately established in the RFM95 modules. Additionally, the data sent from the nodes to the Gateway is visualized on ThingSpeak.

Keywords: LoRa · LoRa Gateway · ThingSpeak · catchment source

1 Introduction

Low-power communications and data transmission are the main objectives of the LoRa Alliance association, which developed and promoted the LoRa technology.

M. Z. Vizuete et al. (Eds.): CI3 2023, LNNS 1040, pp. 138–152, 2024.
https://doi.org/10.1007/978-3-031-63434-5_11

In addition, the use of physical layer techniques allows for coverage of wide areas in terms of kilometers, as reported by [1].

LoRa has been chosen as the technology for designing the wireless sensor network because it provides advantages over its primary rival, SigFox. In contrast to SigFox, LoRa's standard modules serve as both transmitters and receivers without any limitation on data transmission. The acronym LoRa represents 'long-range', indicating one of its primary features. This is achieved through the utilization of a spread spectrum modulation technique known as spread spectrum chirp technology or CSS. LoRa technology is ideal for usage in wireless sensor networks and the Internet of Things due to its ability to implement a wide area network with low power consumption. This enables the maintenance of hundreds of devices connected through a wireless link to a hub or Gateway, which communicates with all network nodes, manages traffic, and shares information to the required destination [2].

A system for measuring the water level of Tulcan City's main water supply source will be developed using wireless communication technology. The system will use a network of wireless sensors connected through LoRa communication technology to identify high and low water levels witnessed in the source's channel.

Awareness of the water level in a water source is vital due to climate change leading to fluctuations in ambient temperature, rainfall frequency, and volume. Knowledge of this physical variable can indicate the potential for floods or droughts and can enable measures to decrease or increase water levels. The water resource level will be measured using an ultrasonic sensor that operates non-invasively without direct water contact. The microcontroller board equipped with a LoRa module processes the sensor signal to facilitate long-range communication with the concentrator node or the dedicated LoRa Gateway board. Processed data will be sent to the cloud and can be accessed via the Internet. The system monitors the water resource level daily, enabling frequent checks to identify changes in water levels and detect increasing or decreasing trends within the flow.

2 Related Work

The paper [3] addresses the problem of efficiently monitoring water levels in natural environments as a disaster prevention measure. Developed by Shinshu University in Japan, the system's objective is to effectively transmit water level, temperature, and humidity data. To achieve this, it employs LoRa technology in the 429 MHz band and a technique called Packet Level Index Modulation (PLIM). The system was implemented in Nagano City, Japan, for real-world testing. The study's conclusions indicate that the system demonstrated stable information aggregation and the ability to avoid packet collisions.

The study [4] tackles the challenge of real-time monitoring and predicting flash floods in Uttarakhand, India, using the Internet of Things (IoT). The aim is to create a low-cost, energy-efficient system that employs a variety of sensors and an Arduino UNO microcontroller. This data is then analyzed using a Long

Short-Term Memory (LSTM) model to issue real-time flood alerts. The system has shown high accuracy, with F1 scores of 97–98% for different alert levels, making it a valuable tool for disaster management.

In [5], the study focuses on addressing the issue of unanticipated flooding by developing an IoT-based monitoring system. Using an ultrasonic sensor and a NodeMCU microcontroller, the system sends data to the ThingSpeak platform for real-time monitoring and alert generation. Although tested in an academic setting, the system has the potential for application in flood-prone areas. The findings indicate that the system is effective in detecting changes in water levels and issuing alerts, which could be crucial for flood preparedness and response.

The authors of the paper [6], addresses the issue of inefficient water level management in dams, which can lead to either man-made or natural disasters. The aim is to develop an IoT-based monitoring system that can automatically route water from dams into canals. This system was tested using a model that includes an Arduino Uno microcontroller and an ultrasonic sensor to measure water levels in real-time. The data is sent to a base station and then stored in the cloud, where a command center makes decisions about opening or closing the dam gates. The findings suggest that the system could be effective in managing water resources and preventing calamities.

The paper [7] addresses the issue of real-time monitoring of water levels in the Al-Gharraf river in Wasit city, Iraq. The aim is to design a low-cost system that uses a microprocessor and an ultrasonic sensor to track water levels and relay the data via GPRS technology. The system was implemented at regulator No.1 in Al-Hay city and utilizes an Arduino Uno board and a JSN-SR04T ultrasonic distance sensor. The data is visualized on a ThingSpeak platform and updates every three minutes. The system achieved an accuracy rate of 96.84% and suggests that it could be improved with an additional network of sensors.

The study [8] tackles the issue of water pollution in the river in Riau Province, Indonesia. It aims to develop a smart sensor node to monitor multiple water parameters such as temperature, pH, dissolved oxygen, and electrical conductivity using Wireless Sensor Networks (WSN). The system was implemented in a laboratory setting and plans to be tested on the actual site. It utilizes a microcontroller and a variety of sensors, and the data is sent to a back-end system for analysis. Initial results show good agreement between theoretical and measured data, suggesting that the system is ready for real-environment testing.

The paper [9] focuses on the issue of water pollution in the Citarum River in Indonesia. The aim is to develop a fuzzy logic-controlled floating robot to monitor water quality in real-time. The system was tested in a controlled environment and uses an Arduino Uno microcontroller and sensors like the MPU6050 for stabilization. The robot can be manually or automatically controlled and sends data via an HC05 Bluetooth module. Although the robot is still not perfect in maintaining its position and has a tendency to sink due to the PLA material used, the results show that the system has a 92.67% accuracy in distance calculation using the Haversine Formula.

3 Development of the Water Level Monitoring System

The project is being developed in the parish of Rio Chico, Tufiño, which corresponds to the city of Tulcán in the province of Carchi. Rio Chico is a tributary of the river of the same name, which is recognised as a source for purifying water due to its special characteristics.

3.1 System Design

The system has been designed with three stages in mind: the sensing stage, the control stage, and the visualization stage. The system comprises three stages. The first stage collects data, specifically the water level variable. The control stage utilizes a control board for data processing while the visualization stage presents detailed information of the measured physical variable.

The system has been designed with three stages in mind: the sensing stage, the control stage, and the visualization stage. The system comprises three stages. The first stage collects data, specifically the water level variable. The control stage utilizes a control board for data processing while the visualization stage presents detailed information of the measured physical variable.

The control stage utilizes a printed circuit controller board and a LoRa module, which establish communication with the sensors. The data collected is subsequently processed by this element and sent to the next respective stage.

The visualization stage involves observing different level measurements obtained from ultrasonic sensors on a web page accessible on the internet. This enables the variation of the water flow level to be viewed, along with other relevant information.

3.2 System Block Diagram

In Fig. 1 describe the system block diagram consists of the layers of the IoT architecture. It describes the essential elements and requirements to be employed in each layer for designing the proposed system in a general manner.

The device layer comprises ultrasonic sensors that acquire data for measurement. These sensors are connected to a micro-controller board, which executes the respective programming. Furthermore, a LoRa module facilitates wireless communication to the LoRa Gateway via this technology. In the Gateway layer, the printed circuit board or controller board with high processing capabilities is present. It is connected to a LoRa module to enable communication with the sensor nodes, forming the central node, or LoRa Gateway. This is the element of the network that receives all data collected from the sensors. The network layer establishes an internet connection, which can be achieved through wired or wireless media. In this case, the data from the LoRa Gateway is transmitted wirelessly to the servers using the MQTT connection protocol. At the Cloud layer or data center, received data is stored before being transmitted to the final layer, i.e., the application layer. Here, the data is displayed using graphical tools, such as a dashboard.

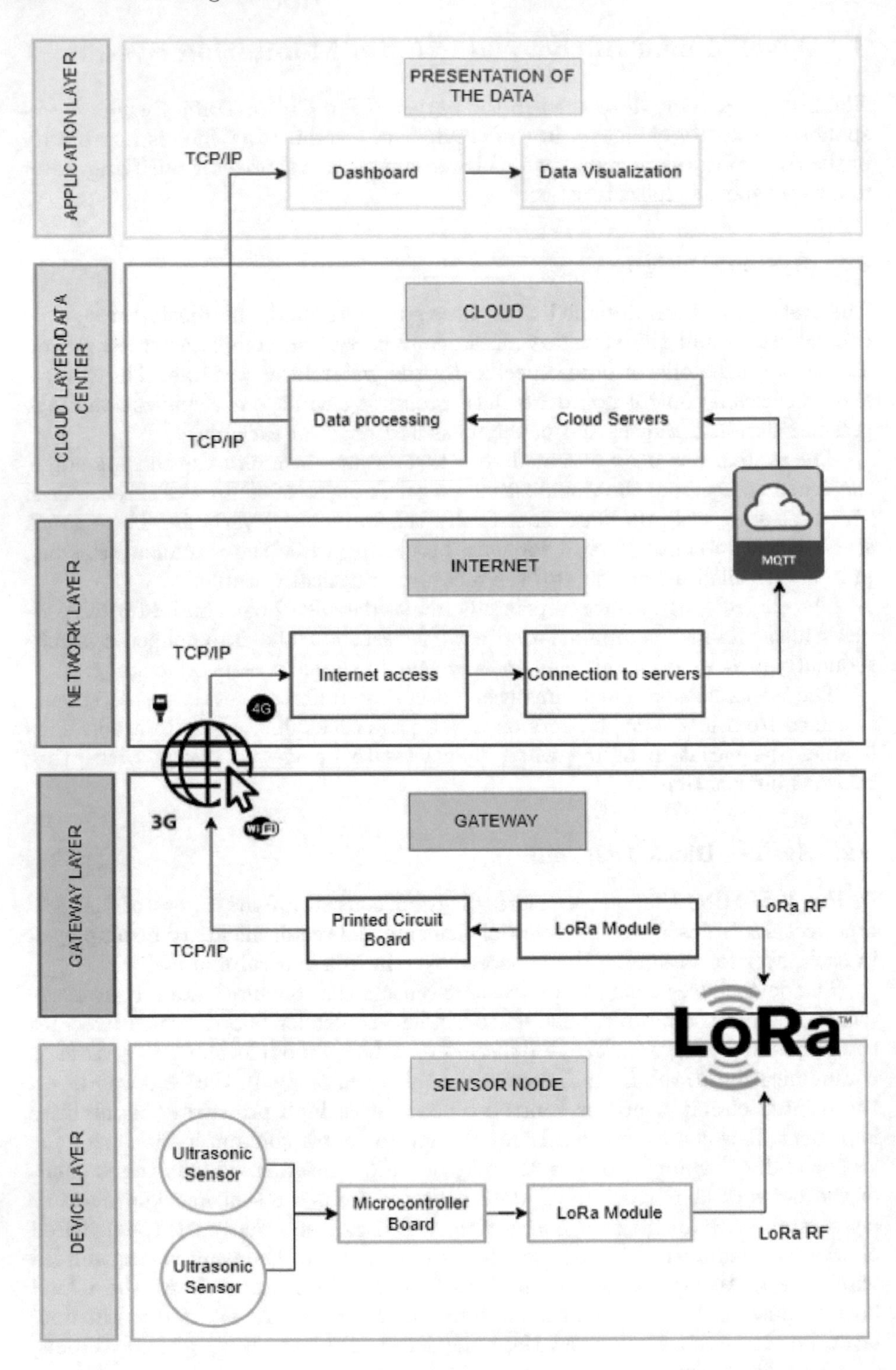

Fig. 1. The block diagram of the system with each layer of the IoT architecture.

3.3 System Architecture of the Water Level Monitoring System

The system architecture, shown in the figure below Fig. 2, represents a star network consisting of the first element, namely the collector nodes, which utilize ultrasonic sensors for measuring water levels. The ESP32 board, acting as a microcontroller, is responsible for the functioning of the sensors and the programming code. For communication between the sensors and the LoRa Gateway, LoRa RF95 modules are employed. The Gateway itself involves communication between an Arduino UNO and a Raspberry Pi 4, linked through a serial port. The final stage consists of processing data in the cloud, after which it is visualized through the use of the IoT platform ThingSpeak and made available for subsequent access by end devices.

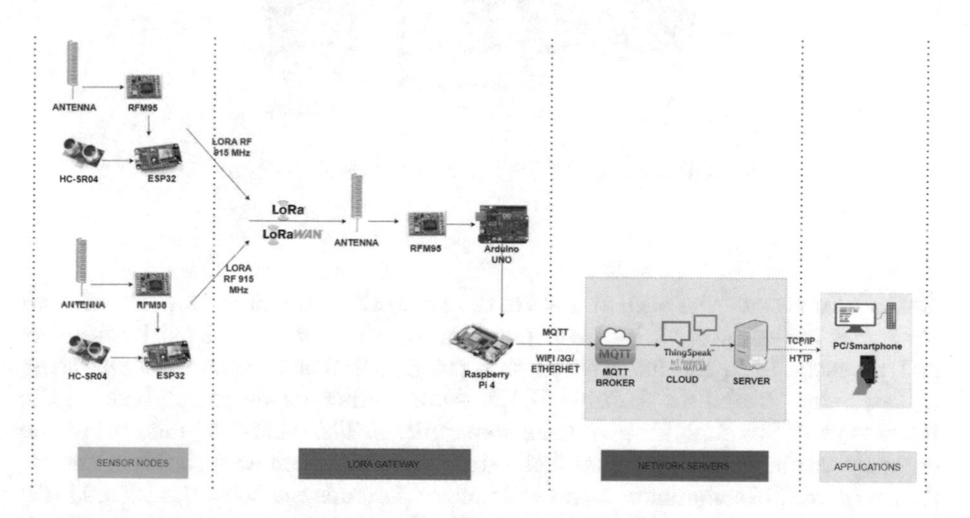

Fig. 2. System Architecture of the Water level monitoring system.

3.4 Interconnection Diagrams

After analyzing the hardware requirements and selecting the components for the sensor network development, connection diagrams for each element are presented to ensure their proper operation, including the sensor node and the LoRa Gateway node.

Sensor Node. The sensor node is composed of an ESP32 microcontroller and an ultrasonic sensor to measure water levels in a flow. The accompanying diagram outlines the connections between the two elements for their respective operation. In Fig. 3, below are the details of how each element is connected to the corresponding pins:

- The Vcc power supply pin of the HC-SR04 sensor is connected to the 5V power supply pin of the ESP32.
- The Trigger pin of the HC-SR04 sensor is connected to pin 16 of the ESP32.
- The Pin Echo of the HC-SR-04 sensor is connected to pin 17 of the ESP32.
- The GND pin, corresponding to the ground of the HC-SR04 sensor, is connected to the GND pin of the ESP32.

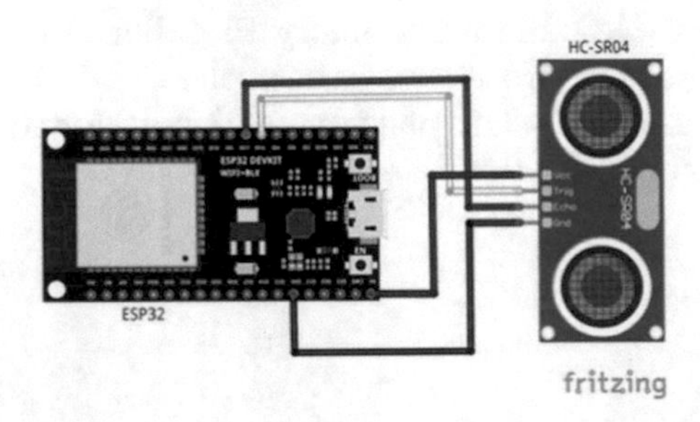

Fig. 3. C-SR04 and ESP32 sensor connection diagram.

LoRa Gateway. The central node is the network component to which all sensor nodes are connected. Its main function is to receive data from the sensors and transmit it to the internet. In this setup, a LoRa Gateway is built using a Raspberry Pi and an Arduino UNO. About other development boards, the Raspberry Pi has superior processing capabilities. This element functions as the gateway within the network that facilitates communication with the sensor nodes that require LoRa communication technology. For this purpose, the RFM95 [13] module is connected to the Arduino UNO as illustrated in the diagram in Fig. 4. The Arduino UNO, in turn, is connected to the Raspberry Pi through a serial connection. The connections made are further detailed below:

- The 3.3 V power pin of the RFM95 module should be connected to the 3.3V pin on the Arduino UNO.
- The ground pin of the RFM95 module must be connected to the ground pin of the Arduino UNO.
- Pin 2 on the Arduino UNO should be connected to the DIO0 pin on the RFM95 module.
- The RST pin of the RFM95 module must be connected to pin 9 on the Arduino UNO.
- Pin 10 on the Arduino UNO should be connected to the NSS pin on the RFM95 module.
- Pin 11 of the Arduino UNO is connected to the MOSI pin of the RFM95 module.
- Pin 12 of the Arduino UNO is connected to the MISO pin of the RFM95 module.

- Pin 13 of the Arduino UNO is connected to the SCK pin, which corresponds to the clock signal of the RFM95 module.
- The antenna is connected to the ANT pin and operates at the required frequency based on its size.
- To read the data via serial communication, the Arduino UNO is connected to the Raspberry PI through a USB cable.

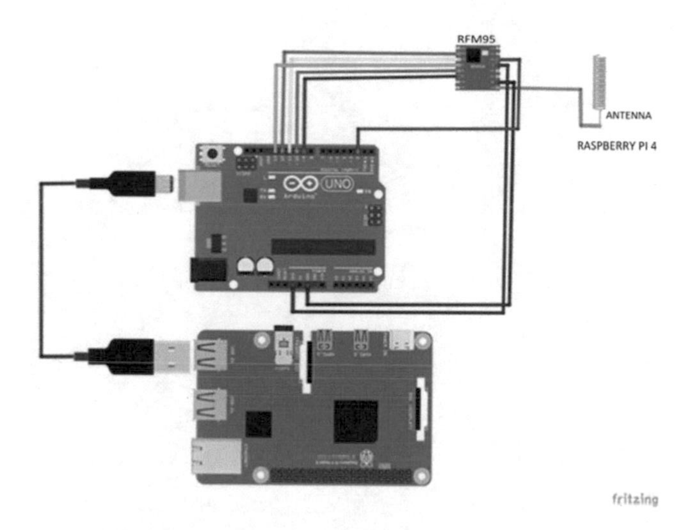

Fig. 4. Connection diagram of LoRa Gateway.

System Flow Diagram. The system flow diagram shows the processes involved and the different functions described in Fig. 5. The process begins with connecting the LoRa RFM95 modules to the ESP32 microcontroller using digital pins to establish wireless communication via LoRa technology. The modules are then verified for proper initialization, which is necessary for establishing a connection to the LoRa Gateway. To measure river flow level, the ultrasonic sensors are connected to the ESP32 microcontroller using its pins. Verification is done to ensure that the sensors are operating correctly, and the data obtained from them is then sent to the LoRa Gateway. Specifically, the LoRa Gateway consists of an Arduino UNO board connected to the Raspberry Pi 4 via serial port. Afterwards, the system confirms that the data is received at this network element and then measures whether the flow level increased or decreased compared to the average value. This information is then used to send out an email alert. Following this, the data is sent to the ThingSpeak platform for cloud storage, which can later be accessed online.

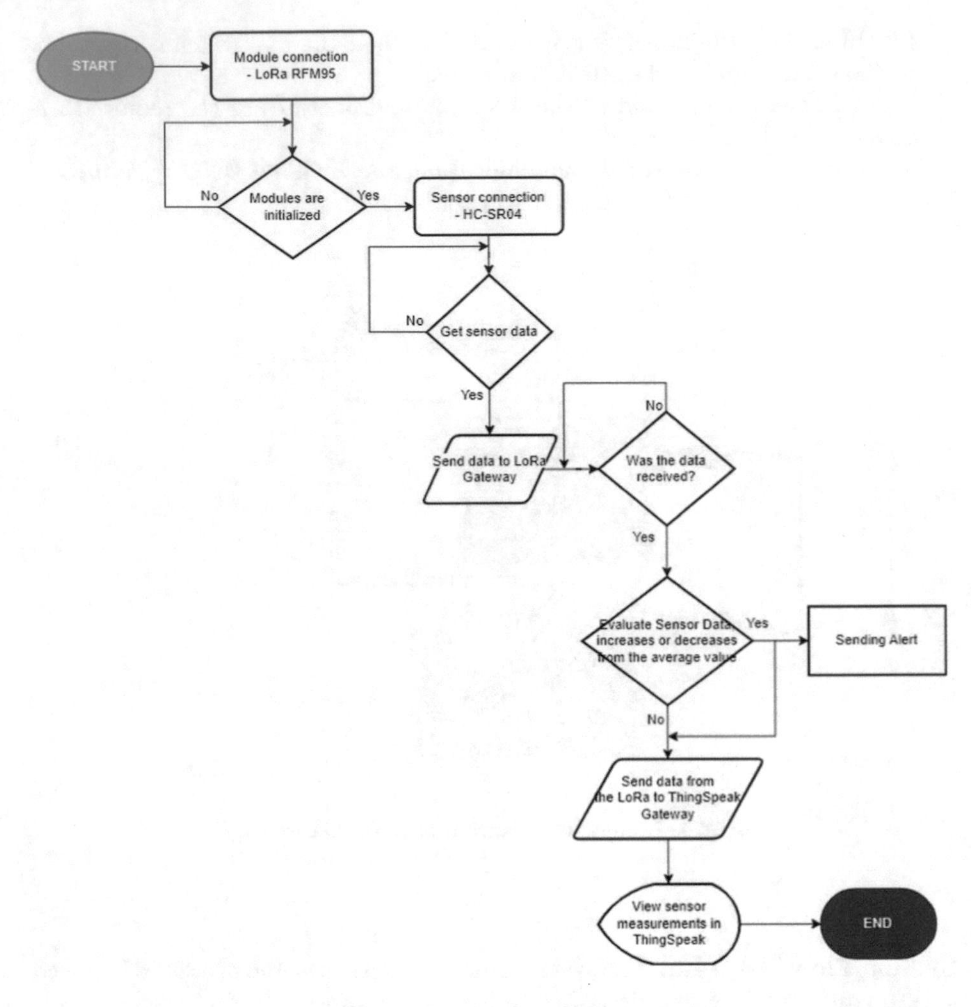

Fig. 5. System flow diagram.

4 Results

4.1 Component Integration

After selecting the appropriate hardware for each network element, the equipment is physically connected according to the established configurations in the previous step.

Peripheral devices are connected to the ESP32 board for the sensor node. Figure 1 illustrates the implementation of the sensor node. It comprises of an HC-SR04 ultrasonic sensor connected to the ESP32 board and a LoRa RFM95 module also connected to the ESP32 board. These are powered by a lithium battery with 5 V input for autonomous operation. The assembled sensor node inside its case looks as Fig. 6:

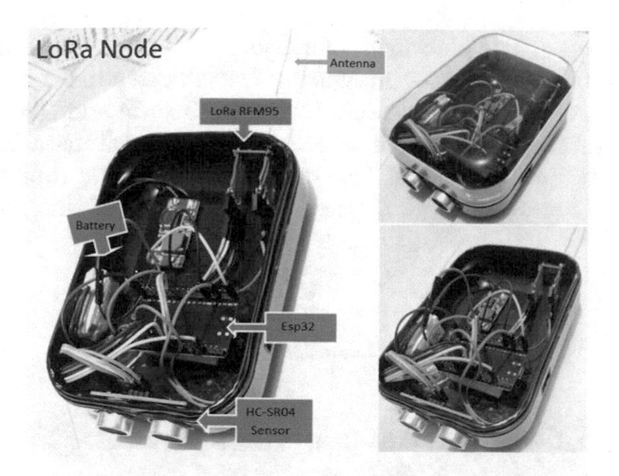

Fig. 6. LoRa sensor node.

The LoRa Gateway obtained the data collected through the RFM95 module using serial communication between the Arduino UNO and the Raspberry PI. The Raspberry PI 4 processed the data, and it was sent to the ThingSpeak platform. The assembled LoRa Gateway, or central node, can be viewed in its casing. Refer to Fig. 7 for details.

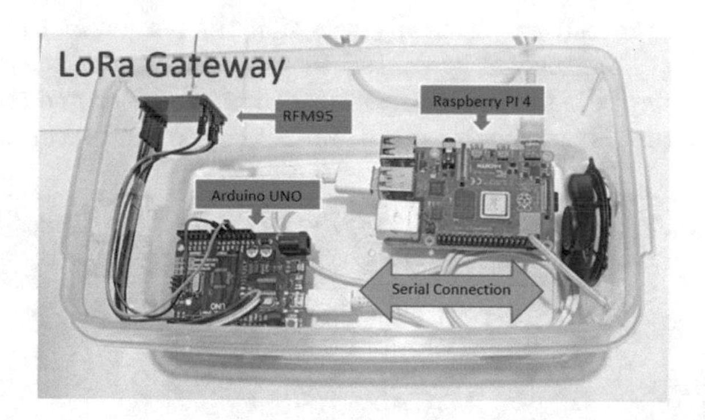

Fig. 7. Gateway LoRa.

4.2 Performance Tests

During the system operation tests, data is obtained from each sensor node, confirming the correct initialization of the LoRa RFM95 module and accuracy of the level value (in centimeters) retrieved from the sensors. Moreover, the LoRa Gateway's capability to receive the data correctly is verified. This ensures the data can be further processed and sent to the ThingSpeak platform without loss.

System Implementation. As previously analyzed in the requirements and system design phase, the project will be implemented in Rio Chico. Two sensor nodes will be placed to measure the level of the river water at two different geographic locations. The placement of sensor nodes takes into account that the water level can be calculated by measuring the cross-section of the river, which includes the distance from the bottom to the surface. In addition, to obtain the river level in centimeters, it is essential to know the total distance between the bottom of the river and the node. This can be calculated by subtracting the two measurements. Figure 8 displays a visual representation of this process.

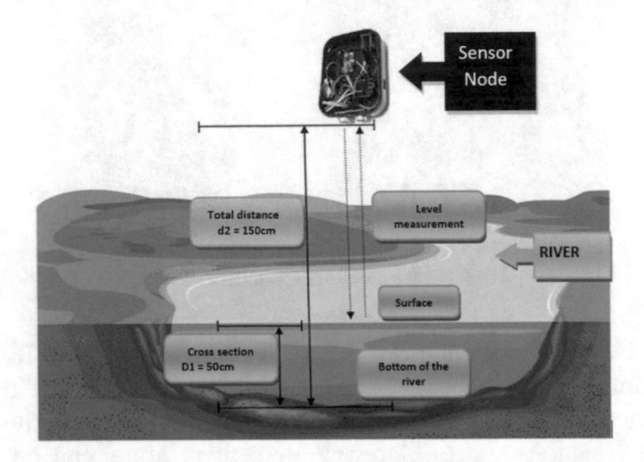

Fig. 8. Sensor node location and measurements.

We will now proceed to physically position the sensor in the chosen location, considering the measures presented in Fig. 8 for proper sensor functionality. Figure 9 displays the sensor node positioned on the river and supported by a PVC pipe structure. This allows for obtaining the required level measurement.

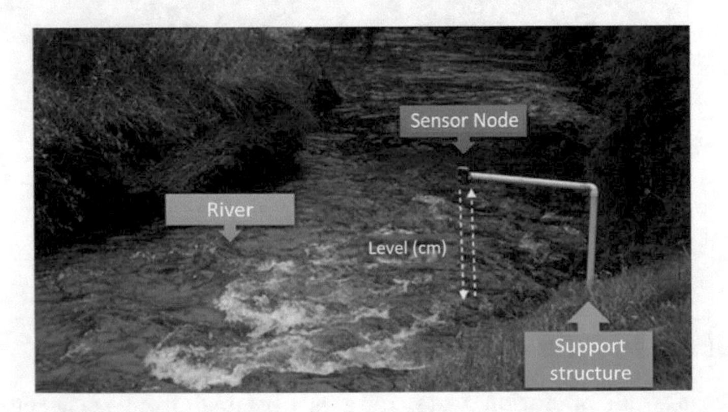

Fig. 9. Location of the sensor node on the river.

Communication Between the Sensor Node and the LoRa Gateway. In this test in Fig. 1, the sensor node is powered by a lithium battery, which allows it to operate autonomously without the need for a USB cable. This feature provides mobility to the node, which can collect and transmit data from its current location (Fig. 10).

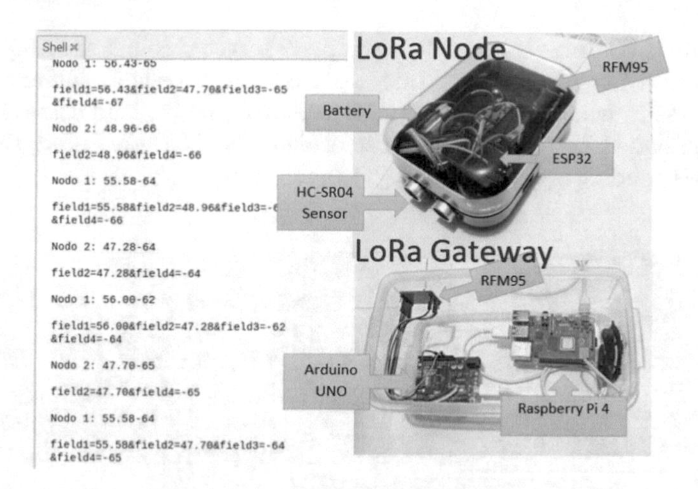

Fig. 10. Receiving data from the sensor node at the Gateway.

Receiving and Displaying Data in the ThingSpeak Platform. Data is sent to the ThingSpeak platform using the MQTT protocol as observed Fig. 11. One can observe the measurements for both Node 1 and Node 2 levels on the platform. Additionally, the RSSI of the signal, expressed in dBm, can be seen to check the signal quality for both nodes.

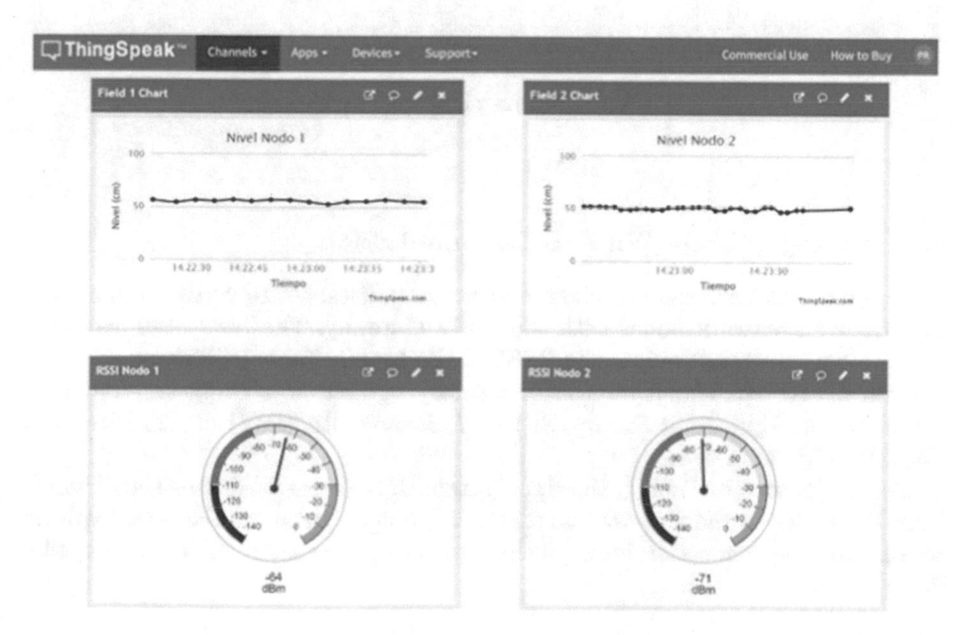

Fig. 11. Visualization platform ThingSpeak.

The data is transmitted to the ThingSpeak server through the MQTT protocol as observed Fig. 11. To verify this, we used the Wireshark sniffer to capture the packets sent, which revealed the used protocol, the payload corresponding to the fields intended for publication in the broker, the TCP port used, the address of the server, and the packet size (Fig. 12).

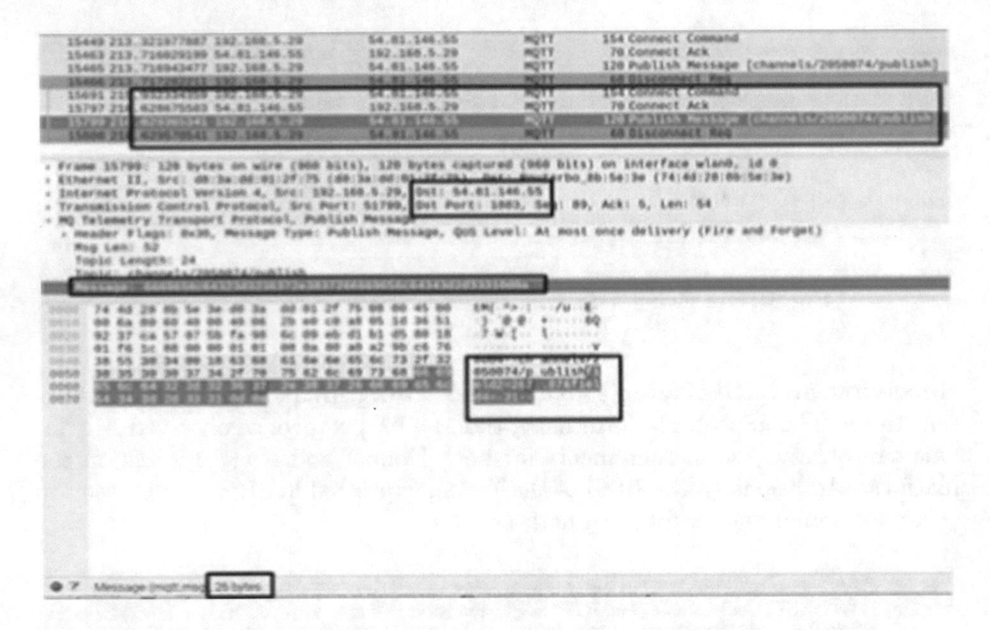

Fig. 12. Visualization platform ThingSpeak.

4.3 Testing of LoRa Wireless Communication

Communication tests were conducted at varying distances to verify communication between network nodes with the LoRa Gateway. The tests obtained data sent by the sensor as well as the RSSI of the signal. Table 1 illustrates a comparison of received signal intensities for varying distances using the following configuration: Spreading Factor (SF) of 7, Bandwidth (BW) of 125 kHz, and Code Rate (CR) of 4/5.

As can be seen in Table 1, the signal strength becomes weaker as the distance from the nodes to the Gateway increases, for longer distances the signal will be weaker and the connection between the nodes and the central node may even be lost.

Table 1. Distance and RSSI Measurements

Distance (m)	RSSI Node 1 (dBm)	RSSI Node 2 (dBm)
5	−59	−61
10	−64	−72
20	−76	−83
30	−84	−86
40	−89	−88
50	−90	−94
70	−110	−117

5 Conclusions

The developed system meets the project's needs by enabling the sensors to read data for water level measurements. Furthermore, LoRa wireless communication technology is used to transmit this data. By analyzing the collected data, it's possible to determine the variation in the water level and relate it to the climatic conditions of the area.

Sensor nodes continually transmit the river level's value to display on the platform. Collecting this data makes it possible to determine if the level is consistent, increasing, or decreasing. From this data, we can analyze whether there is rainfall or drought in the area.

Tests establish correct module configurations according to LoRa requirements since the data collected from sensor nodes can transmit to the Gateway for processing and visualization.

LoRa communication technology widely employed in wireless sensor networks interconnects devices separated over long distances. This feature sets it apart from other wireless communication technologies.

References

1. Narváez Narváez, K.S., Contreras Pérez, V.A.: Design and development of a prototype IoT sensor network using LoRaWAN technology for monitoring indoor and outdoor environmental parameters. Universidad Politécnica Salesiana. Sede Guayaquil, Guayaquil (2020)
2. Maldonado Borda, C.A., Pineda Cusba, L.: Design of a wireless network for the management and control of an integral system of irrigation systems. Universidad Distrital Francisco José de Caldas, Bogotá (2016)
3. Koike, Y., Takyu, O.: Water level monitoring system based on 429MHz LoRa with packet level index modulation. In: 2023 Fourteenth International Conference on Ubiquitous and Future Networks (ICUFN), Paris, France, pp. 76–81 (2023). https://doi.org/10.1109/ICUFN57995.2023.10199961
4. Prakash, C., Barthwal, A., Acharya, D.: FLOODWALL: a real-time flash flood monitoring and forecasting system using IoT. IEEE Sens. J. **23**(1), 787–799 (2023). https://doi.org/10.1109/JSEN.2022.3223671

5. Ashokkumar, N., Arun, V., Prabhu, S., Kalaimagal, V., Srinivasan, D., Shanthi, B.: Monitoring and warning of flooding conditions using IoT based system. In: 2023 9th International Conference on Advanced Computing and Communication Systems (ICACCS), pp. 1199–1203. Coimbatore, India (2023). https://doi.org/10.1109/ICACCS57279.2023.10112930
6. Kumar, M., Sapkal, A., Tiwari, V.: Ganga water quality monitoring system IoT based. In: 2018 IEEE Punecon, Pune, India, pp. 1–5 (2018). https://doi.org/10.1109/PUNECON.2018.8745406.
7. Nawar,p A.K., Altaleb, M.K.: A low-cost real-time monitoring system for the river level in Wasit province. In: 2021 International Conference on Advance of Sustainable Engineering and its Application (ICASEA), Wasit, Iraq, pp. 54–58 (2021). https://doi.org/10.1109/ICASEA53739.2021.9733092
8. Kadir, E.A., Siswanto, A., Rosa, S.L., Syukur, A., Irie, H., Othman, M.: Smart sensor node of WSNs for river water pollution monitoring system. In: 2019 International Conference on Advanced Communication Technologies and Networking (CommNet), Rabat, Morocco, pp. 1–5 (2019). https://doi.org/10.1109/COMMNET.2019.8742371
9. Pratama, R.P., Rusdinar, A., Wibawa, I.P.D.: Floating robot control system for monitoring water quality levels in Citarum River. In: 2019 IEEE International Conference on Internet of Things and Intelligence System (IoTaIS), Bali, Indonesia, pp. 206–211 (2019). https://doi.org/10.1109/IoTaIS47347.2019.8980393
10. Camacho, A., Rodriguez, E., Bajaña, J., Jonathan, P., William, C.: WSN networks for monitoring the banks of the San Pablo River in La Maná Canton. Cotopaxi. Universidad Técnica de Cotopaxi, La Maná (2019)
11. CATSENSORS: LoRa and LoRaWAN technology. Retrieved from CATSENSORS (2021). https://www.catsensors.com/es/lorawan/tecnologia-lora-y-lorawan
12. LoRa Alliance: What are LoRa® and LoRaWAN®? Retrieved from SEMTECH.COM. LoRa Alliance (2022). https://lora-developers.semtech.com/documentation/tech-papers-and-guides/lora-and-lorawan/
13. VISTRONICA S.A.S.: MODULO TRANSCEPTOR LORA RFM95: Article title (2023). https://www.vistronica.com/comunicaciones/wifi/modulo-transceptor-lora-rfm95-868mhz-detail.html

Innovation and Development

A Voice-Based Emotion Recognition System Using Deep Learning Techniques

Carlos Guerrón Pantoja[1], Edgar Maya-Olalla[1](✉), Hernán M. Domínguez-Limaico[1], Marcelo Zambrano[1,2], Carlos Vásquez Ayala[1], and Marco Gordillo Pasquel[1]

[1] Universidad Técnica del Norte, Avenida 17 de Julio 5-21 y General María Córdoba, 100105 Ibarra, Ecuador
eamaya@utn.edu.ec

[2] Instituto Superior Tecnológico Rumiñahui, Avenida Atahualpa 1701 y 8 de Febrero, 171103 Sangolquí, Ecuador

Abstract. The project aims to create an emotion recognition system based on voice using deep learning techniques. The system is based on supervised learning with artificial neural networks, enabling it to accurately predict emotions. The system's potential usage in detecting depression pathologies in the psychological area gives rise to its development. The system is designed using the KDD (Knowledge Discovery in Database) methodology and utilizes an existing database containing audio with various emotions. These audios are subjected to Multilevel Wavelet transform, decomposing the original signal into sub-signals with specific characteristics for each audio to form a training data set that is subsequently normalized, followed by the generation of the LSTM neural network architecture. Performance tests are eventually conducted on patients with depressive pathology, involving the application of the "Beck test", which indicates the severity of depression experienced by the patient. As a result, the individual reads a text that is recorded, followed by the process of feature extraction and emotion recognition performed with the pre-trained neural network. The outcome indicates that 50% of the patients exhibit severe depression, while the remainder displays milder symptoms, which is supported by the emotions detected by the system alongside the administered test.

Keywords: Emotion Recognition · Wavelet Transform · deep learning · emotion recognition

1 Introduction

Emotion recognition can determine human attitude through speech-based emotional cues. Therefore, detecting human emotions is crucial for personal healthcare and mental well-being [1]. Emotions are an innate aspect of human beings, and they reflect a range of moods including sadness, anger, happiness, and neutrality. However, issues may arise due to emotional imbalance, including depression which significantly impacts a person's usual functioning, well-being, and

M. Z. Vizuete et al. (Eds.): CI3 2023, LNNS 1040, pp. 155–172, 2024.
https://doi.org/10.1007/978-3-031-63434-5_12

quality of life [2]. Artificial intelligence (AI) is currently a popular trend, with many research projects utilizing this technique to gain new insights and strategies that enhance people's lives. The research aims to study how people's emotions are expressed through their speech patterns and to create a machine-learning reference criterion that is inspired by the biological neural networks of the human brain. These elements are composed in a way that mimics the biological neurons in their most common functions. These elements are organized in a manner that resembles that of the human brain [3].

To conduct the research, it is important to discuss various concepts that aid in the development of the system. This starts with the voice, which is a sound produced by the vibration of vocal cords in living organisms and has qualities such as timbre, intensity, and quality [4]. The process of extracting acoustic parameters of vocal utterances is based on the idea that variations in speech are caused by different states of excitation or valence in the speaker. These variations can be estimated by different wave parameters [5]. The approach for Speech Emotion Recognition (SER) incorporates two main phases: feature extraction and feature classification [6]. One method of feature extraction is the multi-level modeling method. This method uses the Continuous Wavelet Transform (CWT) to model hierarchical prosodic features, such as an initial signal and energy contour [7].

A neural network is a computational model that emulates the behavior of the human brain. It consists of interconnected artificial neurons, represented by nodes or vertices, and connections, represented by edges. In a neural network, the fundamental element is the artificial neuron. It emulates its natural counterpart found in the brain [8].

The main focus of this research is to create an emotion recognition system using deep learning techniques. This system will detect emotions in patients who are attending consultations. The goal is to implement this system in patient care to provide better support to psychologists and offer improved diagnoses to the patient based on professional judgment.

The cascade model will facilitate the selection of system requirements during the system development process, which comprises several successive phases. For instance, the analysis phase addresses the concepts covered in the project. Another phase of the project is the design and specifications where the programming, training based on the Knowledge Discovery in Databases (KDD) methodology, implementation, functional testing, and finally a verification of the results obtained will be made.

Finally, we obtain the results of the tests conducted with the system on various individuals. We classify the effectiveness of measuring emotions and determine the percentage of emotions measured. We identify the emotion(s) with the lowest percentage of accuracy, precision, and sensitivity, enabling the system administrator to consider these values and use this tool effectively.

2 Development

The creation process of the speech emotion recognition system includes multiple stages such as emotion audio acquisition, signal preprocessing, feature extraction, training data preparation, and neural network training. The entire process is aligned with the Knowledge Discovery in Databases (KDD) methodology, represented in Fig. 1

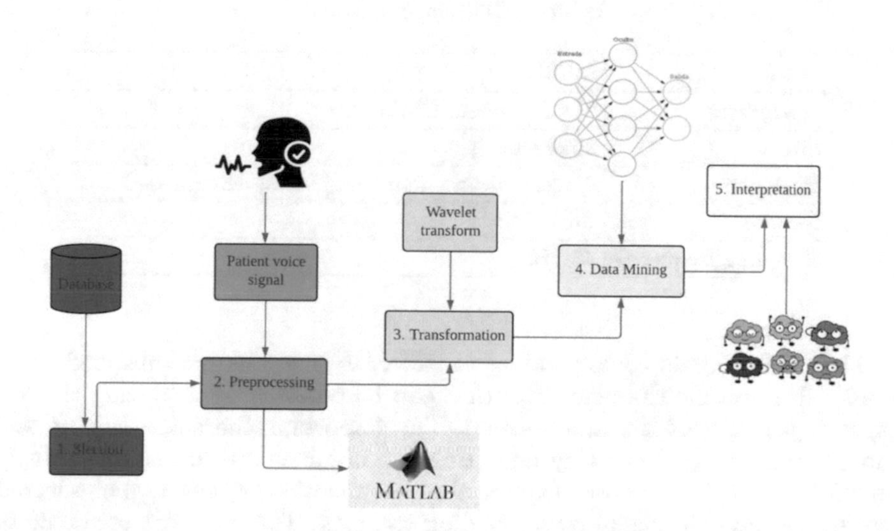

Fig. 1. Generalized KDD process.

2.1 Obtaining Audio of Emotions

Classical data mining tasks require two types of data: a training dataset that trains the algorithm on the problem-specific data and a test dataset that evaluates the quality of the algorithm. Thus, it was necessary to generate at least one audio recording for each human emotion during the data collection phase. However, due to the lack of Spanish audio data, a Mexican audio dataset known as the Mexican Emotion Speech Database (MESD) is utilized.

MESD provides single-word utterances for the affective prosodies of anger, disgust, fear, happiness, neutrality, and sadness with Mexican cultural conformation. The database was retrieved from the Kaggle repository, a scientific community focused on machine learning. The database is licensed for free use, however, proper credit should be given to its author, Saurabh Shahane.

The database consists of single-word utterances expressing affective prosodies of anger, disgust, fear, happiness, neutrality, and sadness. The affective prosodies are selected according to Mexican cultural adaptation. The Multilingual Everyday Speech Dataset (MESD) was recorded by adult actors and non-professional children. Three female, two male, and six child voices are included in the dataset. In this study, a database of 864 voice recordings is evaluated (see Table 1).

Table 1. Database prosodies

Database	MESD
Interpreter	Men, Women, Children
Size	864 recordings
Prosodias	Anger, disgust, fear, happiness, neutral, sadness
License	Free
Sampling frequency	48 kHz

The audio signals corresponding to different emotions are analyzed. Each emotion has specific characteristics that can be reflected by analyzing the corresponding graph of the audio signal. The important characteristics of each emotion can be highlighted by analyzing the obtained raw data. Accordingly, the audio of "Up", which was expressed in the emotion of Anger, was selected. Figure 2 displays the signal corresponding to anger. The first part of the figure shows the signal over time, while the second part shows its frequency spectrum. In Fig. 3, the amplitude of the signal starts with short peaks that are segmented over some samples. This is followed by a gradual increase in amplitude, culminating in very high peaks in the positive and negative axes. This signal is compared to the way anger is expressed vocally, as it begins with a soft tone that ends in a high tone producing a signal as seen in the graph. The signal's energy reaches up to -20dB, as illustrated in the spectrum graph.

2.2 Signal Preprocessing

Preprocessing is a procedure performed on the audio files before analysis, which may include resizing, changing the sample rate, enhancing quality, and extracting features.

It is important to note that the audio files in the database have a bandwidth of 48000 Hz, but their bandwidth is reduced to 16000 Hz after resizing. Thus, all the audio files in the database have a bandwidth of 16000 Hz.

For the audio files obtained in the previous section, the sampling rate of each element is reviewed. This is done to ensure that all the audio files have standardized parameters. To achieve this, the resampling function in MATLAB is utilized. This function alters the sample rate of an audio file to a desired rate. This process is demonstrated in Fig. 4.

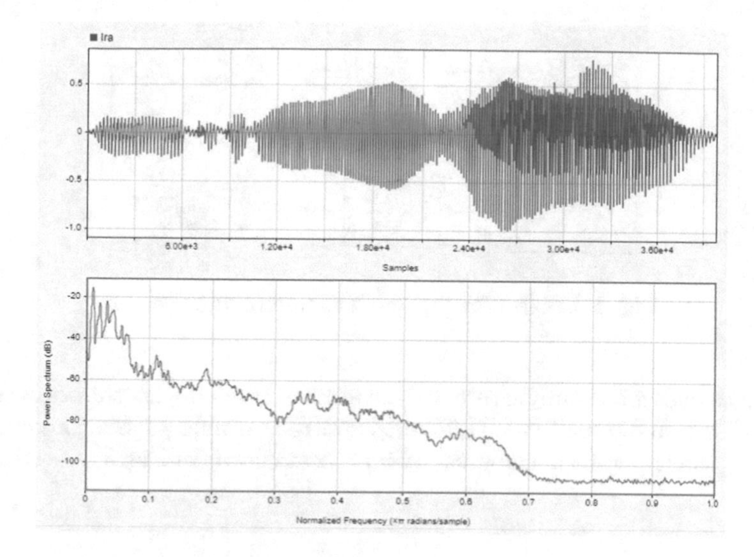

Fig. 2. Plot of anger signal versus time.

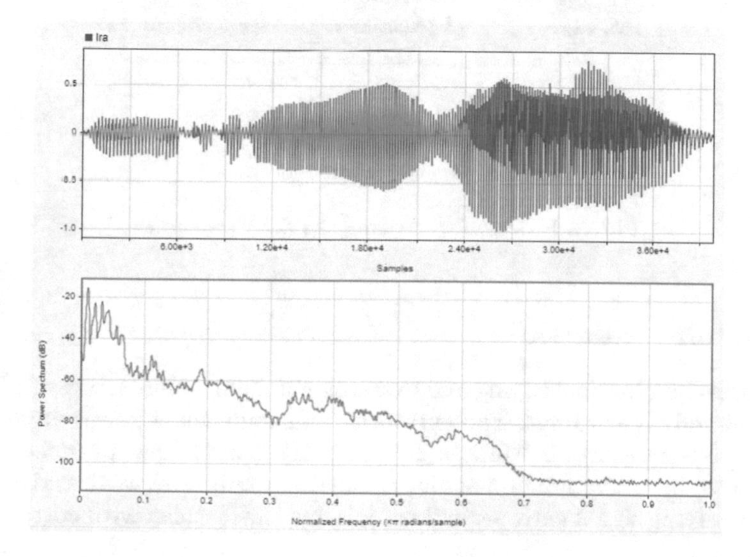

Fig. 3. Plot of the spectrum of the anger signal.

The audio to be processed is read first. Then, it is approximated by rational fractions using the rat function. Next, the sample rate is changed from the current one to the desired one by applying the resample function. Finally, the resulting audio is played back and saved in the desired direction. This process is repeated for each audio file in the folder.

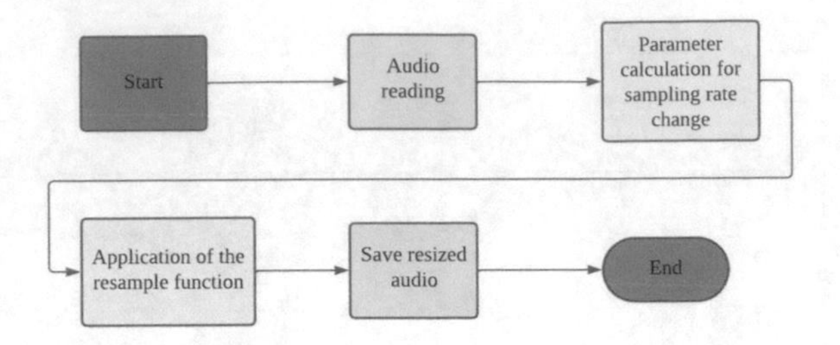

Fig. 4. Process for the audio sampling rate change.

After changing the sample rate, a data store is created with the addresses and names of each audio file. This is done to control variable generation, especially concerning their size in megabytes. The process shown in Fig. 5 is used for this purpose.

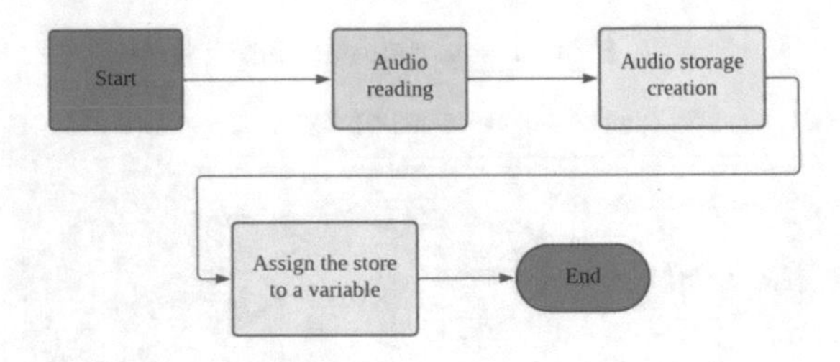

Fig. 5. Process for creating the data warehouse.

2.3 Feature Extraction

This process involves identifying and extracting unique features numerically from emotional audio recordings. The extraction of significant features is performed using wavelet transform in Matlab. To do so, one must follow a logical sequence of five steps, involving reading audio, calculating multilevel wavelet decomposition, quantifying signal entropy, and associating the features with corresponding emotions.

The main concept is to decompose the signal into a sequence of subspaces using chosen wavelet functions (Symmlet and Daubechies). This decomposition is usually accomplished by shifting (right or left) and scaling (amplifying or attenuating) the chosen wavelet function. By doing so, the signal is projected into the area represented by the particular scaling and translation.

In the issue of emotion recognition, the signal energy of the audio signal is recognized as one of the most crucial characteristics. A straightforward approach to extract features would be to employ the signal energy that is available in each subspace or node. Once you acquire each node's feature value, assemble them side by side, and that constitutes your feature vector (refer to Fig. 6).

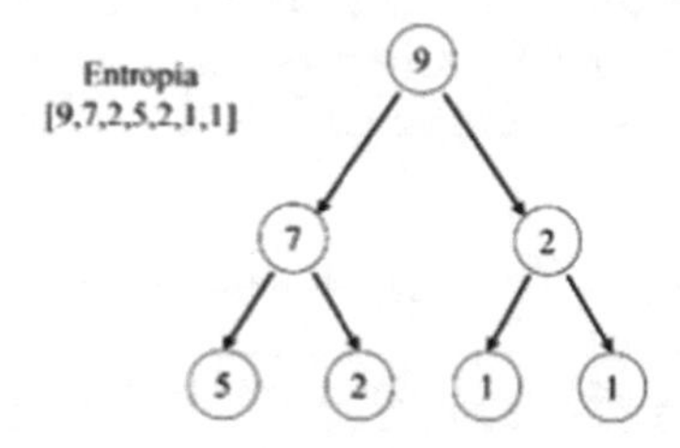

Fig. 6. Example of feature extraction for three levels of decomposition.

To illustrate the process described above, we proceed to calculate the wavelet transform for each considered emotion type. The anger signal is subjected to a wavelet decomposition of the Daubechies-4 family with seven levels of decomposition. As a consequence of using seven levels of decomposition, 255 features are obtained ($2^0 + 2^1 + 2^2 + 2^3 + 2^4 + 2^5 + 2^6 + 2^7 = 255$), as illustrated in Fig. 7.

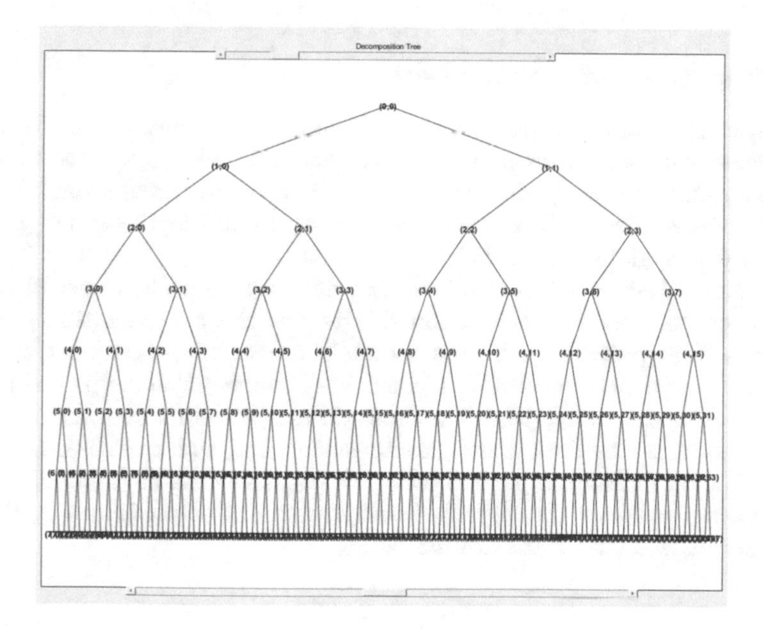

Fig. 7. 7-level wavelet decomposition tree for the emotion of anger.

The first two elements of the final decomposition level are presented, which correspond to the approximation coefficients displayed in Fig. 8. The approximation coefficient exhibits a much clearer signal when compared to the original audio signal, where the amplitude varies constantly.

Emotion audio labeling is the process of assigning emotions to audio. After obtaining the characteristics of the emotion audio, the data is associated with the relevant emotion through a process known as audio labeling. To achieve this, a binary logic, shown in Table 2, is used. Each vector of extracted characteristics is associated with a corresponding binary vector, as shown in Table 2, to label the audio. This labeling process is performed for all available audio in the database.

Table 2. The logic for emotion detection

Ira	Disgust	Fear	Happiness	Neutral	Sadness
1	0	0	0	0	0
0	1	0	0	0	0
0	0	1	0	0	0
0	0	0	1	0	0
0	0	0	0	1	0
0	0	0	0	0	1

2.4 Preparation of Training Data

After processing emotional audio and identifying and extracting signal characteristics, three matrices are created. To accomplish this, the matrix resulting from identifying and extracting characteristics is divided into three groups based on the percentages 80%, 10%, and 10%, which correspond to the training, validation, and testing data, respectively.

The normalization process adjusts the data set to scale values that differ from the original ones. In the obtained data set, the normalization process is performed. This process establishes a certain range for the data set so that it enters the neural network and is trained in a manner that helps the network comprehend the incoming data. The input dataset contains both maximum and minimum values. The data undergoes normalization, which establishes a range of [0,1]. The normalization process takes the negative data as the minimum and the maximum value as the highest. Finally, with the help of Eq. (1), the data can be normalized within the desired range.

$$X' = \frac{X - X_{\text{min}}}{X_{\text{max}} - X_{\text{min}}} \tag{1}$$

where:

- X' represents the normalized value of each datum in the dataset.

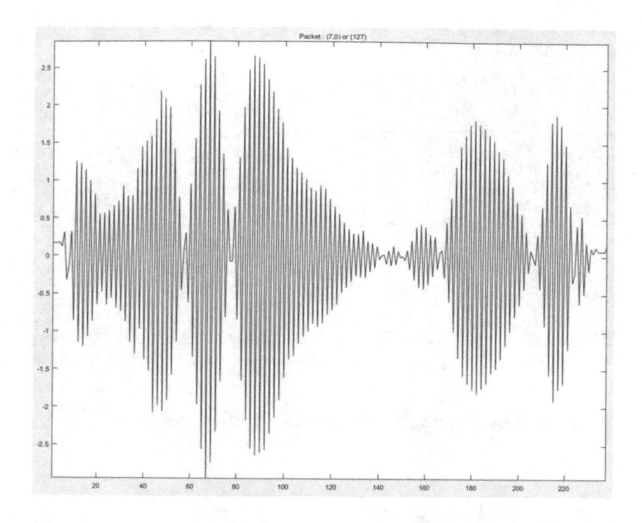

Fig. 8. Approximation coefficient, level 7 associated with the emotion of anger.

In contrast, the detail coefficient indicates that the signal is predominantly present during speech pauses (refer to Fig. 9).

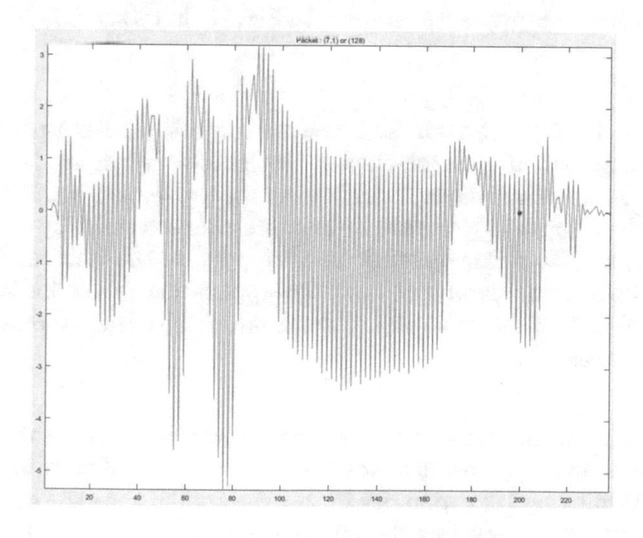

Fig. 9. Detail coefficient, level 7 associated with the emotion of anger.

- The value we wish to normalize is represented by X.
- The value of the minimum data found in the data set is denoted by X_{min}.
- The maximum value found in the data set is represented by X_{max}.

The prepared data set has been normalized, as shown in Fig. 10.

carNorma

868x255 double

	:	244	245	246	247	248	249	250	251	252	253	254	255
856	.2143	0.2117	0.2118	0.2149	0.2132	0.2169	0.2180	0.2168	0.2171	0.2114	0.2126	0.2176	0.2154
857	.2262	0.2241	0.2277	0.2239	0.2264	0.2272	0.2307	0.2316	0.2328	0.2251	0.2288	0.2338	0.2321
858	.2221	0.2240	0.2254	0.2197	0.2229	0.2295	0.2319	0.2311	0.2319	0.2270	0.2288	0.2318	0.2304
859	.1875	0.1845	0.1859	0.1861	0.1872	0.1896	0.1931	0.1942	0.1936	0.1862	0.1893	0.1969	0.1931
860	.1820	0.1870	0.1841	0.1840	0.1847	0.1767	0.1804	0.1816	0.1802	0.1861	0.1839	0.1881	0.1843
861	.2228	0.2219	0.2229	0.2229	0.2217	0.2275	0.2299	0.2305	0.2295	0.2232	0.2252	0.2319	0.2282
862	.2870	0.2895	0.2883	0.2851	0.2863	0.2947	0.2967	0.2967	0.2961	0.2905	0.2912	0.2959	0.2933
863	.3837	0.3755	0.3814	0.3776	0.3806	0.3818	0.3864	0.3840	0.3901	0.3798	0.3845	0.3831	0.3880
864	.0720	0.0712	0.0725	0.0716	0.0721	0.0771	0.0791	0.0806	0.0794	0.0732	0.0758	0.0825	0.0785
865	.2429	0.2444	0.2440	0.2434	0.2434	0.2478	0.2508	0.2512	0.2501	0.2465	0.2476	0.2542	0.2495
866	.2374	0.2370	0.2361	0.2366	0.2347	0.2404	0.2428	0.2436	0.2421	0.2394	0.2407	0.2455	0.2421
867	.2784	0.2788	0.2809	0.2754	0.2781	0.2894	0.2922	0.2918	0.2919	0.2806	0.2831	0.2893	0.2865
868	.2085	0.2020	0.2065	0.2062	0.2084	0.1988	0.2072	0.2079	0.2106	0.2072	0.2115	0.2114	0.2143
869													

Fig. 10. Normalized database with range [0,1].

2.5 Neural Network Training

This is a three-step process: building the LSTM model, parameterizing, and training the neural network. The architecture of the LSTM model is created using MATLAB's Deep Network Designer application, in which two models are built and tested to determine the most optimal model for emotion prediction. The general architecture adopted for the neural network starts with 255 input layer neurons and 6 output layer neurons. In Fig. 11 shows the LSTM architecture used to train the network.

Building the LSTM Model. The neural network models presented in this study are composed of multiple layers, including the input layer, intermediate layers, and output layers. An aspect of particular significance is the activation of the input layer with a data set characterized by dimensions of $255(C) \times 1(B) \times 1(T)$. These dimensions correspond to the features derived from the wavelet transform, signifying that the application of dissimilar multilevel wavelet transforms on an audio input could potentially lead to inaccurate classification outcomes.

Parameterization and Training Process. Upon defining the neural network model, the subsequent phase involves the meticulous configuration of training parameters. While there is variability in the selection of the algorithm and learning rate, certain factors such as the number of epochs and the minimum batch size remain consistent throughout this phase.

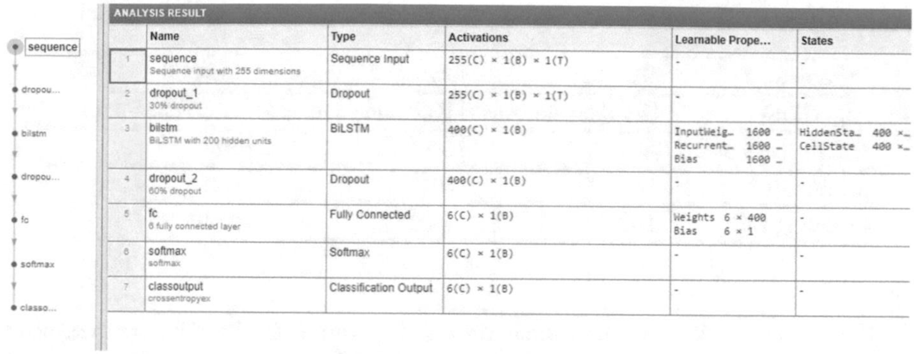

sequence
dropou...
bilstm
dropou...
fc
softmax
classo...

ANALYSIS RESULT

	Name	Type	Activations	Learnable Prope...	States
1	sequence Sequence input with 255 dimensions	Sequence Input	255(C) × 1(B) × 1(T)	-	-
2	dropout_1 30% dropout	Dropout	255(C) × 1(B) × 1(T)	-	-
3	bilstm BiLSTM with 200 hidden units	BiLSTM	400(C) × 1(B)	InputWeig... 1600 ... Recurrent... 1600 ... Bias 1600 ...	HiddenSta... 400 ×... CellState 400 ×...
4	dropout_2 60% dropout	Dropout	400(C) × 1(B)	-	-
5	fc 6 fully connected layer	Fully Connected	6(C) × 1(B)	Weights 6 × 400 Bias 6 × 1	-
6	softmax softmax	Softmax	6(C) × 1(B)	-	-
7	classoutput crossentropyex	Classification Output	6(C) × 1(B)	-	-

Fig. 11. BiLSTM neural network architecture with 7 layers for emotion classification.

The chosen algorithm for this study is the Gradient Descent method. This strategic choice stems from its ability to progressively enhance training efficacy over successive epochs, thereby facilitating the minimization of loss or error functions. The determination of the appropriate number of epochs, as depicted in Table 3, commences from the third epoch onward. This decision is informed by meticulous testing, which indicates that optimal training outcomes are achieved within this timeframe. As such, a total of 50 epochs are judiciously employed for conducting the comprehensive tests mandated by this research.

In establishing the optimal minimum number of batches, a comprehensive consideration of the overall data set is undertaken. In this instance, the dataset comprises a total of 868 instances, representing the maximum number of emotion-laden audio samples available. To facilitate a meticulous division of this dataset, an iterative approach is undertaken. This entails partitioning the data into smaller, manageable batches, each tailored to ensure effective training.

The initial batch size selected is 62, which is subsequently reiterated across 14 epochs. Additionally, a batch size of 217 is implemented, repeated four times throughout the epochs. Finally, the total batch size of 868 is engaged and deployed once across the epochs to encompass the entirety of the dataset.

Regarding the learning rate, a judiciously standardized value of 0.005 is employed. This selection is rooted in its intermediate characteristic, rendering it both compatible and recommended for integration with the Gradient Descent algorithm. Notably, a higher learning rate would impede algorithm convergence, while an excessively low value would significantly prolong the algorithm's weight optimization process.

The graphical representation shown in Fig. 12 corresponds to the results obtained from the first test performed, under the parameterization parameters meticulously outlined in Table 3. A noticeable feature within the graph is the manifestation of overtraining within the neural network architecture. This conspicuous overtraining phenomenon can be attributed to the adoption of remarkably small batch sizes that are insufficient for the effective operation of the system under the given parameter set.

Table 3. Parameters for testing with the LSTM neural network model

Parameter	Test 1	Essay 2	Essay 3
Algorithm	Gradient decrease	Gradient decrease	Gradient decrease
Number of epochs	50	50	50
Minimum lot size	62	217	868
Learning rate	0.005	0.005	0.005

The consequences of using such parameter values go beyond overtraining. They culminate in the prolonged period required for the training process, which significantly lengthens the path to achieving the desired results. Furthermore, the confluence of markedly elevated loss values and concurrently diminished accuracy levels substantiates the limited efficacy of the chosen parameterization. Consequently, the pursuit of optimal results necessitates a recalibration of parameter selection, striking a judicious balance between the architectural nuances of the network and the complexity of the task at hand.

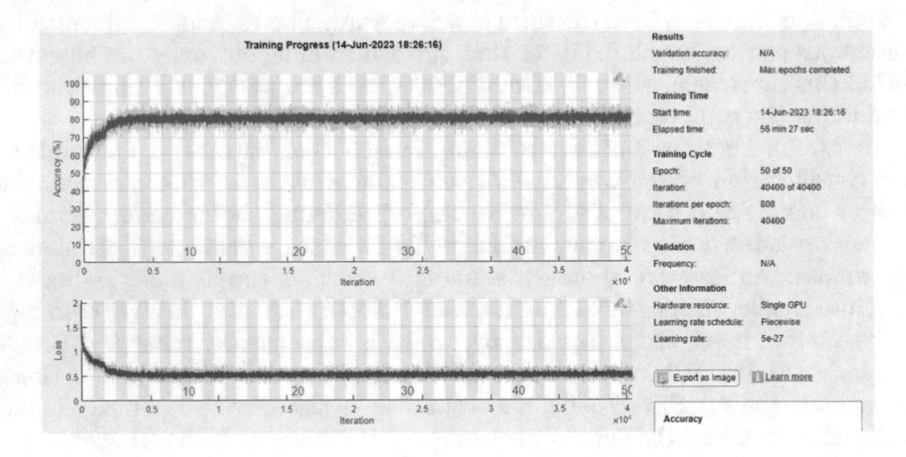

Fig. 12. Accuracy and error result of the first trial applying the LSTM model.

In Fig. 13 gives a visual representation of the training process of the network carried out according to the parameters prescribed in the second test, as described in Table 3. It can be seen that the application of the parameters set for this test leads to consistent results.

In Table 3, the redundancy of the fourth epoch becomes conspicuous, as an observed trend reveals the stability of the accuracy and the parameters from the second epoch onwards. In addition, there is a noticeable reduction in the execution time of the training, depending on the inherent characteristics of the

CPU, which have a significant impact on the temporal dynamics of the training. This correlation confirms the central role of the CPU in influencing the efficiency and duration of the training process.

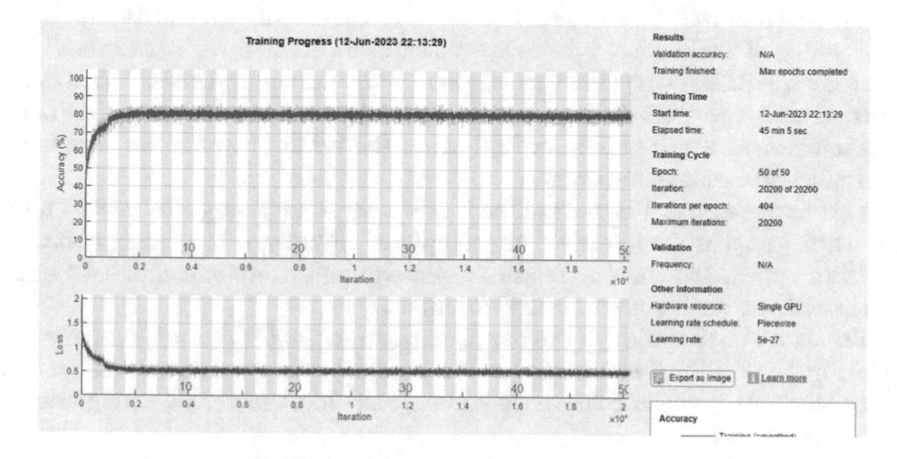

Fig. 13. Accuracy and error result of the second test applying the LSTM model.

In Fig. 14 summarizes the results derived from the final trial, which mirror the results observed in Trial 2. Notably, this final study includes an expanded range of epochs and batch sizes. It is pertinent to emphasize that a consistent pattern similar to that observed in Study 2 is evident, with the stabilization of training occurring from the third epoch onward. The observed stability of training renders further epochs unnecessary, implying that a saturation point has been reached in terms of refining model performance through additional epochs.

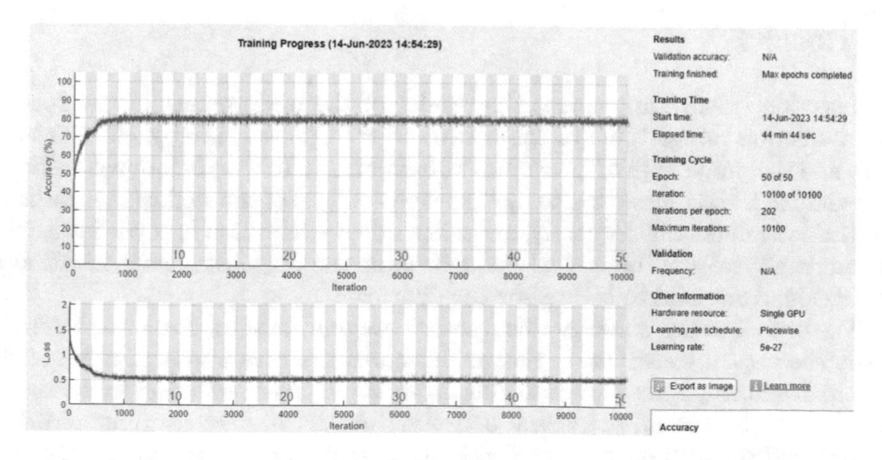

Fig. 14. Accuracy and error result of the third trial applying the LSMT model.

Throughout the neural network training process, careful examination reveals the inherent stability of the learning rate set at 0.005, as opposed to its counterpart at 0.01. As a result, the minimum batch loss exhibits a persistent equilibrium across trials 2 and 3. Significantly, Trial 2 emerges as the optimal training model, characterized by the adoption of a batch size of 217 and the judicious partitioning of the dataset into cohesive segments aligned with the overall size of the dataset. This strategic segmentation avoids the imprudent injection of the entire dataset into the network for training purposes. Conversely, the first batch iteration reveals elevated loss levels, but as the iterations progress, significant loss reduction becomes apparent.

This observed trend bodes well as it anticipates a convergence toward minimal batch losses as the iterations are completed. Finally, it is inferred that after the third epoch, the training reaches a plateau and exhibits consistent performance without further accuracy improvement.

The careful delineation of the parameterization data is comprehensively presented in Table 4. In this tabular format, optimal data parameters are meticulously elucidated, which collectively embody the most favorable configurations for training efforts.

Table 4. Selected training parameters for the neural network

Parameter	Value
Algorithm	Gradient decrease
Number of epochs	3
Minimum lot size	217
Learning rate	0.005

3 Results

This section briefly summarizes the results of the performance of the speech-based emotion recognition system in two different phases. Firstly, the efficiency ratio and the percentage of inaccuracy corresponding to each emotion considered are examined, starting with a comprehensive analysis of the confusion matrix for the trained model. Subsequently, an independent evaluation is carried out with six randomly selected individuals to determine the diagnostic categorization of depression, from mild to severe, for each patient.

The contributions made by this research are multifaceted. First and foremost is the theoretical contribution that underscores the core tenets of this investigation. It encompasses facets ranging from sound representation in speech to sound characterization, the neural networks integral to deep learning and their interconnection with emotion recognition, as well as a juxtaposition with conventional depression detection methods.

Another distinctive contribution is the novel approach to signal characterization. Here, multilevel wavelet decomposition plays a central role. This methodology assigns numerical values to signals depending on the chosen level of decomposition, here at level 7. This complex process results in a comprehensive collection of 255 signal features for each emotional category, meticulously encapsulated in the database.

In addition, this research proposes a linkage between the results derived from the emotion recognition system and the administered depression assessment tests. This synergy complements the psychological domain by providing insights into effective interventions for the treatment of depression.

To critically evaluate the efficiency of the proposed emotion recognition technique, we present the initial model's confusion matrix, as shown in Fig. 15. This analytical lens serves to illuminate the efficiency ratio and the percentage of inaccuracy attributed to each emotion, thus enriching the overall evaluation of the experimental results.

In the first scenario, the accuracy is 73.5%. Although this figure seems relatively modest in the context of the development of neural network systems, it is essential to contextualize this performance. The limited number of audio samples in the database is a prominent factor, limiting the network's ability to reach a superior level of training. In addition, the chosen model and the manner of signal feature extraction inherently influence the accuracy of the results obtained.

The field of emotion recognition is still relatively young. While the earliest investigations date back to 2016, the substantial body of research has proliferated since 2018, highlighting its dynamic and evolving nature. Acknowledging the complex nature of emotion recognition-where variability in emotional expression across individuals poses a challenge literature references are instructive. For example, in [5] reported an effectiveness rate of 56.71% for the neural network used in their research. Similarly, [10] obtained results between 70% and 75%. While the algorithms and databases used are different from the present study, the congruence of the accuracy percentages is noteworthy.

In contrast, the effectiveness of the emotion "anger" is 82.7%, with an associated error rate of 17.3%. Similarly, the emotion "disgust" reaches an effectiveness of 76.1%, paired with a misclassification rate of 23.9%. The emotion "Fear" has a recognition rate of 53.6%, while 46.4% of the cases are misclassified. Similarly, the emotion "happiness" is identified with an efficiency of 50.7% but experiences a misclassification rate of 49.3%. In parallel, the states "neutral" and "sad" manifest themselves as 79.7% and 90.3% correctly classified instances, alongside 20.3% and 9.7% misclassified cases, respectively.

Similarly, the predictive results for the emotion "anger" show an efficiency of 77.6% with an error rate of 22.2%. For the emotion "Fear", the prediction accuracy is 74% with an error rate of 26%. Meanwhile, for the emotion "Disgust", the efficiency rate is 84.4%, coupled with an error rate of 18.6%. "Happiness" has an accuracy of 60%, with an error of 40%. Similarly, "neutral" has an efficiency of 69.2% and an error of 30.8%, while "sadness" has an accuracy of 73.7% and an error of 26.3%.

Confusion Matrix for 10-Fold Cross-Validation
Average Accuracy =
73.5

True Class \ Predicted Class	Anger	Anxiety/Fear	Disgust	Happiness	Neutral	Sadness		
Anger	105	1	1	18			82.7%	17.3%
Anxiety/Fear	7	37	3	5	11	4	53.6%	46.4%
Disgust	2	1	35	1	4	2	76.1%	23.9%
Happiness	21	9	3	36	2		50.7%	49.3%
Neutral		1	1		63	3	79.7%	20.3%
Sadness		1			2	56	90.3%	9.7%
	77.8%	74.0%	81.4%	60.0%	69.2%	73.7%		
	22.2%	26.0%	18.6%	40.0%	30.8%	26.3%		

Fig. 15. Confusion matrix of the LSTM model.

The rigorous testing process includes patient consent, audio analysis, feature extraction, and emotion categorization. A systematic 7-step procedure, as illustrated in Fig. 16, is meticulously followed within this testing framework.

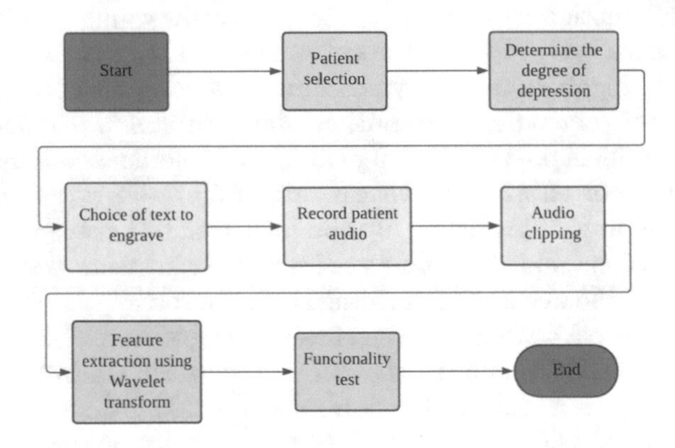

Fig. 16. Confusion matrix of the LSTM model.

The neural network response indicates that the audio input is indicative of the emotion "sadness", as shown in Fig. 17. This result is consistent with the diagnosis of major depression previously confirmed by the Beck test. The schematic shows how the audio labeling was performed. During the prediction phase, the neural network generates values in its six output layers. These values range from 0 to 1, with the value closest to 1 being the expected outcome. To reflect this process, the Simulink diagram uses an adder to effectively round these values to either 1 or 0, thus generating the result graph.

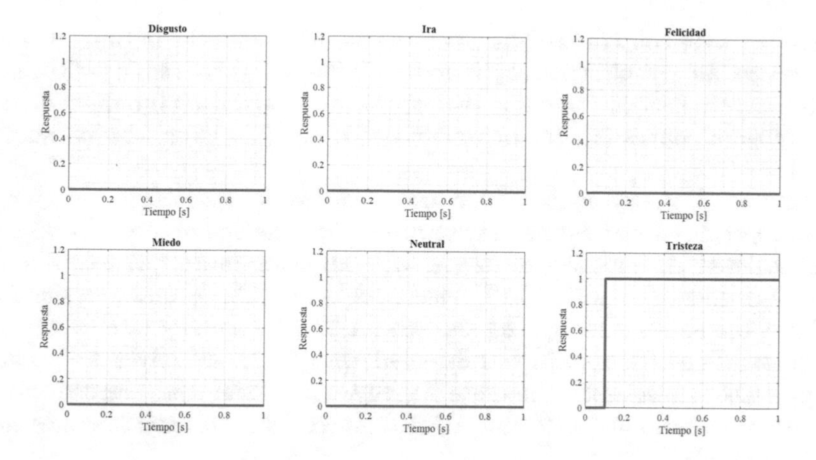

Fig. 17. Response of the emotion recognition system for Individual 1.

Finally, Table 5 is a summary of the results of the emotion detection for each individual. The chart shows the age of the person, the emotion identified by the system, the level of depression determined by the tests administered, and the scores obtained.

Table 5. Summary of the results obtained in the neural network

Individual	Genre	Age	Excitement	Pathology	Test score
Individual 1	Female	18–25	Sadness	Severe Depression	36
Individual 2	Male	18–25	Fear	Mild depression	13
Individual 3	Male	18–25	Fear	Mild depression	16
Individual 4	Female	18–25	Fear	Mild depression	18
Individual 5	Male	18–25	Sadness	Severe Depression	37
Individual 6	Female	18–25	Sadness	Severe Depression	53

4 Conclusions

The voice-based emotion recognition system proves its feasibility by establishing a correlation between the emotions detected by the system and the patient tests administered, thereby facilitating the recognition of the patient's depression level. In this regard, it is estimated that approximately 66.6% of the female participants in the sample are suffering from major depression, while this estimate is 33.3% for the male participants. In addition, the viability of deep learning techniques, particularly neural networks, has been demonstrated.

The Knowledge Discovery in Databases (KDD) architecture was used in the design of the system, including steps such as database acquisition, audio processing, feature extraction, and network training. During neural network training, two models were constructed and analyzed to determine the optimal model for the research. As a result, critical parameters influencing the training of the neural network were identified, including the optimization algorithm (gradient descent), the number of epochs (3), the minimum batch size (217), and the learning rate (0.005).

The overall effectiveness of the neural network was 73.5%. In terms of emotion-specific effectiveness and error rates, the following results were obtained: anger achieved effectiveness of 82.7% with an error rate of 17.3%. Similarly, disgust was identified with 76.1% effectiveness and 23.9% false classifications. Similarly, the network recognized fear with a 53.6% accuracy rate, while 46.4% were misclassified. Happiness was detected with 50.7% effectiveness, accompanied by a 49.3% misclassification rate. Finally, the neutral state and sadness were accurately identified at 79.7% and 90.3%, respectively, with misclassifications at 20.3% and 9.7%.

It is important to note that the effectiveness of the neural network depends significantly on the quality of the extraction of the emotional audio features, as well as the chosen parameterization for the network training. The performance of the network can only be definitively determined through experimentation, using analogous studies from the scientific community as a guide.

References

1. Kaur, G., Pandey, P.S.: Emotion recognition system using IOT and machine learning-a healthcare application. In: Proceeding of The 23rd Conference of Fruct Association, Haryana, 2018, Conference 2016, LNCS, vol. 9999, pp. 1–13. Springer, Heidelberg (2016)
2. Orozco, W.N., Baldares, M.J.V.: Addressing depression: intervention in crisis. Dome J. 19–35 (2012)
3. Olabe Basogain, X..: Artificial neuron networks and their applications, Escuela Superior de Ingeniería de Bilbao, p. 79 (2015)
4. Jahangir, R., Teh, Y.W., Hanif, F., Mujtaba, G.: Deep learning approaches for speech emotion recognition: state of the art and research challenges, vol. 80, no. 16. Multimedia Tools and Applications (2021). https://doi.org/10.1007/s11042-020-09874-7
5. Zloteanu, M., et al.: Speech emotion recognition with deep learning. Sensors **11**(2), 1–18 (2022). https://doi.org/10.1016/j.specom.2021.11.006
6. Hansen, L., Zhang, Y.P., Wolf, D., Sechidis, K., Ladegaard, N., Fusaroli, R.: A general- izable speech emotion recognition model reveals depression and remission. Acta Psychiatr. Scand. **145**(2), 186–199 (2022). https://doi.org/10.1111/acps.13388
7. Sisman, B.: Machine Learning for Limited Data Voice Conversion, National University of Singapore, Singapore (2019)
8. Toshinori, M.: Fundamentals of the new Artificial Intelligence, 2a edn. Springer, Cleveland (2008)
9. Ganapathy, A.: Speech emotion recognition using deep learning techniques. ABC J. Adv. Res. **5**(2), 113–122 (2016). https://doi.org/10.18034/abcjar.v5i2.550
10. Sandoval, V.: Emotion recognition by voice, p. 60 (2019)

Data Mining Applied to the HFC Network to Analyze the Availability of Telecommunication Services

Shirley Alarcón-Loza(✉) and Karen Estacio-Corozo

Instituto Superior Tecnológico ARGOS, Guayaquil, Ecuador
galarcon@tecnologicoargos.edu.ec

Abstract. The failures that affect the telecommunications service in hybrid fiber-coaxial networks (HFC) can be anticipated, through data analysis, to minimize the impact on service availability and users. This work seeks to analyze the causes that affect the performance of HFC networks in Ecuador, through a predictive model for their timely detection and application of anticipated works. The base studied included the components and causes that affected the availability of television and telephony service in a HFC network in 2020, from a sample of 17707 records. For the research, the dependent variable CATEGORY CAUSE is considered, and the objective was to analyze the causes that affect the telecommunications service. The KDD methodology was used and, from the use of WEKA software, classification algorithms were established to determine suitable predictive features, being the causes, the type of event, and the network components, the most significant. The results showed that, for the television service, the best classification algorithm was J48, with a precision value of 68.668% and an area under the curve (ROC Area) of 0.913. For the telephony service, the best classification algorithm was Random Tree, with a precision of 73.666% and an area under the curve of 0.969. The conducted research demonstrated the importance of data mining in the process of analyzing the causes of the impact on television and telephone services.

Keywords: HFC network · Hybrid fiber-coaxial networks · Data Mining · Classification Algorithm

1 Introduction

In recent years, technology has produced a large volume of data worldwide that is stored daily in database servers without being evaluated and then deleted without any action being taken. Today, the trend is to use the data and transform it into valuable information by applying computational models that can predict situations. In this context, data mining is a science that allows problem-solving through data analysis, applying techniques and methods that allow the discovery of patterns [29], which is why it has gained great popularity and is used in various disciplines.

Within the field of telecommunications, there are hybrid fiber optic and coaxial networks (HFC) that provide broadband services to transmit various services such as

M. Z. Vizuete et al. (Eds.): CI3 2023, LNNS 1040, pp. 173–185, 2024.
https://doi.org/10.1007/978-3-031-63434-5_13

telephony, television [8], Internet, and data. In an HFC network, two main sections are distinguished: the inside plant and the outside plant.

The inside plant refers to a segment of the network located inside a building or facility, where the Headend equipment, such as the cable modem termination system (CMTS) and the routing equipment play a fundamental role in the management and distribution of the transmission and reception of the signals of the configured services are located. Figure 1 details the aforementioned scheme [28].

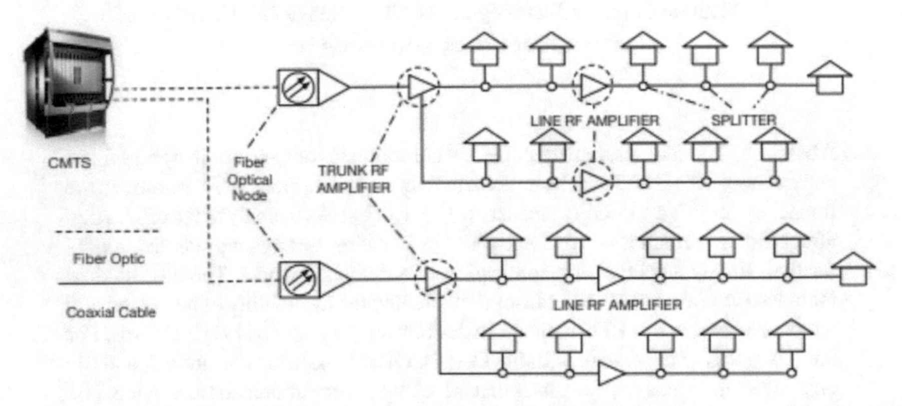

Fig. 1. HFC network architecture [14]

The outside plant represents the area outside the building or facility, where the fiber optic and coaxial cables that connect the network to customer locations are located. This segment is a compound of several sectored areas of optical nodes that are connected through highly reliable fiber optic connections.

In the event of a failure, it is easy to identify the exact point because each fiber node is designed to limit any signal interference. A fiber node, which typically uses multiple line amplifiers, distribution, and possibly splitters, is responsible for connecting the last mile of the network where the customer's cable modem or terminal equipment is installed. The last amplifier located before reaching the end customer is known as the end-line amplifier [26].

1.1 HFC Network Service Failures

In modern HFC networks, covering millions of users, operators must ensure high availability of network access. This need has become very important due to the accelerated growth of activities requiring telecommunication services, making it essential to quickly detect and locate any network faults. However, this task is not easy due to the number of devices present in typical HFC networks [7].

In a conventional HFC network monitoring system [18], transponders are installed to monitor active elements such as optical nodes, amplifiers, power supplies, and transmitters. The transponders allow direct monitoring and control of the aforementioned devices and elements, providing a wide range of detected parameters and service failures related to specific causes.

However, this system has some disadvantages, among which is the cost of monitoring due to the need to use equipment dedicated exclusively to this activity. In addition, the transponders are located in outdoor environments, which exposes them to various environmental conditions that trigger equipment malfunctions or loss of remote management, which generates false alarms and increases operation and maintenance costs by sending technical personnel for review [5].

On the other hand, it is an opportunity and an excellent benefit for Internet Service Providers (ISPs) to create a heat map that shows the devices identified as the root cause of problems affecting services [12]. In this way, if recurring problems are detected in a certain area, technicians attend to the exact spot to perform maintenance work and improve the overall network quality, without the need for it to be related to a specific incident. This approach is useful when faced with problems that are difficult to diagnose accurately. Consequently, by having a broader and longer-term view of problematic devices, ISPs can anticipate problems, optimize the network, and improve the user experience [13]. Table 1, compiles the most relevant cases on reported HFC network failures.

Table 1. Literature review on HFC network failures

Study	Cause	Component	Works related
Noise in upstream network channels	Noise Interference	CMTS	[30]
Cause of network cable failures	Noise	Fiber optic connector and cable	[12, 14]
Network failures	Several	CMTS, Access Point, Amplifier	[26]
RF amplifiers monitoring	Power failure	RF Amplifiers	[22]

The customer-centricity strategy has revolutionized how companies satisfy their customers, as service-oriented companies have recently been given great importance in terms of customer care and satisfaction. Despite this, it is still a challenge to manage the decision-making on what to offer to gain customer loyalty [3].

Based on this foundation, the objective of this article is to analyze the causes that affect the performance of HFC networks in Ecuador, through a predictive model. For this purpose, the data of the chosen sample was organized, the Knowledge Discovery in Databases (KDD) methodology was applied, and the Waikato Environment for Knowledge Analysis (WEKA) program was used to generate decision trees and knowledge rules through classification algorithms, and the results obtained were interpreted.

2 Materials and Methods

In the development of this article, the information was very important to obtain the results, since the data alone did not constitute a clear reality of any situation related to the affectation of television and telephony services in HFC networks. In this sense, the origin of the data, the treatment given to them, the data mining methodology used, the processing in the corresponding software, and the interpretation of the acquired results are explained.

2.1 Database and Resources

The study was focused on various events that affected the television and telephony service of the HFC network, which is managed by a private company. The characteristics of these events include whether they were fortuitous, programmed, or emergent. The causes correspond to failures, electrical interruptions, and physical damage, among others. The cause category includes the components that have failed, and the type of cause refers to failures originating in the telecommunications network itself or outside it.

Table 2 shows the population corresponding to the 50691 events in 2020 that meet the aforementioned characteristics. These events caused the interruption of the telecommunications service, generating discomfort in the users. Based on this aspect, the sample was selected using the non-probabilistic technique by convenience, considering the affectation of television and telephony services in the HFC network.

Table 2. Population distribution and sample used.

Population		Sample	
Events	Telecommunications service	Events	Telecommunications service
50691 service impact records in 2020	Television Telephony Internet Data	17707 service impact records in 2020	Television Telephony

For this study, a database of 17707 cases of damage was considered, according to the sample described, with the characteristics shown in Table 3. The type of event refers to whether the situation was a fortuitous event, scheduled maintenance, or an emergency. The causes were equipment damage, various electrical failures, accidents, redesign due to pole movement or urban regeneration, attenuation, misalignment, noise, intermittency, and signal weakness, among others. The category of the cause considered the affected equipment such as coaxial cable, amplifiers, sources, transmitters, UPS, and couplers, among others. In addition, it is considered whether the causes were due to internal or external factors.

Table 3. Features of the telecommunications, television, and telephony service impact in 2020.

Features	Description	Features	Description
Event Type	Situation in which the service was affected	Cause	Reason for service interruption
Category Cause	Component or device where the failure occurs	Cause Type	Origin of the cause that affects service

For data processing and information generation, WEKA software [23] version 3.8.6 was used, based on a high-level programming language that hosts various machine learning algorithms that allow the analysis of a large set of data. This technological resource is open source and supports data formats including ARFF, CSV, and LibSVM.

2.2 Methodology

The methodology applied in this work is a cross-sectional correlational design, aimed at a process of knowledge discovery in databases (KDD) [11]. It is a methodology for the data mining process that allows data cleaning before processing and also considers the understanding of the problem. This methodology, according to [19], consists of five phases, and its application for this article is described below:

Selection Stage. Data are generated in each of the events that affect telecommunications networks, which are stored in the corresponding software and are collected in a database, so it is of great importance that they receive some treatment to discover patterns associated with the state of HFC networks that provide television and telephony services. By identifying this scenario, the present research was proposed to analyze the causes that affect the operation of HFC networks through a predictive model to identify the cause of the affectation, as well as the faulty component for telecommunications companies to make timely decisions at the time of problems that disturb a large number of users [25]. For this purpose, the data mining classification technique was used and the best algorithms were validated from the participant sample.

Pre-processing Stage. In this phase, the database containing the causes affecting telecommunication services was used, constituting 50691 records of 4 different fields. The data were collected, stored, and filtered in Microsoft Excel, so that 17707 records of 4 fields were organized, shown in Table 3.

Transformation Stage. To determine the causes and their categories, a dependent variable has been chosen, a nominal type, and data mining classification techniques will be applied to this variable in WEKA software. The dependent variable that directly affects the availability of the telecommunications service and, therefore, generates user discomfort, is the variable CATEGORY CAUSE. Table 4 lists the variables that were entered into WEKA.

Table 4. Description of the variables considered in the impact on television and telephony services.

Variables	Meaning	Type
EVENT TYPE	Unintentional Programmed Emergent	Nominal
CAUSE	Accidents: attenuation, cut-off, damage Faults: high resource consumption, electrical, undetermined, mismatch, water or noise intrusion, sulfation, inhibition, signal level, intermittency, breakage, loss of connectivity Scheduled work Theft	Nominal
CATEGORY CAUSE (Dependent variable)	Amplifier, Fiber optic and coaxial cable, UPS, Source, Optical Node, Connector, Coupler, Splice, Receiver, Transmitter, Feeder, CMT, Switch, Servidor, Multiplexer, RF Headend	Nominal
CAUSE TYPE	Internal External	Nominal dichotomous

Data Mining Stage. Based on the test data set that allows analyzing the causes of the impairment of television and telephony service in HFC networks, three experiments were carried out with the dependent variable CATEGORY CAUSE; applying the classification algorithm that considers the best attributes [9] to reduce the processing work.

The classification algorithms [29] used to identify the causes of TV and telephony service impairment in HFC networks were J48 [1], Random Forest, and Random Tree [24] decision trees. In addition, the Cross-Validation [2, 21] evaluation and validation mode were applied, which provides an average accuracy per dependent variable based on N iterations run on the data set. In Table 5, the area under the curve (ROC-Area) [10, 15] and Accuracy [16] have been used as indicators for the quality assessment of the applied classification algorithm.

Table 5. Quality assessment of the applied algorithms.

Algorithm	CATEGORY CAUSE Television		CATEGORY CAUSE Telephony	
	Accuracy	ROC Area	Accuracy	ROC Area
J48	68.668%	0.913	73.636%	0.969
Random Forest	68.661%	0.913	73.606%	0.969
Random Tree	68.661%	0.913	73.666%	0.969

Interpretation Stage. In this paper, the predictive model [4, 20] was formed by the application of classification algorithms known as decision trees [17, 23] in WEKA. Several tests were executed, according to Table 5; the J48 classification algorithm was applied to analyze the causes affecting the availability of television services in HFC networks.

Twenty-eight knowledge rules were generated (see Fig. 2) indicating the causes of service impairment; those that presented more cases are explained below:

A) The CAUSE was a damage in the Fiber Optic Cable and the EVENT TYPE was fortuitous, then there was service affectation and 336 cases were located;
B) The CAUSE was a damage in the HFC Passive Connector 500 and the EVENT TYPE was programmed, then there was service affectation and 154 cases were located;
C) The CAUSE was an electrical failure in the UPS air conditioning and the CAUSE TYPE was external, then there was service affectation and 123 cases were located;
D) The CAUSE was an electrical failure in the Source Amplifiers, then there was an affectation of the service and 183 cases were located;
E) The CAUSE was an accident damage in the Fiber Optic Yarn and the EVENT TYPE was fortuitous, then there was an affectation of the service and 107 cases were located;
F) The CAUSE was a mismatch failure in the PAD Equalizer Amplifier, then there was affectation of the service and 409 cases were located.
G) The CAUSE was a noise ingress failure in the Coaxial Cable Drop, then there was affectation of the service and 111 cases were located;
H) The CAUSE was an inhibition failure in the Television Multiplexer Module, then there was service affectation and 3839 cases were located;
I) The CAUSE was a programmed work in the UPS air conditioning, then there was service affectation and 1025 cases were located;
J) The CAUSE was a programmed work and the EVENT TYPE was emergent in the OTN Node, then there was service affectation and 1902 cases were located;
K) The CAUSE was a failure due to loss of connectivity in the Core Service IPS Switches Interface, so the service was affected and 857 cases were located;
L) The CAUSE was an electrical failure in the Source Batteries, so the service was affected and 217 cases were located; and,
M) The CAUSE was a failure due to the high use of resources in the Television Multiplexor and the EVENT TYPE was fortuitous, so the service was affected and 4798 cases were located.

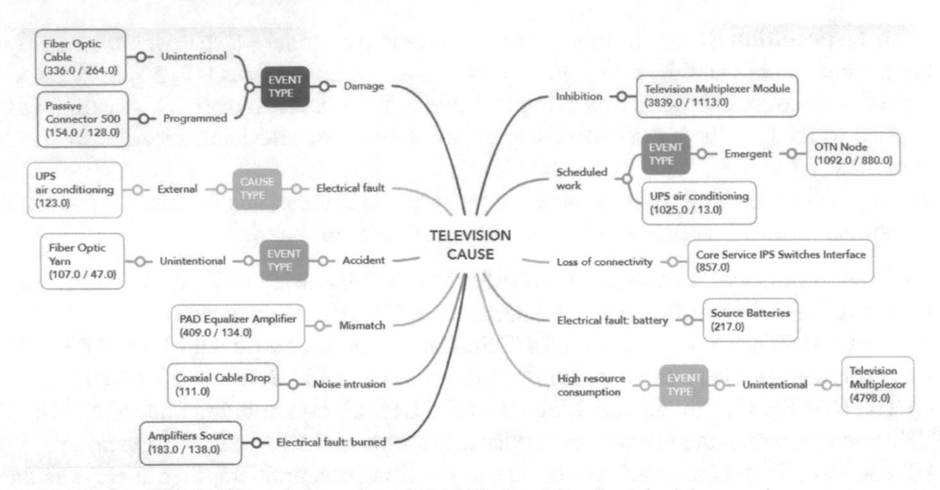

Fig. 2. Knowledge rules of the J48 algorithm for television service.

Similarly, the Random Tree classification algorithm was applied to analyze the causes affecting the availability of telephony service in HFC networks. Thirty knowledge rules were generated (see Fig. 3) indicating the causes of service impairment. Those with the highest number of cases are explained below:

A) The CAUSE was a damage in the fiber optic cable and the EVENT TYPE was fortuitous, then there was service affectation and 335 cases were located;
B) The CAUSE was a damage in the HFC Passive Connector 500 and the EVENT TYPE was programmed, then there was service affectation and 155 cases were located;
C) The CAUSE was an electrical failure in UPS Air Conditioning and the EVENT TYPE was incidental and the CAUSE TYPE was external, then there was service affectation and 307 cases were located;
D) The CAUSE was a burnout electrical failure in Amplifiers Source and the EVENT TYPE was incidental, then there was service affectation and 169 cases were located;
E) The CAUSE was a mismatch failure in PAD Equalizer Amplifier and the EVENT TYPE was programmed, then there was service affectation and 330 cases were located;
F) The CAUSE was a noise ingress failure in the Coaxial Intake, then there was service affectation and 111 cases were located;
G) The CAUSE was an electrical failure in Battery Sources, then there was an affectation of the service and 217 cases were located; and,
H) The CAUSE was a failure due to high use of resources in the Core Service IPS Servers Hardware, then there was an affectation of the service and 1026 cases were located.

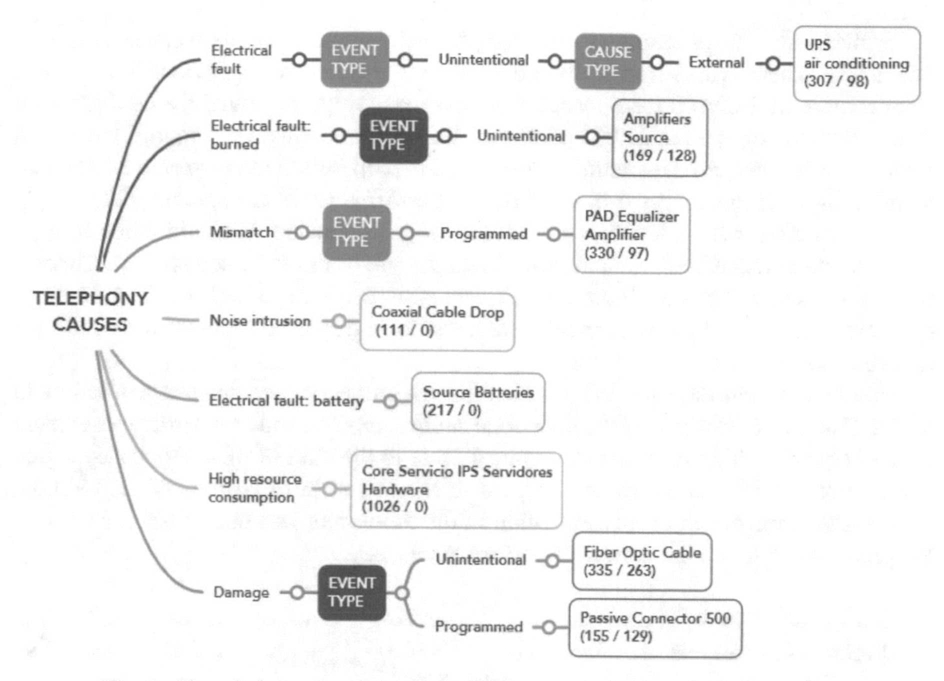

Fig. 3. Knowledge rules of the Random Tree algorithm for telephony service

3 Results

For the dependent variable CATEGORY CAUSE, algorithms that generate decision trees were applied to identify the most influential predictive attributes. In the case of the television service, the J48 algorithm was used and in the case of the telephone service, the Random Tree algorithm was used; both obtained a good percentage of accuracy and quality margin, according to the evaluation of Table 5. In both cases, service impairment was considered by evaluating the most significant causes.

The results of the television service (see Fig. 4), showed as a predominant factor the failures caused by the high use of resources in the multiplexer, causing total or partial inhibition in each of the modules due to traffic overloads, capacity limitations, configuration problems and hardware failures, so that this problem was divided into two independent causes for further treatment.

In the same way, programmed works represented another common cause of service affectation. These works were carried out within the Headend where OTNs, switches, Core routers, and CMTS were installed, which massively converged the configuration of the services of thousands of clients [6]. Likewise, adequate air conditioning of a Headend was considered essential to ensure optimal operation of electronic equipment, maintaining efficient energy use, stability, and performance of the network [27].

The results for the telephony service (see Fig. 5) showed that failure due to high resource usage was a predominant factor in the failure of the HFC network. This occurs when the resources of the IP core servers are saturated due to high traffic volume or excessive load of configured services, which results in degraded performance or service interruption.

Fiber optic cable damage [14] was identified as the second leading cause of failure in the HFC network. This part of the fiber optic cable is represented as the primary segment of the network, which runs from the optical node to the start of radio frequency signal distribution on the coaxial cable and goes to the last mile of customers. In addition, failure due to mismatch or miscalibration in the amplifiers was one of the causes with the greatest impact on the network's performance [22].

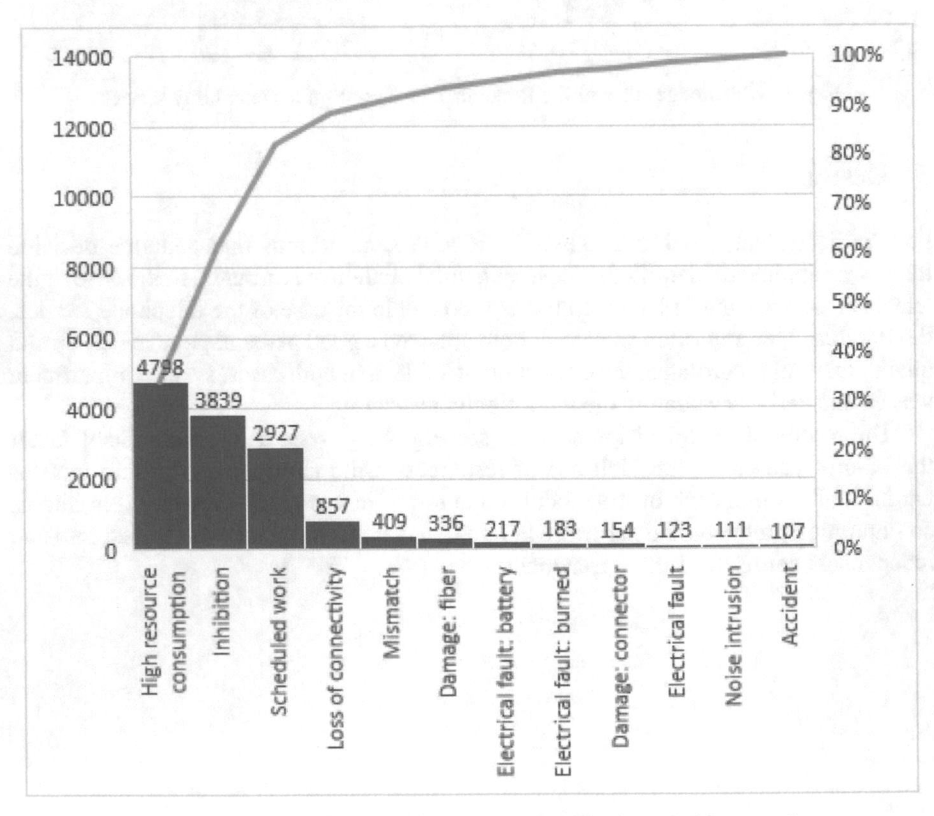

Fig. 4. Causes affecting TV service availability in an HFC network

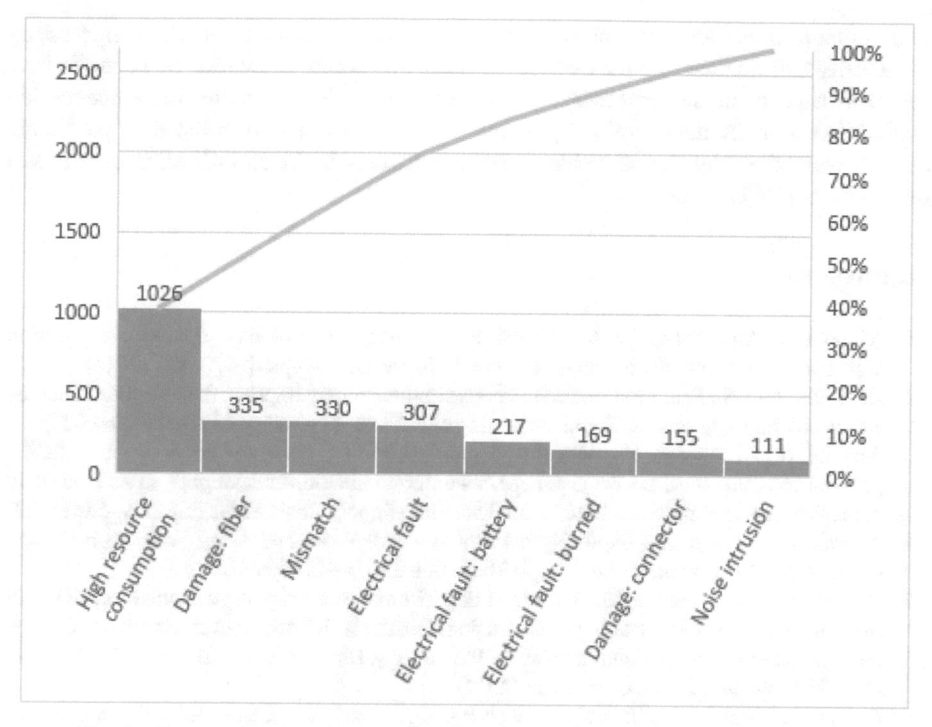

Fig. 5. Causes affecting the availability of telephony service in an HFC network.

4 Conclusions

In this case, it was evidenced that the high use of resources and the poor performance of the HFC network components directly influence the availability of the television service, so public and private companies should take preventive and corrective measures to improve the service. There are a large number of incidents at the Headend since this is the starting point of the network where the services of each of the customers are configured and converged massively, and when a failure occurs in any of its components, it can trigger a generalized service disruption.

Concerning the telephony service, the traffic overload in the IP Services Core Servers is caused by the high use of resources due to the reception and processing of a large amount of data. If the traffic load exceeds the capacity of the hardware to handle it, it can lead to resource saturation, resulting in degraded performance or even equipment inhibition. Core IP equipment can be inhibited to protect itself from complete collapse, this happens when there is a sudden increase in demand for the allocated bandwidth or because of widespread congestion in the network.

On the other hand, the secondary segment of the HFC network, composed of the coaxial cable and amplifiers, can also contribute to failures. The amplifiers are necessary to maintain the quality and strength of the signal as it is distributed over the coaxial cable. If there are power failures, malfunctions, mismatches, or wear, they can cause signal degradation, affect the quality of service, as well as affect the correct operating and signal-to-noise levels in the network terminal equipment. It would be interesting to identify and analyze the data collected from fiber optic networks, to generate new research that considers internet services and examines other factors that influence service availability. For the data mining approach, it is suggested to develop new predictive models with other algorithms, to have a better accuracy in the classification of data and obtain high-quality results.

References

1. Akinola, S., Oyabugbe, O.: Accuracies and training times of data mining classification algorithms: an empirical comparative study. J. Softw. Eng. Appl. **8**, 470–477 (2015)
2. Alabdulwahab, S.: Feature selection methods simultaneously improve the detection accuracy and model building time of machine learning classifiers. Symmetry **12**(9), 03–20 (2020)
3. Alzoubi, H., Alshurideh, M., Al Kurdi, B., Inairat, M.: Do perceived service value, quality, price fairness, and service recovery shape customer satisfaction and delight? A practical study in the service telecommunication context. Uncertain Supply Chain Manage. **8**, 579–588 (2020)
4. Attwal, K., Dhiman, A.: Exploring data mining tool-Weka and using Weka to build and evaluate predictive models. Adv. Appl. Math. Sci. **19**(6), 451–469 (2020)
5. Benhavan, T., Songwatana, K.: HFC network performance monitoring system using DOCSIS cable modem operation data in a 3-dimensional analysis. In: 4th Joint International Conference on Information and Communication Technology, Electronic and Electrical Engineering (JICTEE), pp. 1–5. IEEE, Chiang Rai (2014)
6. Campo, W., Rueda, D., Taimal, I., Arciniegas, J.: Analysis of the DOCSIS protocol for distribution of TDi applications and contents in an HFC network. Revista Avances en Sistemas e Informática **6**(3), 119–132 (2009)
7. Chan, H.N., Martinez, B.C., Gomez, R.M., Granados, J.R.: Integrated multi-purposed testbed to characterize the performance of internet access over hybrid fiber coaxial access networks. In: Gregori, E., Conti, M., Campbell, A.T., Omidyar, G., Zukerman, M. (eds.) NETWORKING 2002: Networking Technologies, Services, and Protocols; Performance of Computer and Communication Networks; Mobile and Wireless Communications, NETWORKING 2002, LNCS, vol. 2345, pp. 996–1007. Springer, Berlin (2002). https://doi.org/10.1007/3-540-47906-6_81
8. Ciciora, W., Farmer, J., Large, D., Adams, M.: The History of Modern Cable Television Technology, 2nd edn. Morgan Kaufmann Publishers, San Francisco (2003)
9. Dinakaran, S., Thangaiah, P.: Role of attribute selection in classification algorithms. Int. J. Sci. Eng. Res. **4**(6), 67–71 (2013)
10. Fawcett, T.: An introduction to ROC analysis. Pattern Recogn. Lett. **27**(8), 861–874 (2006)
11. Fayyad, U., Piatetsky-Shapiro, G., Smyth, P.: From data mining to knowledge discovery in databases. AI Mag. **17**(3), 37 (1996)
12. Heiler, G., Gadermaier, T., Haider, T., Hanbury, A., Filzmoser, P.: Identifying the root cause of cable network problems with machine learning. arXiv abs/2203.06989 (2022)
13. Heiler, G., Gadermaier, T., Haider, T., Hanbury, A., Filzmoser, P.: Improving cable network maintenance with machine learning. arXiv, abs/2203.06989 (2022)
14. Hu, J., Zhou, Z., Yang, X., Malone, J., Williams, J.: CableMon: improving the reliability of cable broadband networks via proactive network maintenance. In: 17th USENIX Symposium on Networked Systems Design and Implementation, pp. 619–632. NSDI 20, Santa Clara (2020)
15. Janssens, A., Martens, F.: Reflection on modern methods: revisiting the area under the ROC Curve. Int. J. Epidemiol. **49**(4), 1397–1403 (2020)

16. Juba, B., Le, H.: Precision-recall versus accuracy and the role of large data sets. In: Proceedings of the 37th AAAI Conference on Artificial Intelligence, pp. 4039–4048. AAAI Press, Washington (2019)
17. Kingsford, C., Salzberg, S.: What are decision trees? Nat. Biotechnoly **26**(9), 1011–1013 (2008)
18. Milan, S., Ina, M., Cica, Z.: Performance monitoring challenges in HFC networks. In: 13th International Conference on Advanced Technologies, Systems, and Services in Telecommunications (TELSIKS), pp. 385–388. IEEE, Serbia (2017)
19. Pareek, M., Bhari, P.: A review report on knowledge discovery in databases and various techniques of data mining. OAIJSE **5**(12), 79–82 (2007)
20. Preet, K., Singh, A.: Exploring data mining tool - WEKA and using WEKA to build and evaluate predictive models. Adv. Appl. Math. Sci. **19**(6), 451–469 (2020)
21. Scholz, M., Forman, G.: Apples-to-apples in cross-validation studies: pitfalls in classifier performance measurement. Int. J. Res. Stud. Sci. Eng. Technol. **6**(11), 40–52 (2010)
22. Sánchez, D., Vega, J., Rueda, D., Rodríguez, A.: Remote monitoring of RF amplifiers in HFC networks: voltage drop detection due to power blackouts. In: Congreso Internacional de Innovación y Tendencias en Ingeniería, pp.1–6. IEEE, Bogotá (2022)
23. Sharma, P.: Analysis of various decision tree classification algorithms using WEKA. Int. J. Recent Innov. Trends Comput. Commun. Comp. **3**(2), 684–690 (2015)
24. Sheela, Y., Krishnaveni, S.: A comparative analysis of various classification trees. In: International Conference on Circuit, Power and Computing Technologies, pp. 1–8. IEEE, Collam (2017)
25. Simakovic, M., Cica, Z., Drajic, D.: Big-data platform for performance monitoring of telecom-service-provider networks. Electronics **11**(14), 2224 (2022)
26. Simakovic, M., Cica, Z.: Detection and localization of failures in hybrid fiber-coaxial network using big data platform. Electronics **10**(23), 2906 (2021)
27. Suárez, I., Escobar-Díaz, A., Vacca, H.: Unidades de climatización para centro de datos. Vínculos **16**, 128–147 (2019)
28. Wauters, T., et al.: HFC access network design for switched broadcast TV services. IEEE Trans. Broadcast. **53**(2), 588–594 (2007)
29. Witten, I., Frank, E., Hall, M.: Data Mining: Practical Machine Learning Tools and Techniques. Elsevier Science & Technology, San Francisco (2011)
30. Zhang, L., Ma, Y., Liu, K., Zeng, Y.: Research of the noise characteristic on the upstream channel for HFC network. In: 2nd International Conference on Signal Processing Systems, pp. 426–430. IEEE, Dalian (2010)

Optimization of the B10s1 Engine for Corsa with Adaptation of the Cylinder Head of a Spark-Ignition Engine and Validation on a Chassis Dynamometer to Verify Its Power

Jose Vicente Manopanta Aigaje[1,2](✉)

[1] Instituto Tecnológico Universitario Ismac, Quito, Ecuador
vmanopanta@tecnologicoismac.edu.ec
[2] Universidad Internacional SEK, Quito, Ecuador

Abstract. In the automotive fleet, the improvement of performance and power have a significant growth, these parameters are evidenced in high-end vehicles with high costs. In common vehicles, these improvements can be obtained with adaptations to the combustion engine. The objective of this research is to improve the performance of the engine by adapting a cylinder head. A Chevrolet Corsa B10S vehicle is selected. The engine power without modification is determined using a chassis dynamometer. The impact on the mechanical performance of the vehicle is given by the adaptation and modification of the cylinder head of an Aveo Activo 16 valves with independent ITBs for the intake manifold, modifications in the exhaust manifold with independent outputs (header), improving the intake and exhaust of gases in the spark-ignition engine. The method used will be the experimental one, making adjustments in the distribution with two camshafts and timing belt, optimizing the maintenance. It is also concluded that as the percentage of air-fuel increases, a better torque and power performance will be obtained in relation to the standard engine.

Keywords: Tuning · Adaptation · Power · Performance

1 Introduction

This theoretical and practical research will take into consideration that motor racing competitions in Ecuador have been practiced since 1930 in cities such as Quito, Guayaquil, Riobamba, Cuenca and Ambato. In 1985 the Tungurahua Automóvil Club (Tungurahua Automobile Club) (TAC) was created and new divisions were formed. One of the categories characterized by the high competitive level in the province, is the 0 to 1150 cc category, dominated by the presence of Suzuki Forsa 1 vehicles [1]. Engine tuning is possible because all manufacturers have over-dimensioned them to obtain safety and reliability margins that ensure their operation in adverse conditions. It is possible to reduce these safety margins to acceptable minimums in order to considerably increase engine performance without jeopardizing the engine service life [2]. In conventional

M. Z. Vizuete et al. (Eds.): CI3 2023, LNNS 1040, pp. 186–198, 2024.
https://doi.org/10.1007/978-3-031-63434-5_14

spark-ignition engines, efficient combustion requires the use of fuels that ensure fast and safe starting of the engine regardless of the outside temperature, guarantee a carbon-free combustion process, reduce the formation of pollutants, and produce minimum deterioration of the main engine parts [3]. The physical-chemical properties of the fuel used must allow the fuel to evaporate to form a very homogeneous fuel-air mixture in the combustion chamber. In spark-ignition engines (SIE) the above process depends on factors such as: the type of fuel, the proper design of the intake system, heat transfer conditions, the presence of waste gases, engine speed, leakage in the chamber due to compression losses and the effects of mixture agitation [4]. The addition of bridges and bores in the cylinder head proves to be very effective as it considerably reduces emissions. In addition, a slight change in the configuration of the bridges and bores can change the flow directions and patterns and vary the way the gases react, which can further reduce emissions [5]. Sha et al. (2015) [6] experimentally investigated the effects of pre-chamber volume and nozzle diameter on the resultant ignition characteristics. It was found that a larger pre-chamber provides higher ignition energy, which results in shortened flame development angle and combustion duration. At a given pre-chamber volume, nozzle diameter mainly affects the combustion duration [7]. The research begins with the aforementioned background, and due to contamination and lack of performance in spark-ignition and compression-ignition engines, the low level of energy consumption of organic materials as energy sources that guarantee an optimal performance in internal combustion engines. The main purpose of this project is the adaptation of the cylinder head of a spark-ignition engine and validation on a chassis dynamometer for the verification of its developed power after the respective modifications of the cylinder head with its respective elements that involve a better performance in the engine.

2 Method

2.1 Applied Test Vehicle

The research will be experimental and a Chevrolet Corsa 1.3 Standard 4-cylinder 1,300 cm^3 gasoline engine, with an indirect injection in the intake manifold, equipped with a three-way catalytic converter of the year 1997 will be used; the vehicle with the lowest fuel consumption and most sold in Ecuador in those years. Table 1 describes the technical specifications of the vehicle.

2.2 Combustion Chamber (Cylinder Head) Features

It is made of aluminum, which helps to dissipate heat. The design of the cylinder head is simple, easy to reproduce and install; given the intake stroke, the mixture reaches the cylinder with little speed, the turbulence is almost null, the combustion is slow and prone to detonation due to the long length of the flame [8]. The measurement of the volume of the combustion chamber can be seen in Fig. 1 with the help of a graduated burette in ml and an acrylic piece.

The mathematical models are generated from the following equations.

Table 1. Chevrolet Corsa 1.3L Standard Technical Data Sheet

Characteristics	Units
Make	Chevrolet
Model	Corsa Wind
Year	1997
Compression Ratio	9.4/1
Torque	111 Nm
Horsepower	60 CV
Bore	77.6 mm
Top Speed	164 km/h
Acceleration, 0–100 km/h	14.8 s
Max Power	60 CV DIN AT 5800 rpm
Displacement	1389 cm^3
Max Torque	111 Nm DIN AT 3400 rpm
Valves	2 valves per cylinder
Fuel	Gasoline
Engine	Inline four
Firing order	1-3-4-2

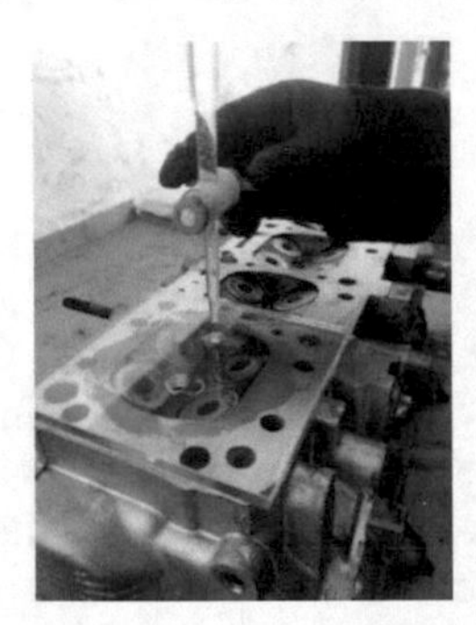

Fig. 1. Measurement of the combustion chamber volume

Piston volume measurement [9]

$$\textit{Compressed Vol.} = \text{Piston } \textit{Vol.} + \textit{Chamber Vol.} + \text{Gasket Vol.} \quad (1)$$

$$\textit{Compressed Vol.} = 14{,}4 + 22{,}2 + 4{,}65 = 41.25\,cm^3$$

Calculation of the compression ratio of the engine

$$Rc = \frac{\textit{Compressed Vol.} + \textit{Displacement Vol.}}{\textit{Compressed Vol.}} \quad (2)$$

$$Rc = \frac{41{,}25 + 343.67}{41.25} = 9.33 : 1$$

With the help of the mathematical models [9] and the corresponding adaptation, calculations are made to obtain the tuned or modified compression ratio that will allow observing the changes in the displacement values, comparing the results of this experiment where a burette is used to have a great accuracy in the liquid measurements and precision instruments to avoid errors in this experiment, as shown in Table 2.

Table 2. Compression Ratio Technical Data

	Engine without Modification	Engine with Modification
Displacement Vol	343.67 cm3	343.67 cm^3
Gasket Vol	4.65 cm^3	3.66 cm^3
Chamber Vol	22.2 cm^3	22 cm^3
Piston Vol	14.4 cm^3	14.4 cm^3
Compressed Vol	41.25 cm^3	40.06 cm^3
Compression Ratio	9.33:1	9.57:1

2.3 Valve Mechanism System

The project involves a traditional OHV type distribution system and a DOHC type system with the variation of two camshafts, improving the engine performance and gas intake and exhaust [10]. The adaptation is based on the use of a cylinder head from a Chevrolet Aveo vehicle in the monoblock of the Corsa vehicle. The DOHV valve system performs an exhaustive work in the gas recirculation. Figure 2 shows several characteristics of the valves both in standard and modified or tuned form, applying an inclination of 30 ° to the valve seat [11].

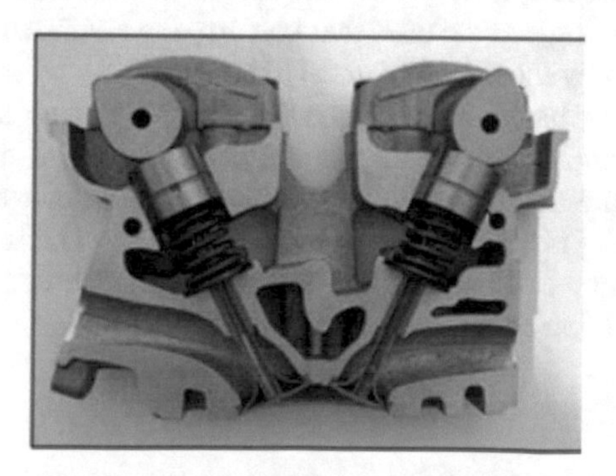

Fig. 2. Camshaft and valve configuration

Table 3 shows the characteristics of the intake and exhaust valves with their standard and modified values for the experimentation with a Corsa Wind vehicle with the engine tuned.

Table 3. Valve System

	Standard Condition	Modified Condition
Intake Valve Measurement		
Retainer Diameter	39.2 mm	28.58 mm
Valve Length	105.1 mm	101.58 mm
Valve Stem	7.1 mm	5.86 mm
Valve Seat	45°	30°
Exhaust Valve Measurement		
Retainer Diameter	31 mm	26 mm
Valve Length	112.5 mm	101.34 mm
Valve Stem	105.1 mm	5.85 mm
Valve Seat	45°	30°

2.4 Intake and Exhaust System of the Spark-Ignition Engine

The cylinder head has the gas intake mainly by means of ITB ducts which are used for the recirculation of gases and managing a proper combustion. Figure 3 shows the ITBs that allow greater air intake. Table 4 details the characteristics and values for standard and tuned measures proposed in this research. In addition, Figs. 2, 3 and 4 show the measurement processes established to obtain the data in the cylinder head [12]. The engines are fed with ethanol-gasoline blends and the users are responsible for performing the tuning, without following a recommended instruction and carrying out the tests in any chassis dynamometer. The service technicians, with their experience rather than formal training, are the ones that perform the engine tunings when operating with fuel not provided by the manufacturer [13].

Intake valve timing has a significant effect on the engine air exchange process. During engine start conditions, the engine valve strategy is to delay the intake valve timing and speed up the exhaust valve timing to ensure the lowest residual exhaust gas and maximum fresh air [14]. The sizing of the manifold nozzles is relevant to appreciate the modification as a consequence of the new performance desired from the vehicle engine, according to Table 5.

Fig. 3. Measurement of the intake manifold nozzle diameter

Table 4. Gas recirculation cylinder head ducts and valve springs

	Engine Without Modification			Engine with Modification		
Cylinder head intake nozzles						
	Diameter (mm)	Length (cm)	Volume (cm^3)	Diameter (mm)	Length (cm)	Volume (cm^3)
CYLINDER 1	33.0	35	500	40.73	18.5	238.7
CYLINDER 2	33.0	35	500	40.73	18.5	238.7
CYLINDER 3	33.0	35	500	40.73	18.5	238.7
CYLINDER 4	33.0	35	500	40.73	18.5	238.7
Cylinder head exhaust nozzles						
	Diameter (mm)	Length (cm)	Volume (cm^3)	Diameter (mm)	Length (cm)	Volume (cm^3)
CYLINDER 1	30.0	22	530	33	26	530
CYLINDER 2	30.0	22	530	33	26	530
CYLINDER 3	30.0	22	530	33	26	530
CYLINDER 4	30.0	22	530	33	26	530
Cylinder head springs						
	Diameter (mm)	Length (mm)	Thickness (mm)	Diameter (mm)	Length (mm)	Thickness (mm)
CYLINDER 1	29.50	30.50	3.05	27.1	40.42	3.48
CYLINDER 2	29.50	30.50	3.05	27.1	40.42	3.48
CYLINDER 3	29.50	30.50	3.05	27.1	40.42	3.48
CYLINDER 4	29.50	30.50	3.05	27.1	40.42	3.48

Fig. 4. Measurement of the ITB diameter

Table 5. Manifold nozzles and cylinder head height

	Engine Without Modification	Engine with Modification
Measurement of the cylinder head height		
	mm	mm
Height	96	130.4
Measurement of the exhaust manifold nozzle diameter		
	Diameter (mm)	Diameter (mm)
CYLINDER 1	30	33
CYLINDER 2	30	33
CYLINDER 3	30	33
CYLINDER 4	30	33
Measurement of the intake manifold nozzle diameter		
	Diameter (mm)	Diameter (mm)
CYLINDER 1	33	40.73
CYLINDER 2	33	40.73
CYLINDER 3	33	40.73
CYLINDER 4	33	40.73

2.5 Determination of the Developed Power

For the measurement process, the drive wheels or front wheels of the car rest on four rollers. The rollers rotate with a known moment of inertia. There is a sensor that registers the rotational speed of the rollers and systematically sends the speed information to the computer. [15] (For this chassis dynamometer, only computerized measurement is admissible). The main test in this study is in the chassis dynamometer, where the SAENZ PERFORMANCE INERTIAL CHASSIS DYNOS-N 08–19 was used, which allows simulating a speed profile as a function of time to perform dynamic tests on diesel and gasoline vehicles. It allows to load several test cycles of the different legislations for which they were designed, determining in an exact way the values of power and torque of the vehicle.

Fig. 5. Dynamometer characteristics

Figure 5 shows a sequential representation of the process and the model of tests performed on the vehicles. This is a chassis dynamometer (Fig. 6).

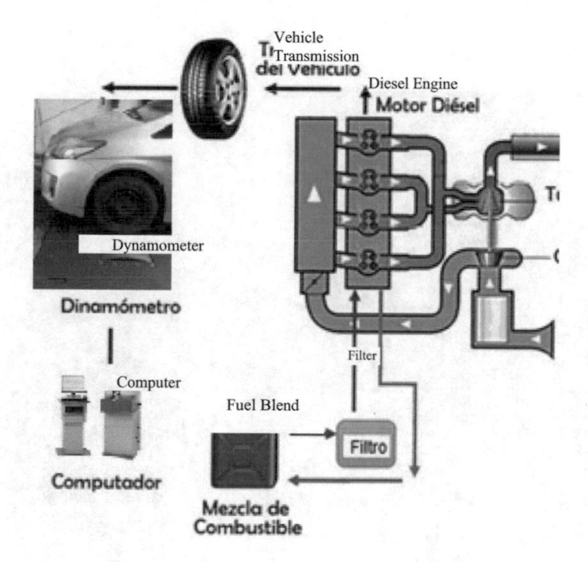

Fig. 6. Chassis dynamometer structure

The power and torque test protocols were performed under the ISO 1585 and ISO 3173 standards; following the test protocol: verify that the diameter of the wheels is equal to or greater than 13-inch rim (Wheels of the Corsa vehicle are 175/70/13). Verify the cleanliness of the test site both in the tire tread and in the dynamometer (Fig. 7).

Secure the vehicle with straps to prevent it from slipping off the rollers; enter the technical data concerning the vehicle to be tested into the software; check the gear and transmission ratio of the unit, which must be 1:1; make sure that the engine temperature is in the range of 85°-90°, otherwise it must pass an engine warm-up period to reach such temperature; start the cooling fan of the dynamometer; start the measurement test; accelerate the vehicle with the pedal fully depressed in the test gear until reaching the

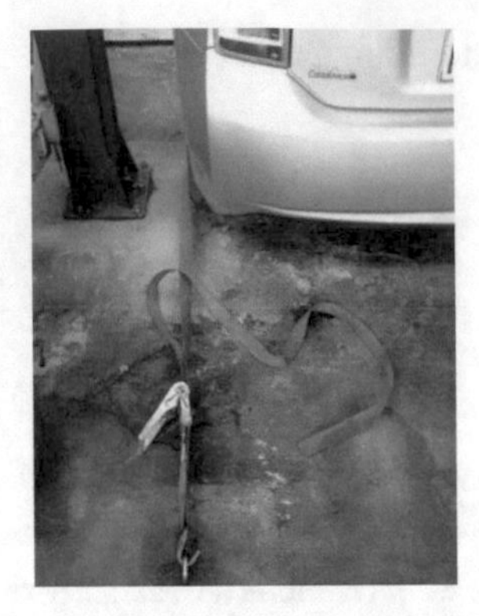

Fig. 7. Security straps and anchoring of the vehicle.

desired speed called "rpm cut" (4500–6000 rpm); when the "rpm cut" has been reached, depress the clutch, leaving the gear engaged. The dynamometer decelerates to a stop [16, 3] (Fig. 8).

Fig. 8. Dynamometer tests

3 Results

3.1 Vehicle Power Without Modification of the Cylinder Head

The power values generated by the engine of the Corsa 1.3 vehicle are obtained with tests. The static test with load was carried out, using the chassis dynamometer, under the ISO 1585 Standard, performing three tests for each case and obtaining the result to be evaluated. Figure 9 shows the results of the tests on the equipment. The results of

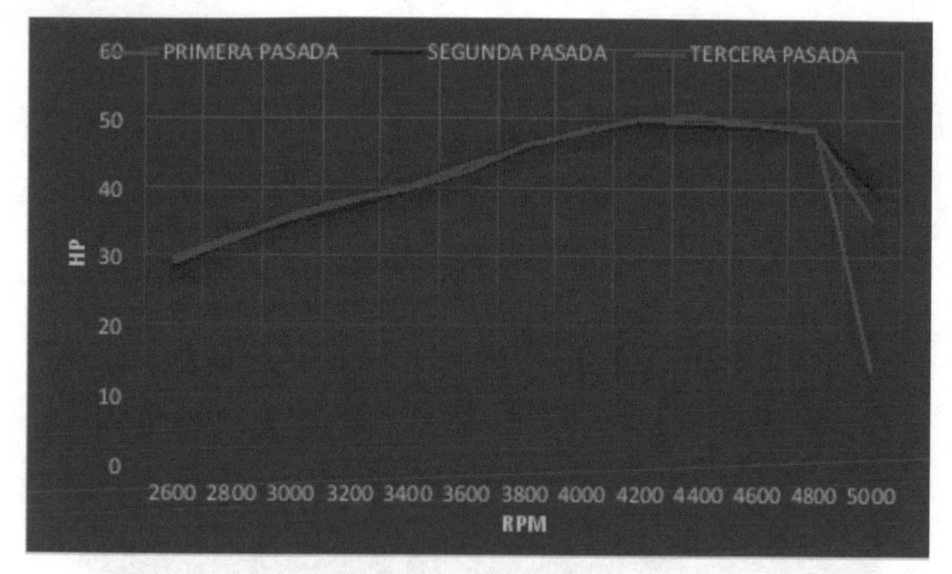

Fig. 9. Vehicle power without the cylinder head modification during three runs

the power values obtained for the vehicle with the dynamometer are as follows. The test was run three times in a row using premium gasoline. In this test, it can be observed that the curves generated do not show a major difference according to the values.

3.2 Vehicle Power with the Cylinder Head Modification

Next, the power parameters generated by the engine of the Corsa 1.3L vehicle are presented. The static test with load was carried out using the chassis dynamometer, under the ISO 1585 Standard. Three consecutive tests were performed obtaining the result shown in Fig. 10, which indicates the power curves obtained with the dynamometer, using premium gasoline. In addition, it is shown that there is no great difference between the values obtained.

Table 6 shows the results obtained in the dynamometer with a power increase of 8 hp, where it is confirmed that the adaptation of the cylinder head of the Aveo vehicle to the engine of the Corsa vehicle has been efficient.

3.3 Discussion of the Results on the Obtained Power

The engine could be fueled by gasoline, ethanol or a blend of both in any proportion. The fuel injection system was controlled by an electronic module for engine development, which allowed optimizing the blend ratio and ignition timing for the whole speed range tested [17].

In Fig. 10, an irregular line representing the torque curve as a function of revolutions per minute of the Corsa Wind 1.3L engine can be seen. The graph starts at 2200 RPM with a torque of 50 hp, when approaching 2300 RPM there is an abrupt change, a rise in the graph curve up to 68 hp and at 5500 RPM rises to almost 69 hp and drops at 2000

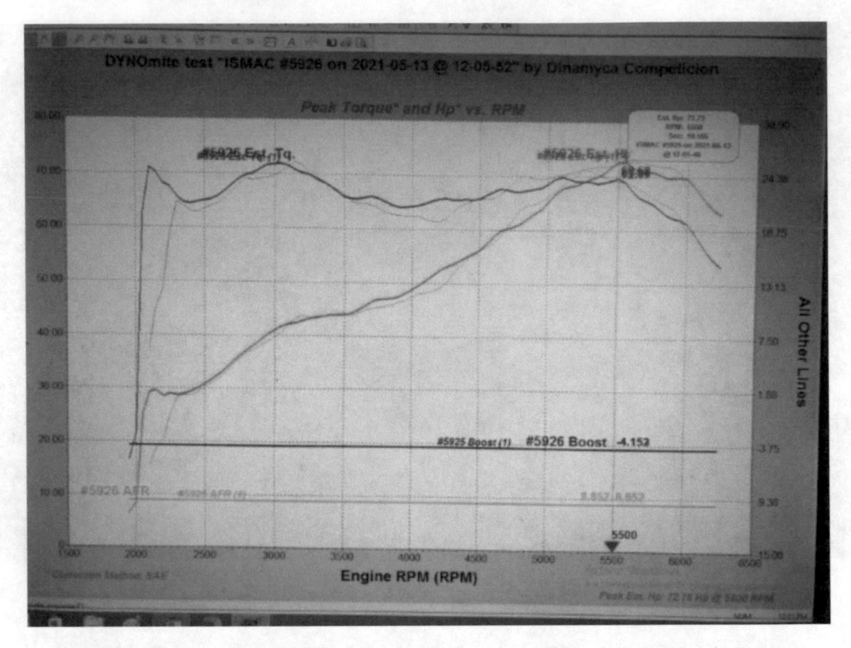

Fig. 10. Vehicle power with the cylinder head modification during three runs

Table 6. Test results of the tuned engine

Dynamometer	Power	RPM
First test	68.58 hp	5500 rpm
Second test	69.58 hp	5500 rpm
Third test	68.58 hp	5500 rpm

RPM to 30 hp, it rises again to almost 50 hp at 2700 RPM and then there is a rise and fall of torque, rises and falls again until it stabilizes from 3500 RPM [18].

The tests carried out and the values obtained are used to analyze the behavior of the vehicle, which will be used to develop the pertinent analysis regarding the performance of the engine. We will proceed as follows, working with the average power values obtained [12]. Figure 9 shows the power variation differences for the standard working tests and the cylinder head modification, clearly evidenced in the chassis dynamometer.

It is then established according to the measurements made that there is an increase of 30.08% in the performance at 4400 rpm with respect to the engine power with the modifications of the cylinder head, and it is clearly seen in Table 6 the difference between the performance of power before and after the modifications of the cylinder head. All the tests were carried out with the premium gasoline that is commercialized in the country. Several authors show similar data, confirming the increased engine efficiency [4] (Fig. 11).

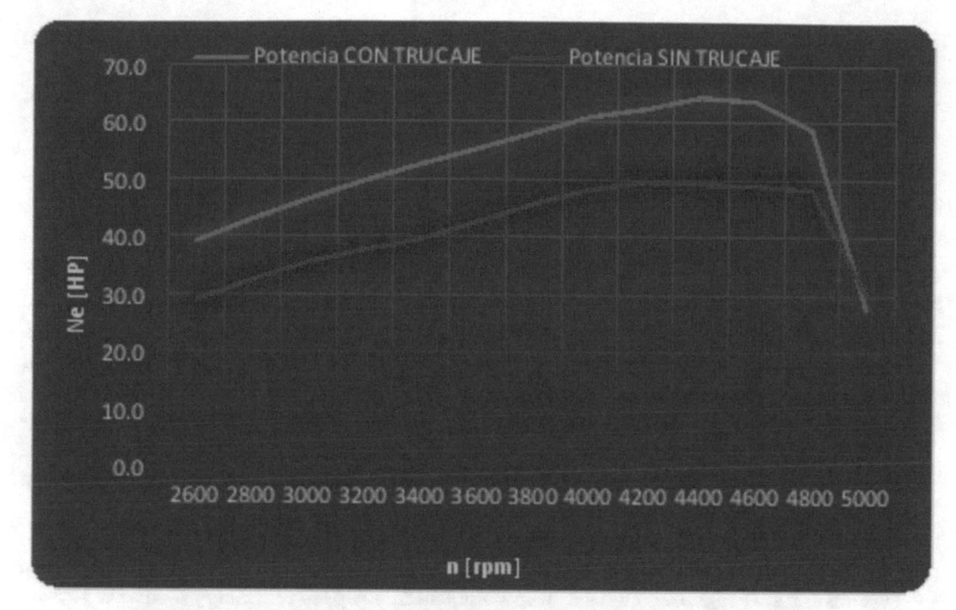

Fig. 11. Developed powers before and after the engine cylinder head tuning.

4 Conclusions

An analysis is made by researching the adaptations and modifications in the cylinder head and engine tuning to find the best performance and yield. In the analyzed studies, it has been found that the best angle of the valve seats is at 30 degrees in the intake and exhaust nozzles and ducts of the cylinder heads, and they are truly relevant and have the greatest influence on the performance of the engine's ITBs (Individual throttle body).

The measurements performed show an increase in power performance of 8 hp in general during the course of the engine operation from idle to maximum power. Therefore, with the modifications and specifically at the maximum power at 4500 rpm, an increase of 13.08% measured in the chassis dynamometer was observed as a result of the modifications of the cylinder head of the Aveo vehicle with respect to the standard cylinder head of the Corsa Wind vehicle.

References

1. Repositorio de la Universidad de Fuerzas Armadas ESPE: Artículo Científico - Estudio y análisis teórico práctico del comportamiento de un motor SUZUKI G10 previo y posterior a su trucaje
2. Herrera Martínez, J.J.: Trucaje de un motor Renault 1022 CC referencia R850 de serie (2016)
3. Jovaj, M.: Motores de automovil, Editorial MIR, Mosc (1982)
4. Lukanin, V.: Motores de combustion interna, Editorial MIR, Mosc (1985)
5. Indudhar, M.R., Banapurmath, N.R., Rajulu, K.G., Patil, A.Y., Javed, S., Khan, T.M.Y.: Optimization of piston grooves, bridges on cylinder head, and inlet valve masking of home-fueled diesel engine by response surface methodology. Sustainability **13**(20), 11411 (2021)

6. Shah, A., Tunestal, P., Johansson, B.: Effect of pre-chamber volume and nozzle diameter on pre-chamber ignition in heavy duty natural gas engines, SAE technical paper (2015)
7. Ge, H., Bakir, A.H., Yadav, S., Kang, Y., Parameswaran, S., Zhao, P.: CFD optimization of the pre-chamber geometry for a gasoline spark ignition engine. Front. Mech. Eng. **6**, 599752 (2021)
8. Kalghatgi, G.T., McDonald, C.R., Hopwood, A.B.: An experimental study of combustion chamber deposits and their effects in a spark-ignition engine, SAE Technical Paper (1995)
9. Kindler, H.: Matemática Aplicada Para La Técnica Del Automóvil. Barcelona España: Reverté (2007)
10. Gillieri, S.: Preparación de Motores de Serie para Competición. Barcelona España: CE (2006)
11. Rosales, J.P.F., Revelo, J.D.G., Cárdenas, E.M.C., Paz, P.A.M.: Determinación experimental de la eficiencia mecánica de un motor de combustión interna de encendido provocado por medio de las curvas de torque y potencia. Polo del Conoc. **7**(7), 1735–1759 (2022)
12. Romero-Piedrahita, C.A., Mejía-Calderón, L.A., Carranza-Sánchez, Y.A.: Estudio de sensibilidad del desempeño durante el calentamiento en vacío de un motor de combustión interna a cambios en su relación de compresión, el contenido de etanol en la mezcla combustible con gasolina y el material de la culata. Dyna **85**(204), 238–247 (2018)
13. Galindo, J., Climent, H., De la Morena, J., Pitarch, R., Guilain, S., Besançon, T.: A methodology to study the interaction between variable valve actuation and exhaust gas recirculation systems for spark-ignition engines from combustion perspective. Energy Convers. Manag. **250**, 114859 (2021)
14. MAUT 014 TRABAJO DE GRADO.pdf (utn.edu.ec)
15. Corsini, A., Di Antonio, R., Di Nucci, G., Marchegiani, A., Rispoli, F., Venturini, P.: Performance analysis of a common-rail diesel engine fuelled with different blends of waste cooking oil and gasoil. Energy Procedia **101**, 606–613 (2016)
16. Yu, S., Zheng, M.: Future gasoline engine ignition: a review on advanced concepts. Int. J. Engine Res. **22**(6), 1743–1775 (2021). https://doi.org/10.1177/1468087420953085
17. Costa, R.C., Sodré, J.R.: Hydrous ethanol vs. gasoline-ethanol blend: engine performance and emissions. Fuel **89**(2), 287–293 (2010)
18. Modificación mecánica y electrónica de un motor Corsa Wind 1300 cc distribución OHC con la implementación de una unidad de control electrónico programable y un sistema de distribución DOHC de 16 válvulas

Analysis of the Reliability of the Calibration of a Camera Through the Knowledge of Its Extrinsic Parameters

Henry Díaz-Iza[1](✉), Harold Díaz-Iza[2], Wilmer Albarracín[1], and Rene Cortijo[1]

[1] Universidad Tecnológica Israel, Quito, Ecuador
hdiaz@uisrael.edu.ec
[2] Universitat Politècnica de València, Valencia, Spain

Abstract. Camera calibration methods support their concepts on pinhole camera models to obtain the intrinsic and extrinsic parameters of a camera. These methods consider the different nonlinearities caused by the lens to determine their results. The different calibration algorithms, to verify whether the parameters of the model are acceptable, use a mathematical factor known as back-projection error that, as much as it approximates to zero, indicates whether the theoretical mathematical model of a camera fits well with the real camera model. In this study, the aim is to analyze the reliability of the calibration results of a camera by comparing the extrinsic parameters (position and orientation) delivered by the algorithms of Tsai, Zhang, and Faugeras with different camera positions and orientations taken using an ABB manipulator robot.

Keywords: Camera calibration · Error analysis · Computer vision

1 Introduction

In computer vision, digital cameras play a significant role in developing vision applications. Digital cameras allow improvements in image definition, resolution, and optical quality. In addition, they have advanced processing capabilities that enable additional features such as fast autofocus, face detection, and burst image capture. These enhancements have significantly impacted various industries and vision applications, expanding possibilities in photography, video surveillance, augmented reality, medicine, robotics, and more.

Calibration of a camera is essential in obtaining accurate measurements of a scene from captured images. For this reason, it is essential to perform a camera calibration with complete assurance that the parameters obtained are as close as possible to values that guarantee an accurate real measure of a scene. Calibration enables a mathematical representation of their physical and geometrical properties and their orientation and position according to a reference plane [1–3].

M. Z. Vizuete et al. (Eds.): CI3 2023, LNNS 1040, pp. 199–210, 2024.
https://doi.org/10.1007/978-3-031-63434-5_15

The basic idea of calibration is to describe a projection model that relates the coordinate systems of the camera and the reference world. In other words, a calibration permits obtaining intrinsic parameters (optical center and focal length) and extrinsic parameters (position and orientation of a camera according to a reference framework) [4,5].

Although there are a variety of calibration methods, all of them use the reprojection error to validate their results [6]; however, in practice, this mathematical factor does not ensure that the camera model is correct due to the coupling between intrinsic and extrinsic parameters that makes the algorithms in one way or another look for a solution to the system that may or not be adequate.

Even though most camera calibration algorithms solve the acquisition of intrinsic and extrinsic parameters, the question arises as to how reliable are the parameters obtained from the calibration. In general terms, this work aims to implement and analyze the most commonly used calibration algorithms in the field of vision and to assess the reliability of the extrinsic parameters by comparing the extrinsic parameter results of the algorithms with the position and orientation values of the camera taken with the help of an IRB140 robot of ABB.

The article is structured as follows: Sect. 2 describes the camera model commonly employed in calibration and the different calibration methods implemented. Section 3 details the results of the experiments performed. Finally, Sect. 4 mentions the conclusions

2 Methods

The camera calibration methodology is divided into two stages. The first stage consists of obtaining the intrinsic and extrinsic camera parameters by applying the most popular calibration algorithms, Zhang and Tsai for 2D and Tsai and Faugeras for 3D, over synthetic data to validate the algorithms. The second stage is responsible for analyzing the reliability of the extrinsic parameters obtained by applying the algorithms over real data and comparing the extrinsic parameter results with the real camera position obtained with the help of the IRB140 robot [7]. The camera model consists of two parameters to consider: intrinsic and extrinsic. The intrinsic parameters model the camera image sensor, i.e., internal geometry and optical characteristics. The extrinsic parameters, on the other hand, measure the position and orientation of the camera with respect to a world coordinate system.

The model of a camera is based on the approximation of the internal geometry, position, and orientation of the camera in the scene. Several camera models depend on the desired accuracy; one of the most widely used is the Pinhole model. In general, this model allows obtaining the necessary data to calculate the geometric information that an image possesses, i.e., they describe a mathematical projection between an object in 3D and 2D image space.

2.1 Pinhole Camera Model

The pinhole model is the most straightforward and practical type of camera model. It operates by projecting a point in the scene onto the image plane through the intersection of a line passing through the point and the center of projection [4,8], as shown in Fig. 1.

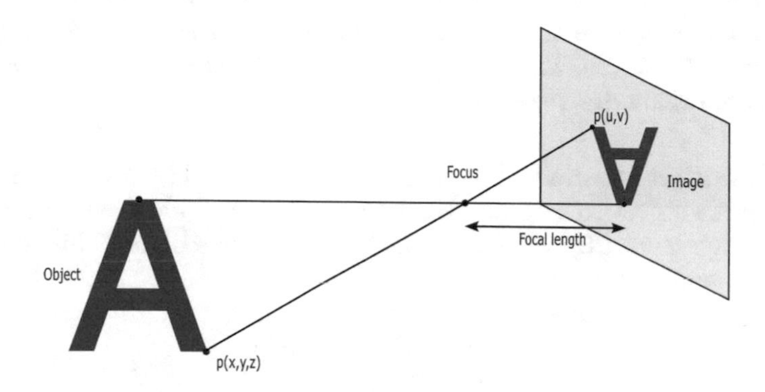

Fig. 1. Pinhole Camera Model [9]

The pinhole model utilizes a projection matrix to convert the three-dimensional coordinates of object points into their corresponding two-dimensional image coordinates, as expressed by Eq. 1.

$$m = P \cdot M \tag{1}$$

where $M = [X_\omega, Y_w, Z_\omega, 1]^T$ represents the vector that contains the coordinates of the world reference system, P is the projection or camera matrix used to transform the world coordinates into the projected point $m = [u, v, 1]^T$ over the image, the units of m are pixels.

2.2 Methods of Calibration

Calibrating a camera is essential for obtaining accurate scene measurements from captured images. The quality of the calibration will have a direct impact on the accuracy of the measurements made based on those images. For this reason, it is necessary to perform a reliable camera calibration and to obtain parameters as close as possible to the valid values. This commitment implies making the right decisions regarding the chosen calibration method and using it correctly.

Calibration involves obtaining the parameters that describe the camera model including their linear and nonlinear components. Several methods are currently available for performing this calibration; some use specific templates, while others dispense with them altogether.

The Hall [10] and Faugeras-Toscani [7] linear calibration methods are commonly used with 3D templates and employ least-squares techniques to obtain the model parameters. By the other hand, nonlinear calibration methods such as Tsai's [11] with 3D or 2D templates and Zhang's [12] with 2D templates employ a two-stage technique. In the first stage, a linear model is calculated based on the calibration data; this provides initial estimates of data for the camera parameters. In the second stage, a nonlinear optimization is performed using the initial estimates parameters as a starting point. This optimization process refines the camera parameters to further improve the calibration accuracy. By combining both linear and nonlinear techniques, the calibration process achieves more accurate and reliable results. The algorithms and their main characteristics used in this work are briefly described below.

Hall Calibration Method. The Hall method [7,10] is based on an implicit calibration whose objective is to find the linear relationship between the 3D points of the scene with the 2D points projected on the image plane. This relationship is shown in the Eq. 2

$$\begin{pmatrix} s^I X_d \\ s^I Y_d \\ s \end{pmatrix} = \mathbf{A} \begin{pmatrix} {}^W X_w \\ {}^W Y_w \\ {}^W Z_w \end{pmatrix} \tag{2}$$

where $({}^W X_w, {}^W Y_w, {}^W Z_w)^T$ expresses a 3D point from the world (scene), $({}^I X_d, {}^I Y_d)^T$ is the 2D point in pixel with respect to the image coordinate system, s is an scale factor, and A is a 3 by 4 transformation matrix proposed by Hall.

Faugeras-Toscani Calibration Method. To obtain a complete model, Faugeras-Toscani [7,13] proposes a slightly different method than that proposed by Hall for the estimation of the calibration matrix. Faugeras-Toscani obtains the intrinsic and extrinsic parameters by equating his calibration matrix with that proposed by Hall. The relationship is shown in the equation

$$\begin{pmatrix} s^I X_d \\ s^I Y_d \\ s \end{pmatrix} = \begin{pmatrix} \alpha_u & 0 & u_0 & 0 \\ 0 & \alpha_v & v_0 & 0 \\ 0 & 0 & 1 & 0 \end{pmatrix} \begin{pmatrix} r_{11} & r_{12} & r_{13} & t_x \\ r_{21} & r_{22} & r_{23} & t_y \\ r_{31} & r_{32} & r_{33} & t_z \\ 0 & 0 & 0 & 1 \end{pmatrix} \begin{pmatrix} {}^W X_w \\ {}^W Y_w \\ {}^W Z_w \\ 1 \end{pmatrix}, \tag{3}$$

then

$$\begin{pmatrix} s^I X_d \\ s^I Y_d \\ s \end{pmatrix} = \mathbf{A} \begin{pmatrix} {}^W X_w \\ {}^W Y_w \\ {}^W Z_w \\ 1 \end{pmatrix} \tag{4}$$

where $({}^W X_w, {}^W Y_w, {}^W Z_w)^T$ expresses a 3D point from the world (scene), $({}^I X_d, {}^I Y_d)^T$ is the 2D point in pixel with respect to the image coordinate system, s is an scale factor, A is a 3 by 4 transformation matrix, $(\alpha_u, \alpha_v, u_0, v_0)$ are the intrinsic parameters and $(r_1, r_2, r_3, t_x, t_y, t_z)$ are the extrinsic parameters.

Tsai Calibration Method. The Tsai method [7,11] models the radial distortion of the lens but assumes that there are camera parameters that the manufacturer provides. This method reduces the number of calibration parameters at the first stage when parameter estimation is started.

The Tsai method has some limitations, among which we can mention that in its model, it only involves radial and not tangential distortions; however, these distortions are often not considered necessary, so the method has been widely extended and used in applications that require higher accuracy.

Zhang Calibration Method. The Zhang method [9,14] is the most widely used in computer vision; in the first instance, the method is based on obtaining an approximation of the camera model by a linear method. This approximation is refined by applying a maximum likelihood criterion iteratively. Calibration with this method is mainly performed with a point plane or a 2D grid. Zhang mentions that for the method to work correctly, it is necessary to take at least three images from different positions and orientations; however, the number of images can be reduced if some intrinsic parameters are fixed.

3 Results and Discussion

A series of images captured from various camera positions were used to extract intrinsic and extrinsic parameters. Different algorithms were implemented to estimate these parameters, firstly using synthetic data to verify the correct functioning of the algorithm to finally using the algorithms with real data. The data obtained from the application of the algorithm over the different images taken were used to verify whether the extrinsic parameters resulting from the calibration algorithms fit the camera positions and orientations by taking the positions acquired with the help of an ABB IRB 140 robot as a reference. The system overview is shown in Fig. 2 and consists of a Canon EOS 700D camera, 2D/3D calibration templates, ABB IRB 140 robot, and measurement pointer. For further reference of the implemented system and acquired data, refer to [15].

3.1 Description of the Experiments Performed

The acquired images have a 5184 × 3456 [pixels] resolution, and the focal distance of the camera is 18 [mm]. The data acquisition was performed in two parts; first, the coordinates of the reference points (corners) of the template were taken with a pointer mounted as an end-effector on the robot. Second, the coordinates and orientations of the different camera locations were taken according to the orientation and position of the robot.

The images were acquired from a camera in the robot as an end-effector and from different camera positions and orientations reached by IRB 140 robot. The robot position data displayed on the Flexpendant were recorded. For the detection of points of the 2D template, necessary for the calibration of the camera, were obtained by the vision tools of Matlab software. Specifically, the

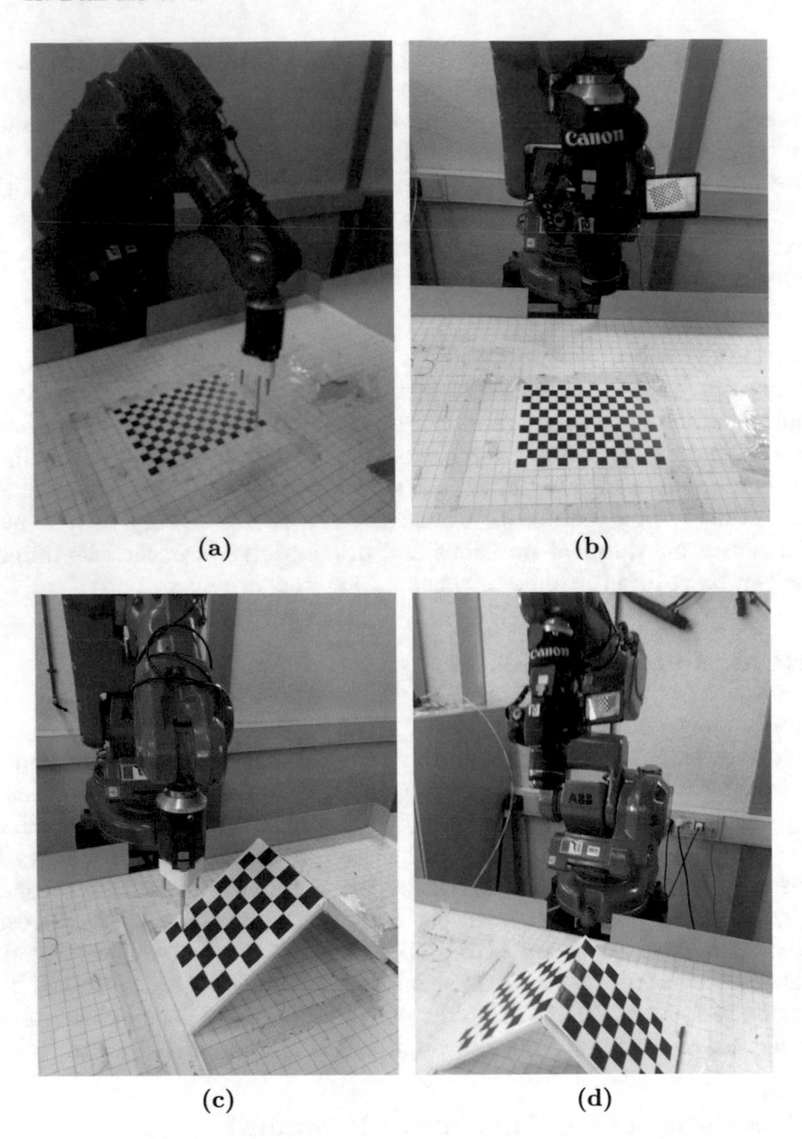

Fig. 2. System used for calibration: pointer and 2D template(a), camera and 2D template(b), pointer and 3D template(c), camera and 3D template(d).

detectCheckerboardPoints function was used to identify the points of the 2D template. On the other hand, the detection of points for a 3D template was performed manually using as a reference the measurement of each point by the robot.

3.2 Tsai and Zhang Method for 2D

Tsai and Zhang's algorithms used a set of 16 images. The extrinsic parameters (camera position and orientation) resulting from the calibration of each image have been compared with the coordinates acquired by the ABB robot. Figure 3 shows the back-projection error of each image used for calibration. It is observed that the maximum back-projection error for Tsai's method is 4.2 pixels and a mean of 3.07 pixels, and for Zhang's method, the maximal back-projection error is 3.7 pixels and a mean of 2.33 pixels. This error means there is no significant difference (depending on the application) between the extrinsic parameters given by both algorithms and the values taken by the robot.

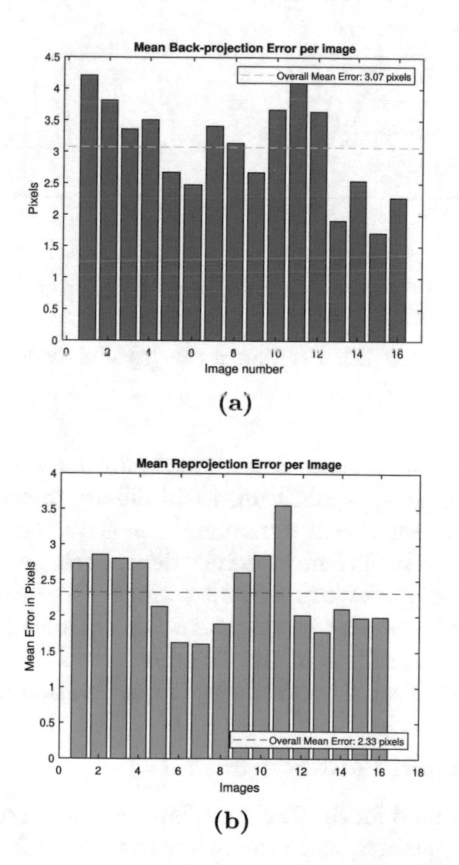

Fig. 3. Back-projection error: Tsai 2D (a) y Zhang (b)

By applying the inverse transform to the extrinsic parameter matrix resulting from the calibration, we can obtain the camera position and orientation that were compared with those acquired by the ABB robot. Figure 4 shows the camera position and orientation error by comparing the robot coordinates as a reference with the extrinsic parameters provided by the Tsai 2D and Zhang 2D algorithms

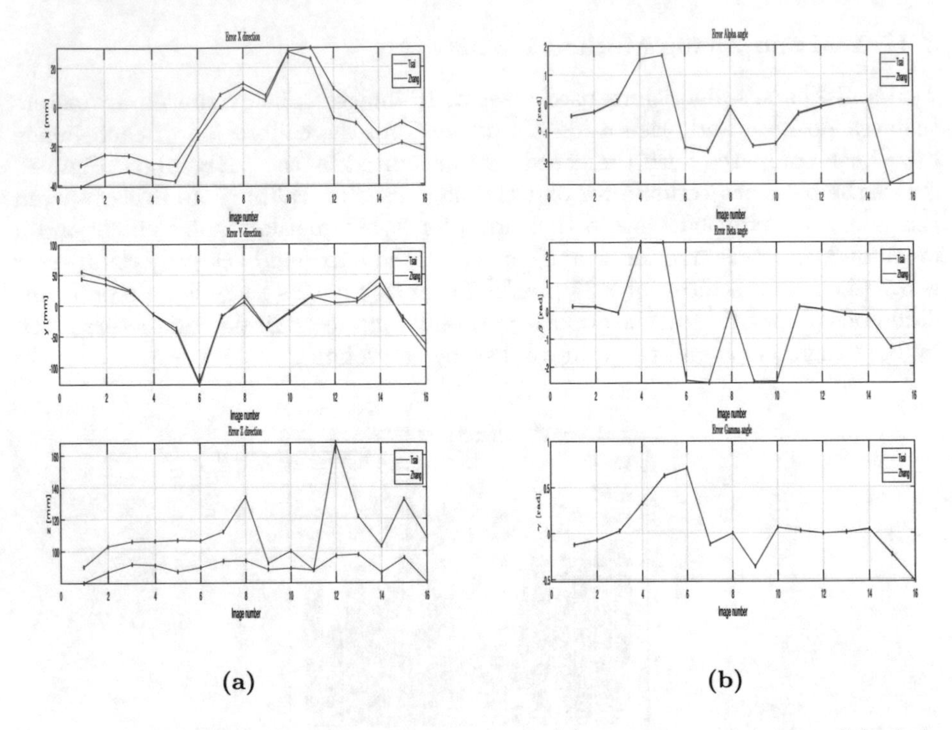

Fig. 4. Tsai 2D and Zhang 2D: Position error (a) and orientation error (b) of the camera.

for each image. The results show that the orientation errors of the camera are similar in both algorithms; in addition, little difference errors can be seen in x and y directions, but a significant difference is shown in the z direction, meaning that the algorithms are inefficient to detect depth, this problem could be due to that the algorithms only take into a count a single camera, to solve the problem with depth the use of a stereo-vision system is recommended.

Figure 5 shows graphically how the different actual camera positions lie over those delivered by the Tsai 2D and Zhang 2D calibration algorithm.

3.3 Tsai and Faugeras Method for 3D

A set of 6 images was used for the Tsai and Faugeras 3D algorithms. The extrinsic parameters (camera position and orientation) resulting from the calibration of each image have been compared with the coordinates acquired by the ABB robot. Figure 6 shows the back-projection error of each of the images used for calibration; it is observed that the average reprojection error of 20.17 pixels for the Tsai algorithm and 11.91 pixels for the Faugeras 3D algorithm. This error means that the difference between the extrinsic parameters given by the algorithms and the values taken by the robot is significant.

Applying the inverse transform to the extrinsic parameter matrix yields the position and orientation of the camera concerning the calibration template; these

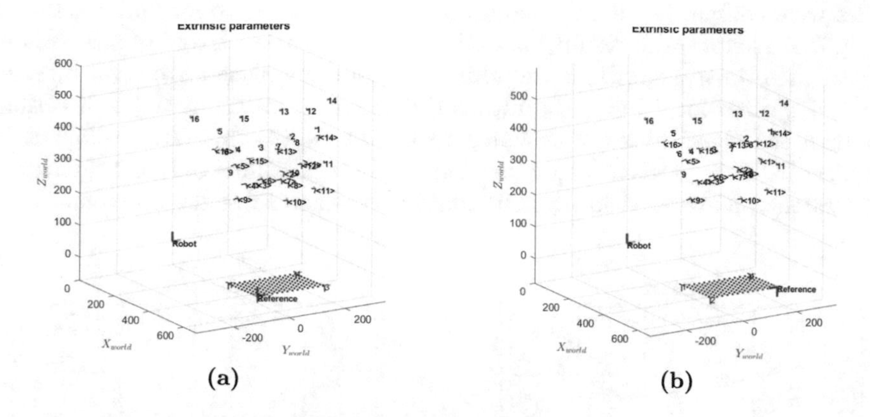

Fig. 5. Calibration comparison of extrinsic parameters.The actual camera positions are above those delivered by the algorithm: Tsai 2D (a) and Zhang (b).

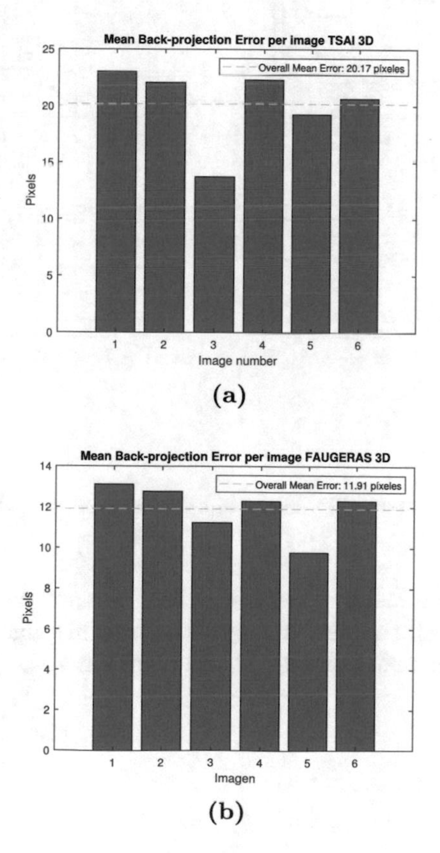

Fig. 6. Back-propagation error: Tsai 3D (a) and Faugeras 3D (b).

results were compared with the values obtained from the robot and are shown in Fig. 7. As before, the results show that the orientation errors of the camera are similar in both algorithms; in addition, little difference errors can be seen in x and y directions, but a significant difference is shown in the z direction. Although the algorithms are designed to detect depth, they are inefficient in detecting it. This problem might be due to the algorithms only considering a single camera, a stereo-vision system could be used to solve the problem.

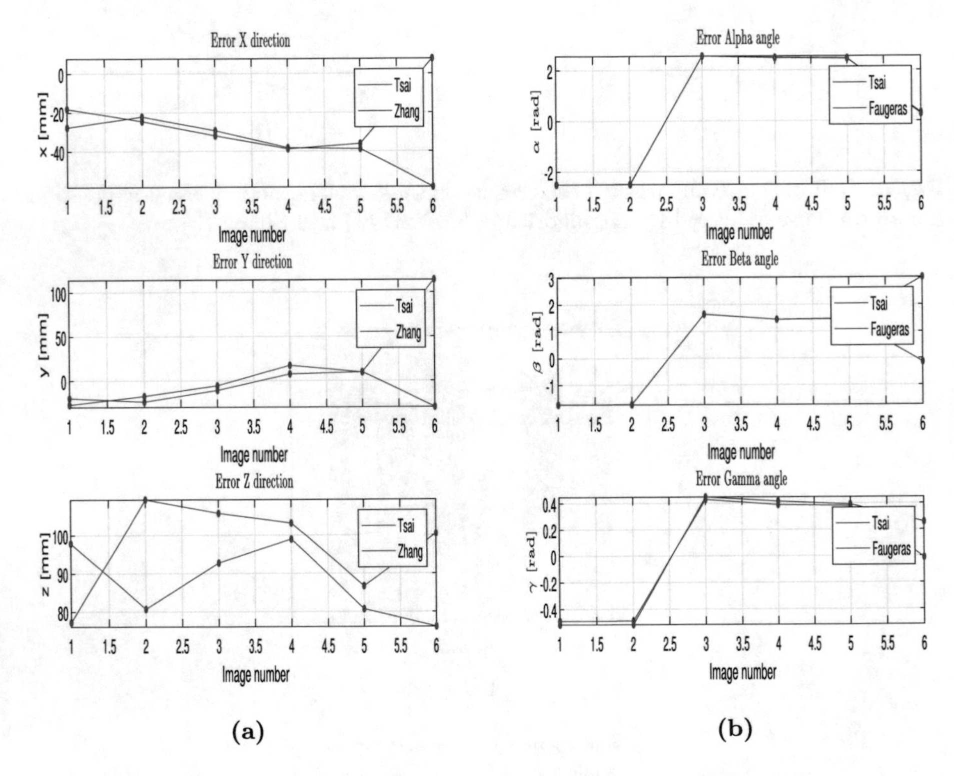

Fig. 7. Tsai 3D and Faugeras 3D: Error of camera position (a) and orientation (b).

Figure 8 shows graphically that the actual camera positions do not match those delivered by both the Tsai 3D and Faugeras 3D algorithms, as the actual position is offset from that resulting in the algorithms.

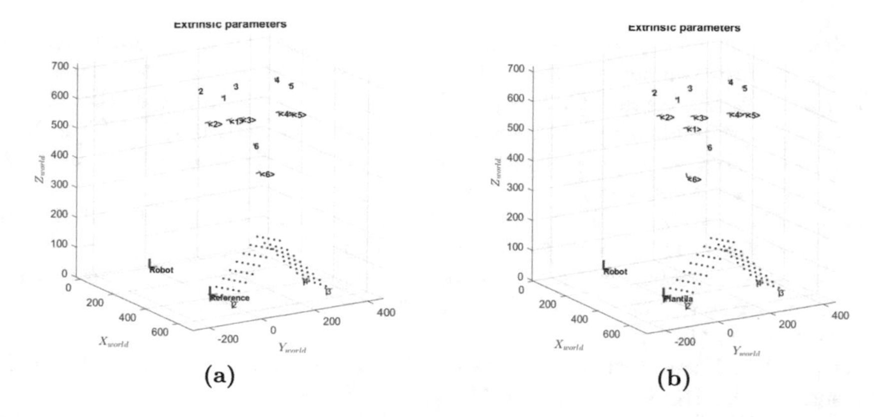

Fig. 8. Comparison of extrinsic parameter calibration: Tsai 3D (a) and Faugeras 3D (b).

4 Conclusions

In our study, the different camera calibration methods showed a specific deviation from the reference values (measurements taken by the robot) in both orientation and translation. However, the 2D Zhang and 3D Faugeras calibration methods had the best approximated extrinsic parameters concerning the reference. Additionally, the results demonstrate that the back-projection error in the 2D calibration methods is small in comparison with the 3D calibration methods; however, this small error does not guarantee that the extrinsic coordinates are the same that the reality coordinates, as shown by the results obtained. Our results underscore that practical experiments provide an essential piece of information to distinguish which algorithm has the best performance at the time to calibrate a camera.

References

1. Viala, C.R.: Caracterización y optimización del proceso de calibrado de cámaras basado en plantilla bidimensional. Ph.D. thesis, Tesis doctoral, Universidad Politécnica de Valencia. Capitulo 2: El modelo de la cámara-2.7 Homografias p. 29 (2006)
2. Förstner, W., Wrobel, B.: Photogrammetric computer vision, pp. 141–159. Chapter robust estimation and outlier detection, Springer, Berlin (2016). https://doi.org/10.1007/978-3-319-11550-4
3. Ricolfe-Viala, C., Esparza, A.: The influence of autofocus lenses in the camera calibration process. IEEE Trans. Instrum. Meas. **70**, 1–15 (2021)
4. Hornberg, A.: Handbook of Machine and Computer Vision: The Guide for Developers and Users. Wiley, Hoboken (2017)
5. Chen, C.H.: Handbook of Pattern Recognition and Computer Vision. World Scientific, Singapore (2015)
6. Chatterjee, C., Roychowdhury, V.P.: Algorithms for coplanar camera calibration. Mach. Vis. Appl. **12**(2), 84–97 (2000)

7. Salvi, J., Armangué, X., Batlle, J.: A comparative review of camera calibrating methods with accuracy evaluation. Pattern Recogn. **35**(7), 1617–1635 (2002)
8. Hui, Z., Hua, W., Huinan, G., Long, R., Zuofeng, Z.: A real-time camera calibration system based on opencv. In: Proceedings of SPIE, vol. 9631, pp. 96311T–1 (2015)
9. Isern González, J.: Estudio experimental de métodos de calibración y autocalibración de cámaras (2003)
10. Hall, E.L., Tio, J.B., McPherson, C.A., Sadjadi, F.A.: Measuring curved surfaces for robot vision. Computer **15**(12), 42–54 (1982)
11. Tsai, R.: A versatile camera calibration technique for high-accuracy 3d machine vision metrology using off-the-shelf tv cameras and lenses. IEEE J. Robot. Autom. **3**(4), 323–344 (1987)
12. Burger, W.: Zhang's camera calibration algorithm: in-depth tutorial and implementation. In: HGB16-05, pp. 1–6 (2016)
13. Faugeras, O.: Three-Dimensional Computer Vision: a Geometric Viewpoint. MIT Press, Cambridge (1993)
14. Zhang, Z.: Flexible camera calibration by viewing a plane from unknown orientations. In: Computer Vision, 1999. The Proceedings of the Seventh IEEE International Conference on, vol. 1, pp. 666–673. IEEE (1999)
15. Díaz Iza, H.J.: Valoración de la fiabilidad de los resultados de calibrado de una cámara conociendo sus parámetros extrínsecos. UPV (2018)

Design and Construction of a Prosthetic Finger with Distal Phalanx Amputation

Enrique Mauricio Barreno-Avila[1](✉), Segundo Manuel Espín-Lagos[2], Diego Vinicio Guamanquispe-Vaca[2], Alejandra Marlene Lascano-Moreta[2], Christian Israel Guevara-Morales[2], and Diego Rafael Freire-Romero[3]

[1] Postgraduate Faculty, Universidad Técnica de Manabí, Av. Urbina y Portoviejo, 130105 Portoviejo, Ecuador
ebarreno3907@utm.edu.ec

[2] Faculty of Civil and Mechanical Engineering/Technology and Transfer Center (CTT), Universidad Técnica de Ambato, Av. los Chásquis, 180207 Ambato, Ecuador

[3] Harbert International Establishment S de RL, de las Magnolias S8-216, Quito, Ecuador

Abstract. There is a large number of workers who have suffered an accident at work in most companies is common phalangeal and transfalangeal amputations is the case of the company Dico-Val, where a few years ago a worker suffered an accident where he lost part of the finger specifically the distal phalanx, in this way, obtaining a prosthesis in order to recover the physical capacity of workers becomes imperative; Therefore, the present research of analysis of mechanism alternatives proposes the design of a mechanically driven prosthesis of the distal phalanx of the hand using bibliographic information and analysing different types of prosthesis; comparative analysis of 3D printing materials was performed, a comparative study between design applications and laminators required for the use of 3D printing was performed, different types of prosthesis were analysed focusing on the selection of alternatives of the type of mechanism, which helps the flexion and extension movements, and the type of material for the design parameters. Finally, the results were positive as the computational verification efforts were obtained; concluding that a functional prosthesis was obtained and the selected 3D printing material (PLA) provided sufficient strength for the prosthesis to perform prehension actions.

Keywords: Distal phalanx · Finger prosthesis · Mechanism · Anthropometry · Kinematic analysis

1 Introduction

Various prosthetic hands (or fingers) have been created for amputees who lost a finger or hand in an accident or complication. These prosthetic hands can be can be classified into three types, depending on the pattern and level of amputation of the patient: Type 1 (middle or proximal phalanx), Type 2 (affects only the wrist) and Type 3 (affects the thumb and part of the metacarpals) [1–3].

M. Z. Vizuete et al. (Eds.): CI3 2023, LNNS 1040, pp. 211–224, 2024.
https://doi.org/10.1007/978-3-031-63434-5_16

Because the proximal interphalangeal (PIP) or metacarpophalangeal (MCP) joints may move, prosthetic hands for Type 1 patients have been created as finger modules that employ a linking mechanism powered by the patient's body [4, 5].

The human hand has a wide range of degrees of freedom, a high force-to-weight ratio (including the source of energy), a low factor of form (compactness), and a complex sensory system. Each of the bases of the fingers has two degrees of freedom, with the exception of the thumb, which has five degrees and two joints that allow for flexion and extension motions. The other degree of freedom (GDL) is located in the palm, which curves the surface where the bases of the toes are located [6, 7].

A prosthesis is a device created to replace a missing part of the human body [8]. Its purpose is to improve the user's quality of life by helping him or her to perform tasks for which the missing body part would normally be needed or by serving as an aesthetic accessory [9]. They say that when choosing the type of prosthesis, several factors have to be taken into account, such as the degree of amputation, the functionality and the economic aspect of the device [10].

The aim of prostheses for distal phalangeal amputations is to restore some (as many as possible) of the lost functions, rather than to replace the missing limb, with the primary goal of improving the quality of life of patients who have undergone some form of amputation [11].

However, the development and fabrication of handmade partial prostheses is not a recent development [7].

This paper will depict the design and explain the construction of a finger prosthesis with distal phalanx amputation. As such, this research will focus on determining a mechanical system to simulate a distal phalanx prosthesis with features that can be attached to the residual limb. Furthermore, to design a model of a distal phalanx prosthesis capable of performing prehension movements of the finger and, finally, to simulate the operation of the prosthesis using software. Finally, to build a finger prosthesis for the distal phalanx using 3D printing techniques [12, 13].

2 Methodology

This section discusses the most important considerations to be taken into account when designing a prosthesis of the distal phalanx and the different working mechanisms to subsequently build a prototype containing suitable and efficient materials and also describes the most appropriate methods to carry out the technical research work.

2.1 Materials

Comparisons were made with other prosthesis models with different types of materials and mechanisms, to find possible advantages and disadvantages that help to make an adequate and improved design for the construction [14].

The materials used for the construction of the 3D printed mechanism are listed in Table 1.

Table 1. 3D printing materials.

Material	Durability	Flexibility	Bed temperature	Waste
PLA (Polylactic Acid)	Good	Little or none	25 °C–60 °C	Biodegradable Recyclable
ABS (Acrylonitrile Butadiene Butadiene Styrene)	High	Little or none	90 °C–100 °C	No biodegradable Recyclable
TPU (Thermoplastic Polyurethane)	High	High	25 °C–60 °C	No biodegradable Recyclable
PETG (Polyethylene Terephthalate)	High	Little or none	70 °C–80 °C	No biodegradable
PC (Polycarbonate)	Very High	None	90 °C–105 °C	No biodegradable Recyclable

The design software and laminators required for the use of 3D printing are currently in use due to their characteristics; either learning type or licensing costs [15, 16]. These aspects are shown in Table 2.

Table 2. Comparative study of design applications and laminators required for the use of 3D printers.

Software	Description	Learning Difficulty	Cost of Licences
Tinkercad	Online 3D modelling software, popular for its simplicity and ease of use	Little or none	Free
Blender	Dedicated to modelling, lighting, animation and the creation of three-dimensional graphics	Very High	Free
Fusion 360	3D CAD product modelling and design, manufacturing, electronics and mechanical engineering.	Medium	$495USD/year No Free
SketchUp	Allows design, 3D modelling for video games or filming movies.	Medium	$299USD/year No Free

(*continued*)

Table 2. (*continued*)

Software	Description	Learning Difficulty	Cost of Licences
Simplify 3D	Printing simulation, support structures, multiple extruder optimisation, mesh generation	Medium	$149USD/Year No Free
Ultimaker Cura	The print parameters can be modified and then converted to G-code	Easy	Free
Matter Control	Allows you to design, cut, organise and manage your 3D prints	Easy	Free
Slic3r	Slic3r is a free 3D cutting engine software for 3D printers. It generates G-code from 3D CAD files	Medium	Free

Prostheses manufactured with 3D printing technologies offer many advantages in all aspects of the human being, in contrast to standard prostheses they have some disadvantages in terms of the following factors: functionality, high prices and unaesthetics are shown in Table 3.

Table 3. Comparative study of design applications and laminators required for the use of 3D printers.

	3D printed prosthetics	Standard prosthesis
Lifetime	They last between 7 and 10 years. It does not require a complete replacement of the prosthesis, it is sufficient to replace the part that is malfunctioning or damaged	The entire prosthesis needs to be replaced after 2 to 3 years depending on your physical activity
Availability factor	The work is carried out together with a doctor and a specialist in the use of 3D printing technology. Orders can also be placed with companies specialising in 3D printing	Only upon request
Couplings	It is made to measure according to the patient's needs. Each piece is modelled according to the specifications of the prototype used	Tailor-made, physical therapy is necessary
Price	From $500 to $700 approx	From $5,000 to $10,000 depending on the amputation, material and functionality required by the patient

The particularities of the prototype finger prosthesis are intended to positively impact the needs of the patient, focusing on the grasping functions of the human hand. Since no patients were involved in the realisation of this project, the design specifications below were based on the improvement needs considered significant (operation, cost and maintenance) compared to other types of prostheses marketed in the country.

2.2 Calculation of Anthropometric Measures

Sample for Anthropometric Measurements. The project focuses on people with disabilities related to distal phalanx dissection; however, some people are more at risk due to the industrial activities they perform. Therefore, the sample was taken from a population of 120 workers. The sample was taken from 60 male workers with the aim of finding out the ergonomic measurements of the right hand.

Before extracting the percentile value, it is necessary to obtain several important data such as: range, class mark and interval factor which are directly proportional to the percentile value. This calculation process is seen in the Fig. 1.

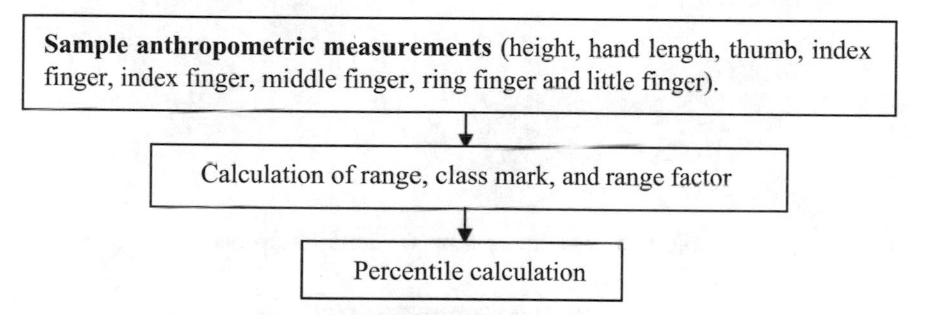

Fig. 1. Anthropometric measurement calculation sequence.

2.3 Selection of the Mechanism

This selection was carried out in three pre-projects. The first one assumes a mechanism when flexing the proximal phalanx, two tensor shafts attached to the sides restrict the movement of the bar behind this phalanx, causing its linear movement in the direction of the joint, as can be seen in Fig. 2.

In the second sketch, the bars are no longer on the sides but on the back of the hand and finger, creating a spider shape in the plan view. As can be seen in Fig. 3.

The third pre-project focuses on the adjustment of the person affected by distal phalanx by means of rings to make them feel comfortable. The aim is to understand the user's perception and to achieve the right position for each joint. This can be seen in Fig. 4.

Finally, the geometry of each phalanx and the anchorage of the prosthesis will be relatively similar for each option, as shown in Fig. 5.

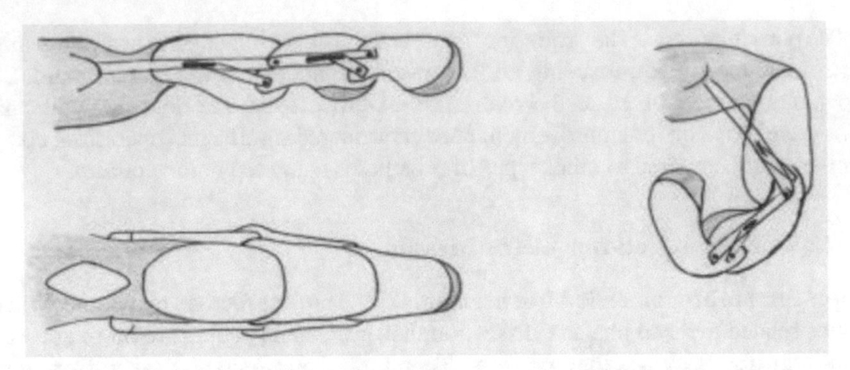

Fig. 2. Free body diagram for the distal phalanx.

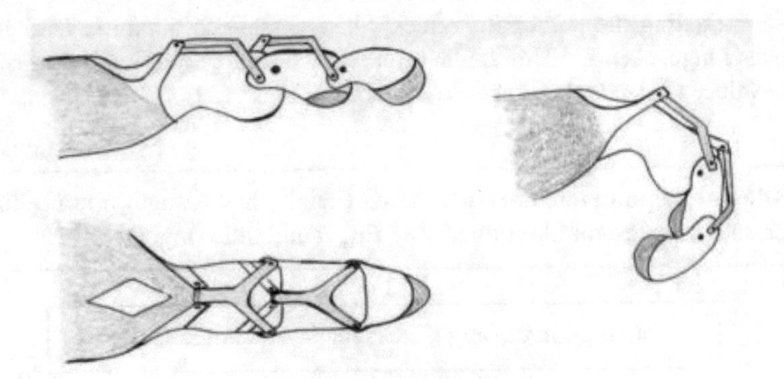

Fig. 3. Y-connection scheme for the distal phalanx

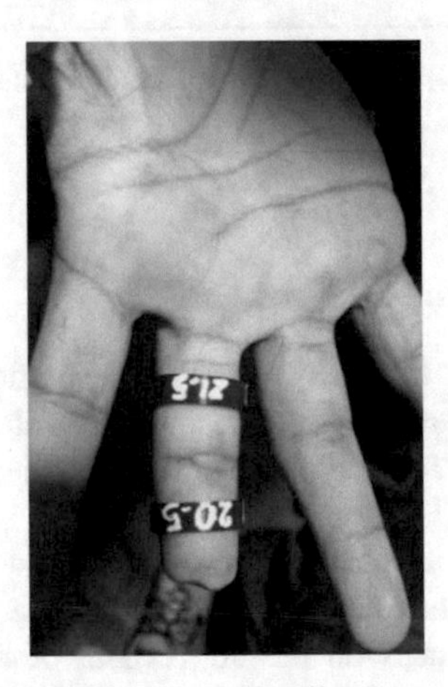

Fig. 4. Suitable dimensions for the prosthesis

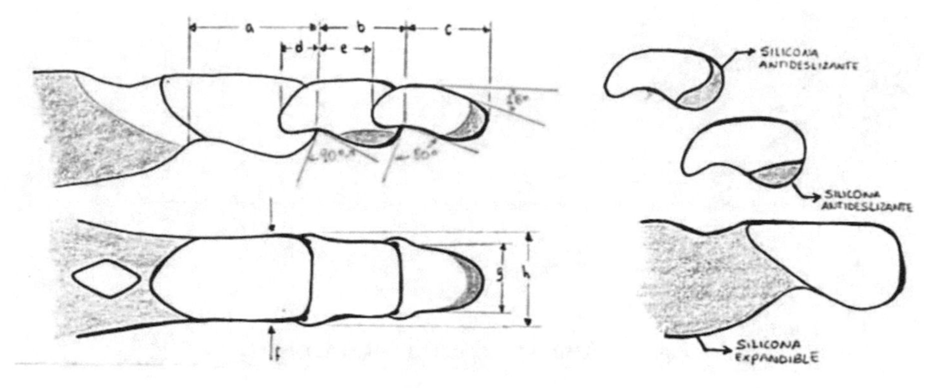

Fig. 5. Genereral scheme for the prosthesis

2.4 Calculation of Forces for Each Phalanx

Free body analysis is structured by body diagram, simulating phalanges as beams, and force of 0.66 kgf or 6.47N.This is illustrated in Fig. 6.

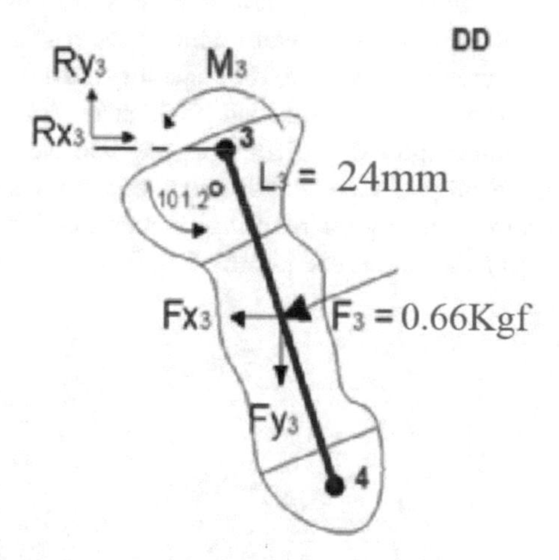

Fig. 6. Free body force diagram for the distal phalanx

The proposed design consists of a four-bar mechanism, activated by the movement of the matrix and the second one generated continuously. The aim is to correct the fourbar mechanism to respect the proposed range of articulation seen in the Fig. 7. The middle finger is chosen for its higher incidence and utility. The design, methodology and calculations can be reproduced for any four-handed finger. The distance $r2$ is the length of the matrix, and the inputs to solve the system are the values of $r2$ and $\theta 2$.

Figure 6 shows how the movement of the phalanx coincides with the movement of the previous one, confirming the complete movement of the joint. For this 4-bar mechanism, points "A" and "B" are fixed, by rotating the first arrangement, $r2$, point "D" will move

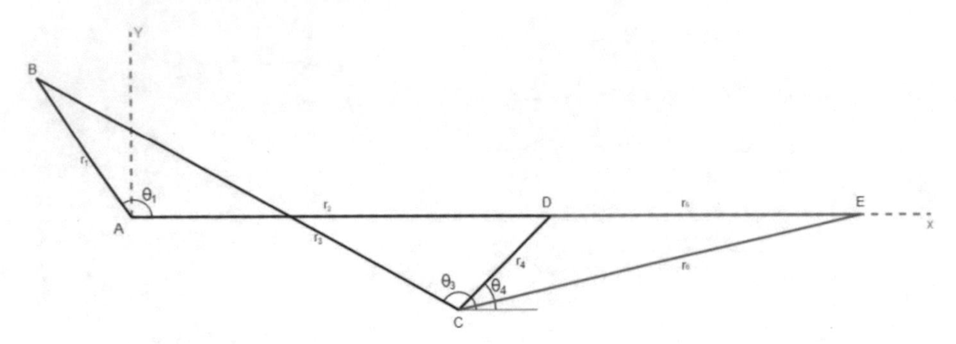

Fig. 7. Schematic diagram of 4-busbar system

an angle θ2; which will produce a rotation of bar $r4$ and it is precisely this rotation that will cause bar $r5$ to no longer be aligned with bar $r2$, that is, by moving the proximal phalanx we have a movement of the middle and distal one.

The distribution of phalanges in the human hand has several positions for grasping objects, and the trajectory is supposed to be circular and have extension movements. The synthesis of the device is presented under two conditions: three trajectory points and two support points ($Q1$ and $Q2$). With these variables, the shape of the link can be defined and compared with ergonomic measurements, as it is seen in the Fig. 8.

The linking mechanism, which consists of the trajectory and its magnitude, is checked by anthropometric measurements. The proximal phalanx is replaced by the transverse mechanism, while the middle and distal phalanges are also part of the link. The link between the middle and distal phalanges is not necessary in the model, as they can be selected in the modelling process by maintaining an angle of inclination of 7 degrees with respect to the neutral axis, as it is seen in the Fig. 9.

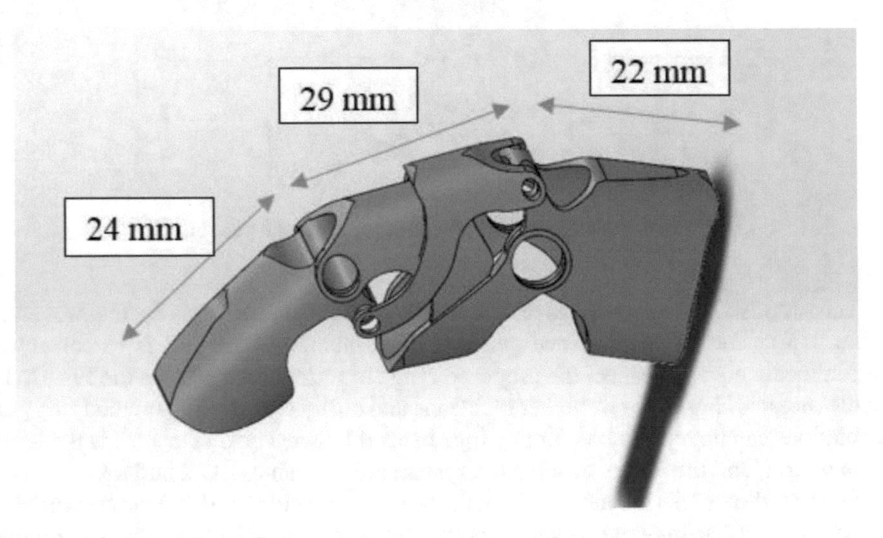

Fig. 8. Distal phalanx mechanism with anthropometric measurements

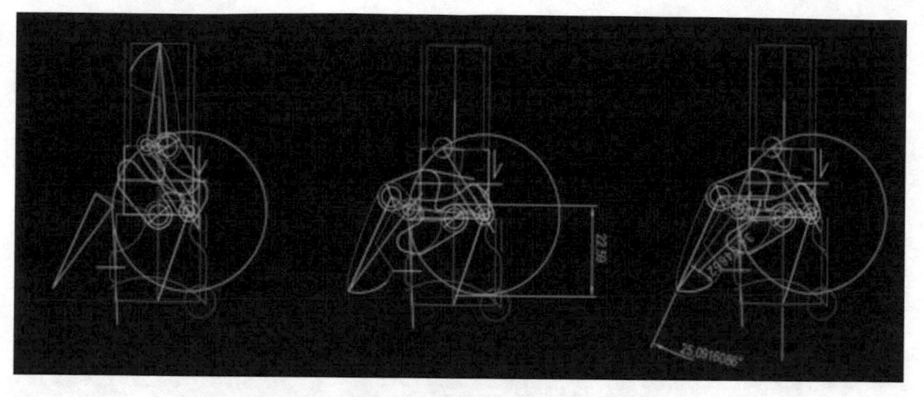

Fig. 9. Movement of the phalanx mechanism

2.5 Movement of the Phalangeal Mechanism.

The values of forces exerted at each support point are shown in Fig. 4. An analysis using 2 mm diameters, from orifices to 23 mm, was used to represent the most important case. Results for Von Mises equivalent effort, displacement, and safety factor for each arrangement are presented. Pines and pines were removed in all pieces to obtain better color gradient images within these regions.

Table 4 presents a detailed report on the accuracy of the convergence analysis, using a method of using smaller elements in possible regions to produce high levels of error.

Table 4. Convergence by Adaptive-h method.

Accuracy level	98%
Accuracy balance	50
Max. no. of loops	5
Mesh thickness application	Disable

The table presents the loads and boundary conditions applied to the proximal link model for the finite element analysis with an accuracy of 98%. Furthermore, It is also necessary to establish mesh parameters so that the analysis can reach a convergence of results seen in the Fig. 10.

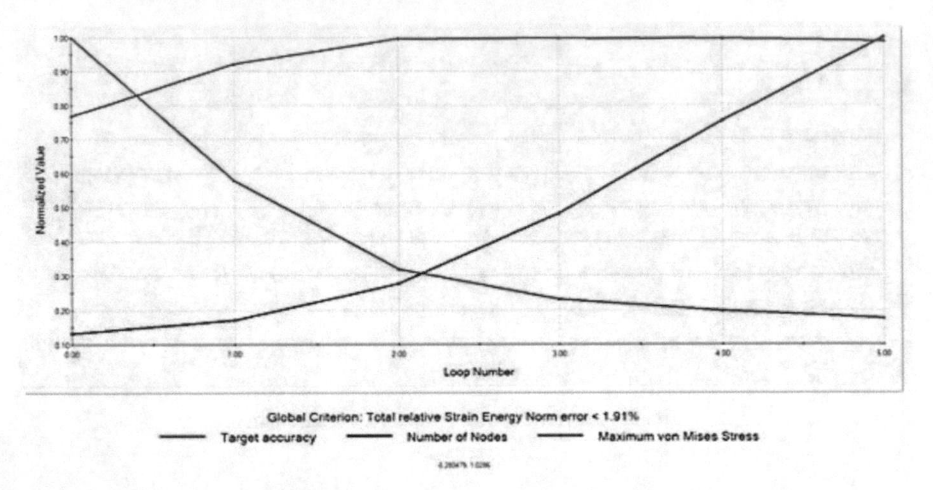

Fig. 10. Convergence of results for the proximal phalanx model.

2.6 Stress Analysis of Components

The stress analysis was carried out in Solidworks software, which had to be analysed part by part since, if it was analysed as an assembly, the software ended up analyzing it as a single part, thus giving misleading results. The values of the forces exerted at each support point were taken from Fig. 4.

Mesh parameters must be established in order for the analysis to get findings that are converging. The details of the mesh characteristics used for the proximal phalanx model are shown in Table 5.

Table 5. Mesh information - Details

Total number of nodes	192286
Total number of elements	122131
Maximum aspect ratio	17,331
% of elements with aspect ratio < 3	94,4
The percentage of elements whose aspect ratio is > 10	0,118
Percentage of distorted elements	0
Time to complete the grid (hh;mm;ss)	00:00:02

3 Results

3.1 Load Analysis Results

Finally, the results of the finite element analysis with the loads and restraints established on the proximal, medial, and distal phalanx model are obtained. Table 6 presents the results of the Von Mises stress, displacement and deformation analyses from the proximal, medial, and distal links shown in the Fig. 10.

Table 6. Mesh information - Details

Type	Proximal link		Medial link		Distal link	
	Min	Max	Min	Max	Min	Max
Stresses by Von Mises	5,916e + 03N/ m^2 Node: 58395	1,119e + 08N/ m^2 Node: 2013	1,351e + 03N/ m^2 Node: 73564	9,650e + 07N/ m^2 Node: 478	8,197e + 02N/ m^2 Node: 74638	1,085e + 0N/ m^2 Node: 273
URES: Resultant displacements	0,000e + 00m m Node: 58	1,102e + 00m m Node: 29963	–	–	–	–
ESTRN: Equivalent strain	5,408e-06 Element: 8735	9,163e-02 Element: 45240	1,896e-06 Element: 62	6,218e-02 Element: 83654	1,000e-10 Element: 62	9,915e-07 Element: 3562

These analyses were performed element by element, therefore, the most representative analysis for this prosthetic element was obtained for the ESTRN: Equivalent strain deformation analyses results from each element. This can be seen in Fig. 11.

Simulated proximal and medial matrix displacement is 0 to 1 mm, and the maximum displacement values in the distal phalanx are 0.009 mm. The Von Mises stress in the medial and proximal phalanges is higher, at 0.99 MPa and 8.92 MPa, respectively. The results represent less than 1% of the yield strength, but the analysis is performed with a load of 6.47N, which generates reactions not exceeding 12N. To know the maximum strength by tensile analysis, analyse with high load.

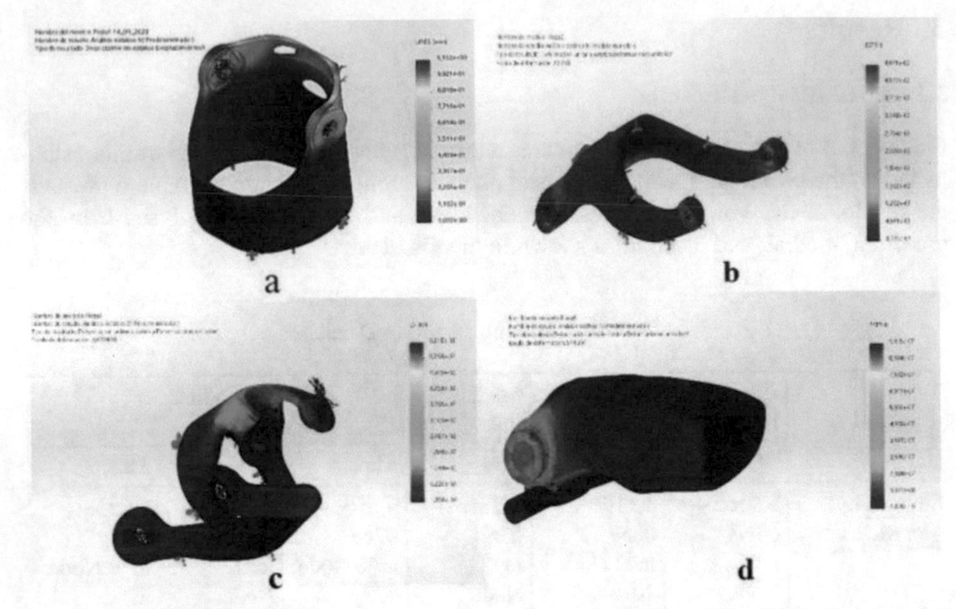

Fig. 11. ESTRN: Equivalent strain deformation analyses results: a) proximal link, b) connecting piece, c) medial link, and d) distal link

Subsequently, from Fig. 10 and Fig. 11 the values are taken to elaborate Table 11 where the comparison of the resulting values for each part with the PLA material is shown (Table 7).

Table 7. Static results generated in Solidworks software for PLA

Von Mises Max. (MPa)				Displacement (mm)			Safety factor Min.		
Phalange	Proximal	Distal	Medial	Proximal	Distal	Medial	Proximal	Distal	Medial
PLA material	0.001	0.965	8.92	0.99	0.000062	0.000647	3	3	3

3.2 Costs Estimation

Table 8 shows the estimated cost values for each part and prosthesis for a distal finger.

Table 8. Estimated cost of finger prosthesis per piece in US dollars.

Part	Printing cost*	Printing time (h)	Filament costs**	Part weight (gr)	Total cost per part***
Proximal	5.54	0.82		0.95	0.74
Medial		0.79		1.14	0.88
Distal		1.6	13.99	4.32	3.36
connecting piece		0.56		0.7	0.54

*The cost of printing on a Stratasys Mojo printer was calculated in USD per hour of printer usage
**Filament cost was evaluated in USD per gram
***Cost per part = printing cost x printing time + filament cost x part weight

Table 9 shows the cost data for printing a prosthesis for an adult.

Table 9. Estimated cost of finger prosthesis per piece in US dollars.

	Costs of production
3D printing	5.54$
Pin and screw	1.74$
Anti-slip silicone	1.00$
Labour	50% surcharge on the total value
Total	16.56$

4 Conclusions

This research aimed to develop a device that can perform flexion and extension movements with an angular amplitude of 32° for the middle phalanx and 30° for the distal phalanx. The device is purely mechanical, requiring no external power source. The four-bar mechanism is influential for designing mechanical prosthetic phalanges, as it responds to prehension, the same movement humans use for certain activities.

Certain 3D printing material (PLA) was selected for its strength, allowing for a cylindrical grip with weights up to 1 kg. The PLA+ case had a maximum Von Mises stress of 9 MPa, representing a safety factor of 3. This technological advance can greatly benefit patients with limb or limb amputation, providing information on designing and manufacturing 3D printed prostheses for patients with distal phalanx damage from work accidents. This low-cost, functional, and aesthetic prosthesis can improve patients' confidence, physical appearance, and self-esteem.

To ensure proper prosthesis manufacturing, consider shrinkage in polymers and a ±2% dimensional tolerance. Explore resistant polymers similar to the human body without affecting user health. Measure cylindrical grip strength and conduct experimental verification with the prosthesis and user for future research.

References

1. Difonzo, E., Zappatore, G., Mantriota, G., Reina, G.: Advances in finger and partial hand prosthetic mechanisms. Robotics **9**(4), Art. n.o 4, dic. (2020). https://doi.org/10.3390/robotics9040080
2. Fajardo, J., Ferman, V., Cardona, D., Maldonado, G., Lemus, A., Rohmer, E.: Galileo hand: an anthropomorphic and affordable upper-limb prosthesis. IEEE Access **8**, 81365–81377 (2020). https://doi.org/10.1109/ACCESS.2020.2990881
3. Ryu, W., Choi, Y.Y., Choi, J., Lee, S.: Development of a lightweight prosthetic hand for patients with amputated fingers. Appl. Sci. **10**(10), Art. n.o 10 (2020). ene. https://doi.org/10.3390/app10103536
4. Koudelkova, Z., et al.: Verification of finger positioning accuracy of an affordable transradial prosthesis. Designs **7**(1) (2023). https://doi.org/10.3390/designs7010014
5. del Piñal, F.: The indications for toe transfer after minor finger injuries. J. Hand Surg. Br **29**(2), 120–129 (2004). https://doi.org/10.1016/j.jhsb.2003.12.004
6. Cuellar, J.S., Plettenburg, D., Zadpoor, A.A., Breedveld, P., Smit, G.: Design of a 3D-printed hand prosthesis featuring articulated bio-inspired fingers. Proc. Inst. Mech. Eng. H **235**(3), 336–345 (2021). https://doi.org/10.1177/0954411920980889
7. Moreira, A.H.J., Queirós, S., Fonseca, J., Rodrigues, P.L., Rodrigues, N.F., Vilaca, J.L.: Real-time hand tracking for rehabilitation and character animation. In: 2014 IEEE 3nd International Conference on Serious Games and Applications for Health (SeGAH), pp. 1–8, May 2014. https://doi.org/10.1109/SeGAH.2014.7067086
8. Bregoli, C., et al.: Osseointegrated metallic implants for finger amputees: a review of the literature. Orthop. Surg. **14**(6), 1019–1033 (2022). https://doi.org/10.1111/os.13296
9. Markus, A.T., Sobczyk, M.R., Perondi, E.A.: Modeling, control, and simulation of a 3-degrees of freedom mechanism actuated by pneumatic artificial muscles for upper limb prosthesis application. J. Mech. Robot. **15**(1) (2023). https://doi.org/10.1115/1.4054084
10. Draycott, J.: Prosthetics and Assistive Technology in Ancient Greece and Rome. Cambridge University Press, Cambridge (2022). https://doi.org/10.1017/9781009168410
11. Cevik, P., Schimmel, M., Yilmaz, B.: New generation CAD-CAM materials for implant-supported definitive frameworks fabricated by using subtractive technologies. Biomed. Res. Int. **2022**, 3074182 (2022). https://doi.org/10.1155/2022/3074182
12. Barreno-Avila, A.F., Monar-Naranjo, M., Barreno-Avila, E.M.: Fusion deposition modeling (FDM) 3D printing parameters correlation: an analysis of different polymers surface roughness. In: IOP Conference Series: Materials Science and Engineering, vol. 1173, no. 1, p. 012071 (2021). ago. https://doi.org/10.1088/1757-899X/1173/1/012071
13. Barreno-Avila, E., Moya-Moya, E., Barreno-Avila, A.: Optimización de parámetros del proceso de corte con láser para Polimetilmetacrilato y Policloruro de vinilo. CienciAmérica **9**(4), 43–50 (2020). dic. https://doi.org/10.33210/ca.v9i4.342
14. Capsi-Morales, P., et al.: Comparison between rigid and soft poly-articulated prosthetic hands in non-expert myo-electric users shows advantages of soft robotics. Sci. Rep. **11**(1), Art. n.o 1, dic. (2021). https://doi.org/10.1038/s41598-021-02562-y
15. Barreno-Avila, E., Moya-Moya, E., Pérez-Salinas, C.: Rice-husk fiber reinforced composite (RFRC) drilling parameters optimization using RSM based desirability function approach. Mater. Today Proc. **49**, 167–174 (2022). ene. https://doi.org/10.1016/j.matpr.2021.07.498
16. Barreno Avila, E., Balladares-Pazmiño, M., Barreno-Avila, A., Espín, S.: Black- berry (Rubus Glaucus) Natural-Fiber Reinforced Polymeric Composites: An Overview of Mechanical Characteristics, pp. 300–315 (2021). https://doi.org/10.1007/978-3-030-72212-8_22

Vehicle Braking and Suspension Systems: Redesign of Preventive and Corrective Maintenance Processes in Automotive Workshops in Southern Quito

Jorge Ramos(✉), Paúl Caza, Cristian Guachamin, Pamela Villarreal, Rodrigo Díaz, Patricio Cruz, and Edisón Criollo

Instituto Tecnológico Universitario Vida Nueva, Quito, Ecuador
jorge.ramos@istvidanueva.edu.ec

Abstract. Mechanical design analysis using technologies such as FEM, CAD (Computer Aided Design) and CAE (Computer Aided Engineering) is a common practice in engineering to determine the behavior of mechanical stresses and deformations in a multifunctional machine. In the specific case of a multifunctional machine for the maintenance of the brake and suspension system of cars, it is important to consider the maximum weight of a car, which in this case is set at 1.5 tons (1.5T). This specification provides a reference for the load and strength analysis of the machine. The use of FEM technology allows a detailed analysis of the structural behavior of the machine. Using the finite element method, the machine is divided into smaller elements to represent its geometry and material. Then, appropriate loading conditions and constraints are applied to simulate the forces and moments acting on the machine during its operation. CAD software is used to create the 3D model of the multifunctional machine, including geometric details, components and assemblies. The CAD model can be imported into the FEM analysis software to generate a finite element mesh and apply load conditions. The CAE software allows the analysis and simulation of the results obtained from the FEM model. It provides tools to visualize and evaluate mechanical force, deformation and other relevant parameters at different points of the machine. It can also help to identify critical areas where high levels of force or deformation occur, allowing design and optimization decisions to be made.

Keywords: Brake system · Suspension system · Vehicle · Maintenance · Pneumatics

1 Introduction

Nowadays, the versatility of the maintenance processes applied to the different systems that make up the automobile is due to the advance of technology both in tools and equipment used by technicians, depending on the useful life of the automobile [1–3]. In the automotive industry market, there is a great variety of this type of resources that seek

M. Z. Vizuete et al. (Eds.): CI3 2023, LNNS 1040, pp. 225–237, 2024.
https://doi.org/10.1007/978-3-031-63434-5_17

to optimize times and reduce labor risk factors [4]; however, the purchasing power of most multi-brand workshops prevents them from having the technologically advanced machines available [5, 6].

The design and simulation processes supported by software of great computation-al capacity help to determine the physical and mechanical behaviors, predicting the maximum and minimum values of forces and displacements to which they will be subjected, which currently minimizes manufacturing costs for companies, but nevertheless, represent components that are difficult to acquire [7]. Under this reality, the manufacture of a multifunctional machine, which gathers the characteristics of several of them in one, turns out to be the answer to this reality, which will allow to have an equipment that provides the reliability to apply preventive and corrective maintenance processes in the brake and suspension systems, minimizing the risk of accidents, optimizing resources and providing high quality technical services [8–11].

2 Brake and Suspension System

Brake and suspension systems are an essential part of the structural geometry of vehicles which, through their components, are exposed to physical and mechanical loads such as exposure to temperature variations, friction, rubbing, shocks, humidity, bending, etc. [12, 13].

Therefore, they must meet a series of mechanical requirements to provide the necessary safety for both the driver and the occupants under any driving conditions, regardless of the different types, marks and models of vehicles in general [14–18].

As mentioned by Bauzá Francisco [19] in his research on braking systems in light vehicles, the objective of the brakes is to stop the car with a very short distance and in the shortest time according to the needs of the driver [20], under the principle of surface friction, dissipating a large amount of energy into the environment in the form of heat [21].

On the other hand, a suspension system has the purpose of absorbing the irregularities of the road on which the vehicle moves and that these are not transmitted to its interior, thus avoiding the corresponding discomfort to its occupants. In this context, Camilo Gavilanes [22] has extended in his research the analysis and importance of the suspension system of light vehicles through a digital model, noting how it influences the behavior under exposure to different loads and how this improves or affects the comfort of its occupants, under simulations in computer programs. It is very common to assimilate the term suspension only to the shock absorbers because they are very important in the geometry of the car and they support the greatest amount of load, however they are a series of elements that achieve the purpose of this system, and it is here where you get the different variants and types of these systems that make up the structure of the car [23, 24].

2.1 Brakes

Mechanical Brake System. This system is used for parking braking (parking brake), using simple mechanisms by a cable by the action of ratchet and locking that has in addition to an adjustment according to the needs of the driver, the following figure shows the characteristics of this system with its constituent components [25] (Fig. 1).

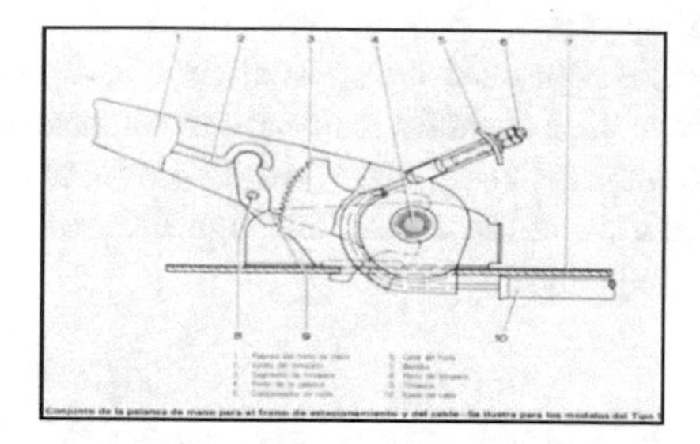

Fig. 1. Mechanical braking or parking system. [2]

Hydraulic Brake System. The brake systems that use a hydraulic fluid and the physical principle of Pascal for its operation are the most used in vehicles today, the hydraulic fluids used in these systems meet the DOT regulations that depend heavily on their boiling points both wet and dry in addition to this they are incompressible that under high loads come to act practically as a solid come to amplify the force transmitted in a circuit formed by pipes that have physical properties to withstand high pressures and temperatures generated in the system [26]; To achieve this multiplication and force is the sequential operation of a series of mechanical components such as a brake servo, a hydraulic pump, pipes, reaching the clamps that through cylinders make contact between sur-faces with the brake discs or pads against the drum, which will depend on the type of circuit that has the vehicle [27].

The figure shows a conventional brake circuit using hydraulic operation (Fig. 2).

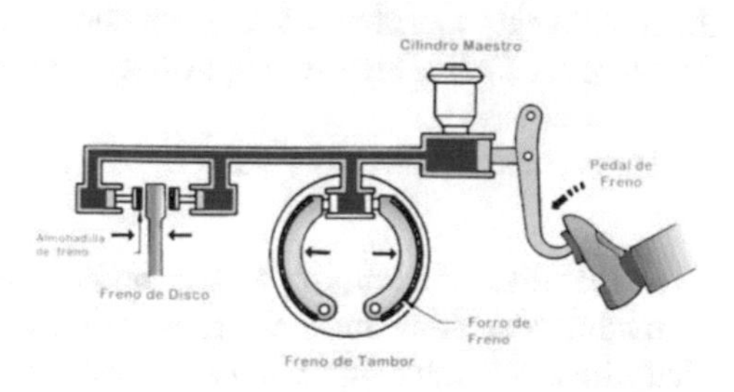

Fig. 2. Hydraulic brake circuit [6]

2.2 Suspension

As mentioned by Mitsubishi Motors [28], the suspension system of a car is a set of elements that have great elasticity capacity with high rates of support deformation loads, which provide stability and control under different driving conditions in addition to providing a high degree of comfort to its occupants. The suspension of an automobile has the function of absorbing the unevenness of the terrain over which it travels.

In a simple suspension system, spring steel linkage components are used in the form of: leaf springs, coil springs, torsion bars, stabilizer bars, etc.

The suspension components must have excellent elastic properties, but their disadvantage is that they have little capacity to absorb mechanical energy, so they can-not be mounted alone in the suspension, requiring the assembly of an element that restrains the oscillations produced in its deformation. The figure shows the constituent elements of a suspension system [29] (Fig. 3).

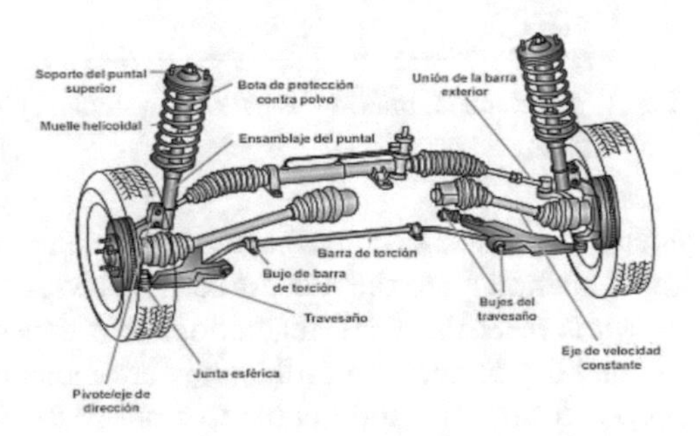

Fig. 3. Suspension system components [8]

2.3 Breakdowns in the Brake and Suspension System

Failures of Brake System. The brake system is extremely important, it is in charge of controlling the dis-placement based on friction, the elements used are subjected to high temperature; each of the components contributes to the reduction of speed. The elements that make up the braking system usually show wear and deterioration that after a certain mileage it is necessary to revise.

The failures that frequently occur are linked to wear and tear, among the most frequent are the following:

- Wear brake linings, whether disc or drum, due to friction between the materials.
- The wear of friction materials such as discs or drums, these materials are made of different materials that withstand high temperatures, however, a thermal shock causes deformation or even breakage.
- High temperature due to emergency braking causes overheating of the friction material, which leads to evaporation of the brake fluid and thus to a constant failure in braking efficiency.
- Low brake fluid level may be an indicator of friction material wear or premature fluid leakage through the brake cylinder seals.
- When the brake pedal is pressed, it moves too low to stop the vehicle, causing a feeling of insecurity.
- The materials produce a scraping noise when braking and even a clanking sound.

- Crystallized brakes due to high temperatures generate a permanent squealing noise or in certain braking conditions.
- The brake servo helps reduce braking force, however, a failure in the vacuum system often results in excessive brake pedal harshness and poor braking.
- Efficiency loss
- Unbalance is another brake system failure generally attributed to the distribution of braking force causing the vehicle to roll sideways under braking.

Suspension System Failures. The elements that make up the suspension system usually show wear and deterioration that must be checked after a certain mileage [30].

The failures that frequently occur are linked to component wear or breakage of elastic elements to the deterioration of shock absorbers. Among the failures considered high risk are the following:

- Excessive elastic oscillation, which increases when passing an irregularity in the roadway and even when crossing a speed breaker.
- The steering axle does not maintain the direction of the vehicle, tending to change the trajectory to one side or the other.
- Irregular tire wear is one of the signs of suspension system failure.
- Suspension squeaking indicates exposure of the articulating elements to oxidizing substances and implies a lack of maintenance and lubrication.
- Excessive vibration of the steering wheel occurs when driving at speeds above 80 km/h.
- If the car is parked and without load, and the car is leaning towards a certain side-wall, an anomaly in the elastic elements is established.
- When going over bumps or irregularities in the road, you feel a strong shock in the area where the tires are located means deterioration in the articulation points of the arms or shock absorbers.
- Shock absorbers with loss of efficiency to reduce the oscillations of the elastic elements.

3 Methodology

According to the question posed at the beginning of the research according to the objective of the study, it has a qualitative approach that allows analyzing the design and modeling of an equipment that allows performing the maintenance of brake and suspension systems of vehicles in multi-brand workshops, considering the operation and applicability that will be developed to generate a redesign of the existing plat-forms, which is intended to reduce the safety risks to which operators are exposed during the various activities, for this reason the following bibliographic, descriptive and experimental research techniques will be used to generate the progress of the investigation [31].

Through corrective and preventive maintenance, it will be possible to analyze the failures or defects that the braking and suspension system of the vehicle may have, by using tools and equipment according to the problems during the inspection of the vehicle.

3.1 Statistical Analysis

The validation of the data under the ANOVA statistical model based on the results obtained in the surveys applied, determine a confidence level of 95% in the feasibility of the construction of this multifunction machine for use in the maintenance process for the brake and suspension systems, for which the following treatments were pro-posed:

Sample 1: PROCESSING 1 (Operation of the mechanical workshop)
Sample 2: PROCESSING 2 (Maintenance of light vehicles)
Sample 3: PROCESSING 3 (Activities performed in the workshop)
Sample 4: PROCESSING 4 (Maintenance of brakes)
Sample 5: PROCESSING 5 (Maintenance of suspension)
Sample 6: PROCESSING 6 (equipment and tools used in the ABC of brakes)
Sample 7: PROCESSING 7 (Tools FOR suspension system maintenance)
Sample 8: PROCESSING 8 (Incidents or occupational accidents have generated)
Sample 1: 45 values in the range de 1,0 a 5,0
Sample 2: 45 values in the range of 5,0 a 15,0
Sample 3: 45 values in the range of 2,0 a 7,0
Sample 4: 45 values in the range of 5,0 a 15,0
Sample 5: 45 values in the range of 2,0 a 7,0
Sample 6: 45 values in the range of 5,0 a 15,0
Sample 7: 45 values in the range of 4,0 a 4,0
Sample 8: 45 values in the range of 2,0 a 2,0

Means with confidence intervals are constructed in such a way that, if two means are equal, their intervals will overlap 95.0% of the time.

It is observed in the graph the intervals selecting, being the representation of the means of and if they are significantly different from others, as it is interpreted in the Table 1.

Table 1. Table of averages with confidence intervals of 95,0%

	Error Est.				
	Cases	*Media*	*(s groups)*	*Lower Limit*	*Upper Limit*
PROCESSING 1	45	4,3111	0,311791	3,87751	4,74472
PROCESSING 2	45	7,3333	0,311791	6,89973	7,76694
PROCESSING 3	45	5,0	0,311791	4,5664	5,4336
PROCESSING 4	45	6,2222	0,311791	5,78862	6,65583
PROCESSING 5	45	5,0	0,311791	4,5664	5,4336
PROCESSING 6	45	6,1111	0,311791	5,67751	6,54472
PROCESSING 7	45	4,0	0,311791	3,5664	4,4336
PROCESSING 8	45	2,0	0,311791	1,5664	2,4336
Total	360	4,9972			

The table shows the means with the 95% confidence level, this method used in the research was used to discriminate between the means is Fisher's Least Significant Difference (LSD) procedure. With this method there is a risk of 5.0% in saying that each pair of means is significantly different, when the real difference is equal to 0.

3.2 Structural Design Analysis

The equipment designed and built is oriented for its application in the automotive field, for preventive and corrective maintenance processes in vehicles [32]. Based on the reference of a vehicle, it is considered that the most relevant part of the multifunction machine represents the structure of the hydraulic lift, which was dimensioned based on its technical characteristics, as shown in Table 2 below.

Table 2. Technical specifications of hydraulic elevator

Technical characteristics of the hydraulic elevator		
Load support	Load support	4 swivel arms, symmetrical
	Capacity	3 T
	length of swivel arm at the front, min	600 mm
	length of swivel arm at the front, max	1070 mm
	length of swivel arm at rear, min	600 mm
	length of swivel arm at rear, max	1070 mm
	distance between supports, max	2300 mm
	Diameter of loading platform	120 mm
Lifting height	lift height	1850 mm
	minimum support height	110 mm
	height adjustment of the silver-loading shape	35 mm
Speed	lifting time	40 s
	descent time (under load)	40 s
Dimensions	dimensions length	2350 mm
	Width	3050 mm
	Height	790 mm
	Input width	2000 mm
	Weight	900 kg

Based on the functionality of the hydraulic elevator, it must withstand loads of great magnitude, so its design and structure must correspond to withstand the influence of these and under different operating conditions.

The simulation of the application of point loads as deformation and displacement on the main arm of the hydraulic elevator is shown below.

In order to generate data on the elevator's behavior under structural loads, the simulation process is generated in the SolidWorks environment, determining stress values and displacements, which are represented in the following figures (Fig. 4).

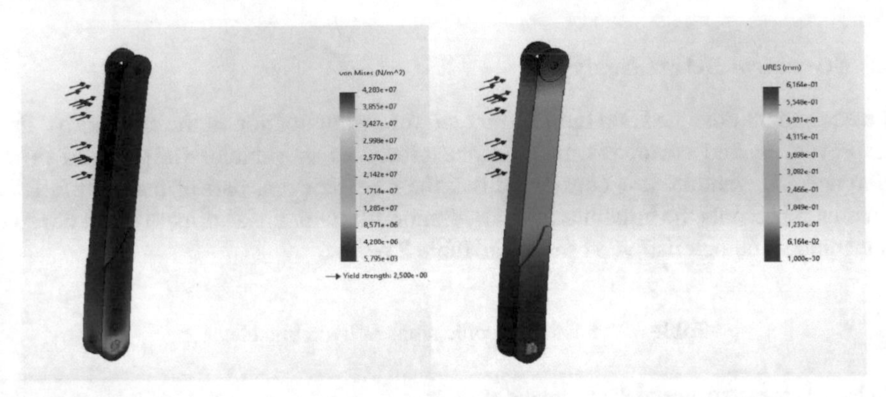

Fig. 4. Simulation of structural loads on the elevator main elevator arm

For the simulation, the point load was applied on the pad considering that it is the element that will support the vehicle load, this magnitude is equivalent to 2500 [N], the same that is distributed over the four pads in the same way.

In the simulation of the extendable arm, the quality of the meshing is another important factor during the finite element analysis, in Fig. 40 you can see the mixed mesh size, which has a value of 0.78 [mm] (Fig. 5).

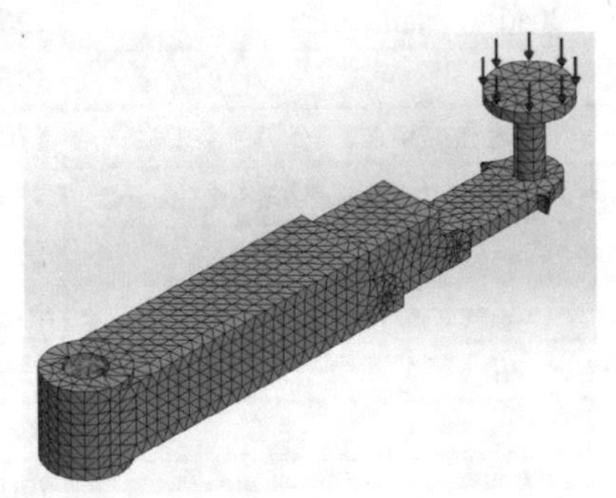

Fig. 5. Application of loads and constraints to analyze the behavior of the main arm pad

Figure 6 shows the results of the maximum and minimum deflection in the analyzed structural system according to the load assigned to the most critical structural element, observing that the maximum deflection is 2.1 [mm] which is a significantly small value in consideration to the magnitude and size of the system, in addition the maximum Von Misses force is presented where a maximum value of 2.51 [N/mm^2] is indicated, if the yield force of the material is considered, it can be observed that the value is lower, so the system cannot fail.

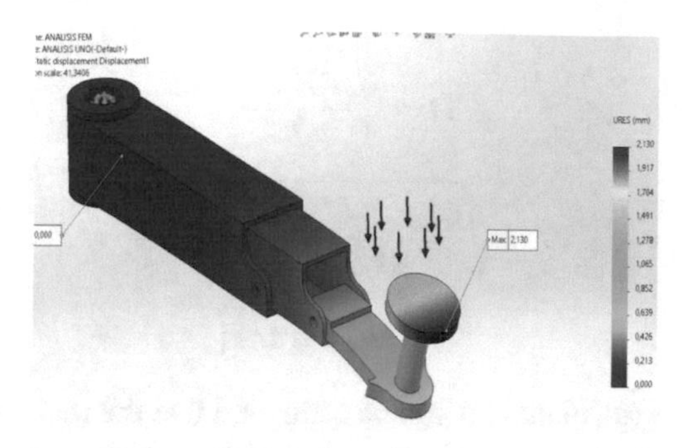

Fig. 6. Von Misses analysis on the extendable arm

3.3 Dimensioning of the Pneumatic and Hydraulic System

Vehicle Lifting System. A hydraulic system is one in which the application of power is generated, transmitted and controlled through the circulation of oil through the circuit. The system as such can be divided into three main parts which are the motor, pump, control valves and cylinder or actuator including the pipes that allow the connection of the system.

Cylinder and pump selection. The parameters to generate the correct selection of the hydraulic equipment are related to three main factors which are:

- Maximum load: 3000 kg
- Stroke: 35.5 cm
- Speed: 6 [cm/s]

In the present investigation the hydraulic cylinder has a stroke of 35.5 cm according to the hydraulic cylinder catalog, based on this selection it can be determined that:

Piston Rod Diameter. The calculation of this parameter is made considering the minimum diameter that the shank must have and is related to the slenderness that the element must have, since it is subjected to compression loads. The safety factor used for this type of buckling case is 3.0.

$$F = \frac{\pi^2 \times I \times E}{L^2 \times n} \tag{1}$$

Considering that it is a circular section, the moment of inertia is calculated with the following equation:

$$I = \frac{D^4 \times \pi}{64} \tag{2}$$

Considering the force of 30,000 [N], with a modulus of elasticity of 206.8 GPa, in conjunction with the force and inertia equations mentioned above, the minimum stem diameter to be considered can be determined.

$$F = \frac{\pi^3 D^4 E}{64 L^2 n} \tag{3}$$

$$D = \sqrt[4]{\frac{64FL^2n}{\pi^3E}}$$

$$D = \sqrt[4]{\frac{64 \times 30000 \times 355^2 \times 3}{\pi^3 \times 206.8x10^3}}$$

$$D = 336.46\,[\text{mm}]$$

According to the calculation, it can be estimated that the minimum diameter to be considered in the stem to avoid buckling failures is 168 [mm], due to the load that the hydraulic element will support.

Hydraulic System Pressure. For the calculation of the system pressure, it is necessary to know the piston area of the cylinder, considering the previous calculation of the piston diameter, we proceed to calculate it with the following equation:

$$P = \frac{F}{A} \tag{4}$$

$$P = \frac{30000[N]}{\frac{\pi \times (0.2)^2}{4}}$$

$$P = 954.92\,[\text{kPa}]$$

Working Fluid Flow Rate. It is also necessary to know the flow rate to determine the required pump power. For which the data of this parameter is related to the elevation speed and the piston area.

$$Q = A \times V \tag{5}$$

Pump Power. The pump power is determined by knowing the previously calculated flow rate and system pressure, with the following equation:

$$P_{ot} = Q \times P \tag{6}$$

The motor to be selected for the construction of the machine must have a power of 2.5 [hp] with a pump flow rate of 1.88 [m^3/s].

4 Discussion and Results

This section presents the results obtained and analyzed during the research, in which it has been possible to study elements that are part of a multifunctional machine used in the maintenance of the brake and suspension system as a preventive part of the mechanical, ergonomic and physical risks for the operators.

In the process of design and construction of the multifunctional machine for the maintenance of the brake and suspension system, an adequate selection of materials and forming processes was carried out, considering that they comply with the technical requirements that allow the handling of the vehicle, which allows reducing accidents or damages caused to the operator's health.

Of the elements that make up the machine, it was determined that the cushion is the most critical component, since it will support the load of the vehicle, so after per-forming a finite element analysis it was observed that the maximum and minimum deflection in the structural system analyzed in terms of the load assigned to the most critical structural element, noting that the maximum deflection is 2.1.

Taking into account the conditions to which the machine will be subjected, which were calculated through finite element analysis, observing the zones of influence and load distribution in the arms that will support the weight of the vehicle during the revision and maintenance of the brake and suspension system.

A key factor in the design of the machine was the ease with which the operator can manipulate the equipment and tools required in the process of maintaining the brakes and suspension of a vehicle, reducing the time factor in the repair of the systems.

5 Conclusions

In this research, a study of the mechanical elements that constitute a multifunctional machine was carried out from the mechanical and safety point of view for the operator in unauthorized mechanical workshops, in which production, work times, maintenance and ergonomic, mechanical and physical risks to which the operator is ex-posed were compared. According to the study carried out, it was identified that the workshops are looking for the implementation of equipment that allows them to better handle the repair and maintenance of the vehicle's brake and suspension system.

Through mechanical design analysis using FEM, CAD and CAE technology, it is possible to evaluate the behavior of mechanical forces and deformations in a multi-functional machine for the maintenance of the brake and suspension system of automobiles. This helps to ensure that the machine is strong and safe enough to withstand the expected loads during operation, considering the maximum weight of a car of 1.5 tons.

It is recommended that future research be carried out to analyze the design of new components that will allow a preventive maintenance analysis of conventional and automatic transmission systems, steering systems and complementary engine systems, taking into account automation and control technology.

References

1. Bonavía, T., Marín, J.A.: An empirical study of lean production in the ceramic tile industry in Spain. Int. J. Oper. Manage. (2006)
2. Bouhmala, J.: Los secretos de una buena suspensión | AMORTIAUTOS. AMORTIAUTOS SAS (2020). https://amortiautos.com/los-secretos-una-buena-suspension/
3. Braglia, M., Carmignani, G., Zammori, F.: A new value stream mapping approach for complex production systems. Int. J. Prod. (2006)
4. Brembo. Mantenimiento del sistema de frenos (n.d.). https://www.brembo.com/es/automóviles/aftermarket/conductores/manutenzione-impianto-frenante. Accessed 20 Mar 2022

5. Gavilánez Endara, C.: Análisis e Importancia de Sistema de Suspensión de Ve-hículos Livianos Mediante Modelo Digital. UNIVERSIDAD SAN FRANCISCO DE QUITO USFQ (2016)
6. Giraldo Gutiérrez, C.A.: Control Predictivo con Restricciones de Suspensión Vehicular Activa. https://www.researchgate.net/publication/320981475_Control_predictivo_con_restricciones_de_suspension_vehicular_activa. Accessed 20 Mar 2022
7. Crabill, J., et al.: Production operations level transition-tolean description manual. Center for Technology, Policy, and Industrial Development. Massachusetts Institute of Technology (2000)
8. Bauzá Fernández, F.J.: Estudio del sistema de frenado en los vehículos ligeros (turismos). Universidad Técnica de Cataluña (2018)
9. García León, R.A., Acosta Pérez, M.A., Flórez Solano, E.: Análisis del compor-tamiento de los frenos de disco de los vehículos a partir de la aceleración del proceso de cor-rosión. Tecnura **19**(45), 53–63 (2015). https://doi.org/10.14483/UDISTRITAL.JOUR.TECNURA.2015.3.A04
10. Gómez, J.: Mantenimiento preventivo de suspensión. Doctor Auto. https://www.doctorauto.com.mx/2019/06/17/mantenimiento-preventivo-del-sistema-de-la-suspension-del-automovil/
11. Rodriguez Pinzon, H.R.: Consideraciones cinematicas y dinamicas para el desarrollo de control a un sistema de suspensión. https://www.researchgate.net/publication/328589913_CONSIDERACIONES_CINEMATICAS_Y_DINAMICAS_PARA_EL_DESARROLLO_DE_CONTROL_A_UN_SISTEMA_DE_SUSPENSION. Accessed 20 Mar 2019
12. Rodríguez Pinzón, H.R.: Diseño y construcción de un sistema para medir el desgaste en forros de frenos de disco. https://www.researchgate.net/publication/332725197_Diseno_y_construccion_de_un_sistema_para_medir_desgaste_en_forros_de_frenos_de_disco. Accessed 20 Mar 2022
13. Correa-Arciniegas, J.: Mapas mecánicos y dinámicos de frenos de disco bajo diferentes condiciones de operación. https://www.researchgate.net/publication/357438518_Propuesta_del_sistema_de_frenos_para_un_vehiculo_Baja_SAE. Accessed 20 Mar 2022
14. Correa-Arciniegas, J.: Propuesta de sistema de frenos para un vehículo Baja SAE. https://www.researchgate.net/publication/357438518_Propuesta_del_sistema_de_frenos_para_un_vehiculo_Baja_SAE. Accessed 20 Mar 2022
15. Rivera-López, J.E.: Análisis térmico y fluido dinámico de un freno de disco automotriz conpilares de ventilación tipo aerodinámico. https://www.researchgate.net/publication/337614379_Thermal_and_fluiddynamic_analysis_of_an_automotive_disc_brake_with_ventilation_pillars_aerodynamic_type. Accessed 20 Mar 2022
16. Gallardo, J.M.: Investigación de la rotura en servicio de amortiguadores de automóviles (2016). https://www.researchgate.net/publication/242230029_INVESTIGACION_DE_LA_ROTURA_EN_SERVICIO_DE_AMORTIGUADORES_DE_AUTOMOVILES. Accessed 20 Mar 2022
17. Obregón Sánchez, L.M.: Diseño y construcción de un módulo didáctico de en-trenamiento para frenos de aire de camiones. UNIVERSIDAD INTERNACIONAL DEL ECUADOR FACULTAD DE INGENIERÍA AUTOMOTRIZ (2013)
18. Mecánica Automotriz, F.: Reparación del sistema de frenos convencionales y ABS (2018). www.mecanica-facil.com
19. Mantenimiento preventivo del sistema de frenos. El Universo. https://www.eluniverso.com/entretenimiento/2018/11/21/nota/7060449/frene-embrague-adecuadamente/
20. Mendez Cuello, A., Velez Cely, M., Monar Monar, W.: Diseño del sistema de freno regenerativo de automóviles híbridos. Rev. Politécnica **37**(2), 10 (2016). https://revistapolitecnica.epn.edu.ec/ojs2/index.php/revista_politecnica2/article/view/451/pdf

21. Trejo, O.: Sistema de inferencia difuso de estructura variable para el control de suspensión mecánica (2016). https://www.researchgate.net/publication/343135582_Sistema_de_inferencia_difuso_de_estructura_variable_para_el_contro_de_suspension_mecanica
22. Padilla, L.: Elaboración de hojas de operación estándar para el mantenimiento del ser-vicio mayor de una empresa automotriz del Sur de Sonora (2018). http://www.ecorfan.org/republicofperu/research_journals/Revista_de_Ingenieria_Industrial/vol2num6/Revista_de_Ingenier%C3%ADa_Industrial_V2_N6_1.pdf
23. Razmi-Ishak, M., Abu-Bakar, A.R., Belhocine, A., Mohd-Taib, J., Wan-Omar, W.Z.: Brake torque analysis of fully mechanical parking brake system: theoretical and experimental approach. Ingeniería Invest. Tecnol. **19**(1), 37–49 (2018). https://doi.org/10.22201/fi.25940732e.2018.19n1.004
24. García-León, R.A.: Análisis del caudal en un disco de freno automo-triz con álabes de ventilación tipo NACA 66-209, utilizando velocimetría por imágenes de partículas (2019). https://www.researchgate.net/publication/342734012_Analisis_metalografico_y_materiales_de_los_frenos_de_disco/related. Accessed 20 Mar 2019
25. García-León, R.A.: Estudio y análisis térmico del diseño de formas de paletas para discos de freno en automoción industria (2020). https://www.researchgate.net/publication/342734012_Analisis_metalografico_y_materiales_de_los_frenos_de_disco/related
26. García-León, R.A.: Análisis de los principales mecanismos de falla en tres frenos de disco automotrices (2020). https://www.researchgate.net/publication/357114147_ANALISIS_DE_LOS_PRINCIPALES_MECANISMOS_DE_FALLA_EN_TRES_FRENOS_DE_DISCO_AUTOMOTRICES
27. Zamora, R.S.: Material de friccion sin asbesto para frenos y embragues con compuesto de baritina (2022). https://www.researchgate.net/publication/287215845_Material_de_friccion_sin_asbesto_para_frenos_y_embragues_con_compuesto_de_baritina
28. Simulación, A.D.: Manual de Mantenimiento de Frenos. Manual De Metodo-logia Da Pesquisa Aplicada À Educação **6**, 1–56 (2019)
29. Systems, B.: What is a braking system ? How does a braking system work ? 1 (n.d.)
30. TEOJAMA COMERCIAL. Recomendaciones para el mantenimiento de la Sus-pensión de Camiones. Blog - Teojama (2018). https://www.teojama.com/blog/recomendaciones-mantenimiento-la-suspension-cami-ones/#:%7E:text=Revisar%20el%20engrase%20de%20chasis,y%20balanceo%20cada%2010.000%20kil%C3%B3metros
31. Jayaraj, T.: Diseño optimizado para freno magnetorreológico usando métodos DOE (2015). https://www.researchgate.net/publication/326106876_Amortiguadores_inteligentes_y_reologia
32. Yogesh Kumar, Estudio sobre el Comportamiento Dinámico del Sistema de Suspensión Wishbone (2022). https://www.researchgate.net/publication/258683057_Study_on_Dynamic_Behaviour_of_Wishbone_Suspension_System

Cylinder Head Tuning of a Spark-Ignition Engine and Validation on a Chassis Dynamometer to Verify Its Power

Jose Vicente Manopanta Aigaje[1,2](✉), Jonathan Daniel Hurtado Muñoz[2], Jefferson Marcelo Peringueza Chuquizan[2], and Fausto Neptali Oyasa Sepa[2,3]

[1] Universidad Internacional SEK, Quito, Ecuador
vmanopanta@tecnologicoismac.edu.ec
[2] Instituto Tecnológico Universitario Ismac, Tumbaco, Ecuador
[3] Universidad Israel, Quito, Ecuador

Abstract. The research is motivated by the high levels of contamination and lack of efficiency, especially due to the exponentially increasing vehicle fleet in Ecuador, the proposal is to modify several components and characteristics of the cylinder head such as the combustion chamber, the valve mechanism system, the intake and exhaust system of the spark-ignition engine, which will result in changes to the power of the engine and thus the vehicle and this was measured in the chassis dynamometer since this result can be obtained directly through the vehicle. With the present research you can achieve a bulwark of machining in the elements of the engine where the calculated theory is reflected in practice and in the chip starting that will be given to the cylinder head linking fuel consumption and compression ratio, the engines look good or bad according to the power sought experimentally. In conclusion, from the measurements made, in general there is an increase in the power performance of 30.08% with the modifications and specifically at the maximum power at 4400 rpm, measured in the chassis dynamometer, which were the result of the modifications of the cylinder head of the engine with respect to the standard cylinder head of the vehicle's engine.

Keywords: Cylinder head tuning · Spark-ignition engine · Compression ratio · Power · Chassis dynamometer

1 Introduction

A small portion of vehicles are replaced by new models every year. Standards for passenger cars, for example, now cover almost two-thirds of their overall energy use, up from 50% a decade ago. For some developing countries that have not yet established standards, the import of second-hand vehicles is an important feature of their market with many of these imports no longer meeting the standards of the exporting country [1].

M. Z. Vizuete et al. (Eds.): CI3 2023, LNNS 1040, pp. 238–250, 2024.
https://doi.org/10.1007/978-3-031-63434-5_18

Each country's economy is directly or indirectly dependent on fossil fuels, gradually shrinking through massive industrialization, transportation and population growth [2]. In addition, concerns about climate imbalances, global warming and the commitments made by developed countries to improve the security of energy supply and encourage the use of renewable energies are just some of the factors that make biodiesel an interesting alternative [3].

The addition of bridges and bores in the cylinder head proves to be very effective as it considerably reduces emissions [4]. In addition, a slight change in the configuration of the bridges and bores can change the flow directions and patterns and vary the way the gases react, which can further reduce emissions [5, 6]. The effect of speed, load, and blend ratio on the competence of a multi-cylinder indirect injecting diesel power unit was investigated by Adam et al. (2015), using statistical tool, Box-Behnken design (BBD) based on RSM to predict and assess their net effects on the responses, such as torque, power, BSFC, and BTE. Blends of 5–20% volume of BDF (prepared from a mixture of palm and rubber seed oils) to diesel fuel were prepared Load was found to be the most effective input, both individually and in combination, in contrast to the blending and speed variables [7]. A strong influence of speed on the results was observed, except for torque, while their combined effect was not vital, except for BSFC and BTE [8].

Sha et al. (2015) [9] experimentally investigated the effects of pre-chamber volume and nozzle diameter on the resultant ignition characteristics. It was found that a larger pre-chamber provides higher ignition energy, which results in shortened flame development angle and combustion duration. At a given pre-chamber volume, nozzle diameter mainly affects the combustion duration [10].

The research was initiated in response to the high levels of contamination and lack of efficiency, especially in the automotive fleet [11], due to the scarce energetic use of organic materials as energy sources that guarantee an optimal performance in internal combustion engines. The main objective of this project is the modification of the cylinder head of a spark-ignition engine and validation in a chassis dynamometer for the verification of its developed power after the adjustments in these variables.

2 Method

2.1 Applied Test Vehicle

For the experimental investigations, a Citroën 1.4 Standard K2d 4-cylinder 1,360 cm^3 gasoline engine, with indirect injection in the intake manifold, naturally aspirated and equipped with a conventional three-way catalytic converter from 1998 was used; selected because it is one of the most sold automobiles in ecuador in those years [12]. Table 1 describes the specifications of the vehicle.

Table 1. Citroën 1.4 Standard K2d Technical Data Sheet

Characteristics	Units
Make	Citroën
Model	Saxo
Year	1998
Compression Ratio	9.3/1
Torque	114 Nm
Horsepower	75 CV
Bore	75 mm
Top Speed	164 km/h
Acceleration, 0–100 km/h	14.8 s
Max Power	76 CV DIN AT 5800 rpm
Displacement	1360 cm^3
Max Torque	111 Nm DIN AT 3400 rpm
Valves	8 valves
Fuel	Gasoline
Engine	Inline four
Firing order	1-3-4-2

2.2 Characteristics of the Cylinder Head for the Combustion Chamber Area

The design of the cylinder head is simple, easy to reproduce and install [13]; given the intake stroke, the mixture reaches the cylinder with little speed, the turbulence is almost null, the combustion is slow and prone to detonation due to the long length of the flame [14]. Figure 1 shows the measurement of the combustion chamber volume, as the main value in this tuning process, in addition to the calculation of the engine compression ratio measured, applying the following equations.

Piston volume measurement

$$\begin{gathered} Piston\ Vol. = 10cm^3 \\ Compressed\ Vol. = \text{Piston } Vol. + Chamber\ Vol. + \text{Gasket Vol.} \\ Compressed\ Vol. = 10 + 25.5 + 6.87 = 42.37\,\text{cm}^3 \end{gathered} \tag{1}$$

Calculation of the compression ratio of the engine

$$\begin{gathered} Rc = \frac{Compressed\ Vol. + Displacement\ Vol.}{Compressed\ Vol.} \\ Rc = \frac{42.37 + 343.67}{42.37} = 9.11\ a\ 1 \\ Chamber\ Vol. = 25.5\,\text{cm}^3 \end{gathered} \tag{2}$$

As a consequence of the calculations, we obtain the tuned or modified compression ratio that will show the results of this experiment [15], as described in Table 2.

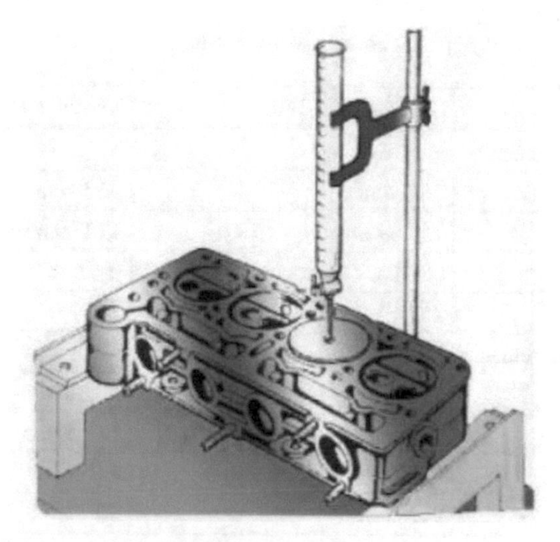

Fig. 1. Measurement of the combustion chamber volume

Table 2. Compression Ratio Calculations

	Engine without Modification	Engine with Modification
Displacement Vol.	343.67 cm^3	343.67 cm^3
Gasket Vol.	6.87 cm^3	6.87 cm^3
Chamber Vol.	25.5 cm^3	24.8 cm^3
Piston Vol.	10 cm^3	10 cm^3
Compressed Vol.	42.37 cm^3	41.67 cm^3
Compression Ratio	9.11 : 1	9.25 : 1

2.3 Valve Mechanism System

The valve system performs an exhaustive work in the recirculation of gases and manages a proper combustion. Table 3 shows several characteristics of the valves both in standard form and in modified or tuned form applied in the study proposed in this research [16].

Table 3. Valve System

	Standard Condition	Modified Ccndition
Intake Valve Measurement		
Retainer Diameter	36.5 mm	36.5 mm
Valve Length	112.69 mm	112.69 mm
Valve Stem	0.7 mm	0.6 mm
Valve Seat	45°	30°
Exhaust Valve Measurement		
Retainer Diameter	29.2 mm	29.2 mm
Valve Length	112.5 mm	112.5 mm
Valve Stem	0.7 mm	0.6 mm
Valve Seat	45°	30°

2.4 Intake and Exhaust System of the Spark-Ignition Engine

The cylinder head has several ducts that are mainly useful for the recirculation of gases and managing a proper combustion. Table 4 shows several features measured for both standard form and modified or tuned form that characterize the study proposed in this research [17]. In addition, Figs. 2, 3 and 4 show the measurement processes established to obtain the data in the cylinder head [18].

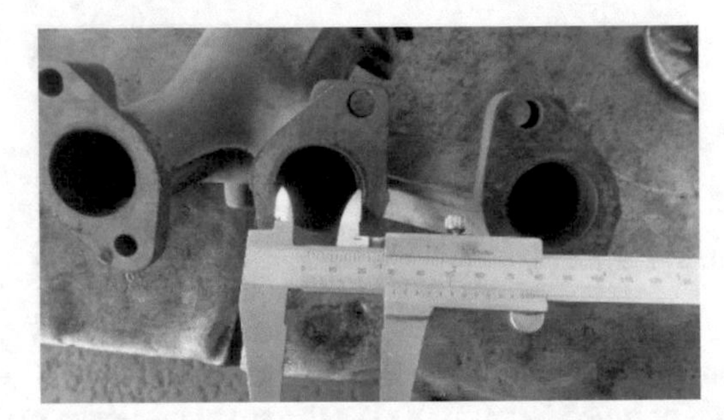

Fig. 2. Measurement of the exhaust manifold nozzle diameter

Fig. 3. Measurement of the intake manifold nozzle diameter

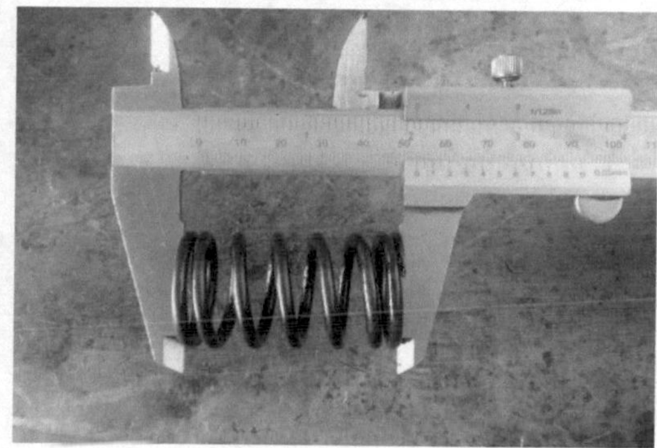

Fig. 4. Measurement of the cylinder head springs

Table 4. Gas recirculation cylinder head ducts and valve springs

	Engine without Modification			Engine with Modification		
Cylinder head intake nozzles						
	Diameter (mm)	Length (cm)	Volume (cm^3)	Diameter (mm)	Length (cm)	Volume (cm^3)
CYLINDER 1	29.40	8.5	66.5	31	8.5	67.2
CYLINDER 2	29.40	8.5	66.5	31	8.5	67.2
CYLINDER 3	29.40	8.5	66.5	31	8.5	67.2
CYLINDER 4	29.40	8.5	66.5	31	8.5	67.2
Cylinder head exhaust nozzles						
	Diameter (mm)	Length (cm)	Volume (cm^3)	Diameter (mm)	Length (cm)	Volume (cm^3)
CYLINDER 1	28.0	6	45.0	33	6	47.0

(*continued*)

Table 4. (*continued*)

	Engine without Modification			Engine with Modification		
CYLINDER 2	28.0	6	45.0	33	6	47.0
CYLINDER 3	28.0	6	45.0	33	6	47.0
CYLINDER 4	28.0	6	45.0	33	6	47.0
Cylinder head springs						
	Diameter (mm)	Length (mm)	Thickness (mm)	Diameter (mm)	Length (mm)	Thickness (mm)
CYLINDER 1	28.80	52.10	3.60	32.40	47.30	4.50
CYLINDER 2	28.80	52.10	3.60	32.40	47.30	4.50
CYLINDER 3	28.80	52.10	3.60	32.40	47.30	4.50
CYLINDER 4	28.80	52.10	3.60	32.40	47.30	4.50

The measurements of the manifold nozzles are relevant to observe its modification as a consequence of the new performance to be obtained in the vehicle's engine, according to Table 5.

Table 5. Manifold nozzles and cylinder head height

	Engine without Modification	Engine with Modification
Measurement of the cylinder head height		
	mm	mm
Height	111.2	110.4
Measurement of the exhaust manifold nozzle diameter		
	Diameter (mm)	Diameter (mm)
CYLINDER 1	31	34.80
CYLINDER 2	31	34.80
CYLINDER 3	31	34.80
CYLINDER 4	31	34.80
Measurement of the intake manifold nozzle diameter		
	Diameter (mm)	Diameter (mm)
CYLINDER 1	27.5	27.5

(*continued*)

Table 5. (*continued*)

	Engine without Modification	Engine with Modification
CYLINDER 2	27.5	27.5
CYLINDER 3	27.5	27.5
CYLINDER 4	27.5	27.5

2.5 Determination of the Developed Power

The main test in this study is the chassis dynamometer, where the chassis dynamometer SAENZ PERFORMANCE INERTIAL CHASSIS DYNOS-N 08-19 was used, which allows simulating a speed profile as a function of time to perform dynamic tests on diesel and gasoline vehicles [19]. It allows to upload different test cycles of the different legislations or self-designed cycles in order to be able to determine exactly the amounts of developed power. Figure 5 shows schematically the model of the testing process performed on the vehicles. The torque and power tests were performed on the chassis dynamometer [20].

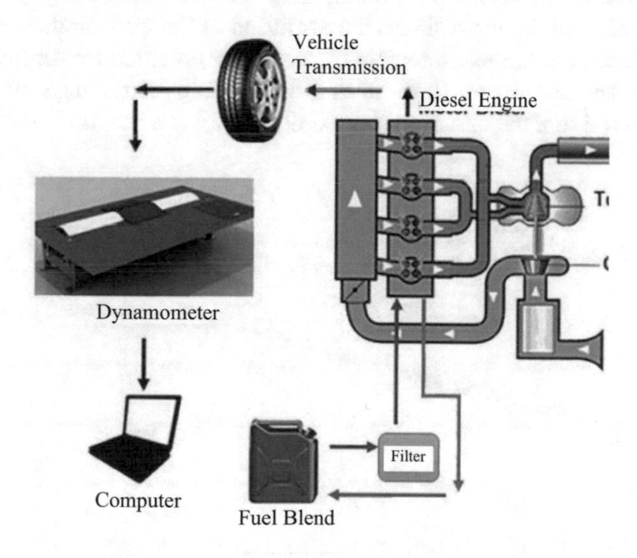

Fig. 5. Test setup on the chassis dynamometer

Power and torque test protocol. The evaluation of the performance obtained on the output shaft through the dynamometer was performed under ISO 1585 and ISO 3173 standards [3]; following the test protocol: verify that the diameter of the wheels is equal to or greater than 13 rim and that they comply with the weight capacity established by the manufacturer; place the test vehicle on the dynamometer rollers; verify that the tire tread is free of stones; lower the lift and leave the wheels resting on the rollers;

check the alignment of the wheel assembly with respect to the dynamometer rollers by rotating the wheels at a maximum speed of 20 km/h; secure the vehicle with straps to prevent it from slipping off the rollers; check the safety of the test area; enter the technical data concerning the vehicle to be tested into the software; check the gear and transmission ratio of the unit, which must be 1:1; make sure that the engine temperature is the normal operating temperature, otherwise it must pass an engine warm-up period to reach such temperature; start the cooling fan of the dynamometer; start the measurement test; accelerate the vehicle with the pedal fully depressed in the test gear until reaching the desired speed called "rpm cut" (4500–6000 rpm); when the "rpm cut" has been reached, depress the clutch, leaving the gear engaged. The dynamometer decelerates to a stop [21].

3 Results

3.1 Vehicle Power Without Modification of the Cylinder Head

The power parameters generated by the engine of the Citroën 1.4 vehicle with the different tests. The static test with load was carried out, using the chassis dynamometer, under the ISO 1585 Standard, performing three tests for each case and obtaining the result to be evaluated. Figure 6 shows the evolution of the tests on the equipment [22]. Here are the results of the measurement of the power obtained for the vehicle with the dynamometer. The test was run three times in a row using premium gasoline. In this test, it can be observed that the curves generated do not show a major difference according to the values.

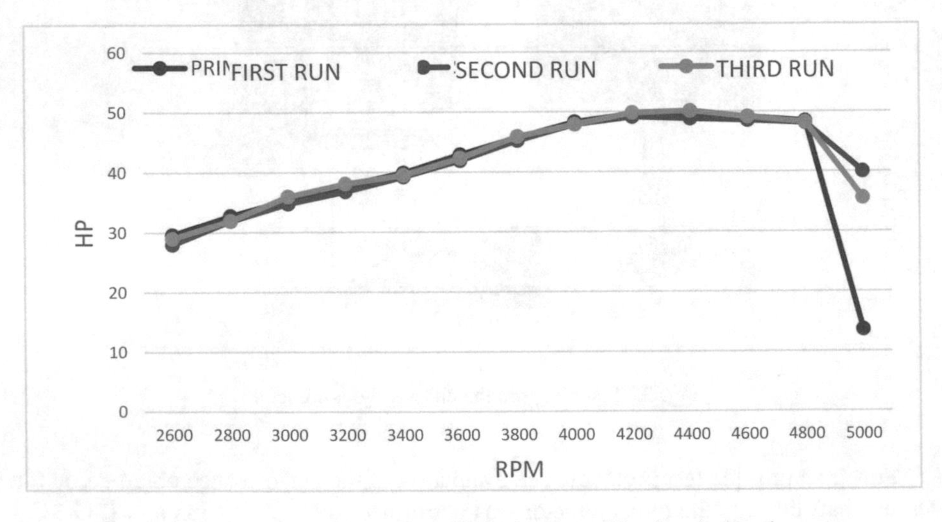

Fig. 6. Vehicle power without the cylinder head modification during three runs

3.2 Vehicle Power with the Cylinder Head Modification

The power parameters generated by the engine of the Citroën 1.4 vehicle with the different tests. The static test with load was carried out, using the chassis dynamometer, under the ISO 1585 Standard, performing three tests for each case and obtaining the result to be evaluated [23]. Figure 7 shows the power curves obtained with the dynamometer. The test was run three times in a row using premium gasoline. In addition, it can be observed that there is not a major difference among the values [24].

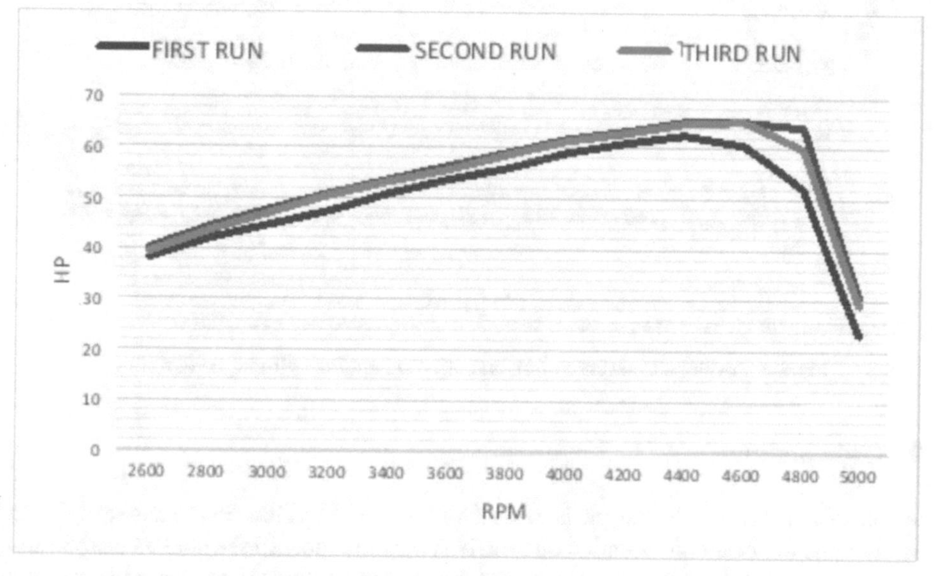

Fig. 7. Vehicle power with the cylinder head modification during three runs

3.3 Discussion of the Results on the Obtained Power

The tests carried out and the values obtained are used to analyze the behavior of the vehicle, which will be used to develop the pertinent analysis regarding the performance of the engine. We will proceed as follows, working with the average power values obtained [25]. Figure 8 shows the power variation differences for the standard working tests and the cylinder head modification, clearly evidenced in the chassis dynamometer.

It is then established according to the measurements made that there is an increase of 30.08% in the performance at 4400 rpm with respect to the engine power with the modifications of the cylinder head, and it is clearly seen in the comparative graph the difference between the performance of power before and after the modifications of the cylinder head. All the tests were carried out with the premium gasoline that is commercialized in the country [26]. Several authors show similar data, confirming the increased engine efficiency [5].

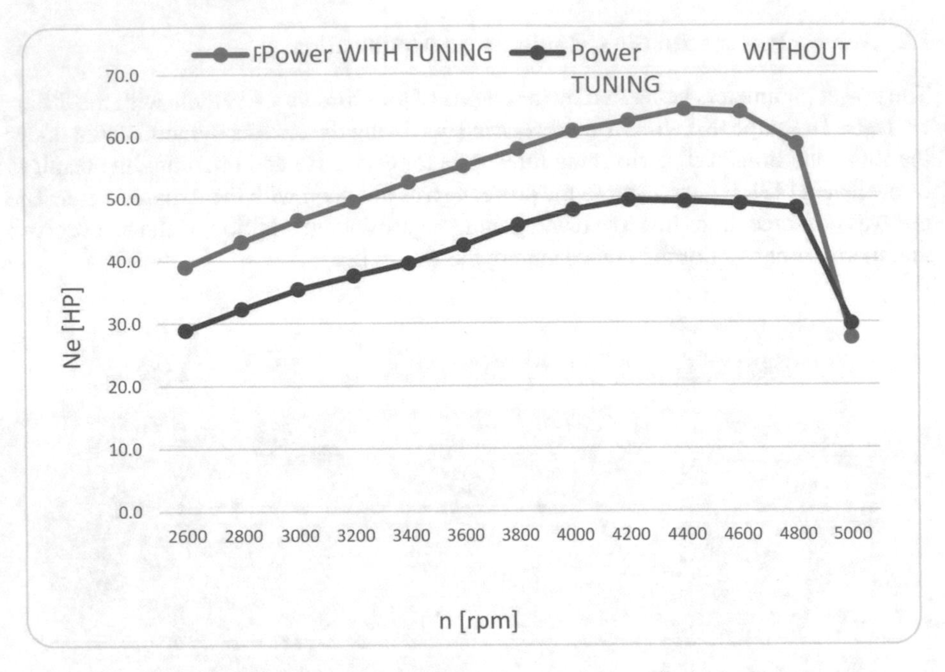

Fig. 8. Developed powers before and after the engine cylinder head tuning.

4 Conclusions

In conclusion, a research analysis is made about the modifications in the engine cylinder head to find the best configuration and the best performance. In the studies analyzed, it has been found that the valve, nozzles and ducts of the cylinder head are truly relevant and have the greatest influence on the engine performance.

The measurements performed show an increase in power performance in general during the course of the engine operation from idle to maximum power. Therefore, with the modifications and specifically at the maximum power at 4400 rpm, an increase of 30.08% measured in the chassis dynamometer was observed as a result of the modifications of the engine cylinder head with respect to the standard cylinder head of the vehicle's engine.

References

1. Cozzi, L., et al.: World energy outlook 2020, pp. 1–461. International Energy Agency Paris, France (2020)
2. Afzal, S., et al.: Exhaust emission profiling of fatty acid methyl esters and NO x control studies using selective synthetic and natural additives. Clean Technol. Environ. Policy **20**, 589–601 (2018)
3. Rocha-Hoyos, J.C., Llanes-Cedeño, E.A., Celi-Ortega, S.F., Peralta-Zurita, D.C.: Efecto de la Adición de Biodiésel en el Rendimiento y la Opacidad de un Motor Diésel. Inf. Tecnol. **30**(3), 137–146 (2019)

4. Menaca, R., Bedoya-Caro, I.D.: Una revisión del uso del hidrógeno en motores de encendido por compresión (diésel) y un análisis de su posible uso en motores duales en Colombia. Rev. UIS Ing. **21**(3), 33–54 (2022)
5. Indudhar, M.R., Banapurmath, N.R., Rajulu, K.G., Patil, A.Y., Javed, S., Khan, T.M.Y.: Optimization of piston grooves, bridges on cylinder head, and inlet valve masking of home-fueled diesel engine by response surface methodology. Sustainability **13**(20), 11411 (2021)
6. Ogden, J.M., Williams, R.H., Larson, E.D.: Societal lifecycle costs of cars with alternative fuels/engines. Energy Policy **32**(1), 7–27 (2004)
7. Shrivastava, P., Verma, T.N., Samuel, O.D., Pugazhendhi, A.: An experimental investigation on engine characteristics, cost and energy analysis of CI engine fuelled with Roselle, Karanja biodiesel and its blends. Fuel **275**, 117891 (2020)
8. Adam, I.K., Aziz, A.R.A., Yusup, S.: Determination of diesel engine performance fueled biodiesel (rubber seed/palm oil mixture) diesel blend. Int. J. Automot. Mech. Eng. **11**, 2675 (2015)
9. Shah, A., Tunestal, P., Johansson, B.: Effect of pre-chamber volume and nozzle diameter on pre-chamber ignition in heavy duty natural gas engines. SAE technical paper (2015)
10. Ge, H., Bakir, A.H., Yadav, S., Kang, Y., Parameswaran, S., Zhao, P.: CFD optimization of the pre-chamber geometry for a gasoline spark ignition engine. Front. Mech. Eng. **6**, 599752 (2021)
11. Demeulenaere, X.: The use of automotive fleets to support the diffusion of alternative fuel vehicles: a rapid evidence assessment of barriers and decision mechanisms. Res. Transp. Econ. **76**, 100738 (2019)
12. Ocaña, B.V., Zavala, A.F., Erazo, C.V.: "Trucaje del cabezote del motor del vehículo Chevrolet Astra para competencias rally", 593 Digit. Publ. CEIT **7**(6), 192–201 (2022)
13. Perumal Venkatesan, E., et al.: Performance and emission reduction characteristics of cerium oxide nanoparticle-water emulsion biofuel in diesel engine with modified coated piston. Environ. Sci. Pollut. Res. **26**, 27362–27371 (2019)
14. Kalghatgi, G.T., McDonald, C.R., Hopwood, A.B.: An experimental study of combustion chamber deposits and their effects in a spark-ignition engine. SAE Technical Paper (1995)
15. Mohiuddin, K., Kwon, H., Choi, M., Park, S.: Effect of engine compression ratio, injection timing, and exhaust gas recirculation on gaseous and particle number emissions in a light-duty diesel engine. Fuel **294**, 120547 (2021)
16. Rosales, J.P.F., Revelo, J.D.G., Cárdenas, E.M.C., Paz, P.A.M.: Determinación experimental de la eficiencia mecánica de un motor de combustión interna de encendido provocado por medio de las curvas de torque y potencia. Polo del Conoc. **7**(7), 1735–1759 (2022)
17. Galindo, J., Climent, H., De la Morena, J., Pitarch, R., Guilain, S., Besançon, T.: A methodology to study the interaction between variable valve actuation and exhaust gas recirculation systems for spark-ignition engines from combustion perspective. Energy Convers. Manag. **250**, 114859 (2021)
18. Romero-Piedrahita, C.A., Mejía-Calderón, L.A., Carranza-Sánchez, Y.A.: Estudio de sensibilidad del desempeño durante el calentamiento en vacío de un motor de combustión interna a cambios en su relación de compresión, el contenido de etanol en la mezcla combustible con gasolina y el material de la culata. Dyna **85**(204), 238–247 (2018)
19. Yang, Z., Deng, B., Deng, M., Huang, S.: An overview of chassis dynamometer in the testing of vehicle emission. In: MATEC Web of Conferences, vol. 175, p. 2015 (2018)
20. Alagumalai, A.: Internal combustion engines: progress and prospects. Renew. Sustain. Energy Rev. **38**, 561–571 (2014)
21. Corsini, A., Di Antonio, R., Di Nucci, G., Marchegiani, A., Rispoli, F., Venturini, P.: Performance analysis of a common-rail diesel engine fuelled with different blends of waste cooking oil and gasoil. Energy Procedia **101**, 606–613 (2016)

22. Chen, L., Wang, Z., Liu, S., Qu, L.: Using a chassis dynamometer to determine the influencing factors for the emissions of Euro VI vehicles. Transp. Res. Part D Transp. Environ. **65**, 564–573 (2018)
23. Bae, C., Kim, J.: Alternative fuels for internal combustion engines. Proc. Combust. Inst. **36**(3), 3389–3413 (2017)
24. Abd-Alla, G.H.: Using exhaust gas recirculation in internal combustion engines: a review. Enery Convers. Manag. **43**(8), 1027–1042 (2002)
25. Yu, S., Zheng, M.: Future gasoline engine ignition: a review on advanced concepts. Int. J. Engine Res. **22**(6), 1743–1775 (2021). https://doi.org/10.1177/1468087420953085
26. Costa, R.C., Sodré, J.R.: Hydrous ethanol vs. gasoline-ethanol blend: engine performance and emissions. Fuel **89**(2), 287–293 (2010)

Additive Manufacturing of the Acceleration Body of the Corsa Evolution Analyzing the Type of Meshing

Denis Ugeño([✉]), Víctor López, Pamela Villarreal, Cristian Guachamin, Jhon Jara, and Jorge Ramos

Instituto Superior Universitario Vida Nueva, Quito, Ecuador
denis.ugeno@istvidanueva.edu.ec

Abstract. Additive manufacturing, also known as 3D printing, has revolutionized the manufacturing industry by enabling the creation of complex parts in an efficient and customized manner. In the case of the Corsa Evolution's throttle body, additive manufacturing has been used to produce this key component of the air intake system. In terms of meshing type analysis, meshing in additive manufacturing refers to the internal structure of the 3D printed part. Meshing can vary in terms of density and design, which directly affects the properties and performance of the final part. The additive manufacturing of the acceleration body of the Corsa Evolution has allowed the production of a customized and efficient component. The analysis of the type of mesh in this process plays a crucial role in optimizing the performance of the final part, taking into account factors such as strength, weight and airflow.

Keywords: Additive manufacturing · 3D printing · Meshing · FEM analysis · CFD analysis

1 Introduction

Reverse engineering has evolved in leaps and bounds these last two decades, enabling a continuous improvement in design processes, therefore, [9] mentions that "in this way designers as architects, engineers will have a much clearer idea of what is being built, with the possibility of minimizing errors when building mechanical elements".

The obtaining of the geometry of the mechanical elements through a 3D scan has gained strength in recent times so [10] affirmed the following:

The function of a 3D scanner is to measure and digitize a physical model, for this it collects samples of the geometry of the object, extracting a cloud of points that will be processed and analyzed to determine its position in space and obtain a three-dimensional model or reconstruction of the object.

For example, you can create replacement parts that correspond to the original design of damaged existing parts, or use reverse engineering processes to integrate complex surfaces of sturdy objects into fasteners with 3D-printable guides, which are useful when you want to modify handmade or mass-produced products [16].

M. Z. Vizuete et al. (Eds.): CI3 2023, LNNS 1040, pp. 251–266, 2024.
https://doi.org/10.1007/978-3-031-63434-5_19

However, despite the advances, the use of composite materials in different automotive parts still has great challenges: how do the mechanical properties of compo-site materials influence the performance and durability of the car's acceleration body?

In this research, we propose to address the question posed through a multidisciplinary approach that combines composite materials theory, solid mechanics and de-sign engineering. Through experiments and theoretical analyses, seeking to obtain accurate and reliable information on how composite materials influence the performance, strength and durability of acceleration bodies.

2 Additive Manufacturing

Additive manufacturing is a technology that is currently being applied for the manufacture of complex parts, [14]. "the technology known as Additive Manufacturing (AM), as it is known internationally, basically consists of manipulating material at the micrometric scale and depositing it very precisely to build a solid" (p. 34).

Taking into account the above, additive manufacturing may vary according to the following parameters:

- Cost
- Type of material
- Maintenance
- Velocity
- Thickness
- Precision

2.1 FMD (Fused Deposition Modeling) Printing

From the point of view of "fused deposition modelling (FDM) is a 3D printing technology based on extrusion. The manufacturing materials used in FDM are thermoplastic polymers and come in filament form", On the other hand, with FDM technology, "a part is manufactured by selectively depositing molten material layer by layer in a path defined by the CAD model. Due to its high precision, low cost and wide selection of materials [17].

According to [14] "FDM technology uses the manufacturing input material in the form of thermoplastic filaments that are liquefied and re-solidified into the de-sired shape according to the defined CAD model (Fig. 1).

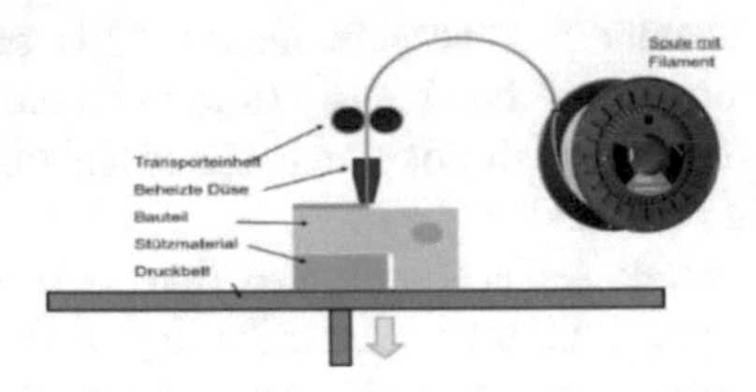

Fig. 1. This graph represents the fundamental elements of FMD printing, taken form [12]

In this type of printing [3] mentions that a very important aspect must be taken into account regarding the filament, so the filament for FMD printing must reach a liquid state so that the individual layers of the model can be added. This must be properly heated and introduced into a thin nozzle. This produces a thin thread of plastic that is used to make layers that overlap to form the 3D component.

Advantages of Fused Deposition 3D Printing. This type of technology, used mainly in sectors such as aeronautics, automotive, health and consumer goods, among others, constitutes a new model in the way parts and products are manufactured. It has numerous advantages that have made it a modality widely used by companies, in addition to any new technology, it also has its limitations, therefore, the following table shows the advantages and limitations of additive manufacturing (Table 1).

Table 1. Advantages and limitations of additive manufacturing

Advantages	Limitations
Geometric complexity, customization	AM Technologies in Development
Creative freedom	Peripheral and auxiliary processes
Adaptation to the market	Lack of knowledge of designers
Access to new market niches	Availability and cost of raw materials
Integrated mechanisms	Surface finish
Weight reduction, lightened products	Manufacturing speed
Reduced time to market	Product quality and process repeatability
Reduction of intermediate process costs	Limited size of parts
Hybrid processes	Cost of machinery

As for the FDM process, these are the most important factors that make it one of the most popular 3D printing technologies.

- FDM is known as one of the cheaper options for 3D printing compared to its counterparts that are more expensive like SLA o SLS
- The manufacturing material is cheaper and widely available which makes printing prototypes and design improvements less expensive than with other technologies.
- The FDM process is fast in terms of printing speed, since the manufacturing filaments are readily available and there is no problem with printing time.

Technical Specifications for FMD. Taking into account the advantages of fused deposition 3D printing, the technical specifications for FMD technology are presented below (Table 2).

Table 2. Technical specifications for FMD technology of 3D printing [17]

Standard delivery time	Minimum 4 days (or 48 h for models using the quick management service), depending on the size of the part, number of components and degrees of finish
Standard accuracy	±0,15% (with a lower limit of ±0,2 mm)
Layer thickness	0.18–0.33 mm (varies depending on material selected)
Minimum wall thickness	1 mm
Maximum dimensions of the construction area	There are no limits to dimensions, as components can be created with different parts. The maximum size is machine is 914 × 610 × 914 mm
Surface structure	Unfinished parts usually have a rough surface, but it is possible to apply any type of surface finish. Parts produced using FDM can be painted and coated

2.2 Filling 3D Printing

Infill is very important in 3D printing, since this filling pattern can be adjusted from 0% to 100% so [2] mentions the following:

- When doing 3D printing there are several parameters that one can control that give the final finish to the piece and one of the most important is the filling
- The filling is how the part to be printed is filled and this parameter determines many characteristics of the part. It influences weight, strength, printing speed, part cost, and even its ability to float.

Fill Density. The density of the filling is a characteristic that 3D printing technology has, so you can modify the percentage of filling material of the piece at the time of printing, con-figured the weight, resistance and time of printing, in Fig. 7 the percentage of filling is shown in a triangular pattern (Fig. 2).

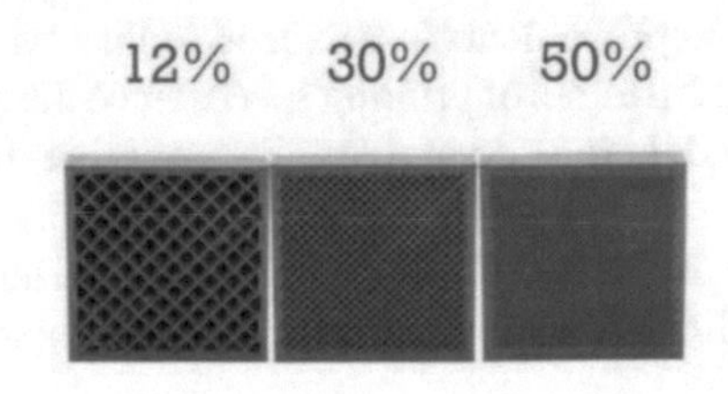

Fig. 2. In the figure you can see the percentage of meshing of a mechanical component [2].

When using software such as Cura or Prusa Slicer, you can choose the percentage of fill of the piece to be printed, default from 20% to 100% fill, so [2] mentions the following:

If you want to use exhibition pieces, the same ones that are going to be exhibited on a shelf or models and do not support loads that may affect their structure, it is recommended to use a filling from 0% to 15%.

If you want to use standard functional parts, the same ones that will be supporting moderate loads such as that of a simple mechanism, it is recommended to use a filling from 15% to 50%, modifying its robustness and printing speed.

If it is going to be used for mechanical parts, a filling percentage from 50% to 100% is recommended, so they will always be supporting combined stresses or as they are commonly called, cyclic stresses, requiring more material [2].

2.3 Materials for Additive Manufacturing

Carbon Fiber. According to the research of [11] mentions that "this fiber is one of the most used in the industrial field so it is composed of carbon atoms, the same that has a diameter of 5 to 10 um", therefore Alegre Gago [5] mentions that "over the years it has gained strength in many sectors because it provides great resistance and can even overcome the resistive stress of the conventional materials", in the same way it offers a low, but compared to other metals for the reason that it provides, it should be noted that carbon fiber in its pure state offers a resistance of 5 times more than that of steel.

The following table shows the characteristics of carbon fiber, highlighting its mechanical properties for printing tensile and bending specimens (Table 3).

Table 3. Properties of carbon fiberglass

Typical Material Properties			
Physical Properties	Unit	Value	Method
Density	g/cm^3	1,35	ISO 1183
Moisture absorption	%	–	
Tensile strength	MPA	76	ISO 527
Tensile elongation @ breakage	%	7,5	ISO 527
Flexible strength	MPA	110	ISO 178
Bending modulus	GPA	6,2	ISO 178
Impact resistance with notches	kJ/m^2	6	ISO 180
Young's module	GPA	220	ISO 178
Impact resistance without notches	kJ/m^2	60	ISO 180
Poisson coefficient		0,25	
Recommended 3D printing temperature	°C	240–260	–

(continued)

Table 3. (*continued*)

Typical Material Properties			
Physical Properties	Unit	Value	Method
Recommended bed temperature	°C	60–70	–
Bed surface	–	modification Glass plate	–
Active cooling fan	%	50	–
Recommended 3D printing speed	mm/w	40–70	–

ABS Plus Fiber. According to the research of [4], mentions that "ABS Plus, is a versatile material that has special properties such as greater resistance and lower processing contraction than standard ABS", therefore Alegre Gago [5] mentions that this fiber "is ideal for mechanical and technical processing. In addition, 3D printing with ABS Plus is faster than with other standard ABS.

On the other hand, [16] mentions that "the possible applications of ABS PLUS are prototypes, electronic components or decorative objects, but mainly the applications of this material find their space in industry, in products that require greater durability".

The following table shows the printing characteristics of ABS fiber BONUS (Table 4).

Table 4. Printing features of ABS Plus fiber

Characteristics	Value
Printing temperature	250–270 °C
Bed temperature	90–110 °C
Closed chamber	recommended
Fan	0–10%
Flowrate	95–105%
Print speed	35–60 mm/s
Surface	glass, Kapton tape, ABS juice
Retraction (direct)	2–3 mm
Retraction (Bowden)	4–6 mm
Retraction speed	20–45 mm/s
Drying conditions	60 °C/4h

3 Methodology

3.1 DOE Experimental Design

The design of experiments (DOE) is a statistical approach that will be applied for the optimization of the research project, that is, it will help to obtain the factorial experimental design, the number of possible combinations or treatments and the number of specimens.

3.2 Factorial Experimental Design

The effects that occur in the study of most experiments that are developed Many experiments, is due to different causes in the research process, so [13] states that "the factorial design is understood as one in which all possible combinations of factor levels are investigated in each complete trial or replica of the experiment".

Other authors mention the following about factorial experimental design:

Two types of elements are present in the object of study. The planning factor (k) is the measurable variable that acts on the object of investigation and takes a certain value in a given test. Planning factors can be quantitative and/or qualitative [15].

The factorial design lies in all possible combinations of levels with factors. According to [15] mentions that "a factorial arrangement is represented by Eq. 1.

$$n^k \tag{1}$$

where:

K = It is the number of factors

n = the number of levels

To apply the factorial design for the calculation of the combinations of the specimens, the type of mesh in 3D printing, the printing speed and printing temperature with their respective levels are taken into account, the same as presented in Table 5.

Table 5. Factors and levels to be applied in the unifactorial design for the calculation of the number of combinations of the specimens

Factors	Levels
Type of meshing	Triangular (A1) Grid (B1) Gromoid (C1)
Print speed	35 mm/s (D1)
Printing temperature	275 °C (E1)

Applying the mathematical model of factorial design yields the following:

$$3^1 \times 1^1 \times 1^1 = 3 \tag{2}$$

Analyzing the value obtained, it can be deduced that 3 treatments need to be performed with their respective combinations, the same as presented in Table 6.

Table 6. Treatments and combinations of test tubes to be performed

Treatment	Combinations	Response variable
T1 (Carbon Fiber)	A1 + D1 + E1	5 units
T2 (ABS PLUS)	B1 + D1 + E1	5 units
T3 (PETG)	C1 + D1 + E1	5 units

Interpreting Table 5, the following can be mentioned:

- Treatment 1 shall be printed 5 units with triangular mesh at the speed of 35 mm/s with a temperature of 275 °C.
- Treatment 2 shall be printed 5 units with the mesh type grid at the speed of 35 mm/s with a temperature of 275 °C.
- Treatment 3 shall be printed 5 units with gyroid meshing at a speed of 35 mm/s with a temperature of 275 °C.

Number of Repetitions of the Experiment. For the results of the experimental design, aspects such as the following variables are taken into account, such as: maximum elasticity force and breaking force, as a background the type of meshing, the printing speed and finally the printing temperature variable, where these variables are intentionally manipulated for the calculation of the number of repetitions, is considered. as visualized in Eq. 3.

$$n = \left(\frac{40 \times \sqrt{n'' \times \sum x^2 - (\sum x)^2}}{\sum x} \right)^2 \quad (3)$$

where:
N″ = Number of preliminary observations
Σ = Sum of values
x = Value of observations
40 = A constant for a 95% confidence level and a 5% error
To apply the mathematical model for the experimental design, the following considerations must be taken into account:
If > 5 the number of missing specimens must be drawn up
If (n < =) 5 will be sufficient with the test specimens.

Evaluation of the Mechanical Properties of the Prototype. Once determined the combinations of the specimens according to the experimental design, we proceed to the design of the tensile and bending specimens for their characterization, using the ASTM D3039-14 and D7264 standards respectively, to guarantee the standardized tests that allow to determine the mechanical properties, in the table the variables that must be considered for the development of the process considered the validation of the mechanical properties of the prototype are presented (Table 7).

Table 7. Dependent and independent variables for prototyping validation considering mechanical properties

Independent variables	Dependent variables
Type of meshing	Tractive effort Stress curve Deformation Deflection effort Deflection curve
Material	Carbon fiber, ABS plus and PETG

Design of Tensile and Flexural Specimens According to ASTM D3039-14 and D7264. The research also focuses on the experimental type for the reason that test tests were carried out through the design and construction of specimens, obtaining the mechanical properties of the same with carbon fiber, ABS PLUS and PETG, these results will be obtained by applying the ASTM 3039–14 standard for tensile tests, the same one that mentions that said specimen must have a thickness of 3 mm, 250 mm in length and 59 mm in width (Fig. 3).

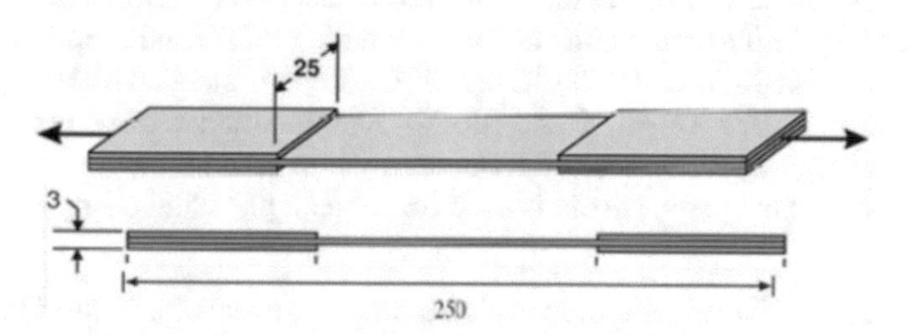

Fig. 3. Tensile test specimen according to ASTM D3039-14

On the other hand, the ASTM-D7264 standard designed for bending tests, which mentions that the specimen must have the measurements: 160 mm long, 13 wide, and 4 thick (Fig. 4).

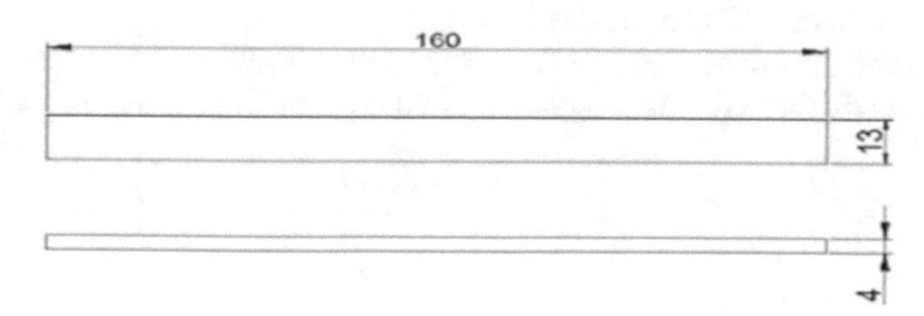

Fig. 4. Specimen for bending test according to ASTM D7264

Once the regulations for the manufacture of specimens have been analyzed, it was designed in SolidWorks, in this way it can be transported in STL format for the configuration of the printing in the Prusa Slicer software, as shown in Fig. 5.

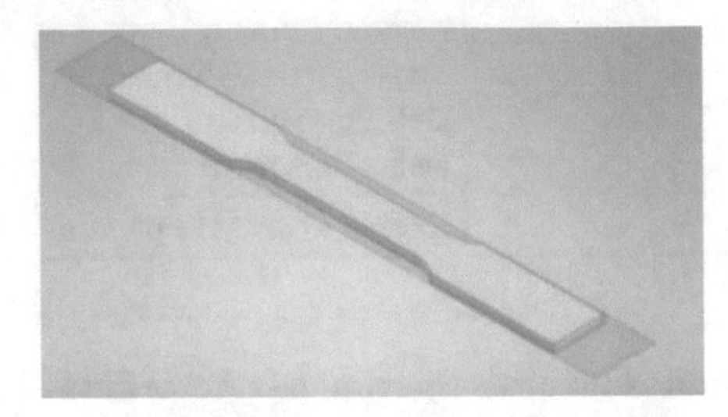

Fig. 5. CAD design of specimen to bend and tensile according to the regulations

Parameters for Printing the Initial Tensile and Bending Specimens. The parameters that are taken into account for the analysis of the printing of the specimens in the Prusa Slicer software with the Infill triangular type, grid and Gyroid, after the CAD design, was the printing speed the same as the value of 35mm/s, also the temperature, taking into account that the printing temperature of the carbon fiber PETG and ABS Plus are almost similar (2400 °C–2600 °C and 2500 °C–2700 °C), Recom-mended bed temperature for each of the fibers is (600 °C–700°C and 900 °C–1000 °C), manufacturing the specimens with carbon fiberglass, ABS Plus and PTG, taking into account that the mesh type grid is used for mechanical parts, gyroid is used for organic parts and triangular is used for mechanical parts.

FEM Simulation. Finite element analysis decomposes a real object into a large number (thousands to hundreds of thousands) of finite elements, such as small cubes. Mathematical equations make it possible to predict the behavior of each element. A computer then adds up all the individual behaviors to predict the actual behavior of the object. It is for this reason that we proceed to perform this type of simulation in the acceleration body of the Chevrolet Corsa, using the SIMSOLID software, considering the mass of the vehicle, its speed to determine the efforts and displacements in real time in case of a frontal impact, modulus of elasticity, Poisson coefficient, density and load, both original material, carbon fiber, PETG and ABS PLUS.

Now to know the impact force of the 2000-kg car traveling at 20 m per second that crashes into a wall, directly affecting the internal combustion engine, Eq. 4 is used.

$$F = \frac{0.5 \times m \times v^2}{d} \tag{4}$$

where:

F = impact force (N)
M = mass of the vehicle (kg)
V = vehicle speed
D = transformation coefficient (0.75)

When replacing the corresponding values, a load of 53.33 KN was obtained, this load will be the one applied in the simulation.

Table 8 shows the general conditions that were taken into account for the development of the first analysis of the acceleration body with the materials mentioned above.

Table 8. General considerations taken into account for the finite element analysis of the acceleration body with different materials

General conditions for FEM-conventional material analysis	
Velocity	72 km/h
Modulus of elasticity	73000 GPA
Poisson coefficient	0,33
Material	Aluminio
Density	2780 kg/m^3
Strength	53,33 kN
General conditions for FEM-carbon fiber analysis	
Velocity	72 km/h
Modulus of elasticity	62 GPA
Poisson coefficient	0,4
Material	Fibra de carbono
Density	1,35 g/m^3
Strength	53,33 kN
General conditions for the EMF analysis of ABS plus	
Velocity	72 km/h
Modulus of elasticity	1800 MPA
Poisson coefficient	0,42
Material	ABS plus
Density	1,07g/cm^3
Strength	53,33 kN
General conditions for FEM-PETG analysis	
Velocity	72 km/h
Modulus of elasticity	2200 MPA
Poisson coefficient	0,42
Material	PETG
Density	12700 kg/m^3
Strength	53,33 kN

Once the general configuration characteristics for FEM simulation in SIMSOLID have been identified, a general environment of that configuration is displayed in Fig. 6.

Fig. 6. Configuration in the SIMSOLID software the FEM simulation of the acceleration body of the Chevrolet Corsa

As can be seen in Fig. 36, the maximum displacement suffered by the acceleration body with the load of 53.33 KN is 0,048 mm right at the top, so in that sector it should be reinforced.

The maximum effort produced by the mechanical element when subjected to the load of 53.33 kN, is 224.8 MPa, and the minimum is 8.210e−5 MPa, so it is very resistant to frontal impacts.

Fig. 7. Analysis of displacements and forces of the acceleration body according to Von Misses with the original material

4 Results and Discussion

4.1 Results

The analysis reveals that the acceleration body experiences a maximum displacement of 157 mm at the top when subjected to a load of 53.33 kN. This information suggests the need to reinforce that specific area of the acceleration body. Reinforcing this area will ensure that the component can withstand the load and prevent possible failures or excessive deformations. The analysis provides a solid basis for taking preventive measures and ensuring safe and reliable performance of the acceleration body.

Automotive auto parts, on the other hand, experience a maximum effort of 217 MPa and a minimum effort of 6.38e−5 MPa. These values indicate a wide difference between maximum and minimum stress, suggesting an uneven distribution of loads along the element. This situation is worrisome, as it can lead to excessive stress concentrations in certain areas, which could lead to unacceptable failures or deformations. It is necessary to address this problem by designing reinforcements or redistribution of loads to ensure a more even distribution of stress along the mechanical element. This will help prevent potential structural failures and ensure safe and reliable operation of the component (Fig. 8).

Fig. 8. Analysis of displacements and forces of the acceleration body with carbon fiber

4.2 Discussion

The analysis reveals that the maximum displacement of the element is 5.41 mm, while the minimum displacement is 8.84e−5 mm. These values indicate a not so significant difference in the displacements along the element, which may be ideal. It is important to consider that such a small minimum displacement may indicate a lack of stiffness or strength in certain areas of the element.

On the other hand, the analysis reveals that the maximum stress on the element is 215 MPa, while the minimum effort is 6.339e−5 MPa. These values indicate a wide difference between the maximum and minimum stresses, suggesting an uneven distribution of loads along the element, taking into account that it can withstand high loads before its deformation (Fig. 9).

Based on the results, it was determined that the minimum deformation in the component is 7,23e−5 mm, while the maximum deformation is 4,43 mm.

These values indicate a significant difference in the deformations experienced along the component. Importantly, such low minimum deformation could be indicative of areas of the component that are experiencing insufficient load or lack the stiffness needed to adequately withstand the applied forces.

Fig. 9. Analysis of displacements and forces of the acceleration body with ABS PLUS

On the other hand, according to the results of the analysis, it was determined that the maximum effort in the component is 2,15e2 MPa, while the minimum effort is 6,339e−5 MPa. These values reveal a wide difference between the maximum and minimum stresses experienced on the component. The existence of such a low mini-mum effort can indicate areas of the component that are subjected to very low loads so it would be a suitable composite material for the manufacture of the acceleration body of the Chevrolet Corsa.

5 Conclusions

In the study of additive manufacturing of the throttle body for the Corsa Evolution, a comprehensive analysis of the meshing technique used in finite element simulations was carried out. The choice of meshing type significantly impacts the accuracy and efficiency of the simulations. The results demonstrated that the selected meshing approach, in combination with the powerful tool SimSolid, allowed for reliable outcomes in a considerably reduced timeframe compared to conventional simulation methods. This optimization of the meshing technique not only enhanced the analysis speed but also contributed to a deeper understanding of the structural behaviors and performance of the additively manufactured throttle body.

The application of 3D printing through fused filament fabrication with carbon fiber filament and ABS Plus has proven to be an effective strategy for enhancing the performance of the Corsa Evolution throttle body. The incorporation of carbon fiber bolstered the structural strength and durability of the component, enabling it to withstand the mechanical and thermal demands it faces under operating conditions. Additionally, the use of ABS Plus as a complementary material provided greater toughness and dimensional stability to the component. This combination of materials and the 3D printing process resulted in a throttle body that not only met performance standards but also provided an advantage in terms of volumetric efficiency.

The additive manufacturing study of the Corsa Evolution throttle body highlights the innovative potential and feasibility of 3D printing technology in the automotive industry. The integration of finite element simulations through SimSolid and the use of advanced materials such as carbon fiber and ABS Plus represent a significant advancement in optimizing the design and manufacturing of automotive components. This considers technical characteristics, geometry, and materials in accordance with mechanical and physical properties to enhance performance, enabling the creation of highly efficient and durable products.

References

1. Alvarez C.K.L., Lagos C.R.F., Aizpun, M.: Influencia del porcentaje de relleno en la resistencia mecánica en impresión 3D, por medio del método de Modelado por Deposición Fundida (FDM). Ingeniare Rev. Chilena Ingeniería **24**(ESPECIAL), 17–24 (2020). https://doi.org/10.4067/S0718-33052016000500003
2. Colombia, B.: Microscopio Trinocular Amscope Soporte de Pluma—3.5X-90X/Luz LED 54. BIOWEB® Colombia (2022). https://colombia.bioweb.co/products/microscopio-trinocular-amscope-boom-stand-3-5x-90x-luz-led-54
3. Dong, C., Davies, I.J.: Flexural properties of macadamia nutshell particle reinforced polyester composites. Compos. B Eng. **43**(7), 2751–2756 (2021). https://doi.org/10.1016/j.compositesb.2021.04.035
4. Fiberlogy. ABS PLUS | Filamento ABS con mayor resistencia. Fiberlogy (2022). https://fiberlogy.com/es/filamentos/abs-plus/. Accessed 21 June 2022
5. Goldschmidt, B.: Infill en Cura: Los mejores patrones de relleno. All3DP (2022). https://all3dp.com/es/2/infill-cura-relleno-impresion-3d/
6. Ikustec. Escaneado 3D de Piezas—Mediciones de precisión. Ikustec (2021). https://ikustec.com/escaneado-3d-de-piezas/
7. Ineo 3D, C. Ineo—Fabricación aditiva—Modelado por deposición fundida (FDM) (2022). https://www.ineo.es/es/tecnologias/modelado-deposicion-fundida-fdm. Accessed 12 July 2022
8. González-Estrada, O.A., Martínez, E.: Evaluación de las propiedades tribológicas de materiales compuestos de matriz metálica (MMCs) procesados por técnicas de fabricación aditiva con haz láser (SLM). Rev. UIS Ingenierías **16**(1), 101–114 (2019)
9. Loste, J., Lopez-Cuesta, J.M., Billon, L., Garay, H., Save, M.: Transparent polymer nanocomposites: an overview on their synthesis and advanced properties. Prog. Polym. Sci. **89**, 133–158 (2019)
10. Miranda, M.F.: Universidad Internacional SEK. Panorama, pp. 5–20 (2020)
11. Möhring, H.C., Brecher, C., Abele, E., Fleischer, J., Bleicher, F.: Materials in machine tool structures. CIRP Ann. Manuf. Technol. **64**(2), 725–748 (2019). https://doi.org/10.1016/j.cirp.2019.05.005
12. Pucha Tambo, M.V.: Caracterización de materiales compuestos con matriz foto polimérica reforzados con fibras de abacá y cabuya mediante impresión 3D (2020). http://localhost:8080/xmlui/handle/123456789/3138
13. Rocha-Hoyos, J.C., Llanes-Cedeño, E.A., Peralta-Zurita, D., Pucha-Tambo, M.: Caracterización mecánica a flexión de materiales compuestos con matriz foto polimérica reforzados con fibras de abacá y cabuya mediante impresión 3D. Ingenius **22**, 100–112 (2020). https://doi.org/10.17163/ings.n22.2020.10
14. Tapia Cabrera, J.E.: Evaluación de las propiedades de amortiguamiento de materiales fabricados por impresión 3D y reforzados con nanotubos y fibras de carbono (2020). https://tesis.pucp.edu.pe/repositorio/handle/20.500.12404/15674
15. Vázquez Núñez, M.: Desarrollo de polímeros conductores para circuitos fabricados mediante impresión 3D (2020). https://ruc.udc.es/dspace/handle/2183/28409

16. Villarreal Bolaños, C.A.: Alternativa para la construcción de autopartes vehicular por medio de la ingeniería inversa e impresión 3D. Caso de estudio tapa de distribución inferior del Chevrolet Spark 2019 (2019). http://localhost:8080/xmlui/handle/123456789/3303
17. Xometry, E.: Tecnología de impresión 3D por deposición fundida. Xometry Europe (2022). https://xometry.eu/es/impresion-3d-por-modelado-por-deposicion-fundida-fdm-vision-general-de-la-tecnologia/
18. 3DWorks. Altura de Capas y Resolución en la Impresión 3D. 3DWorks (2021). https://www.3dworks.cl/post/altura-de-capaseade. Sector automotor en cifras. https://www.aeade.net/wp-content/uploads/2021/04/Sector-en-Cifras-Resumen.pdf. Accessed 25 May 2021

Analysis of Airborne Particles in Powder Coating Process

Caza Paúl(✉), Villarreal Pamela, López Víctor, Díaz Rodrigo, and Cruz Patricio

Instituto Superior Tecnológico Vida Nueva, Quito, Ecuador
paul.caza@istvidanueva.edu.ec

Abstract. This study focuses on the analysis of particles generated during the electrostatic painting process and aims to measure air quality and atmospheric pollution. A study was conducted to evaluate the effectiveness of a cyclone-type extractor system in reducing particle emissions and improving air quality in industrial painting environments. The analysis focuses on understanding the characteristics of particles generated during the electrostatic painting process, including their size, composition, and potential impact on air quality. Various sampling techniques and analytical methods are used to measure and analyze particle concentrations in the ambient air. The effectiveness and efficiency of the cyclone-type extraction system are evaluated, providing insight into its crucial role in reducing particle emissions and improving environmental conditions. This contributes to the development of strategies and technologies aimed at minimizing air pollution in industrial painting operations.

Keywords: Air quality · industrial safety · spay · extraction system · power coating

1 Introduction

The application of electrostatic painting using the spray method is a technique currently being employed in numerous industries to achieve high-quality coatings. This process involves the dispersion of electrically charged paint particles into to the surface to be painted, harnessing the electrostatic energy between the particles and the metallic surface. As the particles move towards the substrate, a uniform layer is formed, providing optimal characteristics and properties such as protection, adhesion, aesthetic appearance, hardness and resistance to environmental and chemical agents.

However, during the electrostatic painting spray process, fine particulate matter is generated and remains suspended in the air, posing challenges in terms of air quality and workplace safety. These suspended particles can vary in size, shape and composition, depending on various factors such as the type of electrostatic paint, spray conditions, equipment configuration and substrate characteristics.

The particulate matter generated in the electrostatic painting application process can have significant consequences for both the coating quality and the health and environment. Suspended airborne particles can negatively affect the appearance and durability

M. Z. Vizuete et al. (Eds.): CI3 2023, LNNS 1040, pp. 267–280, 2024.
https://doi.org/10.1007/978-3-031-63434-5_20

of the final finish, causing defects such as rough texture, thickness irregularities, and poor adhesion. Moreover, exposure to these particles can pose health risks to workers, as some particles may contain toxic or irritant compounds.

2 Contamination

2.1 Air Pollution

It can be stablished that atmospheric pollutions have different emission sources (both natural and anthropogenic) that have a negative effect on the atmosphere. According to these considerations, pollutants can be classified by different aspects such as origin, form of emission, physical state, reactivity and composition, as mentioned above, they can be of natural origin or generated by people and industries [18].

The negative effects generated by these pollutants can be related to premature damage to materials by altering the properties of soil and water, affecting people, causing death due to oxygen substitution, causing skin cancer, destruction of the ozone layer and damage to plantations [12].

Ozone Layer Destruction. Ozone can be considered a reactive gas that cause oxidation in some materials. This damage is generally caused by different pollutants such as NOx, VOCs, CH4 and CO, which are associated with fossil fuels, causing damage to health and the environment due to the level of toxicity and corrosive properties.

It can be mentioned that the ozone layer weakens due to the decrease in the concentration of O3, which causes the formation of ozone holes, this factor is related to diseases related to skin cancer, premature aging of the skin, among others.

Photochemical Pollution. Photochemical pollution is known as acid smog, a mixture of primary and secondary pollutants, which in the presence of sunlight cause severe damage to the biosphere and materials [19]. This process occurs in polluted atmospheres with a high SO2 content, which is composed of high levels of suspended particles, humidity, and cold temperatures, causing it to form a layer of cold air and preventing warmer air from rising. The harmful effects on human health are related to irritation of eyes and mucous membranes, asthmatic attacks, bronchitis, among others.

Air Quality. Air quality can be determined through the analysis of the concentration of different types of pollutants, taking into account the criteria of intake and emission.

For this purpose, a measurement and quantification of the pollutants present in the atmosphere is performed on the sample, which allows a comparison with the legal limit values, thus guaranteeing that the air quality is in accordance with the permitted limits. The procedure to be performed to analyze the air quality considers:

Sampling the same which depends on some factors such as time of day, weather conditions, emission at the site. In accordance with the Ecuadorian Air Quality Standard, Table 1 shows the alert, alarm and emergency levels for the concentration of pollutants to be considered [13].

Table 1. Concentration of pollutant types

Type of pollutants	Alert	Alarm	Emergency
Carbon monoxide 8-h average concentration (µg/m3)	15000	30000	40000
Ozone average concentration 8-h average(µg/m3)	200	400	600
Nitrogen dioxide 1-h average concentration (µg/m3)	1000	2000	3000
Sulfur dioxide 24-h average concentration (µg/m3)	200	1000	1800
Particulate Matter PM10 Concentration in 24 h (µg/m3)	250	400	500
Particulate Matter PM2.5 Concentration in 24 h (µg/m3)	150	250	350

2.2 Mechanical Processes

In fine particle recovery systems, the cyclone is analyzed because it is one of the most efficient equipment for separating and recovering particulate matter, thus helping to reduce the environmental damage caused by industrial processes. The most used control technologies or systems to help reduce particulate matter emission pollution are the cyclone, Venturi scrubber, electrostatic precipitator and baghouse [3].

The main objective of these mechanical processes is to collect as many of the particles emitted in different industrial processes as possible before they are released int the environment [10]. These separators are found in different types considering the separation system to be used, in this research we intend to analyze the efficiency and effectiveness of cyclone type separators.

Cyclone. The cyclone is a device used in systems for extracting and recovering airborne dust particles. Its operation is primarily based on the principle of centrifugal force. The contaminated air, along with the particulate matter, enters the cyclone through a tangential inlet in a spiral form, creating a rotating vortex motion. This rotation pushes the particles towards the cyclone's walls [2].

The particles with higher density separate and fall into the bottom of the cyclone, where they accumulate in a collection container. Meanwhile, the clean air, now free of particles, continues its path towards the cyclone's outlet. It is established that to increase the efficiency of the mechanism, the following should be considered:

- Particulate size
- Particulate properties
- Cyclone design parameters
- Particulate inlet diameter in relation to gas outlet

3 Methodology

Sampling will be carried out using manual equipment that will allow analyzing the concentration of contaminants generated by the dust particles obtained in situ from the analysis of the cyclone-type mechanical system in the electrostatic paint application process.

From the analysis carried out in the sampling, it can be observed that all the values considered in the analysis are related to the pressure generated by the compressor, since fluidization and pulverization depend on it, according to the calibration generated in the equipment. In the manual sampling process, different samples were taken considering the following aspects shown in the table.

3.1 Cyclone Dimensional Analysis

Once the required equations are established, different dimensional analysis techniques can be applied to determine the dimensions and relationships between the various variables of the cyclone. These dimensions are related to the cyclone's diameter, cone height, and the size of the inlet and outlet for air [6].

It is important to highlight that dimensional analysis considers practical limitations in the investigation, such as space constraints, cyclone installation, and the selection of optimal materials that can withstand the working conditions in the electrostatic painting process.

Technical Operating Parameters. To analyze the technical parameters of the cyclone operation in the powder paint application process, it is intended to tabulate the pressure-dependent variables of the electrostatic equipment which generates an optimum performance in the collection of suspended particles in the paint chamber, for which the following parameters are considered [7]:

- The blower motor has an air flow capacity of m3/h 1050. This allows to have higher absorption of dust particles.
- According to the design the diameter of cyclone is 800 mm. Considering the efficiency and efficacy within the collection process which allows to reduce atmospheric pollution.
- The climatic conditions of the painting chamber are to room temperature, which is a suitable factor for electrostatic adhesion on the metal surfaces to be coated.
- According to the implemented system, the vacuum inlet velocity to the cyclone is 190 mbar. Therefore, it is considered an essential factor in the study.

TOPSIS Comparative Analysis. For the multicriteria analysis, a selection is made from several criteria that directly affect the dependent variables such as: environmental pollution, energy consumption of the equipment, work efficiency, equipment manufacturing cost, maintenance cost, construction complexity and operation safety, with all these elements the TOPSIS analysis is developed, which consists of choosing the most efficient option for the analysis of the collection system of suspended particles in the environment. The equipment analyzed arc Venturi scrubber collector, cyclone collector, electrostatic precipitator, and bag filter, as shown in Table 2.

This alternative consists of selecting a minimum distance with respect to a positive ideal distance, considering a greater distance from an anti-ideal alternative. The selection is determined based on seven criteria and four alternatives. The criteria must be reliable and meet the parameters, maximizing the acceptance criteria (Table 3).

Table 2. Criteria for selected alternatives

Criteria	Production capacity (%)	Energy consumption (kW/h)	Environmental contamination (kg)	Production (kg)	Investment	Maintenance cost $/year	Production Volume (gr)
Oven Infrared	90	10	65	120	17000	2500	800
Oven Electrical resistors	90	7,26	10	120	14000	1200	1000
Oven Burners	85	5,2	90	120	20000	2000	500
Oven Convection	70	5,4	85	120	17000	1500	400

Table 3. Weighted table of variables

Criteria	Environmental contamination (%)	Energy consumption (kW/h)	Efficiency	Manufacturing cost (USD)	Maintenance cost $/year	Construction complexity (%)	Handling safety (%)
Venturi Scrubber	65	10	90	7690	2500	90	26
Cyclone	10	5,2	90	6000	1200	85	45
Electrostatic precipitator	90	7,26	90	9500	1200	90	24
Bag filter	85	5,4	90	6300	1500	90	21

In Table 4, the normalized matrix generated in the TOPSIS multicriteria analysis is presented. It is obtained by taking the square root of each criterion and dividing it by the sum of the squares.

Table 4. TOPSIS standardized analysis matrix

Criteria	Production capacity (%)	Energy consumption (kW/h)	Environmental contamination (kg)	Production (kg)	Investment	Maintenance cost $/year	Production Volume (gr)
Oven Infrared	90	10	65	120	17000	2500	800
Oven Electrical resistors	90	7,26	10	120	14000	1200	1000
Oven Burners	85	5,2	90	120	20000	2000	500
Oven Convection	70	5,4	85	120	17000	1500	400

(*continued*)

Table 4. (*continued*)

Criteria	Environmental contamination (%)	Energy consumption (kW/h)	Efficiency	Manufacturing cost (USD)	Maintenance cost $/year	Construction complexity (%)	Handling safety (%)
Venturi Scrubber	0,463	0,6918	0,5	0,512	0,7410	0,5068	0,4264
Cyclone	0,0713	0,3597	0,5	0,3999	0,3557	0,4787	0,7380
Electrostatic precipitator	0,6420	0,5022	0,5	0,6332	0,3557	0,5068	0,3936
Bag filter	0,6063	0,3736	0,5	0,4199	0,4446	0,5068	0,3444

Table 5 displays the weighted normalized decision matrix, where each acceptance criterion is assigned a percentage weight according to a specific percentage ranking. This ranking is determined by multiplying the normalized matrix by the assigned percentage weight.

Table 5. Weighted normalized matrix of the multicriteria selection criteria

Criteria	Production capacity (%)	Energy consumption (kW/h)	Environmental contamination (kg)	Production (kg)	Investment	Maintenance cost $/year	Production Volume (gr)
Oven Infrared	90	10	65	120	17000	2500	800
Oven Electrical resistors	90	7,26	10	120	14000	1200	1000
Oven Burners	85	5,2	90	120	20000	2000	500
Oven Convection	70	5,4	85	120	17000	1500	400
Criteria	Environmental contamination (%)	Energy consumption (kW/h)	Efficiency	Manufacturing cost (USD)	Maintenance cost $/year	Construction complexity (%)	Handling safety (%)
Venturi Scrubber	0,125	0,187	0,095	0,056	0,067	0,025	0,013
Cyclone	0,019	0,097	0,095	0,044	0,032	0,024	0,022
Electrostatic precipitator	0,173	0,136	0,095	0,070	0,032	0,025	0,012
Bag filter	0,164	0,101	0,095	0,046	0,040	0,025	0,010

Finally, the “Euclidean” distance is determined, which represents the distance at which the point of maximum performance in the performance score corresponds to the best alternative in the alternatives ranking from an ideal value to an alternative.

Table 6. Euclide distance or ranking of alternatives

Criteria	Production capacity (%)	Energy consumption (kW/h)	Environmental contamination (kg)	Production (kg)	Investment	Maintenance cost $/year	Production Volume (gr)
Oven Infrared	90	10	65	120	17000	2500	800
Oven Electrical resistors	90	7,26	10	120	14000	1200	1000
Oven Burners	85	5,2	90	120	20000	2000	500
Oven Convection	70	5,4	85	120	17000	1500	400

Positive Ideal Solution	Negative Ideal Solution	Performance Score Pi	Ranking Alternatives Ci
0,140	0,061	0,303	2
0,043	0,179	0,805	1
0,163	0,057	0,260	4
0,149	0,087	0,368	3

In the multicriteria analysis using the TOPSIS method, four teams are presented as options, each fulfilling the best characteristics for collecting airborne dust particles. The evaluation focuses on variables that directly impact the manufacturing of the equipment. The results indicate significant differences among the teams, the TOPSIS analysis shows a high Euclidean distance value for the cyclone system, considering it as the best alternative based on the multi-criteria criterion.

3.2 Factorial Experimental Design

The experimental design is based on two variables such as the particles suspended in the environment considering two different positions inside the paint booth, and two different speeds of the blower in the dust extraction system suspended in the environment are also considered. Table 6 details each of the factors involved in the analysis (Table 8).

Table 7. Variable denomination

Criteria	Production capacity (%)	Energy consumption (kW/h)	Environmental contamination (kg)	Production (kg)	Investment	Maintenance cost $/year	Production Volume (gr)
Oven Infrared	90	10	65	120	17000	2500	800
Oven Electrical resistors	90	7,26	10	120	14000	1200	1000
Oven Burners	85	5,2	90	120	20000	2000	500
Oven Convection	70	5,4	85	120	17000	1500	400

Table 8. The statement of variables used in the experimental design

Variable Designation	
Treatment 1	Treatment 1 Total suspended particulate matter 1 (TSP)/Velocity 1
Treatment 2	Treatment 2 Total suspended particulate matter 1 (TSP)/Velocity 2
Treatment 3	Treatment 3 Total suspended particulate matter 1 (TSP)/Velocity 3
Treatment 4	Treatment 4 Total suspended particulate matter 1 (TSP)/Velocity 4

For the declaration of variables, we start from two very important elements such as the variation of the position of measures or dust collection system inside the cabin and the linear velocity that affects the cyclone is also considered in order to have a higher efficiency factor in the particle collection system. Table 7 details the statement of each of the variables (Table 9).

Table 9. Codification of variables

Criteria	Production capacity (%)	Energy consumption (kW/h)	Environmental contamination (kg)	Production (kg)	Investment	Maintenance cost $/year	Production Volume (gr)
Oven Infrared	90	10	65	120	17000	2500	800
Oven Electrical resistors	90	7,26	10	120	14000	1200	1000
Oven Burners	85	5,2	90	120	20000	2000	500
Oven Convection	70	5,4	85	120	17000	1500	400

Using Stat graphics software, the comparative samples in the results of the statistical summary are presented in Table 10, which shows the values of the number of counts or

Table 10. Coding of variables used in the experimental design

Statement of Variables	
Total suspended particulate matter 1 (TSP)	Particulate matter measured at the back of the spray booth
Total suspended particulate matter 2 (TSP)	Particulate matter measured at the top of the paint booth
Total suspended particulate matter 3 (TSP)	Blower linear velocity (1.43 m/s)
Total suspended particulate matter 4 (TSP)	Blower linear velocity (1.76 m/s)

analyses studied in each treatment, the larger the coefficient of variation, the greater the difference in deviation between the measured values [9] (Tables 11 and 12).

Table 11. Statistical summary

Criteria	Production capacity (%)	Energy consumption (kW/h)	Environmental contamination (kg)	Production (kg)	Investment	Maintenance cost $/year	Production Volume (gr)
Oven Infrared	90	10	65	120	17000	2500	800
Oven Electrical resistors	90	7,26	10	120	14000	1200	1000
Oven Burners	85	5,2	90	120	20000	2000	500
Oven Convection	70	5,4	85	120	17000	1500	400

Table 12. Results obtained using stat graphics software for the average standard deviation between the tests generated

	Count	Average	Standard Deviation	Variation Coefficient	Minimum	Maximum
Analysis I	4	79,135	21,87	27,63%	46,34	90,86
Analysis II	4	66,71	20,12	30,17%	39,69	86,52
Analysis III	4	97,96	81,88	83,58%	40,46	219,11
Analysis IV	4	38,95	10,17	26,11%	30,24	50,59
TOTAL	16	70,69	45,03	63,71%	30,24	219,11

Table 13 shows the results of standardized kurtosis and standardized bias, where information is presented and indicates that the data entered and the measurements obtained come from a normal curve and normal values ranging between –2 and 2, if the data were outside this range they are also known as outliers that should not be considered, therefore, it is considered that the number of tests is correct for this analysis in taking measurements of particulate matter suspended in the air.

In the Anova table, the P-value of the F-ratio is presented, which in this case is 1.27, the estimated coefficient between groups. It can be observed that the P-value is greater

Table 13. Statistical analysis standardized kurtosis and skewness

	Range	Standardized skewness	Standardized kurtosis
Analysis I	44,52	–1,62	1,62
Analysis II	46,83	–0,74	0,24
Analysis III	178,65	1,5	1,42
Analysis IV	20,35	0,25	–1,72
TOTAL	188,87	4,17	6,73

than or equal to 0.05, indicating that there is no statistically significant difference among the four variables. This result is obtained with a confidence level of 95%, as shown in Table 14.

Table 14. Anova table

Source	Sum of squares	Mean square	F Ratio	P-value
Between groups	7352,14	2450,71	1,27	0,3274
Between groups	23014,7	1922,89		
Total (Corr)	30426,9			

Figure 1 shows the response through the block diagram by box and whisker, where the value of the mean is indicated, it is also observed that there are significant differences between the four treatments of analysis I and analysis IV, it is also visualized that analysis II and analysis III have a similar average in the value of the mean, the maximum and the minimum. In addition, in treatment III there is a very high dispersion among the assigned values, since the external values are far from the mean, and it is also observed that in treatment III one of its values has a high dispersion, thus being considered as the most defective treatment in the treatment of samples and reading in the tests.

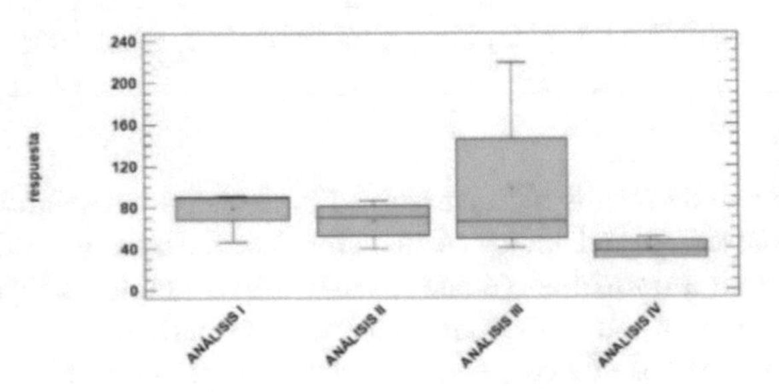

Fig. 1. Box and whisker plot

A very important factor to consider is the dispersion in the reading of the selected measurements, this factor depends on the condition under which each of the analyses were carried out, in case III a very high dispersion in the results is observed, what is recommended is to generate more measurements, on the other hand, in analysis I the reading measurements coincide in the same way, ore readings are recommended in order to have a significant difference, in analysis II it is observed that the samples were taken almost equally, as shown in Fig. 2.

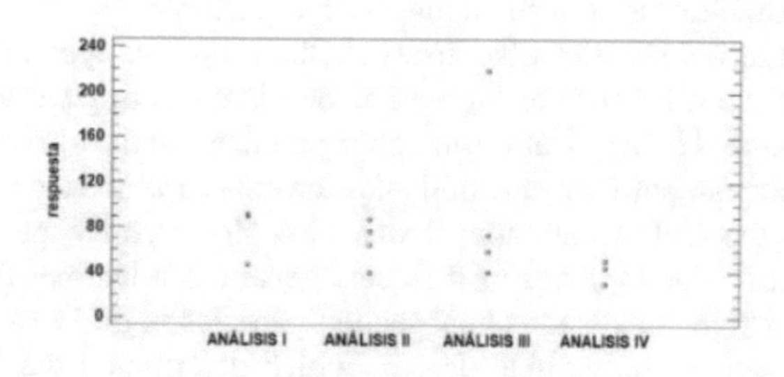

Fig. 2. Sample dispersion box

Finally, Fig. 3 shows the values obtained in the Anova analysis, indicating a dispersion factor P of 0.3274, which is a considerable value for a technical experimental analysis and considering that the sample readings in each of the treatments are similar, which makes it possible to take samples at any internal point of the space analyzed for the collection of suspended particles in the environment.

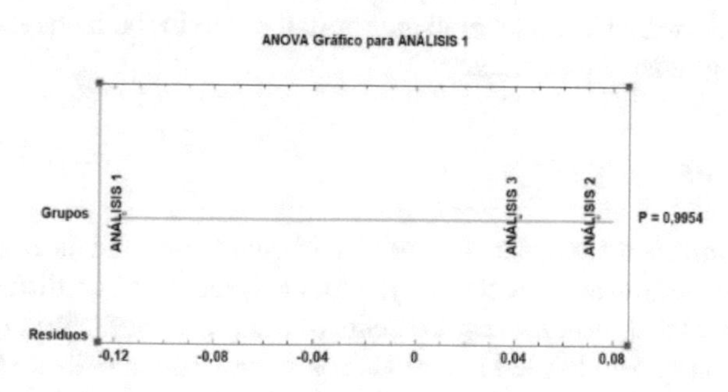

Fig. 3. Anova analysis

4 Results and Discussion

Finally, after conducting various air quality measurements in this research, it has been determined that the technical characteristics of powder coating, such as adhesion, temperature, and humidity, influence the spraying method on the metal surface to be coated, which poses a high risk to operators and the environment.

The ability of powder coatings to adjust their chemical composition according to the needs relies entirely on their components. Therefore, the materials chosen to design the particle collector must meet mechanical, chemical, and physical properties that ensure greater durability of the system used in the electrostatic painting process.

The technical parameters of the electrostatic painting equipment play a crucial role in the application of the coating on the metal component. Poor calibration of pressure, amperage, volumetric amount of powder coating, inadequate or lack of nozzle selection, and circuit closure can result in excessive overspray in the work area, placing maximum demand on the extraction system and application equipment.

From the obtained data, it can be observed that a cyclone type extraction and collection system with a filter ensures high efficiency in collecting suspended particulate matter in the air caused by application and poor practices by the operator or equipment. This, in turn, reduces air pollution and mitigates the risks to which the worker is exposed.

According to the CFD simulation, it was possible to verify and analyze that the cyclone-type system, due to centrifugal forces, generates a laminar flow, allowing for the collection of a high percentage of suspended particles in the working process. This was achieved thanks to the optimal design, which determined the post-construction dimensional analysis parameters, ensuring the proper functioning of the equipment.

Additionally, it has been established that the cyclone system will be driven by a blower that generates centrifugal movement through the use of blades. These blades, powered by the motor's pressure, enable a wide operating range and low energy consumption. This helps reduce the impact of powder coating on both operators and the environment.

The factorial analysis allowed us to determine the Anova graph, which shows a significant distance between the selected equipment, with a minimum distance below 0.05% of the P factor for the cyclone-type equipment selection. This achieves an efficiency and reliability of 95%, which is considered an optimal range in the technical and scientific field for selecting reliable equipment.

5 Conclusions

The present research contributed to the study of air quality through the implementation of a cyclone type extraction and collection system, analyzed based on different acceptance criteria or variables that meet the requirements of design, manufacturing, operation, and maintenance. This system has led to a reduction in ergonomic risks and environmental mitigation. The analysis of variables was performed using the TOPSIS multi-criteria selection method, which yielded results indicating that the cyclone-type system implemented in the electrostatic painting application process is ideal for separating particulate matter, which is collected in a container connected to the collection system, while clean air is recirculated into the environment without contaminants.

The design of the powder particle extraction and recovery system ensured high performance with a 95% efficiency in powder paint recovery, as determined through factorial analysis. Key parameters in the construction process were considered, such as the diameter of the cyclone, cone height, and the size of the air inlet and outlet. The latter is closely related to the airflow rate, density, and particle size. These parameters were carefully evaluated to optimize the system's performance and ensure effective recovery of the powder paint.

The extraction system meets the optimal specifications thanks to the blade design, allowing for the collection of 98% of the suspended powder paint particles in the air during the application in the paint booth. This ensures operator safety and environmental protection, as well as enabling the reuse of the collected paint for coating new metal elements.

It is recommended for future research to conduct a study that analyzes the climatic conditions according to the equipment's location, as they influence the air quality in industrial extraction and collection systems. Understanding the impact of factors such as weather, temperature, humidity, and air composition can provide valuable insights into the performance and efficiency of these systems under different environmental conditions. This information can be used to optimize the design and operation of industrial extraction and collection systems, leading to improved air quality and better environmental outcomes.

References

1. Arredondo, K., Muñoz, J., Carrillo, T.: Proyecto de mejora en los procesos de curado y pintura en una empresa dc productos eléctricos. Rev. Aristas Investig. Básica y Apl. **6**(12), 239–243 (2019)
2. Cabina, T.: Nuevas cabinas de aspiración de pintura en polvo evolucionadas - Tecnicabina S.L. https://tecnicabina.com/nueva-cabina-aspiracion-pintura-polvo/. Accessed 20 June 2021
3. Company Spectris, "Contaminación por partículas y Seguridad Ambiental". https://mx.omega.com/technical-learning/contaminacion-por-particulas-y-seguridad-ambiental.html. Accessed 24 Aug 2021
4. Dalmar protecciones y "Cabinas de pintura en polvo". http://blog.proteccionesypinturas.com/cabinas-de-pintura-en-polvo/. Accessed 28 Sep 2021
5. "Parlamento Europeo P9_TA(2023)0318 Calidad del aire ambiente y una atmósfera más limpia en Europa" (2020)
6. El sitio de la pintura en polvo, "Recuperación de Pintura en Polvo en la Fabricación". http://pinturaenpolvo.org/recuperacion-de-pintura-en-polvo. Accessed 20 Jan 2022
7. He, H., Lou, J., Li, Y., Zhang, H., Wei, X.S.: Effects of oxygen contents on sintering mechanism and sintering-neck growth behaviour of FeCr powder. Powder Technol. **329**, 12–18 (2020). https://doi.org/10.1016/J.POWTEC.2020.01.036
8. Industria elaboradora de pinturas, "Guía para el control y prevención de la contaminación industrial". http://www.ingenieroambiental.com/4014/pinturas.pdf. Accessed 05 July 2020
9. IPM Integraciones y proyectos metálicos S.A. "Cabinas para aplicar la pintura electrostática Metalmecánica, Proyectos Metálicos y Pintura Electrostática". http://ipmsadecv.com/cabinas-para-aplicar-la-pintura-electrostatica. Accessed 20 Jan 2021
10. Kleine Deters, J., Zalakeviciute, R., Rybarczyk, Y.: Modeling PM $_{2.5}$ urban pollution using machine learning and selected meteorological parameters. J. Electr. Comput. Eng. **2020**, 1–14 (2020). https://doi.org/10.1155/2017/5106045
11. Li, Y., Zhang, W., Wang, Y.: Effect of spray powder particle size on the bionic hydrophobic structures and corrosion performance of Fe-based amorphous metallic coatings. Surf. Coat. Technol. **437**, 128377 (2022). https://doi.org/10.1016/j.surfcoat.2022.128377
12. Ministerio para la Transición Ecológica, "Partículas en suspensión". https://www.miteco.gob.es/es/calidad-y-evaluacion-ambiental/temas/atmosfera-y-calidad-del-aire/emisiones/prob-amb/particulas.aspx. Accessed 23 Feb 2020

13. Norma Española, "UNE-EN 50050-1: Equipo manual de pulverización electrostática" (2020)
14. Raoelison, R., Sapanathan, T., Langlade, C.: Cold gas dynamic spray technology: A comprehensive review of processing conditions for various technological developments till to date. Addit. Manuf. **19**, 134–159. https://doi.org/10.1016/J.ADDMA.2020.07.001
15. Rodríguez Guerra, A., Cuvi, N.: Air pollution and environmental justice in Quito, Ecuador. Fronteiras **8**(3), 13–46 (2019). https://doi.org/10.21664/2238-8869.2019v8i3.p13-46
16. Rojas, J.: Reciclado y recuperación de pintura en polvo, Dianetl. https://dialnet.unirioja.es/servlet/articulo?codigo=6733852. Accessed 13 Sep 2020
17. Secretaria del Ambiente, "Norma Ecuatoriana de la Calidad del Aire (2021). http://www.quitoambiente.gob.ec/ambiente/index.php/norma-ecuatoriana-de-la-calidad-del-aire
18. Tapia Núñez, L.: Norma De Calidad Del Aire Ambiente o Nivel De Inmisión Libro Vi Anexo, no. 4 (2019)
19. Valencia, A., Suárez Castaño, R., Sánchez, A., Cardozo, M.B., Buitrago, C.: Gestión de la contaminación ambiental: cuestión de corresponsabilidad Management of Environmental Pollution: a matter of co-responsibility (2019)

Climatic Pattern Analysis Using Neural Networks in Smart Agriculture to Maximize Irrigation Efficiency in Grass Crops in Rural Areas of Ecuador

Evelyn Bonilla-Fonte, Edgar Maya-Olalla, Alejandra Pinto-Erazo, Ana Umaquinga-Criollo, Daniel Jaramillo-Vinueza, and Luis Suárez-Zambrano(✉)

Universidad Técnica del Norte, Avenida 17 de Julio 5-21 y General María Córdoba, 100105 Ibarra, Ecuador
{eamaya,lesuarez}@utn.edu.ec

Abstract. This project describes the development of a system that allows to improve the efficiency of a sprinkler irrigation system based on sensor networks and artificial neural networks. The WSN network is composed of a sensor node, a central node, and an actuator node. The sensor node consists of soil moisture, UV radiation, temperature, and rain sensors. The data collected by the sensor node is sent to the central node or Gateway through LoRa communication where the neural network is trained. The neural network makes predictions of irrigation status it predicts when the crop needs irrigation. Subsequently, the actuator node executes order sent from central node; activates or deactivates the solenoid valve allowing or denying the water flow to the sprinklers. A comparison between the manual irrigation system and the automated system with the neural network, where it is verified that automated system reduces water consumption and contributes to the development grass crop.

Keywords: LoRaWAN · smart irrigation · neural network · precision agriculture

1 Introduction

In [1], a drip irrigation system specifically designed for vegetable crops has been developed, utilizing fuzzy logic as its foundation. This system comprises a Wireless Sensor Network (WSN) that wirelessly collects and transmits data to a control station through Zigbee technology. At the control station, the data is analyzed, and the solenoid valve's opening time for crop irrigation is determined. Parameters such as irrigation durations and solenoid valve operation times are managed. Although this system automates part of the process, it still relies on operator intervention and does not harness IoT-oriented communication technologies.

M. Z. Vizuete et al. (Eds.): CI3 2023, LNNS 1040, pp. 281–298, 2024.
https://doi.org/10.1007/978-3-031-63434-5_21

On the other hand, in [2], a drip irrigation control system applying machine learning techniques to precision agriculture in alfalfa crops is proposed. Data is collected from a Sensor Network, as described in [3], consisting of input parameters processed in an artificial neural network that decides whether irrigation is necessary. Wireless communication is achieved through LoRaWAN modules. However, this system lacks an IoT platform enabling real-time sensor data visualization and access from any location and moment.

Considering the aforementioned projects focused on addressing drip irrigation challenges in greenhouses for various crops, we propose the design of an intelligent irrigation system tailored to open-field agriculture, with a focus on grass cultivation using the sprinkler irrigation method. This system acquires data from a Wireless Sensor Network used as input variables for an Artificial Neural Network that establishes behavioral rules. This enables the system to make predictions and determine whether the crop requires irrigation or not. The algorithm's resulting decision is wirelessly transmitted to an actuator node responsible for activating a solenoid valve controlling water flow in the grass crop via sprinklers. The process continues as described below.

Starting from the design, programming, implementation, and testing of a LoRa-based Wireless Sensor Network (WSN) [4] for a smart irrigation system in a grass cultivation area in the rural zone of the Tocachi parish, Pedro Moncayo canton in Pichincha province, real-time sensor data is collected. This data is visualized and analyzed through an IoT platform. In addition, key parameters of the wireless link such as path loss, receiver power, signal-to-noise ratio, link budget, and Fresnel zone are analyzed. During this phase, artificial intelligence techniques and neural net-works are applied to improve the efficiency of the sprinkler irrigation system in the grass cultivation. Behavior rules are established, and predictions are made to determine the appropriate time for irrigation. The commands are transmitted through the LoRa wireless link to an actuator node with a solenoid valve. Comparative tests are also conducted between the grass cultivation using the automated irrigation system and the cultivation that continues to use manual irrigation. Aspects such as grass growth, leaf coloration, plant height, and root size are evaluated. The document is structured into sections covering the materials and methods used, hardware and software design, results and discussion obtained through the neural network, as well as conclusions and recommendations.

2 Methods

The proposed topology consists of three nodes: Sensor Node, Central Node (Gateway), and Actuator Node [5], as shown in Fig. 1.

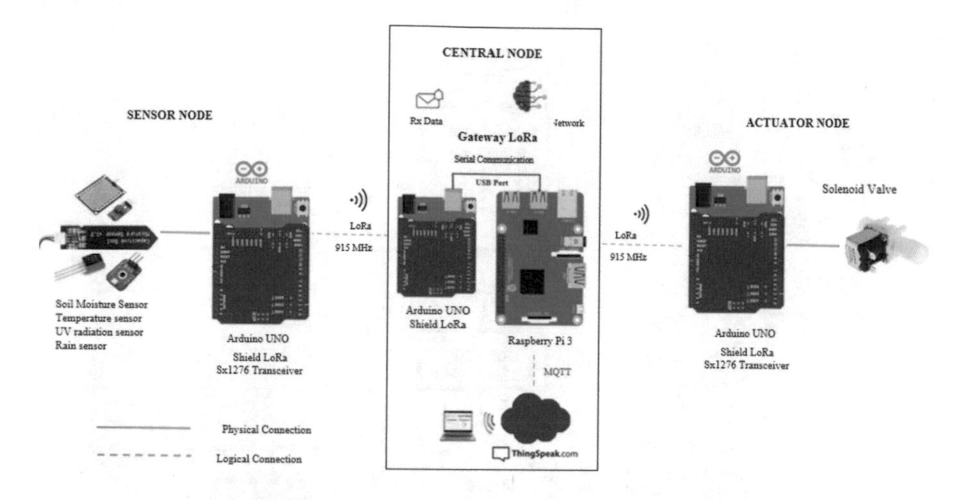

Fig. 1. General System Architecture

2.1 Current Situation

In the first phase, a star-type communication [6] was established using LoRa technology [7] between the Sensor and Actuator nodes towards the Gateway, located in the central node. The data collected by the sensors is transmitted to the central node through the MQTT protocol [8] and displayed on an IoT platform. In the second phase, the neural network is trained to predict irrigation in the grass cultivation. The central node is responsible for sending the command to the actuator node to activate or deactivate the solenoid valve. The execution of the neural network takes place on a Raspberry Pi 3 using Python programming, selected for its low computational load. The actuator node consists of an Arduino UNO and a LoRa transceiver. A 4-channel Relay Shield module is also used to control the opening or closing of the solenoid valve. Once the wireless connection is established, the sensor node's variable data is stored in the Gateway. However, these data cannot be transformed into useful information for the automatic sprinkler irrigation system without proper processing.

2.2 System Design

The soil moisture, temperature, UV radiation, and rain sensors collect data through the sensor node, which is used as a training set for an artificial neural network that makes predictions on when the crop needs to be irrigated and sends a command to the central node to activate or stop irrigation through the solenoid valve. To provide a visual understanding of how the neural network operates in the system, Fig. 2 is shown.

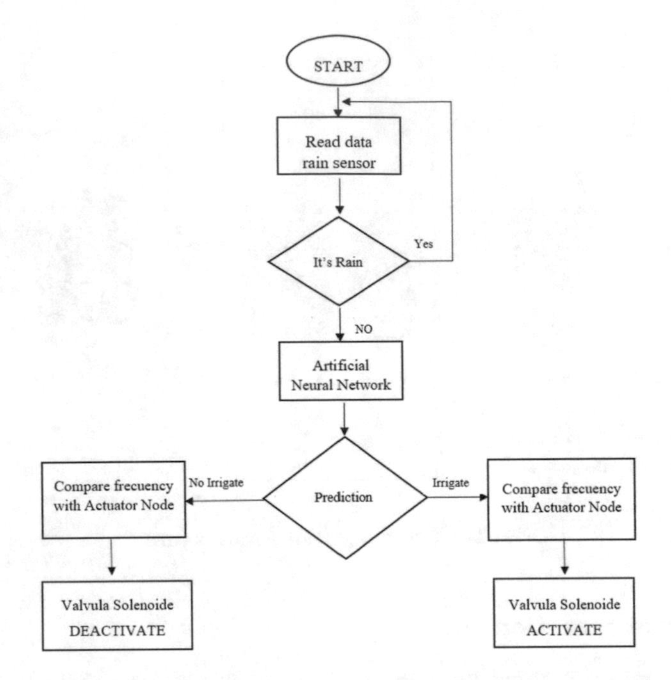

Fig. 2. Flowchart of the neural network application

2.3 Power Supply

The system is powered with 3–5 volts at the sensor node, as well as 3.7–12 V at the actuator node, since each node consumes 173.36 mA and 323.3 mA, respectively, in normal state. The actuator node, like the sensor node, has a lithium battery charging system through a solar panel. The power board includes a TP4056 module for charging the lithium battery and a DC-DC boost converter MT3608 that converts the 3.7V input voltage to 12V to power the solenoid valve.

2.4 System Encoding

The actuator node uses the same LoRa communication parameters as the central node. The training of the neural network is conducted in five stages.

The stages are developed as follows:

Data Collection. It involves gathering data acquired by the sensors in the WSN network, which serves as input for the neural network learning process.

Data Processing. This process involves managing the data obtained from variables such as soil moisture and UV radiation. A total of 8822 data points were obtained, and ranges were assigned to determine a decision based on these two variables: soil moisture and UV radiation.

The node issues a command to not irrigate if the soil is moist or very moist, regardless of the UV radiation. On the other hand, it is decided to irrigate when the soil is dry or very dry, if the UV radiation is low, moderate, or high. If the UV radiation is high, the decision is not to irrigate. Figure 3 presents four ranges for the soil condition and five ranges for UV radiation. The decision to irrigate (1) or not irrigate (0) the grass crop is based on the soil condition and UV radiation.

DATASET

SOIL MOISTURE SENSOR		
Wet	200-298	1
Moist	299-349	2
Dry	350-449	3
Very dry	450-550	4

UV RADIATION SENSOR		
Low	(1-2)	1
Moderate	(3-4-5)	2
High	(6-7)	3
Very high	(8-9-10)	4
Extremely high	11	5

Decision	
Irrigate	1
No Irrigate	0

Soil Moisture			Uv Radiation			Decision	
Very dry	491	4	Moderate	3	2	Irrigate	1
Very dry	489	4	Moderate	3	2	Irrigate	1
Very dry	489	4	Moderate	3	2	Irrigate	1
Very dry	489	4	Moderate	3	2	Irrigate	1
Very dry	483	4	Moderate	3	2	Irrigate	1
Very dry	484	4	Moderate	3	2	Irrigate	1
Very dry	482	4	Moderate	3	2	Irrigate	1
Very dry	481	4	Low	2	1	Irrigate	1

dataset Hoja3 train train_2 Hoja1 Hoja2

Fig. 3. Treatment of soil moisture and UV radiation data

Creation of the Neural Network. The structure of the neural network is defined with input, output, and hidden layers. The number of neurons in each layer is determined. The input layer receives data from the environmental variables, the output layer provides the response of the neural network, and the hidden layer captures specific features of the environment being modeled [9]. In the input layer, two neurons are used for soil moisture and UV radiation variables, while in the output layer, one neuron is used for the irrigation decision (to irrigate or not to irrigate). The number of neurons in the hidden layer is not fixed but can be calculated using the geometric pyramid rule. For a three-layer network with a single hidden layer, Eq. 1 [9] is used.

$$h = \sqrt{m \cdot n} \tag{1}$$

where h is the initial number of neurons in the hidden layer, m is the number of neurons in the output layer, and n is the number of neurons in the input layer, it is obtained that the number of neurons in the hidden layer is 1.4 neurons. However, it would not be feasible to have enough data for prediction. Therefore, the number of neurons in a network with two hidden layers, h1 for the first hidden layer and h2 for the second hidden layer, is determined following Eq. 2 and Eq. 3 [10].

$$h_1 = m \cdot r^2 \tag{2}$$

$$h_2 = m \cdot r \tag{3}$$

where:

$$r = 3\sqrt{\frac{n}{m}} = 3\sqrt{\frac{2}{1}} = 4.24 \tag{4}$$

Replacing the values in Eqs. 2 and 3, we find that the number of neurons in the first hidden layer is 18, and in the second hidden layer it is 4,24.

$$h_1 = 1 \cdot (3\sqrt{2})^2 = 18 \tag{5}$$

$$h_2 = 1 \cdot 3\sqrt{2} = 4.24 \tag{6}$$

To obtain the optimal number of neurons, it is necessary to train and evaluate the network with different numbers of neurons for the hidden layers until achieving satisfactory performance. Therefore, the flow described in Fig. 4 should be followed.

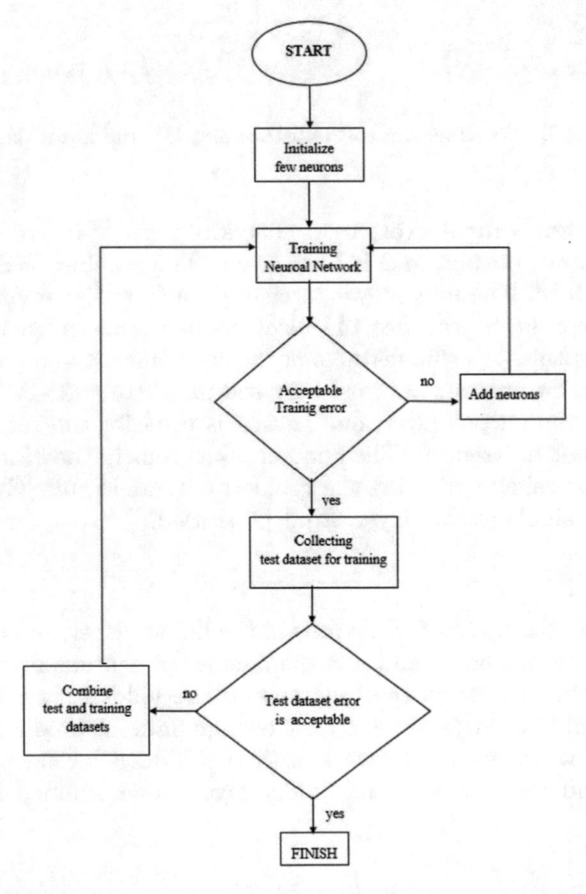

Fig. 4. Flowchart to determine the number of neurons and hidden layers. Adapted from "Design of Multilayer Neural Networks". [11]

Training the Neural Network. The development environment is set up using the Anaconda Navigator interface, which provides access to Jupiter Notebook as a web application. This allows for importing the necessary libraries to execute the neural network training code. The training dataset with 8822 rows and three columns: humidity, radiation, and decision. The data is graphically visualized, where the X-axis represents soil moisture, and the Y-axis represents UV radiation. In the graph, the gray area indicates the irrigation decision made by the system, while the red area indicates that the system should not irrigate (Fig. 5).

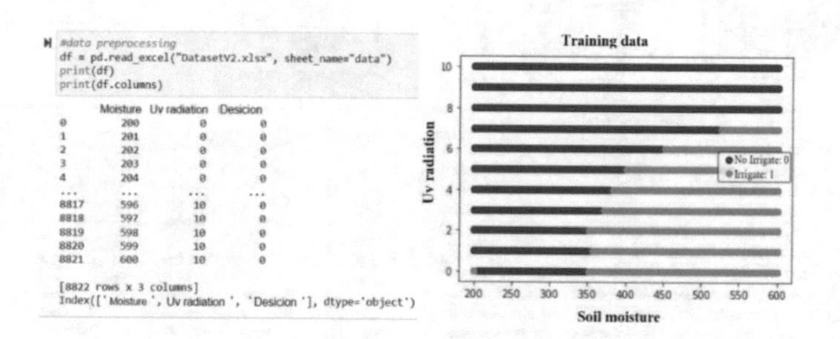

Fig. 5. Data preprocessing and graphical representation of trained data.

The first model initially consists of an input layer, two hidden layers, and an output layer, with eighteen neurons in the first hidden layer and four neurons in the second hidden layer [12]. However, this model fails to achieve the desired prediction and does not fit the training data properly. Therefore, in the second model, it is decided to increase the number of neurons in the hidden layers to twenty neurons. Although this model shows positive results, it does not fulfill the prediction for values according to the training data. Therefore, it is decided to increase the number of neurons in each hidden layer to one hundred neurons, according to the third training of the neural network, which allows for more accurate predictions with respect to the dataset. Hence, this model is selected as the most suitable one, See Fig. 6, Fig. 7, Fig. 8.

Activation Functions: These are used to establish the relationship between nodes when moving from one layer to another. In the hidden layers, the ReLU function is used to transform the input values by nullifying the negatives and preserving the positives. On the other hand, in the output layer, the sigmoid function is used, which produces an output in the range of 0 to 1, suitable for binary classification [13].

Learning Algorithm: The backpropagation algorithm is used, which allows the transmission of an input pattern through the different layers of the neural network to obtain an output. Then, a training stage is conducted where the network weights are adjusted to improve its performance.

Hyperparameters: They allow controlling the training process of this model, which can be divided into two phases: compilation and fitting.

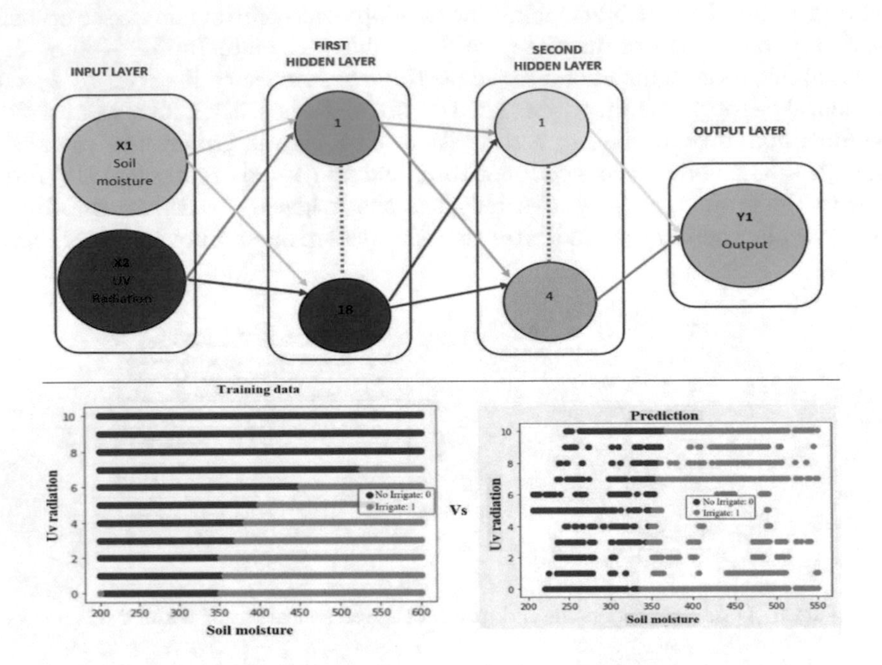

Fig. 6. Training result with the first model.

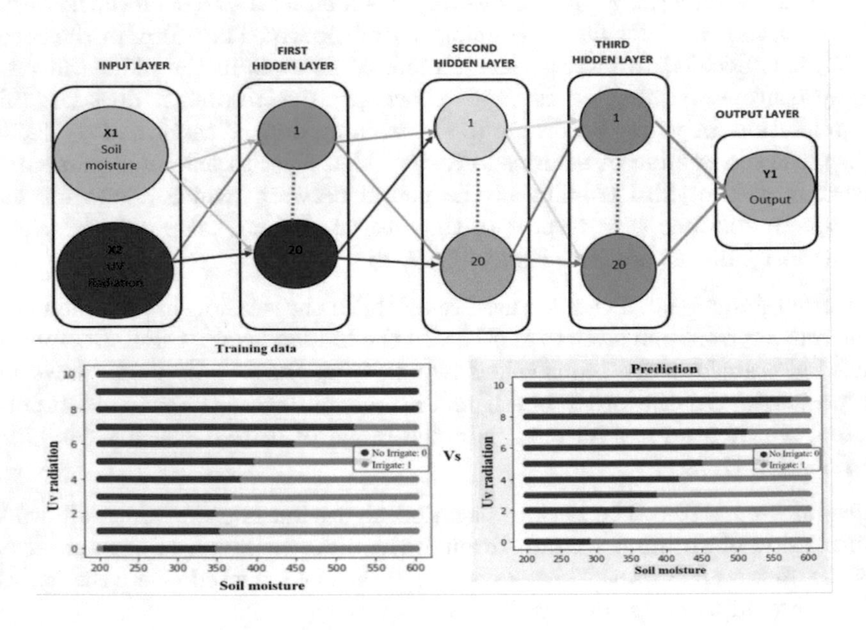

Fig. 7. Training Result with the second model.

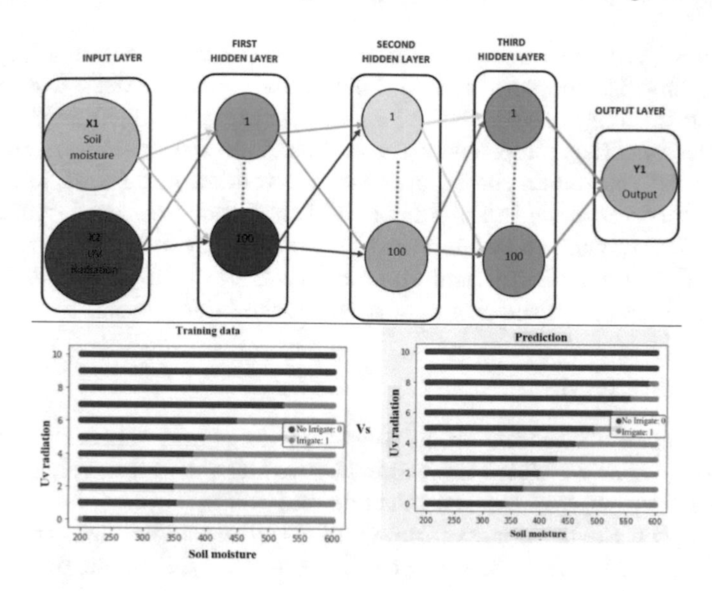

Fig. 8. Training Result with the third model

Compilation: The optimizer, loss function, and list of metrics are defined.

- Optimizer: In this model, the Adam optimizer is used, which adapts the learning rates individually for each parameter. A higher value of this parameter will require more epochs (training iterations) to make subtle changes to the weights in each update.
- Loss function: In this case, the Binary Cross-Entropy loss function is used, which evaluates how accurate the model's predictions are. This function compares the output of the sigmoid function (sigmoid activation function) with the two possible categories: 1 (irrigate) and 0 (do not irrigate).
- List of metrics: Accuracy represents the fraction of correct predictions made by the model, while loss indicates how incorrect the model's predictions have been. If the prediction is perfect, the loss will be zero. These aspects help measure the performance and quality of the model during its training and evaluation.

Fitting: These include the inputs (x) and outputs (y), representing the soil moisture, UV radiation data, and the irrigation results, respectively. The batch size is also included, which defines the number of data points used in each update of the internal parameters. Additionally, the number of epochs is specified, indicating how many times the entire dataset is worked with. Lastly, the validation split is mentioned, determining the percentage of in-put data used exclusively for evaluating the model during training. In this study, it corresponds to 0.3 of the totals provided data.

Testing and Validation of the Neural Network: There are a lot of record of training metrics for each epoch. This includes the model's loss and accuracy during training, as well as the loss and accuracy of the validation dataset. The metric "accuracy" represents the accuracy of the training dataset, while "val_accuracy" represents the accuracy of the validation dataset. On the other hand, "loss" corresponds to the loss of the training dataset, and "val_loss" represents the loss of the validation dataset. The accuracy in both the training and validation datasets indicates that the model has been fully trained, as there is no observed increase in the accuracy trend in the later epochs.

2.5 Actuator Node

The central node utilizes the predictions from the neural network to control the solenoid valve that regulates the water flow to the sprinklers in the grass crop. Communication between the central node and the actuator node is established using LoRa, with a transmission power of 20 dBm, spreading factor SF7 [14], bandwidth of 125 kHz, and error coding rate of 4/5. These parameters are observed in Fig. 9.

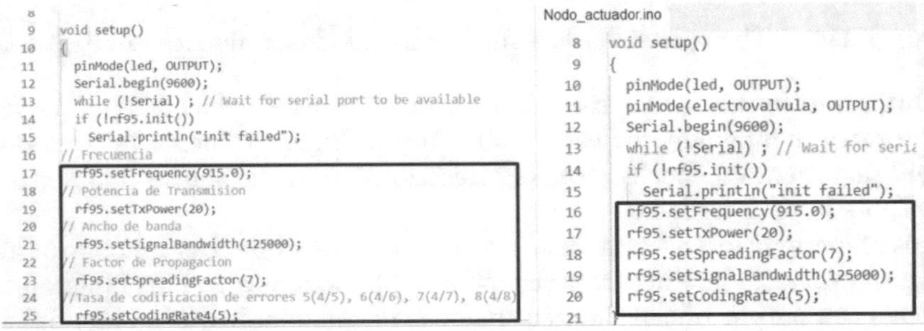

```
8
9   void setup()
10  {
11    pinMode(led, OUTPUT);
12    Serial.begin(9600);
13    while (!Serial) ; // Wait for serial port to be available
14    if (!rf95.init())
15      Serial.println("init failed");
16  // Frecuencia
17    rf95.setFrequency(915.0);
18  // Potencia de Transmision
19    rf95.setTxPower(20);
20  // Ancho de banda
21    rf95.setSignalBandwidth(125000);
22  // Factor de Propagacion
23    rf95.setSpreadingFactor(7);
24  //Tasa de codificacion de errores 5(4/5), 6(4/6), 7(4/7), 8(4/8)
25    rf95.setCodingRate4(5);
```

```
Nodo_actuador.ino
8   void setup()
9   {
10    pinMode(led, OUTPUT);
11    pinMode(electrovalvula, OUTPUT);
12    Serial.begin(9600);
13    while (!Serial) ; // Wait for seri
14    if (!rf95.init())
15      Serial.println("init failed");
16    rf95.setFrequency(915.0);
17    rf95.setTxPower(20);
18    rf95.setSpreadingFactor(7);
19    rf95.setSignalBandwidth(125000);
20    rf95.setCodingRate4(5);
21  }
```

Fig. 9. Basic configurations of the central node and actuator node.

3 Results and Discussion

After designing the sensor network, an electrical test of the system was realized. During this test, sensor information was collected and wirelessly received at the central node using LoRa technology. Additionally, the behavior prediction of the neural network and actuator node was performed.

3.1 Electrical Test

Figure 10 shows the verification of the activation of each device in the system through LED indicators. Functionality tests were performed on the power board, first without the solar panel connection, and then with the solar panel connected.

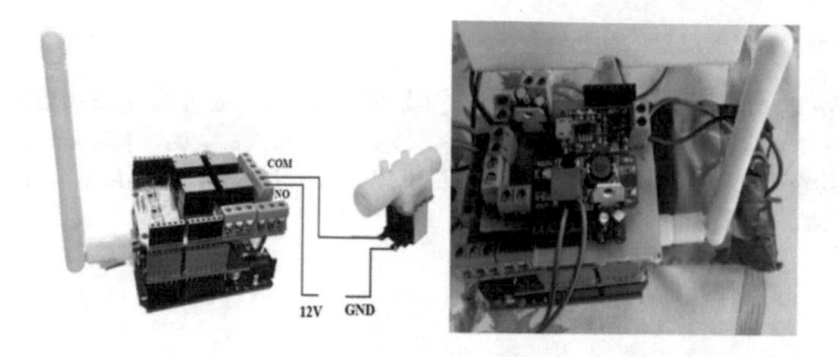

Fig. 10. Actuator node powered by a lithium battery.

3.2 Neural Network Prediction

The environmental variable data captured by the sensors and the predictions from the neural network can be visualized on the central node using the Tonny Python integrated development environment shell. For the neural network training model to run, the rainfall sensor value needs to be above 500, indicating the absence of rain.

In Fig. 11, the following details can be observed: in the first section, the rainfall sensor value is 1023, which activates the model execution. The UV radiation sensor shows a value of 0 (indicating low UV radiation), and the soil moisture sensor exceeds 450 units, indicating that it is very dry. As a result, the neural network prediction is "Irrigate" with a value of 1.0. On the other hand, in the second section, the soil moisture sensor value is below 350, indicating that the soil is moist. In this case, the neural network prediction is "Do not irrigate".

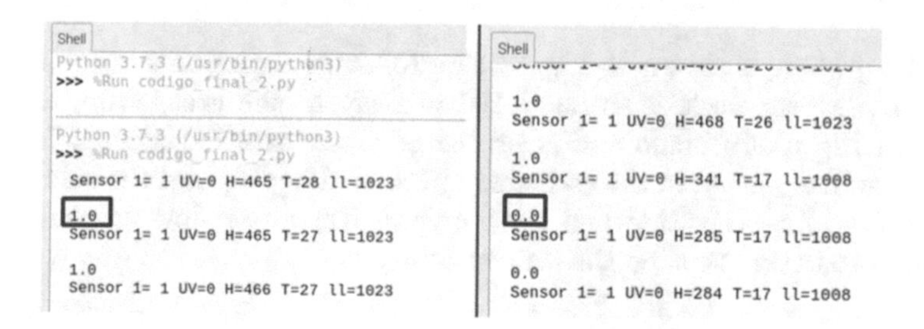

Fig. 11. Neural Network Prediction. Case 1 - Irrigate. Case 2 - Don't irrigate.

3.3 Actuator Node Behavior

The system communicates wirelessly via LoRa from the central node to the actuator node, which is located 160 m away. The system is protected by a sealed casing, shielding the devices from external agents such as water, heat, and humidity, as shown in Fig. 12.

Fig. 12. Sensor, central, and actuator nodes that make up the wireless sensor network (WSN).

In the prototype shown in Fig. 13, the following components are located as follows: the sensor node is situated 160 m away in the grass crop, the central node is in the user's home and is protected from external agents. Lastly, the actuator node is placed in the secondary pipe of the plot, and the solenoid valve is responsible for activating and deactivating the water flow according to the irrigation prediction sent by the central node.

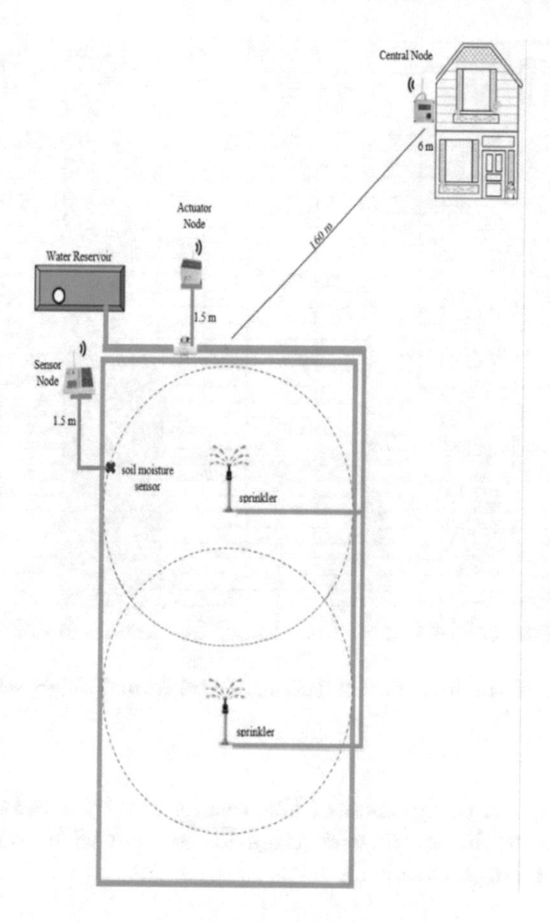

Fig. 13. Location of the nodes in the plot.

3.4 Water Consumption Analysis

The system undergoes testing on a 1000 m^2 plot equipped with two rotating sprinklers. In manual irrigation mode, the sprinklers operate continuously for 3 to 5 h, assuming full irrigation. With the automated system, the time and water consumption are calculated from the moment the neural network predicts the need for irrigation until it stops when the soil reaches the desired moisture level.

In Fig. 14, the system activates irrigation (1) at 18:08 and stops it (0) at 18:46, with an irrigation time of less than one hour on a completely dry grass plot.

Date	Soil moisture	Temperature	UV Radiation	Irrigation Prediction	Rain
2022-10-05T18:08:56	415	21	0	1	0
2022-10-05T18:10:23	416	21	0	1	0
2022-10-05T18:10:48	415	22	0	1	0
2022-10-05T18:11:07	415	23	0	1	0
2022-10-05T18:11:27	414	22	0	1	0
2022-10-05T18:11:48	415	21	0	1	0
2022-10-05T18:12:14	440	22	0	1	0
2022-10-05T18:12:34	453	21	0	1	0
2022-10-05T18:13:08	480	23	0	1	0
2022-10-05T18:15:12	482	22	0	1	0
2022-10-05T18:16:33	483	22	0	1	0
2022-10-05T18:17:59	484	56	0	1	0
2022-10-05T18:19:25	484	21	0	1	0
2022-10-05T18:20:45	480	21	0	1	0
2022-10-05T18:22:12	479	21	0	1	0
2022-10-05T18:25:38	477	21	0	1	0
2022-10-05T18:26:57	451	22	0	1	0
2022-10-05T18:27:17	472	22	0	1	0
2022-10-05T18:29:37	463	21	0	1	0
2022-10-05T18:32:58	439	21	0	1	0
2022-10-05T18:33:19	430	21	0	1	0
2022-10-05T18:33:43	450	22	0	1	0
2022-10-05T18:34:12	467	15	0	1	0
2022-10-05T18:34:32	462	15	0	1	0
2022-10-05T18:35:00	451	15	0	1	0
2022-10-05T18:35:24	440	15	0	1	0
2022-10-05T18:35:51	410	15	0	1	0
2022-10-05T18:36:14	405	15	0	1	0
2022-10-05T18:37:37	400	15	0	1	0
2022-10-05T18:38:04	398	16	0	1	0
2022-10-05T18:39:28	398	21	0	1	0
2022-10-05T18:40:48	395	22	0	1	0
2022-10-05T18:41:23	391	21	0	1	0
2022-10-05T18:43:27	391	18	0	1	0
2022-10-05T18:44:14	385	16	0	1	0
2022-10-05T18:45:34	364	15	0	1	0
2022-10-05T18:46:28	341	16	0	0	0

Fig. 14. Database of the irrigation status, retrieved from ThingSpeak. 1 = Irrigate, 2 = No irrigate

Table 2 presents a comparison of the results obtained between the manual irrigation system vs the automated irrigation system with the neural network over a period of 15 days (Table 1).

3.5 Grass Crop Analysis

An evaluation of the grass crop's development is conducted after the defoliation stage when implementing the automated irrigation system with the neural network. Over a 15-day period, the growth and leaf coloration are observed as the plant recovers its leaf area. Figure 15 presents three images: the first shows the grass during the defoliation stage, the second shows the crop using the automated irrigation system, and the third shows the grass crop using manual irrigation.

The results are remarkable and evident. The automated irrigation system has successfully provided appropriate water dosing, reflected in the emergence of new leaves with elongated growth and vibrant green color. Additionally, it has generated uniform regrowth in the grass crop. In Fig. 16, it can be observed that in the automated irrigation system, the roots are thin, elongated, and abundant, unlike the manual irrigation system where the roots are smaller, thicker, and have not achieved sufficient depth.

Table 1. Water consumption using the manual irrigation system vs the automated irrigation system.

Day	MANUAL IRRIGATION SYSTEM		AUTOMATED IRRIGATION SYSTEM	
	time (hours)	Consumption (m3)	time (hours)	Consumption (m3)
1	5:00	5.4	1:30	1.62
2	3:00	3.24	1:36	1.72
3	0:00	0	0:55	0.99
4	5:00	5.4	1:35	1.7
5	3:00	3.24	1:36	1.72
6	3:00	3.24	1:35	1.7
7	0:00	0	2:11	2.35
8	3:00	3.24	0:00	0
9	5:00	5.4	1:50	1.98
10	0:00	0	2:28	2.66
11	5:00	5.4	0:00	0
12	3:00	3.24	2:04	2.23
13	3:00	3.24	2:17	2.46
14	0:00	0	0:00	0
15	5:00	5.4	1:42	1.83
	TOTAL: 46.44		TOTAL: 22.03	

Fig. 15. a. Defoliation, b. Grass with automated irrigation, c. Grass with manual irrigation.

Fig. 16. Root size with manual irrigation and automated irrigation.

4 Conclusions

- The electrical test demonstrated that the power supply board continuously provides power to the actuator node for approximately 22 h.
- The TP5046 charging module features a micro-USB port to recharge the system's battery with 5 V, if necessary.
- The neural network's response to environmental variables is optimal, activating when the rain sensor indicates the absence of rain, which prevents unnecessary activation of the irrigation system. Additionally, the accuracy of the predictions made by the neural network has been verified.
- The communication between the central node and the actuator node is successful, as they are located at the same distance and use the same LoRa communication parameters, ensuring efficient transmission and reception of information.
- The behavior of the actuator node is correct, activating the solenoid valves if the neural network prediction is "Water" and properly deactivating them if the prediction is "No Water."
- An automated sprinkler irrigation system has been successfully implemented using artificial neural networks to predict irrigation in grass crops.
- The neural network model used consists of an input layer with two neurons, three hidden layers with one hundred neurons each, and an output layer. After experimentation, it was determined that 5000 training epochs are optimal for effective learning without overfitting.
- According to the predictions of the neural networks, it is observed that the sprinkler irrigation system irrigates the crop during dawn and dusk, which promotes water absorption and soil filtration by reducing evaporation compared to direct sun exposure.
- It has been found that the manual sprinkler irrigation system consumes approximately 46.44 mm^3 of water, while the automated system with neural networks consumes around 22.03 mm^3, representing a difference of 15.89 mm^3. These results demonstrate that the automated system significantly improves the efficiency of sprinkler irrigation.
- While the growth and development of a grass crop depend on factors such as soil and nutrients, proper water distribution in the crop contributes positively to its development.

5 Recommendations

- It is crucial to ensure that the gateway has a stable internet connection to achieve efficient packet routing and enable data visualization in the cloud.
- To enhance the precision of irrigation prediction, it is advisable to incorporate supplementary variables, such as wind speed and the occurrence of mist/fog, given their significance in a sprinkler irrigation system. By considering these factors, the accuracy of the irrigation predictions can be improved, resulting in more efficient water distribution and optimized irrigation practices.

- To ensure optimal performance, it is recommended to perform an initial charge of the sensor and actuator nodes' lithium batteries through the micro-USB port. This will help maximize their battery life and overall functionality. By starting with a full charge, the nodes will be better equipped to carry out their tasks effectively and reliably in the automated irrigation system.
- It is important to ensure a clear line of sight between the end node and the gateway, as the presence of obstacles can weaken the transmitted signal. This may require adjustments in the network design during the development process.
- Careful selection of data used to train the neural network is crucial, as the network will learn from this data. The quality and representativeness of the data will influence the network's ability to make accurate predictions.
- Determining the optimal number of layers and neurons in the hidden layer of a neural network can be challenging. It is recommended to perform testing and adjustments to find the appropriate configuration, as this process may require a trial-and-error approach.
- Like the layer and neuron configuration, finding the right number of training epochs or iterations also involves a trial-and-error process. An optimal number of iterations must be found that avoids both overfitting and underfitting of the network.

References

1. Leandro, S.: Diseño de un sistema de riego inteligente para cultivos de hortaliza basado en Fuzzy Logic en la granja La Pradera de la Universidad Técnica del Norte, Universidad Técnica del Norte, 2019
2. Dario, C.: Diseño de un sistema para el control de riego mediante técnicas de aprendizaje automático aplicadas a la agricultura de precisión en la granja La Pradera de la Universidad Técnica del Norte, Universidad Técnica del Norte, 2020
3. Dominguez, A.: Diseño de una red de sensores inalámbricos LPWAN para el monitoreo de cultivos y materia orgánica en la granja experimental La Pradera de la Universidad Técnica del Norte, Universidad Técnica del Norte, 2020
4. Santamaría, F., Archila, D.: Estado Del Arte De Las Redes De Sensores Inalámbricos. Tecnol. Investig. Acad. **1**(2) (2013)
5. Fernández, R., et al.: Redes inalámbricas de sensores: teoría y aplicación práctica, 2009
6. Nayibe, C.C., Tibaduiza, D., Aparicio, L., Caro, L.: Redes de sensores inalámbricas. J. Pediatr. Adolesc. Gynecol. **24**(6), 404–9 (2011)
7. Semtech, Semtech SX1276 (2020). https://www.semtech.com/products/wireless-rf/lora-transceivers/sx1276
8. Poursafar, N., Alahi, M.E., Mukhopadhyay, S.: Long-range wireless technologies for IoT applications: a review. In: Proceedings of the International Conference on Sensing Technology, ICST, vol. 2017-Decem, pp. 1–6 (2017). https://doi.org/10.1109/ICSensT.2017.8304507
9. Zepeda Hernandez, J.Á., Aguilar Castillejos, A.E., Hernandez Sol, Á., Salgado Gutiérrez, M.C.: Metodología para determinar la topología, patrones de parámetros eléctricos y entrenamiento de una Red Neuronal Artificial para un control inteligente. Rev. Tecnol. Digit. **5**(1), 103–113 (2015)

10. Pano Azucena, A.D.: Realización FPGA de una Red Neuronal Artificial para Reproducir Comportamiento Caótico, p. 97 (2015)
11. ESCOM, Diseño de redes neuronales multicapa (2019). https://es.slideshare.net/mentelibre/diseo-de-redes-neuronales-multicapa-y-entrenamiento. Accessed 14 July 2023
12. ¿Qué es una red neuronal? — TIBCO Software. https://www.tibco.com/es/reference-center/what-is-a-neural-network. Accessed 14 July 2023
13. Acevedo, E., Serna, A., Serna, E.: Principios y características de las redes neuronales artificiales. Desarrollo e innovación en ingeniería, no. March 2019, pp. 173–183 (2017)
14. Solera, E.: Modulación LoRa: long range modulation. Medium 1–10 (2018)

Data Lake Optimization: An Educational Analysis Case

Viviana Cajas-Cajas[1](✉), Diego Riofrío-Luzcando[2], Joe Carrión-Jumbo[1], Diana Martinez-Mosquera[3], and Patricio Morejón-Hidalgo[1]

[1] Universidad Internacional SEK, Quito, Ecuador
{viviana.cajasc,joe.carrion,victor.morejon}@uisek.edu.ec
[2] Universidad San Francisco de Quito, Quito, Ecuador
driofriol@usfq.edu.ec
[3] Escuela Politécnica Nacional, Quito, Ecuador
diana.martinez@epn.edu.ec

Abstract. This study focuses on enhancing the performance of Universidad Internacional SEK's (UISEK) Data Lake by addressing challenges in computing resource consumption from a prior data lake implementation. Notably, data has been sourced from the Canvas Learning Management System based on the university's usage since 2019 for both implementations. The restructuring, carried out through three layers, successfully mitigated previous computing resource challenges.

Following the CRISP-DM framework, the new approach exhibited substantial improvements over the previous version. Results include a 73.1% reduction in the estimated size of the last dump, 51.7% more efficient storage utilization in the user behavior table, and a 4.3% improvement in CPU consumption. Additionally, showcased a 50% reduction in ingestion time.

The study emphasizes the significance of a well-organized Data Lake governance structure for streamlined data management. The presented improvements lay a solid foundation for future analyses and machine learning models. Moreover, the study underscores the role of automated processes in maintaining an updated Data Lake, ensuring its relevance for decision-making at UISEK. Overall, this work contributes to advancing the efficiency and performance of educational Data Lakes, providing valuable insights for decision-making and continuous enhancement of the educational experience.

Keywords: API REST · data lake · Data Science · HDFS · LMS Canvas · optimization

1 Introduction

"Data Lake" is a term coined in 2010 by Dixon (2010), sometimes erroneously associated solely with a marketing label. However, it refers to a massively scalable storage repository containing a large amount of raw data in its native format until further processing systems (engine) are needed, capable of ingesting data without compromising data structure.

M. Z. Vizuete et al. (Eds.): CI3 2023, LNNS 1040, pp. 299–309, 2024.
https://doi.org/10.1007/978-3-031-63434-5_22

Additionally, it can integrate SQL and NoSQL database approaches, online analytical processing (OLAP), and online transaction processing (OLTP) capabilities. For this reason, McClure (2016), notes that yesterday's unified storage is today's enterprise data lake.

In this context, the Data Lake serves as a storage system for managing and analyzing data in line with institutional objectives. It addresses the challenge of heterogeneous data volumes within a business that are complex to process for decision-making.

Processes related to teaching and learning generate a vast amount of data with multiple potential applications. Cantabella et al. (2019) suggest that the current trend is towards harnessing Big Data to understand how university students experience and build knowledge, a valuable insight for universities assessing the impact of the teaching-learning process. Analyzing and drawing conclusions through Big Data enables a more accurate understanding of student usage patterns and preferences for virtual platforms, facilitating improvements in both physical and virtual teaching methodologies in real-time.

This work presents the case of the International SEK University (Universidad Internacional SEK Ecuador, 2023), offering undergraduate and postgraduate programs with a student community of 2500 as of 2023. The current issue is that data within Canvas's Learning Management System (LMS) isn't providing value to the administration because the analysis results are missing. Reasons for this include:

- Data not being processed.
- Many data points being forgotten.
- Loss of origin and purpose of the data.

As a partial solution, a proposal was made resulting in an initial architecture (Beltrán, 2022) outlined in a master's thesis. This involved implementing a Data Lake with Canvas LMS data from UISEK, storing it in a single repository with greater flexibility for subsequent data analysis, machine learning, and other applications. In the previous work, complete data was loaded into a Hadoop Distributed File System (HDFS) (Dwivedi & Dubey, 2014) directory created in a Hadoop environment through Canvas's API with Python scripts. Scripts were also generated to create directories in HDFS, tables in Hive, and automate data ingestion. Finally, a scheduled task was set up for daily execution of the described processes.

However, with daily use of the Data Lake, it was observed that computing resources, particularly RAM consumption, reached high limits. This was primarily attributed to a "requests" table, storing all records of each user, i.e., every click made by each student and teacher on the platform. Memory consumption occurred when updating the corresponding file for this table in HDFS, following the process explained in the aforementioned thesis.

Hence, a proposed architecture was introduced, focusing on restructuring the Data Lake into three layers (LT, STI, and CT). HDFS-created paths begin with /LT to identify the Academic Period directory, grouping unprocessed files for the same period, with data arriving in batches daily. Directories named LMS, SocialNetworks, WebServices, AcademisServices, and Sensors group data by source (Martinez-Mosquera et al., 2022).

This work presents the implementation and results of the proposed architecture outlined in the previously mentioned document.

2 Related Work

According to the findings of Romero & Ventura (2020), el Educational Data Mining (EDM) (Ashish et al., 2017) Educational Data Mining (EDM) (Ashish et al., 2017) emerged from the analysis of student-computer interaction log files that prevailed in the early years of the 21st century. The methods evolved from relationship mining to prediction, with a strong emphasis on learning and teaching, as evident in the work of Zaıane and Luo (Zaiane & Luo, 2001). They emphasized that the goal of EDM is to make learners effective, a focus contrasting with the early use of the term learning analytics (LA) by Mitchell & Costello (2000) to refer to business intelligence in e-learning.

Campbell et al. (2007) state that the growth of learning analytics is driven by the need to use insights gained from data analysis to make interventions aimed at improving learning, addressing factors such as the tension related to the impersonality of online systems, as addressed by Mazza & Dimitrova (2007). This sets the stage for a strong line of research in the field, leading to relevant studies related to the present work on data treatment in education:

Gewerc et al. (2016) conduct a case study of students using a social network as a learning context, employing Unicet and NetDraw as analytical tools with metrics such as network density to gauge collaboration intensity and node centrality. Using ADEGA, they extract information to analyze the content of student blog entries and automatically extract terms or words characterizing each entry. The results show interdependence between student-centered teaching and resulting Learning Analytics (LA) indicators.

Cravero et al. (2021) present an architecture for a Chilean university's Data Lake and metadata modeling, consisting of three data access zones, tested with a prototype implemented on AWS. It provides a metadata catalog for data analysts and third-zone (Gold) data for decision-makers, recording management indicators. The Data Lake manages data from survey files, relational databases, processed spreadsheets, and data stored in Data Warehouses. Additionally, applying ontologies highlights relationships between existing data. Future work aims to implement the Data Lake for all available data on a local server.

Oukhouya et al. (2022) propose a new big data architecture for educational systems covering multiple data sources. This architecture organizes data through layers, from managing different data sources to final consumption. The proposed approach includes the use of a data lake to modernize decision-making processes, particularly data warehouses and OLAP methods. It serves as a means of consolidating data for integrating heterogeneous data sources.

Seeling (2022) notes that despite the existence of various open-source Learning Management System (LMS) solutions allowing open access to data for learning analysis stakeholders, commercially prevalent solutions in the U.S. education system are often closed-code and closed-data. Therefore, this work provides an approach to enable research and implementations of open and open-source data within the confined environments of closed-code and closed-data commercial LMS implementations.

Kustitskaya et al. (2023) address the design of an educational database oriented towards educational management, primarily for student attention and learning analytics tasks. Implementing these principles may require significant changes to existing database infrastructures and training for university staff using database-related information systems.

All the aforementioned works underscore the growing importance of learning analytics for research and implementation in educational systems. However, none of them align precisely with the proposed approach. The following section details the methodology of the present work.

3 Methodology

The improvement of performance in data collection, storage, and querying within the Data Lake of Universidad Internacional SEK was addressed following the CRISP-DM framework (Schröer et al., 2021), providing a structured and systematic methodology. This approach breaks down into six interrelated stages, each crucial for the project's success.

The initial project phase focused on thoroughly understanding fundamental objectives, prioritizing the optimization of performance in data collection and storage within the university's Data Lake. A significant aspect was the creation of an on-premise virtualized environment to ensure a scalable and adaptable infrastructure, capable of meeting the institution's changing needs. Notably, the solution was installed on a cluster using commodity hardware with the following minimum specifications:

- Hypervisor Promox Virtual Environment
- 3 DataNodes
- 24 GB RAM
- 4 processors (2 sockets, 2 cores)
- Storage: 300 GB

Subsequently, a comprehensive analysis of data obtained from the Canvas LMS system was conducted. This analysis included transforming records into fact and dimension tables, following a star schema. This structure allowed for a clearer and deeper understanding of the data, especially in terms of academic performance and student engagement.

Data preparation became an essential phase, encompassing the selection of tables, records, and attributes, as well as data transformation and cleaning for use in modeling tools.

In contrast to focusing on model implementation, the project prioritized improving the performance and infrastructure of the Data Lake, emphasizing efficiency in data management over the creation of analysis models.

Initially, data extraction occurred through a REST API (Sohan et al., 2017) connected to Canvas LMS, downloading the complete dump. However, this approach proved inefficient due to the massive consumption of computational resources when bringing in all data with each dump. The proposed solution involved modifying the API to allow incremental data consumption, significantly reducing computational load and improving storage efficiency.

The API now identifies and transmits only the updated data since the last extraction. This approach minimizes resource usage while ensuring the integrity and constant update of information stored in the data lake.

3.1 Implementation Process

Identification of Changes: The REST API now tracks updates in Canvas LMS since the last synchronization.

Incremental Transmission: Only modified or added data is transmitted to the data lake, avoiding redundancy of previously stored information.

Atomized Update: An atomized update process is implemented to ensure the coherence and consistency of data stored in the lake.

Equation 1 reflects the essence of the incremental approach, reducing computational load and optimizing data storage in the data lake.

$$Dt = Dt - 1 \cap D\Delta t \tag{1}$$

where:

Dt corresponds to the total data at time t,

Dt − 1 is the previously stored data, and

DΔt represents changes in data since the last synchronization.

∩ denotes the set intersection operation, ensuring that Dt contains only the updated or added data since the last extraction.

Figure 1 shows the implemented architecture, where a modified REST API allows incremental data consumption, reducing computational load by identifying and transmitting only changes since the last synchronization. This data is stored in a data lake implemented with Hadoop Distributed File System (HDFS) (Dwivedi & Dubey, 2014), from where it is extracted and loaded into Hive internal tables for processing. Finally, visualization is achieved using any presentation tool.

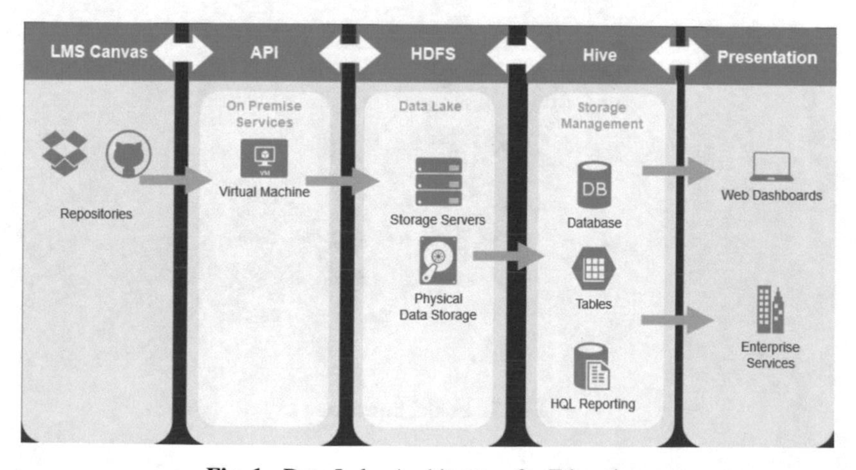

Fig. 1. Data Lake Architecture for Education.

The evaluation was a critical step in the process. Key metrics such as RAM and storage consumption were measured to compare a virtualized environment with the existing system, allowing for the assessment of the impact of improvements.

Finally, the results were deployed to ensure their effective utility. The acquired knowledge was presented clearly and accessibly to UISEK, considering specific needs and client skills to facilitate informed decision-making.

Additionally, an organizational data structure was established within the Data Lake, classifying historical and recent data in an orderly and accessible manner for use and analysis.

Complementary automated processes were implemented to generate the database and for the daily download and upload of the latest data from the Canvas Learning Management System. These automated processes ensure that the Data Lake is always updated and available for decision-making and educational analysis.

The structured approach throughout the different stages of the CRISP-DM framework has generated significant improvements in the infrastructure and performance of UISEK's Data Lake. These improvements ensure greater scalability, flexibility, and efficiency in data management and analysis, providing a valuable tool for decision-making and continuous enhancement of the educational experience.

3.2 Data Lake Governance

Structuring data in the Data Lake of Universidad Internacional SEK has been essential for organizing and classifying information logically and efficiently. A folder hierarchy has been adopted, all generated within the "bin" directory of the Hive installation, to facilitate data management and query. This organizational structure is presented in Fig. 2.

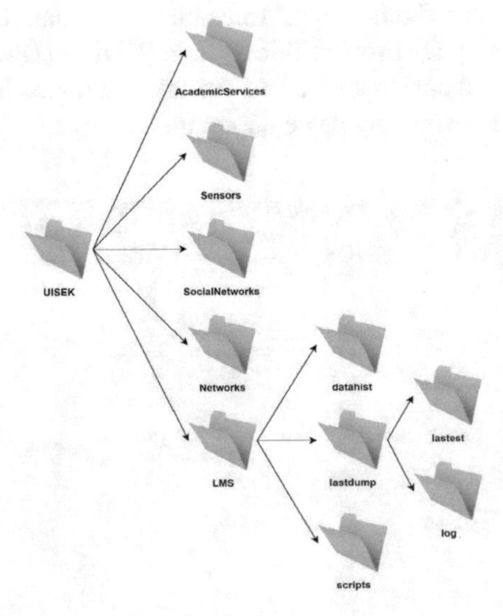

Fig. 2. Folder structure

The "LMS" subfolder within the "bin" folder of Hive contains specific subfolders:

- datahist: Stores historical data from the LMS, including past records and events relevant for the analysis and tracking of the LMS data's evolution over time.
- lastdump: Saves the latest data downloaded from the LMS, providing updated and relevant information for analysis and decision-making based on recent data.
- scripts: Stores codes used in database generation.
- Additionally, within the "lastdump" subfolder, two additional subfolders have been created:
- latest: Contains the most recently downloaded data, allowing quick and direct access to the most current information.
- log: Records events and activities related to the download and update of data in the last dump. This subfolder logs important data and metadata for detailed tracking of operations performed.

Furthermore, an automated process has been implemented for generating the database in the Data Lake, activated in the absence of the database or during the first execution of data download. This process involves a.py file that generates a.hql file with the necessary instructions to create the database and its corresponding tables. These instructions are executed in the Hadoop environment using appropriate commands.

Moreover, an automated process has been developed for the download and upload of the latest data from the Canvas Learning Management System (LMS). A file named "last_dump.py," located within the "lastdump" folder, performs this action, scheduled to run daily at 4 am using the Crontab tool (Xu et al., 2014). This process involves downloading files from the last dump via the Canvas Data API, verifying and copying them into the Hadoop environment (HDFS), maintaining a detailed record of activities and performance metrics. The next section presents the results obtained after applying the proposed approach.

4 Results

The results obtained after implementing the improvement of this Data Lake and automating the download and upload of the latest dump from LMS Canvas were analyzed, focusing on the resources used during the process and corresponding measures to evaluate the system's performance and efficiency on two virtual machines named DataLakeCANVASDev and DATACANVAS. DataLakeCANVASDev represents the previous version with the mentioned issues at the beginning of this article.

For this analysis, the available resources on the respective virtual machines were considered, where DataLakeCANVASDev has 300 GB of storage and 31.25 GB of RAM, while DATACANVAS has 300 GB of storage and 23.44 GB of RAM.

It is important to highlight the two different approaches implemented in the virtual machines DataLakeCANVASDev and DATACANVAS. In DataLakeCANVASDev, the combination of operations related to the download and upload of the lastDump, along with the historical file, posed memory management challenges. The gradual accumulation of data due to integrating the lastDump into the historical file made it difficult to precisely identify the individual size of the lastDump, estimated at approximately 3.4 GB at the end of the DataLakeCANVASDev version.

On the contrary, DATACANVAS implemented a more direct and efficient approach, where the lastDump is incorporated directly into the latest, avoiding additional operations that could impact memory management. This simplified strategy allowed a clear identification of the individual size of the lastDump, estimated at 915 MB, providing an accurate measure of its dimension compared to the complete dataset.

The results showed significant differences in the size of the "request" table between the two versions. In DataLakeCANVASDev, the table exhibited a size of around 282.89 GB, while in DATACANVAS, considering the inclusion of all data from the first dump, it presented a size of approximately 136.79 GB.

The implementation of the DATACANVAS version of the Data Lake revealed substantial improvements in performance and efficiency compared to the DataLakeCANVASDev version. Optimization of CPU and memory usage, along with a significant reduction in runtime, ensured a smoother and more effective experience in data collection, storage, and querying in the Data Lake of Universidad Internacional SEK. These improvements provide a solid foundation for data analysis and informed decision-making in the educational environment of the institution.

Table 1 presents a comprehensive summary of the results obtained, highlighting key similarities and differences in terms of storage, CPU, and efficiency. Both machines exhibit similar storage and CPU configurations, but there is a disparity in RAM, with 31.25 GB used in DataLakeCANVASDev and 23.44 GB in DATACANVAS.

Table 1. Comparison of Resources and Efficiency between DataLakeCANVASDev and DATACANVAS

Virtual Machines	DataLakeCANVASDev	DATACANVAS
Storage (GB)	300	300
RAM (GB)	31.25	23.44
CPU Consumption (%)	17.23	16.5
Ingestion Time (minutes)	20	10
Implementation Approach	Combination of operations	Direct and efficient approach
Estimated lastDump Size (GB)	Aprox. 3.4	0.915
"request" Table Size (GB)	282.89	136.79
Number of Records in "request" Table (millions)	7.32	39499.56

The choice of the implementation approach reveals a crucial point of differentiation: DataLakeCANVASDev opted for a combination of operations, while DATACANVAS implemented a direct and efficient approach. This strategic decision resulted in notable improvements, evidenced by the 73.1% reduction in the estimated size of the lastDump in DATACANVAS, decreasing from 3.4 GB to 0.915 GB.

Efficiency is further accentuated in the size of the "request" table, where DATACANVAS proves to be 51.7% more efficient than DataLakeCANVASDev, with a size of 136.79 GB compared to 282.89 GB, respectively. These results indicate substantial optimization in the system's performance and efficiency with the implementation of the DATACANVAS version of the Data Lake.

Furthermore, other previously unexplained key results are highlighted, where DATACANVAS shows more efficient CPU consumption, with a 4.3% improvement (17.23% vs. 16.5%). The ingestion time also experiences a significant improvement, halving from 20 min in DataLakeCANVASDev to 10 min in DATACANVAS. Additionally, the number of records in the "request" table sees an impressive increase of 538,852% in DATACANVAS, reaching 39,499.56 million records, compared to 7.32 million in DataLakeCANVASDev. These complementary results reinforce the overall conclusion of notable improvements in efficiency and performance with the implementation of DATACANVAS in the Data Lake environment. Finally, conclusions and future work are presented in the next section.

5 Conclusions and Future Work

The adoption of a star schema for data storage proved to be a wise choice, facilitating analysis and access to relevant information. Organizing data into fact and dimension tables allowed a clear separation of fundamental data, simplifying understanding and exploration. Data preparation processes, including selection, transformation, and cleaning, ensured the quality and consistency of stored information.

The differentiated approach between virtual machines in terms of operational efficiency is particularly noteworthy. The old version (DataLakeCANVASDev) involved additional operations such as data decompression, storage in files, and loading into the Hive database. In contrast, the new version (DATACANVAS) stands out for executing exclusively necessary operations, demonstrating a more efficient and results-focused approach.

The Data Lake provides a solid foundation for future analyses and models that can extract valuable insights for the university. Results from the DATACANVAS version exceeded expectations, showing a significant improvement in system performance and efficiency. Reduced memory and storage usage, along with an approximate runtime of 10 min for the download and upload process, ensure fast and up-to-date access to data.

The hierarchical organization of folders in the Data Lake, specifically in the "LMS" section, facilitated the storage and access of historical and latest dump data. This organized structure allows for easier management and intuitive navigation of information. The automation of downloading and loading the latest dump ensures that information is always updated and available for analysis and decision-making. The optimization in resource usage and overall system performance in the DATACANVAS version emphasizes the importance of having appropriate infrastructure and a well-defined strategy for data management.

Furthermore, the implementation of these automated processes ensures that the Data Lake is always up-to-date with the latest data from the LMS Canvas, providing Universidad Internacional SEK with access to current and relevant information for decision-making and educational analysis. This automation enhances the efficiency and reliability of the system, establishing the Data Lake as a powerful source of information for the institution.

As a future endeavor, the implementation of a machine learning model based on the Random Forest Classifier algorithm is proposed. This model aims to facilitate the identification of categories and classes, enabling proactive classification of students, like the approach taken by Rojas Pari (2021) in identifying performance patterns. This approach seeks to effectively anticipate problematic situations, such as low grades or dropouts, with the goal of intervening and providing necessary support in early stages.

Acknowledgments. We appreciate the financial support from Universidad Internacional SEK.

References

Ashish, D., Maizatul, A.I., Tutut, H.: A systematic review on educational data mining. IEEE Access **5**, 15991–16005 (2017)

Beltrán, V.: Implementación de un Data Lake de los datos de uso del LMS Canvas de la Universidad Internacional SEK. Universidad Internacional Sek (2022)

Campbell, J., DeBlois, P., Oblinger, D.: Academic analytics: a new tool for a new era. Educause Rev. **42**(4), 40–57 (2007)

Cantabella, M., Martínez-España, R., Ayuso, B., Yáñez, J.A., Muñoz, A.: Analysis of student behavior in learning management systems through a big data framework. Futur. Gener. Comput. Syst. **90**, 262–272 (2019)

Cravero, A., Lefiguala, I., Tralma, R., Gonzalez, S.: Data Lake para la dirección de análisis universitaria: Arquitectura y Metadata. RISTI: Revista Ibérica de Sistemas e Tecnologias de Informação, Extra **41**, 560–569 (2021)

Dixon, J.: James Dixon's Blog. Pentaho, Hadoop, and Data Lakes (2010)

Dwivedi, K., & Dubey, S. K. (2014). Analytical review on Hadoop Distributed file system. *In 2014 5th International Conference-Confluence The Next Generation Information Technology Summit (Confluence)*, 174–181

Gewerc, A., Rodríguez-Groba, A., Martínez-Piñeiro, E.: Academic social networks and learning analytics to explore self-regulated learning: a case study. IEEE Revista Iberoamericana de Tecnologias Del Aprendizaje **11**(3), 159–166 (2016)

Kustitskaya, T.A., Esin, R.V., Kytmanov, A.A., Zykova, T.V.: Designing an education database in a higher education institution for the data-driven management of the educational process. Educ. Sci. **13**(9), 947 (2023)

Martinez-Mosquera, D., Beltrán, V., Riofrío-Luzcando, D., Carrión-Jumbo, J.: Data lake management for educational analysis. In: 2022 IEEE Sixth Ecuador Technical Chapters Meeting (ETCM), pp. 1–5 (2022)

Mazza, R., Dimitrova, V.: CourseVis: a graphical student monitoring tool for supporting instructors in web-based distance courses. Int. J. Hum. Comput. Stud. **65**(2), 125–139 (2007)

McClure, T.: Yesterday's unified storage is today's enterprise data lake. Yesterday's Unified Storage is Today's Enterprise Data Lake (2016)

Mitchell, J., Costello, S.: International e-VET market research report: a report on international market research for Australian VET online products and services (2000)

Oukhouya, L., El Haddadi, A., Er-raha, B., Asri, H., Laaz, N.: A proposed big data architecture using data lakes for education systems. Emerg. Trends Intell. Syst. Netw. Secur. 53–62 (2022)

Rojas Pari, R.J.: Modelo de Aprendizaje Automático Supervisado para Identificar Patrones de Bajo Rendimiento Académico en los Ingresantes al Instituto de Educación Superior Pedagógico Público–Juliaca. Universidad Peruana Unión (2021)

Romero, C., Ventura, S.: Educational data mining and learning analytics: an updated survey. Wiley Interdiscip. Rev. Data Min. Knowl. Discov. **10**(3) (2020)

Schröer, C., Kruse, F., Gómez, J.M.: A systematic literature review on applying CRISP-DM process model. Procedia Comput. Sci. 526–534 (2021)

Seeling, P.: Enabling open source, open data for closed source, closed data learning management systems. In: ACM Digital Library (Ed.), In Proceedings of the 23rd Annual Conference on Information Technology Education, pp. 124–126 (2022)

Sohan, S.M., Maurer, F., Anslow, C., Robillard, M.P.: A study of the effectiveness of usage examples in REST API documentation. In: 2017 IEEE Symposium on Visual Languages and Human-Centric Computing (VL/HCC), pp. 53–61 (2017)

Universidad Internacional SEK Ecuador. (2023). Universidad Internacional Sek

Xu, L., Cao, S.X., Ma, L.M.: A strategy of database project task based on the Crontab. Appl. Mech. Mater. **610**, 611–614 (2014)

Author Index

M. Z. Vizuete et al. (Eds.): CI3 2023, LNNS 1040, pp. 311–312, 2024.
https://doi.org/10.1007/978-3-031-63434-5

Printed in the United States
by Baker & Taylor Publisher Services